Theory and Observation of Active B-type Stars

Theory and Observation of Active B-type Stars

Theory and Observation of Active B-type Stars

Editors

Lydia Sonia Cidale
Michaela Kraus
María Laura Arias

Basel • Beijing • Wuhan • Barcelona • Belgrade • Novi Sad • Cluj • Manchester

Editors

Lydia Sonia Cidale
Facultad de Ciencias
Astronómicas y Geofísicas
Universidad Nacional
de La Plata
La Plata
Argentina

Michaela Kraus
Stellar Department
Czech Academy of Sciences
Ondřejov
Czech Republic

María Laura Arias
Facultad de Ciencias
Astronómicas y Geofísicas
Universidad Nacional
de La Plata
La Plata
Argentina

Editorial Office
MDPI
St. Alban-Anlage 66
4052 Basel, Switzerland

This is a reprint of articles from the Special Issue published online in the open access journal *Galaxies* (ISSN 2075-4434) (available at: www.mdpi.com/journal/galaxies/special_issues/Btype_Stars).

For citation purposes, cite each article independently as indicated on the article page online and as indicated below:

Lastname, A.A.; Lastname, B.B. Article Title. *Journal Name* **Year**, *Volume Number*, Page Range.

ISBN 978-3-0365-9123-0 (Hbk)
ISBN 978-3-0365-9122-3 (PDF)
doi.org/10.3390/books978-3-0365-9122-3

© 2023 by the authors. Articles in this book are Open Access and distributed under the Creative Commons Attribution (CC BY) license. The book as a whole is distributed by MDPI under the terms and conditions of the Creative Commons Attribution-NonCommercial-NoDerivs (CC BY-NC-ND) license.

Contents

Preface vii

Michel Curé and Ignacio Araya
Radiation-Driven Wind Hydrodynamics of Massive Stars: A Review
Reprinted from: *Galaxies* 2023, 11, 68, doi:10.3390/galaxies11030068 1

Julieta Paz Sánchez Arias, Péter Németh, Elisson Saldanha da Gama de Almeida, Matias Agustin Ruiz Diaz, Michaela Kraus and Maximiliano Haucke
Unveiling the Evolutionary State of Three B Supergiant Stars: PU Gem, ϵ CMa, and η CMa
Reprinted from: *Galaxies* 2023, 11, 93, doi:10.3390/galaxies11050093 34

Juan Zorec
BCD Spectrophotometry and Rotation of Active B-Type Stars: Theory and Observations
Reprinted from: *Galaxies* 2023, 11, 54, doi:10.3390/galaxies11020054 62

Catalina Arcos, Leonardo Vanzi, Nikolaus Vogt, Stefano Garcia, Virginia Ortiz and Ester Acuña
Hidden Spectra Treasures in the Foster Archive: A Pilot Study of the Be Stars α Eri, α Col, ω Car and η Cen
Reprinted from: *Galaxies* 2022, 10, 106, doi:10.3390/galaxies10060106 165

Yanina Roxana Cochetti, Anahi Granada, María Laura Arias, Andrea Fabiana Torres and Catalina Arcos
Infrared Spectroscopy of Be Stars: Influence of the Envelope Parameters on Brackett-Series Behaviour
Reprinted from: *Galaxies* 2023, 11, 90, doi:10.3390/galaxies11040090 185

Anahí Granada, Maziar R. Ghoreyshi, Carol E. Jones and Tõnis Eenmäe
New Method to Detect and Characterize Active Be Star Candidates in Open Clusters
Reprinted from: *Galaxies* 2023, 11, 37, doi:10.3390/galaxies11010037 202

Grigoris Maravelias, Stephan de Wit, Alceste Z. Bonanos, Frank Tramper, Gonzalo Munoz-Sanchez and Evangelia Christodoulou
Discovering New B[e] Supergiants and Candidate Luminous Blue Variables in Nearby Galaxies
Reprinted from: *Galaxies* 2023, 11, 79, doi:10.3390/galaxies11030079 214

Michaela Kraus, Michalis Kourniotis, María Laura Arias, Andrea F. Torres and Dieter H. Nickeler
Dense Molecular Environments of B[e] Supergiants and Yellow Hypergiants
Reprinted from: *Galaxies* 2023, 11, 76, doi:10.3390/galaxies11030076 235

Andrea Fabiana Torres, María Laura Arias, Michaela Kraus, Lorena Verónica Mercanti and Tõnis Eenmäe
New Insight into the FS CMa System MWC 645 from Near-Infrared and Optical Spectroscopy
Reprinted from: *Galaxies* 2023, 11, 72, doi:10.3390/galaxies11030072 258

Tiina Liimets, Michaela Kraus, Lydia Cidale, Sergey Karpov and Anthony Marston
Large-Scale Ejecta of Z CMa—Proper Motion Study and New Features Discovered
Reprinted from: *Galaxies* 2023, 11, 64, doi:10.3390/galaxies11030064 278

Yael Aidelman and Lydia Sonia Cidale
Reddening-Free Q Parameters to Classify B-Type Stars with Emission Lines
Reprinted from: *Galaxies* 2023, 11, 31, doi:10.3390/galaxies11010031 292

Aldana Alberici Adam, Gunther F. Avila Marín, Alejandra Christen and Lydia Sonia Cidale
Synthetic Light Curve Design for Pulsating Binary Stars to Compare the Efficiency in the Detection of Periodicities
Reprinted from: *Galaxies* **2023**, *11*, 69, doi:10.3390/galaxies11030069 314

Paula Esther Marchiano, María Laura Arias, Michaela Kraus, Michalis Kourniotis, Andrea Fabiana Torres and Lydia Sonia Cidale et al.
A Mini Atlas of H-Band Spectra of Southern Symbiotic Stars
Reprinted from: *Galaxies* **2023**, *11*, 80, doi:10.3390/galaxies11040080 354

Olga Maryeva, Péter Németh and Sergey Karpov
Revealing the Binarity of HD 36030—One of the Hottest Flare Stars
Reprinted from: *Galaxies* **2023**, *11*, 55, doi:10.3390/galaxies11020055 366

Preface

Active B-type stars are defined by the presence of emission lines formed in a dense environment, often in the form of a circumstellar disk. Among them are the classical Be stars, which are stars surrounded by an ionized gaseous disk, and B[e] stars, whose disks are so dense that even molecules and dust can form within them. In either case, the disks are the source of characteristic infrared excess continuum emission and linear polarization. Active B-type stars often exhibit signatures of variability on different temporal and spatial scales. Of particular importance are variations associated with the formation and dissipation of the gaseous, molecular and dusty disks around the various types of objects that occur on time scales spanning from years to decades. Although mechanisms such as radiation, rotation, pulsation and binarity and the role they might play in pulling material off the stars have been explored in detail in the past few decades, the relevant physical processes involved in the formation of sustainable disks and their dynamical evolution are still unclear and a matter of debate.

This Special Issue intends to provide a comprehensive update on the state of the art in the field of active B-type stars. It aims to combine reviews and selected contributions on recent groundbreaking advancements in the knowledge of these peculiar objects from both a theoretical and an observational perspective. Latest observations from high-resolution, ground-based facilities and from satellite missions, collected over a wide wavelength range, reveal fascinating details about the shape and structure of circumstellar disks and their interaction with interstellar matter. At the same time, the progress in numerical models combined with ever-increasing computer power facilitates the analysis of complex models for a more realistic representation and treatment of the mechanisms behind mass ejection and the realistic estimation of the amount of mass lost. These new insights, from both theory and observations, provide a basis for exploring complex phenomena such as pulsation–wind connections, the transport of angular momentum, internal rotation law, and mass loss, and for testing the various scenarios proposed for the evolution and variability of stars and their circumstellar environment of the diverse classes of active B-type stars.

Lydia Sonia Cidale, Michaela Kraus, and María Laura Arias
Editors

Review

Radiation-Driven Wind Hydrodynamics of Massive Stars: A Review

Michel Curé [1,*] and Ignacio Araya [2]

1. Instituto de Física y Astronomía, Universidad de Valparaíso, Gran Bretaña 1111, Valparaíso 2340000, Chile
2. Vicerrectoría de Investigación, Universidad Mayor, Santiago 8580745, Chile; ignacio.araya@umayor.cl
* Correspondence: michel.cure@uv.cl

Abstract: Mass loss from massive stars plays a determining role in their evolution through the upper Hertzsprung–Russell diagram. The hydrodynamic theory that describes their steady-state winds is the line-driven wind theory (m-CAK). From this theory, the mass loss rate and the velocity profile of the wind can be derived, and estimating these properly will have a profound impact on quantitative spectroscopy analyses from the spectra of these objects. Currently, the so-called β law, which is an approximation for the fast solution, is widely used instead of m-CAK hydrodynamics, and when the derived value is $\beta \gtrsim 1.2$, there is no hydrodynamic justification for these values. This review focuses on (1) a detailed topological analysis of the equation of motion (EoM), (2) solving the EoM numerically for all three different (fast and two slow) wind solutions, (3) deriving analytical approximations for the velocity profile via the LambertW function and (4) presenting a discussion of the applicability of the slow solutions.

Keywords: stars: massive; stars: mass-loss; hydrodynamics; analytical methods; numerical methods

Citation: Curé, M.; Araya, I. Radiation-Driven Wind Hydrodynamics of Massive Stars: A Review. *Galaxies* 2023, 11, 68. https://doi.org/10.3390/galaxies11030068

Academic Editor: Artemio Herrero

Received: 14 March 2023
Revised: 5 May 2023
Accepted: 9 May 2023
Published: 12 May 2023

Copyright: © 2023 by the authors. Licensee MDPI, Basel, Switzerland. This article is an open access article distributed under the terms and conditions of the Creative Commons Attribution (CC BY) license (https://creativecommons.org/licenses/by/4.0/).

1. Introduction

At the beginning of the XX Century, Johnson [1,2] and Milne [3] argued that the force exerted on ions in the atmosphere of a luminous star could be responsible for the ejection of these ions from the star. They also argued that the ejected ions should carry with them the corresponding number of electrons, and strictly there should be no charge current, but they did not realize at that time that the collisional coupling between ions and protons would drag the rest of the plasma (mostly fully ionized hydrogen) with them as well, at least to supersonic velocities, and this theory was laid aside. It was Chandrasekhar [4,5] who, in the context of globular cluster dynamics, developed the theory of collisions due to an inverse square law, and Spitzer [6] applied Chandrasekhar's theory for collisions between charged particles.

Morton [7] was the first to report far-ultraviolet observations of three OB supergiants from an Aerobee-sounding rocket. After this came *Copernicus*, the first satellite with a telescope on board, and since then it has been possible to obtain stellar spectra in the ultraviolet (UV) region. Morton [7] found that the resonance lines of C IV, N V and Si IV showed the typical P-Cygni profiles (see Lamers and Cassinelli [8], Section 2.2). He found that the displacements in the profiles of C IV $\lambda\lambda$1549.5 and Si IV $\lambda\lambda$1402.8 corresponded to outflow velocities in the range of 1500–3000 km/s.

Snow and Morton [9] showed through a detailed survey that stars brighter than $M_{bol} \sim -6$ have strong P-Cygni profiles in their spectra and therefore *lose mass*. The same conclusion was arrived at by Abbott [10], who compared the radiative force with the gravitational force and concluded that radiative forces could initialize and maintain the mass loss process for stars with an initial mass at the zero-age main sequence (ZAMS) of about 15 $M_\odot$ or greater.

This mass loss process (known as stellar wind), together with supernovae explosions, are the main contributors in supplying the interstellar medium (ISM) with nuclear-processed heavy elements and therefore influence not only chemical evolution (and therefore star formation) but also the energy equilibrium of the ISM and the Galaxy (see [11–13] and the references therein).

Parker [14] was the first to develop the notion of solar wind through a purely gas dynamical theory, which was the only known stellar wind theory until the winds of massive stars were discovered. When this theory was applied to the winds of a typical O-star, the effective temperature necessary to reproduce the observed terminal velocities was of the order of 10^7 K, a value that is completely excluded by the presence of lines such as Si IV, C IV and N V ions, which would be destroyed by collisional ionization at temperatures above 3×10^5 K. It was, therefore, necessary to seek an alternative mechanism to drive the wind. The natural driven mechanism is the force due to the interaction of the radiation field on the wind plasma, and the simplest form is the force due to the continuum, i.e., the Thompson radiative acceleration. This force leads macroscopically to a decrease in the star's gravitational attraction by a constant factor (for O-stars this is between 0.3 and 0.6). It is then clear that the continuum force alone cannot produce a force that exceeds gravity and, therefore, cannot drive these kinds of winds.

Lucy and Solomon [15] resuscitated the proposal of Johnson and Milne and considered the force due to the absorption of spectral lines, but unlike the earlier authors, they considered the flow of the plasma as a whole rather than as the selective ejection of specific ions. They calculated an upper limit on the force on the C IV line $\lambda\lambda 1548$, finding that this exceeds the force of gravity by a factor of approximately a few hundred. Hydrostatic equilibrium in the outermost layers is not possible, and an outflow of material must occur. In their stellar wind model, Lucy and Solomon made a series of assumptions, for instance, that the wind is driven only by resonance lines. They found mass loss rates for O-stars of two orders of magnitudes less than the values obtained from observations.

A significant step in the theory was made by Castor, Abbot and Klein [16] (hereafter CAK), who realized that the force due to line absorption in a rapidly expanding envelope could be calculated using the Sobolev approximation [17,18]. Then, by developing a simple parameterization of the line force using the point star approximation, they were able to construct an analytical wind model. Despite the number of approximations made in that work, e.g., they represented the line force by C III lines and calculated only one model for a typical O5 f star ($T_{eff} = 49,290$ K, $\log g = 3.94$[1] and $R/R_\odot = 13.8$), they obtained a mass loss rate of $\dot{M} = 6.61 \times 10^{-6}\, M_\odot$/year and a terminal velocity of $v_\infty = 1515$ km/s. The value of the mass loss rate was of the same order of magnitude as the values obtained from observations, but the terminal velocity lay below the measured ones. They also gave analytical scaling relations for the mass loss rates and terminal speeds as functions of the stellar parameters. These were widely used to prove (or disprove) the validity of the radiation-driven (or line-driven) wind theory by comparison with the observations.

Abbott [10] improved this theory by calculating the line force using a tabulation of ca. 250,000 lines, which was complete for the elements H to Zn in the ionization states I to VI. Currently, the non-local thermodynamic equilibrium code CMFGEN [20] uses around 900,000 lines and FASTWIND contains 4 million lines [21] (see also [22], which uses ca. 4 million lines). Despite this immense effort to give a more realistic representation of the line force, evident discrepancies with the observations remained. Simultaneously and independently, Friend and Abbott [23] and Pauldrach et al. [24] calculated the influence of the finite cone angle correction on the dynamics of the wind (described in the Appendix from [18]). They found a much better agreement between the improved or modified CAK theory (hereafter m-CAK) and the observations of the mass loss rate and the terminal velocity in a large domain in the Hertzsprung–Russell diagram.

The equation of motion of the m-CAK theory is a highly non-linear differential equation that has singular points, eigenvalues and solution branches (see [16,23–27]). Since it is challenging to solve this differential equation numerically, Pauldrach et al. [24] found that the velocity field, $v(r)$, from the m-CAK theory can be described by a simple approximation, known as the β law approximation (see below). In addition, Kudritzki et al. [28] developed analytical approximations for the localization of the critical point, mass loss rate and terminal velocity with an agreement within 5% for v_∞ and 10% for $\dot{M}$ when compared to the correct numerical calculations.

Radiation-driven stellar winds are hydrodynamic phenomena involving the flow of the outer layers of the atmospheres of massive stars. This review is focused on describing the investigation of the m-CAK hydrodynamic theory, its topology and its three known physical solutions.

Section 2 presents the theory to calculate the radiation (line) force via an analytical description thanks to the Sobolev approximation. Section 3 introduces the m-CAK hydrodynamic theory, and its topological description is given in Section 4. Section 5 shows all three known physical solutions, whilst Section 6 presents analytical approximate solutions based on the LambertW function. Finally, in Section 7, we summarise the main topics of this review and discuss the applicability of slow solutions.

2. The Radiation Force

The exact calculation of the radiation force requires a knowledge of the radiation field (in all the lines and continua) and of the physical processes (scattering, absorption and emission) that contribute to the exchange of energy and momentum throughout the wind. The radiation field is represented by the monochromatic specific intensity $I_\nu(\mu)$, where μ is the cosine of the angle between the incoming beam and the velocity vector of the interacting particles. Thus, the radiation force per unit of volume at a distance r exerted on a point particle per unit of time is equal to the momentum removed from the incident radiation field ($\kappa \rho\, I(\mu)\, \mu/c$) integrated over all the scattering directions. This force is given by

$$\mathbf{F}^{\mathrm{rad}}(r) = \frac{4\pi}{c}\frac{1}{2}\int_0^\infty \int_{-1}^1 \kappa_\nu(r)\, \rho(r)\, I_\nu\, \mu\, d\mu\, d\nu, \quad (1)$$

where the absorption coefficient κ_ν is given in units of $\mathrm{cm}^2\,\mathrm{g}^{-1}$. The net flux density comes from the interaction processes integrated over the whole spectral range between the radiation field emitted by the photosphere and the stellar wind of mass density ρ at the distance r. Here, it is assumed that the emissivity (thermal emission and photon scattering) in the expanding atmosphere is isotropic. Therefore, no net momentum change occurs from this process (see [29], Chapter 20).

The absorption coefficient κ_ν consists of three main contributions:

$$\kappa_\nu = \kappa^{\mathrm{Th}} + \kappa_\nu^{\mathrm{cont}} + \kappa_\nu^{\mathrm{line}}, \quad (2)$$

where κ^{Th} represents the Thomson scattering, $\kappa_\nu^{\mathrm{cont}}$ is the contribution of bound-free and free-free transitions and $\kappa_\nu^{\mathrm{line}}$ is the sum of all line absorption coefficients at frequency ν.

The radiation force can be calculated by state-of-the-art non-local thermodynamic equilibrium (NLTE) radiative transfer codes such as FASTWIND [30,31], CMFGEN [20,32–34] or POWR [35,36], but these calculations depend on the velocity and density profile used to describe the wind.

2.1. Radiative Force Due to Electron Scattering

The interaction between photons and free electrons is described by a Compton process (an excellent review of this process, including Monte Carlo calculations, can be found in [37]). If photons with energy $h\nu \ll m_e c^2$ are scattered by Maxwellian electrons[2] with $kT \ll m_e c^2$, the frequency shift will be very small, but if the scattering process is repeated many times, the small amounts of energy exchanged between the electrons and photons can build up and give rise to substantial effects.

In the non-relativistic limit without the influence of quantum effects ($h\nu \ll m_e c^2$) and neglecting the possible effects described above, the scattering cross-section is frequency independent and called the Thomson cross-section, namely:

$$\sigma^{\text{Th}} = \frac{8\pi}{3} \frac{e^4}{m_e^2 c^4}. \tag{3}$$

The value of this cross-section is $\sigma^{\text{Th}} = 6.65 \times 10^{-25}$ cm^2 and the absorption coefficient is therefore:

$$\kappa^{\text{Th}} \rho = n_e \sigma^{\text{Th}}. \tag{4}$$

Using this value (κ^{Th}) in Equation (2) and integrating Equation (1), we obtain the contribution of Thomson scattering to the radiation force,

$$\mathbf{F}^{\text{Th}} = n_e \frac{\sigma^{\text{Th}} L}{4\pi c\, r^2}, \tag{5}$$

where L is the luminosity of the star. The radiative acceleration on the electrons is then

$$g_e^{\text{Th}} = \frac{1}{m_e} \frac{\sigma^{\text{Th}} L}{4\pi c r^2}. \tag{6}$$

It is useful to define the ratio of the Thomson scattering force and the gravitational force by:

$$\Gamma_e = \frac{g_e^{\text{Th}}}{g^{\text{grav}}} = \frac{1}{m_e} \frac{\sigma^{\text{Th}} L}{4\pi c\, GM_*}, \tag{7}$$

where G is the gravitational constant and M_* is the star's mass. In the standard one-component description of stellar winds, the force over the density of the plasma is given by:

$$g^{\text{Th}} = \left(\frac{n_e}{\rho}\right) \frac{\sigma^{\text{Th}} L}{4\pi c\, r^2} \tag{8}$$

where $\rho = m_p n_p + \Sigma_{\text{ions}}(m_i n_i) + n_e m_e$ is the mass density. The principal contribution of the ions comes from helium, and neglecting the electrons, $n_e m_e$, the density is

$$\rho \simeq m_p n_p (1 + A_{\text{He}} Y_{\text{He}}). \tag{9}$$

Here, A_{He} is the atomic mass of a helium atom, Y_{He} is the relative abundance of helium with respect to hydrogen (the latter being described by the subscript p) and m_p is the proton mass. Based on the conservation of charge, it is possible to express the electron number density as $n_e = n_p(1 + q_{\text{He}} Y_{\text{He}})$, where $q_{\text{He}} = 0, 1$ or 2 depending on the helium ionisation state.

Thus, the ratio n_e/ρ is:

$$\frac{n_e}{\rho} = \frac{1}{m_p}\left(\frac{1 + q_{\text{He}} Y_{\text{He}}}{1 + A_{\text{He}} Y_{\text{He}}}\right) \tag{10}$$

and the acceleration is:

$$g^{\text{Th}} = \frac{1}{m_p}\left(\frac{1 + q_{\text{He}} Y_{\text{He}}}{1 + A_{\text{He}} Y_{\text{He}}}\right)\left(\frac{\sigma^{\text{Th}} L}{4\pi c\, r^2}\right) \tag{11}$$

or

$$\Gamma_e = \left(\frac{1 + q_{\text{He}} Y_{\text{He}}}{1 + A_{\text{He}} Y_{\text{He}}}\right)\left(\frac{\sigma^{\text{Th}} L}{4\pi c\, m_p\, GM_*}\right). \tag{12}$$

Quite often, the canonical value of $\kappa^{\text{Th}} = 0.34$ cm^2 g^{-1} is adopted, which follows from assuming a fully ionised plasma at solar abundance. In addition, since the continuum of OB stars is also optically thin in the lines near its maximum, the contribution of the continuum to the total radiative force is neglected.

The next section provides a general description of the line force based on the Sobolev approximation (see, e.g., Lamers and Cassinelli [8] or Hubeny and Mihalas [29]).

2.2. Radiative Force due to Lines

The contribution to the radiation force due to the spectral lines in the wind of massive stars is provided by the momentum transfer of photons (via absorption and re-emission processes in optically thick lines) mainly from the most dominant ions (i.e., C, O, N and the Fe group). The proper calculation of the line force (per unit volume) is given by:

$$F^{\text{line}}(r) = \frac{2\pi}{c} \sum_l \int_0^\infty \int_{-1}^{+1} \kappa_l(r)\, \rho(r)\, \phi_l(\nu,\mu,r)\, I_\nu(r,\mu)\, \mu\, d\mu\, d\nu \qquad (13)$$

where ϕ is the Gaussian absorption profile. The summation is over all the line transitions (l), assuming non-overlapping lines, for which the wind is optically thick. κ_l is the opacity coefficient (in cm^2 g^{-1}) of lines formed between levels l (lower) and u (upper) with energy h ν_0,

$$\kappa_l \rho = \frac{\pi e^2}{m_e c} f_l n_l \left(1 - \frac{n_u g_l}{n_l g_u}\right). \qquad (14)$$

The number densities n_l and n_u of ions in levels l and u are given in cm^{-3}, g_l and g_u are the corresponding statistical weights and f_l is the oscillator strength of the line. The CAK theory allows us to find an analytical expression for the line force in a moving media with large velocity gradients in terms of the macroscopic variables using the Sobolev approximation. However, this expression only applies to radiating flows in the non-relativistic regime.

2.2.1. The Sobolev Approximation

In a moving plasma such as stellar wind, the interaction of radiation with matter can be better understood as follows. Let us consider a single spectral line thermally broadened with a rest wavelength λ_1. A photon emitted from the stellar surface with wavelength $\lambda_* < \lambda_1$ propagates without interacting with the matter until, due to the Doppler shift, it is scattered at the blue edge of the line in question. Due to the expansion of the wind, the particles viewed from any direction from a certain position always appear to be receding. This means that independent of the scattered direction of the photon (forward or backwards), the distance travelled always causes its comoving wavelength to be red shifted.

After many scatterings, the photon's wavelength has been shifted to the line's red edge, and the interaction of this photon with the line (λ_1) ceases. The region in the wind where an incoming photon can interact with the ions is called the interaction zone. It is also well known that the winds of massive stars reach terminal velocities of several times the sound speed, and the point at which the wind velocity is equal to the sound speed (the sonic point) is very near to the photosphere. This means that almost all the region where stellar winds are found is supersonic.

This description corresponds to the Sobolev approximation [17], where all the relevant physical quantities, such as the opacity, source function, etc., are considered constant in the interaction zone, i.e., the width of the interaction zone is small compared with a characteristic flow length. Thus, for a generic Doppler-broadened line profile, the Sobolev length, L_s, is defined as:

$$L_s = v_{\text{th}}/(dvs./dr), \qquad (15)$$

where T_{eff} is the star's effective temperature, $v_{\text{th}} = \sqrt{2 k_B T_{\text{eff}}/m_p}$ is the thermal speed of the protons and k_B is the Boltzmann constant.

A characteristic length of the flow is

$$L_c \simeq vs./(dvs./dr) \tag{16}$$

Typical values of thermal velocities in OB-type stars are about 7–20 km/s, while terminal velocities are about 1000–3000 km/s (see, e.g., Lamers and Cassinelli [8], Puls et al. [12]). More recent measurements of terminal velocities based on observations performed in the frame of the ULLYSES collaboration [38] have been accomplished by Hawcroft et al. [39].

2.2.2. The Line Force due to a Single Line

Castor [18] analysed the Sobolev approximation in detail in the context of stellar winds and showed that the force produced by the incoming radiation due to a single line can be expressed as[3]:

$$f^{\text{line}} = \left(\frac{F_\nu \Delta \nu_d}{c}\right)\left(\frac{k_l}{\tau_l}\right)(1 - e^{-\tau_l}), \tag{17}$$

where $\Delta \nu_d = v_{\text{th}} \nu/c$ corresponds to the Doppler shift, F_ν is the flux of the radiation field at frequency ν, k_l is the monochromatic line absorption coefficient per unit mass and

$$\tau_l = \int \rho \, \phi(\nu, r) \, k_l \, dr \tag{18}$$

is the optical depth. Evaluating the optical depth for a normalized Gaussian profile and using the Sobolev approximation, we find:

$$\tau_l = k_l \, \rho \, v_{\text{th}}/(dvs./dr). \tag{19}$$

With this expression, we can interpret the RHS of (17) as:

(i) $(F_\nu \Delta \nu_d / c)$ is the rate of momentum emitted by the star per unit area at frequency ν with bandwidth $\Delta \nu_d$;
(ii) $(\tau_l / k_l) = \rho \, v_{\text{th}}/(dv/dr)$ represents the amount of mass that can absorb this momentum;
(iii) $(1 - e^{-\tau_l})$ is the probability that such an absorption occurs.

Then, we define

$$t = \sigma_e \rho \, v_{\text{th}}/(dv/dr), \tag{20}$$

where $\sigma_e = \sigma^{\text{Th}} n_e/\rho$ corresponds to the Thomson scattering absorption coefficient per density. In a moving medium, t represents the optical depth that a line will have if its opacity is equal to its electron scattering opacity. Based on this definition, it is possible to rewrite t as

$$\tau_l = \eta_l \, t \tag{21}$$

where $\eta_l = k_l/\sigma_e$. The first factor in (21) is related only to line properties, and the second only to dynamic variables of the wind.

2.2.3. The Line Force due to a Statistical Distribution of Line Strength

The total line force due to the addition of all the single lines of the ions for a point star approximation and for non-overlapping single lines is given by:

$$f^{\text{line}} = \sum_l \left(\frac{F_\nu \Delta \nu_d}{c}\right)_l \left(\frac{dv/dr}{\rho \, v_{\text{th}}}\right)_l (1 - e^{-\eta_l t}). \tag{22}$$

Expressing (22) in terms of $\Delta \nu_d = \nu \, v_{\text{th}}/c$ and the relation $F = L/4\pi r^2$, we obtain

$$f^{\text{line}} = \frac{L}{c^2}\left(\frac{dv/dr}{4\pi r^2}\right)\sum_l \left(\frac{L_\nu \, \nu}{L}\right)_l (1 - e^{-\eta_l t}) \tag{23}$$

Abbott [10] was the first to compile and publish a list of ca. 250,000 lines for atoms from H to Zn in ionisation stages I to VI. Based on such a line list [22,40,41], it is possible to derive a line strength distribution function [24,42]. This distribution can be described as follows:

$$dN(k_l) = \int_0^N \left(\frac{L_\nu \nu}{L}\right) n(k_l, \nu) d\nu \qquad (24)$$

and represents the number of lines in the line strength interval $(k_l, k_l + \Delta k_l)$ obtained from the total spectrum and weighted by the flux mean of line strength $(L_\nu \nu / L)$. Notice that in Equation (24), the distribution in frequency space of the lines is independent from the distribution in line strength. An alternative formulation of the line statistic is given by Gayley [43] (see also [22]).

The logarithm of the number of lines can be fitted by a linear function, namely:

$$dN(k_l) = N_0 \left(1 - \alpha\right) (k_l / \sigma_e)^{\alpha - 2} d(k_l / \sigma_e) \qquad (25)$$

where N_0 is the number of lines (strong and weak) that effectively contribute to the line force. Typical values of the parameter α are $0.45 \leq \alpha \leq 0.7$ [8,42]. Notice that line force parameters are not free but depend on the transfer problem in each individual star (see [22,41,42,44–46], for a detailed description of the calculation of the line force parameters).

Extending the sum in Equation (23) to an integral, we obtain the line force expression:

$$f^{\text{line}} = \frac{L}{c^2} \frac{(dv/dr)}{4\pi r^2} N_0 (1 - \alpha) \int_{\sigma_e}^{\infty} (1 - e^{-\eta_l t})(k_l / \sigma_e)^{(\alpha - 2)} d(k_l / \sigma_e). \qquad (26)$$

Neglecting the lower limit of the integral, a valid approximation for stars of type OB, and replacing it by zero and integrating, the line force becomes:

$$f^{\text{line}} = \frac{L}{c^2} \frac{1}{4\pi r^2} v_{\text{th}} \sigma_e N_0 (1 - \alpha) \Gamma(\alpha) \left(\frac{dv/dr}{\sigma_e \rho v_{\text{th}}}\right)^\alpha, \qquad (27)$$

where Γ is the Γ-function. Then, dividing by the total density, we obtain the standard form of the line acceleration,

$$g^{\text{line}} = \frac{C}{r^2} \left(r^2 v \frac{dv}{dr}\right)^\alpha, \qquad (28)$$

with

$$C = \Gamma_e\, GM_* \, k \left(\frac{4\pi}{\sigma_e v_{\text{th}} \dot{M}}\right)^\alpha, \qquad (29)$$

where Γ_e is the radiative acceleration due to Thomson scattering in terms of the gravitational acceleration and $\dot{M}$ is the mass loss rate. Here, the continuity equation has been used, and the variables, such as N_0 or $\Gamma(\alpha)$, have been collected into the constant k. Note that this expression for the line force (Equation (28)) only takes interactions between ions and radially emitted photons into account [16,18].

2.2.4. The Correction Factor

Castor et al. [16] (see their appendix) qualitatively discussed the effect on the line force that the proper shape of the star (non-radial incoming photons) would have on the wind kinematics. Later Pauldrach et al. [24] and Friend and Abbott [23] independently investigated the influence of this effect, known as the finite disc correction factor, thereby developing the m-CAK theory.

The expression of the line force for incoming photons from an arbitrary direction for a radial flow velocity field comes from the definition of Equation (20); thus,

$$t_\sigma = \left(\frac{1 + \sigma}{1 + \mu^2 \sigma}\right) \frac{\sigma_e \rho v_{\text{th}}}{dv/dr}, \qquad (30)$$

where

$$\sigma = \frac{d\ln vs.}{d\ln r}. \tag{31}$$

Inserting t_σ into Equation (26) instead of t and integrating, we obtain the following expression for the line force:

$$g^{\text{line}} = \frac{C}{r^2} CF(r, v, v') \left(r^2 \, v \, v' \right)^\alpha \tag{32}$$

where CF is the correction factor, defined as the ratio of the force due to the non-radial contributions to that of a point star approximation, namely:

$$CF = \frac{2}{1 - \mu_c^2} \int_{\mu_c}^{1} \left[\frac{(1 - \mu^2) \, v/r + \mu^2 v'}{v'} \right]^\alpha \mu \, d\mu \tag{33}$$

with $\mu_c = \sqrt{(1 - R_*^2/r^2)}$, where R_* is the stellar radius and $v' = dv/dr$. In Appendix A, we summarised some properties of the correction factor.

2.3. The Ionization Balance

In their work, Abbott [10] assumed a local thermodynamic equilibrium (LTE) and used the modified Saha formula (see Hubeny and Mihalas [29]) to take the dilution of the radiation field and the possible difference between the electron kinetic temperature T_e and the radiation temperature T_r into account. Due to the changes in the ionisation throughout the wind, Abbott fitted the line force not only in terms of $(r^2 \, vs. \, v')^\alpha$ (see Equation (28)) but also as a function of the ratio $n_e/W(r)$, where

$$W(r) = \frac{1}{2}\left(1 - \sqrt{(1 - R_*^2/r^2)}\right) \tag{34}$$

is the dilution factor. They found that the functional dependence of this quotient in the line force is:

$$g^{\text{line}} \propto \left(\frac{n_e}{W(r)}\right)^\delta, \tag{35}$$

where the electron number density, n_e, is given in units of 10^{11}cm^{-3}. This proportionality means that the greater the density, the lower the ionisation level. In view of the fact that the lower ionisation levels have more line transitions, usually at the maximum of the radiation field, the line force increases with increasing density. Values of this δ line force parameter for the fast solution (see below) are in the range $0 < \delta \lesssim 0.2$ [8], but for a pure hydrogen atmosphere, the value is $\delta = 1/3$, as Puls et al. [42] demonstrated.

3. The m-CAK Hydrodynamic Model

The 1D m-CAK stationary model for line-driven stellar winds considers the following assumptions: an isothermal fluid in spherical symmetry and no influence from viscosity effects, heat conduction and magnetic fields.

The stationary continuity and momentum conservation equations are:

$$\dot{M} = 4\pi r^2 \rho \, vs., \tag{36}$$

$$v\frac{dv}{dr} = -\frac{1}{\rho}\frac{dP}{dr} - \frac{GM_*(1 - \Gamma_e)}{r^2} + \frac{v_\phi^2(r)}{r} + g^{\text{line}}(\rho, dv/dr, n_e), \tag{37}$$

where P is the fluid pressure and $v_\phi = v_{rot} R_*/r$, with v_{rot} being the stellar rotational speed at the equator. In addition, $g^{\text{line}}(\rho, dv/dr, n_e)$ corresponds to the acceleration due to an ensemble of lines.

The standard or m-CAK parameterization of the line force [10,23,24] is the following:

$$g^{\text{line}} = \frac{C}{r^2} \, CF(r, v, dv/dr) \left(r^2 v \frac{dv}{dr}\right)^\alpha \left(\frac{n_e}{W(r)}\right)^\delta, \quad (38)$$

where the coefficient C is given in Equation (29).

Substituting the density from the mass conservation equation (Equation (36)) into the momentum equation (Equation (37)), we obtain the equation of motion (EoM).

Transforming to dimensionless variables, that is

$$u = -\frac{R_*}{r}, \quad (39)$$

$$w = \frac{v}{a}, \quad (40)$$

$$w' = \frac{dw}{du}, \quad (41)$$

where a is the isothermal sound speed of an ideal gas, $P = a^2 \rho$.

Using these new variables, the EoM now reads:

$$F(u, w, w') = \left(1 - \frac{1}{w^2}\right) w \frac{dw}{du} + A + \frac{2}{u} + a_{\text{rot}}^2 u - C' \, CF \, g(u)(w)^{-\delta} \left(w \frac{dw}{du}\right)^\alpha = 0, \quad (42)$$

where the constants are the following:

$$A = \frac{GM(1 - \Gamma_e)}{a^2 R_*} = \frac{v_{\text{esc}}^2}{2a^2}, \quad (43)$$

$$a_{\text{rot}} = \frac{v_{\text{rot}}}{a}, \quad (44)$$

$$C' = C \left(\frac{\dot{M}D}{2\pi} \frac{10^{-11}}{aR_*^2}\right)^\delta (a^2 R_*)^{(\alpha-1)}, \quad (45)$$

$$D = \frac{(1 + Z_{\text{He}} Y_{\text{He}})}{(1 + A_{\text{He}} Y_{\text{He}})} \frac{1}{m_p}, \quad (46)$$

where Z_{He} is the number of free electrons provided by helium, v_{esc} is the escape velocity and the function g is defined as:

$$g(u) = \left(\frac{u^2}{1 - \sqrt{1 - u^2}}\right)^\delta \quad (47)$$

In order to find a physical wind solution of the EoM (Equation (42)), i.e., starting from the photosphere with a small velocity and reaching infinity with a supersonic velocity, we first need to understand the topology of this equation.

4. Topological Analysis

As mentioned previously, the first wind model was developed by Parker [14] for the sun. This model possesses a singular point at the sonic point and different solution branches (see Figure 3.1 in [8]). The m-CAK model has a driving force (line force) that depends not only on the radial coordinate r (or u) but also on the velocity and the velocity gradient. These characteristics complicate the study of the EoM's topology that gives rise to the different solutions.

Mathematically, singular points are located where the singularity condition is satisfied, i.e.,

$$\frac{\partial}{\partial w'} F(u, w, w') = 0, \quad (48)$$

and these locations form the *locus* of singular points.

At these specific points, in order to have a smooth wind solution between solution branches, a regularity condition must be imposed, namely:

$$\frac{d}{du}F(u,w,w') = \frac{\partial F}{\partial u} + \frac{\partial F}{\partial w}w' = 0 \qquad (49)$$

Using the following coordinate transformation:

$$Y = w\,w' \qquad (50)$$

$$Z = w/w', \qquad (51)$$

we can now solve Equations (42), (48) and (49), only valid simultaneously at one singular point, obtaining the following set of equations:

$$Y - \frac{1}{Z} + A + 2/u + a_{rot}^2 u \quad - C' f_1(u,Z)\,g(u)\,Z^{-\delta/2}\,Y^{\alpha-\delta/2} = 0, \qquad (52)$$

$$Y - \frac{1}{Z} \qquad\qquad\qquad\quad - C' f_2(u,Z)\,g(u)\,Z^{-\delta/2}\,Y^{\alpha-\delta/2} = 0, \qquad (53)$$

$$Y + \frac{1}{Z} - 2Z/u^2 + a_{rot}^2 Z \quad - C' f_3(u,Z)\,g(u)\,Z^{-\delta/2}\,Y^{\alpha-\delta/2} = 0, \qquad (54)$$

derivation details and definitions of $f_1(u,Z)$, $f_2(u,Z)$ and $f_3(u,Z)$ are summarised in Appendix B.

Solving for Y and C' from the set of Equations (52)–(54), we obtain:

$$Y = \frac{1}{Z} + \left(\frac{f_2(u,Z)}{f_1(u,Z) - f_2(u,Z)}\right) \times \left(A + \frac{2}{u} + a_{rot}^2 u\right) \qquad (55)$$

and

$$C'(\dot{M}) = \frac{1}{gf_2}\left(1 - \frac{1}{YZ}\right) Z^{\delta/2}\,Y^{1-\alpha+\delta/2} \qquad (56)$$

These last two Equations are generalisations of the relations found by [28] (see their Equations (21) and (34) for Y and Equations (21) and (44) for the eigenvalue), but now including the rotational speed of the star.

The Critical Point Function R

The set of Equations (52)–(54) are only valid at the singular point for the unknowns Y_s, C'_s, Z_s and u_s[4]. Due to the fact that there are only three equations and four unknowns, it is not possible to solve them. Nevertheless, from this set of equations, we can derive the function $R(u,Z)$, defined by:

$$R(u,Z) = -\frac{2}{Z} + \frac{2Z}{u^2} - a_{rot}^2 Z + f_{123}(u,Z)\left(A + \frac{2}{u} + a_{rot}^2 u\right) \qquad (57)$$

where $f_{123}(u,Z)$ has the following definition:

$$f_{123}(u,Z) = \frac{f_2(u,Z) - f_3(u,Z)}{f_2(u,Z) - f_1(u,Z)} \qquad (58)$$

The locus of singular points, u_s, is given by the points which are solutions of the following equation:

$$R(u,Z) = 0. \qquad (59)$$

It should be noted that no approximation has been made in the derivation of the above topological equations.

To determine the location of the singular point in the locus of points that satisfies $R(u, Z) = 0$, and therefore determine the values of Y_s, C'_s, Z_s and u_s, we need to set a boundary condition at the stellar surface.

The most used boundary conditions are:

(i) Set the density at the stellar surface to a specific value,

$$\rho(R_*) = \rho_* \qquad (60)$$

Usually this base density is in the range 10^{-8} g cm^{-3} to 10^{-13} g cm^{-3}. For some examples, see the works of de Araujo and de Freitas Pacheco [47], Friend and MacGregor [48], Madura et al. [49], Curé [50] and Araya et al. [51].

(ii) Set the optical depth integral to a specific value, i.e.,

$$\tau_* = \int_{R_*}^{\infty} \sigma_E \rho(r) dr = \frac{2}{3}. \qquad (61)$$

Employing one of these boundary conditions at the stellar surface plus the regularity condition at the singular point, we can solve from the EoM (Equation (42)) the velocity profile, $w = w(u)$, together with the value of the eigenvalue, C', and therefore the mass loss rate, $\dot{M}$.

5. Types of Solutions

We developed a numerical code that discretizes by finite differences the EoM and, using the Newton–Raphson method, iterates to a numerical solution. This code is called HYDWIND and is described in more detail in Curé [50] (see also [52]).

After performing a topological analysis of the EoM, we were able, thanks to HYDWIND, to find the numerical solutions of all three known m-CAK physical solutions: fast, Ω-slow and δ-slow solutions.

5.1. Fast Solution

From the pioneering work of CAK and its improvements from Friend and Abbott [23] and Pauldrach et al. [24], the code HYDWIND is able to obtain the standard solution of the m-CAK theory, hereafter called the fast solution.

To perform our topological analysis, we use a typical O5 V star with the following stellar parameters: $T_{\text{eff}} = 45,000$ K, $\log g = 4.0$, $R/R_\odot = 12$, and line force parameters $k = 0.124$, $\alpha = 0.64$, and $\delta = 0.07$, with the boundary condition $\tau_* = 2/3$.

The function $R(u, Z)$ is shown in Figure 1 for a non-rotating star ($a_{\text{rot}} = 0$). The plane $R(u, Z) = 0$ is plotted in light grey (right panel) and the intersection of both functions, which corresponds to the locus of singular points, is plotted with a black line. The locus of singular points for the fast solution is the one that starts at $Z \sim 0$ and $u = -1$. The other locus of singular points will be discussed below.

Knowing the topology of the m-CAK model, specifically the locus of singular points, we now solve the EoM for the velocity profile, $v(r)$, or equivalently $w = w(u)$, and the eigenvalue C', which is proportional to the mass loss rate, $\dot{M}$. Then, the wind parameters obtained for this model are a terminal velocity (v_∞) of 3467 km/s and a mass loss rate ($\dot{M}$) of $2.456 \times 10^{-6} M_\odot$/yr. Figure 2 shows the velocity profile of this model as a function of $\log(r/R_* - 1)$ (left panel) and as a function of u (right panel). The location of the singular point (r_s) is shown with a red dot, and it is located near the stellar surface ($r_s = 1.029 R_*$ or $u_s = -0.9719$). At this point, the wind velocity is 181.4 km/s, a highly supersonic speed ($a = 24.17$ km/s).

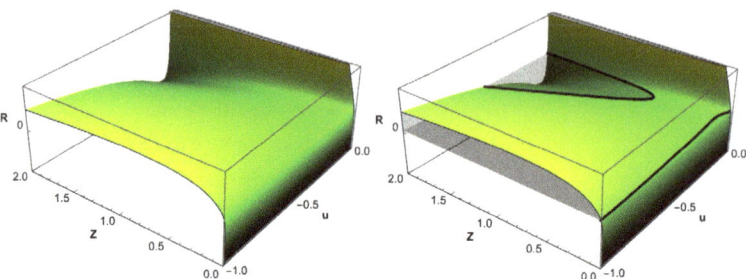

Figure 1. Function $R(u, Z)$ for a typical O5 V star without rotation. The (**left panel**) shows only the function $R(u, Z)$, while the (**right panel**) is similar to the (**left panel**), but the plane $R(u, Z) = 0$ is also plotted in light grey. Furthermore, the intersection of both curves (black solid lines) shows two loci of singular points.

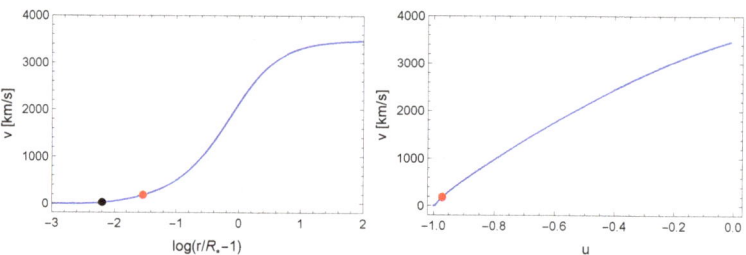

Figure 2. Velocity profile for a typical O5 V star without rotation. The velocity profile is plotted as a function of $\log(r/R_* - 1)$ (**left panel**) and as a function of u (**right panel**). The location of the singular point is shown with a red dot, while the sonic point is in black.

This steep velocity gradient is due to the rapid increase in the line force just above the stellar surface, as shown in Figure 3, where the sound speed is reached at $r = 1.006\, R_*$ and the maximum of g^{line} is reached at $r = 1.3\, R_*$.

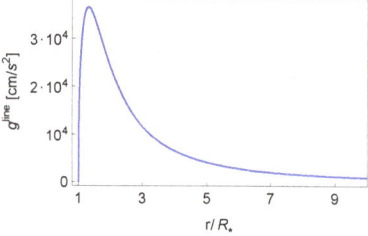

Figure 3. The radiative acceleration, g^{line}, for a typical O5 V star without rotation as function of r/R_* for $r < 10\, R_*$.

As previously mentioned, the wind parameters (v_∞ and $\dot{M}$) must be calculated within the framework of the radiative transport problem. However, to understand the complex non-linear dependence on the wind parameters from the line force parameters, in the following figures, we show how the wind parameters depend on each one of the line force parameters.

Figure 4 shows the dependence of the wind parameters, v_∞ (left panel) and $\dot{M}$ (right panel), as a function of the line force parameter α, using the same stellar parameters and keeping the line force parameters k and δ fixed. There is an increase in the values of both wind parameters as α increases.

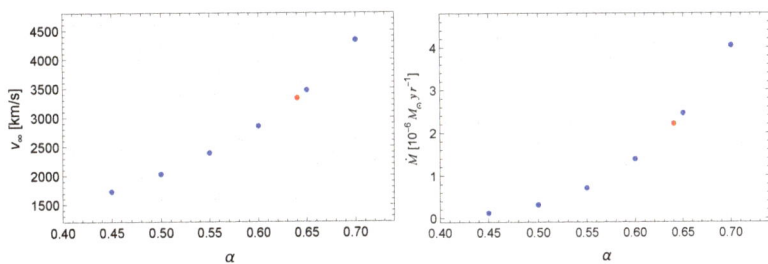

Figure 4. Dependence of the wind parameters as a function of the line force parameter α. Terminal velocity (**left panel**) and mass loss rate (**right panel**). The values obtained for our typical O5 V star without rotation are shown in red.

The dependence of the wind parameters as a function of the line force parameter k is shown in Figure 5. In this case, the wind parameters also increase as k increases. It is clearly seen in Figure 5 that the terminal velocity depends only slightly on the value of k rather than the mass loss rate, which has a significant impact on the value of k [53].

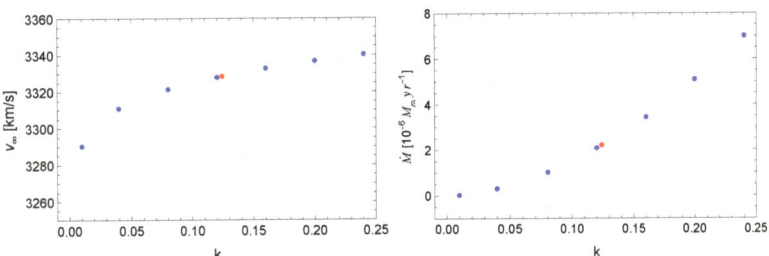

Figure 5. Dependence of the wind parameters as a function of the line force parameter k. Terminal velocity (**left panel**) and mass loss rate (**right panel**). The values obtained for our typical O5 V star without rotation are shown in red.

Finally, the dependence of the wind parameters as a function of the line force parameter δ is shown in Figure 6. We observe that the terminal velocity has a decreasing behaviour when the parameter δ increases, while the mass loss rate can have a decreasing or increasing behaviour. This behaviour depends on the parameter k; for low values of k, the mass loss rate decreases while δ increases, but for larger values, the behaviour is reversed.

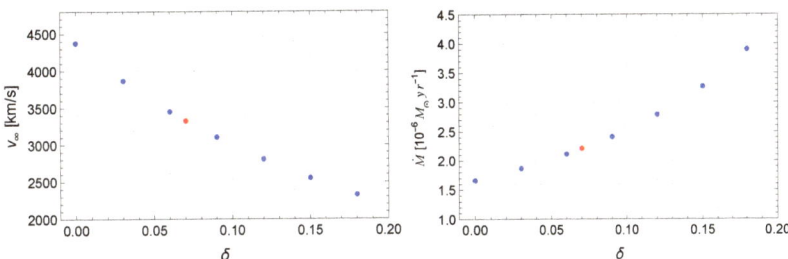

Figure 6. *Cont.*

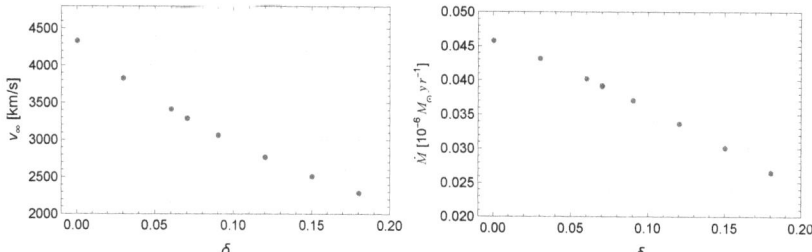

Figure 6. Dependence of the wind parameters as a function of the line force parameter δ. Upper panels are for $k = 0.124$ and lower panels are for $k = 0.0124$. The values obtained for our typical O5 V star without rotation are shown in red.

In the next subsections, we will discuss each of the slow solutions. The combined effects of the line force parameters and the three physical wind solutions are discussed in detail in work by Venero et al. [53].

Overall, the fast solution from the m-CAK theory has been very successful in explaining the terminal velocities and mass loss rates of massive stars (see [8,11–13]).

5.2. Ω-Slow Solution

The original CAK model considered the star point approximation, i.e., all the photons are radially directed over the wind plasma. In that work, CAK only discussed the effect of the finite disc of the star seen by an observer in the wind. Friend and Abbott [23] and Pauldrach et al. [24] implemented the finite disc correction factor and solved the EoM. In both works, they also studied the influence of rotation in the equatorial plane of a rotating star, but they could not obtain solutions for rapidly rotating stars. The reason was found by Curé [50] for values of $\Omega \gtrsim 0.75$, where $\Omega = v_{rot}/v_{crit}$. At this value of Ω, the fast solution ceases to exist, and another type of solution is found. This solution, called Ω-*slow* solution, is characterised by a slower and denser wind in comparison with the fast solution.

It is well known that Be stars are the fastest rotators among stars [54]. Thus, in this section, we will study the topology and the wind solutions for a typical B2.5 V star with the following stellar and line force parameters: $T_{eff} = 20,000 K$, $\log g = 4.11$, $R/R_\odot = 4.0$, $k = 0.61$, $\alpha = 0.5$ and $\delta = 0.0$. The lower (surface) boundary condition is fixed at $\rho_* = 8.7 \times 10^{-13}\,\mathrm{g/cm^3}$. In addition, the distortion of the shape of the star caused by its high rotational speed and gravity darkening effects are not considered.

Figure 7 shows the surfaces $R(u, Z)$ for different values of Ω, together with the plane $R(u, Z) = 0$. The intersection of the surfaces $R(u, Z)$ and $R(u, Z) = 0$ (black lines) correspond to the locus of singular points. We clearly observe two different loci of singular points. The fast solution locus can be observed for $\Omega = 0.3$ (upper left panel), $\Omega = 0.5$ (upper right panel) and $\Omega = 0.7$ (lower left panel). For larger rotational rates ($\Omega \gtrsim 0.75$), the fast solution locus lies completely under the plane $R(u, Z) = 0$, as shown in the lower right panel for $\Omega = 0.9$. Thus, the fast solution *does not exist* for large values of Ω.

Figure 8 shows the velocity profiles, $v(u)$, as a function of u for different values of Ω. All these solutions use the same lower boundary condition. This figure shows (left panel) fast solutions in light grey and Ω-slow solutions in coloured lines. The right panel shows only Ω-slow solutions; the location of the singular point is almost independent of Ω. This is a consequence of the shape of the locus curve of singular points (see Figure 7). This locus is located almost at a constant value of u.

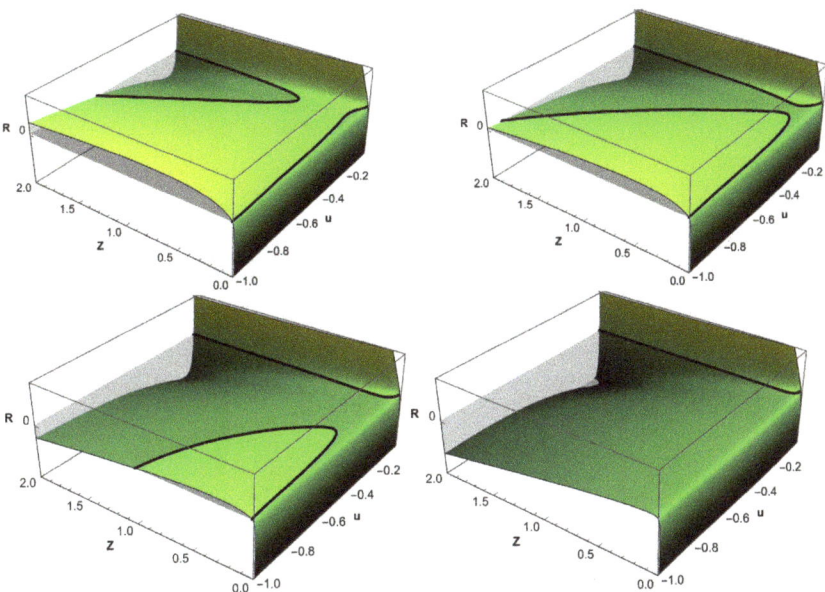

Figure 7. The function $R(u, Z)$ (topology) of the m-CAK theory as function of Ω: $\Omega = 0.3$ (**upper left panel**), $\Omega = 0.5$ (**upper right panel**), $\Omega = 0.7$ (**lower left panel**) and $\Omega = 0.9$ (**lower right panel**). The plane $R(u, Z) = 0$ is shown in light grey, and its intersection with the surface $R(u, Z)$ (locus of singular points) is plotted with a black line.

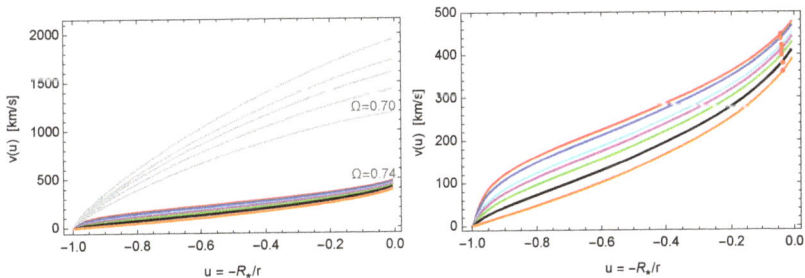

Figure 8. Velocity profiles, $v(u)$, as function of the rotational rate Ω. (**Left panel**): fast solutions ($\Omega \lesssim 0.74$) are plotted in grey lines, while the Ω-slow solutions are in coloured lines: $\Omega = 0.74$ (red line), $\Omega = 0.76$ (blue line), $\Omega = 0.80$ (cyan line), $\Omega = 0.82$ (magenta line), $\Omega = 0.85$ (green line), $\Omega = 0.90$ (black line) and $\Omega = 0.95$ (orange line). (**Right panel**): The same Ω-slow solutions, but zoomed and including the location of the singular points (red dots).

The Ω-slow solutions are only valid in the equatorial plane in this 1D m-CAK model. Notice that this model does not take into account the oblateness and gravity-darkening effects. See Araya et al. [55] for the implementation in the 1D model (equatorial plane) and Cranmer and Owocki [56] for the implementation in the 2D model.

In the equatorial plane, the higher the Ω, the greater the centrifugal force and, consequently, the lower the effective gravity. Therefore, the higher Ω is, the higher the rate of mass loss and, through the continuity equation, the higher the wind density, as shown in Figure 9.

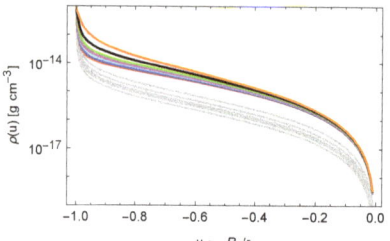

Figure 9. Wind density profiles, $\rho(u)$ (in gr/cm^3) versus u, for fast and Ω-slow solutions. The colour scheme is the same as the one used in Figure 8.

5.3. δ-Slow Solution

The *δ-slow* solution was found numerically by Curé et al. [27]. This solution, based on the m-CAK theory, describes the wind velocity profile when the ionization-related line force parameter δ takes larger values, $\delta \gtrsim 0.28$. These values are larger than the ones provided by the standard m-CAK solution (see [8] and the references therein). Nevertheless, Puls et al. [42] calculated the value of δ for a pure hydrogen atmosphere, finding a value of $\delta \sim 1/3$. These high values of δ are also found in atmospheres and winds with extremely low metallicities (see [57]).

The δ-slow solution, as well as the Ω-slow solution, are characterized by low velocities. This solution could explain the velocities obtained for late-B and A-type supergiant stars and seems to fit well with the observed anomalous correlation between the terminal and escape velocities found in A supergiant stars [27]. Furthermore, in work by Venero et al. [53] (see their Table 2), a gap of solutions between the fast and the δ-slow solutions for different values of the rotational speed was found in the plane δ-Ω.

To present a topological analysis of this type of solution, we adopt the model T19 from Venero et al. [53]. The stellar and line force parameters are $T_{\text{eff}} = 19{,}000$ K, $\log g = 2.5$, $R/R_\odot = 40$, $k = 0.32$ and $\alpha = 0.5$. We use $\tau_* = 2/3$ as a boundary condition at the stellar surface.

In Figure 10, the $R(u,Z)$ function and the plane $R(u,Z) = 0$ are shown for different values of δ. The upper left panel shows the surface $R(u,Z)$ for $\delta = 0.1$. We clearly see that, for this case, the locus of singular points for fast solutions is different from the case of the fast solution shown in Figure 1. Here, this locus is located when $Z \lesssim 0.5$, $\forall u$. The upper right panel shows $R(u,Z)$ for $\delta = 0.12$, where the locus of singular points for fast solutions returns to the behaviour shown in Figure 1. The fast solution is present until $\delta = 0.24$, see the lower left panel. We cannot find fast solutions for slightly larger values of δ until $\delta \sim 0.30$ (lower right panel). For this value of δ, the locus of singular points for fast solutions shifts to slightly larger values of Z for $u \lesssim -1$, and the numerical wind solutions no longer have a singular point in this locus, switching to the other locus of singular points (δ-slow solutions) located at $u \lesssim -0.1$ (or $r \gtrsim 10 R_*$).

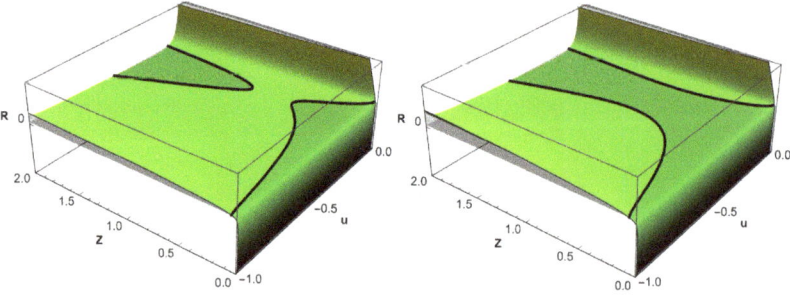

Figure 10. *Cont.*

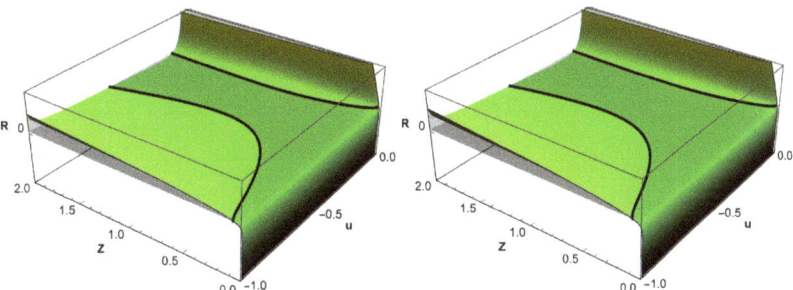

Figure 10. The topological function $R(u, Z)$ of the m-CAK theory as function of δ. (**Upper left panel**): $\delta = 0.1$. (**Upper right panel**): $\delta = 0.12$. (**Lower left panel**): $\delta = 0.24$. (**Lower right panel**): $\delta = 0.3$. The plane $R(u, Z) = 0$ is shown in light grey, and its intersection with the surface $R(u, Z)$ (locus of singular points) is plotted with black lines.

Figure 11 shows the velocity profiles, $v(u)$, as a function of u for different values of δ and the same lower boundary condition. This figure shows (left panel) fast solutions in light grey and δ-slow solutions in coloured lines. The right panel shows only δ-slow solutions. The locus curve of singular points (see Figure 10) is located almost at a constant value of u. Then, the location of the singular point is almost independent of δ, similar to the behaviour of Ω-slow solutions.

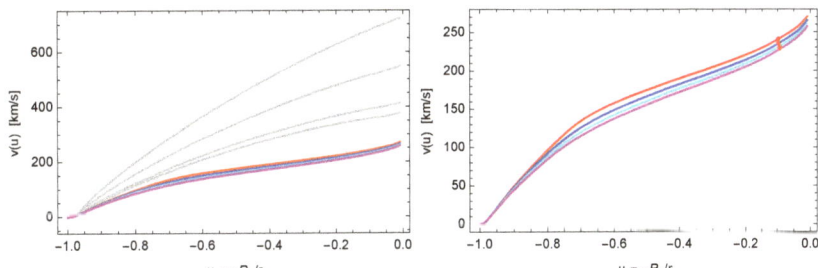

Figure 11. Velocity profiles, $v(u)$, for different values of the line force parameter δ. (**Left panel**): fast solutions are plotted in grey lines for $\delta = 0.0, 0.1, 0.2, 0.24$, while δ-slow solutions are in coloured lines. The red line corresponds to $\delta = 0.3$, the blue line to $\delta = 0.31$, the cyan line to $\delta = 0.32$ and the magenta line to $\delta = 0.33$. The (**right panel**) shows only δ-slow solutions, and the location of the singular points for each solution is shown with a red dot.

5.4. The β-Law Approximation

In the work of Pauldrach et al. [24], after obtaining the numerical solution of the EoM, they assumed a power law approximation to describe the velocity profile only as a function of the radial coordinate r. This approximation is known as the *β-law* approximation and has the following expression:

$$v(r) = v_\infty \left(1 - R_*/r\right)^\beta, \qquad (62)$$
$$= v_\infty \left(1 + u\right)^\beta, \qquad (63)$$

where v_∞ is the terminal velocity and the value of β determines the shape of the velocity profile. In the context of stellar wind diagnostics, these parameters are considered fit parameters that must be determined through spectral line fitting. Usually, the range used for the β parameter is $0.7 \lesssim \beta \lesssim 4$ [11].

Figure 12 shows the velocity profile of the fast solution for the stellar and line force parameters given at the beginning of Section 5.1, together with six different values of β for a β-law velocity profile. From this figure, we can conclude that the fast solution cannot be described properly by a β-law with $\beta > 1.2$.

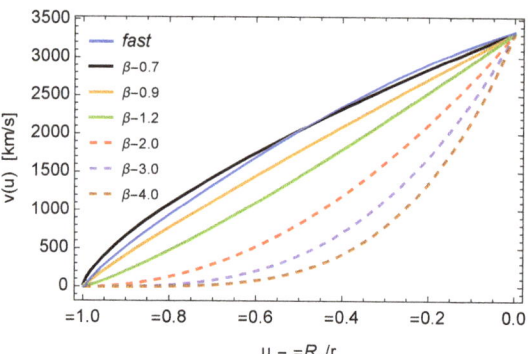

Figure 12. Fast solution velocity profile (solid blue), $v(u)$ vs. u. Six different β-law velocity profiles are also plotted. It is clearly seen that the β-law approximation is a good one for $0.7 \lesssim \beta \lesssim 1.2$. See text for details.

On the other hand, Figure 13 shows the velocity profile for the δ-slow solution for the stellar and line force parameters given at the beginning of Section 5.3, where $\delta = 0.32$. Furthermore, the same β-law profiles of Figure 12 are used, with the proper values of v_∞ for this solution. The same is plotted in Figure 14 for the Ω-slow solution with $\Omega = 0.8$. In the case of the δ-slow solution, the β-law profile cannot fit the m-CAK hydrodynamical solution for any $\beta > 0.7$, while for values around $\beta = 0.7$, the profiles can be considered similar. Finally, from Figure 14, we can definitely conclude that Ω-slow solutions cannot be described properly by any β-law profile.

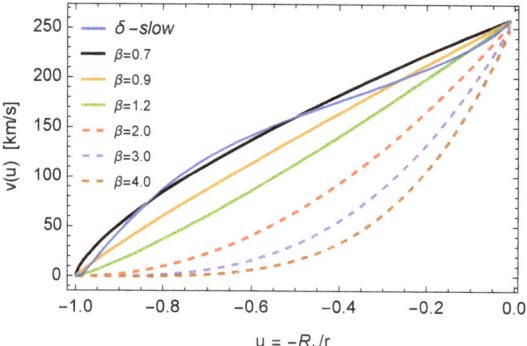

Figure 13. δ-slow solution velocity profile (solid blue), $v(u)$ vs. u. Six different β-law velocity profiles are also plotted. For values around $\beta = 0.7$, the profiles can be considered similar, but it can be clearly concluded that for $\beta > 0.7$, the β-law profile cannot fit the m-CAK hydrodynamical δ-slow solution.

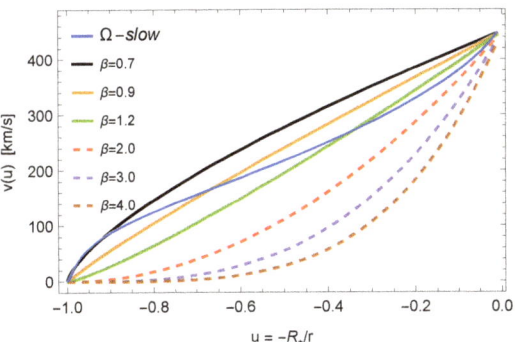

Figure 14. Ω-slow solution velocity profile (solid blue), $v(u)$ vs. u. Six different β-law velocity profiles are also plotted. It can be clearly concluded that no β-law profile can fit the m-CAK hydrodynamical Ω-slow solution.

6. Analytical Wind Solutions

Using an analytical expression to represent the radiation force and solve the equation of motion analytically offers numerous advantages over the numerical integration of the EoM. These formulae can be used in all cases where good first estimates are needed; for example, it gives the advantage of solving the radiative transfer problem for moving media in an easier way.

Kudritzki et al. [28] were some of the first to develop analytical solutions for radiation-driven winds considering the finite disc correction factor in the line force. Based on these solutions, they provided approximated analytical expressions for the terminal velocity and the mass loss rate in terms of the stellar parameters (L, M_* and R_*), the line force parameters (k, α and δ) and the free parameter β from the β-law (they adopted $\beta = 1.0$ for their results).

Other authors have tried to simplify the complicated numerical treatment from the theory. Villata [58], with the purpose of simplifying the integration of the EoM, derived an approximated expression for the line acceleration term, which depends only on the radial coordinate. Müller and Vink [59] presented an analytical expression for the velocity field using a parameterized description for the line acceleration that (as in Villata [58]) also depends on the radial coordinate. These line acceleration expressions do not depend on the velocity or the velocity gradient, as the standard m-CAK description does. Araya et al. [60] proposed to achieve a complete analytical description of the 1D hydrodynamical solution for radiation-driven winds in the fast regime by gathering the advantages of both previous approximations (the use of known parameters and the LambertW function). In addition, Araya et al. [61] developed an analytical solution for the δ-slow regime. To date, no approximation using the LambertW function has been performed for the Ω-slow regime, we expect to do this in the future.

In the following sections, we describe the results and methodology used to analytically solve the equation of motion for fast and δ-slow regimes.

6.1. Solution of the Dimensionless Equation of Motion

In this section, we recapitulated the methodology described by Müller and Vink [59] to obtain the dimensionless equation of motion.

In a dimensionless form, the momentum equation can be expressed as follows,

$$\hat{v}\frac{d\hat{v}}{d\hat{r}} = -\frac{\hat{v}_{\text{crit}}^2}{\hat{r}^2} + \hat{g}^{\text{line}} - \frac{1}{\rho}\frac{d\rho}{d\hat{r}}, \qquad (64)$$

where $\hat{r}$ is a dimensionless radial coordinate $\hat{r} = r/R_*$ and the dimensionless velocities (in units of the isothermal sound speed a) are:

$$\hat{v} = \frac{v}{a} \quad \text{and} \quad \hat{v}_{\text{crit}} = \frac{v_{\text{esc}}}{a\sqrt{2}}, \tag{65}$$

where v_{crit} is the rotational break-up velocity in the case of a rotating star. It is usually determined by dividing the effective escape velocity, v_{esc}, by a factor of $\sqrt{2}$. Similarly, a dimensionless line acceleration can be written as follows:

$$\hat{g}^{\text{line}} = \frac{R_*}{a^2} g^{\text{line}}. \tag{66}$$

Using the continuity equation and the equation of state of an ideal gas, the dimensionless equation of motion reads as follows:

$$\left(\hat{v} - \frac{1}{\hat{v}}\right) \frac{d\hat{v}}{d\hat{r}} = -\frac{\hat{v}_{\text{crit}}^2}{\hat{r}^2} + \frac{2}{\hat{r}} + \hat{g}^{\text{line}}(\hat{r}). \tag{67}$$

Lastly, a 1D velocity profile is derived analytically in terms of the LambertW function [62–64]. See Müller and Vink [59] for a detailed description of the methodology used to arrive at this solution. This analytical solution is expressed as follows:

$$\hat{v}(\hat{r}) = \sqrt{-W_j(x(\hat{r}))}, \tag{68}$$

where

$$x(\hat{r}) = -\left(\frac{\hat{r}_c}{\hat{r}}\right)^4 \exp\left[-2\hat{v}_{\text{crit}}^2 \left(\frac{1}{\hat{r}} - \frac{1}{\hat{r}_c}\right) - 2\int_{\hat{r}_c}^{\hat{r}} \hat{g}^{\text{line}}(\hat{r}) d\hat{r} - 1\right]. \tag{69}$$

In the last equation appears the parameter $\hat{r}_c$, which represents the position of the sonic (or critical) point.

Furthermore, the LambertW function (W_j) has only two real branches, indicated by the sub-index j, where $j = 0$ or -1. These two branches coincide at the sonic point, $\hat{r}_c$, i.e.,

$$j = \begin{cases} 0 & \text{for } 1 \leq \hat{r} \leq \hat{r}_c \\ -1 & \text{for } \hat{r} > \hat{r}_c \end{cases} \tag{70}$$

A regularity condition must be imposed, as in the m-CAK case, since the LHS of the equation of motion (Equation (67)) vanishes at $\hat{v} = 1$ (singularity condition in the CAK formalism). This is equivalent to ensuring that the RHS of Equation (67) vanishes at $\hat{r} = \hat{r}_c$. Therefore,

$$-\frac{\hat{v}_{\text{crit}}^2}{\hat{r}_c^2} + \frac{2}{\hat{r}_c} + \hat{g}^{\text{line}}(\hat{r}_c) = 0, \tag{71}$$

and $\hat{r}_c$ is obtained by solving this last equation. Finally, the velocity profile is derived using the function $x(\hat{r})$, Equation (69), in Equation (68).

6.2. The Fast Regime

Kudritzki et al. [28] analytical study of radiation-driven stellar winds allowed Villata [58] to derive an approximate expression for the line acceleration term. In this case, the line acceleration is only dependent on the radial coordinate, and it reads as follows:

$$g_{V92}^{\text{line}}(\hat{r}) = \frac{G M_* (1 - \Gamma_e)}{R_*^2 \hat{r}^2} A(\alpha, \beta, \delta) \left(1 - \frac{1}{\hat{r}}\right)^{\alpha(2.2\beta - 1)}, \tag{72}$$

with

$$A(\alpha, \beta, \delta) = \frac{(1.76\beta)^\alpha}{1-\alpha} \left[10^{-\delta}(1+\alpha)\right]^{1/(1-\alpha)} \left[1 + \left(\frac{2}{\alpha}\left\{1 - \left[10^{-\delta}(1+\alpha)\right]^{1/(\alpha-1)}\right\}\right)^{1/2}\right]^\alpha. \tag{73}$$

According to Kudritzki et al. [65], the exponent β can be calculated based on the force multiplier parameters and the escape velocity, v_{esc}:

$$\beta = 0.95\,\alpha + \frac{0.008}{\delta} + \frac{0.032\,v_{esc}}{500}, \qquad (74)$$

with v_{esc} in km/s.

Then, using Equation (72) in its dimensionless form (Equation (66)) and inserting it into the dimensionless equation of motion (Equation (67)) yields:

$$\left(\hat{v} - \frac{1}{\hat{v}}\right)\frac{d\hat{v}}{d\hat{r}} = -\frac{\hat{v}_{crit}^2}{\hat{r}^2} + \frac{2}{\hat{r}} + \frac{1}{a^2}\frac{GM_*(1-\Gamma_e)}{R_*\hat{r}^2}\,A(\alpha,\beta,\delta)\left(1 - \frac{1}{\hat{r}}\right)^{\gamma_v}, \qquad (75)$$

with $\gamma_v = \alpha\,(2.2\,\beta - 1)$.

Based on Villata [58] approximation of the line acceleration, this differential equation can be viewed as a solar-like differential equation of motion. Hence, the singular point is the sonic point. Additionally, Villata [58] equation of motion does not have eigenvalues, which means it does not depend explicitly on the star's mass loss rate.

Using a standard numerical integration method, Villata [58] solved the equation of motion and obtained terminal velocities that were within 3–4 % of those computed by Pauldrach et al. [24] and Kudritzki et al. [65].

A parametrized description of the line acceleration was presented years later by Müller and Vink [59] that is dependent on the radial coordinate (similar to Villata [58]). The line acceleration in Müller and Vink [59] was determined using Monte Carlo multi-line radiative transfer calculations [66,67] and a β law. Following this, the line acceleration was fitted using the following formula:

$$\hat{g}_{MV08}^{line}(\hat{r}) = \frac{\hat{g}_0}{\hat{r}^{1+\delta_1}}\left(1 - \frac{\hat{r}_0}{\hat{r}^{\delta_1}}\right)^{\gamma}, \qquad (76)$$

where $\hat{g}_0$, δ_1, $\hat{r}_0$ and γ are the acceleration line parameters.

Then, the solution of the equation of motion, based on their methodology and line acceleration, is:

$$\hat{v}(\hat{r}) = \sqrt{-W_j(x(\hat{r}))}, \qquad (77)$$

with

$$x(\hat{r}) = -\left(\frac{\hat{r}_c}{\hat{r}}\right)^4 \exp\left[-2\hat{v}_{crit}^2\left(\frac{1}{\hat{r}} - \frac{1}{\hat{r}_c}\right)\right.$$
$$\left. -\frac{2}{\hat{r}_0}\frac{\hat{g}_0}{\delta_1(1+\gamma)}\left(\left(1 - \frac{\hat{r}_0}{\hat{r}^{\delta_1}}\right)^{1+\gamma} - \left(1 - \frac{\hat{r}_0}{\hat{r}_c^{\delta_1}}\right)^{1+\gamma}\right) - 1\right]. \qquad (78)$$

As a result of the approximations described above, the velocity profile can be represented analytically, greatly simplifying the solution of the equation of motion.

Furthermore, it is relevant to note that each of the mentioned approximations has its own advantages and disadvantages. Even though Villata's approximation of the radiation force is general and can directly be applied to describe any massive star's wind, the momentum equation still needs to be solved numerically. With Müller and Vink [59] approximation, the equation of motion can be analytically solved based on the $\hat{g}_0$, δ_1, $\hat{r}_0$f and γ parameters of the star. Nevertheless, it is still necessary to perform Monte Carlo multi-line radiative transfer calculations in order to determine these parameters.

This methodology to solve the equation of motion was revisited by Araya et al. [60] in order to derive a fully analytical expression combining Villata [58] expression of the equation of motion with the methodology developed by Müller and Vink [59].

This analytical solution is,

$$\hat{v}(\hat{r}) = \sqrt{-W_j(x(\hat{r}))}, \quad (79)$$

with

$$x(\hat{r}) = -\left(\frac{\hat{r}_c}{\hat{r}}\right)^4 \exp\left[-2\hat{v}_{\text{crit}}^2\left(\frac{1}{\hat{r}} - \frac{1}{\hat{r}_c}\right) - 2\left(I_{gV92}^{\text{line}}(\hat{r}) - I_{gV92}^{\text{line}}(\hat{r}_c)\right) - 1\right], \quad (80)$$

where

$$I_{gV92}^{\text{line}}(\hat{r}) = \left(10^{-\delta}(1+\alpha)\right)^{\frac{1}{1-\alpha}} \left(1 + \sqrt{2}\sqrt{-\frac{(10^{-\delta}(1+\alpha)-1)^{\frac{1}{\alpha-1}}}{\alpha}}\right)^{\alpha}$$

$$\times (1.76\,\beta)^{\alpha}\, G\, M_* \left(\frac{\hat{r}-1}{\hat{r}}\right)^{1+\gamma_v} \frac{\Gamma - 1}{(a^2[\alpha-1](1+\gamma_v)\,R_*)}. \quad (81)$$

As was mentioned in the previous section, $\hat{r}_c$ can be obtained numerically, making the RHS of Equation (75) equal to zero. In order to obtain the terminal velocity in a simpler way, we can use the average value of $\hat{r}_c$ ($\bar{\hat{r}}_c = 1.0026$) obtained by Araya et al. [60]. Note that this value can be used only in the supersonic region.

Equation (79) has the advantage that it is based not only on the LambertW function, but also on stellar parameters and the line force. For a wide range of spectral types, stellar and force multiplier parameters are given (see, e.g., [8,10,22,24,41,44]).

By comparing the analytical solution to the 1D hydrodynamic code HYDWIND, the accuracy of the analytical solution can be tested. Figure 15 compares the results obtained with our analytical approximation to those obtained with the hydrodynamics for four stars taken from Araya et al. [60]. Both solutions have similar behaviour. However, as shown by Araya et al. [60], the analytical approximation close to the stellar surface (subsonic region) is not good enough. In the same way, Figure 16 compares the numerical and analytical velocity profiles near to the stellar surface for ϵ Ori.

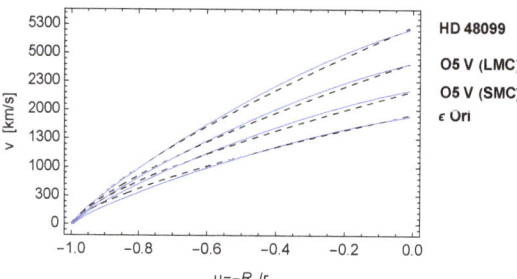

Figure 15. Velocity profiles as a function of the inverse radial coordinate $u = -R_*/r = -1/\hat{r}$ for four models. The hydrodynamic results from HYDWIND are shown in solid blue lines and the analytical solutions are shown by dashed lines. The stellar and line force parameters for the models are given in Araya et al. [60].

There is a limitation to this analytical expression when the line force parameter δ exceeds about 0.3. This is due to the complexity of a term in the proposed line acceleration expression. To obtain an expression with real values, high values of δ would require high values of α. However, such kind of α values would be totally unphysical ($\alpha > 1$). As an illustration of the dependence of this expression on the parameters α and δ, Figure 17 shows the domain of the complex and real regions when this expression is evaluated to the given line acceleration term.

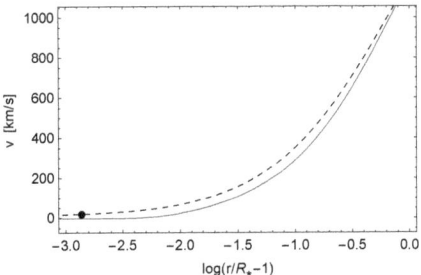

Figure 16. Velocity profiles of ϵ Ori as a function of $\log(r/R_* - 1)$ in a region near to the stellar surface. The solid blue line shows the numerical hydrodynamic result and the analytical solution is shown by a dashed line. The dot symbol indicates the position of the sonic (or critical) point. The difference between both curves is around one thermal speed.

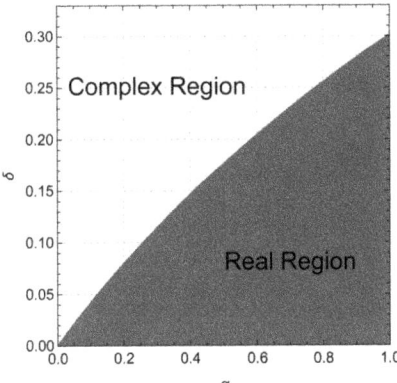

Figure 17. Real and complex regions where the line acceleration expression given by Villata [58] can be found. These regions are delimited by the values of the line force parameters α and δ.

6.3. The δ-Slow Regime

Considering the results obtained when using an approximate description of the wind velocity for the δ-slow case, Araya et al. [61] modified the function of the line acceleration given by Müller and Vink [59] to better describe of δ-slow wind.

As a result, the proposed line acceleration is:

$$\hat{g}^{\text{line}}(\hat{r}) = \frac{\hat{g}_0}{\hat{r}^{1+\delta_1}} \left(1 - \frac{1}{\hat{r}^{\delta_2}}\right)^{\gamma}, \tag{82}$$

where $\hat{g}_0$, δ_1, δ_2 and γ are the new set of line acceleration parameters.

It is notable that the δ_2 parameter, which has been incorporated into this new expression, provides a much better agreement with the numerical line acceleration obtained from the m-CAK model in the δ-slow regime compared with the one from Müller and Vink [59].

Based on this new definition of the radiation force, the new dimensionless equation of motion reads:

$$\left(\hat{v} - \frac{1}{\hat{v}}\right)\frac{d\hat{v}}{d\hat{r}} = -\frac{\hat{v}_{\text{crit}}^2}{\hat{r}^2} + \frac{2}{\hat{r}} + \frac{\hat{g}_0}{\hat{r}^{1+\delta_1}}\left(1 - \frac{1}{\hat{r}^{\delta_2}}\right)^{\gamma}. \tag{83}$$

The LambertW function is used to solve the equation of motion, Equation (83), following the same methodology developed by Müller and Vink [59],

$$\hat{v}(\hat{r}) = \sqrt{-W_j(x(\hat{r}))}, \tag{84}$$

with

$$x(\hat{r}) = -\left(\frac{\hat{r}_c}{\hat{r}}\right)^4 \exp\left[-2\hat{v}_{\text{crit}}^2\left(\frac{1}{\hat{r}} - \frac{1}{\hat{r}_c}\right) - 2\left(I_{\hat{g}^{\text{line}}}(\hat{r}) - I_{\hat{g}^{\text{line}}}(\hat{r}_c)\right) - 1\right],$$

where

$$I_{\hat{g}^{\text{line}}} = \int \hat{g}^{\text{line}}(\hat{r})d\hat{r} = -\frac{\hat{g}_0\,\hat{r}^{-\delta_1}\,{}_2F_1\left[-\gamma,\frac{\delta_1}{\delta_2},1+\frac{\delta_1}{\delta_2},\hat{r}^{-\delta_2}\right]}{\delta_1}, \tag{85}$$

and $_2F_1$ is the Gauss Hypergeometric function. The critical (or sonic) point, $\hat{r}_c$ is obtained numerically, making the RHS of Equation (83) equal to zero.

Ultimately, this expression for the velocity profile is in quite satisfactory agreement with the numerical solution from HYDWIND.

As described in Araya et al. [60], a relationship was established between the Müller and Vink [59] line force parameters ($\hat{g}_0$, δ_1, $\hat{r}_0$ and γ) and the stellar and m-CAK line force parameters. In addition to being easy to use, this relationship provides a straightforward and versatile method of calculating velocity profiles analytically for a wide range of spectral types, since both stellar and m-CAK line force parameters are available (see [8,10,22,24,41,44]).

A similar relationship can be derived for the δ-slow regime using m-CAK hydrodynamic models, that is, creating a grid of HYDWIND models for δ-slow solutions. These models are then analysed using a multivariate multiple regression analysis (MMR [68,69]).

To develop this hydrodynamic grid, the stellar radius is calculated from M_{bol} for each pair of stellar parameters (T_{eff} and $\log g$) by using the flux weighted gravity–luminosity relationship [70,71]. Additionally, a total of 20 stellar radius values were added (ranging from 5 $R_\odot$ to 100 $R_\odot$ in steps of 5 $R_\odot$). The surface gravities are in the range of $\log g = 2.7$ down to about 90% of Eddington's limit in steps of 0.15 dex. The effective temperatures are between 9 000 K and 19, 500 K in steps of 500 K. The range of this grid has been chosen to cover the region of the T_{eff}-$\log g$ diagram that contains B- and A-type supergiants. In Table 1, the m-CAK line force parameters for each set of (T_{eff}, $\log g$) values are listed. For the purpose of obtaining δ-slow solutions, only high values of δ are considered. For the T_{eff}-$\log g$ plane (see Figure 18), we show in blue dots all converged models.

Table 1. m-CAK line force parameter ranges for the grid of models.

Parameter	Range
α	0.45–0.69 (step size of 0.02)
k	0.05–1.00 (step size of 0.05)
δ	0.26–0.35 (step size of 0.01)

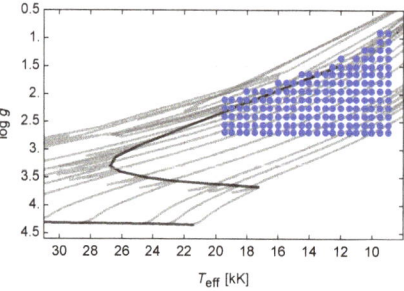

Figure 18. Hydrodynamic models in the T_{eff}-$\log g$ plane. Blue dots represent the converged solutions. Grey solid lines are the evolutionary tracks for stars of $7M_\odot$ to $60M_\odot$ without rotation [72], and black lines represent the zero-age main sequence (ZAMS) and the terminal age main sequence (TAMS).

In order to obtain the new line acceleration parameters ($\hat{g}_0$, δ_1, δ_2 and γ) for each model, the m-CAK line acceleration was fitted, using Least Squares, with the proposed line acceleration expression (Equation (82)). Then, an MMR is applied to the grid of models in order to derive the relationship between the new line acceleration parameters ($\hat{g}_0$, δ_1, δ_2 and γ) and stellar (T_{eff}, $\log g$ and $R_*/R_\odot$) and m-CAK line force parameters (k, α and δ). The estimated parameters are:

$$\hat{g}_0^{0.27} = -4.548 - 1.890 \times 10^{-4} T_{\text{eff}} + \qquad (86)$$
$$4.393 \log g + 3.026 \times 10^{-2} R_*/R_\odot -$$
$$4.802 \times 10^{-3} k + 3.781 \alpha - 3.212 \delta,$$

$$(\delta_1 + 1)^{5.3} = -4.623 - 3.743 \times 10^{-4} T_{\text{eff}} + \qquad (87)$$
$$1.489 \times 10^1 \log g + 1.148 \times 10^{-1} R_*/R_\odot +$$
$$2.415 k + 9.553 \times 10^1 \alpha - 1.320 \times 10^2 \delta,$$

$$\delta_2^{0.45} = 5.359 + 8.262 \times 10^{-5} T_{\text{eff}} - \qquad (88)$$
$$1.327 \log g - 8.327 \times 10^{-3} R_*/R_\odot +$$
$$2.181 \times 10^{-1} k + 9.618 \times 10^{-1} \alpha - 2.296 \delta$$

and

$$(\gamma + 1)^{-3.56} = -1.031 + 7.254 \times 10^{-6} T_{\text{eff}} + \qquad (89)$$
$$2.994 \times 10^{-1} \log g + 3.097 \times 10^{-3} R_*/R_\odot +$$
$$1.836 \times 10^{-1} k - 4.828 \times 10^{-1} \alpha + 1.254 \delta,$$

After the estimated values for each dependent variable ($\hat{g}_0^{0.27}$, $(\delta_1 + 1)^{5.3}$ and $\delta_2^{0.45}$, $(\gamma + 1)^{-3.56}$) are obtained they are transformed into $\hat{g}_0$, δ_1, δ_2 and γ through their respective inverse functions. Finally, we can use these parameters in Equation (84) to calculate the velocity profile.

The velocity profiles obtained via HYDWIND code and the analytical solution are shown in Figure 19 for one model with a δ-slow solution. The model is taken from Curé et al. [27].

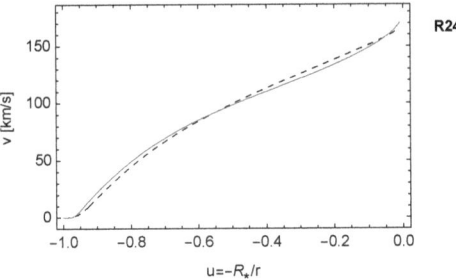

Figure 19. Velocity profiles as a function of the inverse radial coordinate $u = -R_*/r = -1/\hat{r}$ for the model R24 from Curé et al. [27]. The hydrodynamic result from HYDWIND is shown in the solid blue line and the analytical solution is a dashed black line.

Remember, however, that this relationship holds only for δ-slow solutions, especially for values of δ between 0.29 and 0.35. In addition, considering the number of converged models in the grid, the authors recommend using this expression for values of α between 0.45 and 0.55.

Finally, it is important to remark that both analytical solutions for the velocity profile, fast and δ-slow, do depend only on the stellar (T_{eff}, $\log g$ and $R_*/R_\odot$) and m-CAK line force parameters (k, α and δ). Regarding the mass loss rate, Villata [58] proposed an expression to obtain it (in terms of the stellar and m-CAK parameters) following the approximations given by Kudritzki et al. [28]. Furthermore, Araya et al. [61], in the appendix, proposed a method to obtain the mass loss based on Curé [50].

7. Summary and Discussion

Observations in recent decades have shown that the basic wind parameters behave as predicted by theory. This fundamental agreement between observations and theory provides strong evidence that the winds from massive stars are driven by radiation pressure. This has given m-CAK theory a well-established status in the massive star community. However, several issues are contentious and still unclear, such as the calibration of the wind momentum–luminosity relationship (WLR) [73], discs of Be stars, wind parameter determination and the applicability of the so-called slow wind solutions, among others. All these issues are the focus of massive stars research.

This review is focused on the theoretical and numerical research of wind hydrodynamics of massive stars based on the m-CAK theory, with particular emphasis on its topology and hydrodynamic solutions.

We presented a topological analysis of the one-dimensional m-CAK hydrodynamic model and its three known hydrodynamic solutions, the fast, Ω-slow and δ-slow solutions. From a topological point of view, slow solutions are obtained from a new branch of solutions with a locus of singular points far from the stellar surface, unlike fast solutions with a family of singular points near the stellar surface.

We continued analyzing the dependence of the line force parameters (k, α and δ) with the wind parameters (mass loss rate and terminal velocity) in order to understand the complex non-linear dependence between these parameters. In the case of α, there is an increase in both wind parameters as this parameter increases. This behaviour is similar to the k parameter, but the dependence is very slight for terminal velocity, while the mass loss rate has a significant impact. For the δ parameter, the terminal velocity has a decreasing behaviour when this parameter increases, while the mass loss rate can have a decreasing or increasing behaviour, which depends on the parameter k. When k is low, mass loss rates decrease while δ increases, whereas when k is large, the opposite occurs.

In addition, we compared the β-law with the hydrodynamic solutions. We concluded that the fast solution could not be adequately described by a β-law with $\beta > 1.2$, while the δ-slow solution cannot be described by any β-law.

Furthermore, we presented two analytical expressions for the solution of line-driven winds in terms of the stellar and line force parameters. The expressions are addressed to obtain the fast and δ-slow velocity regimes in a simple way. Both solutions are based on the LambertW function and an approximative expression for the wind line acceleration as a function of the radial distance. The importance of an analytical solution lies in its simplicity in studying the properties of the wind instead of solving complex hydrodynamic equations. In addition, these analytical expressions can be used in radiative transfer or stellar evolution codes (see, e.g., [74]).

Concerning the applicability of the slows solutions, in the case of the Ω-slow solution, their behaviour suggests that it can play a paramount role in the ejection of material to the equatorial circumstellar environment of Be and B[e] stars. Be stars are a unique set of massive stars whose main distinguishing characteristics are a rapid rotation and the presence of a dense, gaseous circumstellar disc orbiting in a quasi-Keplerian fashion. There is a long-standing problem in understanding the formation of discs in Be stars; this is one of the major areas of ongoing research in Be stars. The gaseous discs are not remnants of the objects' protostellar environments, nor are they formed through the accretion of material [54,75]. On the contrary, the equatorial gas consists of a decretion disc formed from a material originating from the central star.

As was stated above, attempts to solve this problem have been made without much success, for example, the link between the line-driven winds and these discs, called the wind-compressed discs [76], but the work of Owocki et al. [77] was the first to show this is not a viable mechanism for rapidly rotating stars due to the non-radial line force components. The most accepted model to successfully reproduce many Be star observations is the viscous decretion disc (VDD) model, developed by Lee et al. [78] and examined by Okazaki [79], Bjorkman and Carciofi [25], Krtička et al. [80] and Curé et al. [81]. Currently, how that material is ejected into the equatorial plane and how sufficient angular momentum is transferred to the disc to maintain quasi-Keplerian rotation are among the primary unresolved questions currently driving classical Be star research.

Araya et al. [55] studied the Ω-slow wind solution and its relation with the discs of Be stars. Overall, this work is an extension to the study performed by Silaj et al. [82], where they precisely investigated if the density distribution provided by the Ω-slow wind solution could adequately describe the physical conditions to form a dense disc in Keplerian rotation via angular momentum transfer. They considered a two-component wind model, i.e., a fast, thin wind in the polar latitudes and a Ω-slow, dense wind in the equatorial regions. Based on the equatorial density distributions, Hα line profiles were generated and compared with an ad hoc emission profile, which agreed with the observations. In addition, their calculations assumed three different scenarios related to the shape (oblate correction factor) and the star's brightness (gravity darkening). Finally, they found that under certain conditions (related to the line force parameter of the wind), a significant Hα line profile could be produced, and it may be that the line-driven winds through the Ω-slow solution have an essential role in the disc formation of Be stars.

In addition, Araya et al. [51] studied the zone where the classical m-CAK fast solution ceases to exist, and the Ω-slow solution emerges at rapid rotational speeds. This study used two hydrodynamic codes with time-dependent and stationary approaches. They found that both solutions can co-exist in this transition region, which depends exclusively on the initial conditions. In addition, they performed base density wind perturbations to test the stability of the solution within the co-existence region. A switch of the solution was found under certain perturbation conditions. The results are explained by a possible role in the ejection of extra material into the equatorial plane of pulsation modes, where the Ω-slow solution can play an important role.

A current weakness of this m-CAK model is that it does not consider the role of viscosity and its influence on angular momentum transport. This mechanism might explain the formation of a Keplerian disc.

On the other hand, the δ-slow solution is promising to explain the discrepancies of the wind parameters between observations and theory in late B- and A-type supergiant stars. According to the findings of Venero et al. [53], these suggest that the terminal velocity of early and mid-B supergiants agrees with the results seen from fast outflowing winds. In contrast, the results obtained for late B supergiants and, mainly, those obtained for early A supergiants, agree with the results achieved for δ-slow stationary outflowing wind regimes. Then, the δ-slow solution enables us to describe the global features of the wind quite well, such as mass loss rates and terminal velocities of moderately and slowly rotating B supergiants.

Conversely, Venero et al. [53] stated that the δ-slow solution seems not to be present in stars with $T_{\text{eff}} > 21,000$ K. This restricts the possibility of a switch between fast and slow regimes at such temperatures. Consequently, this would be a physical explanation for why an empirical bi-stability jump can be observed around 21,000 K in B supergiants [83]. From a theoretical perspective, a velocity jump has also been found using Monte Carlo modelling and the co-moving frame method (see, e.g., [13,84,85]).

In addition, it is generally accepted that most O and early B-type stars can be adequately modelled with a β velocity law with $0.7 \lesssim \beta \lesssim 1.2$. However, supergiants A and B exhibit β values that tend towards higher values, often $\beta \geq 2$ (see, e.g., [86–90]). Venero et al. 2023 (in preparation) propose that δ-slow solutions might explain these winds. They show that the δ-slow regime could adequately fit the Hα line profile of B supergiants with the same accuracy as that obtained using a β-law with $\beta \geq 2$, but now with a hydrodynamic explanation of the velocity profile used.

The investigations carried out in the latest works inspired us to go deeper into the possible role of slow wind solutions with respect to the unresolved questions related to massive stars. In view of our results, we are encouraged to develop this line of research further. In the case of the Ω-slow solution and its link to Be stars, or possibly to B[e] stars, it would require 2D/3D models for a better understanding to take into account non-radial forces, the effects of stellar distortion and gravity darkening. These considerations could change, in turn, the nature of the Ω-slow solution or the behaviour regarding the co-existence of solutions and a switch between them.

The δ-slow solution could play an essential role in understanding the winds of B- and A-type supergiants. Moreover, this solution is expected to solve the disagreement between the observations and theory for these stars and, in this way, calibrate the wind momentum–luminosity relationship.

As we mentioned previously, in the standard procedure for finding stellar and wind parameters, the β-law (β and v_∞) and the mass loss rate ($\dot{M}$) are three *'free'* input parameters in radiative transfer codes, comparing synthetic spectra with the observed spectra of a star. The β law comes from an approximation of the fast wind solutions, and the values of β should be in a restricted interval. To improve the hydrodynamic approximation used in this standard procedure, we have developed two hydrodynamic procedures to derive stellar and wind parameters:

- The self-consistent CAK procedure [44], based on the m-CAK model. Here, we iteratively calculate the line force parameters using the atomic line database from CMFGEN, coupled with the m-CAK hydrodynamic until convergence. We obtain the line force parameters and, therefore, the velocity profile and the mass loss rate. Thus, none of the input parameters are *'free'*, but self-consistently calculated.
- The LambertW procedure [45]. In this procedure, we start using a β-law and a value for $\dot{M}$ in CMFGEN. After convergence, we calculate the line acceleration as a function of r, and, using the LambertW function, we obtain a new velocity profile. This is inserted in CMFGEN and the cycle is repeated until convergence. In this LambertW procedure, the only input parameter is the mass loss rate.

We expect that these two alternatives, which reduce the number of input parameters, will in the future have a significant impact on the standard procedures for obtaining stellar and wind parameters of massive stars.

Author Contributions: M.C. and I.A. participated in all aspects related to this work. All authors have read and agreed to the published version of the manuscript.

Funding: We are grateful for the support from FONDECYT projects 1190485 and 1230131. I.A. also thanks the support from FONDECYT project 11190147. This project has received funding from the European Union's Framework Programme for Research and Innovation Horizon 2020 (2014–2020) under the Marie Skłodowska-Curie Grant Agreement No. 823734.

Institutional Review Board Statement: Not applicable.

Informed Consent Statement: Not applicable.

Data Availability Statement: Not applicable.

Acknowledgments: We are grateful for the suggestions and comments from reviewers to improve this work. We would like to thank the continuous support from Centro de Astrofísica de Valparaíso and our colleagues and students from the massive stars group at Universidad de Valparaíso, especially Catalina Arcos. We also thank our long-standing colleagues from other institutes who have contributed to our understanding of the field, especially Lydia Cidale, Diego Rial and Roberto Venero. We also wish to acknowledge the support received from our respective Universities, Universidad de Valparaíso and Universidad Mayor, to continue our research.

Conflicts of Interest: The authors declare no conflict of interest.

Abbreviations

The following abbreviations are used in this manuscript:

ZAMS	Zero-Age Main Sequence
EoM	Equation of Motion
CAK	Castor et al. [16]
RHS	Right Hand Side
MMR	Multivariate Multiple Regression
WLR	Wind momentum–Luminosity Relationship
VDD	Viscous Decretion Disc

Appendix A

In their original work, CAK discussed the stellar point approximation of their model and properly estimated the influence of the disc correction factor (CF) on the wind dynamics, i.e., reducing the line force by about 40% at the stellar surface.

The definition of the CF is:

$$CF = \frac{2}{1-\mu_*^2} \int_{\mu_*}^{1} \left(\frac{(1-\mu^2)v/r + \mu^2 v'}{v'} \right)^{\alpha} \mu \, d\mu \tag{A1}$$

where $v' = dv/dr$ and $\mu_*^2 = 1 - (R_*/r)^2$.

Integrating (A1) and changing the variables from $r \to u = -R_*/r$ and $v \to w = v/a$, where a is the thermal speed, the finite disc correction factor transforms to:

$$CF(u, w/w') = \frac{1}{1-\alpha} \frac{1}{u^2} \frac{1}{(1 + \frac{1}{u}\frac{w}{w'})} \left[1 - \left(1 - u^2 - u\frac{w}{w'}\right)^{(1+\alpha)} \right], \tag{A2}$$

where $w' = dw/dr$. Due to the fact that CF depends on u and the ratio $Z = w/w'$, we can define λ as:

$$\lambda = (u + Z)\, u. \tag{A3}$$

Therefore, $CF(u, Z)$ is re-written as:

$$CF(\lambda) = \frac{1}{(1-\alpha)} \frac{1}{\lambda} \left[1 - (1-\lambda)^{(1+\alpha)} \right]. \tag{A4}$$

The partial derivatives of CF with respect to u, w, w' are then related to $\partial CF/\partial \lambda$ via the chain rule, namely:

$$\frac{\partial CF}{\partial u} = \frac{\partial CF}{\partial \lambda} \times \frac{\partial \lambda}{\partial u}. \tag{A5}$$

$$\frac{\partial CF}{\partial w} = \frac{\partial CF}{\partial \lambda} \times \frac{\partial \lambda}{\partial w}. \tag{A6}$$

$$\frac{\partial CF}{\partial w'} = \frac{\partial CF}{\partial \lambda} \times \frac{\partial \lambda}{\partial w'}. \tag{A7}$$

The function $e(\lambda) = \partial CF/\partial \lambda$ is therefore:

$$e(\lambda) = \frac{(1-\lambda)^{\alpha} - CF(\lambda)}{\lambda}. \tag{A8}$$

Then, (A5)–(A7) are related to (A8) by:

$$e(\lambda) = \frac{1}{2u+Z} \times \frac{\partial CF}{\partial u} = \frac{w'}{u} \times \frac{\partial CF}{\partial w} = -\frac{w'}{uZ} \times \frac{\partial CF}{\partial w'}. \tag{A9}$$

Approximating Z by a β-field ($Z = (1+u)/\beta$), we obtain CF and e *only* as functions of u.

Appendix B

Here, the basic steps toward Equations (52)–(54) are outlined. The reader should keep in mind the original derivation by CAK.

The partial derivatives of $F(u, w, w')$ (Equation (52)), with respect to u, w and w' are:

$$\frac{\partial F}{\partial u} = -\frac{2}{u^2} + a_{rot}^2 - C'\left(\frac{\partial CF}{\partial u} g + CF \frac{\partial g}{\partial u}\right) w^{-\delta} (w\,w')^{\alpha} \tag{A10}$$

$$\frac{\partial F}{\partial w} = \left(1 + \frac{1}{w^2}\right) w' - C'\left(\frac{\partial CF}{\partial w} + \alpha \frac{CF}{w} - \delta \frac{CF}{w}\right) g\, w^{-\delta} (w\,w')^{\alpha} \tag{A11}$$

$$\frac{\partial F}{\partial w'} = \left(1 - \frac{1}{w^2}\right) w - C'\left(\frac{\partial CF}{\partial w'} + \alpha \frac{CF}{w'}\right) g\, w^{-\delta} (w\,w')^{\alpha} \tag{A12}$$

After using the new coordinates $Y = w\,w'$ and $Z = w/w'$, some derivative relation of the correction factor (see Appendix A) and defining:

$$\frac{dg(u)}{du} = g(u) \times h(u) \tag{A13}$$

where

$$h(u) = \delta \left(\frac{2}{u} - \frac{u}{\sqrt{1-u^2}\left(1 - \sqrt{1-u^2}\right)}\right) \tag{A14}$$

the singularity condition ($w'\, \partial F/\partial w' = 0$) now reads:

$$\left(1 - \frac{1}{YZ}\right) Y - C'\, f_2(u, Z)\, g\, Z^{-\delta/2}\, Y^{\alpha - \delta/2} = 0 \tag{A15}$$

where $f_1(u, Z) = CF(u, Z)$,

$$f_2(u, Z) = \alpha\, f_1(u, Z) - u\, Z \times e(u, Z) \tag{A16}$$

and $e(u, Z) = e(\lambda)$ is defined in Appendix A.

The regularity condition ($Z\, dF/du = 0$) now transforms to:

$$\left(1 + \frac{1}{YZ}\right) Y - C'\, f_3(u, Z)\, g(u)\, Z^{-\delta/2}\, Y^{\alpha - \delta/2} = +\frac{2Z}{u^2} - a_{rot}^2 Z \tag{A17}$$

where

$$f_3(u, Z) = (3u + Z)\, Z \times e(u, Z) + f_1(u, Z) \times (h(u)\, Z + \alpha - \delta) \tag{A18}$$

Notes

1. The surface gravity g is given in CGS units, i.e., cm/s^2. The quantity $\log g$ is dimensionless, see Matta et al. [19].
2. Electrons with a velocity distribution function given by the Maxwellian distribution.
3. This equation is for the direct radiation force as no scattering contributions are included within the Sobolev approximation.
4. The subscript s means at the singular point.

References

1. Johnson, M.C. The emission of hydrogen and helium from a star by radiation pressure, and its effect in the ultra-violet continuous spectrum. *Mon. Not. R. Astron. Soc.* **1925**, *85*, 813–825. [CrossRef]
2. Johnson, M.C. The velocities of ions under radiation pressure in a stellar atmosphere, and their effect in the ultraviolet continuous spectrum (Second paper). *Mon. Not. R. Astron. Soc.* **1926**, *86*, 300. [CrossRef]

3. Milne, E.A. On the possibility of the emission of high-speed atoms from the sun and stars. *Mon. Not. R. Astron. Soc.* **1926**, *86*, 459–473. [CrossRef]
4. Chandrasekhar, I.S. The Time of Relaxation of Stellar Systems. *Astrophys. J.* **1941**, *93*, 285. [CrossRef]
5. Chandrasekhar, S. Stochastic Problems in Physics and Astronomy. *Rev. Mod. Phys.* **1943**, *15*, 1–89. [CrossRef]
6. Spitzer, L. *Physics of Fully Ionized Gases*; Courier Corporation: Chelmsford, MA, USA, 1956.
7. Morton, D.C. The Far-Ultraviolet Spectra of Six Stars in Orion. *Astrophys. J.* **1967**, *147*, 1017. [CrossRef]
8. Lamers, H.J.G.L.M.; Cassinelli, J.P. *Introduction to Stellar Winds*; Cambridge University Press: Cambridge, UK, 1999; p. 219.
9. Snow, T.P.J.; Morton, D.C. Copernicus ultraviolet observations of mass-loss effects in O and B stars. *Astrophys. J. Suppl.* **1976**, *32*, 429–465. [CrossRef]
10. Abbott, D.C. The theory of radiatively driven stellar winds. II—The line acceleration. *Astrophys. J.* **1982**, *259*, 282–301. [CrossRef]
11. Kudritzki, R.P.; Puls, J. Winds from Hot Stars. *Annu. Rev. Astron. Astrophys.* **2000**, *38*, 613–666. [CrossRef]
12. Puls, J.; Vink, J.S.; Najarro, F. Mass loss from hot massive stars. *Astron. Astrophys.* **2008**, *16*, 209–325. [CrossRef]
13. Vink, J.S. Theory and Diagnostics of Hot Star Mass Loss. *Annu. Rev. Astron. Astrophys.* **2022**, *60*, 203–246. [CrossRef]
14. Parker, E.N. Dynamics of the Interplanetary Gas and Magnetic Fields. *Astrophys. J.* **1958**, *128*, 664. [CrossRef]
15. Lucy, L.B.; Solomon, P.M. Mass Loss by Hot Stars. *Astrophys. J.* **1970**, *159*, 879. [CrossRef]
16. Castor, J.I.; Abbott, D.C.; Klein, R.I. Radiation-driven winds in Of stars. *Astrophys. J.* **1975**, *195*, 157–174. [CrossRef]
17. Sobolev, V.V. *Moving Envelopes of Stars*; Harvard University Press: Cambridge, MA, USA, 1960.
18. Castor, J.I. On the force associated with absorption of spectral line radiation. *Mon. Not. R. Astron. Soc.* **1974**, *169*, 279–306. [CrossRef]
19. Matta, C.F.; Massa, L.; Gubskaya, A.V.; Knoll, E. Can One Take the Logarithm or the Sine of a Dimensioned Quantity or a Unit? Dimensional Analysis Involving Transcendental Functions. *J. Chem. Educ.* **2011**, *88*, 67–70. [CrossRef]
20. Hillier, D.J. Hot Stars with Winds: The CMFGEN Code. In *From Interacting Binaries to Exoplanets: Essential Modeling Tools*; Richards, M.T., Hubeny, I., Eds.; IAU Symposium; Cambridge University Press: Cambridge, UK, 2012; Volume 282; pp. 229–234. [CrossRef]
21. Pauldrach, A.W.A.; Hoffmann, T.L.; Lennon, M. Radiation-driven winds of hot luminous stars. XIII. A description of NLTE line blocking and blanketing towards realistic models for expanding atmospheres. *Astron. Astrophys.* **2001**, *375*, 161–195. [CrossRef]
22. Lattimer, A.S.; Cranmer, S.R. An Updated Formalism for Line-driven Radiative Acceleration and Implications for Stellar Mass Loss. *Astrophys. J.* **2021**, *910*, 48. [CrossRef]
23. Friend, D.B.; Abbott, D.C. The theory of radiatively driven stellar winds. III—Wind models with finite disk correction and rotation. *Astrophys. J.* **1986**, *311*, 701–707. [CrossRef]
24. Pauldrach, A.; Puls, J.; Kudritzki, R.P. Radiation-driven winds of hot luminous stars—Improvements of the theory and first results. *Astron. Astrophys.* **1986**, *164*, 86–100.
25. Bjorkman, J.E.; Carciofi, A.C. Modeling the Structure of Hot Star Disks. In *The Nature and Evolution of Disks Around Hot Stars*; Ignace, R., Gayley, K.G., Eds.; Astronomical Society of the Pacific Conference Series; Astronomical Society of the Pacific: San Francisco, CA, USA, 2005; Volume 337; p. 75.
26. Curé, M.; Rial, D.F. A new numerical method for solving radiation driven winds from hot stars. *Astron. Nachrichten* **2007**, *328*, 513. [CrossRef]
27. Curé, M.; Cidale, L.; Granada, A. Slow Radiation-driven Wind Solutions of A-type Supergiants. *Astrophys. J.* **2011**, *737*, 18. [CrossRef]
28. Kudritzki, R.P.; Pauldrach, A.; Puls, J.; Abbott, D.C. Radiation-driven winds of hot stars. VI - Analytical solutions for wind models including the finite cone angle effect. *Astron. Astrophys.* **1989**, *219*, 205–218.
29. Hubeny, I.; Mihalas, D. *Theory of Stellar Atmospheres. An Introduction to Astrophysical Non-equilibrium Quantitative Spectroscopic Analysis*; Princeton University Press: Princeton, NJ, USA, 2015.
30. Santolaya-Rey, A.E.; Puls, J.; Herrero, A. Atmospheric NLTE-models for the spectroscopic analysis of luminous blue stars with winds. *Astron. Astrophys.* **1997**, *323*, 488–512.
31. Puls, J.; Urbaneja, M.A.; Venero, R.; Repolust, T.; Springmann, U.; Jokuthy, A.; Mokiem, M.R. Atmospheric NLTE-models for the spectroscopic analysis of blue stars with winds. II. Line-blanketed models. *Astron. Astrophys.* **2005**, *435*, 669–698. [CrossRef]
32. Hillier, D.J. Modeling the extended atmospheres of WN stars. *Astrophys. Journals* **1987**, *63*, 947–964. [CrossRef]
33. Hillier, D.J.; Miller, D.L. The Treatment of Non-LTE Line Blanketing in Spherically Expanding Outflows. *Astrophys. J.* **1998**, *496*, 407–427. [CrossRef]
34. Hillier, D.J.; Lanz, T. CMFGEN: A non-LTE Line-Blanketed Radiative Transfer Code for Modeling Hot Stars with Stellar Winds. In *Spectroscopic Challenges of Photoionized Plasmas*; Ferland, G., Savin, D.W., Eds.; Astronomical Society of the Pacific Conference Series; Astronomical Society of the Pacific: San Francisco, CA, USA, 2001; Volume 247; p. 343.
35. Hamann, W.R.; Schmutz, W. Computed He II spectra for Wolf-Rayet stars—A grid of models. *Astron. Astrophys.* **1987**, *174*, 173–182.
36. Todt, H.; Sander, A.; Hainich, R.; Hamann, W.R.; Quade, M.; Shenar, T. Potsdam Wolf-Rayet model atmosphere grids for WN stars. *Astron. Astrophys.* **2015**, *579*, A75. [CrossRef]
37. Pozdnyakov, L.A.; Sobol, I.M.; Syunyaev, R.A. Comptonization and the shaping of X-ray source spectra—Monte Carlo calculations. *Astrophys. Space Phys. Res.* **1983**, *2*, 189–331.
38. Roman-Duval, J.; Taylor, J.; Fullerton, A.; Fischer, W.; Plesha, R. *The ULLYSES Large Director's Discretionary Program with Hubble: Overview, Status, and Initial Results*; American Astronomical Society: Washington, DC, USA, 2023; Volume 55, p. 223.02.

39. Hawcroft, C.; Sana, H.; Mahy, L.; Sundqvist, J.O.; de Koter, A.; Crowther, P.A.; Bestenlehner, J.M.; Brands, S.A.; David-Uraz, A.; Decin, L.; et al. X-Shooting ULLYSES: Massive stars at low metallicity. III. Terminal wind speeds of ULLYSES massive stars. *arXiv* **2023**, arXiv:2303.12165. [CrossRef]
40. Hillier, D.J.; Miller, D.L. Constraints on the Evolution of Massive Stars through Spectral Analysis. I. The WC5 Star HD 165763. *Astrophys. J.* **1999**, *519*, 354–371. [CrossRef]
41. Noebauer, U.M.; Sim, S.A. Self-consistent modelling of line-driven hot-star winds with Monte Carlo radiation hydrodynamics. *Mon. Not. R. Astron. Soc.* **2015**, *453*, 3120–3134. [CrossRef]
42. Puls, J.; Springmann, U.; Lennon, M. Radiation driven winds of hot luminous stars. XIV. Line statistics and radiative driving. *Astron. Astrophys.* **2000**, *141*, 23–64. [CrossRef]
43. Gayley, K.G. An Improved Line-Strength Parameterization in Hot-Star Winds. *Astrophys. J.* **1995**, *454*, 410. [CrossRef]
44. Gormaz-Matamala, A.C.; Curé, M.; Cidale, L.S.; Venero, R.O.J. Self-consistent Solutions for Line-driven Winds of Hot Massive Stars: The m-CAK Procedure. *Astrophys. J.* **2019**, *873*, 131. [CrossRef]
45. Gormaz-Matamala, A.C.; Curé, M.; Hillier, D.J.; Najarro, F.; Kubátová, B.; Kubát, J. New Hydrodynamic Solutions for Line-driven Winds of Hot Massive Stars Using the Lambert W-function. *Astrophys. J.* **2021**, *920*, 64. [CrossRef]
46. Poniatowski, L.G.; Kee, N.D.; Sundqvist, J.O.; Driessen, F.A.; Moens, N.; Owocki, S.P.; Gayley, K.G.; Decin, L.; de Koter, A.; Sana, H. Method and new tabulations for flux-weighted line opacity and radiation line force in supersonic media. *Astron. Astrophys.* **2022**, *667*, A113. [CrossRef]
47. de Araujo, F.X.; de Freitas Pacheco, J.A. Radiatively driven winds with azimuthal symmetry: Application to be stars. *MNRAS* **1989**, *241*, 543–557. [CrossRef]
48. Friend, D.B.; MacGregor, K.B. Winds from rotating, magnetic, hot stars. I. General model results. *Astrophys. J.* **1984**, *282*, 591–602. [CrossRef]
49. Madura, T.I.; Owocki, S.P.; Feldmeier, A. A Nozzle Analysis of Slow-Acceleration Solutions in One-dimensional Models of Rotating Hot-Star Winds. *Astrophys. J.* **2007**, *660*, 687–698. [CrossRef]
50. Curé, M. The Influence of Rotation in Radiation-driven Wind from Hot Stars: New Solutions and Disk Formation in Be Stars. *Astrophys. J.* **2004**, *614*, 929–941. [CrossRef]
51. Araya, I.; Curé, M.; ud-Doula, A.; Santillán, A.; Cidale, L. Co-existence and switching between fast and Ω-slow wind solutions in rapidly rotating massive stars. *MNRAS* **2018**, *477*, 755–765. [CrossRef]
52. Curé, M. Multi Component Line Driven Dtellar Winds. Ph.D. Thesis, Luwdig-Maximillians Universität, Munich, Germany, 1992.
53. Venero, R.O.J.; Curé, M.; Cidale, L.S.; Araya, I. The Wind of Rotating B Supergiants. I. Domains of Slow and Fast Solution Regimes. *Astrophys. J.* **2016**, *822*, 28. [CrossRef]
54. Rivinius, T.; Carciofi, A.C.; Martayan, C. Classical Be stars. Rapidly rotating B stars with viscous Keplerian decretion disks. *Astron. Astrophys.* **2013**, *21*, 69. [CrossRef]
55. Araya, I.; Jones, C.E.; Curé, M.; Silaj, J.; Cidale, L.; Granada, A.; Jiménez, A. Ω-slow Solutions and Be Star Disks. *Astrophys. J.* **2017**, *846*, 2. [CrossRef]
56. Cranmer, S.R.; Owocki, S.P. The effect of oblateness and gravity darkening on the radiation driving in winds from rapidly rotating B stars. *Astrophys. J.* **1995**, *440*, 308–321. [CrossRef]
57. Kudritzki, R.P. Line-driven Winds, Ionizing Fluxes, and Ultraviolet Spectra of Hot Stars at Extremely Low Metallicity. I. Very Massive O Stars. *Astrophys. J.* **2002**, *577*, 389–408. [CrossRef]
58. Villata, M. Radiation-driven winds of hot stars—A simplified model. *Astron. Astrophys.* **1992**, *257*, 677–680.
59. Müller, P.E.; Vink, J.S. A consistent solution for the velocity field and mass-loss rate of massive stars. *Astron. Astrophys.* **2008**, *492*, 493–509. [CrossRef]
60. Araya, I.; Curé, M.; Cidale, L.S. Analytical Solutions for Radiation-driven Winds in Massive Stars. I. The Fast Regime. *Astrophys. J.* **2014**, *795*, 81. [CrossRef]
61. Araya, I.; Christen, A.; Curé, M.; Cidale, L.S.; Venero, R.O.J.; Arcos, C.; Gormaz-Matamala, A.C.; Haucke, M.; Escárate, P.; Clavería, H. Analytical solutions for radiation-driven winds in massive stars - II. The δ-slow regime. *MNRAS* **2021**, *504*, 2550–2556. [CrossRef]
62. Corless, R.M.; Gonnet, G.H.; Hare, D.E.G.; Jeffrey, D.J. Lambert's W function in Maple. *Maple Tech. Newsl.* **1993**, *9*, 12–22.
63. Corless, R.M.; Gonnet, G.H.; Hare, D.E.G.; Jeffrey, D.J.; Knuth, D.E. On the LambertW function. *Adv. Comput. Math.* **1996**, *5*, 329–359. [CrossRef]
64. Cranmer, S.R. New views of the solar wind with the Lambert W function. *Am. J. Phys.* **2004**, *72*, 1397–1403. [CrossRef]
65. Kudritzki, R.P.; Pauldrach, A.; Puls, J. Radiation driven winds of hot luminous stars. II - Wind models for O-stars in the Magellanic Clouds. *Astron. Astrophys.* **1987**, *173*, 293–298.
66. de Koter, A.; Heap, S.R.; Hubeny, I. On the Evolutionary Phase and Mass Loss of the Wolf-Rayet–like Stars in R136a. *Astrophys. J.* **1997**, *477*, 792. [CrossRef]
67. Vink, J.S.; de Koter, A.; Lamers, H.J.G.L.M. On the nature of the bi-stability jump in the winds of early-type supergiants. *Astron. Astrophys.* **1999**, *350*, 181–196.
68. Rencher, A.; Christensen, W. *Methods of Multivariate Analysis*; Wiley Series in Probability and Statistics; Wiley: Hoboken, NJ, USA, 2012.

69. Mardia, K.V.; Kent, J.T.; Bibby, J.M. *Multivariate Analysis (Probability and Mathematical Statistics)*; Academic Press: Cambridge, MA, USA, 1980.
70. Kudritzki, R.P.; Bresolin, F.; Przybilla, N. A New Extragalactic Distance Determination Method Using the Flux-weighted Gravity of Late B and Early A Supergiants. *Astrophys. J.* **2003**, *582*, L83–L86. [CrossRef]
71. Kudritzki, R.P.; Urbaneja, M.A.; Bresolin, F.; Przybilla, N.; Gieren, W.; Pietrzyński, G. Quantitative Spectroscopy of 24 A Supergiants in the Sculptor Galaxy NGC 300: Flux-weighted Gravity-Luminosity Relationship, Metallicity, and Metallicity Gradient. *Astrophys. J.* **2008**, *681*, 269–289. [CrossRef]
72. Ekström, S.; Georgy, C.; Eggenberger, P.; Meynet, G.; Mowlavi, N.; Wyttenbach, A.; Granada, A.; Decressin, T.; Hirschi, R.; Frischknecht, U.; et al. Grids of stellar models with rotation. I. Models from 0.8 to 120 $M_\odot$ at solar metallicity (Z = 0.014). *Astron. Astrophys.* **2012**, *537*, A146. [CrossRef]
73. Kudritzki, R.; Lennon, D.J.; Puls, J. Quantitative Spectroscopy of Luminous Blue Stars in Distant Galaxies. In *Proceedings of the Science with the VLT*; Walsh, J.R., Danziger, I.J., Eds.; Springer: Berlin/Heidelberg, Germany, 1995; p. 246.
74. Gormaz-Matamala, A.C.; Curé, M.; Lobel, A.; Panei, J.A.; Cuadra, J.; Araya, I.; Arcos, C.; Figueroa-Tapia, F. New self-consistent wind parameters to fit optical spectra of O-type stars observed with the HERMES spectrograph. *Astron. Astrophys.* **2022**, *661*, A51. [CrossRef]
75. Porter, J.M.; Rivinius, T. Classical Be Stars. *Publ. Astron. Soc. Pac.* **2003**, *115*, 1153–1170. [CrossRef]
76. Bjorkman, J.E.; Cassinelli, J.P. Equatorial disk formation around rotating stars due to Ram pressure confinement by the stellar wind. *Astrophys. J.* **1993**, *409*, 429–449. [CrossRef]
77. Owocki, S.P.; Cranmer, S.R.; Gayley, K.G. Inhibition FO Wind Compressed Disk Formation by Nonradial Line-Forces in Rotating Hot-Star Winds. *Astrophys. J.* **1996**, *472*, L115. [CrossRef]
78. Lee, U.; Osaki, Y.; Saio, H. Viscous excretion discs around Be stars. *Mon. Not. R. Astron. Soc.* **1991**, *250*, 432–437. [CrossRef]
79. Okazaki, A.T. Viscous Transonic Decretion in Disks of Be Stars. *Publ. Astron. Soc. Jpn.* **2001**, *53*, 119–125. [CrossRef]
80. Krtička, J.; Owocki, S.P.; Meynet, G. Mass and angular momentum loss via decretion disks. *Astron. Astrophys.* **2011**, *527*, A84. [CrossRef]
81. Curé, M.; Meneses, R.; Araya, I.; Arcos, C.; Peña, G.; Machuca, N.; Rodriguez, A. Revisiting viscous transonic decretion disks of Be stars. *Astron. Astrophys.* **2022**, *664*, A185. [CrossRef]
82. Silaj, J.; Curé, M.; Jones, C.E. Line-Driven Winds Revisited in the Context of Be Stars: Ω-slow Solutions with High k Values. *Astrophys. J.* **2014**, *795*, 78. [CrossRef]
83. Lamers, H.J.G.L.M.; Snow, T.P.; Lindholm, D.M. Terminal Velocities and the Bistability of Stellar Winds. *Astrophys. J.* **1995**, *455*, 269. [CrossRef]
84. Vink, J.S. Fast and slow winds from supergiants and luminous blue variables. *Astron. Astrophys.* **2018**, *619*, A54. [CrossRef]
85. Vink, J.S.; Sander, A.A.C. Metallicity-dependent wind parameter predictions for OB stars. *Mon. Not. R. Astron. Soc.* **2021**, *504*, 2051–2061. [CrossRef]
86. Stahl, O.; Wolf, B.; Aab, O.; Smolinski, J. Stellar wind properties of A-type hypergiants. *Astron. Astrophys.* **1991**, *252*, 693–700.
87. Lefever, K.; Puls, J.; Aerts, C. Statistical properties of a sample of periodically variable B-type supergiants. Evidence for opacity-driven gravity-mode oscillations. *Astron. Astrophys.* **2007**, *463*, 1093–1109. [CrossRef]
88. Markova, N.; Puls, J. Bright OB stars in the Galaxy. IV. Stellar and wind parameters of early to late B supergiants. *Astron. Astrophys.* **2008**, *478*, 823–842. [CrossRef]
89. Haucke, M.; Cidale, L.S.; Venero, R.O.J.; Curé, M.; Kraus, M.; Kanaan, S.; Arcos, C. Wind properties of variable B supergiants. Evidence of pulsations connected with mass-loss episodes. *Astron. Astrophys.* **2018**, *614*, A91. [CrossRef]
90. Rivet, J.P.; Siciak, A.; de Almeida, E.S.G.; Vakili, F.; Domiciano de Souza, A.; Fouché, M.; Lai, O.; Vernet, D.; Kaiser, R.; Guerin, W. Intensity interferometry of P Cygni in the H α emission line: Towards distance calibration of LBV supergiant stars. *MNRAS* **2020**, *494*, 218–227. [CrossRef]

Disclaimer/Publisher's Note: The statements, opinions and data contained in all publications are solely those of the individual author(s) and contributor(s) and not of MDPI and/or the editor(s). MDPI and/or the editor(s) disclaim responsibility for any injury to people or property resulting from any ideas, methods, instructions or products referred to in the content.

Article

Unveiling the Evolutionary State of Three B Supergiant Stars: PU Gem, ϵ CMa, and η CMa

Julieta Paz Sánchez Arias [1,*], Péter Németh [1,2], Elisson Saldanha da Gama de Almeida [3], Matias Agustin Ruiz Diaz [4], Michaela Kraus [1] and Maximiliano Haucke [5]

Citation: Sánchez Arias, J.P.; Németh, P.; de Almeida, E.S.G.; Ruiz Diaz, M.A.; Kraus, M.; Haucke, M. Unveiling the Evolutionary State of Three B Supergiant Stars: PU Gem, ϵ CMa, and η CMa. *Galaxies* 2023, 11, 93. https://doi.org/10.3390/galaxies11050093

Academic Editor: Jorick Sandor Vink

Received: 17 March 2023
Revised: 19 August 2023
Accepted: 23 August 2023
Published: 29 August 2023

Copyright: © 2023 by the authors. Licensee MDPI, Basel, Switzerland. This article is an open access article distributed under the terms and conditions of the Creative Commons Attribution (CC BY) license (https://creativecommons.org/licenses/by/4.0/).

[1] Astronomical Institute, Czech Academy of Sciences, Fričova 298, 25165 Ondřejov, Czech Republic
[2] Astroserver.org, Fő tér 1, 8533 Malomsok, Hungary
[3] Instituto de Física y Astronomía, Universidad de Valparaíso, Av. Gran Bretaña 1111, Casilla 5030, Valparaíso, Chile
[4] Instituto de Astrofísica de La Plata, CONICET-UNLP, Paseo del Bosque s/n, La Plata 1900, Argentina
[5] Instituto de Ingeniería y Agronomía, Universidad Nacional Arturo Jauretche, Av. Calchaquí 6200, Florencio Varela 1888, Argentina
* Correspondence: julieta.sanchez@asu.cas.cz

Abstract: We aim to combine asteroseismology, spectroscopy, and evolutionary models to establish a comprehensive picture of the evolution of Galactic blue supergiant stars (BSG). To start such an investigation, we selected three BSG candidates for our analysis: HD 42087 (PU Gem), HD 52089 (ϵ CMa), and HD 58350 (η CMa). These stars show pulsations and were suspected to be in an evolutionary stage either preceding or succeeding the red supergiant (RSG) stage. For our analysis, we utilized the 2-min cadence TESS data to study the photometric variability, and we obtained new spectroscopic observations at the CASLEO observatory. We used non-LTE radiative transfer models calculated with CMFGEN to derive their stellar and wind parameters. For the fitting procedure, we included CMFGEN models in the iterative spectral analysis pipeline XTGRID to determine their CNO abundances. The spectral modeling was limited to changing only the effective temperature, surface gravity, CNO abundances, and mass-loss rates. Finally, we compared the derived metal abundances with prediction from Geneva stellar evolution models. The frequency spectra of all three stars show stochastic oscillations and indications of one nonradial strange mode, $f_r = 0.09321\ d^{-1}$ in HD 42087 and a rotational splitting centred in $f_2 = 0.36366\ d^{-1}$ in HD 52089. We conclude that the rather short sectoral observing windows of TESS prevent establishing a reliable mode identification of low frequencies connected to mass-loss variabilities. The spectral analysis confirmed gradual changes in the mass-loss rates, and the derived CNO abundances comply with the values reported in the literature. We were able to achieve a quantitative match with stellar evolution models for the stellar masses and luminosities. However, the spectroscopic surface abundances turned out to be inconsistent with the theoretical predictions. The stars show N enrichment, typical for CNO cycle processed material, but the abundance ratios did not reflect the associated levels of C and O depletion. We found HD 42087 to be the most consistent with a pre-RSG evolutionary stage, HD 58350 is most likely in a post-RSG evolution and HD 52089 shows stellar parameters compatible with a star at the TAMS.

Keywords: stars: massive; stars: supergiants; stars: winds; outflows

1. Introduction

Massive stars are one of the most important objects in the universe due to their key role in the enrichment of interstellar medium with metals for future star generations, as well as in the evolution of host galaxies [1]. Nevertheless, understanding their evolution is challenging due to the significant changes they experience at different evolutionary stages, especially during the post-Main Sequence (MS). Therefore, any inaccuracy in the input parameters results in large uncertainties of the evolutionary models [2,3]. Fortunately, there are many sophisticated stellar evolutionary models calculated for massive stars e.g., [4,5]. However, they depend on internal parameters such as rotation, chemical mixing, and angular momentum transport, for which no decisive observational constraints are available. In addition, massive supergiant stars undergo mass loss via line-driven winds, and their rates are far from being firmly established, adding even more uncertainties to the evolution of these stars.

BSGs comprise extreme transition phases, in which the stars shed huge amounts of material into their environments. During the evolution of massive stars, those stars with masses between 20 and 40 $M_\odot$ evolve back to the blue supergiant state after a Red Supergiant (RSG) state, either as a Blue Supergiant (BSG) or in a follow-up Wolf-Rayet phase. On the other hand, massive stars with <20 $M_\odot$ may experience "blue loops", where the star changes from a cool star to a hotter one before cooling again. Therefore, hot BSGs can be found at the pre-RSG stage, burning only H in a shell or at the post-RSG during the He-core burning. The exact reason why some massive stars experience "blue loops" and others do not still remains unknown, although it is known that extra mixing processes within the layers surrounding their convective core, along with mass-loss events during their evolution, play an essential role [6–8].

In addition, BSGs can show an extremely rich spectrum of stellar oscillations. Since these oscillations depend on the internal structure of the star, their analysis paves the way to understanding phenomena such as the occurrence of "blue loops" or the discernment between the different evolutionary stages in which blue supergiant stars can exist. The observed oscillation modes in B supergiants are driven by different excitation mechanisms, such as the classical κ mechanism, stochastic wave generation caused by the presence of convective layers in the outer regions of the star or in their envelope, and tidal excitation causing the so-called Rossby modes and strange modes. Strange modes are nonlinear instabilities that require a luminosity over a mass ratio of $L_\star/M_\star > 10^4 L_\odot/M_\odot$ to be excited [9]. Their existence has been related to the variable mass loss these stars experience [10–12].

Recent studies on stellar oscillations of individual BSGs [10,12,13] improved our understanding of the complex variability shown by these stars. Furthermore, considerable progress has been made in theoretical studies on the stellar oscillations of massive stars at different evolutionary stages focusing on specific mass ranges. For example, ref. [14] investigated stellar oscillations in stars at pre- and post-RSG stages for masses between 13 and 18 solar masses for different physical parameters such as metallicity and overshooting. In [15], a thorough study of stellar oscillations is presented for a broader range of masses, revealing that the pulsation properties in pre- and post-RSG evolutionary phases are fundamentally different: stars in the post-RSG stage pulsate in many more modes, including radial strange modes, than their less evolved counterparts, although these authors did not inspect the effect of different wind efficiency on the stability of the modes.

Despite the efforts accumulated over years in the study of the stellar pulsations of these objects in different evolutionary stages, the correct identification of the evolutionary stage of BSGs additionally requires a detailed analysis of spectral observations due to the large uncertainties these stars have in astrophysics parameters such as the mass and the radii.

Therefore, we started a comprehensive study of these objects with the aim of gaining insights into the evolutionary state of BSG stars. Combining information about the pulsation behavior extracted from photometric lightcurves with newly determined stellar parameters and precise chemical abundances obtained from the modeling of acquired spectroscopic data, we strive to find clear evidence for either a pre- or a post-RSG state of the objects. The current paper presents our methodology and first results on a small sample of objects.

This paper is organized as follows: in Section 2, we summarize the main parameters for these stars found in the literature; Section 3 describes the spectroscopic and photometric observations employed in this work. In Section 4, we present the analysis of the light curves and the frequency spectra. Section 5 is devoted to the spectral analysis of the selected objects.

We describe the numerical tools employed, including brand new capabilities implemented in XTGRID [16] for this work and the results obtained. Finally, Sections 6 and 7 are devoted to the discussion and conclusions.

2. Target Selection and Parameters Values from Literature

Haucke et al. [11] presented the most recent comprehensive analysis of the wind properties of 19 pulsating BSGs. We focus on three objects from their sample, for which we have obtained new spectroscopic observations. These are the stars HD 42087 (PU Gem), HD 52089 (ϵ CMa), and HD 58350 (η CMa). Table 1 summarizes the values for these stars derived by Haucke et al. [11]. They serve as reference (or starting) values for our analysis. In Figure 1, we show the position of the selected stars in an HR diagram for the parameters in Table 1. We notice that these hot BSG stars could be either at the immediate post-Main Sequence or at the post-Red Supergiant stage.

The three selected objects have been studied extensively in the literature, and a compilation of literature values for their stellar and wind parameters can be found in Haucke et al. [11]. Here, we present a brief overview highlighting a variety of stellar and wind parameters.

2.1. HD 42087

Searle et al. [17] derived the following parameters for this star using spectra from October 1990: T_{eff} = 18,000 ± 1000 K, $\log g$ = 2.5, $\log(L_\star/L_\odot)$ = 5.11 ± 0.24, $R_\star$ = 36.6 $R_\odot$, and $v \sin i$ = 71 km s^{-1}. They employed CMFGEN complemented with TLUSTY to derive T_{eff} and $\log(L_\star/L_\odot)$, $\log g$, along with the CNO abundances, which were $\epsilon(C)$ = 7.76, $\epsilon(N)$ = 8.11, and $\epsilon(C)$ = 8.80. For the wind parameters, they obtained $\dot{M}$ = 5.0 × 10^{-7} $M_\odot$ yr^{-1}, β = 1.2, and v_∞ = 650 km s^{-1}. Morel et al. [18] showed that HD 42087 has a high Hα variability with a spectral variability index (as defined by these authors) of ∼91% in this line (see Section 4 of [18]), evidencing a cyclic behaviour with P ∼25 d. We also mention that the values derived for this object in Haucke et al. [11] were obtained with TLUSTY using an optical spectrum covering only the Hα region. These spectra from 2006 showed a P Cygni feature with a weak emission and a strong absorption component.

2.2. HD 52089

In Morel et al. [19], this star was studied as a slowly rotating B-type dwarf star. They derived T_{eff} = 23,000 K and $\log g$ = 3.30 ± 0.15 using spectroscopic data from April 2005. By using the DETAIL/SURFACE code, they determined the non-LTE abundances $\epsilon(C)$ = 8.09 ± 0.12, $\epsilon(N)$ = 7.93 ± 0.24, and $\epsilon(O)$ = 8.44 ± 0.18. Fossati et al. [20] derived an effective temperature and surface gravity of T_{eff} = 22 500 ± 300 K and $\log g$ = 3.40 ± 0.08. They also obtained updated values for the surface abundances by analyzing a FEROS spectra from 2011 and found $\epsilon(C)$ = 8.30 ± 0.07, $\epsilon(N)$ = 8.16 ± 0.07 and $\epsilon(O)$ = 8.70 ± 0.12. Additionally, they estimated a 12.5 $M_\odot$ for this object.

2.3. HD 58350

Lefever et al. [21] derived T_{eff} = 13,500 K, $\log g$ = 1.75, $R_\star$ = 65 $R_\odot$, $\log(L_\star/L_\odot)$ = 5.10, v_∞ = 250 km s^{-1}, $\dot{M}$ = 1.4 × 10^{-7} $M_\odot$ yr^{-1} and β = 2.5 for the stellar and wind parameters of this star. In Searle et al. [17], they derived T_{eff} = 15,000 ± 500 K, $\log g$ = 2.13, $R_\star$ = 57.3 ± 2.64 $R_\odot$, and $\log(L_\star/L_\odot)$ = 5.18 ± 0.17. For the CNO abundances, they obtained $\epsilon(C)$ = 7.78, $\epsilon(N)$ = 8.29, and $\epsilon(O)$ = 8.75.

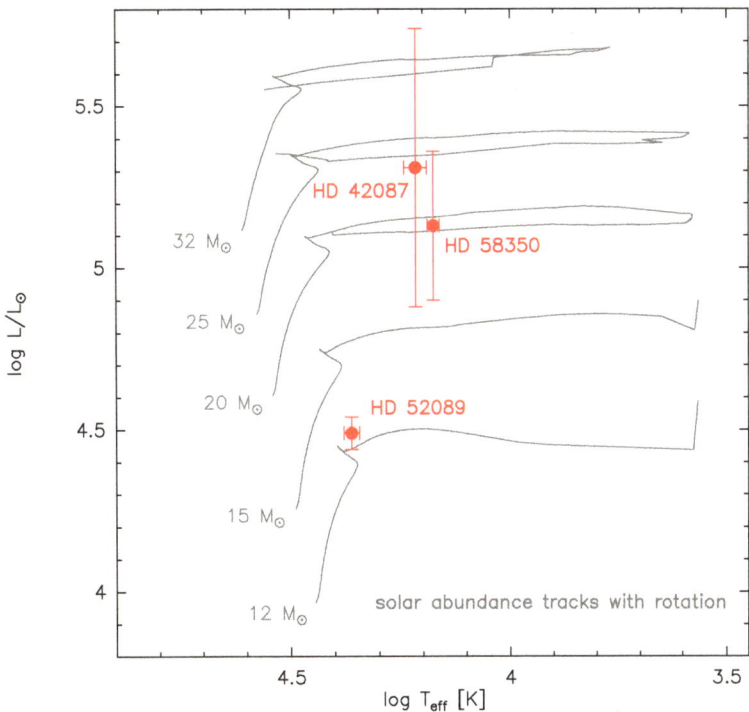

Figure 1. Our three BSG stars in the HR diagram according to the values derived in [11]. The evolutionary tracks are taken from [22].

Table 1. Stellar and wind parameters derived by Haucke et al. [11] for our star sample.

Parameter	HD 42087	HD 52089	HD 58350
T_{eff} [K]	$16{,}500 \pm 1000$	$23{,}000 \pm 1000$	$15{,}500 \pm 700$
$\log g$ [cgs]	2.45 ± 0.10	3.00 ± 0.10	2.00 ± 0.10
$\log L_\star$ [$L_\odot$]	5.31 ± 0.43	4.49 ± 0.05	5.18 ± 0.32
$R_\star$ [$R_\odot$]	55	11	54
$v \sin i$ [km s^{-1}]	80	10	40
$\dot{M}$ [$M_\odot$ yr^{-1}]	$(5.7 \pm 0.5) \times 10^{-7}$	$(2.0 \pm 0.6) \times 10^{-8}$	$(1.4 \pm 0.2) \times 10^{-7}$
v_∞ [km s^{-1}]	700 ± 70	900 ± 270	200 ± 30
β	2.0	1.0	3.0

3. Observations

With the aim of shedding light on the evolutionary state of these objects, we analyze new spectroscopic observations and combine these results with information extracted from their photometric light curves.

3.1. Spectra

The spectra we employed in this work to derive new parameters were taken on 23 and 24 January 2020 for HD 42087 and HD 58350, respectively, and 14 February 2015 for HD 52089. All of them cover the wavelength range from 4275 Å up to 6800 Å with a signal-to-noise ratio S/N of 140, 140, and 130 for HD 42087, HD 52089, and HD 58350, respectively.

We utilized the REOSC spectrograph attached to the Jorge Sahade 2.15 m telescope at the Complejo Astronómico El Leoncito (CASLEO), San Juan, Argentina. The resolving power at 4500 Å and 6500 Å is $R\sim$12,600 and $R\sim$13,900, respectively. The spectra were reduced and normalized following standard procedures using IRAF[1] routines.

In order to show the Hα variability in our targets, we have collected previous observations in CASLEO, depicted in Figures 2–4; from 15 January 2006 for HD 42087; 5 February 2013 for HD 52089; and 15 January 2006 and 5 February 2013 for HD 58350.

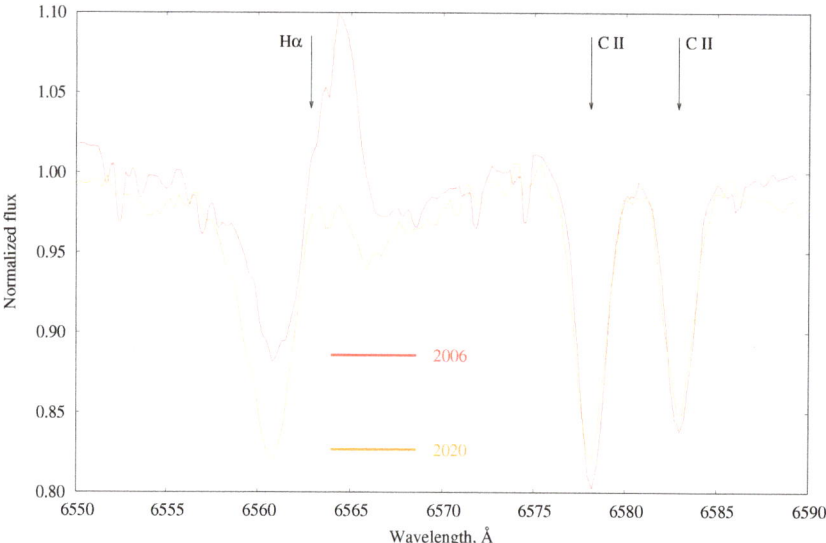

Figure 2. Evolution of the Hα line profile of HD 42087 between 2006 and 2020. The emission component weakened by 2020.

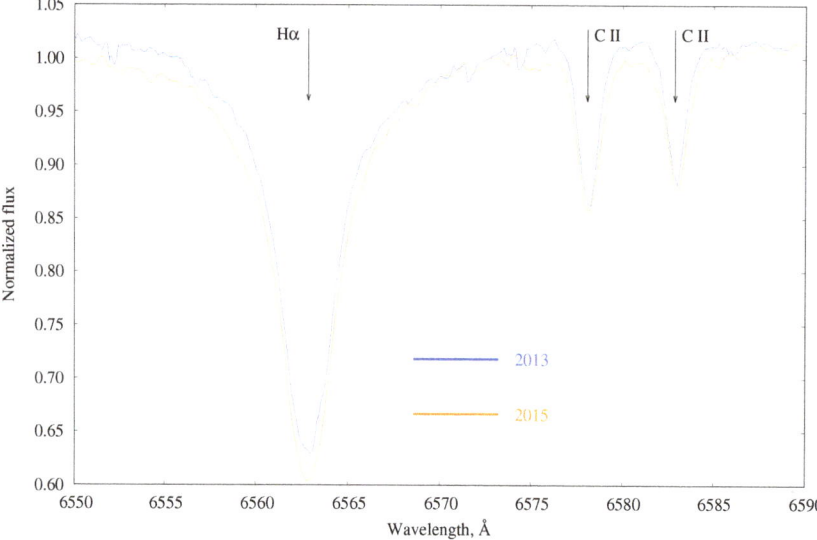

Figure 3. Evolution of the Hα line profile of HD 52089 between 2013 and 2015.

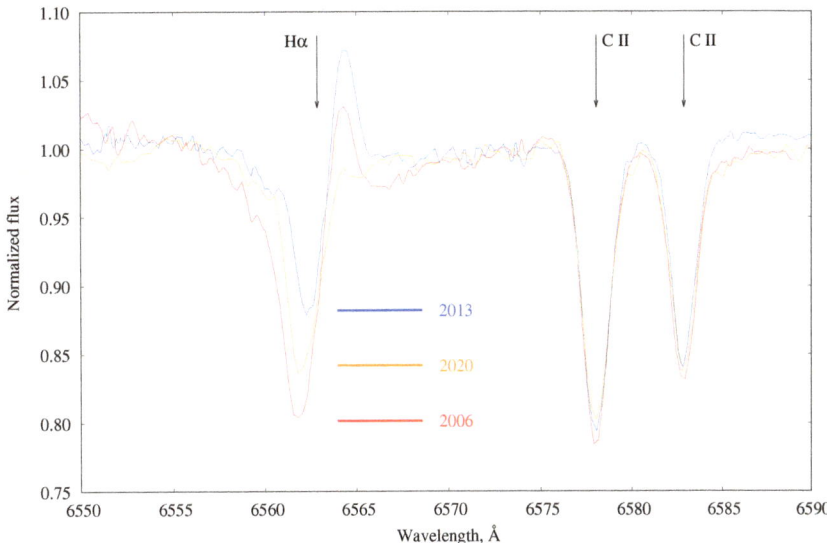

Figure 4. Evolution of the Hα line profile of HD 58350 between 2006 and 2020. The emission component has strengthened between 2006 and 2013 and became noticeably weaker by 2020.

3.2. Photometric Light Curves

We complemented the optical spectra with space photometry collected with the Transiting Exoplanet Survey Satellite (TESS) mission [23,24]. We retrieved a high cadence (120 s) PDCSAP light curve for each object using Astroquery [25] and cleaned the light curves using the Lightkurve package [26]. For this, we first selected only the data points with a quality flag equal to 0, meaning they are not cosmic rays or data from bad pixels. Next, we carried out sigma clipping to remove outliers, using a value of 6σ for all light curves, following a similar procedure as in Garcia et al. [27]. The resulting light curves were transformed to display the variation of the magnitude (Δm) from the mean magnitude:

$$\Delta m = -2.5 \log(\text{pdcsap\_flux}) + 2.5 \log(\overline{\text{pdcsap\_flux}}); \tag{1}$$

where pdcsap_flux is the individual flux at each exposure and $\overline{\text{pdcsap\_flux}}$ is the mean flux of the entire light curve. In this way, we obtain a normalized light curve, whose amplitude can be expressed in units of mag. The time axis is in units of Barycentric Julian Date, which is the Julian Date corrected for differences in the earth's position with respect to the barycenter of the solar system.

4. Frequency Analysis

We employed the Fourier Transform with Period04 [28]. For each star and sector, the frequencies were searched in the interval [0;50] d^{-1}, widely covering their frequency content. No frequency beyond 2 d^{-1} was found for any star. The amplitude and phase were calculated using a least square sine fit for each detected frequency. After obtaining the first frequency, the analysis was performed on the residuals. The Fourier analysis was stopped after obtaining 15 frequencies. Once we derived the frequencies following this procedure, we dismissed those frequencies below 0.1 d^{-1} since TESS data of a single sector do not allow us to derive periods higher than ∼10 d. Additionally, we discard those frequencies with a separation of less than 2/T, where 1/T is the Rayleigh resolution and T is the time span of the observations. The S/N ratio was computed for the derived frequencies along with the uncoupled uncertainties in the frequencies and amplitudes using a Monte Carlo simulation.

When available, we also analyzed the frequency content of the combined, consecutive sectors, since the longer time baseline would facilitate the detection of longer periods from radial modes possibly connected to strange modes. We considered the values recommended in Baran and Koen [29] for the S/N when dealing with individual TESS sectors and combined sectors, which resulted in 5.037 for our individual TESS sectors, and of 5.124 and 5.194 for the combined sectors of HD 52089 (2 sectors combined) and HD 42087 (3 sectors combined), respectively. Nevertheless, we took into account frequencies with lower S/N whenever they turned out to be interesting for the analysis (see below).

Finally, when comparing frequencies from different sectors, we adopted the separation criterion for the combined sectors.

Next, we provide the details of the light curves and the frequencies extracted for each star.

4.1. HD 42087

This star was observed in 3 consecutive sectors: Sector 43, in the period from 16 September to 10 October 2021; Sector 44, from 10 October to 6 November 2021; and Sector 45, from 6 November to 2 December 2021, covering an observation time span of 24.287 days, 24.156 days and 24.551 days, respectively. The light curve and the amplitude spectra for the individual sectors and all sectors combined are displayed in Figure 5.

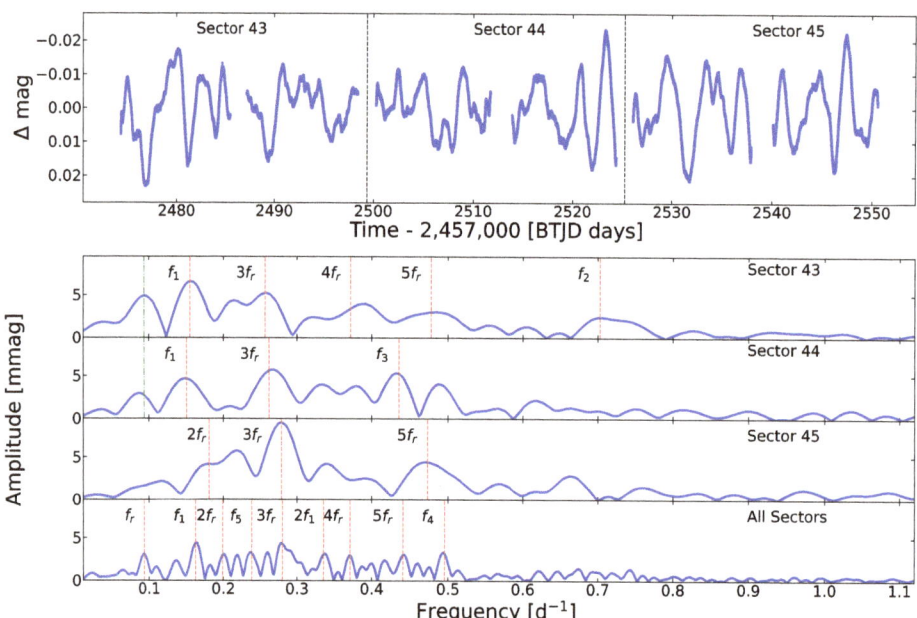

Figure 5. Top panel: Light curve of HD 42087 corresponding to Sectors 43, 44, and 45. **Lower panels**: Amplitude spectra for each sector and all sectors combined.

The frequencies derived for sectors 43, 44, and 45 are listed in Table 2 and the threshold considered for the frequency separation is 0.0823 d^{-1}, 0.0828 d^{-1} and 0.0815 d^{-1}, respectively. For the analysis of all sectors combined, the time span is $T = 72.994\ d$; therefore, the separation criterion we employed is $2/T = 0.0274\ d^{-1}$. We adopted this value as a conservative threshold when comparing frequencies from different sectors.

Table 2. List of frequencies, their amplitudes, and S/N ratios, found for HD 42087.

Sector	Frequency [d^{-1}]	$3\sigma_f$ [d^{-1}]	Amplitude [mmag]	$3\sigma_A$ [mmag]	S/N	Id
43	0.15543	0.00044	7.7615	0.15	10.65	f_1
	0.25729	0.00063	5.7467	0.14	8.41	$3 \times f_r$
	0.47759	0.00064	4.7663	0.13	7.79	$5 \times f_r$
	0.37169	0.00090	3.5347	0.13	5.45	$4 \times f_r$
	0.70257	0.00037	2.9734	0.14	5.41	f_2
44	0.26268	0.00068	5.5317	0.16	7.68	$3 \times f_r$
	0.15072	0.00082	5.2203	0.17	6.75	f_1
	0.43536	0.00090	4.9832	0.19	7.45	f_3
45	0.27928	0.00009	9.2074	0.10	10.65	$3 \times f_r$
	0.18110	0.00059	5.0188	0.13	5.57	$2 \times f_r$
	0.47295	0.00159	4.0068	0.21	5.07	$5 \times f_r$
All	0.28022	0.00019	4.4715	0.12	8.48	$3 \times f_r$
	0.16305	0.00021	3.9728	0.11	7.05	f_1
	0.49515	0.00022	3.4982	0.11	7.44	f_4
	0.23874	0.00027	3.2680	0.11	6.04	f_5
	0.44102	0.00023	3.2321	0.10	6.67	$5 \times f_r$
	0.33531	0.00030	2.9450	0.11	5.74	$2 \times f_1$
	0.19899	0.00029	2.8086	0.10	5.07	$2 \times f_r$
	0.37073	0.00033	2.6635	0.10	5.29	$4 \times f_r$
	0.09321	0.00032	2.4597	0.12	4.20	f_r

The pulsational content of HD 42087 lies at low frequencies below 0.8 d^{-1} and the amplitudes are between 2.5 and 9.5 mmag, approximately. When analyzing the combined sectors, we also searched for frequencies down to 0.025 d^{-1} due to the extended length of the observations. This allowed us to find the frequency f_r and its five harmonics. Some of these harmonics appear in the individual sectors. Due to the short time span for individual sectors, frequencies below $\sim$0.1 d^{-1} are not reliable; nevertheless, f_r seems to appear for sectors 43 and 44 (indicated with green lines in Figure 5). Besides f_r and its harmonics, five frequencies appear randomly over the four sets of observations, which are likely related to stochastic oscillations (see Section 6).

4.2. HD 52089

This star was observed in 4 TESS sectors: Sector 6, in the period from 15 December 2018 to 6 January 2019; Sector 7, in the period from 8 January to 1 February 2019; Sector 33 from 18 December 2020 to 13 January 2021; and Sector 34 in the period from 13 January to 8 February 2021. The total time spans for Sectors 6, 7, 33, and 34 are 21.771 d, 24.454 d, 25.839 d, and 24.962 d, respectively. Figure 6 displays the light curves and amplitude spectra for Sectors 6 and 7, and Figure 7 displays the same information for Sectors 33 and 34. Due to the large time gap between sectors 7 and 33, we decided to evaluate the sectors in pairs; i.e., together with the analysis of sectors 6 and 7, we studied the frequencies of these sectors combined, and the same with sectors 33 and 34. For the combined sectors 6 and 7, the time span is 46.225 d, and for sectors 33 and 34 together, the time span is 50.801 d.

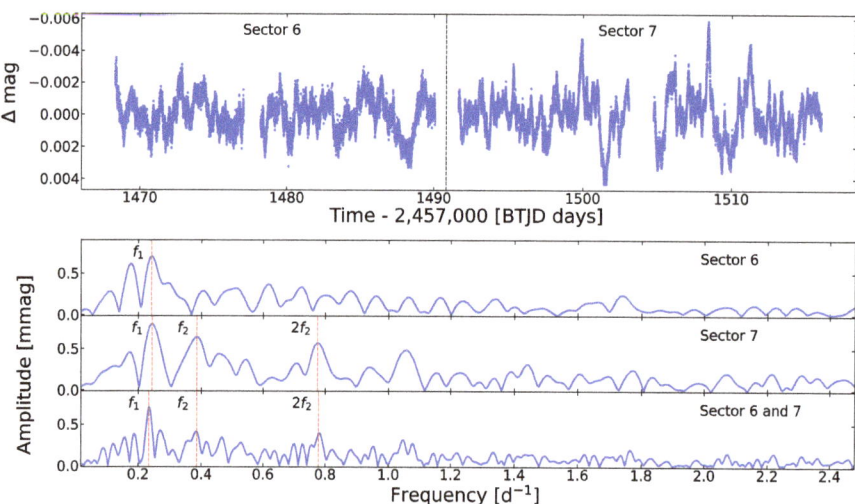

Figure 6. Top panel: Light curve of HD 52089 corresponding to Sectors 6 and 7. **Lower panels**: Amplitude spectra for each sector and all sectors combined.

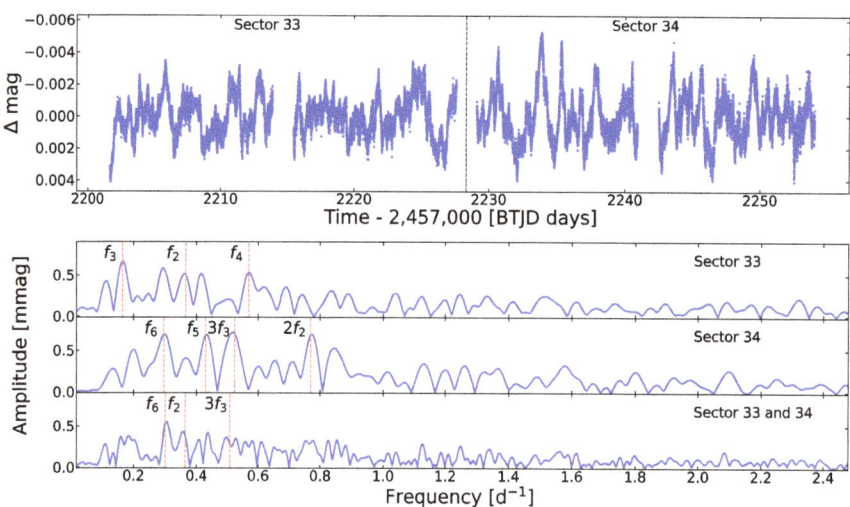

Figure 7. Top panel: Light curve of HD 52089 corresponding to Sectors 33 and 34. **Lower panels**: Amplitude spectra for each sector and all sectors combined.

The derived frequencies, their amplitudes and identifications are listed in Table 3. The separation criteria we employed are 0.0918 d^{-1}, 0.0817 d^{-1}, 0.0774 d^{-1}, and 0.0801 d^{-1} for sectors 6, 7, 33, and 34, respectively. For sectors 6 and 7 combined and 33 and 34 combined the separation criteria are 0.0432 d^{-1} and 0.0393 d^{-1}, respectively and the lowest detectable frequency for the sectors combined is ∼0.04 d^{-1}. We adopted conservative separation criteria with 0.0393 d^{-1} to compare frequencies from different sectors. We find only one significant frequency, f_1, in Sector 6, which is also in Sector 7. In Sector 7, we found another frequency, f_2, which stays present also in Sector 33, and its harmonic, which appears in Sector 34. We searched for lower frequencies in both combined light curves, but we did not find any. After the gap of ∼2 years in the observations, new frequencies appear randomly in the new Sectors, which indicates they are not related to stellar oscillations due to their short lifetime (f_3, f_5, f_6). We note that $f_2 - f_3 \sim f_4 - f_2$ in Sector 33, indicating the presence of a triplet centred in f_2 with a possible rotation frequency of 0.20 d^{-1}. Nevertheless, the observations' short time span makes precise classification difficult. We did not find any significant frequency at lower ranges in the combined sectors.

Table 3. Sector, Frequency, Amplitude, and Identification for HD 52089.

Sector	Frequency [d^{-1}]	$3\sigma_f$ [d^{-1}]	Amplitude [mmag]	$3\sigma_A$ [mmag]	S/N	Id
6	0.24083	0.00115	0.7830	0.05	6.15	f_1
7	0.24316	0.01540	0.8982	0.47	6.14	f_1
	0.38381	0.00192	0.7286	0.05	5.15	f_2
	0.77252	0.00051	0.5888	0.02	4.37	$2 \times f_2$
6 & 7	0.23321	0.00072	0.7736	0.07	7.28	f_1
	0.38342	0.02775	0.4930	0.17	4.79	f_2
	0.77679	0.00073	0.4769	0.04	4.83	$2 \times f_2$
33	0.16358	0.00070	0.7716	0.02	6.868	f_3
	0.36366	0.00089	0.5784	0.02	5.42	f_2
	0.56843	0.00089	0.5083	0.02	4.72	f_4
34	0.42781	0.00107	0.7556	0.03	4.98	f_5
	0.52280	0.00128	0.7113	0.03	4.74	$3 \times f_3$
	0.76818	0.00127	0.7038	0.03	4.75	$2 \times f_2$
	0.29483	0.00147	0.6389	0.03	4.35	f_6
33 & 34	0.29937	0.00167	0.6627	0.12	5.01	f_6
	0.36220	0.00331	0.5806	0.34	4.39	f_2
	0.50772	0.00158	0.4714	0.09	4.09	$3 \times f_3$

4.3. HD 58350

This star has been observed in Sector 34 in the period from 14 January to 8 February 2021, covering a total timespan of $T = 24.96$ days. The light curve for Sector 34 is displayed in the top panel of Figure 8, and we show the amplitude spectra in the lower panel.

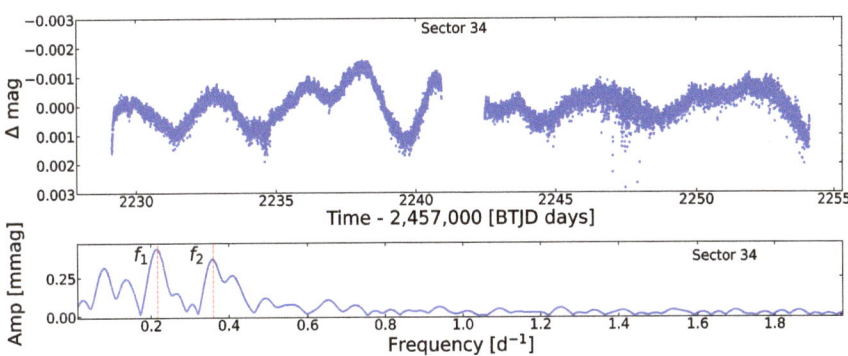

Figure 8. Top panel: Light curve of HD 58350 acquired during Sector 34. **Lower panel:** Amplitude spectrum for the same sector.

The resulting list of frequencies for HD 58350 is shown in Table 4. The separation criterion is $2/T = 0.0801\ d^{-1}$. The pulsational content for this BSG lies below $0.5\ d^{-1}$. We found two independent frequencies with similar amplitudes.

Table 4. Sector, Frequency, Amplitude, and Identification for HD 58350.

Sector	Frequency $[d^{-1}]$	$3\sigma_f$ $[d^{-1}]$	Amplitude [mmag]	$3\sigma_A$ [mmag]	S/N	Id
34	0.21533	0.00069	0.4392	0.01	8.54	f_1
	0.35927	0.00088	0.3339	0.01	7.07	f_2

5. Modeling Tools for Spectral Analysis

5.1. The Code CMFGEN

CMFGEN [30] is one of the state-of-the-art 1-D non-LTE radiative transfer codes for studying the physical and chemical properties of massive hot stars. It has been successfully used for reproducing different types of observables of massive stars, such as in the case of OB supergiants: photometry, spectroscopy, and interferometry, e.g., [17,31,32].

CMFGEN solves the radiative transfer and statistical equilibrium equations for the photosphere and wind under the assumption of radiative equilibrium and considers a spherically symmetric stationary wind. For the photospheric region, CMFGEN solves the radiative transfer and hydrostatic equations (describing the state of radiation and gas in the photosphere) in a self-consistent way. Further details on this procedure can be found in Section 3.1 of Bouret et al. [33].

The wind velocity, $v(r)$, and density, $\rho(r)$, profiles are related to each other as follows, based on the assumption of a stationary and spherical symmetric wind:

$$\dot{M} = 4\pi r^2 \rho(r) v(r) f(r), \qquad (2)$$

where r is the distance from the center of the star and $\dot{M}$ is the wind mass-loss rate, which is constant at any location of the wind. Here, the volume filling factor $f(r)$ parameterizes the inclusion of micro-clumping; e.g., [34], that is, inhomogeneities that are found in the winds of massive stars, e.g., [35,36]. In CMFGEN, this parameterization is performed as follows:

$$f(r) = f_\infty + (1 - f_\infty) e^{-\frac{v(r)}{v_{\text{initial}}}}, \qquad (3)$$

where f_∞ is the filling factor value at $r \to \infty$, and v_{initial} is the onset velocity of clumping in the wind. Since our initial guesses in the fitting procedure (as described in Section 5.2) for the stellar and wind parameters are based on the results from Haucke et al. [11], we set $f_\infty = 1.0$ in our default models. This means that our default CMFGEN models are calculated considering a homogeneous wind.

Furthermore, the wind's velocity profile is parameterized in CMFGEN by the so-called β-law approximation, as follows:

$$v(r) = v_\infty \left(1 - \frac{R_\star}{r}\right)^\beta, \qquad (4)$$

where v_∞ is the wind terminal velocity and $R_\star$ is the stellar radius (r higher than $R_\star$). For OB supergiants, values of β are usually found to be as high as $\sim$2.0–3.0, e.g., [37]. In short, Equations (2) and (4) sets the relation between the most important and fundamental physical wind parameters: the wind mass-loss rate and the terminal velocity.

Despite its high computational cost, when compared to other non-LTE codes, such as FASTWIND, one of the biggest advantages of CMFGEN relies on its allowance to set a complex chemical composition, with the inclusion of energy levels of different ions from hydrogen up to nickel. For instance, Haucke et al. [11] studied our star sample of three BSGs employing radiative transfer models calculated with FASTWIND considering hydrogen, helium, and silicon with an approximation for the treatment of the line transfer of iron-group elements. In fact, the main difference between these two codes relies on an "exact" treatment of line-blanketing that is performed by CMFGEN (e.g., see [38] and references therein). Table 5 summarizes the atomic species, the number of energy levels, and bound–bound transitions that are included as the default in our models. Our models are calculated considering a robust atomic model for studying massive hot stars, including species of hydrogen, helium, carbon, nitrogen, oxygen, neon, magnesium, silicon, sulfur, iron, and nickel.

Finally, due to the complexity of the code and its computational cost, a very common approach when using CMFGEN is to vary its parameters manually in order to find an acceptable "by eye" fit to the observations, e.g., [32,39–42]. However, in this paper, we implement an automatic fitting procedure with CMFGEN to find the best-fit models for the observed spectrum of each star, as described below.

5.2. Spectral Analysis with XTgrid

XTGRID [16] is a steepest-descent iterative χ^2 minimizing fit procedure to model hot star spectra. The procedure was developed for the model atmosphere code TLUSTY [43–45] and was previously applied to ultraviolet and optical spectral observations of O and B-type stars [10], Horizontal Branch stars [46], hot subdwarfs [47–49], and white dwarfs [50,51]. It was designed to perform fully automated or supervised spectral analyses of massive data sets. The procedure starts with an input model; by applying successive approximations along a decreasing global χ^2, it iteratively converges on the best solution. All models are calculated on the fly, which is the main advantage of the procedure. XTGRID does not require a precalculated grid; with its tailor-made models—although at a high computational cost—it is able to address nonlinearities in a multidimensional parameter space. After the fitting procedure has converged for a relative change less than 0.5%, parameter errors are evaluated in one dimension, changing each parameter until the χ^2 variation corresponds to the 60% confidence limit. Parameter correlations are evaluated only for the effective temperature and surface gravity. If the procedure finds a better solution during error calculations, it returns to the descent part using the previous solution as the initial model.

Instead of using TLUSTY (plane-parallel model, no wind), we used CMFGEN, since these stars have non-negligible stellar winds with mass-loss rates in the order of 10^{-8}–10^{-7} $M_\odot$ yr^{-1}. Using TLUSTY, instead of CMFGEN, would prevent us to address the wind variability and perform homogeneous modeling for BSGs. In addition, beyond the short-term photometric variability that is typical for pulsating stars and discussed in Section 4, gradual spectral variations may occur due to inhomogeneities in the wind density structure. These, together, require re-evaluating the surface and wind parameters for each observation. Therefore, we decided to proceed with CMFGEN models and start out from the results of Haucke et al. [11]. We updated XTGRID to apply CMFGEN and minimize the wind properties along with stellar surface parameters. Then, we performed a new analysis of the most recent CASLEO spectra to measure the CNO abundances in each of the three stars.

The focus of our analysis was on the CNO abundances. Therefore, we kept the abundances of all elements heavier than O at their solar values [52], and we adopted He abundances $nHe/nH = 0.2$, based on the analysis of Searle et al. [17]. To maintain an approximate consistency with the results of Haucke et al. [11], we kept the stellar radii and turbulent velocity fixed at the values determined in the previous analysis (see Table 1), and we applied unclumped wind models. Adopting clumping in our models resulted in different mass-loss rates. We neglected macroturbulence in the current analysis because it shows a degeneracy with the projected rotation at low spectral resolutions. We note that our goal was to measure the CNO abundances and not to make a comparison with the analysis of Haucke et al. [11]. Our method is not suitable for such a comparison; as we used different model atmosphere codes, our observations were taken at different epochs and we used a different fitting procedure.

Table 5. Summary on the atoms and ionization states of our default CMFGEN models, including number of energy levels, super-levels, and bound–bound transitions for each atomic species.

Ion	Full-Levels	Super-Levels	b-b Transitions
H I	30	30	435
He I	69	69	905
He II	30	30	435
C II	322	92	7742
C III	243	99	5528
C IV	64	64	1446
N II	105	59	898
N III	287	57	6223
N IV	70	44	440
N V	49	41	519
O II	274	155	5880
O III	104	36	761
O IV	64	30	359
O V	56	32	314
Ne II	48	14	328
Ne III	71	23	460
Ne IV	52	17	315
Mg II	44	36	348
Si II	53	27	278
Si III	90	51	640
Si IV	66	66	1090
S III	78	39	520
S IV	108	40	958
S V	144	37	1673
Fe II	295	24	2135
Fe III	607	65	6670
Fe IV	1000	100	37,899
Fe V	1000	139	37,737
Fe VI	1000	59	36,431
Ni II	158	27	1668
Ni III	150	24	1345
Ni IV	200	36	2337
Ni V	183	46	1524
Ni VI	182	40	1895

Finally, our best-fit CMFGEN models to the CASLEO spectra of HD 42087, HD 52089, and HD 58350, are shown, respectively, in Figures 9–11, and the derived parameters are listed in Table 6. The luminosity and mass in this table were derived using $L_\star = 4\pi R_\star^2 \sigma T_{\text{eff}}^4$ and $g = GM_\star/R_\star^2$.

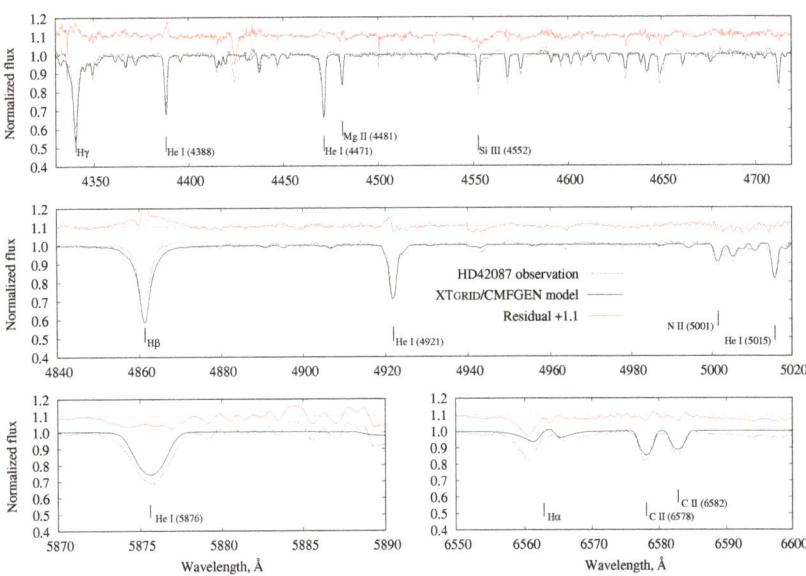

Figure 9. Best-fit XTGRID/CMFGEN model for HD 42087. In each panel, the CASLEO observation is in grey, the CMFGEN model in black, and the residuals, shifted by +1.1 for clarity, are in red.

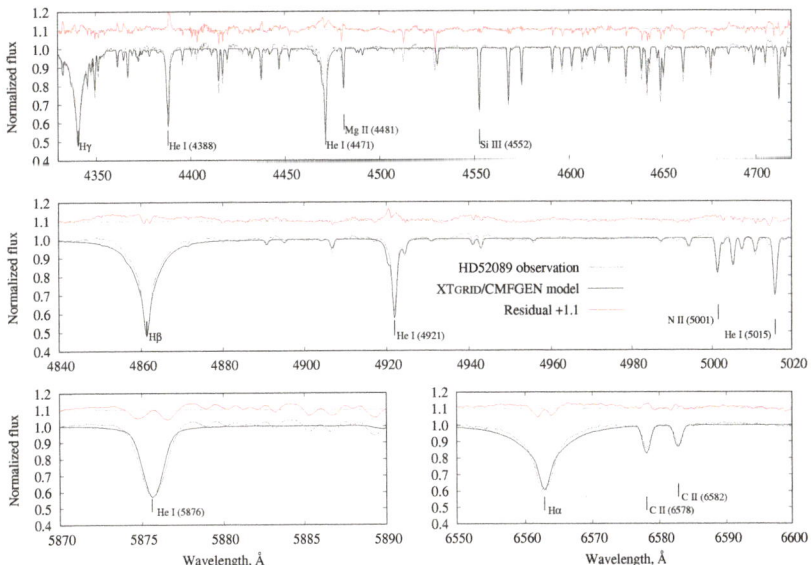

Figure 10. Best-fit XTGRID/CMFGEN model for HD 52089. In each panel, the CASLEO observation is in grey, the CMFGEN model in black, and the residuals, shifted by +1.1 for clarity, are in red.

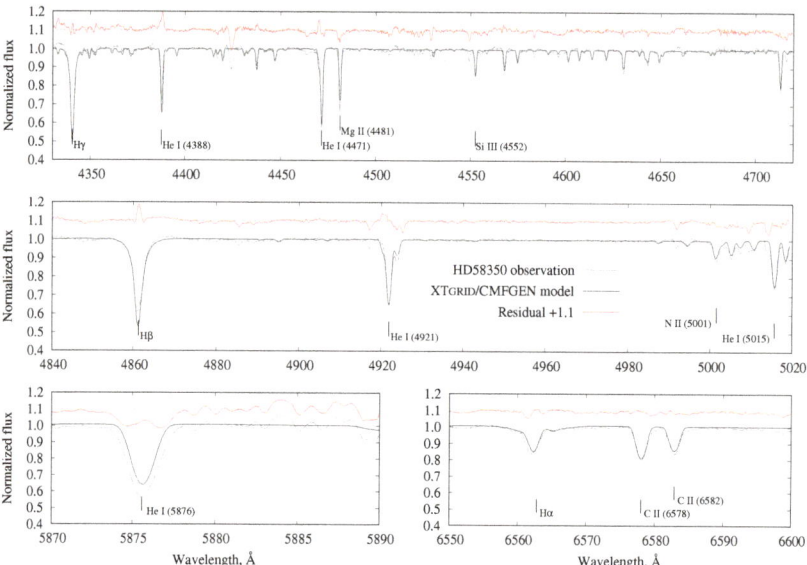

Figure 11. Best-fit XTGRID/CMFGEN model for HD 58350. In each panel, the CASLEO observation is in grey, the CMFGEN model in black, and the residuals, shifted by +1.1 for clarity, are in red.

Table 6. Summary of the spectroscopic results from unclumped CMFGEN models using the XTGRID fitting procedure. Abundances are reported on the 12 scale: $\epsilon = \log N_x/N_H + 12$. Fixed parameters are marked with an "x" prefix. Metals not listed here were included at their solar metallicity from [52].

Parameter	HD 42087		HD 52089		HD 58350	
$T_{\rm eff}$ (K)	$18{,}400^{+1000}_{-200}$		$23{,}800^{+3900}_{-1400}$		$15{,}800^{+100}_{-400}$	
$\log g$ (cm s^{-2})	$2.34^{+0.01}_{-0.17}$		$3.40^{+0.01}_{-0.60}$		$1.95^{+0.02}_{-0.03}$	
$v \sin i$ (km s^{-1})	73.4 ± 8.0		38.4 ± 5.0		51.5 ± 5.0	
$v_{\rm turb}$ (km s^{-1})	x10		x10		x12	
$\dot{M}$ ($M_\odot$ yr^{-1})	$(2.3 \pm 1.0) \times 10^{-7}$		$(1.9 \pm 0.2) \times 10^{-8}$		$(6.2 \pm 2.0) \times 10^{-8}$	
v_∞ (km s^{-1})	x700		x900		x230	
β	x2		x1		x3	
$L_\star$ ($L_\odot$)	$312{,}700^{+74000}_{-13000}$		$35{,}000^{+29200}_{-7500}$		$163{,}800^{+4200}_{-15900}$	
$M_\star$ ($M_\odot$)	24.3		11.1		9.5	
$R_\star$ ($R_\odot$)	x55		x11		x54	
$\log L_\star/M_\star$	4.1		3.5		4.2	
Mean atomic mass (a.m.u.)	1.4490		1.5097		1.5095	
Distance (pc)	2470^{+420}_{-290}		124 ± 2		608^{+148}_{-148}	
$E(B-V)$ (mag)	0.4		0.005		0.03	
Element	ϵ	mass fr.	ϵ	mass fr.	ϵ	mass fr.
Hydrogen	12	5.89×10^{-1}	12	5.52×10^{-1}	12	5.52×10^{-1}
Helium	x11.23 ± 0.10	4.01×10^{-1}	x11.30 ± 0.17	4.41×10^{-1}	x11.31 ± 0.12	4.41×10^{-1}
Carbon	8.31 ± 0.08	1.37×10^{-3}	8.19 ± 0.15	1.04×10^{-3}	8.07 ± 0.08	7.75×10^{-4}
Nitrogen	8.12 ± 0.06	1.09×10^{-3}	7.97 ± 0.06	7.25×10^{-4}	8.21 ± 0.12	1.25×10^{-3}
Oxygen	8.60 ± 0.08	3.75×10^{-3}	8.30 ± 0.13	1.78×10^{-3}	8.19 ± 0.09	1.38×10^{-3}
Abundance ratios	[N/C] 0.41	[N/O] 0.38	[N/C] 0.38	[N/O] 0.53	[N/C] 0.74	[N/O] 0.88

6. Discussion

We used the evolutionary sequences from Ekström et al. [22] to explore the evolutionary stage of our star sample. The main physical ingredients of these sequences relevant to our analysis include initial abundances of H, He, and metals set to X = 0.720, Y = 0.266, and Z = 0.014 with chemical initial abundances of C = 2.283× 10^{-3}, N = 6.588× 10^{-4}, and O = 5.718× 10^{-3}, in the mass fraction. We considered evolutionary tracks with differential rotation at two different rates: Ω/Ω_{crit} = 0.568 and Ω/Ω_{crit} = 0.4, employing for the latter an interpolation of the models with Ω/Ω_{crit} = 0.568 and 0, provided in Ekström et al. [22]. The mass-loss recipes employed in these sequences are those of Vink et al. [53] for initial masses above 7 $M_\odot$. For initial masses above 15 $M_\odot$ and $\log(T_{\rm eff}) > 3.7$, de Jager et al. [54], the recipe was adopted and the correction factor for the radiative mass-loss rate from Maeder and Meynet [55] was implemented in these rotating models (see Equation (10) from Ekström et al. [22]). Detailed descriptions on the microphysics and mass-loss recipes employed in these evolutionary sequences can be found in Ekström et al. [22] and Yusof et al. [4]. Our selected stars with the derived $T_{\rm eff}$, $\log L_\star$ and mass, along with the evolutionary tracks, are depicted in Figure 12. The errors in this Figure correspond to an error of 10% in the radii [11], which in turn result in a 21% error in the mass and luminosity.

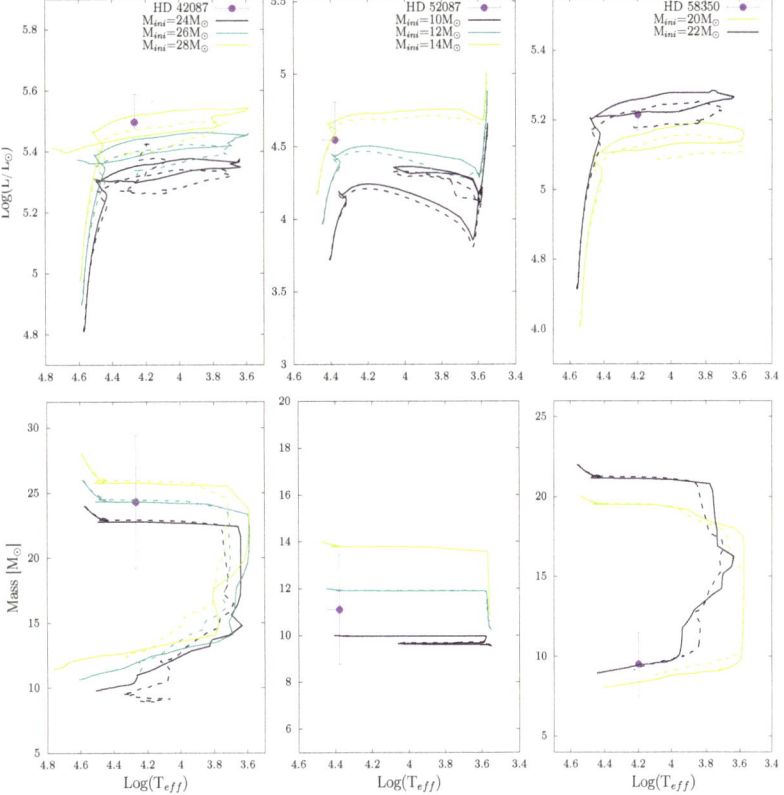

Figure 12. Left panels: HR diagram (**top**) and mass evolution (**bottom**) showing the position of HD 42087 and evolutionary tracks with initial masses of 24, 26, and 28 $M_\odot$. Middle panels: The same as in the left panels for HD 52087 and initial masses of 10, 12, and 14 $M_\odot$. Right panels: the same as before for HD 58350 and initial masses of 20 and 22 $M_\odot$. Solid and dashed lines represent models with $\Omega/\Omega_{crit} = 0.568$ from [22] and the interpolated ones for $\Omega/\Omega_{crit} = 0.4$, respectively.

6.1. HD 42087

In the left panel of Figure 12, we show the new position of HD 42087 in an HR diagram along with evolutionary tracks corresponding to initial masses of 24, 26, and 28 $M_\odot$. Continuous and dashed lines indicate rotation velocities of Ω/Ω_{crit} = 0.568 and Ω/Ω_{crit} = 0.4, respectively. In the lower panels, we include a diagram showing the total mass evolution. The derived values for T_{eff} and $\log(L_\star/L_\odot)$ suggest an initial mass of $\sim 26\ M_\odot$; our value obtained for the current mass indicates that this star is at the pre-RSG stage, which is in agreement with the mass evolution diagram for stars with the mentioned initial mass.

Our analysis for this object resulted in a higher T_{eff} and lower $\log g$ than those from Searle et al. [17] and Haucke et al. [11]. We emphasize that Haucke et al. [11] did not have the Si lines covered by their spectra; in principle, our new values would be more reliable. Additionally, with our values, HD 42087 lies in the linear relation ($\log T_{eff} - \log g$) found in Searle et al. [17] for Galactic BSGs (their Figure 6). In Figure 13, we show the theoretical SED using the D = 2470^{+420}_{-289} pc Gaia EDR3 distance and $E(B-V)$ = 0.4 mag extinction adopted from the STILISM maps [56], along with the binned IUE spectrum in black. Our procedure uses the fit formula from Cardelli et al. [57] based on the extinction data from Fitzpatrick and Massa [58,59] with R_v = 3.1. The theoretical SED fits the photometric measurements, including the IUE spectrum, and the resulting SED and CMFGEN model masses agree within error bars, with 22.5 $M_\odot$ being the mass derived from the SED. However, we were unable to match the Hα and Hβ profiles with our homogeneous wind models. Considering the reported line variability and asymmetries by Morel et al. [18], which is also obvious from Figure 2, we conclude that our steady and smooth wind model is inadequate for the 2020 CASLEO spectrum of HD 42087.

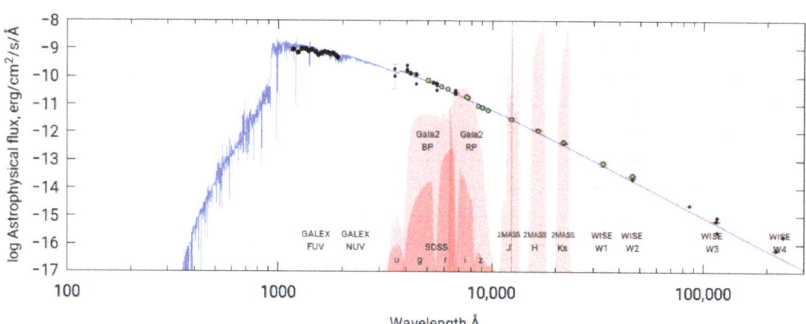

Figure 13. SED of HD 42087. All data points were taken from the VizieR Photometry Viewer service. The photometric data were de-reddened using $E(B-V)$ = 0.4 mag. The green points were used to match the slope of the passband convolved CMFGEN fluxes to the observations, and the model was normalized to the observed SED in the 2MASS/J band. The binned IUE spectrum is included with black dots.

Nevertheless, we were able to derive a mass-loss rate of $\dot M = 2.3 \times 10^{-7}\ M_\odot\ \mathrm{yr}^{-1}$ for this star with our procedure. This value is close to the value found by Haucke et al. [11] of $\dot M = 5.7 \times 10^{-7}\ M_\odot\ \mathrm{yr}^{-1}$, being lower by a factor of ~ 2.6. From looking at their observed Hα line profile (taken in 2006, Figure 2), we see a stronger emission component in Hα compared to our data (2020). This is in line with a higher mass-loss rate (in 2006) as found by these authors.

The frequency analysis performed for HD 42087 shows three frequencies, f_1, f_2, and f_3, which appear randomly over the three sectors, with two more detectable in the long time span of the three consecutive sectors (f_4 and f_5). The short lifetime for these variations prevents us from associating them with stellar pulsations but rather with stochastic variations at the stellar surface [60]. Additionally, the analysis over the combined sectors allowed us to find one stable frequency, f_r, possibly associated with stellar pulsations. The nature of this mode is not certain. Its frequency lies in the usual range ([10, 100] d) of strange modes that are known to facilitate the mass loss in massive stars [12]. These modes can be radial and non-radial [61], and they appear trapped in the strongly inflated envelopes of highly non-adiabatic stars, usually with $\log L_\star/M_\star > 4$. The low amplitude of f_r and the new value found for $\log L_\star/M_\star = 4.1$ indicate that this can be a non-radial strange mode. If true, it would explain the high variability in the wind lines (Figure 2) and why the spectrum can not be modelled with a smooth wind approach. Nevertheless, the only way to determine whether this mode facilitates the mass loss in HD 42087 is to perform a nonlinear stability analysis and to check whether the mode velocity on the surface can exceed the stellar escape velocity. We did not detect any frequency corresponding to the ∼25 d period observed for the variability in the Hα line. We stress here that the detection of any periodicity from the Hα profile in the light curves is unlikely to be observed. Any perturbation on the stellar surface produced in the large wind volume where the Hα line is formed may lead to new perturbations (for example, due to densities and inhomogeneities in the wind), making its detection difficult. Furthermore, the Hα profile, in most cases, is composed of absorption and emission components simultaneously, also hindering its detection. However, as shown in Krtička and Feldmeier [60], wind variations can cause stochastic light variations if the base perturbations are sufficiently large. The lack of multiple independent frequencies identified as stellar pulsations modes in this object would indicate that this star belongs to the pre-RSG stage, considering that massive post-RSGs have in general more excited modes than pre-RSGs, as shown in Saio et al. [15]. This is in agreement with our values for $T_{\rm eff}$, $\log g$, $M_\star$ (Figure 12).

6.2. HD 52089

The new position of HD 52089 in the HR diagram, along with evolutionary tracks for initial masses 10, 12, and 14 $M_\odot$ and $\Omega/\Omega_{\rm crit}$ = 0.568 and 0.4 (continuous and dashed lines, respectively), are depicted in Figure 12. The HR diagram suggests initial masses between 12 and 14 $M_\odot$ at the TAMS in agreement with our derived value for the current mass (∼11$M_\odot$) in this stage, considering the adopted errors in the mass. However, we cannot dismiss a merger scenario for this object that would explain the measured luminosity, which is slightly high considering an 11 $M_\odot$ object at the TAMS. Such an event can lead to rejuvenation and an excess in the luminosity of the merger remnant due to the energy injection from the secondary star during the merging of the two components. The merger product might hence appear as if it would have a significantly higher mass. The best known such case is the B[e] supergiant star R4 in the Small Magellanic Cloud [62,63]. If true, then HD 52089 would be a highly important object to study merger remnants, and it would be interesting to search for possible remnants of ejecta from the past merger event.

Additionally, this star showed an inconsistency in the $T_{\rm eff} - \log g$ distribution of [11], with a higher $\log g$ for its temperature than other stars in their sample. Our new fit to the 2015 CASLEO spectrum confirmed the earlier results and shows a discrepancy among the stellar mass, luminosity, and surface gravity, given its high effective temperature. With its close distance of D = 124 ± 2 pc and moderate interstellar extinction of $E(B - V)$ = 0.005 mag, adopted from [64] and STILISM, respectively, we find that its spectroscopic mass is in good agreement with the mass derived from the SED (9.8 $M_\odot$) depicted in Figure 14. We also notice the model from the optical fit matches the slope of the binned IUE spectrum, but there is an offset, possibly due to the low metallicity found for this object.

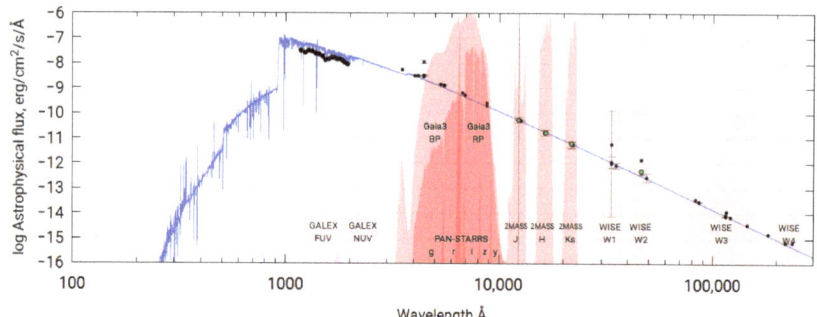

Figure 14. SED of HD 52089. Same as Figure 13, but de-reddening was conducted using $E(B-V) = 0.005$ mag.

The frequency analysis for HD 52089 over the 4 TESS sectors revealed one triplet centred in f_2 in Sector 33, corresponding to a rotational period of 5 d ($f_{rot} = 0.2\ d^{-1}$). The short baseline observations provided by TESS single sectors translate into high Rayleigh frequency separation, preventing making precise mode identification for short-term variabilities. In the case of f_3, f_2, and f_4 being members of a triplet, f_2 would be an $\ell = 1$ nonradial mode [15]. In this case, the star would have a surface rotation velocity of $\sim$110 km s^{-1}, considering the radius and rotation period derived for this object. Moreover, with the value of v sin i obtained from our analysis, we can derive the inclination of the star, which would be the $\sim$20 degree. Besides the triplet, we found f_2 in Sectors 7 and 33, and signatures of this mode in Sectors 6 and 34 with very low amplitude (not included in Table 3) along with its harmonic supporting the hypothesis of f_2 being a non-radial mode due to its apparently extended lifetime. As in the case of HD 42087, we found low-frequency signals (f_1, f_5, and f_6) randomly excited over the observed sectors, which are probably related to convective variabilities stochastically excited at the stellar surface. These observations confirm the results of [65].

Our determination of $\dot{M}$ for HD 52089, $\dot{M} = 1.9 \times 10^{-8}\ M_\odot\ yr^{-1}$ agrees very well with the one reported by Haucke et al. [11]: $\dot{M} = 2.0 \times 10^{-8}\ M_\odot$. From comparing our observed Hα line profile of this star, we do not find any significant morphological difference between the observed spectrum shown in Haucke et al. [11] and our data: a pure absorption Hα line profile with flux at the core of the line of $\sim$0.6 (normalized flux) (see Figure 3).

6.3. HD 58350

The right panel of Figure 12 shows the position of HD 58350 in the HR diagram and the evolution of the total mass along with evolutionary tracks for 20 and 22 $M_\odot$ with different rotational velocities. We found an excellent agreement between our derived values for T_{eff}, $\log(L_\star/L_\odot)$ and its current mass for a star evolution model with initial mass $\sim$22 $M_\odot$ at the post-RSG stage, after losing a considerable amount of mass during its evolution. We notice as well that our derived values for the T_{eff} and $\log g$ are in good agreement with the linear relation found for galactic BSGs in Searle et al. [17].

The theoretical SED fits the optical photometric measurements for HD 58350 (Figure 15). We notice a slightly different slope for the IR photometric data, possibly pointing towards a time-variable wind in data taken at different dates. Considering its distance $D = 608 \pm 148$ pc [11], we derived an extinction $E(B-V) = 0.18$ mag. We also found a discrepancy between the stellar masses obtained from the SED modeling ($\sim$5.5 $M_\odot$) and the best-fitting CMFGEN model ($\sim$9.5 $M_\odot$) of unclear origin. We have found different values for the effective temperature (and extinction) in the literature (see Section 2) for this object, ranging from 13,500 K to 16,000 K, possibly due to a combination between the use of different methodologies to derive it and due to stellar oscillations. This hampers a reliable comparison between our SED model and the photometric data, possibly leading to this mass discrepancy. Spectroscopic time-series observations analyzed homogeneously can help to place reliable constraints on the effective temperature.

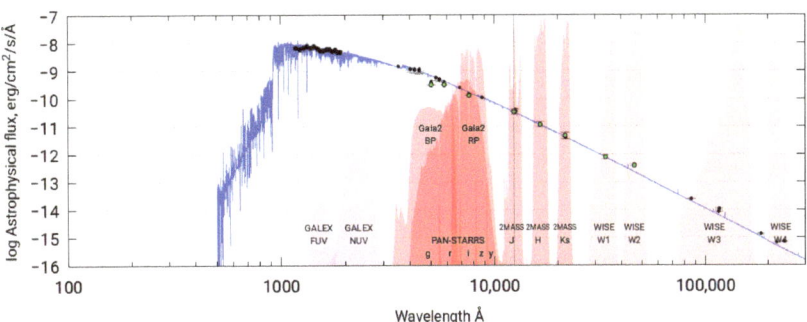

Figure 15. SED of HD 58350. Same as Figure 13, but de-reddening was conducted using $E(B-V) = 0.18$ mag (see main text).

For HD 58350, our derived $\dot{M}$ of $6.2 \times 10^{-8}\ M_\odot\ \mathrm{yr}^{-1}$ is lower than the one from Haucke et al. [11] by a factor of $\sim$2.4. This can be understood since the observed Hα line profile reported in Haucke et al. [11] (taken in 2006 and 2013) shows a more intense emission component in comparison with our data (very weak emission in Hα).

The frequency content for this star lies below $0.4\ d^{-1}$. The only sector provided by TESS at the current time allowed us to find two independent frequencies. We were not able to find the frequency discovered in Lefever et al. [21], with $f = 0.1507\ d^{-1}$, indicating that this mode is not an opacity-driven mode. From the analysis of one single TESS sector, we cannot suggest that this star is a non-radial oscillator. However, the marked Hα line profile variation (see Figure 4) suggests that these frequencies are connected to line-driven wind instabilities [60]. The analysis in [15] demonstrated that a BSG in a post-RSG state should undergo multiple pulsations. It is therefore unfortunate that HD 58350, which is the best candidate in our sample for a post-RSG star, was observed only in one TESS sector. Only with multiple, and in particular consecutive TESS observations, would it be possible to properly analyze the frequency spectrum of this object and to confirm the theoretical predictions of Saio et al. [15].

6.4. Surface Abundances

With the aim of framing these discrepancies with our derived values for surface abundances, we compared our results with those from different authors and Geneva evolutionary tracks. Figure 16 shows our measured CNO abundances for the selected stars along with the solar abundances and the average CNO abundances derived in Searle et al. [17] for their Galactic BSG sample.

In general, for all three stars, we found well-constrained CNO abundances, all showing a slight depletion of C and O, and a mild overabundance of N when compared to the solar mixture. This pattern agrees with the previously reported CNO abundance profiles in Searle et al. [17]; however, at the same time, we found higher C and slightly lower O abundances. We notice that some systematic differences are expected from the analysis itself. The global spectral modeling applied in XTGRID is fundamentally different from the methodology of Searle et al. [17], who used different diagnostic lines for T_{eff}, $\log g$, and abundance determinations. Searle et al. [17] noted that their CMFGEN models, with derived C abundances from the C II 4267 Å line, overestimated the C II 6578 and 6582 Å lines. In contrast, our C abundance analysis was based on the strongest C features in the CASLEO observations (C II 6578 and 6582 Å lines), but the global analysis is also sensitive to variations in all other spectral lines due to a change in the C abundance. This difference in diagnostics is applied consistently for all surface and wind parameters in the XTGRID. Furthermore, Nieva and Przybilla [66] showed that LTE analyses could result in discrepant abundances based on different carbon lines, and the C II 4267 multiplet tends to underestimate the carbon abundance. Further possible sources of a discrepancy may be the differences in atomic data used in the analyses and the different fit procedures. Additionally, we note that our sample of three stars is too small to match Searle et al. [17] population averages, and the individual objects in our selection may show large deviations. To uncover such systematics, one needs to process larger, homogeneously modeled datasets and multi-epoch observations, which are beyond the scope and limitations of our current work.

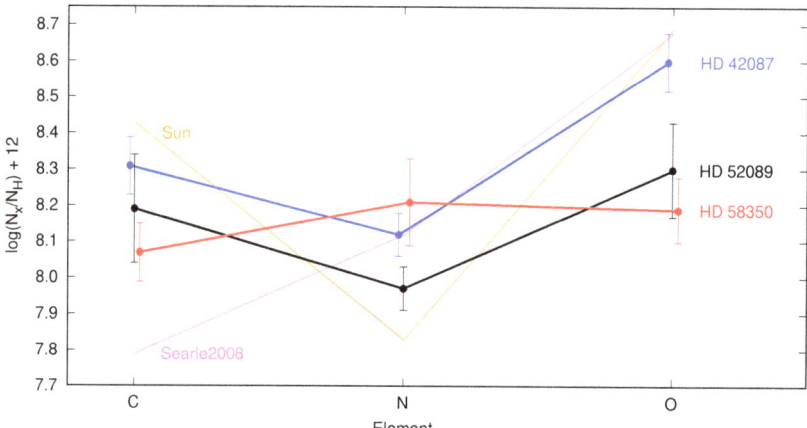

Figure 16. The measured CNO abundance patterns in all three stars compared to the solar pattern from Asplund et al. [52]. In all three stars, C and O are depleted and N is overabundant compared to the solar mixture. The mean CNO abundances found by Searle et al. [17] for Galactic BSGs are also shown for reference.

At the same time, we noticed that the total amount of metals we derived (see Table 6) correspond to subsolar metallicities, being approximately 70% $Z_\odot$, 60% $Z_\odot$, and 50% $Z_\odot$ for HD 42087, HD 52089, and HD 58350, respectively, which is possibly due to the adopted fixed value for $n\text{He}/n\text{H} = 0.2$. We have also adopted fixed solar abundances for all metals beyond oxygen and did not include elements that do not contribute to the opacity, such as Ar, K, Ti, V, Mn, and Co. All these features contribute to reducing the metal mass fractions.

In Figure 17, we compare the N/C and N/O abundance ratios from different star samples with Geneva stellar evolution models for different masses. We included the sample of A and B supergiants studied in [17,67] and the B-type main sequence (MS) stars analyzed in [68] as a reference, along with our three stars. Three well-defined groups can be identified in the panels of Figure 17. B-type MS stars between 5 and 10 $M_\odot$ form a clear sequence in temperature and show the lowest abundance ratios, which marks that these stars have low N abundances relative to C and O. Below ~13,000 K, cooler A-type supergiant stars show up as a compact group, while the hot side of the observed BSG abundance ratios shows a much larger scatter. The larger scatter can be interpreted in several ways. Krtička et al. [69] have shown that winds are driven mostly by C, Si, and S for hot BSGs and iron for cooler BSGs, implying a decreasing mass-loss rate for temperatures lower than 15,000 K. A lower mass-loss rate might operate in favor of a better determination of N/C and N/C ratios. The observed compact group could also be a signature of the fast post-RSG evolution through the cool BSG stage. Additionally, atmospheres of cool supergiants usually exhibit more lines, helping to obtain precise values in their abundances. However, it is clear that none of the measurements reach the predicted high N/C and N/O ratios for the post-RSG domain.

Based on the abundance ratios alone, we find that all our stars are more consistent with a pre-RSG stage of evolution; we do not observe the predicted and very high N/C and N/O ratios. This is in contrast with the spectroscopic mass of HD 58350, which suggests that it is in a post-RSG stage. Saio et al. [15] showed that the N/C and N/O ratios are increased mainly by the mass loss. However, they found N/C and N/O abundances consistent with models at the pre-RSG stage, in contradiction with the position in the HR diagram for Deneb and Rigel, which arrive at the same conclusion.

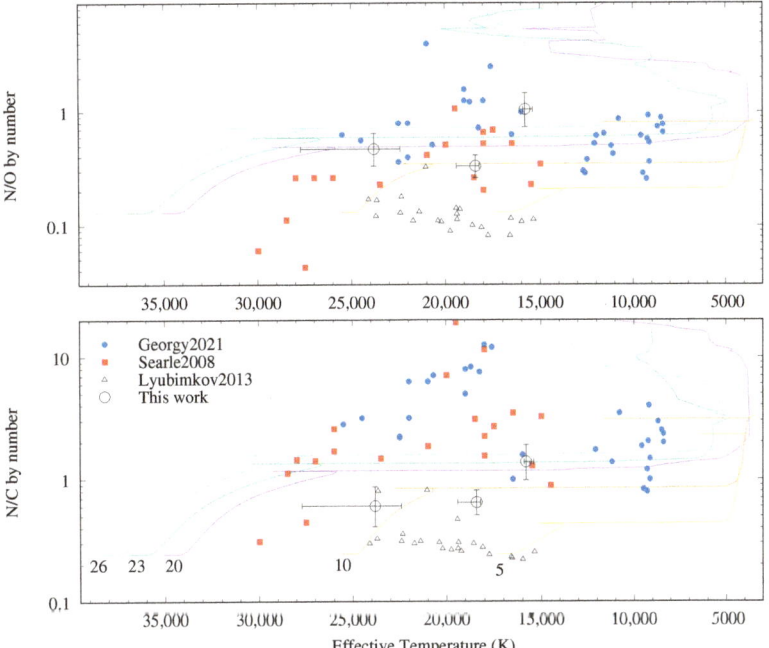

Figure 17. N/C (bottom) and N/O (top) abundance ratios by number from Searle et al. [17] and Georgy et al. [67] compared to our measurements, as well as Geneva solar metallicity (Z = 0.014), $\Omega/\Omega_{crit} = 0.4$ interpolated evolutionary tracks for 5, 10, 20, 23, and 26 $M_\odot$. For reference, we also show B-type main sequence stars from Lyubimkov et al. [68].

There is a discrepancy in the abundance ratios between evolutionary model predictions and spectroscopic measurements. The observed ratios remain significantly lower than predictions. Much larger samples, multifaceted efforts, and homogeneous modeling will be necessary to address this issue statistically.

Figure 18 shows the correlations between the N/O and N/C ratios, which is analogous to the distribution found for O-type stars in Martins et al. [70]. The offset between theoretical predictions and the measurements implies that some systematics may exist in the C and O abundances. A systematically underestimated C or overestimated O abundance can produce the observed offset. It is unlikely that all the methodologies used to analyze the surface abundances in these stars are biased in the same way; therefore, the origin of the offset remains unclear. It may be related to stellar variability as well as to atomic data or shortcomings in the surface abundance predictions. Martins et al. [70] demonstrated an anticorrelation between the N/C ratio and $\log g$; the lower the gravity, the larger the N/C ratio. Meanwhile, Saio et al. [15] concluded that recent developments in the modeling of RSG [71] make these stars more compact for a given luminosity. The combination of the two trends acts towards decreasing the offset with respect to the Geneva models in Figure 18.

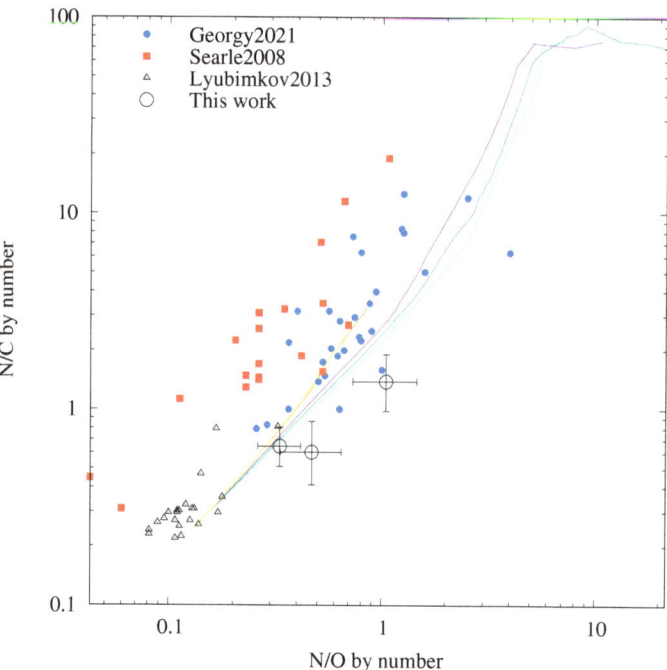

Figure 18. N/C and N/O abundance ratio correlations for our sample and the same tracks as in Figure 17. We included Searle et al. [17] and Georgy et al. [67] samples and the B-type main sequence stars from Lyubimkov et al. [68].

7. Conclusions

This paper reports our first step towards a comprehensive study of BSG stars, taking into account their photometric and spectroscopic variabilities.

The evolutionary tracks predict HD 42087 to be a pre-RSG star for our derived values of $\log(L_\star/L_\odot)$, $T_{\rm eff}$, and $M_\star$, in agreement with our values for the abundance ratios. HD 52089 is most likely an 11 $M_\odot$ star at the TAMS. However, we noticed that the derived luminosity is slightly high for an 11 $M_\odot$ star, and a binary merger scenario is plausible. Finally, for HD 58350, the evolutionary tracks for our derived values indicate that this star is at the post-RSG. However, the derived surface abundance ratios are compatible with those at the pre-RSG stage, finding the same discrepancies mentioned in Saio et al. [15] for Deneb and Rigel. The rather short sectoral observing windows of TESS are insufficient to cover the low frequencies usually present in these stars. However, we were able to detect a frequency splitting allowing us to infer a rotational period of ~5 d for HD 52089 and to find a low frequency, $f_r = 0.093\ d^{-1}$, related possibly to strange mode instabilities in HD 42089 in agreement with its new derived value for the $\log(L_\star/M_\star)$.

We found lower values for mass-loss rates of HD 42097 and HD 58350 compared with those derived by Haucke et al. [11], in agreement with the detected variability of the Hα emission component, related to changes in the wind mass-loss rate in these stars at these different epochs. The finding of low-frequencies stochastic oscillations in these objects is possibly connected with such variations.

Despite the numerous and valuable efforts to study BSG stars, there are still many issues to address; here, we highlight the most important ones:

- A large sample of BSGs needs to be studied with homogenously modeled datasets of multi-epoch observations, with the aim of uncovering systematic deviations of surface abundance ratios from evolutionary models. Additionally, multi-epoch observations

will allow us to place constraints on the current uncertainties observed in the radii of these objects and to identify changes in radii caused by radial pulsations.
- To frame the current studies for the needed M_{core}/M_r ratio for massive stars to evolve back towards the blue region of the HR diagram, such as the effect of stellar rotation, convective boundaries criteria, mixing length theories, overshooting that is adopted for the stellar interior, and evolutionary models, in terms of the CNO abundances. This will allow us to untackle the observed indetermination in the evolutionary stage of these objects with precise values for their CNO surface abundances and, in turn, will help to set the needed constraints to the current poorly established theoretical mass-loss recipes in the diverse evolutionary states and mass ranges.
- In addition, we emphasize that stellar pulsations play a key role in the analysis of BSGs, not only as a test to infer their evolutionary stage, as proposed in Saio et al. [15], but also as a mechanism that facilitates the mass loss in massive stars, as suggested in Kraus et al. [10] and theoretically confirmed in Yadav and Glatzel [72], affecting, therefore, their surface abundances. The systematic differences noticed when comparing evolutionary tracks with surface abundance measurements for BSGs (Figure 18) should be discussed, considering the effect of stellar pulsations over their evolution. Furthermore, short-term mass-loss variabilities should be contemplated in detailed evolutionary sequences, as they can act as an additional source for the discrepancies found with evolutionary models.

From a spectral analysis software development point, we have added CMFGEN modeling capabilities to the automatic spectral analysis procedure XTGRID[2]. The first results shown here demonstrate its feasibility for processing spectra of massive stars and deriving homogeneous parameters from diverse data. The next steps will include improving the accuracy of spectral inference to reduce the observed discrepancies in mass and surface abundances. Additionally, we will work on better optimizing the calculations by recycling previously calculated models. Future applicability for large datasets relies on the development of efficient methods to search the parameter space, either by utilizing large grids [73] and/or machine learning techniques.

We would like to mention, as well, that we found diffuse interstellar bands (DIB) in the spectra of HD 42087 and HD 58350. The 4428 and 6613 Å DIB bands are clearly present in both stars. Neither of the two bands is visible in the spectrum of the relatively nearby HD 52089. Although the two stars with DIBs are at larger distances, and they have very different extinction values, the DIBs show very similar strengths. The lack of DIBs in HD 52089 is likely due to its short distance and the very low interstellar extinction in its direction. In addition, HD 52089 is not only the hottest star in our sample, but also the strongest ultraviolet source in the night sky [74], which may be able to photodissociate DIB carriers.

Author Contributions: Conceptualization, J.P.S.A., P.N. and M.K.; methodology, J.P.S.A., P.N. and E.S.d.G.d.A.; software, P.N.; validation, P.N. and E.S.d.G.d.A.; formal analysis, P.N.; investigation, J.P.S.A., P.N., E.S.d.G.d.A., M.K. and M.A.R.D.; resources, M.A.R.D. and M.H.; writing—original draft preparation, J.P.S.A., P.N., M.A.R.D., M.K. and E.S.d.G.d.A.; writing—review and editing, J.P.S.A., P.N., E.S.d.G.d.A. and M.K.; visualization, J.P.S.A., P.N., M.K. and M.A.R.D.; supervision, J.P.S.A.; project administration, J.P.S.A. and P.N. All authors have read and agreed to the published version of the manuscript.

Funding: This research received funding from the European Union's Framework Programme for Research and Innovation Horizon 2020 (2014–2020) under the Marie Skłodowska-Curie Grant Agreement №823734 (POEMS project). The Astronomical Institute in Ondřejov is supported by the project RVO:67985815. J.P.S.A. and M.K. acknowledge financial support from the Czech Science foundation (GAČR 20-00150S). E.S.G.A. has been financially supported by ANID Fondecyt postdoctoral grant folio № 3220776. P. N. acknowledges support from the Grant Agency of the Czech Republic (GAČR 22-34467S). M.A.R.D. acknowledges support from CONICET (PIP 1337).

Data Availability Statement: The data underlying this article will be shared on reasonable request to the corresponding author.

Acknowledgments: We are grateful to the referees for their thoughtful reports and useful suggestions that helped us improve the manuscript. Based on observations taken with the J. Sahade Telescope at Com-plejo Astronómico El Leoncito (CASLEO), operated under an agreement between the Consejo Nacional de Investigaciones Científicas y Técnicas de la República Argentina, the Secretaría de Ciencia y Tecnología de la Nación, and the National Universities of La Plata,Córdoba, and San Juan. This research has used the services of www.Astroserver.org under reference M1DE05 (accessed on 20 August 2023).

Conflicts of Interest: The authors declare no conflict of interest. The funders had no role in the design of the study; in the collection, analyses, or interpretation of data; in the writing of the manuscript; or in the decision to publish the results'.

Notes

[1] IRAF is distributed by the National Optical Astronomy Observatory, which is operated by the Association of Universities for Research in Astronomy (AURA) under a cooperative agreement with the National Science Foundation.

[2] https://xtgrid.astroserver.org/, (accessed on 20 August 2023).

References

1. Abbott, D.C. The return of mass and energy to the interstellar medium by winds from early-type stars. *Astrophys. J.* **1982**, *263*, 723–735. [CrossRef]
2. Martins, F.; Palacios, A. A comparison of evolutionary tracks for single Galactic massive stars. *Astron. Astrophys.* **2013**, *560*, A16. [CrossRef]
3. Agrawal, P.; Szécsi, D.; Stevenson, S.; Eldridge, J.J.; Hurley, J. Explaining the differences in massive star models from various simulations. *Mon. Not. RAS* **2022**, *512*, 5717–5725. [CrossRef]
4. Yusof, N.; Hirschi, R.; Eggenberger, P.; Ekström, S.; Georgy, C.; Sibony, Y.; Crowther, P.A.; Meynet, G.; Kassim, H.A.; Harun, W.A.W.; et al. Grids of stellar models with rotation VII: Models from 0.8 to 300 $M_\odot$ at supersolar metallicity (Z = 0.020). *Mon. Not. RAS* **2022**, *511*, 2814–2828. [CrossRef]
5. Maeder, A.; Meynet, G. Evolution of massive stars with mass loss and rotation. *New Astron. Rev.* **2010**, *54*, 32–38. [CrossRef]
6. Wagle, G.A.; Ray, A.; Dev, A.; Raghu, A. Type IIP Supernova Progenitors and Their Explodability. I. Convective Overshoot, Blue Loops, and Surface Composition. *Astrophys. J.* **2019**, *886*, 27. [CrossRef]
7. Bowman, D.M.; Burssens, S.; Pedersen, M.G.; Johnston, C.; Aerts, C.; Buysschaert, B.; Michielsen, M.; Tkachenko, A.; Rogers, T.M.; Edelmann, P.V.F.; et al. Low-frequency gravity waves in blue supergiants revealed by high-precision space photometry. *Nat. Astron.* **2019**, *3*, 760–765. [CrossRef]
8. Langer, N. Presupernova Evolution of Massive Single and Binary Stars. *Annu. Rev. Astron Astrophys.* **2012**, *50*, 107–164. [CrossRef]
9. Glatzel, W. On the origin of strange modes and the mechanism of related instabilities. *Mon. Not. RAS* **1994**, *271*, 66. [CrossRef]
10. Kraus, M.; Haucke, M.; Cidale, L.S.; Venero, R.O.J.; Nickeler, D.H.; Németh, P.; Niemczura, E.; Tomić, S.; Aret, A.; Kubát, J.; et al. Interplay between pulsations and mass loss in the blue supergiant 55 Cygnus = HD 198 478. *Astron. Astrophys.* **2015**, *581*, A75. [CrossRef]
11. Haucke, M.; Cidale, L.S.; Venero, R.O.J.; Curé, M.; Kraus, M.; Kanaan, S.; Arcos, C. Wind properties of variable B supergiants. Evidence of pulsations connected with mass-loss episodes. *Astron. Astrophys.* **2018**, *614*, A91. [CrossRef]
12. Aerts, C.; Lefever, K.; Baglin, A.; Degroote, P.; Oreiro, R.; Vučković, M.; Smolders, K.; Acke, B.; Verhoelst, T.; Desmet, M.; et al. Periodic mass-loss episodes due to an oscillation mode with variable amplitude in the hot supergiant HD 50064. *Astron. Astrophys.* **2010**, *513*, L11. [CrossRef]
13. Saio, H.; Kuschnig, R.; Gautschy, A.; Cameron, C.; Walker, G.A.H.; Matthews, J.M.; Guenther, D.B.; Moffat, A.F.J.; Rucinski, S.M.; Sasselov, D.; et al. MOST Detects g- and p-Modes in the B Supergiant HD 163899 (B2 Ib/II). *Astrophys. J.* **2006**, *650*, 1111–1118. [CrossRef]
14. Ostrowski, J.; Daszyńska-Daszkiewicz, J. Pulsations in B-type supergiants with masses M < 20 $M_\odot$ before and after core helium ignition. *Mon. Not. RAS* **2015**, *447*, 2378–2386. [CrossRef]
15. Saio, H.; Georgy, C.; Meynet, G. Evolution of blue supergiants and α Cygni variables: Puzzling CNO surface abundances. *Mon. Not. RAS* **2013**, *433*, 1246–1257. [CrossRef]
16. Németh, P.; Kawka, A.; Vennes, S. A selection of hot subluminous stars in the GALEX survey - II. Subdwarf atmospheric parameters. *Mon. Not. RAS* **2012**, *427*, 2180–2211. [CrossRef]
17. Searle, S.C.; Prinja, R.K.; Massa, D.; Ryans, R. Quantitative studies of the optical and UV spectra of Galactic early B supergiants. I. Fundamental parameters. *Astron. Astrophys.* **2008**, *481*, 777–797. [CrossRef]
18. Morel, T.; Marchenko, S.V.; Pati, A.K.; Kuppuswamy, K.; Carini, M.T.; Wood, E.; Zimmerman, R. Large-scale wind structures in OB supergiants: A search for rotationally modulated Hα variability. *Mon. Not. RAS* **2004**, *351*, 552–568. [CrossRef]

19. Morel, T.; Hubrig, S.; Briquet, M. Nitrogen enrichment, boron depletion and magnetic fields in slowly-rotating B-type dwarfs. *Astron. Astrophys.* **2008**, *481*, 453–463. [CrossRef]
20. Fossati, L.; Castro, N.; Morel, T.; Langer, N.; Briquet, M.; Carroll, T.A.; Hubrig, S.; Nieva, M.F.; Oskinova, L.M.; Przybilla, N.; et al. B fields in OB stars (BOB): On the detection of weak magnetic fields in the two early B-type stars β CMa and ϵ CMa. Possible lack of a "magnetic desert" in massive stars. *Astron. Astrophys.* **2015**, *574*, A20. [CrossRef]
21. Lefever, K.; Puls, J.; Aerts, C. Statistical properties of a sample of periodically variable B-type supergiants. Evidence for opacity-driven gravity-mode oscillations. *Astron. Astrophys.* **2007**, *463*, 1093–1109. [CrossRef]
22. Ekström, S.; Georgy, C.; Eggenberger, P.; Meynet, G.; Mowlavi, N.; Wyttenbach, A.; Granada, A.; Decressin, T.; Hirschi, R.; Frischknecht, U.; et al. Grids of stellar models with rotation. I. Models from 0.8 to 120 $M_\odot$ at solar metallicity (Z = 0.014). *Astron. Astrophys.* **2012**, *537*, A146. [CrossRef]
23. Ricker, G.R.; Winn, J.N.; Vanderspek, R.; Latham, D.W.; Bakos, G.Á.; Bean, J.L.; Berta-Thompson, Z.K.; Brown, T.M.; Buchhave, L.; Butler, N.R.; et al. Transiting Exoplanet Survey Satellite (TESS). In *Space Telescopes and Instrumentation 2014: Optical, Infrared, and Millimeter Wave*; Oschmann, J.M., Clampin, M., Fazio, G.G., MacEwen, H.A., Eds.; Society of Photo-Optical Instrumentation Engineers (SPIE) Conference Series; SPIE: Bellingham, WA, USA, 2014; Volume 9143, p. 914320. [CrossRef]
24. Ricker, G.R.; Winn, J.N.; Vanderspek, R.; Latham, D.W.; Bakos, G.Á.; Bean, J.L.; Berta-Thompson, Z.K.; Brown, T.M.; Buchhave, L.; Butler, N.R.; et al. Transiting Exoplanet Survey Satellite (TESS). *J. Astron. Telesc. Instruments Syst.* **2015**, *1*, 014003. [CrossRef]
25. Ginsburg, A.; Sipőcz, B.M.; Brasseur, C.E.; Cowperthwaite, P.S.; Craig, M.W.; Deil, C.; Guillochon, J.; Guzman, G.; Liedtke, S.; Lian Lim, P.; et al. astroquery: An Astronomical Web-querying Package in Python. *Astron. J.* **2019**, *157*, 98. [CrossRef]
26. Lightkurve Collaboration; Cardoso, J.V.d.M.; Hedges, C.; Gully-Santiago, M.; Saunders, N.; Cody, A.M.; Barclay, T.; Hall, O.; Sagear, S.; Turtelboom, E.; et al. *Lightkurve: Kepler and TESS Time Series Analysis in Python*; Astrophysics Source Code Library: College Park, MD, USA, 2018. Available online: https://arxiv.org/pdf/2111.14278.pdf (accessed on 16 March 2023).
27. Garcia, S.; Van Reeth, T.; De Ridder, J.; Tkachenko, A.; IJspeert, L.; Aerts, C. Detection of period-spacing patterns due to the gravity modes of rotating dwarfs in the TESS southern continuous viewing zone. *Astron. Astrophys.* **2022**, *662*, A82. [CrossRef]
28. Lenz, P.; Breger, M. Period04 User Guide. *Commun. Asteroseismol.* **2005**, *146*, 53–136. [CrossRef]
29. Baran, A.S.; Koen, C. A Detection Threshold in the Amplitude Spectra Calculated from TESS Time-Series Data. *Acta Astron.* **2021**, *71*, 113–121. [CrossRef]
30. Hillier, D.J.; Miller, D.L. The Treatment of Non-LTE Line Blanketing in Spherically Expanding Outflows. *Astrophys. J.* **1998**, *496*, 407–427. [CrossRef]
31. Bouret, J.C.; Hillier, D.J.; Lanz, T.; Fullerton, A.W. Properties of Galactic early-type O-supergiants. A combined FUV-UV and optical analysis. *Astron. Astrophys.* **2012**, *544*, A67. [CrossRef]
32. de Almeida, E.S.G.; Hugbart, M.; Domiciano de Souza, A.; Rivet, J.P.; Vakili, F.; Siciak, A.; Labeyrie, G.; Garde, O.; Matthews, N.; Lai, O.; et al. Combined spectroscopy and intensity interferometry to determine the distances of the blue supergiants P Cygni and Rigel. *Mon. Not. RAS* **2022**, *515*, 1–12. [CrossRef]
33. Bouret, J.C.; Lanz, T.; Martins, F.; Marcolino, W.L.F.; Hillier, D.J.; Depagne, E.; Hubeny, I. Massive stars at low metallicity. Evolution and surface abundances of O dwarfs in the SMC. *Astron. Astrophys.* **2013**, *555*, A1. [CrossRef]
34. Hillier, D.J.; Davidson, K.; Ishibashi, K.; Gull, T. On the Nature of the Central Source in η Carinae. *Astrophys. J.* **2001**, *553*, 837–860. [CrossRef]
35. Eversberg, T.; Lépine, S.; Moffat, A.F.J. Outmoving Clumps in the Wind of the Hot O Supergiant ζ Puppis. *Astrophys. J.* **1998**, *494*, 799–805. [CrossRef]
36. Bouret, J.C.; Lanz, T.; Hillier, D.J. Lower mass loss rates in O-type stars: Spectral signatures of dense clumps in the wind of two Galactic O4 stars. *Astron. Astrophys.* **2005**, *438*, 301–316. [CrossRef]
37. Martins, F.; Marcolino, W.; Hillier, D.J.; Donati, J.F.; Bouret, J.C. Radial dependence of line profile variability in seven O9-B0.5 stars. *Astron. Astrophys.* **2015**, *574*, A142. [CrossRef]
38. Sander, A.A.C. Recent advances in non-LTE stellar atmosphere models. In *Proceedings of the The Lives and Death-Throes of Massive Stars*; Eldridge, J.J.; Bray, J.C.; McClelland, L.A.S.; Xiao, L., Eds., 2017; Volume 329, pp. 215–222. Available online: https://www.cambridge.org/core/journals/proceedings-of-the-international-astronomical-union/article/recent-advances-in-nonlte-stellar-atmosphere-models/B993A4F409FCE7BACF9DAE43CCCF908B (accessed on 16 March 2023). [CrossRef]
39. Martins, F.; Schaerer, D.; Hillier, D.J.; Meynadier, F.; Heydari-Malayeri, M.; Walborn, N.R. On stars with weak winds: The Galactic case. *Astron. Astrophys.* **2005**, *441*, 735–762. [CrossRef]
40. Marcolino, W.L.F.; Bouret, J.C.; Martins, F.; Hillier, D.J.; Lanz, T.; Escolano, C. Analysis of Galactic late-type O dwarfs: More constraints on the weak wind problem. *Astron. Astrophys.* **2009**, *498*, 837–852. [CrossRef]
41. de Almeida, E.S.G.; Marcolino, W.L.F.; Bouret, J.C.; Pereira, C.B. Probing the weak wind phenomenon in Galactic O-type giants. *Astron. Astrophys.* **2019**, *628*, A36. [CrossRef]
42. Rivet, J.P.; Siciak, A.; de Almeida, E.S.G.; Vakili, F.; Domiciano de Souza, A.; Fouché, M.; Lai, O.; Vernet, D.; Kaiser, R.; Guerin, W. Intensity interferometry of P Cygni in the Hα emission line: Towards distance calibration of LBV supergiant stars. *Mon. Not. RAS* **2020**, *494*, 218–227. [CrossRef]
43. Hubeny, I.; Lanz, T. Non-LTE Line-blanketed Model Atmospheres of Hot Stars. I. Hybrid Complete Linearization/Accelerated Lambda Iteration Method. *Astrophys. J.* **1995**, *439*, 875. [CrossRef]

44. Lanz, T.; Hubeny, I. A Grid of NLTE Line-blanketed Model Atmospheres of Early B-Type Stars. *Astrophys. J. Suppl.* **2007**, *169*, 83–104. [CrossRef]
45. Hubeny, I.; Lanz, T. TLUSTY User's Guide III: Operational Manual. *arXiv* **2017**, arXiv:1706.01937. https://doi.org/10.48550/arXiv.1706.01937.
46. Lin, J.; Wu, C.; Wang, X.; Németh, P.; Xiong, H.; Wu, T.; Filippenko, A.V.; Cai, Y.; Brink, T.G.; Yan, S.; et al. An 18.9 min blue large-amplitude pulsator crossing the 'Hertzsprung gap' of hot subdwarfs. *Nat. Astron.* **2023**, *7*, 223–233. [CrossRef]
47. Lei, Z.; He, R.; Nemeth, P.; Zou, X.; Xiao, H.; Yang, Y.; Zhao, J. Mass distribution for single-lined hot subdwarf stars in LAMOST. *arXiv* **2023**, arXiv:2306.15342. https://doi.org/10.48550/arXiv.2306.15342.
48. Németh, P.; Vos, J.; Molina, F.; Bastian, A. The first heavy-metal hot subdwarf composite binary SB 744. *Astron. Astrophys.* **2021**, *653*, A3. [CrossRef]
49. Luo, Y.; Németh, P.; Li, Q. Hot Subdwarf Stars Identified in Gaia DR2 with Spectra of LAMOST DR6 and DR7. II.Kinematics. *Astrophys. J.* **2020**, *898*, 64. [CrossRef]
50. Wang, K.; Németh, P.; Luo, Y.; Chen, X.; Jiang, Q.; Cao, X. Extremely Low-mass White Dwarf Stars Observed in Gaia DR2 and LAMOST DR8. *Astrophys. J.* **2022**, *936*, 5. [CrossRef]
51. Vennes, S.; Nemeth, P.; Kawka, A.; Thorstensen, J.R.; Khalack, V.; Ferrario, L.; Alper, E.H. An unusual white dwarf star may be a surviving remnant of a subluminous Type Ia supernova. *Science* **2017**, *357*, 680–683. [CrossRef]
52. Asplund, M.; Grevesse, N.; Sauval, A.J.; Scott, P. The Chemical Composition of the Sun. *Annu. Rev. Astron Astrophys.* **2009**, *47*, 481–522. [CrossRef]
53. Vink, J.S.; de Koter, A.; Lamers, H.J.G.L.M. Mass-loss predictions for O and B stars as a function of metallicity. *Astron. Astrophys.* **2001**, *369*, 574–588. [CrossRef]
54. de Jager, C.; Nieuwenhuijzen, H.; van der Hucht, K.A. Mass loss rates in the Hertzsprung-Russell diagram. *Astron. Astrophys. Suppl.* **1988**, *72*, 259–289.
55. Maeder, A.; Meynet, G. Stellar evolution with rotation. VI. The Eddington and Omega -limits, the rotational mass loss for OB and LBV stars. *Astron. Astrophys.* **2000**, *361*, 159–166. [CrossRef]
56. Capitanio, L.; Lallement, R.; Vergely, J.L.; Elyajouri, M.; Monreal-Ibero, A. Three-dimensional mapping of the local interstellar medium with composite data. *Astron. Astrophys.* **2017**, *606*, A65. [CrossRef]
57. Cardelli, J.A.; Clayton, G.C.; Mathis, J.S. The Relationship between Infrared, Optical, and Ultraviolet Extinction. *Astrophys. J.* **1989**, *345*, 245. [CrossRef]
58. Fitzpatrick, E.L.; Massa, D. An Analysis of the Shapes of Ultraviolet Extinction Curves. I. The 2175 Angstrom Bump. *Astrophys. J.* **1986**, *307*, 286. [CrossRef]
59. Fitzpatrick, E.L.; Massa, D. An Analysis of the Shapes of Ultraviolet Extinction Curves. II. The Far-UV Extinction. *Astrophys. J.* **1988**, *328*, 734. [CrossRef]
60. Krtička, J.; Feldmeier, A. Stochastic light variations in hot stars from wind instability: finding photometric signatures and testing against the TESS data. *Astron. Astrophys.* **2021**, *648*, A79. [CrossRef]
61. Saio, H. Linear analyses for the stability of radial and non-radial oscillations of massive stars. *Mon. Not. RAS* **2011**, *412*, 1814–1822. [CrossRef]
62. Zickgraf, F.J.; Kovacs, J.; Wolf, B.; Stahl, O.; Kaufer, A.; Appenzeller, I. R4 in the Small Magellanic Cloud: A spectroscopic binary with a B[e]/LBV-type component. *Astron. Astrophys.* **1996**, *309*, 505–514.
63. Wu, S.; Everson, R.W.; Schneider, F.R.N.; Podsiadlowski, P.; Ramirez-Ruiz, E. The Art of Modeling Stellar Mergers and the Case of the B[e] Supergiant R4 in the Small Magellanic Cloud. *Astrophys. J.* **2020**, *901*, 44. [CrossRef]
64. van Leeuwen, F. Validation of the new Hipparcos reduction. *Astron. Astrophys.* **2007**, *474*, 653–664. [CrossRef]
65. Burssens, S.; Simón-Díaz, S.; Bowman, D.M.; Holgado, G.; Michielsen, M.; de Burgos, A.; Castro, N.; Barbá, R.H.; Aerts, C. Variability of OB stars from TESS southern Sectors 1-13 and high-resolution IACOB and OWN spectroscopy. *Astron. Astrophys.* **2020**, *639*, A81. [CrossRef]
66. Nieva, M.F.; Przybilla, N. C II Abundances in Early-Type Stars: Solution to a Notorious Non-LTE Problem. *Astrophys. J. Lett.* **2006**, *639*, L39–L42. [CrossRef]
67. Georgy, C.; Saio, H.; Meynet, G. Blue supergiants as tests for stellar physics. *Astron. Astrophys.* **2021**, *650*, A128. [CrossRef]
68. Lyubimkov, L.S.; Lambert, D.L.; Poklad, D.B.; Rachkovskaya, T.M.; Rostopchin, S.I. Carbon, nitrogen and oxygen abundances in atmospheres of the 5–11 $M_\odot$ B-type main-sequence stars. *Mon. Not. RAS* **2013**, *428*, 3497–3508. [CrossRef]
69. Krtička, J.; Kubát, J.; Krtičková, I. New mass-loss rates of B supergiants from global wind models. *Astron. Astrophys.* **2021**, *647*, A28. [CrossRef]
70. Martins, F.; Hervé, A.; Bouret, J.C.; Marcolino, W.; Wade, G.A.; Neiner, C.; Alecian, E.; Grunhut, J.; Petit, V. The MiMeS survey of magnetism in massive stars: CNO surface abundances of Galactic O stars. *Astron. Astrophys.* **2015**, *575*, A34. [CrossRef]
71. Davies, B.; Kudritzki, R.P.; Plez, B.; Trager, S.; Lançon, A.; Gazak, Z.; Bergemann, M.; Evans, C.; Chiavassa, A. The Temperatures of Red Supergiants. *Astrophys. J.* **2013**, *767*, 3. [CrossRef]
72. Yadav, A.P.; Glatzel, W. Stability analysis, non-linear pulsations and mass loss of models for 55 Cygni (HD 198478). *Mon. Not. RAS* **2016**, *457*, 4330–4339. [CrossRef]

73. Zsargó, J.; Fierro-Santillán, C.R.; Klapp, J.; Arrieta, A.; Arias, L.; Valencia, J.M.; Sigalotti, L.D.G.; Hareter, M.; Puebla, R.E. Creating and using large grids of precalculated model atmospheres for a rapid analysis of stellar spectra. *Astron. Astrophys.* **2020**, *643*, A88. [CrossRef]
74. Gregorio, A.; Stalio, R.; Broadfoot, L.; Castelli, F.; Hack, M.; Holberg, J. UVSTAR observations of Adara (epsilon CMa): 575-1250 Å. *Astron. Astrophys.* **2002**, *383*, 881–891. [CrossRef]

Disclaimer/Publisher's Note: The statements, opinions and data contained in all publications are solely those of the individual author(s) and contributor(s) and not of MDPI and/or the editor(s). MDPI and/or the editor(s) disclaim responsibility for any injury to people or property resulting from any ideas, methods, instructions or products referred to in the content.

Review

BCD Spectrophotometry and Rotation of Active B-Type Stars: Theory and Observations

Juan Zorec

Institut d'Astrophysique de Paris, Sorbonne Universités, CNRS, UPMC, UMR7095, 98bis Bd. Arago, F-75014 Paris, France; zorec@iap.fr

Abstract: This review has two parts. The first one is devoted to the Barbier–Chalonge–Divan (BCD) spectrophotometric system, also known as the Paris spectral classification system. Although the BCD system has been applied and is still used for all stellar objects from O to F spectral types, the present account mainly concerns normal and 'active' B-type stars. The second part treats topics related to stellar rotation, considered one of the key phenomena determining the structure and evolution of stars. The first part is eminently observational. In contrast, the second part deals with observational aspects related to stellar rotation but also recalls some supporting or basic theoretical concepts that may help better understand the gains and shortcomings of today's existent interpretation of stellar data.

Keywords: Techniques: spectrophotometric; Stars: activity; Stars: emission-line, B, Bn, Be, B[e], Bm; Stars: rotation; Stars: evolution; Stars: chemically peculiar; Stars: magnetic field

1. Introduction

This review is dedicated to the BCD spectrophotometric system (Barbier–Chalonge–Divan; spectral classification system of Paris) applied to normal and active B-type stars and their rotation. This paper presents observational and theoretical aspects related to the topics covered.

Although the BCD spectrophotometry and rotation may seem to be disjoint themes, the main reason for their joint presentation resides in the fact that BCD spectroscopy has allowed to obtain the apparent astrophysical parameters of rapidly rotating B-type stars with and without emission lines (Be, B[e]), whose correction for the effects induced by the rapid rotation on their emitted energy distribution gave access, for the first time, to the knowledge of their true mass and evolutionary state. Be stars have been long considered to be in the phase of the secondary contraction, which favors the increase in the surface rapid rotation, an outstanding characteristic of these objects.

Citation: Zorec, J. BCD Spectrophotometry and Rotation of Active B-Type Stars: Theory and Observations. *Galaxies* **2023**, *11*, 54. https://doi.org/10.3390/galaxies11020054

Academic Editor: Oleg Malkov

Received: 20 December 2022
Revised: 24 February 2023
Accepted: 4 March 2023
Published: 10 April 2023

Copyright: © 2023 by the authors. Licensee MDPI, Basel, Switzerland. This article is an open access article distributed under the terms and conditions of the Creative Commons Attribution (CC BY) license (https://creativecommons.org/licenses/by/4.0/).

Among the main advantages of the BCD spectrophotometry over spectroscopy are: the deduced parameters refer to stellar layers that are on average deeper than those producing the spectral lines, and thus, they can provide a more faithful description of the central body of the object; these parameters are not affected by the interstellar extinction nor by the disturbing emissions and absorptions produced in the circumstellar media that characterize active B-type stars, such as Be and B[e] stars.

With the exception of $V\sin i$, the first corrections that were made for effects due to rapid rotation on the stellar observational parameters were made on the astrophysical parameters ($T_{\rm eff}$, $\log g$) obtained with the BCD system. With them, it has been possible to demonstrate with certainty that the Be phenomenon can appear at any moment in the evolution of B-type stars in the main sequence, between the Zero-Age-Main-Sequence (ZAMS) and the Terminal-Age-Main-Sequence (TAMS). The ages of isolated individual stars obtained in this way take into account the evolutionary changes carried by the rotation. On the other hand, they are complementary to those deduced by the method of isochrones applied to normal and active stars in galactic clusters.

The BCD system is easy to use, and only requires spectra with low spectral resolution which are obtained for a short spectral domain, between the near ultraviolet and the visible. Moreover, their reduction is easily automatable. With the help of large telescopes and modern detectors, its use can be easily generalized to characterize stellar populations in other galaxies and possibly in environments of cosmological interest.

Rotation is a complex phenomenon that induces changes in the stellar geometry, determines non-uniform distributions of the surface temperature and gravity, and induces an important series of internal stellar instabilities. These instabilities produce a redistribution of the angular momentum stored by the star during its formation phases. This redistribution leads to internal and surface differential rotation, destabilizes stellar internal regions to convection and favors the creation of magnetic fields. Due to the rotation, a star evolves as having a smaller effective mass and the characteristics of the emitted spectrum depend on the angle of observation.

The amount of angular momentum, its internal distribution and the initial chemical composition, determine the observed spectral characteristics of rotating stars. It is then important to have some information on the circumstances related to stellar formation, and on the theoretical bases that help understand the internal structure of rotating stars that lead to the spectrum finally emitted by the star, particularly the Spectral-Energy-Distribution (SED) studied with the BCD system.

It is worth noting that, when speaking of active B-type stars, we consider objects where observational characteristics indicating some specific physical properties are enhanced. These properties may concern line emission, flux excess, abundance peculiarities and outbursts. The division between active and nonactive objects may be artificial. Stars that are in the pre-main sequence (pre-MS) evolution phases are currently considered in the category of active objects. They are, however, progenitors of what will probably be inactive stars, which they will remain during all of their life in the MS, or a large fraction of it. Likewise, objects apparently inactive during a considerable period of their life in the MS phase may at some point acquire the conditions to manifest particularities, as it happens with Be and B[e] stars. This means that, in the account of stellar activities, it is equally important to consider normal states that can, if certain conditions are met, evolve into properties labeled as active. Evolution in binary systems can trigger such conditions, particularly to display the Be and B[e] phenomena. Then, in this review, we deal with both types of objects, normal and active.

Section 2 is dedicated to the presentation of the BCD spectrophotometric system, and encompasses the basic definitions, its use for field and cluster B-type stars, normal, Bn, Be, helium-weak and helium-strong B-types star and B[e] stars. A short account is given on its development to date. Although the application of this system to objects with complex spectra can lead to some uncertainties, it is nevertheless a method that in many circumstances can provide valuable complements to the spectroscopic results. In this section, emphasis is placed on the need for the detailed modeling of BCD parameters, particularly the second Balmer discontinuity in Be stars, because it can carry important information about the structure of the circumstellar disc (CD) zones close to the central object, where the transfer of the angular momentum is organized to the rest of the CD.

The effects induced by the rotation on the spectra emitted by stars are discussed in Section 3. Some theoretical generalities are exposed about the possible content and distribution of the angular momentum in a dynamically stable self-gravitating object. They allow us to visualize the limits of current rotating star models and think about the improvements that could be made. Herein, some elements of the astrophysical parameter correction technique are given the for rotational effects that ultimately allow us to determine the true mass and evolutionary state of a rapidly rotating star. The mixing of the chemical composition and the redistribution in stellar interiors is shortly discussed in Section 4. In Section 5, a detailed presentation of the uncertainties that affect the $V\sin i$ parameter is given, including the effect of differential rotation on the stellar surface. The important contribution of interferometry to the study of rotating normal active and B-type stars is exposed in Section 6. Section 7 is devoted to the origin of the stellar angular momentum and its internal evolution. Debates are exposed which consider fast rotators as isolated objects, but also as possible results of the phenomenon of merging in binary systems. Observational results are shown that confirm the fast rotation of the Be stars, but not the generality of their quasi-critical rotation. The evolution of rotational velocities in single tars is presented in Section 8. In Section 9, it is shown that the evolution of B-type stars may imply phenomenological kinship between Bn and Be stars. The rotation of magnetic B-type stars is presented in Section 10 and that of B[e] stars in Section 11. A discussion and the concluding remarks are presented in Section 12.

2. B-Type Stars Observed with the BCD Spectrophotometric System

2.1. Brief Historical Account

Daniel Chalonge (1895–1977), one of the most talented students of the renowned physicist Charles Fabry (1867–1945), started studying the continuum energy distribution of stars in wavelength regions around the Balmer discontinuity during two observation missions in 1934 at the Swiss stations of Arosa and Jungfraujoch. The first results were published by Chalonge [1], Arnulf et al. [2], wherein the main interest was to measure the Balmer discontinuity (BD), the color gradients of the energy distribution on both sides of the BD and to standardize the correction for atmospheric absorption, particularly that due to the ozone at $\lambda < 3350$ Å. Furthermore, in 1938, D. Chalonge introduced the micro-photometer, especially designed to record the spectra he was to use for their research. The calibration lamp that enabled converting the spectral micro photograms into spectrophotometric intensities was then described by Guérin [3].

The Chalonge spectrograph gave definite characteristics to the Paris stellar spectrophotometric classification system which was presented by Baillet et al. [4]. The spectra obtained with this instrument are 15 mm long from 3100 to 6100 Å, their dispersion varies from 580 Å/mm at 6000 Å, to 220 Å/mm at Hγ and 78 Å/mm at 3100 Å. On each 6×6 cm photographic plate, 20 stellar spectra were recorded with intensity of 13 and wavelength calibration spectra interspersed in the middle of the plate. All these spectra have a triangular shape due to an oscillating chassis that enables to collect a more significant number of ultraviolet photons, which are strongly absorbed by the atmosphere, and disperse over a wider surface of the plate, the blue–red photons avoiding thus over-exposures.

2.2. Basics of the BCD system

The BCD stellar classification scheme is based on three parameters: the logarithmic flux drop, D dex, of the BD; the wavelength of the midpoint of the intensity decrease, commonly presented in the reduced form $\lambda_1 - 3700$ Å, where λ_1 is the actual wavelength of the midpoint of the discontinuity; the blue spectral gradient between 4000 and 4600 Å, Φ_b μm. Usually, two other gradients are also given, Φ_{rb} defined over the 4000 and a 6100 Å wavelength interval and generally determined only for stars hotter than the A0 spectral type, and Φ_{uv} defined between 3150 and 3700 Å (see the definition of these gradients in Equations (3) and (5)).

The value of D is calculated at $\lambda = 3700$ Å, as $D = \log_{10} F^+_{3700}/F^-_{3700}$, where F^+_{3700} is the Paschen side of the flux and F^-_{3700} is the flux in the Balmer continuum. The value of F^+_{3700} is obtained by the extrapolation of the rectified Paschen continuum to $\lambda = 3700$ Å, for which a relation such as $\log F_\lambda/B_\lambda = p \times (1/\lambda) + q$ is used because this relation as a function of $1/\lambda$ is nearly a straight line. B_λ can be the flux of a comparison star or simply the Planck function, calculated for a higher effective temperature than that expected for the studied star.

$$D = \log\left[\frac{F_{3700+}/B_{3700}}{F_{3700-}/B_{3700}}\right] \text{ dex}. \quad (1)$$

In the original application of the BCD method, the normalization of intensities is obtained with standard stars. Some of these standard stars were published in Divan [5]. The zero-scale value of the BD determinations in the BCD system was determined using the discontinuity of the B0-type supergiant ϵ Ori [6], whose D was determined in an absolute way with a laboratory black body. The values of D are determined within a typical error $\Delta D \lesssim 0.015$ dex.

At first glance, Equation (1) seems to define a value perfectly independent of the interstellar extinction. However, while F^{-}_{3700} can be identified without proceeding to any energy distribution extrapolation, F^{+}_{3700} is necessarily extrapolated and thus depends on the Paschen energy distribution over a larger wavelength region, which can be affected (reddened) by the interstellar extinction. At the same time, the error on D carried by this extinction is relatively small

$$\begin{aligned} D_{\text{cor}} &= D_{\text{obs}} - 0.004\, e_b \\ e_b &= \Phi_b - \Phi_b^* \end{aligned} \quad (2)$$

where Φ_b^* is the stellar intrinsic gradient and e_b is the gradient excess due to the interstellar extinction which, translated into the UBV Johnson–Cousins' photometric system is $e_b = 1.61 E(B-V)$ [7,8]. This means that, for $E(B-V) \lesssim 2.3$ mag, the discontinuity D can be considered unperturbed by the interstellar extinction. Otherwise, a short iteration using Equation (2) may rapidly converge to the correct estimate of D.

Because the coalescence of the higher lines of the Balmer series shifts the apparent BD to longer wavelengths, to avoid uncertainties on the identification of the last recorded merged member, which depends on the spectral resolution power and the plate photographic optical sensitivity, Barbier and Chalonge [9] introduced the parameter λ_1. It is determined by the intersection of the pseudo-continuum that joins the overlapping wings of the coalescent Balmer lines, with the $\log F_\lambda - D/2$ line (see Figures 1 and 2). This parameter is currently given as $\lambda_1 - 3700$ Å. It can be determined within an error not larger than $\Delta \lambda_1 \sim 1$ to 3 Å.

The value of D roughly ranges from 0.035 dex for O4-5 type stars to nearly 0.500 dex for A2-type stars, and then back for cooler spectral types to 0.035 dex in F9-type stars. The $\lambda_1 - 3700$ goes roughly from 70 Å for dwarfs to $\lambda_1 - 3700 \simeq -5$ Å for supergiants, which indicates that λ_1 is a well-resolved quantity useful for the stellar luminosity class classification. The unreddened Φ_{rb} gradient varies in a continuous way from $\Phi_{rb} = 0.60$ μm in O-type stars passing by $\Phi_{rb} \simeq 1.0$ μm in early A-type stars, and attains $\Phi_{rb} \simeq 2.6$ μm at spectral type F9. This gradient can, in some cases, help determine whether a given value of D corresponds to the hot or cold section of the BCD spectral classification system. In general, Φ_{rb} is affected by the interstellar reddening, and an iterated value of the color excess $E(B-V)$ can eventually disentangle the right spectral classification side. Figure 1 shows the various BCD parameters determined from the spectrum of a B-type star around the BD. In Figure 2, the determination of the BCD parameters in the Be stars is shown. Specific details on the BCD parameters of Be stars are given in Section 2.5.1.

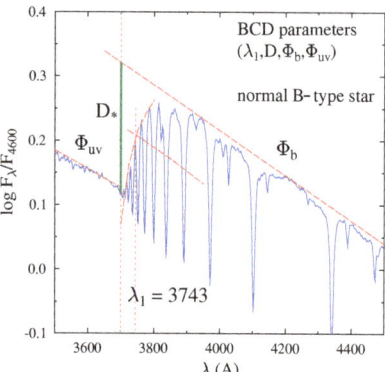

Figure 1. Graphical explanation of the BCD spectrophotometric parameters ($D, \lambda_1, \Phi_b, \Phi_{uv}$) determination for a B-type normal star. The photospheric D_* (heavy full line) and the λ_1 parameters are indicated. The slopes that define the gradients Φ_{uv} and Φ_b are also indicated. The spectrum was obtained in the Complejo Astronómico El Leoncito (CASLEO), San Juan, Argentina. Adapted from Aidelman et al. [10].

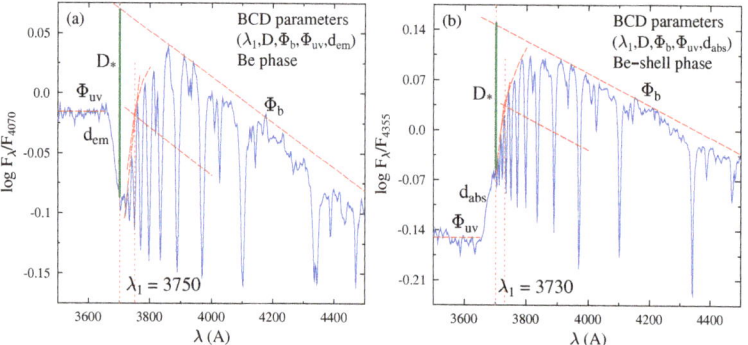

Figure 2. Same as in Figure 1, but in (**a**) is shown the energy distribution of a Be star, whose second component of the BD is in emission (flux excess). In (**b**), the spectrum is for a Be star in shell phase, where the second component of the BD appears in absorption (flux deficiency). As for Figure 1, spectra were obtained in the Complejo Astronómico El Leoncito (CASLEO), San Juan, Argentina, with the Boller and Chivens spectrograph mounted on the J. Sahade 2.15 m telescope. Adapted from Aidelman et al. [10].

When carrying out the spectrophotometric study of the energy distribution near the Balmer discontinuity, two other parameters were obtained: the color gradient Φ_{uv}, given in μm and defined for the 3200–3700 Å spectral region, and the Paschen gradient defined in two versions: the already mentioned Φ_b valid for the spectral regions 4000–4800 Å, and Φ_{rb} defined over the 400–6700 Å spectral region; both are given in μm. The determination of Φ_{uv} requires determining the amount of ozone absorption in the near UV spectral region. Using wavelength units in μm, a color gradient is defined as [11]:

$$\Phi = 5\lambda - \frac{d \ln F_\lambda}{d(1/\lambda)}, \qquad (3)$$

which for a black body at temperature T becomes :

$$\Phi(T) = (C_2/T)\left(1 - e^{-C_2/\lambda T}\right), \qquad (4)$$

where $C_2 = hc/k = 1.4388$ cm·deg is the radiation constant. Assuming that, for a given stellar energy distribution F_λ, Φ can be considered being constant between two wavelengths λ_a and λ_b of the chosen continuum energy distribution, and the expression for Φ becomes:

$$\Phi = \ln\left[\frac{\lambda_a^5 F_{\lambda_a}}{\lambda_b^5 F_{\lambda_b}}\right] / (1/\lambda_a - 1/\lambda_b). \quad (5)$$

As the local temperature of the formation region of the Paschen continuum is close to the stellar effective temperature, from Equations (4) and (5), we note that stars with the same effective temperature but different surface gravity define a common region in the plane (λ_1, D). This fact was used by Barbier and Chalonge [12] and by Chalonge and Divan [13] to determine the curvilinear quadrilaterals that characterize the BCD classification system.

The spectral classification system developed in 1943 at Yerkes Observatory by W. W. Morgan, Philip Childs Keenan, and Edith Marie Kellman, widely known as the MK (or MKK) classification or the Yerkes system, retained the sequence of stellar spectral types named O, B, A, F, G, K, M. They also introduced a range of luminosity classes which indicate whether the star is a supergiant, giant, dwarf or some intermediate class. Barbier and Chalonge [12], Barbier [14] and Chalonge [15] linked the MK classification to the BCD classification based on the (λ_1, D) parameters, but only using the MK classification of stars made by Keenan and Morgan [16] themselves. To this end, the authors of the BCD classification system simply delimited the common region occupied by stars having the same MK spectral type with curves of intrinsic constant Φ_{rb} parameters. In the same way, they have drawn the 'horizontal' lines that separate the MK luminosity classes. The BCD authors attempted to keep inside a common strip of stars, which are all spectral types having the same MK luminosity class label assigned by Morgan and Keenan. Each spectral type obtained in the BCD system and labeled with a given MK designation thus represents a rather wide range of (λ_1, D) parameters, characterizing a curvilinear quadrilateral. The 'vertical' curves of constant gradients Φ_b or Φ_{rb} separate the spectral types and the 'horizontal' curves separate the luminosity classes. Chalonge and Divan [13] is considered today as the founding publication of the BCD stellar classification system.

The largest collection of (λ_1, D) parameters of early-type stars can be found in Chalonge and Divan [13], Underhill et al. [17], Zorec and Briot [18], Underhill et al. [17] and Zorec et al. [19].

A graphical presentation of BD calculated with the LTE model atmospheres is presented in Underhill et al. [17], and a comparison between BD determinations with the LTE and non-LTE model of non-extended atmospheres was made in Zorec [20], where it is shown that differences are of the order of some $\delta D \sim 0.001$ dex, which is much smaller than the measurement uncertainties. Rough estimates of δD using the LTE models of extended atmospheres with $T_{\rm eff} \leq 12{,}000$ K [21] show that the differences are of the same order of magnitude as given above.

2.3. Relation between the BCD Quantities and the Physical Parameters of Stars

A very useful relation implying that the (λ_1, D) parameters are the calibration of the visual absolute magnitude $M_V = M_V(\lambda_1, D)$ obtained with stars in the solar neighborhood [8]. This calibration enabled the estimation of the distance modulus $V_o - M_V$ (V_o is the apparent visual magnitude corrected for interstellar extinction) of several Galactic clusters using only a small number of cluster members. As it is known, regions with different metallicities are characterized by different absolute magnitude scales. Relations $M_V = M_V(\lambda_1, D)$ must then be determined with stars that belong to regions characterized by the specific metallicity. The above calibration of M_V determined for the solar metallicity when used for stars in the Magellanic Clouds produces deviations of $V_o - M_V$ to the expected values [22,23].

In the (λ_1, D) diagram, the parameter s defined as the value of D taken at $\lambda_1 - 3700 = 60$ Å is considered the spectral type classification parameter in the BCD system. Chalonge and Divan [24] showed the high sensitivity of the BD to the stellar effective temperature.

As noted above, the color gradient Φ_b was introduced in 1955 in the BCD system to distinguish F-type stars from B-type stars having the same (λ_1, D) pairs. Later, for stars with spectral types later than A3, Chalonge and Divan [25] have shown the deviation $\Phi_b - \Phi_n$, where Φ_n is the curve joining the lowest points in a $\Phi_b = \Phi_b(s)$ relation, which is strongly correlated with the [Fe/H] abundance ratio.

An observed value of Φ_b or Φ_{rb} also provides an estimation of the ISM extinction as follows [7,8]

$$A_V = 3.1\, E(B-V) = 1.9\,(\Phi_b - \Phi_b^*) = 1.7\,(\Phi_{rb} - \Phi_{rb}^*) \tag{6}$$

where Φ_b and Φ_{rb} are the intrinsic gradients of stars corresponding to their (λ_1, D) parameters. Relations similar to those in Equation (6) must be redefined each time the gradients characterize spectral regions that are not the same as those that define the original BCD Φ_b and Φ_{rb} gradients [10,26].

The great advantage of the BCD system is that it can be used for a wide range of spectral types going from mid-O-type to late F-type stars, and it relies on low-resolution spectra obtained over a relatively short wavelength range (3500–4500 Å), which automated reduction codes can treat. It might then be a valuable tool to characterize faint stars in clusters, or belonging to stellar populations in distant regions of our Galaxy, or those of other more or less neighboring galaxies, using the multi-object spectrographs and/or spectro-imaging devices of large modern telescopes.

The $M_V(\lambda_1, D)$ and $T_{\text{eff}}(\lambda_1, D)$ calibrations for normal B-type stars were then revisited by Zorec and Briot [18] and Zorec et al. [19], respectively. The $M_{\text{bol}}(\lambda_1, D)$ and $\log g(\lambda_1, D)$ calibrations were presented in Zorec [20] and partially in Divan and Zorec [27]. These calibrations are reproduced in Figure 3.

2.4. The BCD System Today

The original Challonge spectrograph has not been in service since December 1988. Other instruments have then been employed to perpetuate the BCD system today. Recently, the most frequent use of the BCD system was made in Argentina, where low-resolution spectra are obtained at the Complejo Astronómico El Leoncito (CASLEO), San Juan, with the Boller and Chivens spectrograph mounted on the J. Sahade 2.15 m telescope. The instrumental configuration consists of a 600 l/mm grating (# 80), a slit width of 250 μm and a CCD detector of 512 × 512 pixels. The spectra cover the 3500–4700 Å wavelength range with a 2-pixel of 4.53 Å, or R = 900. A standard reduction procedure is applied to the spectroscopic images using over-scan, bias- and flat-field corrections. When needed for faint stars, dark-frame subtractions are carried out. The He-Ne-Ar comparison and spectrophotometric flux standard star spectra are regularly obtained, respectively, for wavelength and flux calibrations. The low-resolution spectra for the BCD spectrophotometry are obtained with the widest possible opening of the slit, which thus minimizes light losses and enables to obtain spectra to carry out correct spectrophotometric measurements. Observations are made at the lowest possible zenith angle to minimize refraction effects due to the Earth's atmosphere and thus to avoid other light losses as a function of wavelength. Generally, observations are reduced with the IRAF software package, and all spectra are fully corrected for atmospheric extinction and calibrated with flux standard stars regularly observed during the same run, e.g., [10,28].

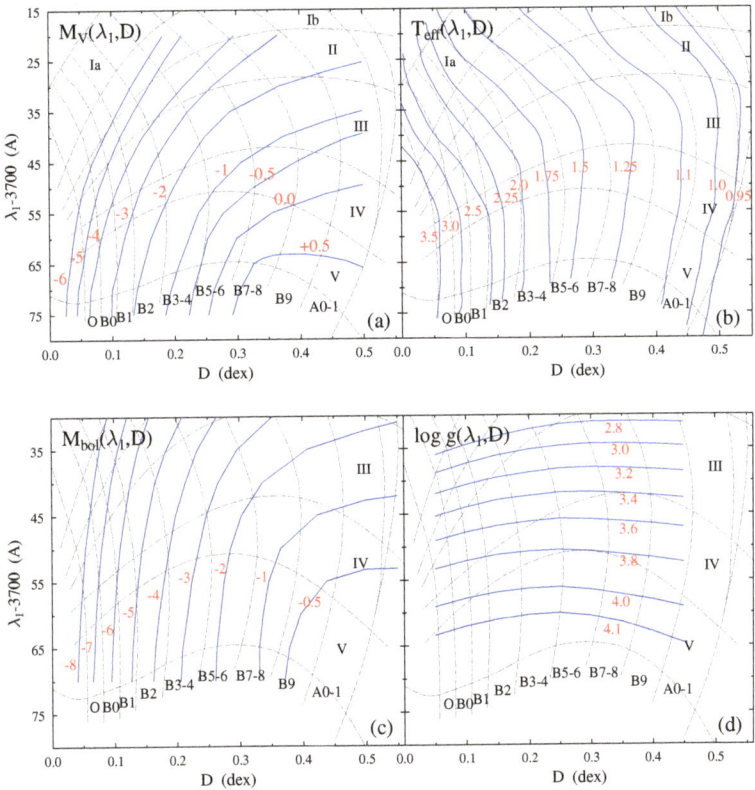

Figure 3. Empirical calibrations of absolute magnitudes M_V and M_{bol} and parameters T_{eff} and $\log g$ as a function of the BCD parameters (λ_1, D). The horizontal black dashed lines separate the MK luminosity classes indicated on the left side of each panel. The vertical black dashed lines separate the groups of the MK spectral types specified at the bottom of each panel. The blue lines represent the calibrations of the indicated astrophysical parameters, respectively, labeled in red (in units of magnitude for M_V and M_{bol}, $T_{eff}(K)/10^4$ for effective temperatures and dex for $\log g$. Figures adapted: (**a**) from Zorec and Briot [18], (**b**) from Zorec et al. [19], (**c**,**d**) from Zorec [20].

Mennickent et al. [29] conducted spectroscopic observations at Cerro Tololo Inter-American Observatory, Chile (CTIO), Las Cumbres Observatory, USA (LCO), and La Silla European Southern Observatory (ESO) Chile Observatory in 2003 to produce BCD classifications of double-periodic blue variables in the Magellanic Clouds observed in the OGLE survey [30]. At CTIO, the 1.5 m telescope was used with the Cassegrain spectrograph and the Loral 1-K detector. Grating N°26 tilted at 15.95° and a slit width of 2 arcsecs yielded a spectral range of 3500–5300 Å, with a resolution of 2 Å. Wavelength calibration functions with a typical standard deviation of 0.1 Å were obtained with around 30 He-Ar lines for the comparison spectra. Observations of the standard stars were used for flux calibrations. They used the NTT (ESO New Technology Telescope) with the EMMI blue arm at La Silla in medium dispersion grating mode, with grating No 4. This instrumental setup and a 1 arcsec slit yielded a spectral range of 3500–5050 Å and a resolution of 5 Å. The wavelength calibration functions have a standard deviation of 0.2–0.3 Å. The standard stars were observed with a 5-arcsec slit for flux calibrations. At LCO, the Irénée du Pont 2.5 m telescope was used with the modular spectrograph and the SiTe2 detector. The combination of grating N° 600 blazed at 5000 Å, with a slit width of 1.5 arcsec, yielded a spectral range of

4000–6050 Å with a resolution of 2.5 Å. About 60 He-Ar-Fe lines in the comparison spectra enabled wavelength calibration functions with a standard deviation of 0.2–0.3 Å.

Gkouvelis et al. [26] studied a large population of Be stars photometrically detected by the IPHAS survey [31] and used them as galactic structure tracers. The study was based on the analysis of the follow-up spectroscopy of stars performed during the period 2005–2012 at the 1.5 m Fred Laurence Whipple Observatory (FLWO) Tillinghast telescope on Mount Hopkins in Arizona, using the Fast Spectrograph for the Tillinghast Telescope (FAST) spectrograph [32]. The data were taken with the 300 l/mm grating and a projected slit width of 3". The data were in a wavelength range from 3500 to 7500 Å at a spectral resolution of $\Delta\lambda \simeq 6$ Å. Since the BCD requires flux-calibrated spectra each night, they selected calibration spectra from the FAST archive to ensure that all spectra were calibrated with flux standards observed the same night. The calibration has been performed using standard image reduction and IRAF analysis facility routines. The authors selected sources with spectra having SNR $\gtrsim 30$ around the Balmer discontinuity. Gkouvelis et al. [26] developed a semi-automatic procedure to obtain the fundamental parameters and distances of stars based on the BCD system.

Additional mid-resolution (2–4 Å) and high SNR (30–100 at 3700 Å) spectra of a number of stars were obtained at the Roque de Los Muchachos Observatory in La Palma, Canary Islands, Spain, to transform the resolution of spectra into that currently used in the BCD system since the astrophysical parameters of the studied stars were obtained using the calibrations established with the original BCD (λ_1, D) quantities, the details of which can be found in [26]. For this task, the telescopes and instruments used were the Isaac Newton Telescope (INT) equipped with the Intermediate Dispersion Spectrograph (IDS), and the Nordic Optical Telescope (NOT), using the Andalucía Faint Object Spectrograph and Camera (ALFOSC).

A series of B emission-line stars for BCD classification were acquired with the VLT/X-shooter instruments [33]. X-shooter is a multiwavelength medium-resolution spectrograph mounted at the Cassegrain focus of UT2 of the VLT at ESO Paranal with a mirror diameter of 8.2 m. The X-shooters' three arms are UVB, covering 300,550 nm; VIS, covering 550–1010 nm; and NIR, covering 1000–2500 nm. The resolution depends on the chosen slit width. It ranges from R = 1890 to 9760 in the UVB, from 3180 to 18,110 in the VIS, and from 3900 to 11,490 in the NIR arm, respectively, [34]. The spectra were reduced with the ESO Recipe Flexible Execution Workbench (REFLEX) for X-shooter [35], a workflow environment to run ESO VLT pipelines. This workflow provides an interactive way to reduce VLT science data. The steps executed by the ESO X-shooter pipeline include bias subtraction, flat fielding, wavelength and flux calibration, and order merging. The authors did not give details on the spectrophotometric reliability of their spectra.

Another series of spectroscopic observations to carry out BCD classification was selected by Shokry et al. [36] from the NOAO Indo-U.S Archive of Coudé Feed Stellar Spectra [37]. The spectra of 1273 stars were carried out using the 0.9 m Coudé Feed telescope at Kitt Peak National Observatory in the spectral range of 3460–9464 Å at a low-resolution of 1.2 Å FWHM [37]. Nearly 140 B-type star spectra exist in this archive, but only 83 spectra were suitable to apply the BCD method because their spectra were either badly calibrated or were not obtained in the wavelength domain required for BCD.

2.5. Be Stars Observed with the BCD System

2.5.1. Main Characteristics of the Balmer Discontinuity in Be stars

Be phenomenon is defined in B-type stars, which are not supergiants and have shown some emission in their hydrogen Balmer line spectrum at least once [38,39]. The numerous known properties and reviewed definitions of classical Be stars are nicely summarized in Rivinius et al. [40]. They have two outstanding characteristics: they are rapid rotators and, on average, are the fastest among the non-degenerate stellar population; and they show spectral and photometric variability in all spectral domains.

It is known that supergiants can have emission components in their Balmer line spectrum due to their extended atmosphere. Moreover, they may have some emission in the hydrogen Paschen lines due to non-LTE effects. The Be phenomenon is thus conceptually associated with the capability of a star to create by its own a circumstellar disc (CD) or envelope where line emissions are formed. The Be phenomenon can be present in stars of mid-to-late O spectral type, B-type objects and early A-type stars.

Barbier and Chalonge [41] noticed that the continuum spectrum of ζ Tau (HD 37202) presented a peculiar BD. From the confluence point of the last seen members of the Balmer line series, towards shorter wavelengths, there is a stall of the level of the energy distribution, called the 'second BD component in absorption' (scBD in absorption), which attains its maximum absorption at the theoretical limit of the Balmer line series (λ 3648 Å). Barbier and Chalonge [41] attributed this phenomenon to a circumstellar medium having low gas pressure. During the first observation missions in the BCD system, roughly from 1941 to 1948 at the Swiss Jungfraujoch station and the Institut d'Astrophysique in Paris, where the main objective was to obtain a large enough number of stars of all classes to develop their spectrophotometric stellar classification system, Barbier, Chalonge and their colleagues incidentally observed also a series of B-type stars that displayed the above-noted behavior of the BD, but also other B-type stars where the stall corresponds to a flux excess that steadily increases from roughly 3650 Å to a maximum at 3648 Å. This flux excess was called the 'second BD component in emission' (scBD in emission).

Be stars with strong emission lines generally display a second BD component in emission, while those with a shell-line spectrum have a second BD component in absorption. They can lose their line and continuum emission or absorption characteristics to acquire a transitory B-normal aspect. Since the IAU Coll. No 98 [42], it is customary to call each of these spectroscopic aspects a 'phase' of the Be phenomenon. In Figure 2, examples of Be stars are shown with the second BD component in emission and in absorption. The total BD of Be stars is written as

$$D = D_* + d \qquad (7)$$

where D_* is the stellar proper or photospheric BD component, and d ($d > 0$ absorption; $d < 0$ emission) is the second BD component due to the circumstellar gaseous environment. Many Be stars have undergone either B⇌Be (59 Cyg [43,44]), or B⇌Be-shell (88 Her [45]; θ CrB [46]) phase transitions. Some of them had shown both types of transitions as 59 Cyg [44], γ Cas [47]; however, Pleione (28 Tau) is the most typical example [48,49]. These transitions are detected in the behavior of the Balmer line profiles as well as by the changes of the continuum spectrum around the BD Zorec [20], de Loore et al. [50], Divan et al. [51].

One of the first interpretations of the continuum energy distributions such as those displayed in Figure 2 in terms of a circumstellar gaseous envelope was made by Barbier [52]. He used the BCD spectra of γ Cas (HD 5394) obtained from 1934.6 until 1943.5 [12,53–56]. The spectra obtained from 1934.6 to 1935.5, before the huge stellar outburst in 1937.5, characterized by a brightening of nearly two magnitudes, were considered to represent the star during a normal or emission-less phase. Considering a spherical circumstellar envelope, the observed BCD parameters were interpreted using the following expression for the observed radiation flux

$$F_\lambda = F_\lambda^* \exp(-\tau_\lambda) + S\,E\,[1 - \exp(-s\tau_\lambda)] \qquad (8)$$

where F_λ^* is the underlying photospheric stellar flux, S is the effective emitting surface of the circumstellar envelope, E is the source function of this envelope, τ_λ is the non-LTE bound-free opacity of the envelope made up of hydrogen atoms, and s is a parameter defined as $\exp(-s\tau) = 2E_3(2\tau)$ with E_3 being the exponential integral of the third order. With Equation (8), rarely cited, but many times re-invented in the specialized literature of Be stars, D. Barbier described the normal, Be and Be-shell phases of γ Cas. He concluded that these phase changes occur with opacities $\tau_{\text{visible}} \lesssim 0.4$ and $\tau_{\text{uv}} \simeq 1-6$. The estimated geometrical dilution factor characterizing the extent of the circumstellar envelope producing emission and absorption in the analyzed wavelength region is $W(R) \simeq 0.04$, which implies relatively short distances from the stellar surface.

There is a list of approximately 50 Be stars in the original BCD archives which were observed several times with the original Chalonge spectrograph, although quite irregularly until 1983. The individual observations were almost never published, except for γ Cas, X Per (HD 24534) [50], 59 Cyg (HD 200120) [43], HD 60848 [51] and 88 Her (HD 162732) [45,57]. The photospheric BCD (λ_1, D) parameters for nearly 50 Be stars were published in Zorec and Briot [18], Zorec [20].

2.5.2. The Apparent HR Diagram of Be Stars

Only when there is a powerful emission in the scBD does the scBD overlap the photospheric component of the BD and makes its determination somewhat uncertain. Additionally, the Paschen continuum is perturbed by the radiation from the circumstellar medium, which introduces a change in its distribution and makes the determination of D_* more difficult. Nevertheless, the reddening or bluing of the gradient due to the circumstellar environment can be interpreted as an increased ISM reddening because, in the short wavelength range concerned by these gradients, the circumstellar and ISM reddening laws are barely distinguished. The perturbed value of the BD can then be corrected as indicated in Equation (2) to obtain the genuine D_*. In the remaining cases, the value of D_* is constant within the limits of the uncertainty of its determination, even though the intensity of the line and continuum circumstellar emission may change. It is remarkable that during the 'shell' phases, very frequently Φ_b remains almost unperturbed, and the determination of D_* does not offer any additional difficulty. When the presence of the circumstellar medium rather strongly perturbs the determination of D_*, the value of λ_1 can be slightly uncertain as well.

Excluding the very extreme cases of the presence of strong emission, corrections on D_* and λ_1 are small or negligible. We can then consider the pair (λ_1, D_*) as parameters characterizing the photosphere of the observed stellar hemisphere of the Be star. Following these considerations, Divan [58] proposed a spectral classification of Be stars in terms of (λ_1, D_*).

An in-depth study based on a new data set of BCD parameters added to the existing collection allowed Zorec [20] to present a diagram where, for the first time, it is shown with unperturbed parameters by the circumstellar medium that the Be phenomenon can appear at any moment of the stellar evolution in the main sequence (MS), as well as a little later. This diagram is shown in Figure 4, where l in Å is the value of $\lambda_1 - 3700$ Å, determined at $D_* = 0.2$ dex and can be considered as the BCD continuous luminosity class parameter. Let us note that, until the 1980s, the HR diagrams of Be stars were constructed with photometric data not corrected for the effects due to the circumstellar emission or absorption. The Be phenomenon was then believed to likely appear in the late evolutionary phases of OBA stars on the MS [59]. It was thus suggested that the occurrence of the Be phenomenon could be related to the short secondary contraction period that follows the hydrogen exhaustion in the core [60], where the increase in the stellar rotational rate favors the reaching of its critical limit.

In the diagram of Figure 4, we note that Be stars earlier than the spectral type B3-4 appear over an extensive range of luminosity classes. The tendency for the remaining stars is that the phenomenon tends to be present at more evolved evolutionary stages on the MS with a later B spectral type. Later studies on the evolutionary stages of Be stars in the solar neighborhood have confirmed this result [61].

Finally, it is worth noting that the (λ_1, D) or (l, D) parameters reflect the photospheric characteristics of the stellar hemisphere distorted by rotation and projected towards the observer. They are then considered as apparent parameters that need to be corrected for the effects induced by the rapid rotation to obtain a more realistic view of the real physical properties and evolutionary status of Be stars.

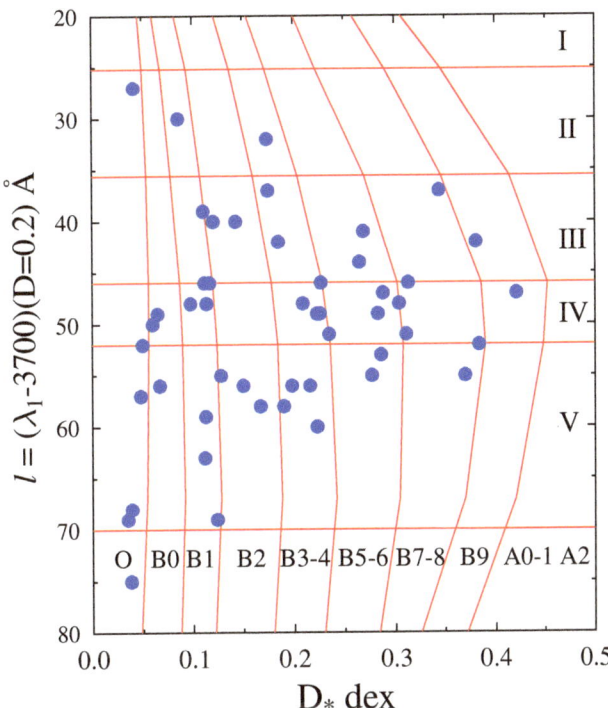

Figure 4. BCD spectral classification diagram for Be stars in terms of luminosity class parameter l against the photospheric BD, D_*. This is one of the first HR-type diagrams showing that the Be phenomenon can appear at whatever evolutionary phase because (l, D_*) characterizes the stellar photosphere of the central star [20].

2.5.3. Correlations of the BCD Parameters with the Emission Characteristics of Be Stars

It has been known for a long time that one of the main characteristics of Be stars is their line and continuum variations [62–64]. As much as it concerns, the variations in the continuum spectrum have very different time scales. Among them, long-term variations deserved attention in the BCD system because they have a clear impact on the values of the gradient Φ_b (or Φ_{rb}) and on the scBD $d = D - D_*$ (see Figure 5).

Moujtahid et al. [65] interpreted the long-term spectrophotometric variations of Be stars in terms of sporadic mass ejections from which their CD is formed and dissipated. According to the dissipating disc model, these authors predicted 'loop-shaped' relations between the flux excess in the V magnitude (ΔV) and the emission excess in the second component of BD (ΔD). Similar loop-shaped behavior in a color-magnitude diagram of Be stars of the Small Magellanic Cloud (SMC) (EROS microlensing experiment, e.g., [66]) was sometime later discussed by de Wit et al. [67] who also based their interpretation on a simple-time-dependent model, where the bound–free and free–free emissions is produced by an outflowing CD. These authors also discussed the correlation between the optical and the near-IR flux excess, from which they concluded that their outflowing CD model could provide reasonable explanations for the observations.

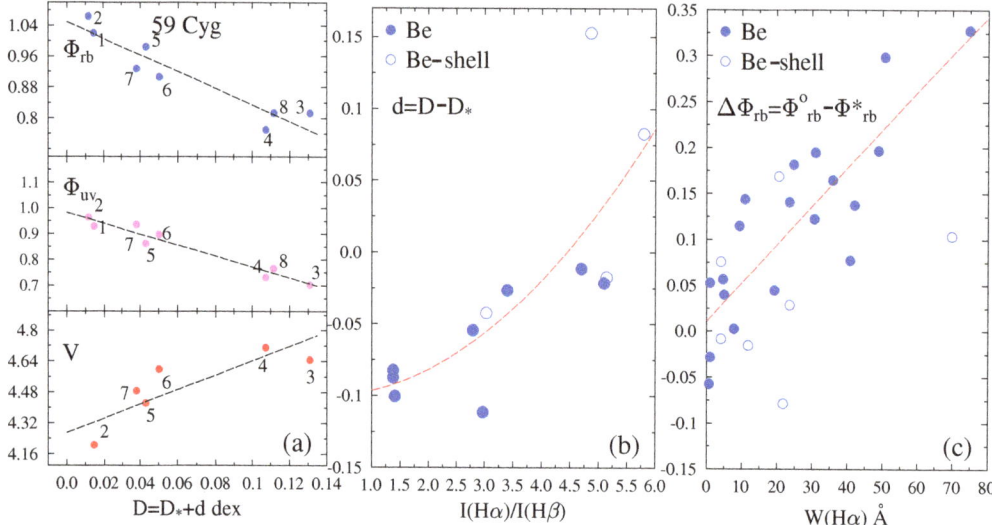

Figure 5. (**a**) Variation of gradients Φ_{rb}, Φ_{uv} and magnitude V against the total BD of 59 Cyg from 1948 to 1981 (1: Spt.48, 2: Oct.65, 3: Jul.77, 4: Nov.77, 5: Jun. 79, 6: Nov.79, 7: Jul.80, 8: Spt.81). (**b**) Second component d of the BD against the Balmer line decrement $I(H\alpha)/I(H\beta)$. (**c**) Reddening and bluing of the gradient Φ_{rb} corrected from ISM extinction against the equivalent width of the emission component of $H\alpha$. In (**b**,**c**), filled circles are for Be phases and open circles for Be-shell phases of Be stars.

The short- and long-lived outbursts discovered by Hubert and Floquet [68] were also reported for some Be stars by Mennickent et al. [69] (MACHO microlensing experiment, e.g., [70]) and Keller et al. [71] (OGLE, e.g., [72]). These authors definitively provided the required observational proof to realize that the long-term photometric behaviors of Be stars are related to the formation of CDs by sudden huge mass ejections, their subsequent dissipation and partial re-accretion onto the star.

2.5.4. Disc-shaped Envelopes in Be Stars, Just a Few Reminders

In the approaches mentioned above by Moujtahid et al. [65] and de Wit et al. [67], the authors used simplified heuristic representations of the radiation produced by a CD as a function of its optical depth changes during its expansion and dissipation phases. New models have since been proposed, which are worth recalling because they suggest a deeper physical perception of the CD structure in Be stars, its formation and evolution. However, this reminder aims to focus on the observational aspects put forward by the BCD system on the spectrophotometric behavior of Be stars that today's most accepted theories of the formation and evolution of a CD should take into account. In fact, BCD data refer to the structure of the CD in regions close to the star, where interactions between an already existent environment, discrete ejecta and a variable stellar wind take place. In these regions, angular momentum transport is organized over the entire extent of the disc. This finally enables the disc to acquire a Keplerian rotation and a subsonic expansion over a large part of its structure.

The envelopes of Be stars deserved attentive discussions since Limber and Marlborough [73] concluded that their support in the equatorial plane must be centrifugal and that viscosity (turbulent or magnetic) should be the agent to provide the required transfer of angular momentum from the star to the envelope. They did not elaborate their suggestion. A steady-state solution for a CD formed through mass ejections forced by rotation was obtained by Limber [74], provided the functional dependence of the azimuth velocity, $u_\phi(R, \theta = \pi/2)$, with R in the disc was specified. In this solution, the radial velocity,

$u_R(R, \theta = \pi/2)$ is subsonic near the star and supersonic far from the star. Based on a formulation by Limber and Marlborough [73], Marlborough [75] proposed a model for the Be star's CD density in hydrostatic equilibrium in the zdirection. This model was then used by [76–79] to produce extensive predictions of the emission profiles in the Balmer lines, continuum flux excesses and line and continuum polarization.

It could also be mentioned that in the 1970s, a series of ad hoc models were proposed for the Be star circumstellar envelopes (or discs) that aimed at inferring from observations the likely dynamical conditions controlling their structure, as can be seen in the reviews by [49,80–82]. Traces of the extensive discussions and considerable efforts provided to disentangle the physical nature of envelopes around Be stars can be found in Slettebak and Snow [83], Waters and Marlborough [84], where step by step, the idea of disc-shaped envelopes has been progressively adopted against spheroidal envelopes [85,86]. There is also the problem of the structure of the stellar atmospheres underlying these circumstellar environments. In the future, much attention must be paid to the suggestions made by Thomas [87] and the references therein concerning the properties of atmospheres of active stars having, in addition to the atmospheric structures modified by the rapid rotation, a more or less permanent supply of non-thermal energy by the sources of hydrodynamic and magnetic nature, to better understand the internal structure of these objects. Up to now, instead, few authors have studied the consequences produced by the deviations from the classical thermal structure of stellar atmospheres on the emitted spectra, e.g., [88–96].

The two main observational constraints finally establish the limits for a strict modeling frame of CDs in Be stars. According to the correlation between the FWHM (full width at half maximum) of emission lines with the $V \sin i$ parameter of Be stars [97,98], discs have rotational support, which implies that there is a mechanism that transports angular momentum to the disc and makes it in Keplerian rotation. On the other hand, there is no evidence of supersonic radial velocities in the discs.

Apart from the mechanisms that produce the discrete mass ejection, which remain widely unknown, the disc formation and its further dissipation are now nicely explained by the viscous decretion disc model (VDD). The disc-shaped envelopes start to take hold with Lee et al. [99], who, following the analysis of the dynamics of viscous discs by Shakura and Sunyaev [100], Pringle [101], suggested that angular momentum is transferred from the stellar interior to the equatorial surface. The matter, thus supplied to the inner edge of the disc, drifts outward thanks to the angular momentum transfer through the viscous stress. Following further developments in the frame of viscous accretion discs [102], accretion and excretion discs [103], Okazaki [104,105] proposed the model of 'viscous decretion discs' that respects the observation constraints mentioned above. Among the first explorations of the dynamical properties of VDD were those carried out by Haubois et al. [106], while in the frame of these models, the line emission, polarimetric and photometric behaviors of Be stars were carried out by, e.g., Ghoreyshi et al. [107] and references therein.

According to VDD models, the particle density structure in a disc that is in hydrostatic equilibrium in the vertical direction is described by the following relation [108]

$$\begin{aligned} \rho(R,z) &= \rho_0 (R_*/R)^n \exp\{-[(R/r)-1]/H(R)^2\} \\ r &= (R^2+z^2)^{1/2} \\ H(R) &= R(V_s/V_k) = V_s^2 R^3 / G M_* \end{aligned} \quad (9)$$

where R is the radial cylindrical distance in the disc, z is the vertical coordinate, ρ_0 is the base disc density, the exponent n is a free parameter or determined by the VDD model, V_s is the sound speed, V_k is the Keplerian velocity, and $H(R)$ is the vertical disc height scale, G is the gravitational constant, and M_* is the stellar mass. Haubois et al. [106] noted that the exponent n in the radial density function is a function of R, $n = n(R)$ as it was also noted empirically by Zorec et al. [96]. The relation $n = n(R)$ changes during the disc evolution from formation to dissipation. Generally, the variable average value of n is used in the literature, which changes as the CD structure evolves, i.e., from $n \simeq 4$ at the disc formation phases, $n \simeq 3$ near its steady state, to $n \simeq 2$ when it dissipates. However, at the disc formation phases, it can locally be $n \lesssim 0$ near the star and $n > 0$ away from the star. When a steady state of the disc is temporarily attained, it has the average value $n = 3.5$ [109].

Detailed studies on the temperature distribution in the CD of Be stars show that the temperature distribution is highly non-uniform. Its global behavior depends on the physical inputs and on the assumptions made in the models [109–114]. The CD temperature decreases in the equatorial region near the star and increases somewhat at larger distances and higher zcoordinates. On the contrary, by including viscous heating, Kurfürst et al. [108] found that the temperature grows in the equatorial regions and decreases towards larger distances in both R and z coordinates.

It has long been known that the source function of Balmer lines, in its representation according to an equivalent two-level atom with continuum, is strongly dominated by photoionization processes [115–119]. The stellar radiation field thus dominates the production and destruction of line photons, and the source function is dissociated from local temperature and density in the line formation region. Therefore, Balmer lines are blind to the temperature structure of the CD. This phenomenon does not concern FeII lines where the temperature dependence of the source function is high [96,120], as well as for the bound-free and free-free continuum radiation.

2.5.5. What Can BCD-like Data Contribute to the Study of Be Star CDs?

Moujtahid et al. [65] have shown that the photometric behavior of a Be star in the visible spectral range is a function of the structure of the CD near the central object as well as of the aspect angle under which the star+CD system is seen. This dependency is also valid for the scBD. Balmer lines, such as Hγ and Hδ, and FeII emission lines also form in the CD regions which are very close to the central object. The aforementioned VVD models should also be applied to the study of changes in these spectral and spectrophotometric characteristics to disentangle the properties of the CD regions, which make the transition between the star and the CD layers that organize the angular momentum transport and are responsible for the Keplerian rotation of the Be star CD.

The mentioned transition region may have particular dynamic characteristics. Stellar winds, mostly when they are massive as predicted by Curé [121], can contribute to the global dynamics of the mass around the star, not only with an added huge amount of mass but also with momentum and energy. The massive discrete ejecta and the fraction of the already existent environment, ablated by winds, produce a mass-loaded flux that can take the aspect of an expanding windblown bubble. These types of phenomena were studied by Hartquist et al. [122], Dyson and Hartquist [123] and Arthur et al. [124]. According to these authors, the most simplified structure of the circumstellar environment in the stationary snowplow phase encompasses three dynamically distinct regions: a wind expansion region; a decelerated, subsonic wind momentum-dominated core; a pressure-dominated supersonic expanding halo. Such a phenomenological picture may then characterize the transition region to other regions where the viscous transport of angular momentum and mass can occur.

The reasons that underlie the triggering of sporadic massive mass ejections are still unknown. Without specifically mentioning them, Kroll and Hanuschik [125] have modeled some characteristics of the ejection proper and its consequences that can lead to the formation of CDs. Kee et al. [126] have proposed that the combination of prograde g-modes of stellar non-radial pulsation could lead to sporadic mass ejections, provided that stars are at nearly critical rotation, which seems far from being a generalized phenomenon [127]. Krtička et al. [128] examined the nature of the mass loss via an equatorial decretion disc in massive stars with near-critical rotation induced by the evolution of the stellar interior. This mechanism stems from the angular momentum loss needed to keep the star near but below the critical rotation. This suggestion suffers from two shortcomings: there is no evidence for the near critical rotation to exist in all Be stars, and the predicted mass-loss rates do not conform with the estimates obtained by modeling the photometric behavior of Be stars with the VVD models [129].

Although the MiMeS experiment yielded no detection of an organized magnetic field with dipole field components weaker than 100 G in Be stars [130,131], random bipolar spot distributions may exist [132] with magnetic fields compatible with the MiMeS non-detections. Hope for its possible existence should not be lost because the average solar magnetic field on the surface does not exceed 2–3 G. This experiment would probably consider the Sun a star without a magnetic field.

Magnetic fields in Be stars can, however, be entertained by sub-photospheric convection as predicted by Cantiello et al. [133], Cantiello and Braithwaite [134,135], as well as by deeper convective regions induced by the rapid rotation as suggested by Maeder et al. [136] and Clement [137]. Hard X-ray emission in γ Cas and in γ Cas class of stars [138] is supposed to be a phenomenon connected with surface magnetic field excluding the effects related with an accreting companion [139]. The X-ray emissions observed in γ Cas could hardly be explained with the possible soft X-ray emissions predicted by the radiation-hydrodynamics simulations of the nonlinear evolution of instabilities in radiatively driven stellar winds [140].

The subphotospheric differential rotation [141] associated with the deep envelope convection [136,137] can also favor the creation of magnetic fields and subsequent magneto-hydrodynamical instabilities that end up with sporadic mass ejections, e.g., [142].

The low probability that Be stars are critical rotators (see discussion on this issue in Section 7.4), the simultaneous presence of rapid and slow stellar winds, discrete mass ejections, and the existence of surface 'hidden' magnetic fields in Be stars justifies the above short recall of the CD models, which aims to focus attention on the scBD. Their correct physical modeling might provide clues to understanding the complex physics that organizes the formation of CDs in Be stars.

2.6. BCD Parameters of B and Be Stars in Clusters

Spectroscopic investigations of open galactic clusters are generally scarce and limited to a reduced sample of cluster members. To study the physical parameters of galactic clusters and their individual members, observations of B and Be stars were made in the BCD spectrophotometric system at the CASLEO observatory in Argentina. These studies were published in Aidelman et al. [10,143] and Aidelman et al. [144]. Data for the following open clusters were obtained: Collinder 223, Hogg 16, NGC 2645, NGC 3114, NGC 3766, NGC 4755, NGC 6025, NGC 6087, NGC 6250, NGC 6383 and NGC 6530. The BCD parameters were derived using the interactive code MIDE3700 [10]. For all the studied stars, the astrophysical parameters T_{eff}, $\log g$, M_V and M_{bol} were determined. They enabled determining the ISM extinction in the direction of each studied cluster and to provide distances and cluster age estimates, stellar masses ($M/M_\odot$) and ages of individual stars. They were obtained by interpolation in models of stellar evolution. The relation between the red-blue gradient of energy distribution and the classical color excess $E(B-V)$ was redetermined according to the spectral wavelength interval $\lambda\lambda 4000 - 4600$ Å used to define the BCD-like color gradients.

From a sample of 230 B stars in the direction of the 11 open clusters studied, six new Be stars were found, including four blue straggler candidates, and 15 B-type stars, called Bdd, which have a double Balmer discontinuity. Neither of these show line emission features or previously been reported as Be stars. The spectra of these Bdd may perhaps indicate the presence of circumstellar envelopes. These data enabled to discuss the distribution of the fraction of B, Be, and Bdd star cluster members per spectral subtype. The authors concluded that the majority of the Be stars are dwarfs and their distribution against the spectral type presents a maximum at the spectral type B2-B4 in young and intermediate-age open clusters (<40 Myr). There is another maximum of Be stars at spectral type B6–B8 in open clusters older than 40 Myr, where the population of Bdd stars also becomes relevant. In conclusion, these results support previous statements that the Be phenomenon is present along the whole MS phase, from Zero Age Main Sequence (ZAMS) to Terminal Age of Main Sequence (TAMS). There is clear evidence for the augmentation of stars with circumstellar envelopes as the cluster age increases. The Be phenomenon reaches its maximum in clusters of intermediate age (10–40 Myr), and the number of B stars with circumstellar envelopes (Be plus Bdd stars) is also high for the older clusters (40–100 Myr).

Figures 6–8 show spectra observed in the CASLEO of cluster Be stars having the scBD in absorption, cluster Be stars displaying an scBD in emission, and the spectra of cluster B stars here named Bdd. Figure 9 illustrates the HR diagram obtained using the BCD parameters of the B and Be stars observed in the cluster NGC 3114.

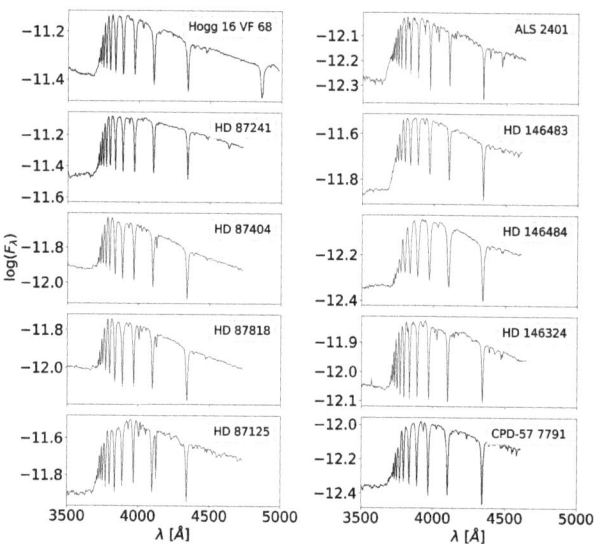

Figure 6. Low-resolution flux-calibrated spectra taken in CASLEO for Be stars in clusters with the scBD in absorption. Adapted from Aidelman et al. [144].

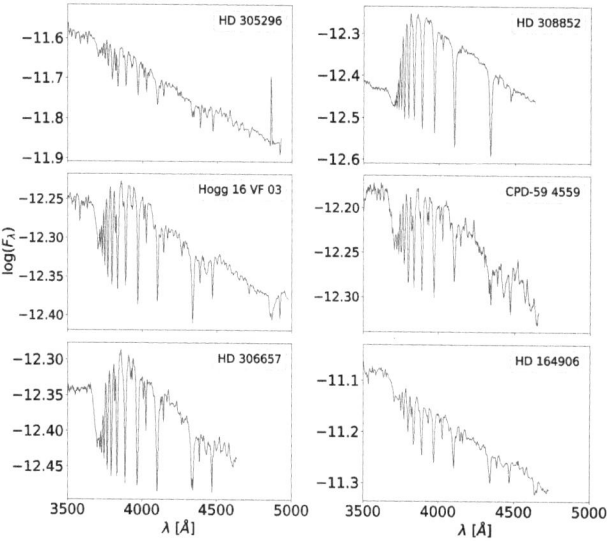

Figure 7. Low-resolution flux-calibrated spectra taken in CASLEO for Be stars in clusters with the scBD in emission. Adapted from Aidelman et al. [144].

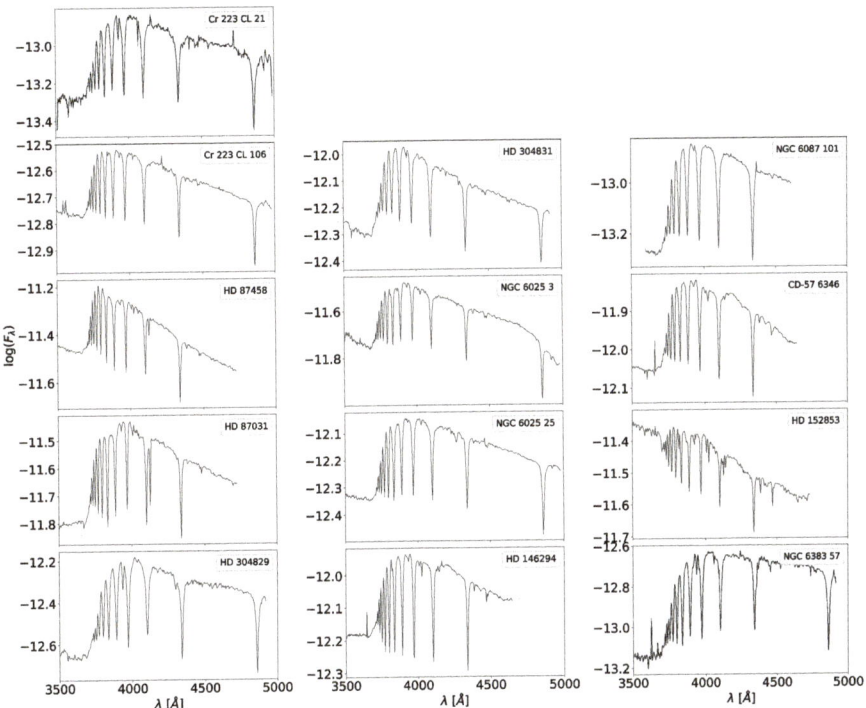

Figure 8. Low-resolution flux-calibrated spectra taken in CASLEO for Bdd stars in clusters. Adapted from Aidelman et al. [144].

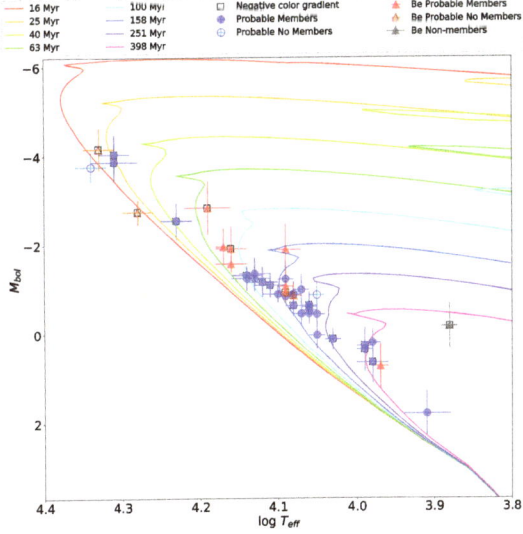

Figure 9. HR diagram of B and Be stars (members or probably not members) in NGC 3114 whose parameters were determined using the BCD system. The isochrones are from Ekström et al. [145]. Adapted from Aidelman et al. [143].

In the left panel of Figure 10, the $\log g$ parameter values obtained with the BCD calibrations and those derived using models of stellar evolution are compared. There is an unexplained difference between both estimates in the $2.7 \lesssim \log g(\lambda_1, D) \lesssim 3.5$ interval that implies overestimated $\log g_{evol}$ values by 0.5 dex. However, except for only two objects, the agreement between the M_{bol} magnitudes is excellent (right panel of Figure 10).

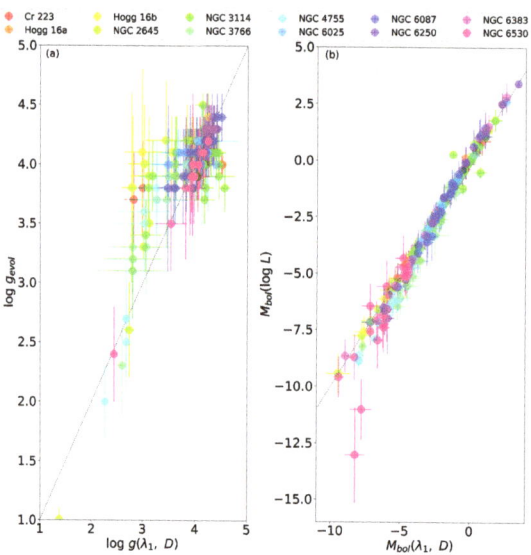

Figure 10. (a) Comparison of $\log g$ parameters of cluster stars obtained using evolutionary tracks without rotation with those derived from the BCD $\log g(\lambda_1, D)$ calibration. (b) Comparison of M_{bol} parameters obtained using the evolutionary tracks of stars without rotation with those derived from the BCD $M_{bol}(\lambda_1, D)$ calibration. Adapted from Aidelman et al. [144].

2.7. Bn Stars Observed with the BCD System

As noted by, e.g., Cochetti et al. [146], a significant fraction of stars have a spectral type sub-tagged by the letter 'n' or 'nn' for showing the presence of 'nebulous' or 'very nebulous' metal lines in their spectrum. Given the temperature range covered by most of these, the rotational origin of their line-broadening is a logical explanation. In particular, Bn stars are mostly B spectral type MS stars that display broad absorption spectral lines and broad hydrogen lines in absorption. Because a high rotation rate is an essential factor in the development of the Be phenomenon, the possible link between Bn and Be stars deserves a thorough exploration. Mechanisms related to the distribution of the internal angular momentum can be present with different degrees of intensity in Bn stars, which at some point of the stellar evolution may enable the star to display the Be phenomenon.

Taking the magnitude-limited sample of well-known bright O-, B-, and A-type stars in the *Bright stars catalog* [147], *Supplement to the bright stars catalog* [148] and errata to both of them published since then, Zorec [149] attempted to determine the possible difference existing between the Initial Mass Functions (IMF) of B stars with and without emission lines. To this end, the assumption was made that: the distribution in the space of Be stars mirrors that of other B-type stars; the relation between the visible absolute magnitude and mass is the same for both groups of objects; their main sequence lifetime is, on average, not strongly different; the star formation rate is constant for each class of objects. These simplifications imply that the ratio between the IMF of the B and Be stars is determined by the ratio of the respective present-day mass functions, which are proportional to the respective counts of main sequence stars [150,151], i.e., $\ln[\Psi(M|Be)/\Psi(M|B)] \propto \ln[N(Be)/N(B)]$, where M is the stellar mass in the interval $3 \lesssim M \lesssim 20 M_\odot$, $\Psi(M)$ is the IMF, $N(Be)$ and $N(B)$ are the counts of main sequence Be and B stars, respectively. The function $\ln[N(Be)/N(B)]$ against $\ln M$ is shown in Figure 11. The extrapolation of the regression line obtained for stars from spectral types B0–B7 and the stellar counts of stars cooler than B7 readily shows that there may be approximately 150 missing, still undiscovered, or 'latent' Be stars. This estimate approaches the number of counted Bn stars in the same magnitude-limited counting volume.

To the above IMF determination, corrections for the over luminosity of Be stars [18] and the blurring of absolute magnitude vs. mass relations for fast rotation should be taken into account [152]. Nevertheless, once these corrections are made, the final estimate of "missing" Be stars does not changes sensitively [153]. Figure 11a compares B and Bn star countings. In Figure 11b is shown the (λ_1, D) diagram of Be and Bn stars studied in the aforementioned magnitude-limited volume by Cochetti et al. [146], where it appears that Bn stars are more frequently found among the late B spectral types.

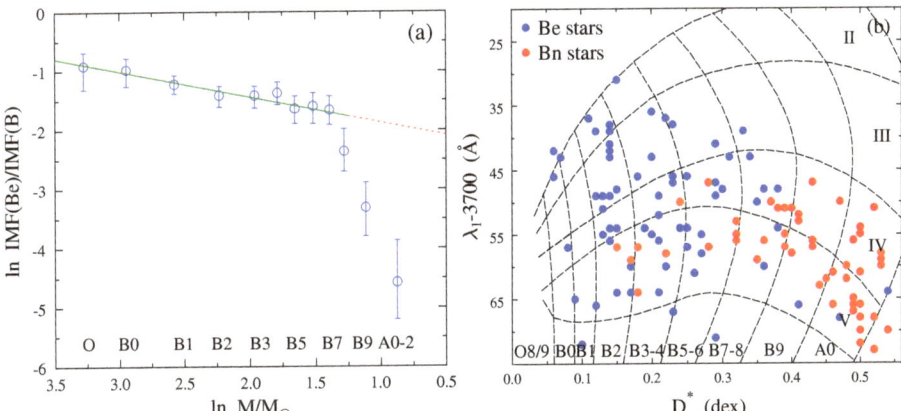

Figure 11. (a) Fraction of Be stars compared to the entire number of B-type stars; adapted from Zorec and Briot [18]; and (b) BCD parameters of Be and Bn stars; adapted from Cochetti et al. [146].

Among the reasons that may explain the lack of Be stars with spectral types cooler than B7 is that, for effective temperatures, $T_{\rm eff} \lesssim 12{,}500$ K, the average CD temperature in the effective emitting volume, drops to $T_{\rm CD} \sim 0.6\, T_{\rm eff} \lesssim 9000$ K [109], so that hydrogen atoms are almost neutral and the recombination rates very low. Thus, there may be potential Be stars among stars having a late B spectral types which, at the moment, are considered Bn. On the other hand, B-type stars, having seen pole-on, show apparently thin absorption lines, but they can be rapid rotators indeed. The actual number of late B-type stars with high rotation should then be higher than actually found. Finally, it has been shown that the time scales of internal angular momentum transfer to the external layers responsible for a more effective reach of the surface near-critical rotation rates are longer for late B-type stars [154] than for the earlier ones.

Some Bn stars are classified as such in the *Bright stars catalog* and its *Supplement* and have been found to have emission lines, and have since been considered to be Be stars, e.g., [146,155,156].

Of particular interest is the observation made by De Marco et al. [157] which was also extensively reported by Aidelman et al. [144] and Cochetti et al. [146], where late B-type stars have a larger BD than expected for their $(T_{\rm eff}, \log g)$ parameters. As much as it concerns this kind of star in the Cochetti et al. sample, no Balmer line emission has been seen. In all these cases, the authors suggest the possible existence of a CD. Using the models of rapidly rotating photospheres, Porter and Townsend [158] leans towards an explanation involving the gravitational darkening effect induced by a rapid stellar rotation. Nevertheless, their calculations produce a tiny scBD in absorption, whose magnitude is far from the observed stall in absorption. As shown in Figure 8, there is also a case with an apparent scBD in emission (HD 152853). Perhaps the spectra of a higher resolution than those shown in this work may indicate whether there is a genuine scBD in emission or a conjunction of strong spectral lines that give the spectrum such an appearance.

In general, the scBD of Bn stars are not large and show up in absorption. They look like those in Figure 6 or Figure 8. A discussion on Bn stars using BCD parameters, but considering these objects as rapid rotators, is postponed to Section 9.

2.8. He-Weak and He-Strong Stars Observed with the BCD System

As compared to the so-called 'normal' upper main sequence stars of approximately the same effective temperature and surface gravity, those having abnormal abundances of some chemical species are called Ap/Bp stars or chemically peculiar (CP) stars [159]. Moreover, among CP stars, class subdivisions are made: CP1 or Am stars (metallic-line A-type) which have weak CaII and strong heavy metal lines. These stars do not have large-scale organized magnetic fields. Many of them may not be magnetic at all. They are called CP2 when it comes to Si, Cr and SrCrEu Ap stars, CP3 or HgMn stars (CP3), there are also CP4 or B-type objects with excessively weak or abnormally strong lines of He I. According to the strength of the HeI lines, they are called He-weak and He-strong stars, which represent the high-temperature tail of Ap/Bp stars [160]. The interaction of variable multipolar magnetic fields with the gravitational and radiative diffusion processes induces non-homogeneous distributions of different chemical elements. On the contrary, CP1 and CP3 do not have such strong magnetic fields.

Knowing that the determination of astrophysical parameters (effective temperature, surface gravity, visual and bolometric absolute magnitudes) for these stars is often much more complex than for normal stars due to their abundance anomalies. Several methods were employed by Cidale et al. [161] to study the evolutionary status of stars and the physical processes that take place in their atmospheres and interiors. Among these is the BCD spectrophotometric method carried out on low-resolution spectra obtained at the CASLEO Observatory in Argentina. The parameters thus obtained were compared to those estimated through integrated fluxes, which enable simultaneously obtaining effective temperatures and angular diameters and to parameters drawn from the fitting of the observed energy distributions with non-LTE models atmosphere calculations for different He/H abundance ratios. The non-LTE synthetic spectra were obtained with the TLUSTY and SYNSPEC codes and references therein [162], assuming the following He/H ratios: 0.1, 0.2, 0.5 and 1.0. From the TLUSTY website, the atomic models were taken, i.e., 9 levels for HI, 20 individual levels for HeI and 20 levels for HeII. The stark broadening of HeI lines are from Dimitrijevic and Sahal-Brechot [163,164] or computed with approximate relations [165]. A microturbulence velocity of 2 km s^{-1} was used. All spectra were reduced to the resolution of the spectra observed in CASLEO.

Cidale et al. [161] concluded that the effective temperatures, surface gravities, and the bolometric absolute magnitudes of He-weak stars estimated with the BCD system agree well with those issued from the integrated flux method and with other estimates previously derived based on several different methods found in the literature. There are, however, discrepancies between the absolute visual magnitudes derived using the HIPPARCOS parallaxes and the BCD values by approximately ±0.3 mag for He-weak stars and ±0.5 mag when it comes to He-strong stars. For He-strong stars, we note that the BCD calibration, based on stars in the solar environment, leads to overestimated values of T_{eff}. Using model atmosphere calculations with enhanced He/H abundance ratios, it was noted that the larger the He/H ratios, the smaller the BD, which explains the T_{eff} overestimation. Nevertheless, these calculations enabled introducing a method to estimate the He/H abundance ratio in He-strong stars based on the BD discrepancy δD

$$\delta D = -0.056\left[1 - 0.233\left(\frac{T_{eff}}{10^4}\right)^{0.974}\right]\left(\frac{He}{H}\right) \qquad (10)$$

It is worth noting that the behavior of HD 37479 was observed at different epochs and showed near-UV flux, Balmer jump and line intensity variations, while the Paschen continuum does not seem to undergo detectable changes [161]. The intensity of H lines increased when that of He I lines decreased, and the near-UV flux was lower. A noticeable difference was detected in the equivalent widths of spectral lines (∼30%) and on the He/H line ratio as measured in short periods. There was also some flux excess in the Balmer continuum near the BD, reminiscent of the scBD seen in Be stars, although such a flux excess in He-strong stars needs to be studied to see whether it is a matter of emission. On the contrary, it was proven that the stellar T_{eff} remains relatively unchanged when there are substantial variations of He and H lines. Figure 12 shows the difficulties of obtaining the best fit with the model atmosphere calculations of energy distributions of He-strong and He-weak stars.

As much as it concerns the evolutionary status of He-weak and He-strong stars, both types of He-peculiar stars seem to be in the MS evolutionary phase. The He-strong stars are situated roughly within the $T_{\text{eff}} \gtrsim 19{,}000$ K region of the HR diagram, and the He-weak are in the $T_{\text{eff}} \lesssim 19{,}000$ K zone. However, the M_{bol} determinations for a much larger number of He-strong and He-weak stars are desirable to obtain more significant insights into their evolutionary status.

2.9. B[e] Stars Observed with the BCD System

The B[e] stars are identified according to criteria used in previous studies, such as Allen and Swings [166,167] and Zickgraf [168]. They encompass the four main characteristics that can be expressed in terms of physical conditions that characterize the circumstellar medium (CM) around the stars [169]. They are: (1) strong Balmer emission lines; (2) low excitation permitted emission lines of predominantly low ionization metals in the optical spectrum, e.g., Fe II; (3) forbidden emission lines of [Fe II] and [O I] in the optical spectrum; and (4) a strong near- or mid-infrared excess due to hot circumstellar dust.

Lamers et al. [169] reviewed the classification criteria of the B[e]-type stars in terms of physical characteristics of the stars and of the circumstellar matter (CM). According to their physical characteristics, Lamers et al. [169] suggested that instead of the name 'B[e] stars', that the term 'B[e] phenomenon' be used. These authors identified five different classes of stars which show the B[e] phenomenon: (a) B[e] supergiants, named in short 'sgB[e] stars'; (b) pre-main sequence B[e]-type stars, or 'HAeB[e] stars'; (c) compact planetary nebulae B[e]-type stars, or 'cPNB[e] stars'; (d) symbiotic B[e]-type stars, or 'SymB[e] stars'; and (e) unclassified B[e]-type stars, or 'unclB[e]' stars. Several classification criteria for each group have been specified to more clearly describe their characteristics. In some cases, stars can satisfy more than one of these criteria; their evolutionary phase is not obvious. If so, the stars are said of unclear type, or 'unclB[e]' class.

The sgB[e] stars form the most homogeneous group of B[e] stars. It is formed by B-type supergiants, and most of them were discovered in the Large Magellanic Cloud (LMC) and Small Magellanic Cloud (SMC) [170–176]. Objects with similar characteristics were also identified in the Galaxy, e.g., [177–179]. The properties of sgB[e] stars were reviewed by Zickgraf [168] and Kraus [180]. The main criteria defining them are: (a) that the stars show the B[e] phenomenon; (b) they should be supergiants with $\log L/L_\odot \gtrsim 4.0$; (c) there must be indications of mass loss in the optical spectrum, e.g., P Cygni profiles of the Balmer lines, or double-peaked Balmer emission lines with violet shifted central absorption; (d) hybrid spectra, i.e., simultaneous presence of narrow low excitation emission lines and of broad absorption features of higher-excitation lines; (e) enhanced N-abundance with abundance ratio of $N/C > 1$ or an enhanced He/H ratio, which indicates that the star, indeed, is in an evolved evolution stage where the products of the CN-cycle have reached the stellar surface; (f) in the Galaxy, they have a high extinction with $A_V \gtrsim 3$ mag confirmed by the presence of strong interstellar bands, because they are probably massive stars located at large distances in the galactic plane; and (g) generally, the photometric variations of B[e] supergiants are minor, roughly from 0.1 to 0.2 mag.

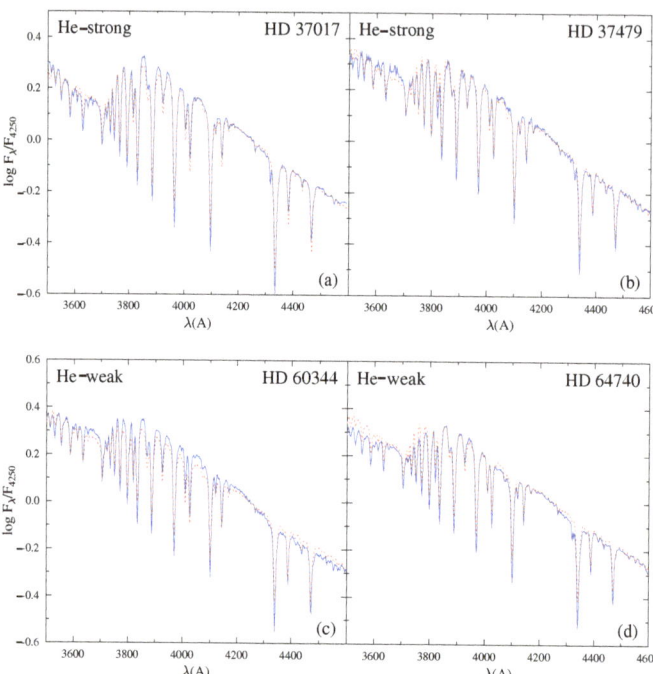

Figure 12. Examples of energy distributions of He-strong (**a,b**) and He-weak stars (**c,d**). Fluxes are normalized to the flux at $\lambda 4250$ Å. Observed fluxes are marked with solid blue lines, while models are shown with red dotted lines. The figure is adapted from Cidale et al. [161].

In the case of HAeB[e] stars, apart from showing the B[e] phenomenon, these are: (a) associated with star-forming regions, although in some cases they seem to be isolated; (b) they show spectroscopic evidence of accretion or infall of matter on the star, evidenced by the presence of inverse P Cygni line profiles; (c) these objects have $\log L/L_\odot \lesssim 4.5$, suggesting that they are probably progenitors of stars in the mass range from 2 to about 15 $M_\odot$ [181]; (d) these stars show large irregular photometric variations on time scales from some days to 103 days, usually characterized by a variable extinction [182]; and (e) the energy distribution shows the presence of warm and cool dust.

Ciatti et al. [183] suggested that some objects designated as BQ[ ] stars are of low mass, probably evolving into a planetary nebula. Apart from showing the B[e] phenomenon, objects designated as cPNB[e] stars have: (a) spectra which indicate they are possibly nebulae; (b) their luminosity is $\log L/L_\odot \lesssim 4.0$; (c) in addition to the forbidden low ionization lines, there are forbidden emission lines of high excitation, such as [O III], [S III], [Ne III], [Ar III] and [Ar V]; (d) the spectrum can show evidence for N enhancement due to an evolved evolutionary phase; and (e) their energy distribution can show the presence of cold dust ($T_d < 100$ K) as a possible remnant due to the AGB wind.

Symbiotic stars are interacting binaries with a cool giant and a compact hot object [184]. They can present the B[e] phenomenon and are named SymB[e] stars if: (a) the visual spectrum shows evidence for a cool star, in particular the TiO band, unless the cool star is heavily obscured; (b) the presence of a late-type stellar spectrum in the near-infrared is noted.

There are objects showing the B[e] phenomenon, such as HD 45677, HD 50138, HD 87643, and MWC 349A, which cannot be classified into one of these groups because they do not fit the criteria for a given class or they can share more than one class. The cited objects and roughly half of those in the original list of B[e] stars were considered unclassified B[e] stars or unclB[e].

Miroshnichenko [185], Miroshnichenko et al. [186] suggested a new classification for stars with the B[e] phenomenon based on the time of dust formation in their CM. The properties of the unclassified Galactic B[e] stars were then analyzed again, which enabled to conclude that these objects are binary systems that are currently undergoing or have recently undergone a phase of rapid mass exchange associated with a strong mass loss and dust formation. Since then, the name FS CMa stars (HD 45677) was suggested, and classification criteria were proposed for the unclassified B[e] stars. However, this does not mean that the newly presented class contains all the unclB[e] stars. In this context, the star MWC 137 can be mentioned, whose evolutionary state has long been badly defined. The object was sometimes considered HAeB[e] because of a massive star with an optical nebula embedded in a dusty and cold molecular gas component. The finding of ^{13}CO, a sign of processed matter around the central object, excludes the pre-main-sequence nature of MWC 137, and suggests that the star must be classified sgB[e] [187]. In Kraus et al. [188], a long account of classifying difficulties of MWC 137 can be found. In this work, a period of 1.93 days was also found, probably due to stellar pulsation but not to a binary. Let us also mention that CI Cam, a star with an orbital period of 19.41 days [189], which was first considered unclB[e], and then the FS CMa candidate, which is now classified as a Galactic sgB[e], which underwent a dramatic X-ray outburst in 1998 [190]. The classification into the FS CMa may sometimes be controversial because the binary nature is not straightforward to determine and the presence of warm dust cannot only be explained by its creation during a mass transfer episode but also by wind–wind and wind–atmosphere interactions. Moreover, some FS CMa stars are now considered post-merger objects, as IRAS 17449+2320 [191], where the Hα line of IRAS 17449+2320 shows night-to-night variability. This is an atypical phenomenon for FS CMa stars, but notwithstanding present in some of them.

As it comes from the above description of objects with the B[e] phenomenon, one of the most subtle challenges concerning the knowledge of their nature is to determine their right evolutionary state. Unfortunately, because the spectra of these objects are strongly marred by line emissions, flux excesses and high ISM extinctions, it is difficult to determine the fundamental parameters that can unambiguously reveal their evolutionary state. Because the BCD system provides parameters that are free from or can be easily corrected for the ISM extinction and perturbation due to their circumstellar environment, a series of low-resolution spectra were obtained at the CASLEO Observatory in Argentina of several stars presenting the B[e] phenomenon. The purpose was to obtain effective temperatures, surface gravities, and photospheric visual and bolometric absolute magnitudes and determine their position in the evolutionary diagram. Up to now, some recent papers were published presenting the BCD parameter determinations of stars with the B[e] phenomenon by Cidale et al. [28], Aidelman et al. [192] and Arias et al. [193]. However, the first observations in the BCD system of a star presenting the B[e] phenomenon were carried on HD 45677 from Dec. 1963 to Jan. 1967, during which the star underwent huge flux variations in the blue and visible spectral region. Burnichon et al. [194] determined the photospheric (λ_1, D) parameters of this object and classified it as B2IV, in agreement with older classifications dating back to Merrill [195] and Swings and Struve [196,197], who assigned the B2 spectral type according to the line spectrum. Using observations in the original BCD system from 1959 to 1980, Zorec et al. [198] updated the spectral classification, putting the star in a wider range of luminosity classes, from B2V to B2III-IV. On the other hand, the spectroscopic distance derived thanks to the BCD parameters (λ_1, D) and the conservative estimation of the amount of energy absorbed by the circumstellar dust in the far-UV and re-emitted in the far-IR imposed the choice of a B2V type, which agreed with the HIPPARCOS parallax [199]. In this work, the BCD classification of HD 50138 was given, for which these authors assigned the B6III-IV spectral type. Thanks to the characterization of the photosphere of HD 45677, Moujtahid et al. [200] studied the far-UV ISM extinction towards this object. They disentangled the interstellar and the circumstellar absorption components from each other.

Both objects, HD 45677 and HD 50138, are considered today of FS CMa class [185]. A discussion is nevertheless presented in Lamers et al. [169] on whether these stars could be regarded as of the HAeB[e] class because they are projected towards the southern filament of the Orion and Monoceros system of molecular clouds. While according to their distance estimates, they can, in principle, be considered as an isolated HAeB[e] class [201,202] located in the Gould belt plane between the Orion and Vela complexes, they have large spatial velocities so that it should not be excluded, they escaped from the Monoceros molecular clouds. Jeřábková et al. [203] preferred the merger solution for HD 50138, although its binary nature is not sure [204].

Low-resolution spectra in the $\lambda\lambda 3500 - 4600$ Å wavelength interval of 23 stars presenting the B[e] phenomenon were obtained at CASLEO, Argentina, to determine their BCD (λ_1, D) parameters and subsequent $(T_{\text{eff}}, \log g, M_{\text{bol}})$ physical parameters [28]. It was possible to clearly see the BD in only 15 stars of the observed set. For the remaining stars, other methods were used to estimate their effective temperatures. In some cases, significant differences between the effective temperatures derived using the BCD classification system and those obtained elsewhere, either based on photometric or spectroscopic analysis, imply spectral-type classification disagreements that range from 2 to 3 up to 6 B sub-spectral types. It clearly shows the difficulty of determining the photospheric characteristics of these types of objects and assigning them an evolutionary status. For HD 53179, a double star system, and for HD 45677 and HD 50138, which are also suspected of being binaries, Cidale et al. [28] predicted the characteristics of the components that are consistent with the observed (λ_1, D) parameters.

In the study carried out on B[e] stars, the observed stars in the BCD system were reported by Arias et al. [193]. In this work, authors attempted to improve the spectral classification and infer the properties of the circumstellar environments in a selected group of five bright IRAS sources (IRAS 02155+6410, MWC 728, AS 119, MWC 819, and IRAS 07080+0605). One of them, IRAS IRAS 07080+0605, displays a large infrared excess; four are unclB[e] stars or FS CMa stars. It was only possible to determine the BCD parameters for the star AS 119, its spectral type, and the $(T_{\text{eff}}, \log g, M_V)$ quantities that enabled to estimate its spectroscopic distance, which agrees with the HIPPARCOS parallax. Their medium-resolution spectra in the region of 2.17–2.39 μm and in the L band were shown for the first time. Infrared features, such as the absorption molecular CO bands and metallic lines, reveal the presence of a late-type companion in MWC 728 and suggest the possible binary nature in AS 119. The binary nature of MWC 728 was previously suggested by Miroshnichenko et al. [205], and that of AS 119 by Miroshnichenko et al. [206]. The star IRAS 07080+0605 shows evidence of a surrounding cool CO molecular cloud. According to the L-band spectra, CS molecular emission may be present in MWC 819, which suggests that the object is a protoplanetary nebula. However, Polster et al. [207] considers that the molecular lines and the late-type spectrum can be produced by a disc. This assumption seems to be confirmed by the polarimetric observations by Zickgraf and Schulte-Ladbeck [208].

In Figures 13 and 14, there are examples of energy distributions of B[e] stars observed in CASLEO, Argentina, to determine their astrophysical parameters according to the BCD system. In these figures, some spectral line identifications are shown. Let us comment on the spectrum of AS 202 in Figure 13. Miroshnichenko [185] has classified this object as an FS CMa star and considered it as a binary with an A0-2-type and K-type components. Other references on this object can be found in Condori et al. [209]. According to this stellar classification, the features at 4200 and 4686 Å that could belong to HeII are dubious and require further investigation to decide whether they are lines or spectral glitches. However, in other spectra of AS 202 and obtained in other recent epochs, not shown here, Balmer lines have nascent emissions, and some metallic lines have P-Cyg profiles. They can be taken as signatures of some kind of activity or interaction between components.

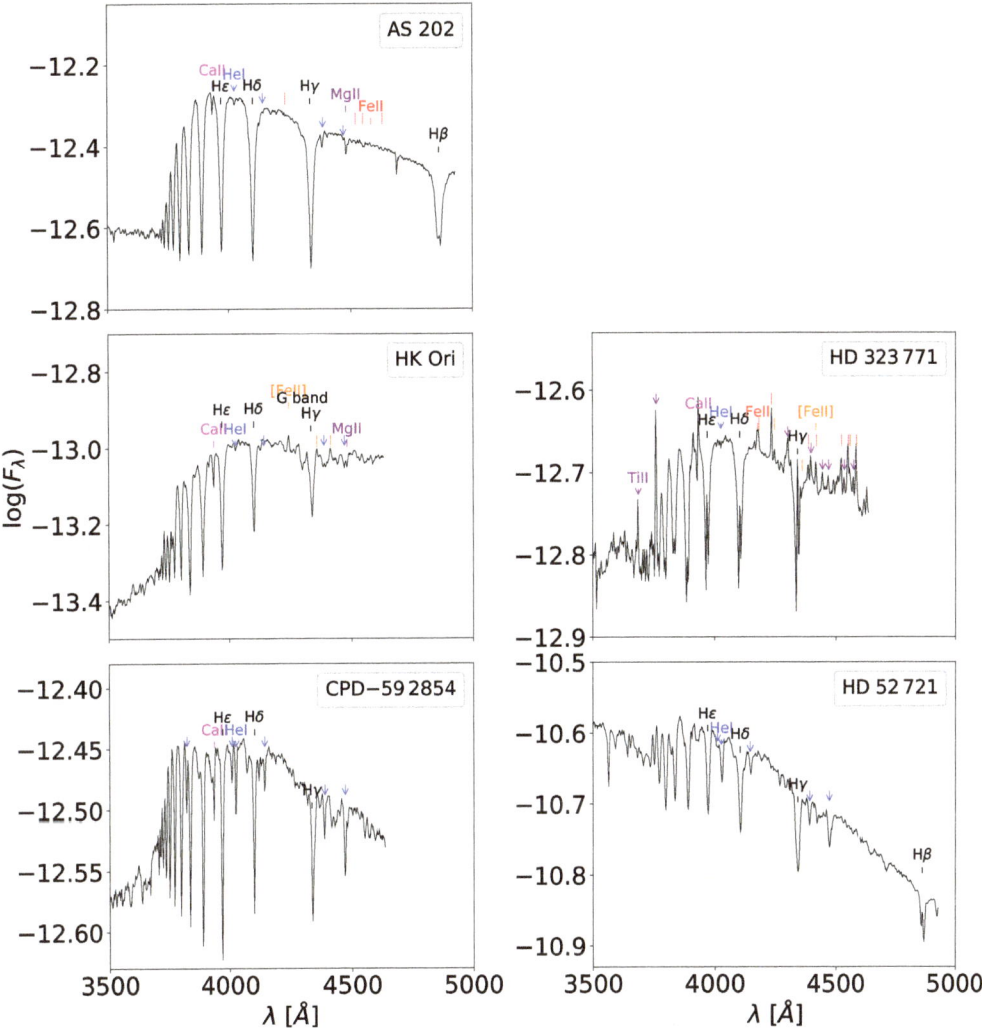

Figure 13. Examples of the energy distributions of B[e] stars without a second component of the BD in which some spectral line identifications are shown. The spectra were observed at CASLEO, Argentina. Figure adapted from Aidelman et al. [192].

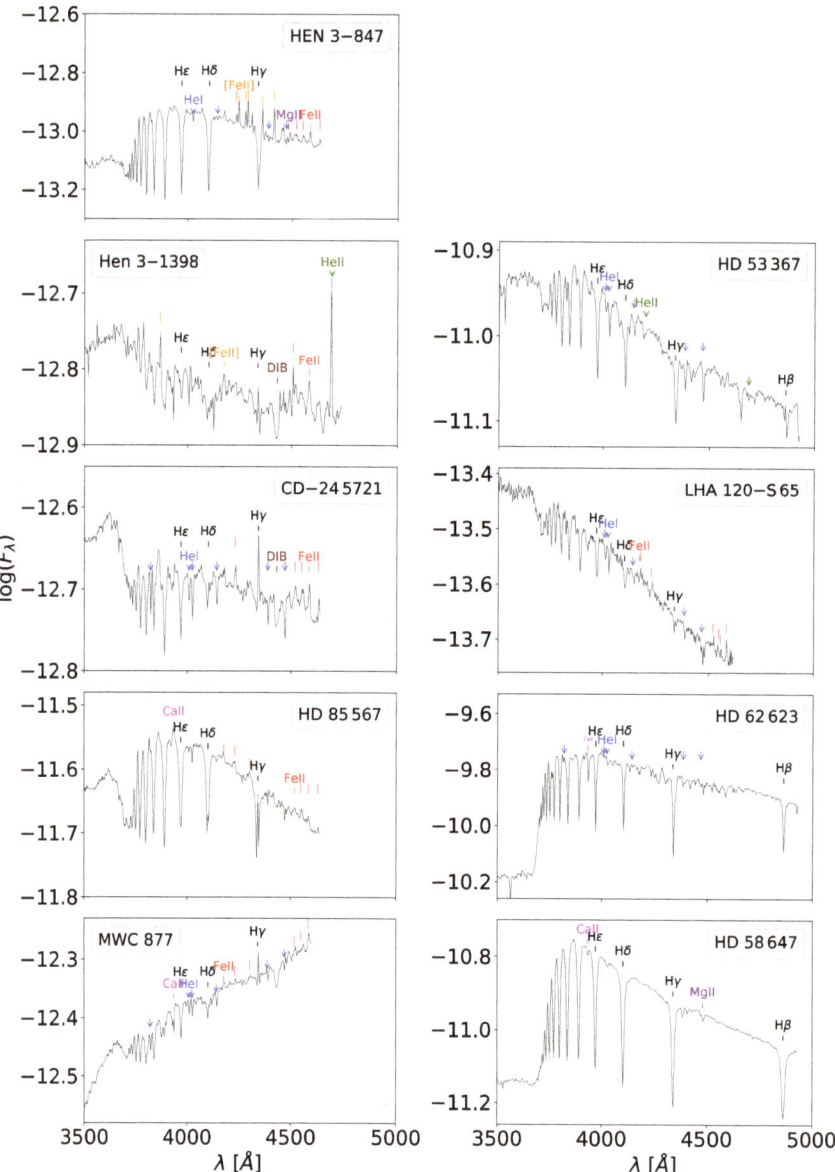

Figure 14. Examples of energy distributions of B[e] stars with a second component of the BD in which some spectral line identifications are shown. The spectra were observed at CASLEO, Argentina. Figure adapted from Aidelman et al. [192].

Stars in transition phases means that they sometimes show the B[e] phenomenon, and sometimes the characteristics of luminous blue variables (LBVs) undergo strong and irregular mass ejection events. At the moment, these changes seem impossible to predict with stellar evolution models. Moreover, since these stars are deeply embedded in their circumstellar environments, only a few of them were studied in some detail. To bring new elements to describe the central objects of these kinds of stars, low-resolution spectra were obtained for a sample of 14 stars with the B[e] phenomenon and characteristics of LBVs to measure their (λ_1, D) BCD parameters [192]. Aidelman et al. [192] were thus able to determine a complete set of physical parameters and related quantities such as luminosity, distance modulus and distance for each studied star. In this group, there are two pre-main sequence HAeB[e] stars, seven main-sequence B[e]-type stars, two B[e] supergiants and three LBV candidates. With the help of the size–luminosity relation, estimates of the inner radius of the dusty disc around the pre-MS and MS B[e] stars were obtained. The use of the BCD system showed its utility in deriving stellar parameters and physical properties of B-type stars in transition phases. Once this method is combined with near-IR color–color diagrams, additional elements can help specify the evolutionary stage and character of B[e] stars. Figure 15 summarizes the results concerning the evolutionary status of stars studied by Aidelman et al. [192] in two HR diagrams. Figure 15Left displays classical evolutionary tracks ($\log L/L_\odot$, $\log T_{\rm eff}$) for rotating and not rotating stars. Figure 15Right shows a spectroscopic HR diagram (sHR) defined according to Langer and Kudritzki [210].

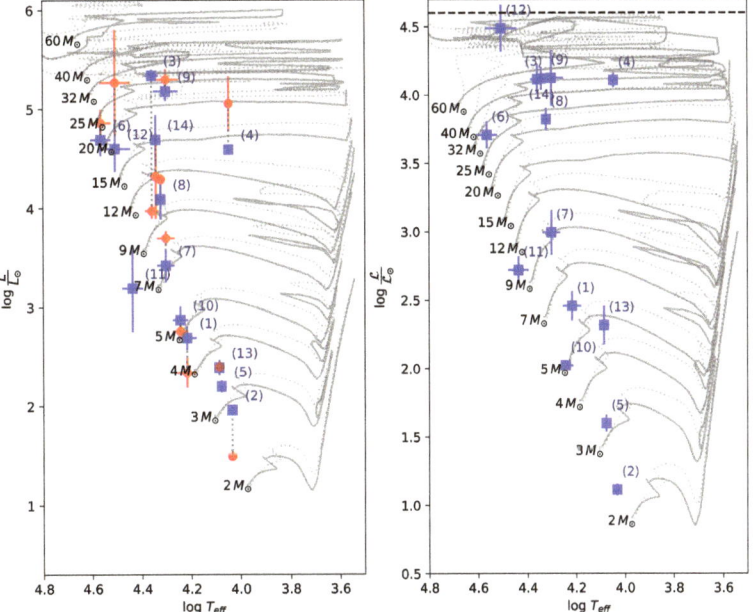

Figure 15. *Left*: The HR diagram of the stars studied in Aidelman et al. [192] where the distances have been calculated using the $M(\lambda_1, D)$ absolute magnitudes (blue squares). The red circles indicate the positions of stars as if they were located at distances given by Gaia EDR3 (Bailer-Jones et al., 2021). Evolutionary tracks with and without rotation ($\Omega/\Omega_c = 0.4$ and $\Omega/\Omega_c = 0.0$ are from Ekström et al. [145]). *Right*: The spectroscopic HR (sHR) diagram as defined by Langer and Kudritzki [210]. The dashed black line at $\log L/L_\odot = 4.6$ corresponds to the Eddington luminosity limit for a hydrogen-rich composition. The numbers identify the studied stars and the B[e] class classification is indicated in braces: 1 = Hen 3-847 (msB[e]), 2 = FX Vel (msB[e]), 3 = Hen 3-1398 (sgB[e]), 4 = HD 62,623 (sgB[e]), 5 = HK Ori (HAeB[e]), 6 = CPD-24 2653 (msB[e]), 7 = HD 85,567 (evolved msB[e]), 8 = CPD-59 2854 (LBV), 9 = MWC 877 (LBV), 10 = HD 323,771 (HAeB[e]), 11 = HD 52,721 (msB[e]), 12 = HD 53,367 (msB[e]), 13 = HD 58647 (evolved msB[e]), 14 = LHA 120-S 65 (LBV). Adapted from Aidelman et al. [192].

The determination of the astrophysical parameters of B[e] stars is highly challenged by the physics of these stars and their complicated spectra. The structure of the outermost atmospheric layers of these objects can be perturbed not only by their possible (interacting) binary nature but also by the presence of circumstellar gaseous and dusty envelopes or discs, whose structure is still relatively unknown. In B-type stars with high effective temperatures, the BD is small and subjected to relatively high uncertainties, which also affect the estimate of the λ_1 parameter. However, the BCD parameters depend on the continuum energy distribution and are thus related to the physical properties of atmospheric layers that on average are deeper than those in which spectral lines are formed.

The main problem in determining the astrophysical parameters of B[e] stars using high-resolution spectra is the identification of genuine photospheric lines. Moreover, the circumstellar environments affect the depth of these lines. These changes can translate in different ways, into increased absorptions, partial filling up by some emission, and by the veiling effect. It may then not be surprising that the astrophysical parameters of some B[e] stars had fundamental parameters that differed according to the methods used to determine them. It is interesting to note the case of HD 62,623, where the BD and the λ_1 are unambiguously determined. According to them, Aidelman et al. [192] obtain $T_{\text{eff}} = 11,140 \pm 300$ K, while Chentsov et al. [211] and Miroshnichenko et al. [212] using spectroscopy determine $T_{\text{eff}} = 8500 \pm 000$ K.

Although establishing the correct evolutionary phase of a star shows that the B[e] phenomenon cannot avoid the use of well-determined stellar physical parameters, as those determined by Aidelman et al. [192] with the BCD system, still other criteria must be called upon for help. On the one hand, to decide whether the phenomenon concerns the evolution of a single star or the consequence of merger phenomena implies stellar evolution in a binary system [213–216]. On the other hand, well-determined rotational parameters and surface abundances of chemical elements are required, as is the case of sgB[e] stars. These determinations are needed to decide whether they are in a post-MS evolutionary phase or share this same place in the HR diagram because they are in the blue loop after the red supergiant phase [213,217–219]. Similar questions may be raised regarding the B[e]⇌LBV transitions.

3. Rotation of B-Type Stars: Theory and Observations

3.1. Introduction

It is not the purpose of the present review to make an exhaustive report on the structure of rotating stars, but it is worth recalling some concepts that underlay models currently used to interpret observations, to grasp possible difficulties and needed theoretical improvements more easily.

In the classical modeling frame of stellar structure, the Vogt–Russell theorem stipulates that *'the structure of a star in hydrostatic and thermal equilibrium is uniquely determined by the total mass $M/M_\odot$ and the run of the chemical composition throughout the stars, provided that the total pressure P and the specific internal energy, the constituent equations are functions only of the local density ρ, temperature T and chemical composition'* [220]. In a more simplified way and for a common user mode, a non-rotating star is mainly characterized by the following reduced sets of parameters, the mass $M/M_\odot$, the radius $R/R_\odot$ and the chemical composition (X, Y, Z), with the possibility of using whatever possible combination of two independent quantities from the set $(T_{\text{eff}}, \log g, L/L_\odot, M/M_\odot, R/R_\odot)$ instead of $(M/M_\odot, R/R_\odot)$, knowing that the luminosity $L = S\sigma T_{\text{eff}}^4$ and the surface gravity $g = GM/R$ (S is the area of the stellar surface; σ is the Štefan-Boltzmann constant; G is the gravitational constant).

Rotation upsets this modeling scheme in the sense that its action on the stellar structure depends on the amount of the global angular momentum stored in the star, J/M (J total angular momentum) and its internal distribution, i.e., distribution of the specific angular momentum $j(\varpi, z) = \Omega(\varpi, z)\varpi^2$ (Ω is the angular velocity; ϖ and z are the cylindrical coordinates).

Rotation generates important effects on the structure and evolution of stars. There are first hydrostatic structural changes because of the centrifugal force, which causes a thermal imbalance that leads to meridional circulation currents [221], which in turn induce several hydrodynamical instabilities [222,223] and magnetic instabilities [224–229]. These instabilities produce turbulent motions which trigger the mixing of chemical composition and the redistribution of the internal angular momentum.

It is well known that, given a total angular momentum, the motion implying the least energy is that of uniform rotation [230]. Although the physical circumstances in the stellar interiors created by the evolution of stars might then be thought to favor a tendency to recover uniform rotation, other internal rotation laws can prevail over a long interval of time, provided they satisfy some general dynamic stability criteria as the Solberg–Høiland stability criteria against axisymmetric perturbations [229,231].

3.2. Rigid against Differential Rotation; Some Theoretical Principles

Since, in a gaseous rotating body, the specific angular momentum ranges from pole to equator as $j(\varpi = 0, z = R_p) \leq j \leq j(\varpi = R_e, z = 0)$ (where R_p is the polar radius; R_e is the equatorial radius), this body can no longer be spherical, but does acquire either an axisymmetric or triaxial equilibrium configuration, which depends on its internal structure, amount and distribution of the angular momentum [231,232].

3.2.1. Rigid Rotation of Homogeneous Bodies

In a homogeneous body, the internal density distribution corresponds to a polytrope with the index $n = 0$. The solution of hydrostatic equations describing its equilibrium conditions at rigid rotation produces two kinds of rotational sequences. From the virial theorem, the results show that the value of the ratio of rotational kinetic energy K to the gravitational potential W in this case is limited to the range of values $0 \leq \tau = K/|W| \leq 0.5$. The first sequence of equilibrium configurations is called Maclaurin's solution and represents a sequence of axisymmetric spheroids. $\tau = K/|W| \simeq 0.14$ branches off Jacobi's sequence of triaxial ellipsoids, whose mechanical energy is lower than for Maclaurin's spheroids and which are thus more stable. However, the analysis of sectorial modes of oscillation shows that homogeneous Maclaurin's spheroids and Jacobi's ellipsoids become dynamically unstable for $\tau_{max} = 0.27$.

3.2.2. Rigid Rotation of Polytropes with Index $n \neq 0$

For centrally condensed bodies, i.e., self-gravitating systems with a polytropic index $n \neq 0$, there is a maximum value of $\tau_{critical} = (K/|W|)_{max}$ that depends on the polytropic index n at which the rotating configurations terminate, because the surface effective gravity becomes zero: $\tau_{crit}(n=0) \simeq 0.5$, $\tau_{crit}(n=0.81) \simeq 0.14$, $\tau_{crit}(n=3) \simeq 0.01$ [231]. Thus, stars in rigid rotation in the Zero Age Main Sequence (ZAMS) and beyond can only be characterized with very low rotational energies, i.e., $\tau_{crit}(n \gtrsim 3) \lesssim 0.01$, so that they can never acquire a global triaxial ellipsoidal configuration. If the mechanisms of angular momentum redistribution led to a huge concentration of angular momentum in the stellar core during the stellar evolution, the core could in principle attain the condition $\tau_{crit} \simeq 0.14$, until dissipation takes over and prevents the formation of a triaxial core [233].

3.2.3. Differential Rotation of Polytropes with Index $n \neq 0$

If for some reason, a self-gravitating object stores more rotational energy than allowed by τ_{crit} at rigid rotation, i.e., the object is necessarily in differential rotation with $\tau_{crit} \lesssim \tau \lesssim 0.14$, the angular momentum must be redistributed in the stellar interior in such a way that the Solberg–Høiland stability criterion is satisfied [231], which in a very simplified way, implies that the Rayleigh dynamic stability condition $\partial j / \partial r > 0$ (where r is the radial coordinate) is obeyed over the entire internal structure of the object.

The dynamical instability of rotating homogeneous spheroids and ellipsoids against sectorial modes of oscillation noted in Section 3.2.1 also appears in differentially rotating polytropes ($n \neq 0$), but for values of $0.24 \lesssim \tau_{max} \lesssim 0.27$, where the value of τ_{max} depends on the index n and on the law of internal differential rotation [231].

It is worth noting that the amount of total specific angular momentum a star can store depends on the energy ratio τ and it is roughly $(J/M) \simeq 1.83 \, 10^{18} (MR)^{1/2} \tau^{0.57}$ cm^{2}s^{-1}.

3.2.4. Barotropes and Baroclines

An object rotating with a conservative rotation law, i.e., the angular velocity depends only on the distance ϖ to the rotation axis ($\partial \Omega / \partial z = 0$, i.e., angular velocity constant over cylinders) is said barotrope. According to Poincaré–Wavre's theorem, as can be seen in [231] (p. 78), in barotropes, the shape of the star and the effective gravity can be derived from a potential, the effective gravity is perpendicular to surfaces of constant density (isopycnic surfaces) and the surfaces of equal pressure (isobars) and the isopycnic surfaces coincide. Otherwise, when the internal angular velocity is a function of both cylindrical coordinates, $\Omega = \Omega(\varpi, z)$, the rotation law is said non-conservative and the stellar structure is considered barocline. In that case, the above listed properties derived from Poincaré–Wavre's theorem are not obeyed anymore.

Rigid rotation is very frequently used to construct models of rotating stars, and it is a particular case of barotropes characterized by the special conservative law Ω = constant that holds over the entire stellar structure.

3.3. Geometry of Rotating Stars

Assuming that stars have genuine axial rotation, their geometry is shaped by the equilibrium established between the gravitational and the inertial centrifugal force. Since the centrifugal acceleration is larger, the larger the distance of regions from the rotational axis is, the more stellar layers are enlarged steadily from the pole towards the equator. Moreover, because rotating stars behave as having lower effective masses (see Section 3.5), the lower the effective mass becomes, the larger the core density is [234,235]. As polar regions undergo lesser centrifugal accelerations, the increased gravitational force in these directions and the increased central density produce a polar flattening or dimple, whose depth depends on the rotational law and the amount of angular momentum stored in the star [236].

The calculation technique of the geometry of a rotating star depends on whether the rotation law is conservative, $\Omega = \Omega(\varpi)$ or not conservative $\Omega = \Omega(\varpi, z)$. When conservative rotation laws are assumed, the effective gravity g_{eff} at any point in the star is given by [231]

$$\mathbf{g}_{eff} = -\nabla \Phi(\varpi, z) \qquad (11)$$

where $\Phi(\varpi, z)$ is the total potential function

$$\Phi(\varpi, z) = \Phi_G(\varpi, z) - \int_0^\varpi \Omega(\varpi') \varpi' \, d\varpi' \qquad (12)$$

where the rotational potential is added to the gravitational potential $\Phi_G(\varpi, z)$. The rotational potential exists only if the rotation law is conservative. The gravitational potential is related to the density $\rho(\varpi, z)$ through the Poisson equation

$$\nabla^2 \Phi_G(\varpi, z) = 4\pi G \rho(\varpi, z) \qquad (13)$$

with G as the constant of gravitation. Equations (12) and (13) can be solved simultaneously using the iterative self-consistent field (SCM) method introduced by Ostriker and Mark [237], largely used in the literature to study stellar structure and evolution with rotation in the original form, or with some modifications, e.g., [234,235,238–244]. A somewhat different method to calculate rotating stellar structures was introduced by Endal and Sofia [245], where the equipotential surfaces have the role of an independent parameter. Iterative methods based on the integrated form of the Poisson equation were introduced by Eriguchi and Mueller [246], Hachisu [247] and Eriguchi and Mueller [248] (see also references in those papers). An elegant two-dimensional finite-difference iterative technique applied to the problem of finding the gravitational potential associated with an axisymmetric density distribution was introduced by Clement [137,249].

Methods to calculate baroclinic stellar structures (i.e., stellar structures with non-conservative rotation laws) were developed by Uryu and Eriguchi [250], Uryu and Eriguchi [251] and Fujisawa [252]. To date, the most extensively used baroclinic models are those calculated by Meynet and Maeder [253] and Meynet and Maeder [254], and with some modifications by Heger et al. [255].

The turbulence in the stellar interior related to the differential rotation produced by the evolution and transport of the angular momentum is anisotropic and characterized by a strong horizontal transport that favors the establishment of a constant angular velocity on isobars, called a 'shellular' angular velocity distribution. On account of this specific differential rotational law, in Meynet and Maeder [253], the following auxiliary function was defined

$$\Psi(\varpi, z) = \Phi(\varpi, z) - \frac{1}{2}\Omega(\varpi, z)\varpi^2, \tag{14}$$

which has the property of determining surfaces $\Psi(\varpi, z) = $ constant that are parallel to isobaric surfaces, which is also useful to obtain relations that enable prediction the observational parameters affected by rotation.

The calculation of parameters that can be tested with observations also needs specific expressions describing the geometry of the stellar surface. In general, it is accepted that the gravitational potential of rotating centrally condensed objects can be given by a multipole expansion [256]

$$\Phi_G = -\frac{GM}{R(\theta)}\left[1 - \sum_{n=1}^{\infty}\left(\frac{R_o}{R(\theta)}\right)^{2n} J_{2n} P_{2n}(\cos\theta)\right], \tag{15}$$

where R_o is the radius of the undistorted star by rotation, J_{2n} are the zonal harmonic coefficients and $P_{2n}(\cos\theta)$ are the Legendre polynomials. To calculate the geometry of a rotating stellar surface, Equation (15) is considered with all coefficients $J_{2n} = 0$, which is then known as the Roche approximation. When the stellar surface is assumed to rotate as a rigid body, $\Omega = $ constant, Equation (12) is considerably simplified. Considering $\Phi_G = $ constant, from Equation (12), the equation giving the radius vector $R(\theta)$ for any azimuthal angle θ of the rotating stellar surface is given by

$$\frac{GM}{R(\theta)} + \frac{1}{2}\Omega^2 R(\theta)^2 \sin^2\theta = \frac{GM}{R_p(\eta)} = \frac{GM}{R_e(\eta)} + \frac{1}{2}\Omega^2 R_e(\eta)^2 \tag{16}$$

where R_p and R_e are the polar and the equatorial radii, respectively, and $\eta = Fc/Fg$ is the force ratio of the centrifugal acceleration Fc to the gravitational acceleration Fg

$$\eta = (V/V_c)^2 (R/R_c) \tag{17}$$

where V is the equatorial linear velocity and V_c is the equatorial critical velocity. Equation (16) admits the widely known analytical solution for the radius vector $R = R(\theta)$, e.g., [257], which describes the shape of the stellar surface at rigid rotation

$$\frac{R(\theta)}{R_e(\eta)} = \left(\frac{3}{1 + \frac{1}{2}\eta}\right)\left(\frac{1}{\epsilon\sin\theta}\right)\left\{\cos\frac{1}{3}[\pi + \arccos(\epsilon\sin\theta)]\right\} \tag{18}$$

with the additional definition

$$\epsilon = \left(\frac{3}{2+\eta}\right)\left(\frac{3\eta}{2+\eta}\right)^{1/2} \tag{19}$$

Using the algorithm by Clement [249] and barotropic representations $P = P(\rho)$ adapted for the stars of masses $M = 5M_\odot$ and $M = 15M_\odot$ at evolutionary stages near the ZAMS and the TAMS, rotating either as rigid or differential rotators with conservative rotation laws of the form (see details in [141])

$$\Omega(\varpi) = \frac{\Omega_o}{1 + \beta(\varpi/R_e)^n}, \qquad (20)$$

where β is a free parameter, and the obtained iso-density surfaces are shown in Figures 16 and 17. Let us note that horizontal shear instabilities will probably destroy the cylindrical differential rotation enforcing a 'shellular' rotation law [222,223,229]. According to these assumptions, the polar dimples have a low probability of existing in 'normal' stars. However, it has been shown in this review that new approaches on the rotating stellar structures suggest that the internal rotational profile should neither be rigid, cylindrical or shellular. In their discussion of stellar shapes with polar dimples, Smith and Collins [236] concluded that the predicted photometric effects are not present in the observational literature, which suggests that strong differential rotation is not common among stars. In a previous work by Collins and Smith [258] also dedicated to the photometric effect of cylindrical differential rotation in the A stars, concluded that "photometry alone can only put rather weak constraints on the degree of differential rotation within stars".

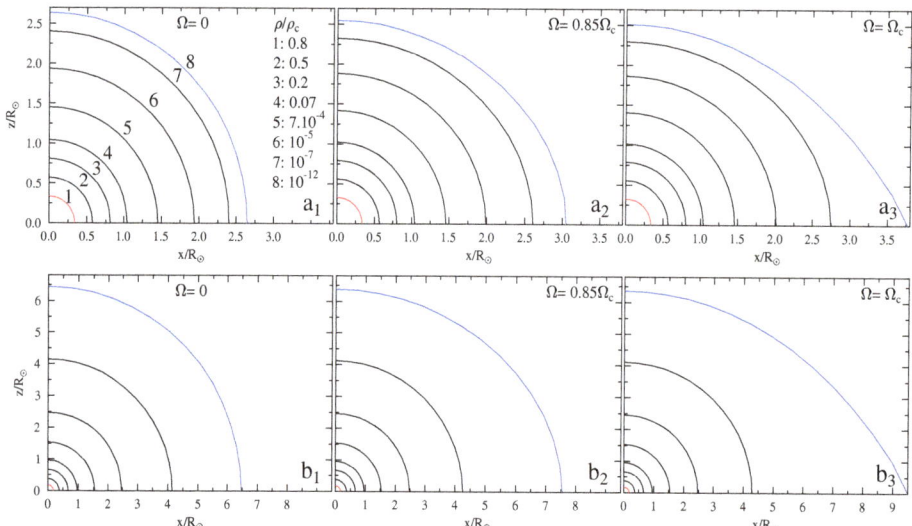

Figure 16. Rigid rotators. (**a**): Iso-density surfaces in model stars of $M = 5M_\odot$ in the ZAMS rotating at (**a$_1$**) $\Omega = 0$; (**a$_2$**) $\Omega = 0.85\Omega_c$; (**a$_3$**) $\Omega = \Omega_c = 1.92 \times 10^{-4}$ s^{-1}; (**b**): Iso-density surfaces in stars of $M = 5M_\odot$ in the TAMS rotating at (**b$_1$**) $\Omega = 0$; (**b$_2$**) $\Omega = 0.85\Omega_c$; (**b$_3$**) $\Omega = \Omega_c = 4.79 \times 10^{-5}$ s^{-1}. The iso-density surfaces are labeled with the corresponding density ratios ρ/ρ_c, which are the same in all panels of the figure. Adapted from Zorec [20], Zorec et al. [141]

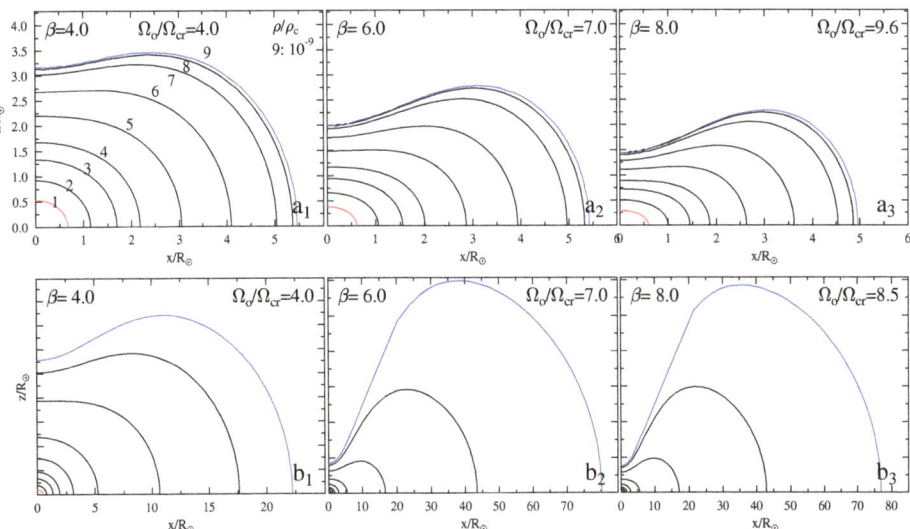

Figure 17. Differential rotators. (**a**): Iso-density surfaces in model stars of $M = 15M_\odot$ in the ZAMS having the internal rotation law (20) whose parameters Ω_o/Ω_{cr} and β are indicated: (**a1**) $\beta = 4.0$, $\Omega_o/\Omega_{cr} = 4.0$; (**a2**) $\beta = 6.0$, $\Omega_o/\Omega_{cr} = 7.0$; (**a3**) $\beta = 8.0$, $\Omega_o/\Omega_{cr} = 9.6$. Ω_{cr} is for the critical rigid rotation. The iso-density surfaces are labeled with the corresponding density ratios ρ/ρ_c, which are the same as in Figure 16 and for all panels of this figure. (**b**): Iso-density surfaces in the model stars of $M = 15M_\odot$ in the TAMS having the internal rotation law (20), whose parameters Ω_o/Ω_{cr} and β are indicated: (**b1**) $\beta = 4.0$, $\Omega_o/\Omega_{cr} = 4.0$; (**b2**) $\beta = 6.0$, $\Omega_o/\Omega_{cr} = 7.0$; (**b3**) $\beta = 8.0$, $\Omega_o/\Omega_{cr} = 8.5$. Ω_{cr} is for the critical rigid rotation. In each panel, the ordinates are in the same scale as the abscissas, but they differ from one to the other panel. Adapted from Zorec et al. [141].

When it comes to non-conservative rotation laws, it is no longer possible to define a rotational potential. In such cases, Maeder [229] suggested that the surface of a star can be identified as the region where an arbitrary displacement $d\mathbf{s}$ does not imply any work performed by the effective gravity, i.e., $\mathbf{g}_{\rm eff}.d\mathbf{s} = 0$. It can then be shown that the effective gravity can be expressed in the form

$$\mathbf{g}_{\rm eff} = -[\nabla \Psi(\varpi,z) + \frac{1}{2}\varpi^2 \nabla \Omega^2] \,, \tag{21}$$

where the function Ψ is defined in Equation (14) so that that the condition $\mathbf{g}_{\rm eff}.d\mathbf{s} = 0$ can be given the form

$$d\Psi(\varpi,z) + \frac{1}{2}\varpi^2(\nabla \Omega^2.d\mathbf{s}) = 0 \,, \tag{22}$$

whose integration over a meridian curve leads to the relation

$$\Phi_G(\theta) - \frac{1}{2}\Omega^2(\varpi,z)\varpi^2 + \frac{1}{2}\int_0^\theta \varpi^2(\nabla \Omega^2.d\mathbf{s}) = \Phi_G(0) \,, \tag{23}$$

which is the equation required to calculate the shape of a star having non-conservative rotational laws. However, with the following change in coordinates describing the stellar surface $\varpi_s(\theta) = R_s(\theta)\sin\theta$ and $z_s(\theta) = R_s(\theta)\cos\theta$, whilst the integration by parts of Equation (23) produces

$$\Phi_G(\theta) - \frac{1}{2}\int_0^\theta \Omega_s^2(\theta')\left(\frac{d\varpi^2}{d\theta'}\right)d\theta' = \Phi_G(0) \,. \tag{24}$$

that with the definition $\Phi_G(\theta) = -GM/R_s(\theta)$ can be put in the following more workable form

$$\frac{R_s(\theta)}{R_e} = \frac{1}{1+\eta_o[I(\pi/2)-I(\theta)]}$$

$$I(\theta) = \frac{1}{2}\int_0^\theta \left[\frac{\Omega_s(\theta')}{\Omega_o}\right]^2 \left(\frac{d\varpi^2}{d\theta'}\right)d\theta'.$$

(25)

In relation (25), the function $R_s(\theta)$ that describes the geometry of the stellar surface also appears in the integrand in terms of ϖ, $R_s(\theta)$ is obtained by iteration. Using the Maunder representation of the surface angular velocity as a function of the azimuthal angle θ given by

$$\Omega(\theta) = \Omega_o(1+\alpha(\cos^2\theta)), \quad (26)$$

geometrical shapes of stellar surfaces such as those shown in Figure 18 can be easily obtained. Since in Equation (25), it is $I(0) = 1/2$, for rigid rotation, this relation acquires the known form

$$\frac{R_e}{R_p} = 1 + \frac{1}{2}\eta \quad (27)$$

The value of R_p for rigid rotation can be estimated using the interpolation expression obtained in Frémat et al. [259] by interpolation in models of rotating stars calculated by Bodenheimer [235], Clement [137], and Zorec [20]

$$\frac{R_p}{R_o} = 1 - P(M/M_\odot)\tau$$

$$P(M/M_\odot) = 5.66 + \frac{9.43}{(M/M_\odot)^2} \quad (28)$$

$$\tau(\eta) = K/|W| = [0.0072 + 0.008\eta^{1/2}]^{1/2}.$$

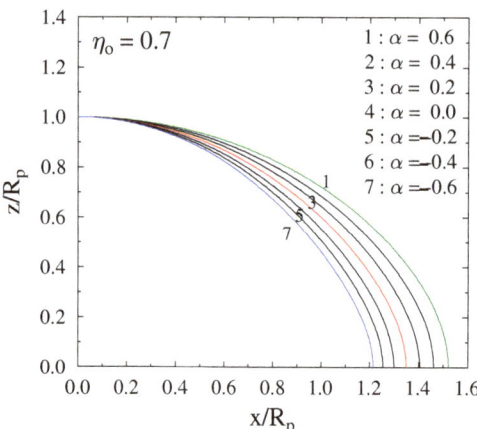

Figure 18. Geometrical shape of stars having the average surface rotational parameter $\eta_o = 0.7$, and latitudinal differential rotations given by Equation (26) for several values of the parameter α. The rigid rotation ($\alpha = 0$) is indicated by the red line.

3.4. On the Critical Rotation Rates

From the relations in Equations (27) and (28), the radii of the stars at critical rigid rotation $R_c(M,t)/R_o(M,t)$ can be calculated as R_o is the stellar radius without rotation. Thus, the equatorial critical linear velocity is [127]

$$V_c = 436.7 \left[\frac{M/M_\odot}{R_c(M,t)/R_\odot}\right]^{1/2} \text{ km s}^{-1}, \quad (29)$$

When characterizing stars rotating at critical rotation rates, it may be worth recalling the following relations between the ratio of angular velocities Ω/Ω_c, the ratio of linear velocities V/V_c at the equator and the equatorial acceleration η ratio

$$\begin{aligned} V/V_c &= (\Omega/\Omega_c)(R/R_c) \\ \eta &= (\Omega/\Omega_c)^2[R(\Omega/\Omega_c)/R_c]^3 \end{aligned} \quad (30)$$

where (Ω, Ω_c) and $(R(\Omega/\Omega_c), R_c)$ are the angular velocities and equatorial radii, with the subindex 'c' indicating the respective parameters at critical rotation. The parameter η and the velocity ratios compare to each other as

$$\eta \leq V/V_c \leq \Omega/\Omega_c, \quad (31)$$

which is worth recalling because we frequently refer to a near-critical rotation when $\Omega/\Omega_c \simeq 0.95$ actually corresponds to $V/V_c \simeq 0.83$ and to $\eta = 0.60$, which cannot be considered as a critical equilibrium condition in terms of forces. In Figure 19, these quantities are compared, where the error bars indicate the differences due to stellar masses and ages on the MS of early-type stars.

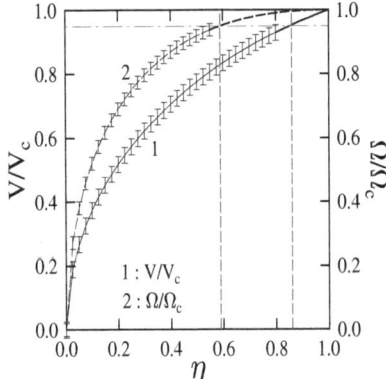

Figure 19. Angular velocity ratio Ω/Ω_c as a function of η. The uncertainty bars indicate differences for masses and ages in the main-sequence phase. It is shown that the $\Omega/\Omega_c \simeq 0.95$ (red line) identifies the force ratio $\eta \simeq 0.6$. The ratio $V/V_c \simeq 0.95$ (red line) corresponds to $\eta \simeq 0.86$. Adapted from Zorec et al. [260].

3.5. Notes on the Stellar Effective Mass, Total Energy Production and Age

An important effect related to the reduction in the local gravitational force due to the centrifugal force is that it makes a star behave as an object with a lower effective mass

$$\begin{aligned} M_{\text{eff}} &= M - \Delta M \\ \Delta M/M &= (2/3)\langle Fc/Fg \rangle_P \end{aligned} \quad (32)$$

where $\langle Fc/Fg \rangle_P$ represents the pressure-averaged ratio of the centrifugal to gravitational acceleration [261,262]. A larger $\Delta M/M$ ratio means a lower stellar mass, but the global lowering of the stellar effective mass, proportional to the stellar mass, is larger the higher the total angular momentum content is. This effect also controls the radiative energy production, e.g., [137,235,244,263,264]. The lowering of the emitted bolometric luminosity calculated by Clement [137], Bodenheimer [235], Sackmann [262] can be represented with the following interpolation relation [259]

$$\begin{aligned}
L(\eta, M/M_\odot) &= L_o(M/M_\odot) f_L(\eta, M/M_\odot) \\
f_L(\eta, M/M_\odot) &= A(M/M_\odot) + [1 - A(M/M_\odot)] \exp[B(M/M_\odot)\tau(\eta)] \\
A(M/M_\odot) &= 0.675 + 0.046(M/M_\odot)^{1/2} \\
B(M/M_\odot) &= 0.52.71 + 20.63(M/M_\odot)^{1/3}
\end{aligned} \qquad (33)$$

where $L_o(M/M_\odot)$ is the bolometric luminosity emitted by the parent non-rotating star, and $\eta = Fc/Fg$ is defined in Equations (17).

Due to a lower effective mass, rotating stars begin evolving in the ZAMS at a somewhat lower effective temperature and bolometric luminosity than corresponding to a non-rotating star having the same mass. As evolution goes on, chemical mixing provides new opacity conditions to the stellar envelope and the bolometric luminosity augments so that the evolutionary paths in the MS run over those of stars of the same mass and without rotation [229]. Since rotating stars evolve as having a lower effective mass, they evolve with longer evolutionary time scales, which, additionally, are sensitively prolonged due to the hydrogen dredged into the extended convective core by overshooting, as provided by the meridional circulation [265]. In massive stars, the mass-loss phenomenon through stellar winds also has a significant impact on the stellar evolution in the MS, e.g., [145,229]. To the mass-loss phenomena, the sporadic huge mass ejections should also be added, whose triggering mechanisms and total amount of ejected mass are still not well explained. These take place in Be stars, which are rapid rotators (see Section 2.5.3).

3.6. Effects of the Rotation on the Emitted Radiation Field

Rotating stars of high and intermediate mass can be described assuming that they have a photosphere in radiative and hydrostatic equilibrium. According to the radiative equilibrium, the temperature gradient is proportional to the emitted radiation flux and the pressure gradient is proportional to the effective gravity

$$\begin{aligned}
\nabla T &\propto F_{rad} \\
\nabla P &\propto g_{eff}.
\end{aligned} \qquad (34)$$

The detailed analytical treatment of these equations led to a relation that describes the emergent radiative flux as a function of the latitudinal coordinate θ and is commonly known as the gravity darkening relation

$$F_{rad}(\theta) = \sigma_{SB} T_{eff}^4(\theta) = \frac{L}{4\pi GM} f(\theta) g_{eff}(\theta)^\beta \qquad (35)$$

called the von Zeipel theorem [266], which describes the non-uniform distribution of the emitted radiation flux or the surface effective temperature as a function of the stellar latitude. In Equation (35), σ_{SB} is the Štefan–Boltzmann constant, L is the stellar bolometric luminosity, G is the constant of gravitation, M is the stellar mass, $f(\theta)$ is the gravitational darkening function and g_{eff} is the surface effective gravity. Physically, von Zeipel's theorem can be understood in terms of a radiation field flowing out of the star isotropically. As a consequence of the geometrical deformation of the star, the stellar surface covering a given solid angle gains less energy per unit of surface in the equator than for the same solid angle does in the polar region.

The theorem raised the known von Zeipel's paradox, which stipulates the impossibility of having stable barotropic stellar models in radiative equilibrium, e.g., [267]. This produces the setting up of a permanent meridional circulation due to the latitudinal thermal imbalance between the pole and the equator.

In the original von Zeipel's formulation for stars in rigid rotation, the gravity darkening term $f(\theta)g_{eff}(\theta) = $ is written as $c(\eta)g_{eff}^\beta$, where $c(\eta)$ and β are independent of θ, and $\beta = $ constant $= 1$ for radiative envelopes, while $\beta = $ constant $\simeq 0.32$ for convective envelopes [268,269]. The original von Zeipel gravity-darkening law holds only for stars with a conservative rotation law and radiative atmospheres in which the transfer of energy is approximated by a diffusion equation. Using radiation transfer equations, Smith and Worley [270] showed that, if meridional circulation is neglected, the gravity darkening (GD) relation is approximately $F_{\rm rad}(\theta) \propto g_{\rm eff}(\theta)^{1/2}$ and it recovers its original form $F_{\rm rad}(\theta) \propto g_{\rm eff}(\theta)$ if meridional circulation is taken into account. However, according to these authors, for non-conservative rotation laws, there is no simple relation between $F_{\rm rad}(\theta)$ and $g_{\rm eff}(\theta)$.

The GD relation in Equation (35), in spite of the noted θ−dependencies, is still frequently written as

$$\sigma_{\rm SB} T_{\rm eff}^4(\theta) = \kappa(\eta) g_{\rm eff}(\theta)^{\beta(\eta)} \qquad (36)$$

where $\beta(\eta)$ is considered independent of the stellar colatitude θ (for the definition of η see Equations (17) and (30)). In such a case, we have [127]

$$\beta = 1 - \frac{2}{3}\left[\frac{\ln(1-\eta) + \eta(1+\eta/2)}{\ln(1-\eta) - 2\ln(1-\eta/2)}\right]^{-3}.$$

This relation shows that $\beta=1$ for $\eta = 0$ and $\beta=1/3$ when $\eta = 1$. The current interferometric imaging and modeling of rapidly rotating atmospheres produce inferences of the gravity-darkening exponent in intermediate-mass stars $\beta_{\rm pbs} \lesssim 1$, e.g., [271–276].

Kippenhahn [277] studied the GD relation for non-conservative rotation laws and showed that $\beta \leq 1$, whose value depends on the specific rotation law. Maeder [278] considered the case of a rotating star with the non-conservative "shellular" profile of the form $\Omega = \Omega(r)$, where r is the radial coordinate of an associated spherical star. In this case, the star is baroclinic and the properties valid for barotropic structures cannot be applied. He considered that all intervening quantities can be developed linearly around their average on an isobar. For chemically homogeneous stellar surfaces in radiative equilibrium where horizontal flux contributions are ignored, he obtained the expression for the emerging bolometric radiation flux

$$F(\theta) \propto g_{\rm eff}[1 + \zeta(\theta)] \qquad (37)$$

where the additional term favors a flux increase in the polar regions and a decrease in the equatorial zone. In more recent formulations of von Zeipel's theorem, also useful for stellar surfaces in differential rotation, the gravity-darkening function in Equation (35) is no longer a constant but a complex function of the surface angular velocity $\Omega(\theta)$ that, following Maunder and Maunder [279], can be written as [260,280]

$$\Omega(\theta) = \Omega_o\left\{1 + \alpha[\cos^k(\theta); \sin^k(\theta)]\right\} \qquad (38)$$

where α and $k \geq 0$ are fitting parameters. The sign of α indicates whether the angular velocity $\Omega(\theta)$ is accelerated towards the pole or towards the equator. Espinosa Lara and Rieutord [280] and Zorec et al. [260] showed that, if the original form of the gravitational darkening relation is conserved, due to the dependence of the function f with $(\theta, \eta, \alpha, k)$, the exponent β averaged over the stellar surface becomes a strong function of (η, α, k).

The most outstanding consequence of the von Zeipel theorem and the geometrical deformation of the star induced by the rotation is that the spectrum emitted by a rotating stars is dependent on the aspect angle under which the object is observed, e.g., [281,282]. This happens in particular for the observed absolute bolometric luminosity, and consequently, also for the visual absolute magnitude.

Due to the lowering of the production of the stellar radiative energy, its geometrical changes and the gravitational darkening effect, the position of a rotating star in the HR diagram depicted just with fundamental parameters averaged over the surface, i.e., $\langle T_{\rm eff}\rangle$; $\langle L\rangle = \sigma_{\rm SB} S \langle T_{\rm eff}\rangle^4$ with respect to another object without rotation, will tend to be at a lower bolometric luminosity and at a lower effective temperature [235,283,284]. However, the detailed position of a star in the HR diagram still depends on the precise values of the apparent quantities ($\log L(i)$, $T_{\text{eff}}(i)$), which in turn rely on the internal rotation law on the amount of stored angular momentum, and on the inclination angle i of the rotation axis with respect to the observer [285–287].

3.7. 'Apparent' and 'Parent Non-Rotating Counterpart' Parameters

The spectrum of a rotating star can be represented as the integrated flux of radiation emitted per unit of wavelength interval per steradian in the direction towards the observer defined by the aspect angle i (inclination angle between the stellar rotation axis and the line of sight). It is given by [281,282,288]

$$L_\lambda(\eta, i) = 2 \int_0^{\pi/2} d\phi \int_0^\pi I_\lambda R^2(\theta, \eta) \frac{|\mu|}{\cos \delta} \sin \theta \, d\theta \tag{39}$$

where $R(\theta, \eta)$ is the co-latitude θ–dependent radius vector to the surface of the star distorted by rotation; $\mu = \mu(\phi, \eta) = \hat{n}.\hat{\imath}$ ($\hat{n}$ is the unit vector normal to the stellar surface and $\hat{\imath}$ is the unit vector representing the direction of the line of sight); $\cos \delta = \cos \delta(\theta, \eta) = -\hat{n}.\hat{r}$ ($\hat{r}$ is the unit vector in the direction of $R(\theta, \eta)$); $I_\lambda = I_\lambda(\mu, \eta)$ is the μ–dependent monochromatic specific radiation intensity calculated for the local effective temperatures $T_{\text{eff}}(\theta, \eta)$ and surface effective gravity $g_{\text{eff}}(\theta, \eta)$.

To study the effects of fast rotation on the fundamental stellar parameters it is useful to use two types of fundamental parameters. The reference model parameters of stellar atmospheres in radiative and hydrostatic equilibrium at each point of the observed hemisphere to calculate Equation (39), are called 'parent non-rotating counterpart' (pnrc) fundamental parameters and are noted as $T_{\text{eff}}^{\text{pnrc}}$, $\log g^{\text{pnrc}}$ and L^{pnrc} (bolometric luminosity). The fundamental parameters obtained by fitting the spectral energy distribution in a given spectral domain $\lambda_1 \leq \lambda \leq \lambda_2$ using classic plane-parallel model atmospheres in radiative and hydrostatic equilibrium take the name of 'apparent fundamental parameters', $T_{\text{eff}}^{\text{app}}$, $\log g_{\text{eff}}^{\text{app}}$ and L^{app}. The apparent, hemisphere-dependent bolometric luminosity, effective temperature, effective gravity and rotation parameter (τ or η, which are synonymous for models that are considered rigid rotators in the ZAMS) are then functions of at least seven unknowns: mass $M/M_\odot$, age t/t_{MS} (t_{MS} is the time a star can spend in the MS evolutionary phase), initial metallicity Z, equatorial rotation parameter η, surface differential rotation (Maunder) parameter α and power β_{GD} in the von Zeipel relation and the inclination angle i

$$\begin{aligned}
T_{\text{eff}}^{\text{app}} &= T_{\text{eff}}^{\text{pnrc}}(M, t) \, C_T(M, t, \eta, Z, i) \\
g_{\text{eff}}^{\text{app}} &= g_{\text{eff}}^{\text{pnrc}}(M, t) \, C_G(M, t, \eta, Z, i) \\
L^{\text{app}} &= L^{\text{pnrc}}(M, t) \, C_L(M, t, \eta, Z, i) \\
\frac{(V \sin i)_{\text{app}}}{V_c(M, t)} &= \left[\frac{\eta}{R_e(M, t, \eta, Z)/R_c(M, t, Z)}\right]^{1/2} \sin i - \frac{\Sigma(M, t, \eta, Z, i)}{V_c(M, t)},
\end{aligned} \tag{40}$$

where $R_e(M, t, Z, \eta)$ and $R_c(M, t, Z)$ stand for the actual and critical stellar equatorial radii, generally determined using the 2D models of rigidly rotating stars as in Zorec et al. [141] and Zorec and Royer [289]. The functions $C_T(M, t, \eta, Z, i)$, $C_G(M, t, \eta, Z, i)$, and $C_L(M, t, \eta, Z, i)$ carry all the information relative to the geometrical deformation of the rotating star and of its GD over the observed hemisphere. The last equation of the system Equation (40), which determines the equatorial linear rotational velocity, contains the correction term Σ that takes the Stoeckley effect into account (see Section 5.4). The dependence on the β_{GD} exponent of the gravitational darkening is also included in $\log g_{\text{app}}$ and $(V \sin i)_{\text{app}}$, because their prediction is based on spectral lines affected by the GD. The above five-fold parametric dependency is also valid for the equivalent widths W_λ of spectral lines, the line FWHM, the zeroes q_n of the Fourier transform of spectral lines, as well as the energy distributions and spectrophotometric parameters such as (λ_1, D) of the BCD system, which are highly sensitive to ($\log g^{\text{app}}$, T^{app}). For completeness, it must be noted that the functions C_T, C_G, and C_L also depend on the differential rotation law and on the specific relation representing the von Zeipel gravitational darkening. However, the system of Equation (40) must be completed with relations giving the surface-averaged fundamental parameters, because they are required as entry quantities to the models of stellar evolution with rotation to infer the mass M and the age t [145,254,290]

$$\begin{aligned} \langle T_{\text{eff}} \rangle &= T_{\text{eff}}^{\text{pnrc}}(M,t,Z) F_T(M,t,\eta,Z) \\ \langle g_{\text{eff}} \rangle &= g^{\text{pnrc}}(M,t,Z) F_g(M,t,\eta,Z) \end{aligned} \quad (41)$$

In Figure 20, the functions C_L, C_T and C_G calculated with the code FASTROT [61,127,259] are shown. In these references, the details concerning the solution by iteration of Equations (40) and (41) can also be seen. From Figure 20, it is apparent that functions C_L, C_T and C_G can in some cases be weakly dependent on η over the interval $0.8 \lesssim \eta \lesssim 1.0$, which introduces a source of indeterminacy of the physical and fundamental parameters. This can be particularly sensitive when it comes to the estimate of the inclination angle i. As i is determined once $\sin i$ has been determined as a solution of Equation (40), the error on the estimate of i runs as $\Delta i \simeq \tan i [\Delta (V \sin i)/V \sin i] + \Delta V/V$, which could be a significant source of the lack of estimated angles approaching 90° [291]. The distributions of inclination angle estimates are strongly asymmetrical. An in-depth statistical processing of these estimates makes it possible to correct the under-estimates that may occur when $i \gtrsim 70^\circ$ [292]. Apart from the interferometric inferences of i, the system Equation (40) thus remains a useful method to determine this quantity in B stars with and without emission, in contrast to that in Sigut and Ghafourian [291], which can only be applied to Be stars.

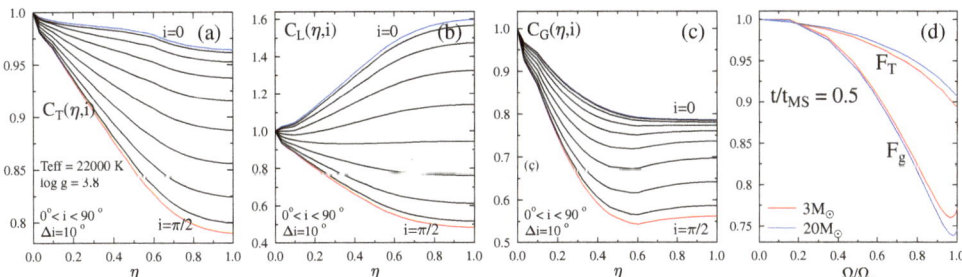

Figure 20. (a) Function $C_T(M,t,\eta,i)$; (b) function $C_L(M,t,\eta,i)$; and (c) function $C_g(M,t,\eta,i)$. The pnrc parameters used here are $T_{\text{eff}}^{\text{pnrc}} = 20{,}000$, $\log g_{\text{pnrc}} = 3.8$ and inclination angles in the interval $0 \leq i \leq \pi/2$ at steps $\Delta i = 10^\circ$. (d) Functions F_T and F_g are for $t/t_{\text{MS}} = 0.5$ and two stellar masses. Adapted from Zorec et al. [61,127].

Since a significant part of the present review is devoted to the BCD parameters, it can perhaps be timely to also show the effects due to the rapid rotation directly on the apparent BCD parameters (λ_1, D), which is given in Figure 21.

Finally, it is worth noting that FASTROT is a calculation code adapted for stellar atmospheres where the plan-parallel approximation is valid. For stellar sufficiently extended atmospheres, where the mean free path of photons is of the same order as the size of the atmosphere, curvature effects must be taken into account, because they can modify the source functions of lines and continua by orders of magnitude, as may be the case for some B[e] stars. The case of extended spherical atmospheres has already deserved some attention, as can be seen in, e.g., [293–297]. However, the effect of the atmospheric curvature also needs to be studied in the frame of rotationally deformed atmospheres, where the gravitational darkening effect can produce changes in the internal atmospheric temperature, as briefly mentioned in Section 5.5.1.

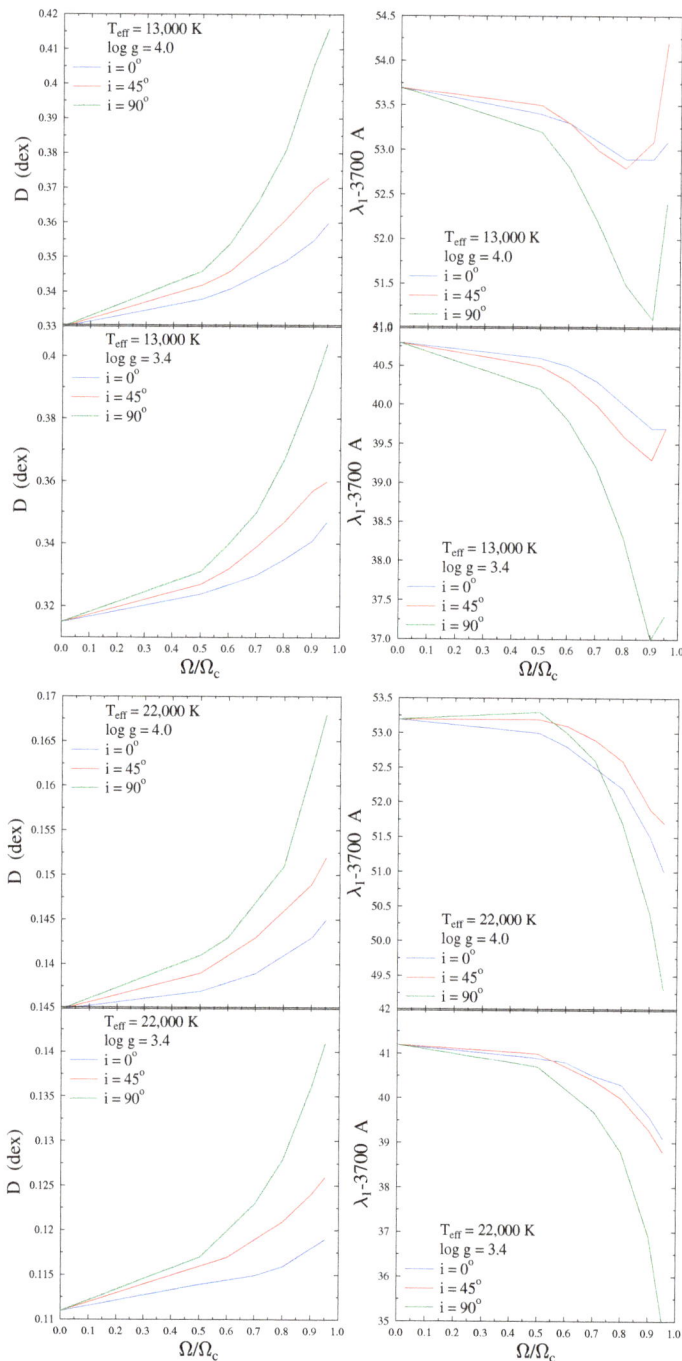

Figure 21. Apparent parameters (λ_1, D) modified by the stellar rapid rotation in stars having the indicated pnrc fundamental parameters and for several inclination angles. Adapted from Zorec et al. [61].

4. CNO Abundances and Stellar Rotation

As already noted in the introduction, the geometrical structural distortions induced by the rotation cause a thermal imbalance and lead to meridional circulation currents [221] that in turn induce a number of hydrodynamic [222,223] and magnetic instabilities [226,228]. These instabilities produce turbulent motions which trigger the mixing of chemical composition and the redistribution of the internal angular momentum. The predictions of quantities such as luminosity, stellar lifetimes, surface chemical abundances and mass-loss rates depend not only on the masses, initial rotation laws in the ZAMS and metallicity, but also on the prescriptions used to calculate the multiple diffusion coefficients that characterize the chemical and angular momentum diffusion. Extensive tests of surface chemical abundances of He and CNO abundances synthesized in the rotating stellar core were carried out in B and Be stars in the Milky Way (MW), SMC and LMC to validate the models of rotating stars, e.g., [298–307] and the references therein. In these contributions, contrasting results can be found, mainly on the surface N abundance, which according to models is meant to be enriched in rapid rotators. Thus, an unexpected significant enrichment was detected for MS stars in LMC with $V\sin i \lesssim 50$ km s^{-1}, and with and without overabundance for LMC B-type stars with $V\sin i \gtrsim 100$ km s^{-1}. This enrichment was attributed to merger phenomena in binary systems. No significant N enrichment seems to exist among rapidly rotating MW B-type stars. It was also noted that the N enrichment of B and Be stars was quite similar and that observed abundances in Be stars in the LMC agree with those predicted with population synthesis models calculated by Brott et al. [308] assuming that Be stars do not rotate faster than $V \simeq 0.4 V_c$.

Since the spectral signatures of the CN elements are rather small, their atmospheric abundances can be more easily detected in stars with higher metallicity, and are in general better determined in stars with $V\sin i \lesssim 100$ km s^{-1}, which implies pole-on seen fast rotators. Among somewhat evolved MS fast rotators, the enrichment of N is expected to be larger in pole-on stars than in those with $V\sin i \gtrsim 100$ km s^{-1}. This phenomenon is due to two concomitant effects that the radiation pressure tends to produce, namely CN elements being synthesized in the stellar core and raised to the surface being drifted towards gravity darkened and thus cooler equatorial regions [309], where their lines cannot be excited enough to produce observable spectral signatures. Except for Dunstall et al. [306], other cited contributions did not correct spectra for veiling effects in Be stars. The correction for the veiling effect sensitively increases the estimate of their chemical abundances. According to Potter et al. [307], the models of rotating stars need refined tuning of several diffusion constants, because abundance predictions strongly depend on their values. Uncertainties affecting the values of these constants make it so that, at the moment, the comparison of theory with observations is rather uncertain. Finally, the explanation of abnormal abundances stemming from single rapidly rotating stars is challenged by the possibility that the observed objects could be members of evolved close binary systems, where the spin-up towards the near critical rotation of one of the components and the N enrichment are both due to the accreted mass during a mass-transfer phenomenon [215,310].

5. Measurement of the Rotational Velocity in Early-Type Stars

The most straightforward though apparent information regarding a star as a rotator is obtained through the parameter $V\sin i$, where V is the true equatorial linear velocity and i is the inclination angle of the stellar rotation axis with respect to the observer. For a given star, V and $\sin i$ cannot be disentangled straightforwardly, unless a complementary analysis is performed on the stellar spectra emitted by the hemisphere deformed by rotation and projected towards the observer (see Section 3.6).

Shajn and Struve [311] discussed the broadening effect that rotation can produce in spectral lines, and two years later Struve [312] pledged that the rotation should be considered as an independent stellar fundamental parameter. Among the first determinations of $V\sin i$ for early-type stars can be cited those of Elvey [313], who used a graphical method where the observed MgII 4481 line profiles were compared to spun-up thin line profiles. Carroll [314] introduced a method based on the Fourier transform (FT) of line profiles that is today still currently used to determine the $V\sin i$. The method introduced by Elvey [313] and previously suggested by Shajn and Struve [311] was systematized by Slettebak [315], who determined the $V\sin i$ parameter for 123 early-type stars with and without emission lines.

Since these pioneering works on the stellar rotation, a large series of studies based on different methods have been dedicated to the determination of the $V\sin i$ of early-type stars, whatever their normal or active character. Following Howarth [316], we classify these methods into the following broad groups: (1) Methods based on the FWHM of observed line profiles that are compared with model FWHM calculated as in Slettebak et al. [317], where the limb-darkening (LD) in the line profiles and the GD effect are taken into account. In the old days, the comparison of observed FWHM was performed with observed or synthetic line profiles enlarged by a rotational broadening function; (2) Methods based on the FT, where neither the wavelength-dependent limb-darkening within the line profiles nor the gravitational darkening effect can be considered; (3) Methods that determine the $V\sin i$ parameter by fitting the observed line profiles with synthetic line profiles calculated for a given spectral domain with static model atmospheres, which are broadened with a rotational broadening function that does not consider the change in line profiles due to wavelength-dependent limb-darkening, nor the effects of the GD, e.g., [318,319]. The $V\sin i$ determination by the cross-correlation of observed high-resolution spectra against an observed or synthetic spectral template where lines are not considered broadened by rotation, e.g., [320] can also be associated to this category. Measurements based on the calibration of the FWHM lines as in Daflon et al. [321] are considered in this category; (4) Methods where observed line profiles are fitted with spectra calculated with model atmospheres that take into account the limb-darkening varying with wavelength in the interval of line profiles, but using not deformed stars by rotation neither affected by the GD effect [322]; and (5) Methods that use synthetic spectra calculated for stars deformed by rotation and gravity darkened, where the limb-darkening in the local line profiles is consistently taken into account, e.g., [259,323]. In Table 1, which is not intended to be exhaustive, the $V\sin i$ determinations of normal and active B-type stars are listed.

Table 1. Non-exhaustive table of $V\sin i$ determinations for early-type stars.

Source	Environment	Stars	Method	Number
Balona [324]	MW	O, Oe, B, Be	1	585
Conti and Ebbets [325]	MW	O, Oe	3	205
Slettebak [326]	MW	Oe, Be, (A, F)$_{shell}$	1	183
Wolff et al. [327]	Mw	B, Be	1	306
Abt and Morrell [328]	MW	A, Ashell, (B, Be)	1	1700
Halbedel [329]	MW	B.Be	1	164
Penny [330]	MW	O, On, Oe	3	177
Brown and Verschueren [331]	MW, assoc.	B, Be	1, 3	156
Howarth et al. [320]	MW	O, B, (Be)	3	373
Steele et al. [332]	MW	Be	1	58
Chauville et al. [333]	MW	O, B, A, Be	4	233
Abt et al. [334]	MW	B, Be	1	1092
Royer et al. [335]	MW	B, A	2	525
Royer et al. [336]	MW	B, A	2	249
Keller [337]	LMC	B	3	100
Penny et al. [338]	MW, SMC, LMC	O	3	56
Frémat et al. [259]	MW	B, Be	1	233
Glebocki and Gnacinski [339]	MW	all, (B, Be)	(1)	28,179
Strom et al. [340]	MW, cluster, field	B, Be	1	216
Dufton et al. [318]	MW, cluster	O, B, Be	3	234
Wolff et al. [341]	MW, cluster	O, Oe	1	44
Frémat et al. [342]	MW	Be	5	64
Mokiem et al. [343]	SMC	0, B	3	31
Levenhagen and Leister [344]	MW	Be	2	141

Table 1. Cont.

Source	Environment	Stars	Method	Number
Huang and Gies [323]	MW, cluster	B, Be	5	496
Martayan et al. [345]	LMC, cluster, field	B, Be	5	202
Hunter et al. [302]	SMC, LMC	B	3	50
Martayan et al. [346]	SMC, cluster, field	B, Be, (O, A)	5	346
Trundle et al. [305]	MW, SMC, LMC	B	3	61
Wolff et al. [347]	MW, cluster	B, Be	1	168
Wolff et al. [348]	LMC, cluster	OB	3	34
Huang and Gies [349]	MW, field, cluster	B, Be	5	108
Hunter et al. [350]	SMC, LMC	B, Be	3	407
Penny and Gies [351]	MW, SMC, LMC	O, B	3	258
Fraser et al. [352]	MW, supergiant	B	2	57
Huang et al. [353]	MW, field, cluster	B, Be	5	634
Díaz et al. [319]	MW	A	2	251
Marsh Boyer et al. [354]	MW, cluster	B, Be	3	104
Bragança et al. [355]	MW	B, Be	3	350
Dufton et al. [356]	LMC	O, B, Be	2, 3	334
Ramírez-Agudelo et al. [357]	LMC	O	2, 3	216
Simón-Díaz and Herrero [358]	MW	O, B	2, 3	203
Garmany et al. [359]	MW	B, Be	3	130
Zorec et al. [127]	MW	Be	4	233
Holgado et al. [360]	MW	O	2, 3	285
Solar et al. [361]	MW	Be	2	57
Xiang et al. [362]	MW	O, B, A	3	132, 548
Gaia DR3 [1], vbroad	MW	late-B to early-M	3	33, 812, 183
Gaia DR3 [1], vsini_esphs	MW	O, B, A	3	780, 461

Notes: MW (Milky Way); LMC (Large Magellanic Cloud); SMC (Small Magellanic Cloud); B can also imply also Bn; 'all' indicates all spectral types. [1] see Section 5.1.

5.1. Rotational Broadening in Gaia DR3

The third release of the *Gaia* catalog *Gaia* DR3, ref. [363], provides what is, to date, the largest survey of rotational line-broadening measurements (i.e., for $T_{\text{eff}} \geq 3100$ K). It is based on the analysis of the data obtained by the Radial Velocity Spectrometer (RVS), ref. [364]. Two measurements of rotational line-broadening are available. The first value was obtained by taking the median of the measurements performed on epoch/transit RVS spectra, prior to the radial velocity (RV) determination. It is named vbroad in the gaia_source table. The second estimate was directly measured on the co-added RVS data by the ESP-HS module of the Apsispipeline [365]. Its determination was inserted in the process that led to the computation of the astrophysical parameters (APs: T_{eff} and $\log g$) and the interstellar reddening of/towards O-, B-, and A-type stars from the simultaneous fitting of the *Gaia* BP/RP and RVS spectra with synthetic ones. The value is saved in the field vsini_esphs of the *Gaia* DR3 astrophysical_parameters table. Both estimates were made by ignoring any broadening mechanisms other than rotation (e.g., they assumed the absence of macroturbulent motions, as can be seen in Section 5.6) and were based on the maximization/optimization of the cross-correlation between the observed and theoretical spectra (which we assimilated to method 3 in Table 1).

It is worth reminding that the RVS wavelength domain ($\lambda\lambda$ 846–876 nm) is centered on the NIR Ca II triplet lines, which reach their maximum strength in F-, G- and K-type stars. In B-type stars (Figure 22), the higher members of the Paschen series dominate the whole spectrum and carry most of the spectroscopic information available to measure the line-broadening. This partly explains the behavior of the vbroad measurements, as reported by Frémat et al. [366]. In particular, at the B-type stars' temperature range, the values are usually expected to be underestimated and strongly affected by the noise at

$G_{RVS} > 10$. Furthermore, because of issues related to the RV determination of O-, B-, and A-type stars [366,367], all the vbroad measurements obtained for targets hotter than $T_{eff} = 14,500$ K were filtered out during the post-processing [368]. Finally, no vbroad measurement is available in the catalog when, in more than 40% of the epoch RVS spectra, the emission in the Paschen lines is rising above the local (pseudo-)continuum [368].

While the vsini_esphs estimate is expected to suffer from the same limitations, its value for the hot stars was kept as being part of the AP determination carried out by the ESP-HS algorithm. We compare in Figure 23 the values found in *Gaia* DR3 to those published in some of the most recent studies listed in Table 1 a similar comparison is presented for vbroad in Figure 11 of [366]. Although the studies we consider in this plot have analyzed stars brighter than $G_{RVS} = 10$, (with the exception of [362], whose study includes mainly fainter stars), there is little overlap with *Gaia* DR3 vsini_esphs measurements. This is due: (1) to the filters applied during the post-processing and based on the analysis of the goodness-of-fit distributions in both BP/RP and RVS domains; and (2) to the degeneracy between the APs at increasing interstellar reddening, as can be seen in, e.g., Figure 2 of [369] during the selection of the target based on the spectral type tagging (spectraltype_esphs). As can be seen in Figure 23, a significant scatter is observed relative to ground-based measurements (which are usually obtained at higher resolving power and in better suited wavelength domains). However, the measurement remains fairly representative of the rotational broadening, as can be measured in the RVS spectra.

Similarly to what was performed for the determination of vbroad, the ESP-HS avoided using the RVS data that could be affected by line emission. In these cases, the APs were derived by analyzing the BP/RP data only, and no line-broadening was derived. The decision to flag the spectrum for possibly having emission lines was based on the pseudo-equivalent width of the Hα line directly measured on the RP spectrum [370]. Because of the low-resolving power of the data, of the steep decrease in the instrument response in the blue wing of Hα, and of the method adopted, a fraction of weak Hα emittors were not flagged [371,372] and still have their vsini_esphsvalue published (see lower left panel of Figure 23, where we plot the estimates obtained by [127] for a sample of Be stars). In these cases, a representative broadening value may still be expected, as long as the spectrum of the star does not show any characteristics of stars with shell-like absorption.

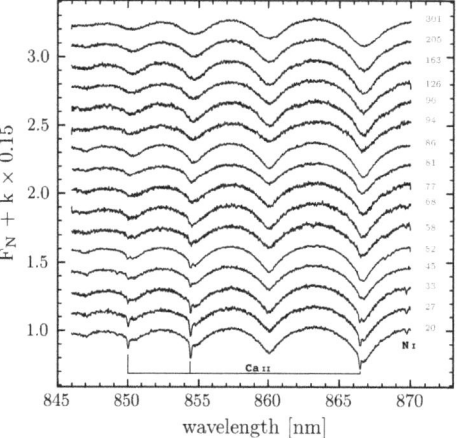

Figure 22. Variation with vsini_esphs (noted in blue, in km s^{-1}) of published *Gaia* DR3 RVS spectra with SNR > 200, $11,500$ K $\leq$ teff_esphs $< 12,500$ K, $3.7 \leq$ logg_esphs < 4.2. The spectra were normalized to the pseudo-continuum by the *Gaia* pipeline (F_N) and vertically shifted relatively to each other.

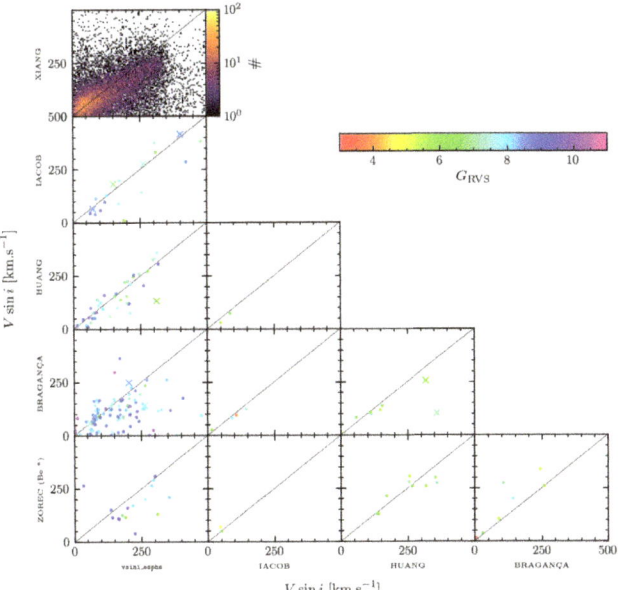

Figure 23. Comparison of the *Gaia* DR3 vsini_esphs estimates obtained for B- and O-type stars with measurements found in various catalogs as noted by the axis labels: XIANG [362], IACOB [358,360], HUANG [353], BRAGANÇA [355], and ZOREC [127]. The color code usually follows the G_{RVS} (see corresponding color bar), except for the XIANG vs. vsini_esphs panel where the data were sampled into 5 km s^{-1} bins. In the IACOB, HUANG, and BRAGANÇA plots, the crosses highlight those targets known, according to Simbad (CDS), for showing emission-lines. In these comparisons, we only took into account those targets with $T_{eff} \geq 10,000$ K.

5.2. Caveats on the FT and 'Spun-Up'-Based Methods

When Carroll [314] introduced their method to determine the $V\sin i$ parameter based on the Fourier transform (FT) of spectral lines, he clearly warned that an easy solution would be to represent the effect of the limb-darkening on the intensity of a spectral line non-affected by rotation by splitting the line profile into a product of line profile proper and a limb-darkening function, whose objective was to have a function to carry out the Fourier transform analytically

$$I(\lambda, \mu) = I(\lambda)_1 \, a(\mu) \qquad (42)$$

where $a(\mu)$ is a polynomial, or whatever other function of μ (μ, is the cosine of the angle between the normal to the stellar surface and the line of sight), and $I(\lambda)_1$ is the line profile seen at $\mu = 1$. Thus, this form does not take into account the effect of the limb-darkening in the line profile itself, which at Carroll's time, was already known to have the following approximate analytical aspect [373]

$$\frac{I(\lambda, \mu)}{I_c(\mu)} = \frac{1 + A(\lambda)\,\mu + \frac{B(\lambda)}{1+C(\lambda)\,\mu}}{1 + a(\mu)} \qquad (43)$$

where $A(\lambda)$, $B(\lambda)$ and $C(\lambda)$ depend on the line absorption coefficient. This form is far from that described in Equation (42), where the function $a(\mu)$ depends only on the constants related to the continuum spectrum, but not concerning the studied spectral line proper. This inconsistency is inherent to all applications of this FT methods [374,375], even though complicated forms of $a(\mu)$ are proposed, thinking that the $V\sin i$ parameters will then be more precise. This same comment is also valid for all methods where spectra are first calculated for non-rotating stellar atmospheres and then spectral lines are broadened according to different projected velocities $V\sin i$ using rotational broadening functions based on limb-darkening expressions like in Equation (42). Finally, Collins and Truax [376] notes that these methods do not take into account the effects carried on the spectral lines by the gravity darkening induced by rotation, although these authors based their criticisms online profile representations resembling that of Carroll.

5.3. Uncertainties on the $V\sin i$ Related to the Non-Consideration of Velocity Fields in the Calculations of Radiation Transfer in Spectral Lines

The line profiles used for the further broadening by rotation are obtained by radiative transfer calculations that consider static stellar atmospheres. Already known are the asymmetries of spectral lines produced by velocity fields in non-rotating stellar atmospheres, as can be seen in e.g., [119] and references therein. However, line asymmetries can be particularly significant when they are calculated using radiative transfer codes that take into account velocity fields in rotating stellar atmospheres, as shown in Korčáková and Kubát [297,377,378]. Nevertheless, further progress on the calculation of broadened line profiles by rotation can be expected if the aforementioned codes will also consider the non-spherical geometries of stellar atmospheres whose temperature structure in depth is calculated consistently with the deviations introduced by the gravity darkening, discussed in Section 5.5.1. Another discussion of asymmetrical rotational broadened line profiles is given in Section 5.10.

Many B-type stars have spectra marred by the presence of circumstellar envelopes or discs. To this end, CMFGEN is a code that was developed to model the spectra of a variety of objects with moving extended atmospheres: O stars, Wolf–Rayet stars, luminous blue variables, A and B stars, central stars of planetary nebula and supernovae, as seen in [379–381] and the references therein. Spectral lines predicted by this code, submitted to rotational broadening, can be of particular interest for studying B[e] stars, mainly in helping to identify genuine photospheric signatures in their observed spectra, otherwise strongly perturbed by the signatures of complicated gaseous and dusty circumstellar environments.

5.4. Uncertainty on the $V\sin i$ Related with the Gravity Darkening

As noted in Section 3.6, in rapid rotators, the emitted bolometric flux weakens from the pole to the equator as a function of the surface effective gravity [260,266,278,280]. The contribution of the radiation to the total λ-dependent flux in a spectral line broadened by the rotation is thus less effective from the equatorial regions, which consequently translates into an underestimated $V\sin i$ parameter. As this effect was first studied by Stoeckley [382], in what follows, we refer to it as Stoeckley's effect. As already quoted in Equation (40), the expected value of $V\sin i$ corrected from the Stoeckley's effect can be written as

$$V\sin i = (V\sin i)_{\text{obs}} + \Sigma(\eta, i, M, t) \qquad (44)$$

where $(V\sin i)_{\text{obs}}$ is the apparent or observed projected rotational, and Σ is the correction for the underestimation induced by the effects associated with the rapid rotation that hereafter, and depending on the circumstances, we call Stoeckley's correction or Stoeckley's underestimation. After Stoeckley [383], a number of authors have calculated Σ, in particular Townsend et al. [384], Cranmer [385] and Frémat et al. [259], who calculated this correction in the frame of a strict surface rigid rotation with the classical formulation of the GD effect by von Zeipel [266]. A schematic representation of the relation between the true $V\sin i$ and the observed one is shown in Figure 24.

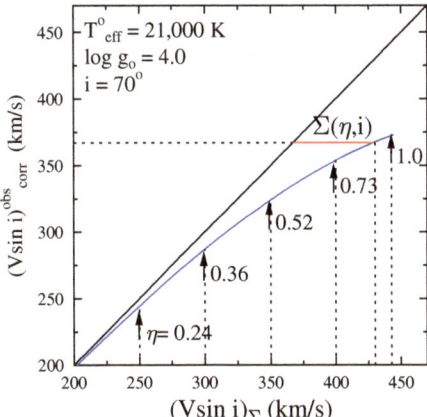

Figure 24. The blue line represents the relation between the observed $(V\sin i)^{obs}_{corr}$ corrected from observational uncertainties (ordinates) against the $V\sin i$ corrected from Stoeckley's underestimation (abscissas). This relation was calculated with He I 4471 line profiles for models with *pnrc* parameters $T_{eff} = 21{,}000$ K and $\log g = 4.0$, inclination angle $i = 70°$, and η ratios ranging from 0 to 1.0. The red bars indicate Stoeckley's correction Σ for an arbitrary near critical $(V\sin i)^{obs}_{corr}$ parameter. The η values indicated in the figure are for the actual $(V\sin i)_\Sigma = V(\eta)\sin i$ parameters in abscissas. Adapted from Zorec et al. [127].

A particular underestimation of the $V\sin i$ parameter can be produced if no attention is paid to the spectral lines that are formed in the polar regions of rapidly rotating stars. Heap [386,387] discovered that the $V\sin i$ parameter of ζ Tau determined using the classical spun-up measurement methods of photospheric lines in the spatial ultraviolet (far-UV) observed with the satellite Copernicus were twice as low than in the visible, and that the effective temperature required to model the far-UV lines was higher than for those in the visible spectral range. She concluded that this discrepancy is due to the gravity darkening, according to which the hot polar regions contribute with smaller line broadening linear velocities than the somewhat cooler equatorial regions, which contribute with the largest line broadening radial velocities. This hypothesis was confirmed with model calculations by Hutchings [388], Hutchings and Stoeckley [389] and Sonneborn and Collins [390]. The rotation of a larger set of rapid rotators using far-UV lines $\lambda 1299$ of SiIII and $\lambda 2756$ of FeII observed by the satellite IUE was studied by Carpenter et al. [391], who made models of gravity-darkened stellar atmospheres and concluded that the obtained $V\sin i$ parameters agree with those obtained from spectral lines in the visible, which is another way to confirm the gravity-darkening effect in rapid rotators and warns of the necessity of using other than simple FT analysis or spun-up profiles to determine the $V\sin i$ parameter from lines whose formation is sensitive to the local formation conditions of temperature and gravity, particularly when it comes to rapid rotators.

5.5. Effects on the $V\sin i$ Related with the Structure of Rotating Stellar Atmospheres

Stress is put here on the formation of spectral lines in gravity-darkened stellar atmospheres. The phenomena concern: (1) the thermal structure in the depth of stellar atmospheres as a function of the colatitude θ and its incidence on the line source function; and (2) the sensitivity of line source functions to local physical conditions.

5.5.1. The Thermal Structure in Depth of Rotating Stellar Atmospheres

Very early, it was recognized that the rotation and simultaneous hydrostatic and radiative equilibrium contradict each other, and that the latitudinal variation of the temperature in the stellar atmosphere produces a non-vanishing θ-dependent divergence of radiation flux [392], which unleashes the horizontal diffusion of light that has to be treated properly to obtain a stable atmospheric thermal structure. Two-dimensional radiation transfer calculations in stars deformed by rotation, similar to those started by Pustyl'Nik [393] and Hadrava [394], might perhaps be able to tackle these questions and predict the reliable dependencies of the line source functions with θ as well as GD relations that can be considered consistent with the right thermal structures of rotating atmospheres. Unfortunately, to date, there are no such detailed models. The current approximation used to model the spectra emitted by rotating objects is to assume that at each latitude θ, the structure of the atmosphere as a function of the optical depth corresponds to that of a classical plane-parallel model atmosphere characterized by the local parameters $[T_{\rm eff}(\theta), \log g_{\rm eff}(\theta)]$, where $T_{\rm eff}(\theta)$ is inferred with the adopted GD law. The thermal structures of rotating atmospheres used to date have never been demonstrated to be consistent, because the horizontal radiation flux was never taken into account. It is worth mentioning that the improbable physical structure of the atmosphere in the equatorial regions of a fast rotator, that the relationship in Equation (35) leads to, implies that, for $\lim g_{\rm eff} \to 0$, the radiative flux is canceled so that $\lim T_{\rm eff} \to 0$. The low temperature condition is not a sufficient condition to make the atmosphere convective, a state that actually requires the specific entropy to decrease towards the surface of the star, which imposes conditions not only on the temperature gradient but also on the density gradient [229]. By simulating the effects produced by some extreme physical circumstances masking the above outlined situation, predictions show that there can be significant changes in the line intensities and their widths, which may lead to estimates of $V\sin i$ that sensitively depend on the chosen spectral lines [260].

5.5.2. The Line Source Functions in a Gravity Darkened Atmosphere

It has long been known that due to the non-LTE effects and atomic structures proper, the source function of spectral lines have selective sensitivities to collisional and radiative processes that dominate the population of atomic levels [87,116,119]. Because the gravity-darkened atmospheres in rapidly rotating stars induce a wide range of electron temperatures and densities over the hemisphere projected toward the observer, spectral lines with different sensitivities to local formation conditions do not reflect the same physical properties prevailing in the observed hemisphere. Owing to these differences, the spectra of rotating stars interpreted with models calculated for non-rotating atmospheres can produce not only different values of apparent $(T_{\rm eff}, \log g_{\rm eff})$-parameters, but they also produce different estimates of the $V\sin i$ (the details of which can be found in [260]).

5.6. *Effects of the Microturbulence and the Macroturbulence on the Determination the $V\sin i$.*

Microturbulence and macroturbulence hide a series of motions that are not yet well identified. It was recognized very early on that spectral lines undergo broadening due to the random motions of eddies in stellar atmospheres [395,396]. Underhill [397,398], and Rosendhal [399] reported that these effects should be present mainly in early-type stars with low surface gravity. Random motions of eddies of the order or lower than the mean free-path of photons were called *microturbulence* and can affect the line absorption coefficient producing the so-called 'second-class' line broadening [396]. Movements of the eddies implying distance scales larger than the photon mean that free-paths were called *macroturbulence*, but they do not change the effective mean atomic-line absorption coefficient. In this case, the line broadening is simply produced by the Doppler effect associated with the macroscopic motion of eddies and was called 'first class'. More recently, Cantiello et al. [133] and Grassitelli et al. [400] evoked the subphotospheric convection to account for the microturbulent motions, while Aerts et al. [401,402] suggested that, in B-type stars, macroturbulence could be ascribed to the low-amplitude gravity modes of non-radial pulsations. The errors affecting the $V\sin i$ and macroturbulence velocity ($v_{\rm macro}$) determinations over the time series of line variations are star-dependent. Aerts et al. [402] find that the maximum error on the $V\sin i$ is $\delta_{V\sin i} \lesssim 60 \ {\rm km\,s^{-1}}$ and the averaged error is $\langle \delta_{V\sin i} \rangle \lesssim 8 \ {\rm km\,s^{-1}}$. Regarding the maximum error on the $v_{\rm macro}$ estimate is also $\delta_{v_{\rm macro}} \lesssim 60 \ {\rm km\,s^{-1}}$, while its averaged error is $\langle \delta_{v_{\rm macro}} \rangle \lesssim 17 \ {\rm km\,s^{-1}}$. Similar conclusions were also put forward by Simón-Díaz [403]. Howarth et al. [320] claimed that macroturbulence can be an important line-broadening mechanism in O- and early B-type supergiants, which adds to rotation. Nevertheless, its determination is not unique, because the same broadening can be produced by combining large ranges of macroturbulent velocities v_{mt} and $V\sin i$ values [404].

To describe the line broadening carried by macroturbulent motions, two models are used: isotropic Gaussian and anisotropic with radial and tangential components, where each has a Gaussian-dependent velocity distribution [405,406]. The anisotropic model depends on at least three free parameters, but they are reduced to only one so that both models are characterized by a unique dispersion of macroturbulent velocities σ_{mt}. The anisotropic model generally produces the more effective fits of spectral line profiles, leading to slightly larger values of v_{mt} than the isotropic model does [358,407,408]. However, they both mask still unidentified line-broadening phenomena.

Macroturbulence was mostly explored in early-type stars, mainly supergiants, which all have low values of $V\sin i$ cf. [320,352,404,407–409].

However, macroturbulence was also suggested for later B-type supergiants [408]. As found in Simón-Díaz and Herrero [358] and Sundqvist et al. [409], macroturbulence should not be entirely negligible in early-type dwarfs and giants. Using the rotational period of two magnetic O-type stars (HD 108 and HD 191612), [409] found that the apparent rotational velocity of the star is $V\sin i \lesssim 1$ km s^{-1}. This same parameter redetermined with the Fourier transform method becomes $V\sin i \sim 40 - 50$ km s^{-1}. It can then be expected that macroturbulent-like movements also exist in the atmospheres from B-type dwarfs to the giants of lower effective temperatures than those explored in the above cited works. This can be the case of all those stars that undergo non-radial pulsations and a wide funnel of instabilities induced by the rotation. The photospheric spectral lines currently used to determine the $V\sin i$ parameters of Be stars can then legitimately also be expected to be affected by macroturbulence. The macroturbulent motions are detected when apparent rotational velocities are $V\sin i \leq 150$ km s^{-1}, while lines broadened by $V\sin i \gtrsim 150$ km^{-1} seem to remain fairly insensitive to the broadening induced by macroturbulence. However, this does not mean that, in these cases, it does not exist, as in Be stars, where atmospheres can undergo significant upheavals maintained by non-radial pulsations, a large spectrum of instabilities induced by the rapid rotation and possible disordered magnetic fields generated by under-photospheric convection [133,134,136,137].

5.7. Effects Carried by the Differential Rotation on the Value of $V\sin i$.

The existence of a surface differential rotation in B and Be and other massive stars was speculated or predicted with models by a number of authors cf. [20,260,289,410–416], who derived parameters indicating its possible existence. To support this hypothesis, two different theoretical approaches to the stellar structure can be mentioned, wherein the results indicate that there are mechanisms capable of inducing differential rotation at the surface of early-type stars. Using conservative rotation laws [137] or non-conservative 'shellular' rotation laws [136] for models of stellar structures, it was shown that the two convection zones in the envelope associated with increased opacity due to He and Fe ionization can be considerably enlarged in depth by the rapid rotation. These regions together establish an entire convective zone beneath the surface that spreads out over a non-negligible region: from 1/8 of the stellar radius at the pole to nearly 1/4 at the equator [136]. However, differential rotation in convective zones can be induced in the same way it is induced in the Sun [417]. Following the solar rotational picture, where the convective layers have differential rotation, we may expect that some coupling can exist between convection and rotation beneath the surface in rapidly rotating massive and intermediate mass stars, whose imprint on the outermost stellar layers will translate into an angular velocity dependent on the latitude.

Solving the baroclinic balance relation obtained with the curl of the time-independent momentum equation of an inviscid, axisymmetric rotating star without magnetic fields, Zorec et al. [141] obtained solutions for the angular velocity distribution in the envelope under several conditions: (1) the surfaces of specific entropy S are parallel to the surfaces of specific angular momentum j, $S = S(j^2)$; (2) the surfaces of specific entropy and angular velocity coincide, i.e., $S = S(\Omega^2)$; and (3) the surfaces of specific entropy S are parallel to the surfaces of constant specific rotational kinetic energy, $S = S(\varpi^2 \Omega^2)$. In Figure 25, some examples of curves $\Omega(r, \theta) = const$ that obey these assumptions are shown. Rieutord [415], Espinosa Lara and Rieutord [418], Rieutord et al. [419] and Gagnier et al. [420] presented a new consistent way to compute a two-dimensional model of a fast rotating star including the large-scale flows, where among many other improvements, the calculated internal and surface differential rotation have some similarities to those shown in Figure 25.

The curves $\Omega(r, \theta) = const$ can have surface imprints as an atmospheric latitudinal differential rotation that can be sketched using the following relation

$$\Omega(\theta) = \Omega_p (1 + \zeta \varpi^\kappa)^\gamma, \qquad (45)$$

where $\varpi = r(\theta) \sin\theta$ is the distance to the rotation axis; $r(\theta) = R(\theta)/R_e$ is the normalized radius vector that describes the stellar surface, Ω_p is the polar angular velocity, ζ, κ and γ are constants. It follows that from Equation (45) that the equatorial angular velocity is $\Omega_e = \Omega_p (1 + \zeta)^\gamma$. The Doppler displacement $\Delta\lambda$ due to radial velocity at a point $\vec{R}$ in the stellar surface is

$$\Delta\lambda = (\lambda/c)(\vec{\Omega} \wedge \vec{R})_z = (\lambda/c)\Omega(\theta) x \sin i, \qquad (46)$$

where the reference system (x, y, z) is centered on the star, with the z axis is positively directed toward the observer and the x and y axes are in the plane of the sky. To more clearly characterize the Doppler displacements described by Equation (46), this relation is normalized by the displacement produced in the limb at the equator

$$\mathscr{C} = \frac{\Delta\lambda}{\Delta\lambda_e} = \left[\frac{\Omega(\theta)}{\Omega_e}\right] \frac{x}{R_e}, \qquad (47)$$

where Ω_e is the angular velocity at the equator and $\mathscr{C}$ is a constant independent of $\sin i$ that represents an isoradial velocity curve. Figure 26a,b show $\mathscr{C} = $ constant for angular velocities accelerating from the pole toward the equator, while Figure 26c shows a case where $\Omega(\theta)$ is accelerated from the equator towards the pole. In cases 'a' and 'b', the maximum Doppler displacement is produced at the stellar limb in the equator, while in case 'c', this maximum is produced in the middle of the closed circles ('owl' eyes) on the visible part of the meridian contained in the $(x\hat{\imath}, \vec{\Omega})$-plane. In this case, the curve $\mathscr{C} = 1$ is also closed and contains the equatorial limb point, while all the remaining curves for $\mathscr{C} > 1$ are closed. In Figure 26b, two curves are shown (red dotted and dashed curves) that produce the same Doppler displacement, i.e., $\mathscr{C}_r = \mathscr{C}_d$, where $\mathscr{C}_r$ denotes the radial velocity for rigid rotation (red dotted line) and $\mathscr{C}_d$ is for differential rotation (red-dashed curve). Both isoradial velocity curves coincide at the equator, nevertheless, the $\mathscr{C}_d$ curve is longer than for $\mathscr{C}_r$. This is the main source for the difference detected in the residual intensities of absorption lines broadened by rigid and differential rotation laws.

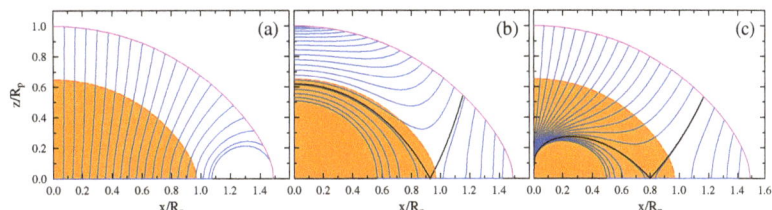

Figure 25. Curves of internal constant angular velocity $\Omega(r,\theta) = const$ in the stellar envelope (blue lines). (**a**) Curves corresponding to the condition $S = S(j^2)$; (**b**) curves obtained with $S = S(\Omega^2)$; and (**c**) curves calculated for $S = S(\varpi^2\Omega^2)$. The solutions are valid in the convective envelope, i.e., above the shaded central region [S = specific entropy; j = specific angular momentum; and ϖ = distance to the rotation axis]. Ordinates coincide with the rotation axis; abscissas are in the equatorial plane. Adapted from Zorec et al. [141].

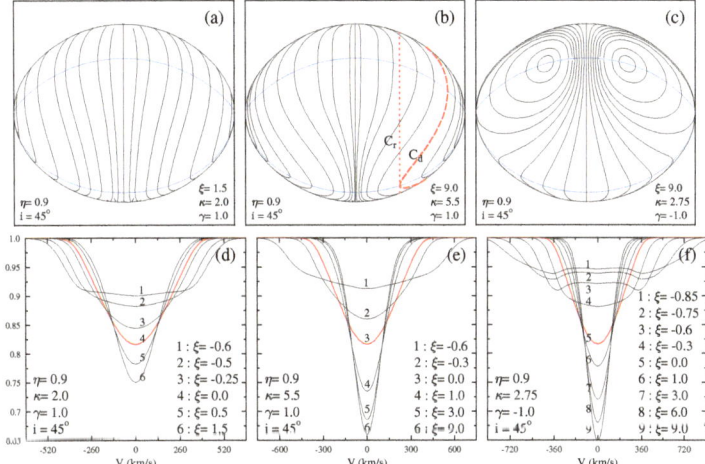

Figure 26. Upper series: Curves of constant radial velocity $\mathscr{C}$ contributing to the rotational Doppler broadening of spectral lines. (**a**) $\mathscr{C}$ curves for a Maunder-like surface angular velocity law with $\partial\Omega(\theta)/\partial\theta > 0$; (**b**) $\mathscr{C}$ curves for a law given by Equation (45) with $\gamma > 0$, so that $\partial\Omega(\theta)/\partial\theta > 0$; (**c**) $\mathscr{C}$ curves for a law given by Equation (45) with $\gamma < 0$, so that $\partial\Omega(\theta)/\partial\theta < 0$. In cases (**a**,**b**), the largest Doppler displacement is produced in the limb at the equator, while in case (**c**), the largest Doppler displacement is produced in the center of the upper closed curves. Case (**b**) identifies the lines producing the same Doppler displacement, $\mathscr{C}_r = \mathscr{C}_d$, where $\mathscr{C}_r$ (red dotted line) indicates a rigid surface rotation, and $\mathscr{C}_d$ (red-dashed line) a differential surface rotation. In all cases, the parent Doppler displacement from rigid rotation is determined by the straight line touching the corresponding $\mathscr{C}_d$ curve at the equator. Lower series: Gaussian line profiles broadened by rotation, whose intensity and equivalent width depends on the effective temperature and gravity such as a HeI 4471 transition: (**d**) line profiles for the curves of constant radial velocities of the type shown in (**a**) for several parameters ξ in Equation (45); (**e**) line profiles are for the constant radial velocities shown in (**b**) for several parameters ξ; (**f**) line profiles are for the constant radial velocities shown in (**c**) for several parameters ξ. Calculations include gravitational darkening. The parent non-rotating object has $T_{\text{eff}} = 22,000$ K and $\log g = 4.0$. The angular velocity law is given by Equation (45) with the same parameters (κ, γ, η, i) as in the corresponding upper blocks, except that ξ takes a wider range of values. In all cases, $\xi = 0$ represents the rigid rotation and the corresponding line profile is red-highlighted. In all blocks, the ordinates are the same. Adapted from Zorec et al. [127].

Putting $k = 2$ and $\gamma = 1$ in Equation (45), the known Maunder relation is recovered and used here below to produce examples of effects due to the surface differential rotation on the value of $V \sin i$

$$\Omega(\theta, \alpha) = \Omega_e(1 + \alpha \cos^2 \theta), \qquad (48)$$

In Figure 27, the $V \sin i$ parameters in km s^{-1} obtained are shown with the FT technique of the HeI 4471 and MgII 4481 lines broadened by differential rotation. In the ordinates, the measured apparent parameter is given, while in the abscissas is given the nominal rotational quantity that would correspond to the true equatorial velocity V_e. Thus, if the stars had their atmosphere in differential rotation, the parameters $V \sin i$ obtained with the currently used measurement methods would depend on the degree of differential rotation and sensitively differ from the expected values if they were rigid rotators. The difference is due, on the one hand, to the geometry of the constant radial-velocity curves contributing to the individual monochromatic Doppler displacements that contribute to the broadening of a spectral line, and on the other hand, to the inadequacy of the measurement methods, which are not adapted to the physical circumstances imposed by the differential rotation. In stars rotating with an angular velocity accelerated from the pole toward the equator ($\alpha < 0$), the measured $V \sin i$ implies that $V < V_{eq}$, where V_{eq} is the actual linear rotational velocity of the equator. Conversely, a differential rotation accelerated from the equator toward the pole ($\alpha > 0$) leads to a $V \sin i$ where $V > V_{eq}$. It also follows from these results that, for a given absolute value $|\alpha|$, the differences $|V - V_{eq}|$ are larger for $\alpha < 0$ than for $\alpha > 0$.

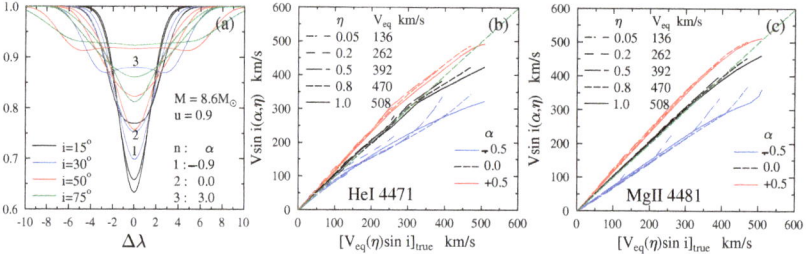

Figure 27. (a) Gaussian absorption lines broadened by rotation, whose intensity and equivalent width depend on the effective temperature and gravity, as does the HeI 4471 transition. The gravitational darkening is calculated for an object of mass $M/M_\odot = 8.6$ and fractional age $t/t_{MS} = 0.5$ rotating at the equator with $u = V/V_c = 0.9$. The angular velocity law used in the calculations is given by Equation (45). The colors indicate inclination angles, but in this figure, only the $i = 30^\circ$ (blue) corresponding to different values of α are explicitly identified. (b) Measured $V \sin i$ parameters in km s^{-1} by means of the FT technique of the HeI 4471 line broadened by differential rotation characterized by the different values of the parameter α in an object having $T_{eff} = 23{,}000$ and $\log g = 4.1$ rotating at different rates η. In abscissas, there are the true equatorial $V \sin i$ for the same rates η. (c) Idem for the MgII 4481 line. Adapted from Zorec et al. [127].

5.8. Line Profiles Produced by Near-Critical Rigid Rotators

Jeffery [421] pointed out that the line profiles emitted by rapidly rotating B-type stars may have flattened cores or even present emission-like reversals. This phenomenon is partially due to the GD effect that makes the local absorption contributions from the equatorial regions to the broadened profile reduced. On the other hand, due to the stellar rotational flattening, the iso-radial velocity curves $V_r = $ constant $\simeq 0$ can be shorter than some of them for slightly larger radial velocities, thus favoring a less efficient absorption that translates into an apparent reversal or emission in the broadened line core. This phenomenon was widely studied by Takeda et al. [422,423] in the metal lines of Vega (α Lyr,

HD 172167). The formation of emission-like reversals due to similar physical circumstances, but in objects with surface differential rotation was evoked by Zorec [424].

5.9. Double Valued V sin i.

Objects with angular velocity laws accelerated toward the pole may present polar dimples [235,236,258,260]. In such configurations, the largest contribution to the Doppler broadening of spectral lines comes from a point situated on an intermediate latitude on the stellar meridian facing the observer. For large inclination angles, the contribution to the line broadening produced by the closed curves corresponding to $\mathscr{C} > 1$ (see Figure 26c) can be partially hidden in the troughs, which, in addition, can be more or less dimmed by the limb-darkening. The relation between the FWHM of spectral lines and the $V\sin i$ then becomes double valued, as shown in Zorec [20,153,424]. Photometric, spectroscopic and interferometric measurements have not yet been able to signal the existence of 'normal' stars with polar dimples, as mentioned in Section 3.3.

5.10. Asymmetrical Rotational Broadening Functions: Expanding Layers and tidal Interactions

When a common rotational broadening function (RBF) is determined by the deconvolution of all spectral lines present in a given spectral range, there is only a small chance that its shape is entirely symmetric [425,426]. This is because lines due to a variety of elements and their different ionization states are formed in atmospheric layers having different physical conditions and perhaps characterized by different velocity fields entertained by non-radial pulsations, convection movements, more or less expanding layers of nascent stellar winds, etc. It can be shown that, independently of the sign of the skewness of the RBF, the derived $V\sin i$ parameters are somewhat underestimated [260].

During the 1970s and 1980s, a large amount of literature dealt with the spectral line formation problem in moving stellar atmospheres (as can be seen in [86,87,119,427–437]). The typical outbursts and fadings in Be stars, which are thought to be the consequences of huge discrete ejections of mass [68,69,71,438], also imply that there are periods with outward accelerated layers in the stellar atmospheres. The combination of expanding velocities with rotation produces broadened and blue-shifted spectral lines that can be schematically described with rotation-expansion broadening functions (REBF). These functions are asymmetric, and depending on the ratio of characteristic velocities V_{exp}/V_{rot}, they can become truncated. This phenomenon was analytically studied in the frame of uniform expansions by Duval and Karp [439] and numerically in Zorec et al. [260]. Their results suggest that: (1) the underestimation of $V\sin i$ cannot be negligible if the studied B and Be stars were observed at phases when their atmospheres were driven by expanding velocity fields; and (2) the blue-shifts of lines can be more or less chaotic; they could partially explain irregular radial velocity drifts, which are sometimes suspected due to undisclosed binaries.

6. Interferometry of Rotating Stars

Because optical and IR long baseline interferometry is sensitive to the stellar geometry and to its brightness distribution projected onto the sky, this technique has become an important tool to study the effects of stellar rotation all over the HR diagram. The mathematical implement to study stars by interferometry is the CittertZernike theorem, which establishes a relation between the complex visibility and the Fourier transform of the brightness distribution on the object. Moreover, the spectro-interferometry (differential interferometry) is a new technique combining high spectral and angular resolution, which enables the study of phenomena that induce chromatic signatures produced by spots, large-scale mass motions such as differential rotation, zonal currents triggered by the hydrodynamic instabilities and non-radial pulsations. In this technique, the phase of the interference fringes is proportional to the barycenter of the stellar brightness distribution [440–443]. Among the most important contributions with these modern techniques were those made with interferometers including: (1) the Center for High Angular Resolution Astronomy (CHARA) array, e.g., [444], formed by six telescopes of 1 m distributed in a y-shaped configuration having 15 baselines ranging from 34 to 331 m, and joined by several beam combiner instruments operating from the visible to near-IR in the San Gabriel Mountains of southern California; (2) the Navy Prototype Optical Interferometer (NPOI), e.g., [445] in Arizona, which is a long-baseline optical interferometer with subarrays for imaging and for astrometry. The imaging subarray consists of six movable siderostats of 50 cm with baseline lengths from 2 to 437 m; (3) the Palomar Testbed Interferometer (PTI), e.g., [446], a long-baseline infrared interferometer (operating in the H and K bands) installed at Palomar Observatory having baselines up to 110 m; and (4) the Very-Large Telescope Interferometer (VLTI), e.g., [447,448]) on Cerro Paranal (Chile). This European Southern Observations (ESO) interferometer includes four unit telescopes (UTs) of 8.2 m and four auxiliary telescopes (ATs) of 1.8 m that can form six baselines, all with different lengths and orientations. The baselines (AT and UT) range from 16 to 140 m. The light beams collected by the individual telescopes are brought together in the VLTI using a complex system of mirrors in underground tunnels. At the beginning, the beam combiner instruments active at near and mid-infrared wavelengths were AMBER, MIDI, PRIMA and VINCI. Today, the operating instruments are GRAVITY, MATISSE, and PIONIER (https://www.eso.org/sci/facilities/paranal/telescopes/vlti.html). Details on the properties of the visibility curves calculated for rapid rotators that can be detected with these instruments may be found in Domiciano de Souza [449,450].

The first although unsuccessful attempt to measure the geometrical deformation of a rapidly rotating star was carried on Altair (HD 187642, an A7-type star), which dates back to Brown [451]. Domiciano de Souza et al. [452] were the first who determined the apparent rotational flattening of Achernar (HD 10144, a B5IIIe star) using the VINCI instrument at the VLTI. They obtained a flattening ratio much higher than the limit imposed by the Roche approximation ($R_{eq}/R_p = 1.5$). Vinicius et al. [453] discussed the fundamental parameters of this star and suggested that a circumstellar disc may contribute with additional IR radiation to produce an apparently excessively large flattening. This possibility was lately studied by Kanaan et al. [454] who concluded that the disc should be smaller than that foreseen by Vinicius et al. [453]. Kervella and Domiciano de Souza [455] found that the Be star Achernar also presents a polar wind. Later interferometric observations of this star during an emissionless phase by Domiciano de Souza et al. [275] lead the authors to determine the actual flattening of the star $R_{eq}/R_p = 1.352$, which corresponds to a critical velocity ratio $V/V_c \simeq 0.883$ and a measured GD exponent $\beta = 0.56$ in units of β used in this review (i.e., in the present review, we used the exponent β related with the bolometric flux: $T_{eff}^4 \propto g_{eff}^\beta$, while other authors may consider this exponent related with the effective temperature: $T_{eff} \propto g_{eff}^{\beta'=\beta/4}$). Technical aspects and interferometric results on B-type stars as rotators are summarized in van Belle [456].

The study of the photocenter displacement using interferometric data obtained for Regulus (α Leo, a bright B7V star) with AMBER at high spectral resolution across the $Br\gamma$ spectral line enabled Hadjara et al. [457] to determine the stellar flattening $R_{eq}/R_p = 1.31 - 1.35$, the critical rotational velocity ratio $V/V_c = 0.88 - 0.89$, the inclination angle $i = 86°$ and the GD exponent $\beta = 0.66 - 0.75$ in units used herein. Vega (α Lyr), the second brightest star in the northern hemisphere after Arcturus, has been long suspected of being a rapidly rotating star seen nearly pole-on, e.g., [458], deserving attentive interferometric studies [459–462]. Concerning Vega, the authors concluded that the star rotates at $V/V_c \simeq 0.75$ ($\Omega/\Omega_c \simeq 0.92$) and that the inclination angle of its rotation axis is $\simeq 4.8°$. The B-type star δ Per was observed with the VEGA/CHARA interferometer by Challouf et al. [463], and the data were analyzed using a code of stellar rotation, CHARRON [464,465], in order to derive the stellar physical parameters. The estimated stellar flatness is $R_{eq}/R_p = 1.121$, the inclination angle $i = 85°$ and the velocity ratio $V/V_c \simeq 0.57$.

As noted in Section 5.7, the iso-radial velocity curves in the observed hemisphere of a star having surface differential rotation are not straight lines but curves. These curved lines induce distortion effects on the spectral lines and thus on the position of the wavelength dependent position of the brightness distribution barycenters that could in principle be exploited by the spectro-interferometric techniques [466] to derive the inclination angle of stars and the Maunder differential rotation parameter (as can be seen in Section 5.7). Differential rotation also induces changes on the stellar geometry that can be distinguished from those induced by the rigid rotation which could also be studied by interferometry [141]. In order to constrain the parameters empirically characterizing the surface differential rotation, instruments operating in the visible range are needed to combine high spatial resolution (∼0.2 mass) and spectral resolution as high as 100,000 in order to reach 0.1 Å in spectral lines which are sensitive to this effect, in particular the HeI lines. It is estimated that with spectral resolutions of 20,000, it would perhaps be possible to have access to parameters of the differential rotation if the measurements of radial velocities are made with precision $15 \lesssim \Delta Vr \lesssim 30$ km s^{-1} and phase shifts with $2^o \lesssim \Delta\phi \lesssim 5^o$ [467,468].

Apart from the above cited works, a significant number of Be stars were observed by interferometry with VLTI/AMBER, VLTI/MIDI (mid-IR) and CHARA/VEGA (visible) that enable carrying out statistical studies on the geometrical and kinematic properties of their CD [467,469–478]. However, as the inclination angles of the rotation axis can be estimated by the interferometric methods, in some of the above cited works, the critical velocity ratio V/V_c has also been estimated for many of the studied Be stars. Meilland et al. [469] concluded that κ CMa has a rotation ratio no larger than $V/V_c \simeq 0.52$. In Meilland et al. [472], it is shown that δ Sco must have $V/V_c \simeq 0.7$. From a study of eight Be stars [473], it comes that they have on average $V/V_c = 0.83 \pm 0.08$, with extreme values $V/V_c = 0.72 \pm 0.20$ for μ Cen and $V/V_c = 0.95 \pm 0.23$ for α Col. Stee et al. [476] concluded that γ Cas is a critical rotator. An interesting discussion was presented by Delaa et al. [467] on the Be star α Cephei, whose previous estimates of Ω/Ω_c range from $0.084^{+0.026}_{-0.049}$ [479] to 0.941 ± 0.020 [273] and the exponents of the von Zeipel relation ranging from $0.34^{+0.10}_{-0.20}$ to 0.22 ± 0.02, respectively. Delaa et al. [467] obtained that, in this star, $\beta < 1$ (as an exponent of the von Zeipel relation), and demonstrated that a differential rotation at the surface of a fast rotating star may affect the brightness distribution over the stellar disc and significantly modify the value of the β exponent. Accordingly, other parameters, particularly inclination angle, can also be affected. Recently, Cochetti et al. [478] observed 26 Be stars and determined the critical velocity ratio for seventeen of them, the average value of which is $\langle V/V_c \rangle = 0.71 \pm 0.06$.

7. The Origin of the Stellar Angular Momentum

7.1. First Findings and Discussions

The rotation velocities of young stars are related to the mechanisms of star formation by means of angular momentum dissipation processes in protostellar clouds. Apart from the highly puzzling questions associated with the problem of how stars dispose of angular momentum and how the angular momentum transfer proceeds between discs and protostellar clouds [480–482], observational studies have inquired into the nature of the relation between stars and protostellar clouds.

Leaning on the idea by Larson [483], where it is suggested that the processes of star formation through the dynamics characterizing the collapse of protostellar clouds are controlled by the rate of dissipation of angular momentum, it was postulated that the rotation of stars may also depend on their location in the Galaxy. Accordingly, Burki and Maeder [484] showed that the size of young open clusters (age < 15 Myr) in the solar neighborhood increases with the galactocentric distance. They also noted that, within 1 kpc from the Sun, the averages $\langle V \sin i \rangle$ of B and Be stars of spectral type B0–B4 increase towards the galactic center, while the fraction of binaries does not seem to depend on the galactocentric distance in the studied environment. Their conclusion was that the noted gradient of rotational velocities can be due to the collapsing time of protostellar clouds, where a higher content of heavy elements can accelerate the collapsing process with a consequent lesser dissipation of angular momentum, thus forming stars with higher rotational velocities.

Using a representative sample of rather early-type stars (spectral types B0–B5), Wolff et al. [327] noted quite a large number of slow rotators that they could not account for neither by magnetic braking, loss of angular momentum through tidal interaction in close binaries, nor by evolution from the ZAMS in the frame of angular momentum conservation. They suggested that the stars were formed with initial low angular momentum and that their observed rotational velocity, as well as the random orientation of inclinations, are the consequence of gravitational interactions in the proto-cluster clouds.

Guthrie [485] tried to clarify the picture foreseen by Wolff et al. [327], where the sample of stars with luminosity class V seemed to be underrepresented and that there is a bias against Be stars. To this end, Guthrie [485] studied the distribution of $V\sin i$ parameters of B-type stars in young and older subgroups of OB associations and field stars. For B0–B5 type stars, in both subgroups, the distributions of $V\sin i$ are quite similar for $V\sin i \gtrsim 100 \text{ km s}^{-1}$, but in the older subgroups, there is an excess of objects with $V\sin i \lesssim 50 \text{ km s}^{-1}$. A noticeable difference exists between the distributions of $V\sin i$ of B0–B5 compared with those of B6–A0 stars in young subgroups. The latter have clear bimodal distributions. When it comes to B0–B5 field stars, those of luminosity class V do not reveal a striking difference in the $V\sin i$ distributions between fields and older subgroups of associations. B0–B5 field stars of luminosity classes III–V have a similar distribution to those above, but they have a high fraction of rotators with $V\sin i \lesssim 100 \text{ km s}^{-1}$. These results suggest that there may be a difference between the rotational properties of stars in tightly bound clusters as compared to those of field stars which probably originated in loose stellar systems.

Knowing that the ratio between the specific angular momentum in an interstellar cloud and that of an average star is approximately $(J/M)_{cloud}/(J/M)_{star} \sim 10^7$, there must exist an efficient mechanism of angular momentum dissipation. When this dissipation proceeds through successive collapses, theoretical predictions say that the rotation axes should tend to be perpendicularly aligned to the galactic plane [481,486]. According to Guthrie [485] and the references therein, the $V\sin i$-distribution of late B-type stars in tightly bound clusters has a bimodal distribution of true rotational velocities V, where it is then speculated that the group of larger values of V could be broad and the axes of rotation being rather aligned, i.e., while there is a paucity of $V\sin i$ from roughly 80 to 160 km s^{-1}, the larger values of V correspond to a distribution of $V\sin i$ from 160 to 400 km s^{-1} with a maximum at $V\sin i = 230$ km s^{-1}. The distribution of V for all B-type field stars is consistent with a Maxwellian law, in accordance with which was predicted by Deutsch [487] using methods of statistical mechanics and assuming random orientations of the axes of rotation. Guthrie [485] concluded that the older subgroups originally had a similar distribution of rotational velocities to those of young subgroups, but the excess of slow rotators was probably formed at a later epoch, out of materials with low or exhausted angular momentum.

Maeder et al. [488] used the fraction of Be stars in different galactic regions and Magellanic Clouds as a surrogate indicator for the influence of the metallicity on the rapid rotation of stars. The results show that the lower the metallicity the faster is, the stellar rotation, a phenomenon that can be easily explained assuming that, under similar distributions of initial angular momentum, stars having lower metallicity rotate faster because their radii at the ZAMS are smaller than those of stars with higher metallicity. Keller [337] also studied the effect of metallicity on the rotation of B-type stars in the LMC. He concluded that stars in clusters rotate more rapidly than in the surrounding field regions, and that both field stars and stars in clusters rotate more rapidly in the LMC than in the Galaxy.

7.2. Stellar Axial Rotation of Field and Cluster B and Be Stars

Strom et al. [340] formulated the hypothesis that large differences can exist among the rotational velocities of stars born in forming regions having large densities compared to those formed in regions with low densities. The hypothesis is that differences in accretion rates should leave imprints characterizing the axial rotation of stars, because higher accretion rates in denser regions produce a higher rotation favored by the initial radii along the birth line, which are larger among stars having masses in the range 4.5–9$M_\odot$. These objects should then show the largest differences in the rotation speeds, which is a direct result of a greater spin-up during the contraction period towards the ZAMS. However, at even higher masses, the difference in rotation speeds between the cluster and field stars can be smaller, because the initial radii along the birth line are similar, or they can even be identical if the birth line joins the ZAMS.

Strom et al. [340] have compared the rotational velocities of B0–B9 stars in the dense double cluster h and χ Persei with those in the nearby field having roughly same ages ranging from 12 to 15 Myr. The age group represented by masses in the range 3.5–5$M_\odot$ have twice as large average rotational velocities $\langle V\sin i\rangle$ than nearby field stars do. This ratio is about one and a half that of the field for objects with masses from 5 to 9$M_\odot$, and velocities are just slightly different between the younger groups with 9–15$M_\odot$, as foreseen in the above formulated hypothesis. This hypothesis was studied in two other contributions by Wolff et al. [347] and Wolff et al. [348]. In the first one, using a large sample of B0–B3 stars in nine clusters and respective neighborhood fields, the authors confirmed that the rotation velocities of stars in clusters are larger than for field stars, and that there is a lack of slow rotators among stars in the clusters. Moreover, the velocities of stars with masses from 6 to 12$M_\odot$ do not seem to change significantly over an initial evolutionary period from the ZAMS up to 12 Myr. In the second contribution, it is shown that rotation velocities in the compact R136 LMC cluster are larger than those in the nearby field stars, and also that the LMC stellar rotation velocities both in the R136 LMC cluster and field are larger than those in the MW stars in similar clusters and fields. In these last comparisons, the difference between LCM metallicity and the average one in the MW plays a significant role in the establishment of the noted axial rotation velocity difference.

In a study of rotational velocities of B stars in the Galactic clusters NGC 3293, NGC 4755 and NGC 6611, Dufton et al. [318] note that, in the two older clusters NGC 3293 and NGC 4755, the true rotational velocities of all B-type stars have a Gaussian distribution, and that the $V\sin i$ are larger than those of field stars, which agrees with the conclusions by Strom et al. [340]. They also find a spin-down effect for stars with masses $15 \lesssim M/M_\odot \lesssim 60$, which is probably due to angular momentum loss by strong stellar winds.

Huang and Gies [323] studied the rotation of B-type stars in 19 MW open clusters with ages spanning from 6 to 73 Myr. As compared to nearby field stars, clusters have fewer slow rotators, which in a way agrees with the findings in the above cited works by Strom et al. [340] and Wolff et al. [347]. At ages > 10 Myr, the number of rapid rotators increases. However, contrarily to Strom et al. [340] and Wolff et al. [347], Huang and Gies [323] speculated that the field stars rotate more slowly than their counterpart in clusters because they are slightly more evolved. In Huang and Gies [489], the authors addressed the study of the evolution of the axial rotation of OB stars in clusters with masses in the range from 8.5 to 16$M_\odot$ and find that the rotation of these objects slows down during the MS phase. The small number of rapid rotators found in their sample are supposed to be spun up in the short secondary contraction phase, or by mass-transfer in close binaries. Their sample has many He-weak and He-strong objects having $T_{\rm eff} < 23{,}000$ K. For stars with $T_{\rm eff} < 18{,}000$ K, the He abundance is characterized by a broadly scattered distribution, which impedes to differentiate between the He-weak, He-normal, and He-strong stars. This makes it impossible to study the He abundance changes due to evolution. However, the He abundance augments at advanced evolutionary stages in more massive and highly rotating stars, which supports the theoretical prediction of mixing induced by rotation. By comparing the rotational velocities of the field and the cluster samples, Huang and Gies [349] concluded that the main reason for the overall slower rotation of the field sample is that it contains a larger fraction of older stars than found in clusters. Since in the approach by Huang and Gies [349], the estimated $\log g$ parameters are better surrogates of stellar ages than the spectral-type groups defined in Strom and Wolff et al., where they considered the spectral classes as indicators of groups having distinct evolutionary stages, their conclusions cast some doubts on the incidence of higher densities in the formation environments to produce faster rotators.

The interesting subject of the stellar rotation census of B-type stars from ZAMS to TAMS was addressed by Huang et al. [353], where the results indicate that the rotational distribution functions of $V/V_{\rm crit}$ for the least evolved B stars show that those with lower masses are born with a larger proportion of rapid rotators than higher mass B stars. Nevertheless, the upper limit of $V/V_{\rm crit}$ separating highly rotating B stars from B stars without emission lines is smaller among the higher mass B stars. The spin-down rates observed in stars with masses $\sim 9M_\odot$ agree rather well with theoretical predictions, but the rates are larger for the low-mass group (mass $\sim 3M_\odot$). The faster spin-down in the low-mass B stars agrees well with the predictions based on differential rotation if the angular momentum is conserved by individual shells. The results also suggest that, among the fastest rotators, which are probably Be stars, the most massive ones have probably been spun up by evolution. It is not excluded that the merger phenomenon in binaries may explain the rapid rotation in many cases over the full range of B star masses.

Using the data published in all the above cited works, the quoted results are summarized in a graphical way in terms of the true rotational velocities V and ratios $V/V_{\rm c}$, and ($V_{\rm c}$ is the critical equatorial velocity) as a function of the stellar mass $M/M_\odot$ and fractional age $t/t_{\rm MS}$ ($t_{\rm MS}$ is the time spent by a star in the MS) in Figures 28 and 29. The distributions of velocities and velocity ratios are histograms smoothed using the methods specified in Zorec et al. [127]. From Figure 28, it can be concluded that field stars are slightly more evolved than those in clusters, as claimed in Huang and Gies' papers. It is also obvious that in clusters, there is a good probability of finding a larger fraction of more rapid stars, as suggested by Strom et al. [340] and Wolff et al. [347]. Figure 29 resumes the evolution of field and cluster stars, which shows that within the margins of uncertainties, there is a small difference among the field stars with mass $M \lesssim 5M_\odot$, as found in the first quarter of the MS lifespan. The diagram suggests a slight acceleration of the rotation for the low-mass stellar group that passes from the fractional age intervals $0 \lesssim t_{\rm MS} \lesssim 0.25$ to $0.25 \lesssim t_{\rm MS} \lesssim 0.5$. They show no particular evolution until the last age interval $0.75 \lesssim t_{\rm MS} \lesssim 1.0$, except that, in this last case, the distribution of $V/V_{\rm c}$ against the mass is more scattered. Concerning the cluster stars, it is apparent that the less massive stars have on average larger values of $V/V_{\rm c}$ than the more massive ones, but for all masses and fractional age intervals, they have on average larger ratios $V/V_{\rm c}$ than field stars. In the age interval $0.75 \lesssim t_{\rm MS} \lesssim 1.0$, there seem to be a lowering of ratios $V/V_{\rm c}$ among the less massive cluster objects. The acceleration noted for stars with $M \gtrsim 15M_\odot$ should not be taken at face value, since there are very few objects entering the calculation of $\langle V/V_{\rm c} \rangle$. The behavior of the rotational velocities of field stars with masses $M \lesssim 3M_\odot$ is described in some detail in Section 8.4.

7.3. What About Be Stars?

In the preceding studies, no particular attention was paid to Be stars for reasons that can lead them to become rapid rotators. There are at least three reasons that were observationally put to test in order to try to understand why they are extreme rotators: (1) the Be star may have been born as a rapid rotator; (2) the mass transfer can produce efficient spin up; and (3) evolutionary factors can produce spin up during the MS evolution of B stars.

Fabregat and Torrejón [490] concluded that the highest fractions of Be stars in clusters having spectral types B0–B2 appear more often at ages from 13 to 25 Myr, which could be interpreted that they have undergone evolutionary spun up by the end of a B star's MS lifetime. Keller et al. [491] also finds that the fraction of Be stars attains a maximum at the end of the MS. Similar behavior can be seen in the Be population of young clusters in the Magellanic Clouds. These results may suggest that there is an evolutionary enhancement of the frequency of the Be phenomenon toward the end of the MS lifetime as a consequence of a spin-up produced by angular momentum redistribution phenomena related to the stellar evolution.

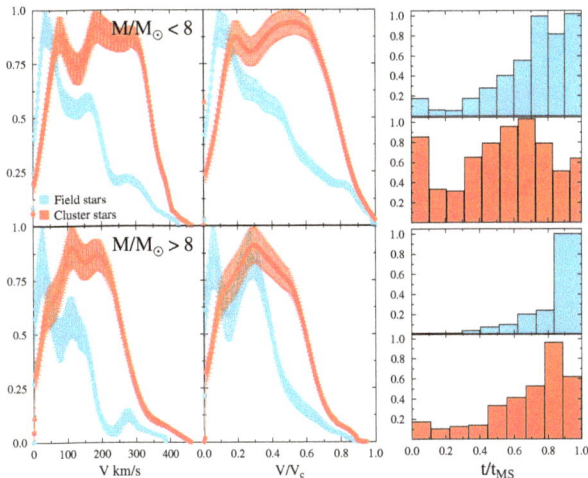

Figure 28. Two left blocks: smoothed and normalized distributions of the true V and ratios V/V_c of the B-type star rotational velocities with masses $M < 8M_\odot$ (upper row) and with masses $M > 8M_\odot$ (lower row). Right blocks: histograms of fractional age distributions t/t_{MS} of stars in the respective left blocks.

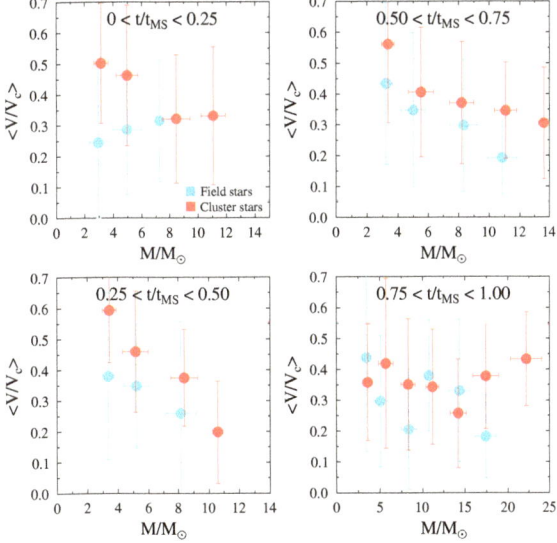

Figure 29. Average ratios of true velocities V/V_c of field and cluster B-type stars against the stellar mass $M/M_\odot$ per fractional age t/t_{MS} intervals.

From a photometric study of Be stars in more than 50 MW open clusters, McSwain and Gies [492] noted that the frequency of Be stars correlates neither with the gradient of the metallicity in the Galaxy, as could be expected from the discussion by Maeder et al. [488], nor with the cluster density proposed by Strom et al. [340] and Wolff et al. [347]. On the contrary, McSwain and Gies [492] found that the fraction of Be stars increased with age until 100 Myr, and Be stars are most common among the most massive B-type stars above the ZAMS. They showed that a spin-up phase at the TAMS cannot produce the observed distribution of Be stars, but up to 73% of the detected Be stars must have been spun up by mass transfer in binary systems. The remaining Be stars could be rapid rotators at birth.

In the above-cited photometric studies, no correction of magnitudes and photometric color indices were carried for overluminosty and reddening induced by the circumstellar discs of Be stars [18,152], nor were they corrected for rapid rotation effects [259], which can introduce significant age overestimates [493]. Using BCD parameters, which avoid the perturbation effects due to circumstellar emissions, Zorec [20] showed that Be stars appear throughout the MS evolution lifespan. Correcting fundamental parameters for rotational effects, Zorec et al. [61] brought additional proof that the Be phenomenon can appear at any evolutionary stage from ZAMS to TAMS.

As noted in the introductory paragraph of this section, stars can gain angular momentum thanks to mass transfer phenomena in binary systems [494,495]. Klement et al. [496] confirmed the existence of far-infrared SED turndowns in a series of 26 Be stars, which can be due to the presence of companion stars that influence the structure of the Be star CD. The confirmation of the mass transfer phenomenon as responsible for the rapid rotation of these Be stars is now suspended to the confirmation of the subdwarf nature of their companions. A further discussion on the merger phenomenon is performed in Section 8.1.

7.4. Distribution of the Rotational Velocities of B and Be Stars

Among the first statistical studies carried out on the distribution of $V\sin i$ parameters can be noted the discussion by Huang [497]. He speculated on the possibility that the spin vector of the star follows a Maxwellian distribution, as previously suggested by van Dien [498] for B-type stars. Assuming that the rotational velocity decreases as the space velocity increases when stars become older, Huang [497] also expected to find some correlation between the rotational velocity and the space velocity, but the results were not conclusive.

Since the $\sin i$ projection factor obliterates the actual information carried by the true stellar axial rotational velocity V, Slettebak [499] and Slettebak and Howard [500] used a homogeneous sample of $V\sin i$ parameters to study the distribution of the true velocity V of B-type stars gathered in two sub-spectral type groups: B2–B5 and B8–A2. To this end, he employed the analytical representation of the true rotational velocities suggested by van Dien [498]

$$\Psi(V) = (J/\sqrt{\pi})\left\{\exp[-J^2(V-V_1)^2] + \exp[-J^2(V+V_1)^2]\right\} \qquad (49)$$

where J and V_1 are free parameters which can be determined using the Abel-like integral equation studied by Kuiper [501] and Chandrasekhar and Münch [502], which transforms $\Psi(V)$ into the distribution of apparent rotational distribution $\Phi(V\sin i)$ assuming that the inclination angles are distributed at random. Slettebak concluded that the mean true rotational velocity of the main-sequence stars with types B2–B5 is approximately 200 km s^{-1}, which is a little larger than the corresponding one for the B8–A2 stars. In both groups, no distinction was made between B-types stars with and without emission. Moreover, the subdivision of main-sequence stars into B1–B3 and B5–B7 groups shows that the B5–B7 stars have the greatest axial rotation, and the stars of intermediate luminosity have a smaller axial rotation than the main-sequence stars. A detailed study of true rotational velocities of stars in groups of spectral types O9–B1.5 and B2–B5 was carried out by Balona [324]. The $V\sin i$ parameters were transformed into the system established by Slettebak et al. [317] and the distributions of $V\sin i$ were corrected for measurement uncertainties using the method of Eddington [503]. The stellar sample was carefully divided into dwarfs, giants, supergiants, but also dwarf and giant emission-line stars, and binaries. Balona [324] used polynomials to represent the distribution functions $\Phi(V\sin i)$ and $\Psi(V)$ and concluded that B stars probably do not rotate as solid bodies. Because of difficulties to decide whether the studied Be stars were dwarfs or giants, no conclusion was drawn on their critical rotation. Nevertheless, the distributions of the Be star true rotational velocities show that almost all are far from being rotating near their break-up velocity.

From a compilation of rotational parameters, in Figure 30, the distributions of true rotational velocities of B stars are shown with and without emission lines listed in the Bright Stars Catalog [147,148]. These distributions were calculated using the Lucy–Richardson method [504,505]. Corrections for observational uncertainties as detailed in Zorec et al. [127] were taken into account to transform the distributions of $V\sin i$ into distributions of true rotational velocities. These enable having a first overview of the effect of the evolution of rotational velocities, as well as differences between their values for B and Be stars. The population of B stars without emission lines encompasses objects of different classes, Bn, B magnetic, etc., which can be readily realized due to the irregular aspect of their distributions. The distributions of true rotational velocities are shown in Figure 30, where it can be seen, mostly for Be stars, that the lower the maxima of distributions are, the more evolved the stars are. The evolution of rotational velocities in Be stars has also been studied by Yudin [506], who shows that the average of ratios V/V_c per spectral type classes increases as stars evolve in the MS.

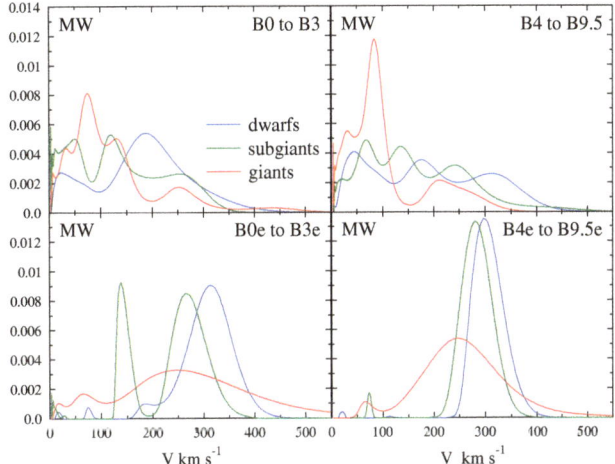

Figure 30. Normalized distributions of true rotational velocities of Milky Way (MW) B and Be stars separated by spectral type groups and luminosity classes. Velocities are not corrected for the gravitational darkening effect. Adapted from Zorec [153].

In the above discussions, the overestimation of the $V\sin i$ parameters due to the GD effect in stars with rapid rotation was not taken into account. For the first time, this was performed by Stoeckley [382] in a study of Be stars. As already noted in Section 5.4, the GD effect induced an underestimation of the $V\sin i$ parameters in rapid rotators, which encouraged Stoeckley [382] and Townsend et al. [384] to conclude that Be stars rotate close to the equatorial breakup velocity. Cranmer [385] and Zorec et al. [127] produced distributions of V/V_c ratios with $V\sin i$ parameters corrected for GD effects which enable the calculation of the fraction of Be stars which are critical rotators in the studied sample. This correction uses the theorem of von Zeipel with $\beta = 1$, which overestimates the GD effect [260,280]. Cranmer [385] and Zorec et al. [127] conclude that just a tiny fraction of Be stars are perhaps critical rotators. The distribution of their V/V_c is near symmetrical, it covers the range of velocity fraction $0.3 \lesssim V/V_c \lesssim 1.0$, and the maximum of their frequency is at $V/V_c \simeq 0.65$ (as can be seen in Figure 31). The small fraction of stars in the $0.95 \lesssim V/V_c \lesssim 1.0$ interval could have these characteristics only temporarily. This could be due to the prograd velocities of non-radial pulsations that add to the stellar rapid rotation, and put the object in conditions of critical rotation that help trigger discrete mass ejections [126]. This could also correspond to evolutionary phenomena, as discussed by

Krtička et al. [128]. Nevertheless, the large majority of Be stars do not rotate at their critical limit. Marsh Boyer et al. [354] also concluded that Be stars in the h and χ Per clusters are not critical rotators, but no clear correction of the measured $V \sin i$ parameters for the GD effect is apparent in this work.

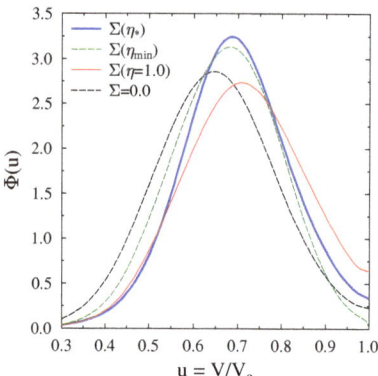

Figure 31. Distributions $\Phi(u)$ obtained with Stoeckley's corrections $\Sigma(\eta)$ calculated for different values of η: $\eta = \eta_*$; $\eta = \eta_{\min}$; $\eta = 1.0$ and $\eta = 0.0$. Adapted from Zorec et al. [127].

Because the distribution $\Phi(u)$ of ratios $u = V/V_c$ depends on Stoeckley's correction Σ given in Equation (44), it is worth showing its dependence on the adopted value of the acceleration ratio η. In Figure 31, four extreme cases are shown: (a) all stars have the lower possible rotational parameter $\eta_{\min}$, i.e., the value derived when we consider that $V(\eta_{\min}) = V \sin i$ (dashed green line); (b) Stoeckley's corrections estimated for the critical rotation, i.e., $V(\eta = 1) = V_c$ (red line); (c) neglecting Stoeckley's correction, i.e., $\eta = 0.0$ in all stars (black dashed line); (d) Stoeckley's corrections estimated for $\eta = \eta_*$ derived for each star individually (blue full line). As expected, approximation (a) produces a slight excess of slow rotators (dashed green line); solution (b) is characterized by an excess of rapid rotators; (c) the distribution is shifted to smaller values of $u = V/V_c$.

In this section, we mentioned several methods that were used by different authors to obtain the distribution of the true rotational velocities of stars, particularly those of Be stars used to test their critical or non-critical character as rotators. There exist, however, other methods to deconvolve the known Fredholm integral that relates the distribution of apparent rotational velocities with the probability distribution of true rotational velocities, although so far, they have not been applied to discuss the specific cases of stellar rotational velocities. Curé et al. [507] developed a method to deconvolve this inverse problem through the cumulative distribution function for true stellar rotational velocities, which completes thus the work of Chandrasekhar and Münch [502]. Christen et al. [508] discretized the Fredholm integral using the Tikhonov regularization method to directly obtain the probability distribution function for stellar rotational velocities. Orellana et al. [509] proposed a method based on the maximum likelihood (ML) estimate to determine the true rotational velocity probability density function expressed as the sum of known distribution families.

7.5. Rotational Velocities of B and Be Stars as a Function of the Metallicity

Stars with low metallicities undergo less violent nuclear reactions and are consequently more compact. Given a similar distribution of total angular momenta in media of different metallicities, stars will on average have faster rotational velocities the lower the metallicity is. Motivated by this fact, much effort has then been devoted to comparing the rotation rates of OB stars to the center of our Galaxy and as a function of the metallicity in our neighboring SMC and LMC dwarf galaxies. These comparisons have been made using Be stars as testimonials and as rotators, knowing that they are indeed very fast rotators. Thus, Maeder et al. [488] found that the fraction of number of Be stars $N(Be)/N(Be + B)$

increases if the metallicity decreases. On the other hand, Martayan et al. [345,346] studied the distribution of true velocity ratios V/V_c as a function of the Be star mass and as a function of the metallicity, using MW, SMC and LMC stars. Their results, summarized in Rivinius et al. [40], show that the Be stars of all masses have on average relations V/V_c closer to 1 the lower the metallicity of the medium is where they were formed. The angular velocities of SMC Oe/Be stars are on average $\langle \Omega/\Omega_c \rangle \simeq 0.95$ in the ZAMS, which are in the theoretically predicted zone for the Long-Gamma-Ray-bursts (LGRBs) progenitors [510]. Martayan et al. [511] discussed this occurrence and considered that low-metallicity early Be and Oe stars can be potential LGRB progenitors. They estimated the yearly rate per galaxy of LGRBs and the number of LGRBs produced in the local universe up to $z = 0.2$, which agreed with the observed rate.

Recently, Ramachandran et al. [512] compared the distributions of $V\sin i$ parameters of OB stars in the MW [318,355,358], SMC [346,356,357] and LMC [350,513]. These distributions concern all OB stellar masses together, without particular distinction regarding the object peculiarities. From these comparisons, rough relations were obtained showing the variations of peaks of $V\sin i$ distributions and fractions of stars having $V\sin i \gtrsim 200$ km s^{-1} as a function of metallicity. The relations obtained are

$$\begin{aligned}\langle V\sin i\rangle_{\text{peak}} &= -121 \times \log(Z/Z_\odot) + 22 \text{ km s}^{-1} \\ \text{fraction} &= -0.2 \times \log(Z/Z_\odot) + 0.1 ,\end{aligned} \quad (50)$$

which clearly show that the lower the metallicity is, the larger the rotational velocities are. These relations depend on the characteristics of the evolution of the stellar rotational velocities in environments having different metallicities. In fact, apart from of the stellar size dependent on the metallicity of the medium where they are formed, which may characterize the initial stellar rotational velocity, the further evolution of the stellar surface rotational velocity is also determined by the mass-loss rate. These rates are significant in massive stars and they depend on the metallicity. They are stronger the higher metallicity is [514].

7.6. Two Different Visions on the Origin of the Stellar Angular Momentum

Krumholz [515] summarized and discussed the two main different classes of models proposed for the formation of massive stars. The first postulates that massive stars are assembled in a similar way to their lower mass counterparts, i.e., from in-falling material located in a rotating protostellar core, channeled by a magnetic field and rooted in the star through a circumstellar accretion disc, called the 'magnetospherically mediated accretion' (MMA). According to Bonnell et al. [516], massive stars should form preferentially in cluster-forming environments. The second class of models stipulates that massive stars start their formation as low-mass protostellar cores in clusters and then grow up via mergers of cores located in regions where the density of cores is high (as can be seen in [517] and references therein). They can be referred to as 'core merger' models (CM). Wolff et al. [341] tackled the problem of determining through observational tests which of these contrasting models—MMA or CM—dominates in nature. Data for O-type stars combined with data in the literature show that over a mass range from 0.2 to $50 M_\odot$ the specific angular momentum J/M of stars varies slowly and continuously as $J/M \propto M^{0.3}$. Moreover, nearly all stars in this mass range rotate along the birth line at rates that are not larger than $\langle V/V_{\text{crit}}\rangle \sim 0.3$. These findings lead the authors to conclude that a single mechanism must be at work to keep the rotation rates low and at similar values for stars of all masses at birth, and that their results seem to rule out CM models, or at best, they must be the exception. The continuity of angular momentum properties across the whole mass range from M stars to O stars argues for a common formation mechanism through a disc.

Because a high fraction of massive stars, perhaps the majority, are born as members of close binary systems, the populations of early-type MS stars may contain stellar mergers and several products of mass transfer in binaries. de Mink et al. [215,216] have undertaken the simulation of a massive binary-star population typical for our Galaxy, assuming continuum star formation. de Mink et al. [215,216] have found that 4–17% of early-type stars can be the products of mergers in binary systems and that 15–40% of MS massive stars are the products of binary interactions. Due to these binary interactions, 10–25% of all massive MS stars have rotational velocities in excess of 200 km s^{-1}. In spite of several uncertainties related to the mass transfer efficiency and the detailed treatment of magnetic braking, the fraction of rapid rotators these authors derive is quite similar to the observed one. This casts some doubts on whether single stars can be formed as rapid rotators. Predictions allow for the possibility that all early-type Be stars result from binary interactions. Finally, the interpretation of anomalies of chemical abundances induced by rotational mixing is also strongly challenged by this working frame.

8. Understanding the Evolution of Rotational Velocities in Single Early-Type Stars
8.1. Models of Stellar Evolution with Rotation

In spite of the numerous efforts made to understand the evolution of rotating stars e.g., [154,255,290,518,519], the results obtained until now cannot be entirely validated, because they depend on several strong assumptions which are difficult to confirm, and because the model predictions have not been confronted with observations in all its aspects.

One of the greatest unknowns concerns the internal angular momentum transport. On the one hand, models by Maeder and Meynet [154] are based on diffusive and advective transport of angular momentum, while those by Heger et al. [255] consider only diffusive transports of angular momentum. More recently, Potter et al. [307,519,520] considered 'shellular' rotational velocity distribution as in the aforementioned models, but have taken into account both diffusive and advective transports of angular momentum, and tested several prescriptions for the many intervening diffusion coefficients. Moreover, they did not impose rigid rotation in the stellar convective core. On the other hand, all these models make, however, evolve stars assuming rigid rotation at the Zero-Age-Main-Sequence (ZAMS). This implies that, in all cases, stars are assumed to have the minimum possible content of total angular momentum, which is limited to a ratio of kinetic rotation energy, K, to gravitational potential, W, not exceeding the ratio for a rigid critical rotation $K/|W| \simeq 0.01$. This rotation law switches in quite short time scales to a differential rotation law, in some 10^4 y [254,521]. Internal differential rotation in stars has then every chance to be present before the stellar ZAMS phase. Nevertheless, before the ZAMS each star had a long pre-main sequence (PMS) evolutionary history, where, in any of its stages, it does not necessarily acquire a strict rigid rotation. Rigid rotation in the ZAMS is mainly supposed to be justified because: (a) dynamic stability against axisymmetric perturbations could be warranted for rigid stellar rotators [230]; and (b) the redistribution of angular momentum can be promoted by the turbulent viscosity [136]. It is then generally assumed that in the PMS full convection phase, stars become rigid rotators. However, Wolff et al. [522] concluded from the comparison of rotational rates on both sides of the convective–radiative boundary that stars do not rotate as solid bodies during the transition from the convective to radiative evolutionary tracks.

On theoretical grounds, it can be argued that, in the regions unstable to convection, the redistribution of the specific angular momentum $j = \Omega \varpi^2$ can be controlled by the convective plumes [523], which favor the existence of rotation laws characterized by $j(\varpi) \simeq$ constant. Moreover, 2D hydrodynamic calculations show that convection does not maintain rigid rotation, but it produces an internal angular velocity distribution profile $\Omega(\varpi) \propto \varpi^{-p}$ (ϖ = distance to the rotation axis), with a value of p that corresponds to an intermediate configuration between the complete redistribution of specific angular momentum and rigid rotation, $0 \lesssim p \lesssim 2$ ($p = 0$ is for rigid rotation) [263,524,525]. It was rather recently shown that, in the stellar convective cores of rotating stars, thanks to the Reynolds stresses, there is a continuous redistribution of angular momentum that leads to a kind of cylindrical differential rotation because of a readjustment that tries to establish a constant specific angular momentum distribution, far from a rigid rotation condition [526–529]. According to these comments, initial (or ZAMS) internal rotation laws can be differential, which may enable, in principle, to concentrate larger amounts of angular momentum towards the stellar core than that enabled by the rigid rotation at critical regimes, i.e., global energy ratios that could attain $0.01 \lesssim K/|W| \lesssim 0.14$. Finally, it can be noted that the solar convective regions, which are not only characterized by significant turbulence, but rotate differentially, do not conform to any intermediate rotation law between the above and extreme frames [417]. This rotation law may finally look like one of those that are inferred if it is assumed that the baroclinic balance relation is obeyed [141].

Undoubtedly, knowledge of the way that the internal angular momentum distribution is built up during the stellar formation processes may carry important consequences for the further evolution of stars on the MS. The magnetic coupling between the accretion disc and the star is believed to spin down PMS objects below the critical surface rotation. However, Rosen et al. [530] showed that such magnetic torques are insufficient to spin down massive PMS stars due to their short formation times and high accretion rates. The mechanism can only be effective if the disc lifetime of massive stars is longer than believed, or the magnetic fields are stronger than suggested by observations. Haemmerlé et al. [531] have tackled this problem assuming that the accreted matter has an angular velocity equal to that of the accreting stellar surface. These authors considered stellar masses ranging from 2 to 22 $M_\odot$ and found that the stars more massive than approximately 8 $M_\odot$ cannot reach the ZAMS with velocity rates higher than $V/V_{\text{crit}} \simeq 0.45$, and that for a given ratio V/V_{crit}, the internal differential rotation is of approximately 80% for objects with 2 $M_\odot$, but decreases for more massive stars, reaching the 20% in stars with 14 $M_\odot$. Moreover, the studied mechanism does not enable one to obtain rapidly rotating massive stars in the ZAMS. In another contribution, Haemmerlé et al. [532] studied several accretion scenarios and noted that smooth angular momentum accretion ends up with an angular momentum barrier that impedes the formation of massive stars. They observed that because shear instabilities and meridional circulation are quite ineffective to transport angular momentum during the accretion phase, the internal rotational profile reflects the angular momentum accretion history. This transport is more efficient by convection. However, as during the accretion phase, large radiative zones are formed and the angular momentum is conserved locally, which can lead to significant internal differential rotation. Whatever the angular momentum accretion scenario during the PMS period, the internal rotation acquires its initial MS structure in a short fraction of the MS life span and depends on the stellar mass, its chemical composition and the total angular momentum stored by the star. This leaves the open question of what the right accretion scenario is?

8.2. Rigid or Differential Rotation: First Observational Guesses

The effects carried by the rapid rotation on the stellar fundamental parameters scale according to the ratio of centrifugal to gravitational accelerations, $\eta = F_c/F_g = V^2 R/GM$. From η, it is also apparent that, for a given total angular momentum $\eta \propto M^{-3}$, which implies that the lower the stellar mass, the larger the effects induced by the rotation can be. This can be one of the reasons why, in the past, more attention was paid to studying the internal rotational properties in late-type stars. Knowing that, for inclination angles distributed at random, it is $\langle (V\sin i)^2 \rangle = (2/3)\langle V^2 \rangle$, the displacement in absolute magnitude ΔM_V from the zero rotation main sequence (ZRMS), that can be studied using the $V\sin i$ parameters by adopting the following expression [257,267,533,534]

$$\Delta M_V = k V^2 \qquad (51)$$

with k being a constant over large MS spectral type domains. In this expression, the value of k can testify to the nature of the internal stellar rotation: rigid or differential. In a first attempt, Maeder [535] considered that the expression in Equation (51) should be written as $\Delta M_V = k V^\alpha$, since, from observations, both quantities (k, α) vary as a function of the spectral type. However, in Maeder and Peytremann [281] and Maeder and Peytremann [285], the relation in Equation (51) was adopted to conclude that stars in the interval B6–A7 are nearly rigid rotators, while the stars of later spectral types do not rotate as rigid bodies.

In contrast with these results and also basing their research on the use of Equation (51), Smith [536] came to the conclusion that, in spite of the uncertainties marring the estimate of k, particularly the empirical definition of the ZRMS, late-type stars in the Praesepe and Hyades clusters with absolute magnitudes in the range $1.5 \lesssim M_V \lesssim 4.5$ likely have internal differential rotation, although the rotation profile could not be determined. Some time after, Cotton and Smith [537] inquired into the values that k may have according to the models of stars rotating with different internal rotation laws.

Following Oke and Greenstein [538] and Sandage [539], who investigated how rotation varied in groups of stars across the HR diagram, Danziger and Faber [540] addressed the question of the distribution of angular momentum in evolving A and F stars by considering two limiting points of view: (a) all stellar shells in a star are completely coupled, so that they rotate as rigid bodies; and (b) all stellar shells in a star are completely uncoupled from each other, so that they are differential rotators.

Danziger and Faber [540] concluded that stars with radii expanded by the evolution of a factor of no more than 2 seem to rotate as solid bodies, but at more evolved stages, observations suggest differential rotation. Using similar arguments, Balona [324] concluded that B-type stars are likely differential rotators. In the same line of thought, and respecting the progenetic relations of spectral types according to the average luminosity classes of stars classified in the BCD system, Zorec [20,153] noted that the B and Be stars of luminosity classes IV and III should behave as having internal rotation laws intermediate between rigid and completely differential, as shown in Figure 32. According to this figure, Be stars seem to have a more rigid rotation than B stars without emission, but the separations of points in the figure are within the uncertainty bars.

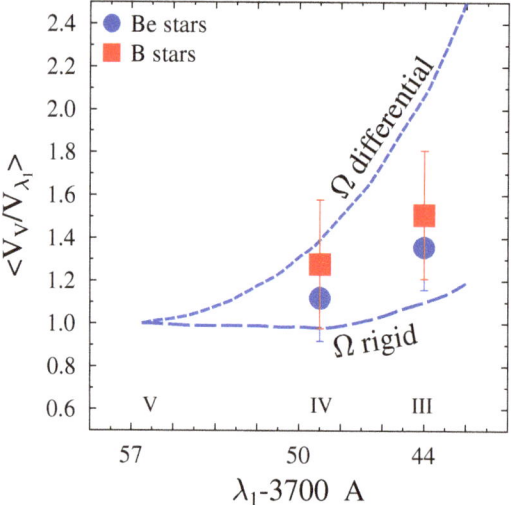

Figure 32. Average true rotational velocity ratios $\langle V_V/V_{\lambda_1}\rangle$ ($\langle V_V$ = true rotational velocity of stars of luminosity class V (dwarfs); V_{λ_1} = true rotational velocity of stars of other luminosity classes identified by an average λ_1 parameter marked in the abscissas) compared to the sequences of evolution of model ratios of true rotational velocities assuming rigid rotation over the MS evolutionary life span and complete differential rotation. Adapted from Zorec [20,153].

8.3. Rigid or Differential Rotation: Extreme Behaviors

More detailed studies of the evolution of late B- and early A-type stars were made by Zorec and Royer [289], where the observed distributions of rotational velocities from a statistically reliable sample were compared with the models of stellar evolution with rotation calculated by Ekström et al. [541]. To facilitate the understanding of the observed evolution of rotation velocities, let us again consider the already mentioned extreme cases of their evolution. If it is assumed that late B- and early A-type stars evolve to conserve their angular momentum, and that, in the ZAMS, they are rigid rotators, as assumed by the existent models of stellar evolution, then stars on the ZAMS should not have amounts of total angular momentum larger than they can bear at critical rigid rotation in the terminal age main sequence (TAMS), i.e., $J \leq J_{\text{crit}}^{TAMS}$. Within this restrictive condition, the evolution of the surface rotational velocities may respond to either an entire redistribution of the angular momentum at any instant, or to a total lack of exchange of angular momentum among the stellar shells, i.e., conserving the specific angular momentum. For the rigid rotation case, the equatorial velocity at a time t elapsed from the ZAMS ($t = 0$) is roughly given by

$$\frac{V(t)}{V_{\text{ZAMS}}} = \frac{R(t)}{R_{\text{ZAMS}}} \frac{I_{\text{ZAMS}}}{I(t)}, \tag{52}$$

where $[V(t), V_{\text{ZAMS}}]$, $[R(t), R_{\text{ZAMS}}]$ and $[I(t), I_{\text{ZAMS}}]$ are the stellar equatorial linear rotational velocity, the radius and the moment of inertia at time t and at the ZAMS, respectively. The moment of inertia is calculated using 2D models of stellar structures with rotation [141].

In the second case, the specific angular momentum will become increasingly centrally condensed as evolution goes on, so that the amount left on the surface will decrease strongly as the star evolves from ZAMS to TAMS. For simplicity, we neglect the geometrical deformation of the stellar surface due to the rotation. The angular momentum of a spherical shell of width dr is then given by

$$dJ(t) = \frac{8\pi}{3}\rho(r,t)r^4\Omega(r,t)dr, \tag{53}$$

where $\rho(r,t)$ and $\Omega(r,t)$ are the density and the angular velocity at radius r and time t. Ω can be assumed to be uniform over each shell, i.e., a 'shellular' distribution of the angular velocity. If r and r' are the radii at times t and t' of the same shell, whose mass $dM(r) = 4\pi\rho(r)r^2 dr$ and angular momentum are conserved, we have

$$\Omega(r')/\Omega(r) = (r/r')^2 \tag{54}$$

which, for the equatorial velocity at t and t' in the stellar surface, implies that

$$V(r')/V(r) = R_e(t)/R_e(t') \tag{55}$$

where $R_e(t)$ and $R_e(t')$ are the equatorial radii of the star at times t and t', respectively. The details of calculations of the surface stellar rotational velocities in these two extreme cases can be seen in Zorec and Royer [289].

For the sake of comparing the evolution of rotational velocities with model predictions, the observed $V\sin i$ parameters in Zorec and Royer [289] were grouped by mass and age intervals so as to obtain averages $\langle V\sin i\rangle$ that were converted into $\langle V\rangle$ assuming a random distribution of the inclination angles. In this section, we compared the observed true velocity ratios against the fractional time t/t_{MS} (t_{MS} which is the life span of a star from the ZAMS to the TAMS) for stars with masses ranging from $\langle M/M_\odot\rangle = 2.5$ and 3.0, with the calculated evolution of velocity ratios according to Equations (52) and (54). This comparison is shown in Figure 33. Figure 33a shows the ratios $\langle V/V_{\text{ZAMS}}\rangle$ against the fractional time t/t_{MS}, and in Figure 33b, the ratios $\langle V/V_{\text{crit}}\rangle$ are compared. The width of the shaded curves corresponds to the mass range $1.5 \leq M/M_\odot \leq 3.0$ used for the calculation. The outstanding characteristic of $\langle V/V_{\text{ZAMS}}\rangle$ curves is that they increase fast in the first third of the MS phase, and they decrease more or less monotonically in the second half of the MS phase. Owing to the shown uncertainties, it is not possible to say that the slopes reveal a slightly faster decrease than predicted for differential rotators. If it were actually the case, this decrease could imply some redistribution of angular momentum toward the center of the stars. In Figure 33b, the evolution of ratios $\langle V/V_c\rangle$ is shown against the fractional time t/t_{MS}, where the calculated evolution of velocity ratios according to Equation (54) was omitted. Even if the evolution of differential rotators is not shown in the figure, we notice that the shape of the curve $\langle V/V_c \rangle$ differs strongly from what is predicted from the two mentioned extreme possibilities of angular momentum redistribution. For both mass-groups, in the first third of the MS, the ratio $\langle V/V_c \rangle$ increases faster than suggested by the theoretical predictions. The $\langle V/V_c \rangle$ reaches a maximum at $t/t_{MS} \approx 0.3$. Then, there is a decrease that lasts roughly $\Delta(t/t_{MS}) \approx 0.2$, which is followed by a rather uniform value of $\langle V/V_c \rangle$ until the TAMS. In Figure 33, the error bars represent the uncertainties associated with the averages of velocity ratios and fractional ages.

8.4. Rigid or Differential Rotation: Comparison of Models with Observations

The aim of this section is to shortly review some aspects of the evolution of the observed surface rotational velocities already used in the preceding section of late B- and early A-type stars that we take as witnesses of what can be expected as possible trends or behaviors in more massive stars, i.e., 'normal' or 'active'. Here, the comparison is made with detailed theoretical predictions, but only for stellar masses $M = 3M_\odot$ because, at the time of Zorec and Royer [289], it was the only one at disposal.

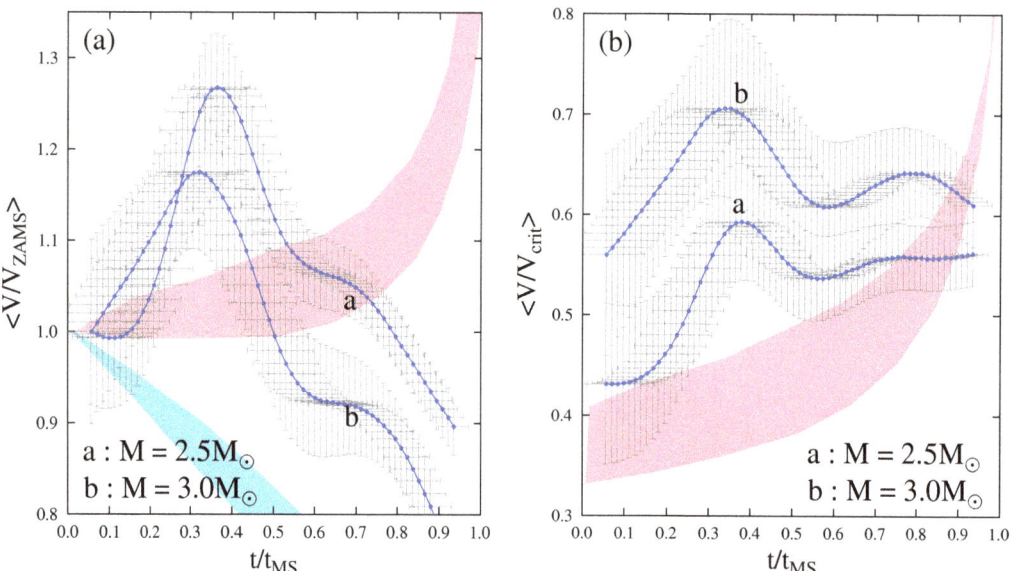

Figure 33. (a) Evolution of the averaged equatorial velocity ratio $\langle V/V_{ZAMS} \rangle$ in the MS lifetime span, where V_{ZAMS} is the true equatorial velocity of stars on the ZAMS having masses 2.5 and $3M_\odot$. The shaded region corresponds to the theoretical evolution of the V/V_{ZAMS} ratios for strict rigid rotators (pink) and differential rotators (blue). (b) Evolution of the equatorial velocity ratio $\langle V/V_{crit} \rangle$ in the MS life time span. The ratios of true equatorial velocities are calculated for two masses: 2.5 and $3M_\odot$. The shaded regions correspond to the theoretical evolution of the $\langle V/V_{crit} \rangle$ ratio of strict rigid rotators whose total angular momentum is $J = J_{TAMS}^{crit}$ (pink). Let us note that, in this figure, the ratio V/V_{ZAMS} was used, while in Figure 32, we have V_V/V_{λ_1}, where V_V replaces V_{ZAMS} and V_{λ_1} indicates the mean velocities per luminosity class. Adapted from Zorec and Royer [289].

Figure 34 shows the comparison of the evolution of observed velocity ratios $\langle V/V_{ZAMS} \rangle$ and $\langle V/V_{crit} \rangle$ presented in Zorec and Royer [289] with those obtained by Ekström et al. [541] for model stars with mass $M = 3M_\odot$ and different initial $\langle V/V_{ZAMS} \rangle$. The respective model V_{ZAMS} values correspond to the initial velocities V in the ordinates in Figure 34b, where velocities are presented as ratios $\langle V/V_{crit} \rangle$. The differences between observations and theoretical predictions shown in Figure 34a are eloquent enough to see that, in the first half of the MS evolutionary phase, the angular momentum in actual stars may undergo different redistribution processes than those considered in the theoretical predictions. In Figure 34b, the theoretical $\langle V/V_{\text{crit}} \rangle$ ratios reveal an initial short period where a star redistributes its internal angular momentum passing from a rigidly rotating object to a differential rotator at a low-energy regime [154], but the remaining predicted evolution does not resemble the one inferred from the observed rotation velocities.

A different insight into the evolution of rotational velocities can be obtained by comparing the minimum value of the total angular momentum $(J/M)_{\text{min}}$ in rigid rotators required to account for the equatorial velocities, as shown in Figure 33. Because the mass loss rate in stars with masses $M < 3M_\odot$ is very low, they can be considered to evolve as conserving their total angular momentum all along the MS phase. According to Figure 33, four evolutionary epochs can be distinguished: there is the first time interval $t/t_{\text{MS}} \approx 0.07$, which represents the ZAMS or the near ZAMS stage, then $t/t_{\text{MS}} \approx 0.3 - 0.4$ corresponds to the epoch when V is maximum, followed by the epoch $t/t_{\text{MS}} \approx 0.65$ characterized by the inflection of V, and finally, $t/t_{\text{MS}} \approx 0.9$, which corresponds to the evolution approaching the TAMS. For each evolutionary stage the angular momentum $(J/M)_{\text{min}}$ required by the rigid rotation associated with the observed velocity V can be estimated, as well as the angular momentum that corresponds to the respective critical rigid rotation $(J/M)_{\text{crit}}$. These values of J/M are given in Table 2, together with the observed rotational velocity V at the indicated evolutionary stages. Thus, we note: (i) in all cases, it is $J/M < (J/M)_{\text{crit}}^{\text{ZAMS}}$; (ii) for evolutionary stages from $t/t_{\text{MS}} \approx 0.3$ to 0.6, it is $J/M > (J/M)_{\text{ZAMS}}$, which implies that, to explain the observed rotational velocities, more rotational kinetic energy would be required than the inferred for rigid rotators on the ZAMS; (iii) the condition $J/M < (J/M)_{\text{crit}}^{\text{ZAMS}}$ for all evolutionary stages means that the evolution of rotational velocities in the MS can be consistent with the low regimes of rotational energies, i.e., $\tau = K/|W| < \tau(\text{ZAMS})_{\text{crit}}^{\text{rigid}}$; (iv) knowing that, for all evolutionary stages, including the ZAMS, the estimated total specific angular momentum is for rigid rotators, and the values $J(t)/M > (J/M)_{\text{ZAMS}}$ at any $t/t_{\text{MS}} > 0$ imply that there must be some transfer of angular momentum towards the stellar surface. However, because rigid rotation in the initial ZAMS epoch demands $\Omega_{\text{core}} = \Omega_{\text{envelope}}$, later evolutionary phases where $J/M > (J/M)_{\text{ZAMS}}$ would imply $\Omega_{\text{core}} < \Omega_{\text{envelope}}$, which is nonsensical. An increase in Ω_{envelope} could then be possible if stars actually evolve as differential rotators having $\Omega_{\text{core}} > \Omega_{\text{envelope}}$. However, the condition $J/M < (J/M)_{\text{ZAMS}}^{\text{crit}}$ suggests that they could be differential rotators with a low energy regime ($\tau \lesssim \tau_{\text{crit}}^{\text{rigid}}$) on the ZAMS; and (v) because the total angular momentum is conserved during the MS for the studied masses, in stages where $J/M > (J/M)_{\text{TAMS}}^{\text{crit}}$, stars have to redistribute their total angular momentum and end up behaving as neat differential rotators in their MS phase ($\tau \gtrsim \tau_{\text{crit}}^{\text{rigid}}$).

Yang et al. [542] discussed the results by Zorec and Royer [289] and showed that they can be reproduced by making evolved models with differential rotation since the ZAMS, where the angular momentum transport is caused by hydrodynamic instabilities during the MS phase. In their picture, the observed initial acceleration should result in an effect of outward transport of angular momentum stored in the stellar interior, while in the last evolutionary stages in the MS, the core and the radiative envelope become uncoupled with the envelope being uniformly rotated. However, the rapid decrease in the equatorial velocity during the middle MS phase remains unexplained.

The bi-modality of rotational velocities of stars with masses $2 \lesssim M/M_\odot \lesssim 4$ put forward by Zorec and Royer [289] was revisited by Sun et al. [543] with a huge stellar sample from the LAMOST medium-resolution survey [544]. Minor differences in the distributions of rotation velocity ratios V/V_{crit} seem to be looming, but in a sample having approximately 14,000 stars, a finer selection of binary candidates and chemically peculiar stars might be required to definitely assess the new results. These authors have, however, found that a bimodal rotation distribution, composed of two branches of slowly and rapidly rotating stars, emerges for more stars with masses $M/M_\odot \simeq 2.5$, whereas for those with

$M/M_\odot \gtrsim 3.0$, the gap between the bifurcated branches becomes prominent. They also find that metal-poor ($[M/H] < -0.2$ dex) objects only exhibit a single branch of slow rotators, while metal-rich ($[M/H] > 0.2$ dex) stars clearly show two branches. The difference could perhaps be due to some unknown high spin-down mechanism, while in the metal-poor sub-sample, this can partly be produced by magnetic fields.

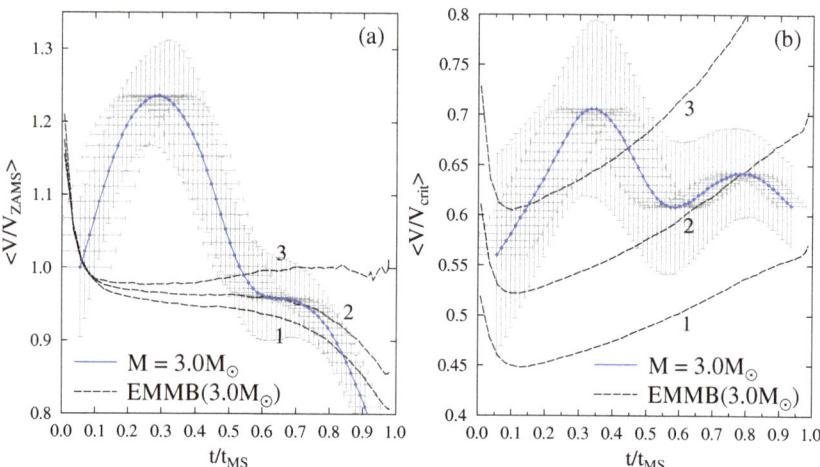

Figure 34. Comparison of the observed evolution of equatorial velocity ratios in the MS lifetime span for stars with $M = 3M_\odot$ inferred from observations with theoretical ones calculated by Ekström et al. [541] (EMMB) for different V/V_{crit} on the ZAMS. (**a**) The evolution of V/V_{ZAMS} ratios; (**b**) evolution of V/V_{crit} ratios. Curves 1, 2 and 3 refer to the different values of V_{ZAMS}, which correspond to the values of V in the initial ratios V/V_{crit} of panel **b**. Adapted from Zorec and Royer [289].

Table 2. Synopsis of the average evolution of equatorial velocities of $2.5\,M_\odot$ and $3.0\,M_\odot$ stars.

t/t_{MS}	V km s^{-1}	$(J/M)_{min}$ (10^{17} cm^2 s^{-1})	V_{crit} km s^{-1}	$(J/M)_{crit}$ (10^{17} cm^2 s^{-1})
		$M = 2.5\,M_\odot$		
0.07	182	1.086	422	1.974
0.36	231	1.270	391	1.842
0.62	195	1.154	361	1.652
0.90	167	0.875	299	1.201
		$M = 3\,M_\odot$		
0.07	249	1.558	439	2.389
0.29	290	1.725	410	2.243
0.62	229	1.419	376	2.017
0.90	192	1.157	310	1.431

8.5. Rigid or Differential Rotation: New Observational Inferences

The knowledge of the internal angular momentum distribution in a star may give significant clues to the understanding of its origin and evolution. After the method developed by Rhodes et al. [545] and applied by Deubner et al. [546] to study the subphotospheric rotation of the Sun, to our knowledge, Ando [547] was one of the first to propose a method to estimate the inner rotational angular velocity in stars, which is based on the rotational frequency splitting of non-radial pulsations. The method was meant to produce reliable results provided that the identification of pulsation modes is correct and that the mass and evolutionary stage of stars are correctly estimated. The method uses the one-zone approximation to calculate the frequency splitting. This zone is identified with an effective radius that eventually enables determining its angular velocity. Unfortunately, the application of the method to three B-type stars (53 Per, β Cep, 12 Lac) was not conclusive on their differential rotation character due to the uncertainties marring the observational quantities.

Improved methods to infer the internal rotation of stars based on asteroseismic measurements were developed since then [548]. Their application revealed internal differential rotation and put constraints on the internal rotation of some main-sequence B stars, e.g., [549–552]. Detailed developments in this field are summarized and discussed in Aerts et al. [553], where from a stellar sample of more than 1200 stars with masses $0.8 \lesssim M/M_\odot \lesssim 8$, the authors concluded that: 'single stars rotate nearly uniformly during the core-hydrogen and core-helium burning phases; stellar cores spin up to a factor of 10 faster than the envelope during the red giant phase; the angular momentum of the helium-burning core of stars is in agreement with the angular momentum of white dwarfs; observations reveal a strong decrease of core angular momentum when stars have a convective core that current theories of angular momentum transport fail to explain'. More recently, from a compilation of 52 slowly pulsating B stars for which the internal rotation has been measured asteroseismically, Pedersen [554] also concluded that the core rotation frequencies decrease as a function of age and put forward indications of angular momentum transport taking place on the main sequence.

Incompletely and sparsely concerning different B- and near B-type objects, as well as methods and models used to interpret asteroseismic data, let us quote some recent works. Based on a simple three-zone modeling of the internal rotation Hatta et al. [555] inferred via the interpretation of asteroseismic data of KIC 11145123, a $2M_\odot$ MS star, that its radiative region rotates almost uniformly, while the convective core may be rotating about 6 times faster than the radiative region above. They also found some evidence for a latitudinal differential rotation in the outer envelope. The study of gravity modes probed the stellar interior near the convective core and enabled the calibration of internal mixing processes in 26 rotating stars with masses between 3 and 10 solar masses covering the entire slowly pulsating B stars (SPB) instability strip and having rotation rates $0.0 \lesssim \Omega/\Omega_c \lesssim 1.0$. The study revealed a wide range of internal mixing profiles, which could provide further guidance for three-dimensional hydrodynamic simulations of transport processes in the deep interiors of stars. In particular, Burssens et al. [556] found that the rotational frequencies in HD 192575 (B0.5 V) are different at radii $r/R = 0.1$ and r/R, namely $f(r/R = 0.1) \sim 5f(r/R = 0.7)$, implying differential rotation. HD 129929 is a massive ($M \sim 9.4 M_\odot$) slowly rotating β Cephei pulsator (surface rotation $V {\sim} 2 \text{ km s}^{-1}$), where the interpretation by Salmon et al. [557] of its rich spectrum of detected oscillations revealed the presence of radial differential rotation with the stellar core rotating $\sim$3.6 times faster than the surface. From a test of hydrodynamic and magnetic instability transport processes of angular momentum, it was realized that the impact of the Tayler magnetic instability on the angular momentum transport is insufficient to account for the asteroseismic inferences, but they can be accounted for with hydrodynamic processes. Long ground-based photometric surveys, e.g., [558] and the recent 4-year data provided by the Kepler mission [559] and the references therein have enabled the identification of rapid rotators among classical B-star pulsators (considered SPB), where the properties of their internal structure and pulsations might perhaps become an approximation for Be stars. This is the case of KIC 7760680 [560], which is found to be a nearly rigid rotator with $V/V_c \sim 0.26$.

The surface differential rotation could be considered as an imprint of the internal rotation profile [141]. Departures of line profiles broadened by rotation from theoretical predictions due to a surface rigid rotation were studied by Stoeckley [410], Stoeckley and Morris [561], Buscombe and Stoeckley [562], Stoeckley et al. [563] and Stoeckley and Buscombe [412]. In these studies on the surface differential rotation, the interpretation of line profiles based on the specific models of the surface rotation lead the authors to suggest that most of the studied O-, B- and A-type stars with and without emission lines on the MS may have surface differential rotation with the angular velocity accelerated towards the pole. The shapes of spectral lines varying with the stellar latitude of α Lyr were studied by Elste [564] using the classical methods of line broadening. The raised uncertainties on their interpretation stem on several more or less competing effects, i.e., limb-darkening on gravitational darkened atmospheres due to rapid rotation, macro-turbulent motions, and differential rotation.

The existence of zonal currents in the atmospheres of rapidly rotating stars analogous to those found in planetary atmospheres was suggested by Cranmer and Collins [414]. The induced Doppler displacement resulting from such zonal wind belts can distort the line profiles broadened by rotation leading to significant departures from the line profiles predicted by the classical model of rotating stars, which may be interpreted as components of differential rotation, but they do not necessarily act as imprints of the stellar subphotospheric rotation. These zonal wind belts also lead to changes in the photospheric polarization from those characteristic models with uniform rotation. In any case, Collins and Truax [376] concluded that the effects due to limb-darkening, differential rotation in rapidly rotating stars cannot be disentangled using classical methods to measure line rotational line broadening. Interferometric methods to study the surface differential rotation were proposed by Domiciano de Souza et al. [466] and Zorec et al. [141].

9. Be and Bn Stars, Rapid Rotation Siblings?

9.1. Evolutionary Status of Be Stars

Zorec et al. [61] studied a sample of approximately 100 galactic field Be stars by taking into account the effects induced by the fast rotation on their astrophysical parameters. The program stars were observed in the BCD spectrophotometric system, which enabled to minimize the perturbations produced by the circumstellar environment on the spectral photospheric signatures. This approach is one of the first attempts at determining stellar masses and ages by using simultaneously model atmospheres and evolutionary tracks, both calculated for rotating objects. The stellar fractional ages t/t_{MS} (t_{MS} is the amount of time a rotating star spends in the main sequence evolutionary phase), as revealed by the mass-dependent trend as shown in Figure 35. This trend shows that there are Be stars spread over the whole interval $0 \lesssim t/t_{MS} \lesssim 1$ of the main sequence evolutionary phase, and that the distribution of points in the (t/t_{MS}, $M/M_\odot$) diagram indicates that among massive stars ($M/M_\odot \gtrsim 12$) the Be phenomenon is present at smaller t/t_{MS} age ratios than for less massive stars ($M/M_\odot \lesssim 12$). Such a distribution can be due, on the one hand, to higher mass-loss rates in massive objects, which can reduce the surface fast rotation and prevents the phenomenon occurring at later stages of the MS phase. On the other hand, the meridional circulation time scales to transport angular momentum from the core to the surface are longer the lower the stellar mass is, which then demands the Be phenomenon to appear at later epochs of the MS evolutionary phase.

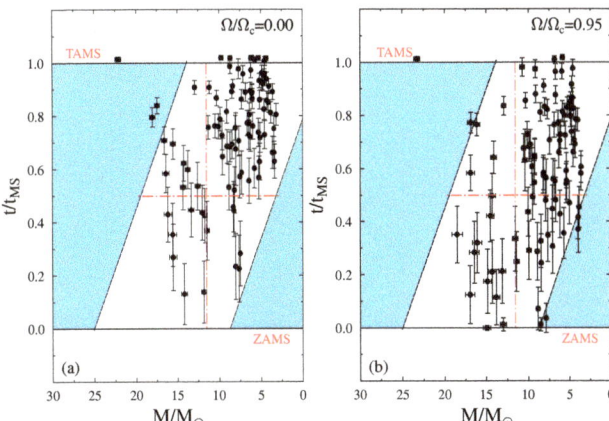

Figure 35. Age ratios t/t_{MS} against the mass of Be stars studied in Zorec et al. [61]. (**a**) Parameters derived using evolutionary tracks without rotation; and (**b**) parameters derived using evolutionary tracks with rotation. Adapted from Zorec et al. [61].

9.2. Distribution of Rotational Velocities of Bn Stars

Herein, the results obtained in Cochetti et al. [146] are reported, where the stellar samples are marred by some selection effects. Bn stars are currently identified as such when they are seen rather equator-on, otherwise, the lack of emissions in their spectra does not enable us to identify those which rotate rapidly but are seen pole-on. Figure 36a,b show the distributions of the observed $V\sin i$ of their program Be and Bn stars corrected for the GD effect. In these figures, histograms correspond to the velocity ratios $v = V\sin i/V_c$. The superimposed green curves $\Psi(v)$ describe the smoothed distributions of the ratios $v = V\sin i/V_c$ corrected for observational uncertainties. The cited authors preferred to use the ratios $V\sin i/V_c$ instead of the $V\sin i$ parameters because the critical equatorial velocity V_c was consistently estimated with the mass and evolutionary state of each star, which then minimizes somewhat mass- and evolution-related effects on the distributions [127].

Because pole-on Bn stars are unavoidably missing in this discussion, the distributions of their rotational velocities do not respect the randomness of inclination angles. Nonetheless, the distributions of the ratios of true velocities $u = V/V_c$ for both Be and Bn stars were calculated as the inclination angles were distributed at random. These are the distributions $\Phi(u)$ shown in Figure 36a,b (blue curves). The lack of randomness of the inclination angles in the distributions of Be star rotational velocities in this work is impossible to correct. However, they attempted to account for this lack in the distributions of Bn rotational velocities. To this end, the transformation of the distribution of ratios $v = V\sin i/V_c$ into $u = V/V_c$ ratios of true velocity ratios was carried out with a probability density distribution $P(i) = \sin i$ of inclination angles that was obliterated from $i = 0°$ up to some limiting inclination i_L using a 'guillotine' function $G(i)$, details of which can be found in [146]. The most outstanding result that can be drawn from Figure 36b,c is that changing the value of i_L affects neither the skewness of the distribution nor its mode. The only characteristic that seems to vary a little more concerns the number of the fastest rotating stars: the higher the limiting inclination angle i_L is, the lower the number of fast rotators is, and accordingly, the larger the number of objects just behind the mode.

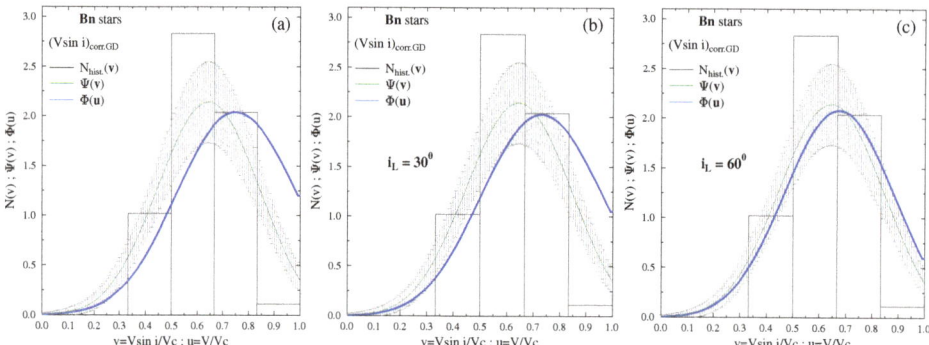

Figure 36. Distributions of ratios of rotational velocities of Bn stars correction for GD effects. (**a**) Histograms of velocity ratios v= $V\sin i/V_c$; functions $\Psi(v)$ are the smoothed histograms after correction for observational uncertainties (green curves); functions $\Phi(u)$ are distributions of true velocity ratios $u = V/V_c$ (blue curves). (**b**,**c**): histograms, functions $\Psi(v)$ from (**a**) and distributions of true velocity ratios $u = V/V_c$ corrected for GD effect and for the density probabilities of inclination angles bridled at inclinations $i_L = 30°$ and $i_L = 60°$, respectively. Adapted from Cochetti et al. [146].

9.3. Evolutionary Status of Bn Stars

Since Bn and Be stars are both rapid rotators, we can ask whether they have similar structures or other common properties that would enable us to consider them both as members of a single population and, in particular, to think of Bn stars as potential Be stars. A first insight into the apparent differences between Bn and Be stars was shown in Figure 11b using the HR diagram in terms of the BCD (λ_1, D) parameters. When (λ_1, D) are transformed into fundamental quantities $(T_{\rm eff}, \log L/L_\odot)$, an HR diagram can be obtained as shown in Figure 37, which concerns the samples of Bn and Be stars studied in Cochetti et al. [146]. In Figure 37a, parameters are not corrected for the effects of rapid rotation, while in Figure 37b they are corrected for rotational effects assuming $\Omega/\Omega_c = 0.95$ for all plotted stars. The main difference seen in Figure 37 is that Bn stars have masses $M \lesssim 9M_\odot$, while Be stars are present in the whole mass interval $3M_\odot \lesssim M \lesssim 20M_\odot$.

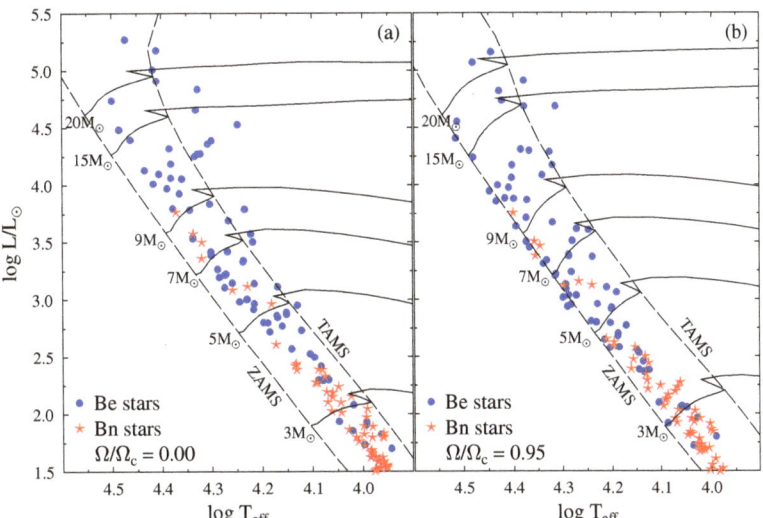

Figure 37. *Panel a*: Hertzsprung–Russell diagram of Be (blue points) and Bn stars (red stars) studied by Cochetti et al. [146] where $(logL/L_\odot, T_{eff})$ are not treated for rotational effects. *Panel b*: same as in Panel a, but parameters are corrected for rotational effects, assuming that for all stars, it is $\Omega/\Omega_c = 0.95$. The evolutionary tracks are from Ekström et al. [145] without rotation (Panel **a**), and with rotation (Panel **b**) where Vo = 300 km s^{-1} in the ZAMS. Adapted from Cochetti et al. [146].

Another way of presenting the evolutionary stage of Bn and Be stars is using masses and fractional ages $(M/M_\odot, t/t_{MS})$, as shown in diagrams of Figure 38. In Figure 38a,b parameters $(M/M_\odot, t/t_{MS})$ are plotted without and with correction for rotation effects, respectively. Figure 38c,d are zooms of Figure 38a,b for masses $M/M_\odot \lesssim 10$, respectively, in order to better see the behavior of Bn stars (red stars) from that of Be stars (blue points). These $(M/M_\odot, t/t_{MS})$ diagrams suggest that Bn stars seem to fill up an apparent gap of Be stars nested in the mass range $2 \lesssim M/M \lesssim 4$, which encompasses a more or less large evolutionary span. When rotational effects are not accounted for, Be and Bn stars are likely located in the second half of the MS. Otherwise, corrections carried on the estimates of $(M/M_\odot, t/t_{MS})$ for rotational effects considering $\Omega/\Omega_c = 0.95$, Be stars are redistributed from the ZAMS to the TAMS, while Bn are scattered over a rather large evolutionary time interval downwards to the ZAMS.

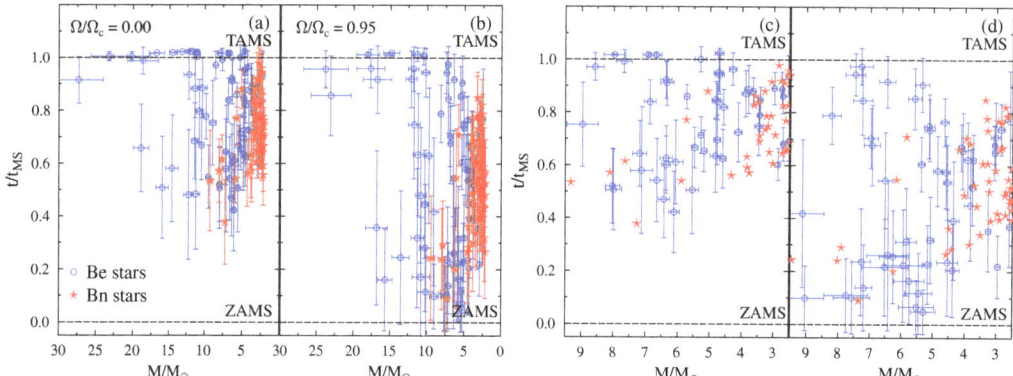

Figure 38. Fractional age t/t_{MS} vs. stellar mass $M/M_\odot$ not treated for rotation (panels **a**,**c**), and using the parameters corrected for rotational effects (panels **b**,**d**). Panels c and d are the zooms of a and b for masses $M/M_\odot < 10$, respectively. Error bars indicate uncertainties affecting the t/t_{MS} and $M/M_\odot$ determinations. Adapted from Cochetti et al. [146].

9.4. Rotational Velocities of Be and Bn Stars in the ZAMS

Having a rather small number of B and Be stars in their discussion, Cochetti et al. [146] could not determine the true rotational velocities V for a statistically significant sample having fractional ages in a reduced interval near the ZAMS, i.e., $0 \lesssim t/t_{MS} \lesssim 0.1$, in order for them to be considered witnesses of the respective distribution of rotational velocities in the initial phases of their MS evolution. Such distributions could provide information to answer the question as to why some B-type rapid rotators in ZAMS can or cannot display the Be phenomenon. To infer vague insights into these distributions, they used the model predictions of changes of the surface rotational velocity [145], which somewhat depends on the prescribed mass-loss rates. Mass loss is indeed a key phenomenon that controls the evolution of stars and their surface rotational velocity, so that high uncertainties may also affect predictions accordingly. In Ekström et al. [145], two series of models were considered for stars with metallicity $Z = 0.02$. The first series of models depends on the mass-loss rates given by de Jager et al. [565] and Kudritzki and Puls [566], here called EMMB; these rates are specified as $\dot{M}$(EMMB) mass-loss rates. The second series of models relies on mass-loss rates suggested by Vink et al. [567], referred here to as $\dot{M}$(VdKL). These two mass-loss prescriptions differ somewhat for $t/t_{MS} \gtrsim 0.5$ and carry differences in the evolution of the V/V_c velocity ratios by the end of the MS phase, mostly for stars with masses $M \gtrsim 9 M_\odot$. By interpolating in the curves of the theoretically predicted evolution of ratios V/V_c, the velocities for $t/t_{MS} \sim 0.01 - 0.02$, the epoch at which objects acquire a stabilized rotational law [254], Cochetti et al. [146] obtained the distributions of true equatorial velocities V of Be and Bn stars in the ZAMS shown in Figure 39.

Apart from a slight shift in the points dependent on $\dot{M}$(EMMB) toward slightly lower VZAMS values than those calculated using models with $\dot{M}$(VdKL) mass-loss rates for stellar masses $M \gtrsim 12 M_\odot$, the effect that seems to be more marked among Be stars, probably because, for Bn stars, the mass-loss rates are lower or zero, no other significant difference is apparent in this figure. Then, since both types of objects begin their MS evolutionary phase with similar rotational velocities and only a fraction of these objects at some moment display the Be phenomenon, it is worth inquiring a little further into the distributions of $(V/V_c)_{ZAMS}$ to uncover whether they can reveal other properties that may distinguish Bn and Be stars in the ZAMS. In Figure 40, it is seen that, independently of the mass-loss rate prescription, the distributions of $(V/V_c)_{ZAMS}$ have different skewness signs when it comes to Bn or to Be stars, and that the $(V/V_c)_{ZAMS}$ of Be stars can attain lower values than in the Bn ones. At the moment, it is not possible to say anything more that would enable to conclude that these characteristics are responsible for only some B stars becoming Be.

As already noted, the models on which these conclusions are based use specific receipts for the diffusion coefficients. They are not all the same in the models by Maeder and Meynet [154] and Heger et al. [255], and are still different in those by Potter et al. [519]. From the results obtained by Potter et al. [519] concerning the TAMS rotational velocities, it comes that they are different according to the used prescription for the diffusion coefficients. The conclusions reported here on the V_{ZAMS} velocities of Be and Bn stars must then be considered as a mere first approach.

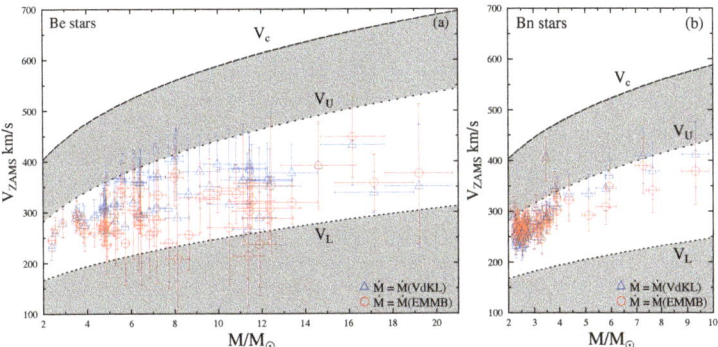

Figure 39. Equatorial rotational velocities V_{ZAMS} of Be (panel **a**) and Bn (panel **b**) stars deduced using the models of stellar evolution by Ekström et al. [541] with mass-loss rates $\dot{M}$(EMMB) (red circles) and $\dot{M}$(VdKL) (blue triangles). In the figure, the curve of V_c in the ZAMS is shown as a function of the stellar mass and the limiting curves V_L and V_U that surround the deduced V_{ZAMS} values. Panel (**b**) is reduced to the mass interval $M \leq 10 M_\odot$ concerning the program Bn stars. Adapted from Cochetti et al. [146].

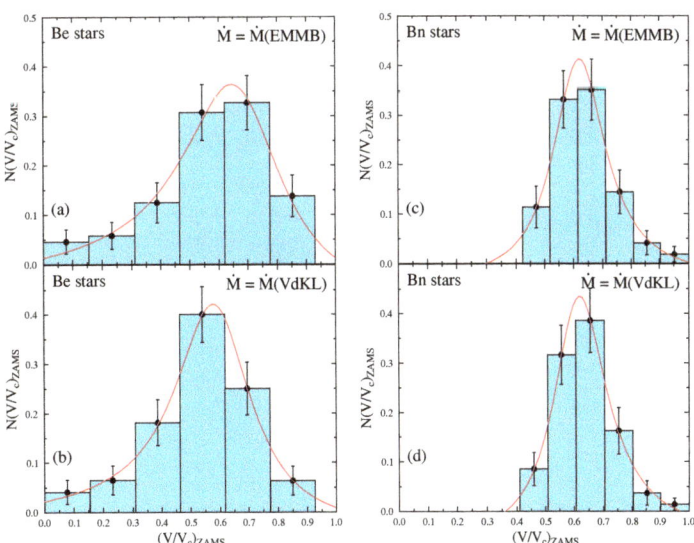

Figure 40. Normalized histograms showing the distribution of V/V_c ratios in the ZAMS derived for Be (panels **a,b**) and Bn (panels **c,d**) stars. The results shown in panels a and c were obtained with EMMB mass-loss rates, while in panels **b,d**, the results are for VdKL mass-loss rates. Error bars indicate sampling uncertainties according to the errors associated with the measured $V \sin i$ parameters. Fitted Pearson distributions are superimposed (red curves) to better perceive the differences in the asymmetries (skewness) of distributions for each type of stars. Adapted from Cochetti et al. [146].

10. Rotation of Magnetic B-Type Stars

Magnetic fields can be detected in almost all types of objects over the entire HR-diagram, cool and hot stars assumed to be single or in binary systems. The presence of magnetic fields may carry significant consequences on the internal stellar rotation, its structure and evolution [568], as well as shaping their circumstellar discs or envelopes. Most often, magnetic fields are measured using spectropolarimetric techniques, but their presence can also be suggested from photometric and spectroscopic studies. Directly, magnetic fields are detected through the Zeeman effect that splits spectral lines into several distinctly polarized components. The easiest detection of the magnetic field signatures is from the Stokes V polarized vector, whose measurement is frequently performed with the Least-Squares-Deconvolution method (LSD) [569,570].

Stellar photospheric magnetic fields are basically of two types, fossil magnetic fields that concern mainly early-type stars which have radiative envelopes, and dynamo-driven fields in stars with convective envelopes. In hot stars, magnetic fields are thought to be frozen in the radiative envelope during the early stages of the stellar evolution in the MS. They frequently have few kG, are rather stable [571], and detected in quite few cases, no more than approximately 10% of mid- to early-type stars [572].

Among the youngest and most rapidly rotating magnetic OB-type stars, the Hα line can be seen in emission, whose origin is assigned to a 'magnetosphere' [573–575]. According to Petit et al. [575], there are two types of magnetospheres: the dynamic magnetosphere and centrifugal magnetosphere. In a dynamic magnetosphere, the stellar rotation is not fast enough for the radiation-accelerated wind plasma to be slowed down by gravity, and eventually pulled back to the star. The presence of dynamic magnetospheres is generally more detected through far-UV lines and by Hα emission, but apparently, this only happens in O-type stars. The centrifugal magnetosphere exists when, in the magnetosphere, the gravitational infall is prevented by the centrifugal force [576,577]. Plasma densities in centrifugal magnetospheres can be significantly higher than in dynamic magnetospheres, which enable the Hα emission to become strong enough to be detected. Using the Kepler corotation radius R_K, where centrifugal and gravitational forces are equivalent, and the Alfvén radius R_A, which indicates the extent of close magnetic loops, Hα emission is detected when $R_K \lesssim 2R_*$ (R_* stellar radius) and $R_A \gtrsim 8R_K$ [578,579]. Moreover, there is a correlation between the Hα emission strength and radio luminosity [580]. Magnetic stars with Hα emission are generally in the first quarter of their MS evolutionary phase [578,581].

Hot magnetic stars can be distinguished from the non-magnetic ones because: (a) they have peculiar abundances, mainly of He, Fe and Si, either over- or under-abundant that currently appear in He-strong or He-weak stars, which are thought to arise due to radiative diffusion in strongly magnetized atmospheres. As these abundance irregularities cover some fraction of the stellar surface, spectroscopic and photometric variability modulated by the rotation can be detected. It is then possible to determine the reliable rotational periods P_{rot}; (b) Magnetic stars are on average less rapid rotators than the non-magnetic ones, which is attributed to a loss in angular momentum through a magnetically confined magnetic field [582,583]. This last property also explains that, among them, the more rapidly rotating stars appear at evolutionary stages with $t/t_{MS} \lesssim 0.25$ [578]. Extensive compilations and the newly determined astrophysical parameters of hot magnetic stars, in particular $V\sin i$ and P_{rot}, can be found in Shultz et al. [580,584,585]. Data suggest that the magnetic star HD 142990 might undergo an irregular rotational evolution [586].

Several surveys aiming to measure magnetic fields in early-type stars have taken mainly advantage of instruments such as ESPaDOnS, FORS2, Narval, and HARPSpol spectropolarimeters (3.6 m CFHT telescope, French 2 m TBL, 3.6 m ESO telescope): MiMeS collaboration [130,587,588], BinaMIcs for magnetism in hot and cool binary stars [589], BOB (B fields in OB stars) collaboration as an ESO large survey [588,590–595], and LIFE collaboration dedicated to the magnetic field and evolution of hot stars [596,597]. Thanks to the TESS and BRITE space missions, it was also possible to find many magnetic hot stars, including pulsating ones. Using TESS data, the MOBSTER collaboration has shown that the rotational ratio V/V_c decreases with stellar age, consistently with the magnetic braking observed in a population of magnetic chemically peculiar stars [598]. A number of known magnetic B-type stars were also observed by BRITE and BRITEpol, and the results were commented by Shultz et al. [599].

Since magnetic B-type stars are also considered chemically peculiar, most often classified as He-strong and He-weak because their abundances do not agree with those in normal stars having the same $(T_{\text{eff}}, \log g)$ parameters, it could also be interesting to see in what way they can be distinguished according to the physical parameters, particularly as rotators. In Figure 41, the He-weak (red points), He-strong (blue points), and those that have not yet been classified are clearly separated as a function of certain parameters, particularly as rotators, and the intensity of their surface magnetic field (here, we consider the maximum value measured of the longitudinal magnetic field $\langle Bz \rangle$ taken in absolute value $|\rangle Bz \rangle|$). Data are from Shultz et al. [580,584,585]. According to the existing data, He-weak stars have $\log L/L_\odot \lesssim 3.25$, $T_{\text{eff}} \lesssim 21{,}000$ K, $\log g \gtrsim 4.0$, and true rotational velocities $V \lesssim 150$ km.s^{-1}. No distinction between He-strong and He-weak stars can be established in the $(\langle|Bz|\rangle, V)$ diagram. He-weak stars follow tight $(\langle|Bz|\rangle, \log L/L_\odot)$ and $(\langle|Bz|\rangle, T_{\text{eff}})$ relations, but there is no clear distinction between He-strong and He-weak objects in the $(\langle|Bz|\rangle, \log g)$ diagram. Probably more detailed relations could be established using the quantitative indices of their chemical peculiarity, namely the index (He/H) or the discrepancy of the Balmer discontinuity δD given in Equation (10).

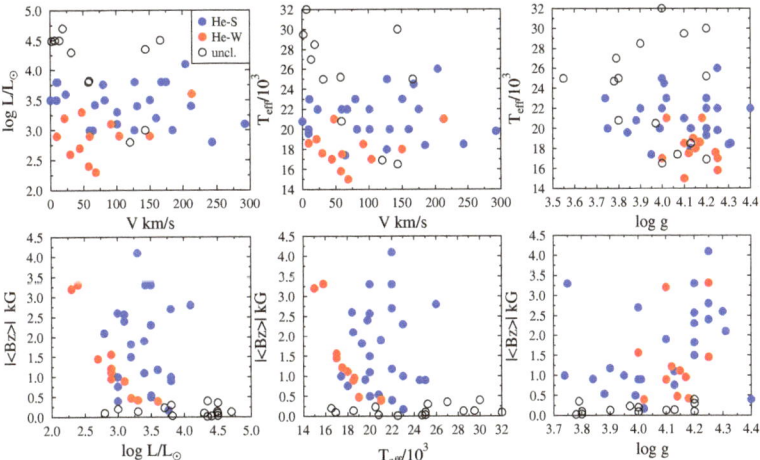

Figure 41. Magnetic B-type stars differentiated into He-weak (red points), He-strong (blue points) and He-unclassified (open black circles).

11. Rotation of B[e] Stars

Excluding cPNB[e] and SymB[e] stars that identify very specific evolutionary stages or environmental conditions in binary systems, the stellar classes HAeB[e], sgB[e] and FS CMa stars are of particular interest for studies of stellar formation and acquisition of angular momentum, as well as of advanced evolutionary properties associated with rotation. However, these phenomena can be understood either as concerning single stars, or objects whose rotational properties respond to evolution in close binary systems [213,215]. The characterization of these objects is not only dependent on their aspects as rotators, but also on the formation mechanisms of their circumstellar discs or envelopes. Both characterizations can be formation- and stellar evolution-related. Unfortunately, the spectra of B[e] stars rarely display features that could be attributed to the photosphere of the central object, which explains the lack of $V \sin i$ determinations for this class of stars. Moreover, the fraction of confirmed binaries among them is still not sufficiently high to decide whether their characteristics are related to phenomena taking place in single stars, or if they obey evolution phenomena in a close binary system.

Whatever the possibility, near-critical rotation is frequently evoked to create their circumstellar environment. This is particularly the case for sgB[e] objects considered single-evolved stars. If stars are very massive, the Eddington factor can approach $\Gamma \to 1$, which enables the ratio $V/V_c \to 1$ by the end of the MS evolutionary phase [213]. Since, in post-MS phases, the surface velocity of stars evolves as $V \propto R^{-1}$, the critical velocity ratio can then be attained only in the blue loop passing from the red-supergiant to the blue-supergiant phase [213]. Because such an evolutionary phase is rather short, it may explain the scarcity of sgB[e] stars. The critical rotation is accompanied by a strong gravity darkening, which can favor the existence of fast polar winds, as well as slow and dense equatorial winds [121]. They may account for the characteristics of the two-component wind in sgB[e] proposed by Zickgraf et al. [170] and Zickgraf [168] for the hypergiant R126 star in the LMC, where rough estimates suggest an Eddington factor $\Gamma \sim 0.7$.

The rotational velocity was estimated for HD 45677 (prototype of FS CMA class stars) and HD 50138, and according to their astrophysical parameters, they have rotational velocities that are far from being critical today. However, their nature as critical rotators lasts a rather short time. Assuming that these stars today have the observed characteristics because they underwent a merger phenomenon, theoretical predictions by Schneider et al. [600] suggest that their initial possible critical rotator nature should only last for a while, because the magnetic field created in the 'merging' process slows down the stellar surface rapid rotation. During this process, a dense envelope prevents the direct observation of such objects.

The B[e] stars HD 45677 and HD 50138 were considered by some authors as unclB[e], because they may be escaped from the background molecular clouds on which they appear to overlap [169].

12. Concluding Remarks

The main conclusions drawn from the first part of this report is that the BCD spectrophotometric system is a powerful observational method to study normal and 'active' (or abnormal) stars over a large spectral type interval. Its qualities are:

(1) The BCD system produces parameters that depend on the continuum energy distribution and are thus related to the physical properties of atmospheric layers that are on average deeper than those where the spectral lines are formed;

(2) The reduction in low-resolution spectra required in the BCD system can be easily automated which can provide the rapid physical characterization of stellar objects not only in deep regions of our Galaxy, but also in other galactic and more or less distant extra-galactic systems;

(3) The rather short wavelength interval that is required to be observed ($\lambda\lambda 3500 - 5500$) does not impose long exposure times;

(4) Excepting for some stellar examples where the extreme spectral peculiarities cannot be apprehended straightforwardly (mainly stars of B[e] class), the spectral signatures from the photospheric regions and those emitted by their close circumstellar regions can be clearly separated;

(5) The BCD (λ_1, D) spectrophotometric quantities are independent, or can easily be rendered independent of the interstellar and circumstellar perturbing extinctions, which helps obtain astrophysical parameters [$T_{\text{eff}}, \log g$] free from the marring effects of these media;

(6) The second component of the Balmer discontinuity, not yet studied in detail in a theoretical approach of Be stars, keeps precious information on the transition region between the star and the circumstellar disc, where the transfer of angular momentum to the disc is organized;

(7) Discrepancies of the Balmer discontinuity reveal abnormal surface chemical abundances, mainly of He, which can be controlled either by phenomena related with rotation or with stellar magnetic fields.

Although the BCD system was devised to classify the stars of spectral types between O and F, it provides important apparent quantities for both normal and active stars, which can be easily corrected by eliminating effects induced by the rapid rotation. These then enable to estimate the true stellar masses and determine the corresponding evolutionary states. This characteristic of the BCD system inspired us to jointly present two topics, BCD and rotation, which at first glance, appear to be disjoint.

In the second part of the text, a short review was presented on the study of normal and active B-type stars as rotators. Reminders of some supporting theoretical concepts were added. The line of thought of this review was related to the angular momentum of stars, its origin and internal redistribution, although without discussing the hydrodynamic and magneto-hydrodynamic instabilities that control its evolution. We addressed the problem of uncertainties marring the $V\sin i$ parameter determination and treated the calculation of the geometrical deformation of stars with surface differential rotation. The interpretation of astrophysical parameters and the determination of fundamental parameters in rotating stars was discussed. Observational and theoretical approaches concerning the origin of the stellar angular momentum in field and cluster stars were shortly described. There are still dark areas concerning the mass and angular momentum accretion during the stellar formation phases, so that it is today impossible to assert how much angular momentum stars can or cannot store. The modeling and measurement of quantities related to the stellar structure considered as physical laboratories to probe rotation and the related phenomenology show that there are still open questions concerning the actual content of angular momentum and its internal redistribution. It is thus worth mentioning that the existent models of stellar evolution systematically impose rigid rotation in the ZAMS, while many theoretical approaches of rotating stellar structures suggest that stars are differential rotators already at their initial MS evolutionary phase. Differential rotation enables the star to have larger amounts of angular momentum than supported by the rigid rotation. Even though stars may try to attain a state of minimal rotational energy (i.e., rigid rotation) during their evolution in the MS, it should not be excluded that some of their activities are related to phenomena where they try to get rid of a certain amount or 'excess' of angular momentum. In the most currently calculated models, the 'shellular' distribution of angular velocity is assumed to prevail and control the evolution of rotating stars. However, new approaches regarding the stellar structure suggest that the internal rotational profile is neither rigid, cylindrical nor 'shellular'. Moreover, it was shown in Clement [137] and Maeder et al. [136] that rapid rotation can destabilize radiative zones against convection. This convection should happen in deeper and extended zones than just in a small subphotospheric region. It can produce differential rotation in depth and generate magnetic fields. Activities related to magnetic instabilities could then trigger mass ejection phenomena to build up the circumstellar discs in Be stars. To this purpose, non-radial pulsation was evoked many times. Unfortunately, to be effective, this mechanism requires stars to rotate near critical rates, a phenomenon which, among Be stars, is more exception than the rule. Differences between the predicted evolution of surface rotational velocities with the observed one of intermediate-mass stars were pointed out, which testify the existence of unsolved problems concerning the mechanisms that control the internal angular momentum redistribution. The rapid rotation of Be stars and of Bn stars was studied and their evolution compared during the MS evolutionary phase. A short account is given of magnetic B-type stars, where rotation is nothing but a phenomenon that helps the magnetosphere in some of these objects have enough circumstellar matter to produce a detectable line emission. A kind of distinction is also noted between He-weak (apparently slow rotators) and He-strong stars (apparently rapid rotators) in diagrams ($\log L/L_\odot$, V) and ($\langle|Bz|\rangle$, $\log L/L_\odot$). Regarding the stars presenting the B[e] phenomenon, a lot of work is still required to identify the genuine photospheric spectral lines that will enable determining their actual rotation rate and identifying what phenomena underlie the origin of a sgB[e] star.

Radiation transfer codes adapted for extended moving media, some of which have been cited in this review, can certainly provide valuable tools for studying the complex spectra of B[e] stars. However, the extended atmospheres of these objects probably have non-spherical geometry, whose shape depends on their rotation law. In addition, the extended atmosphere could be affected by gravitational darkening, which induces a thermal structure controlled not only by the radial radiative flux but also by a horizontal one. The horizontal radiative flux has not yet even been considered in the plane-parallel atmospheres of rapidly rotating stars. As B[e] stars can also undergo mass loss phenomena through winds and discrete ejections, their extended atmospheres can be in non-radiative equilibrium, which increases the non-LTE effects. The classical methods used to calculate the radiative transfer in extended atmospheres to date may be insufficient to tackle the described phenomenological pattern. In these cases, Monte Carlo simulations might perhaps be appropriate to carry out the modeling of the radiation transfer in the mentioned complex atmospheres.

Funding: This research received no external funding.

Institutional Review Board Statement: Not applicable.

Informed Consent Statement: Not applicable.

Data Availability Statement: Data in Figure 41: Reduced ESPaDOnS spectra are available at the CFHT archive maintained by the CADC5, where they can be found via standard stellar designations. ESPaDOnS and Narval data can also be obtained at the PolarBase archive6. Kepler-2 and TESS data are available at the MAST archive7. Hipparcos data are available through SIMBAD and VizieR. DAO and uGMRT observations are available from the authors upon request. Data in all the remaining figures are available through SIMBAD and VizieR.

Acknowledgments: With these few words, I would like to pay tribute to the memory of Lucienne Divan who introduced me to the secrets of the BCD system. Due to the many contributions cited in this paper that we have written together, Yves Frémat should be listed as a co-author of the present review. I am particularly grateful to him for his contribution to Section 5.1. I cannot find the right words to thank him for his great friendship, wise advice, productive discussions and his precious scientific complicity. The long and permanent support from Anne-Marie Hubert has marked my scientific career, which I wish to note and thank warmly. I am deeply grateful to the Argentinian scientists, colleagues of my Alma Mater, with whom the collaboration has continued to grow over the years as evidenced by the documents we signed with Lydia Cidale and her team, notably Maria-Laura Arias, Yael Aidelman, Janina Cochetti and René Rohrmann. My foray into the world of interferometry was only possible thanks to the friendship and complicity of Aramando Domiciano de Souza, Philippe Stee and Omar Delaa. I would like to thank the guest editors Lydia Sonia Cidale, Michaela Kraus, and María Laura Arias for their kind invitation to write this review. I am deeply grateful to the anonymous referees for their numerous and valuable suggestions, grammatical corrections and corrections to the content that helped to substantially improve this paper.

Conflicts of Interest: The authors declare no conflict of interest.

References

1. Chalonge, D. Étude du spectre continu de quelques étoiles entre 3100 et 4600 Å. *J. Obs.* **1936**, *19*, 149.
2. Arnulf, A.; Barbier, D.; Chalonge, D.; Safir, H. Étude du rayonnement continu de quelques étoiles entre 3100 et 4600 Å (I). *Annales d'Astrophysique* **1938**, *1*, 293.
3. Guérin, P. Réalisation et étude d'une lampe luminiscente étalon secondarie de répartition spectrale. In *Contributions de l'Institut d'Astrophysique de Paris*; CNRS; 1954; Volume 162, p. 1.
4. Baillet, A.; Chalonge, D.; Divan, L. Quartz spectograph with oscillating plateholder for stellar spectrophotometry. Review of some applications. *Nouvelle Revue d'Optique* **1973**, *4*, 151–158. [CrossRef]
5. Divan, L. List of Spectrophotometric Standards in Use at the Institut d'ASTROPHYSIQUE of Paris. In *Spectral Classification and Multicolour Photometry: Proceedings of the IAU Symposium No. 24, Saltsjöbaden, Sweden, 17–21 August 1964*; Loden, K., Loden, L.O., Sinnerstad, U., Eds.; Academic Press: London, UK, 1966; Volume 24, p. 311.

6. Divan, L. Discussion: Notes on the definition of the Balmer discontinuity in the BCD system. In Proceedings of the Problems of Calibration of Absolute Magnitudes and Temperature of Stars, Geneva, Switzerland, 12–15 September 1972; Reidel: Dordrecht, The Netherlands; Boston, MA, USA, 1973; Volume 54, p. 267.
7. Divan, L. Recherches sur la loi d'absorption de la poussière interstellaire et sur le spectre continu des étoiles O et B. *Annales d'Astrophysique* **1954**, *17*, 456.
8. Chalonge, D.; Divan, L. La classification stellaire BCD: Paramètre caractéristique du type spectral calibration en magnitudes absolues. *Astron. Astrophys.* **1973**, *23*, 69.
9. Barbier, D.; Chalonge, D. On the Continuous Spectrum of Stars with Extended Atmospheres. *Astrophys. J.* **1939**, *90*, 627–629. [CrossRef]
10. Aidelman, Y.; Cidale, L.S.; Zorec, J.; Arias, M.L. Open clusters. I. Fundamental parameters of B stars in NGC 3766 and NGC 4755. *Astron. Astrophys.* **2012**, *544*, A64. [CrossRef]
11. Allen, C.W. *Astrophysical Quantities*, 4th ed.; Springer: Berlin/Heidelberg, Germany, 2002.
12. Barbier, D.; Chalonge, D. Étude du rayonnement continu de quelques étoiles entre 3 100 et 4 600 Å (4e Partie-discussion générale). *Annales d'Astrophysique* **1941**, *4*, 30.
13. Chalonge, D.; Divan, L. Recherches sur les spectres continus stellaires. V. Etude du spectre continu de 150 etoiles entre 3150 et 4600 A. *Annales d'Astrophysique* **1952**, *15*, 201.
14. Barbier, D. Introduction to the stellar classification based on the Balmer discontinuity. In Proceedings of the Principles of Stellar Classification, Paris, France, 29 June–4 July 1953; CNRS, 1955; p. 47.
15. Chalonge, D. Classification using two or three parameters of early type stars. In Proceedings of the Principles of Stellar Classification, Paris, France, 29 June–4 July 1953; CNRS, 1955; p. 55.
16. Keenan, P.C.; Morgan, W.W. Classification of Stellar Spectra. In *Proceedings of the 50th Anniversary of the Yerkes Observatory and Half a Century of Progress in Astrophysics*; Hynek, J.A., Ed.; McGraw-Hill: New York, NY, USA, 1951; p. 12.
17. Underhill, A.B.; Divan, L.; Prevot-Burnichon, M.L.; Doazan, V. Effective temperatures, angular diameters, distances and linear radii for 160 O and B stars. *Mon. Not. R. Astron. Soc.* **1979**, *189*, 601–605. [CrossRef]
18. Zorec, J.; Briot, D. Absolute magnitudes of B emission line stars: Correlation between the luminosity excess and the effective temperature. *Astron. Astrophys.* **1991**, *245*, 150.
19. Zorec, J.; Cidale, L.; Arias, M.L.; Frémat, Y.; Muratore, M.F.; Torres, A.F.; Martayan, C. Fundamental parameters of B supergiants from the BCD system. I. Calibration of the (λ_1, D) parameters into T_{eff}. *Astron. Astrophys.* **2009**, *501*, 297–320. [CrossRef]
20. Zorec, J. *Thèse d'État, Structure et Rotation difféRentielle dans les Étoiles B avec et sans Émission*; Université Paris VII: Paris, France, 1986.
21. Husser, T.O.; Wende-von Berg, S.; Dreizler, S.; Homeier, D.; Reiners, A.; Barman, T.; Hauschildt, P.H. A new extensive library of PHOENIX stellar atmospheres and synthetic spectra. *Astron. Astrophys.* **2013**, *553*, A6. [CrossRef]
22. Divan, L. Quantitative Spectral Classsification in the BCD System for LMC Supergiants. In Proceedings of the Spectral Classification and Multicolour Photometry, Villa Carlos Paz, Argentina, 18–24 October 1971; Reidel: Dordrecht, The Netherlands; Boston, MA, USA, 1973; Volume 50, p. 27.
23. Divan, L. Calibration en Magnitudes Absolute de la Classification BCD. Application à la DéTERMINATION du Module de Distance du Grand Nuage de Magellan. In Proceedings of the Problems of Calibration of Absolute Magnitudes and Temperature of Stars, Geneva, Switzerland, 12–15 September 1972; Reidel: Dordrecht, The Netherlands; Boston, MA, USA, 1973; Volume 54, p. 78.
24. Chalonge, D.; Divan, L. BCD classification: Relation between the spectral type and the effective temperature. *Astron. Astrophys.* **1977**, *55*, 121–124.
25. Chalonge, D.; Divan, L. BCD system of stellar classification and chemical composition. *Astron. Astrophys.* **1977**, *55*, 117–120.
26. Gkouvelis, L.; Fabregat, J.; Zorec, J.; Steeghs, D.; Drew, J.E.; Raddi, R.; Wright, N.J.; Drake, J.J. Physical parameters of IPHAS-selected classical Be stars. I. Determination procedure and evaluation of the results. *Astron. Astrophys.* **2016**, *591*, A140. [CrossRef]
27. Divan, L.; Zorec, J. Absolute magnitudes and other basic parameters of O and B stars. In *Proceedings of the The Scientific Aspects of the Hipparcos Space Astrometry Mission*; Perryman, M.A.C., Guyenne, T.D., Høg, E., Jaschek, C., Lacroute, P., Eds.; ESA: Paris, France, 1982; Volume 177, pp. 101–104.
28. Cidale, L.; Zorec, J.; Tringaniello, L. BCD spectrophotometry of stars with the B[e] phenomenon. I. Fundamental parameters. *Astron. Astrophys.* **2001**, *368*, 160–174. [CrossRef]
29. Mennickent, R.E.; Cidale, L.; Díaz, M.; Pietrzyński, G.; Gieren, W.; Sabogal, B. Revealing the nature of double-periodic blue variables in the Magellanic Clouds. *Mon. Not. R. Astron. Soc.* **2005**, *357*, 1219–1230. [CrossRef]
30. Zebrun, K.; Soszynski, I.; Wozniak, P.R.; Udalski, A.; Kubiak, M.; Szymanski, M.; Pietrzynski, G.; Szewczyk, O.; Wyrzykowski, L. The Optical Gravitational Lensing Experiment. Difference Image Analysis of LMC and SMC Data. The Catalog. *Acta Astron.* **2001**, *51*, 317–329.
31. Witham, A.R.; Knigge, C.; Drew, J.E.; Greimel, R.; Steeghs, D.; Gänsicke, B.T.; Groot, P.J.; Mampaso, A. The IPHAS catalogue of Hα emission-line sources in the northern Galactic plane. *Mon. Not. R. Astron. Soc.* **2008**, *384*, 1277–1288. [CrossRef]
32. Fabricant, D.; Cheimets, P.; Caldwell, N.; Geary, J. The FAST Spectrograph for the Tillinghast Telescope. *Publ. Astron. Soc. Pac.* **1998**, *110*, 79–85. [CrossRef]

33. Shokry, A.; Rivinius, T.; Mehner, A.; Martayan, C.; Hummel, W.; Townsend, R.H.D.; Mérand, A.; Mota, B.; Faes, D.M.; Hamdy, M.A.; et al. Stellar parameters of Be stars observed with X-shooter. *Astron. Astrophys.* **2018**, *609*, A108. [CrossRef]
34. Vernet, J.; Dekker, H.; D'Odorico, S.; Kaper, L.; Kjaergaard, P.; Hammer, F.; Randich, S.; Zerbi, F.; Groot, P.J.; Hjorth, J.; et al. X-shooter, the new wide band intermediate resolution spectrograph at the ESO Very Large Telescope. *Astron. Astrophys.* **2011**, *536*, A105. [CrossRef]
35. Freudling, W.; Romaniello, M.; Bramich, D.M.; Ballester, P.; Forchi, V.; García-Dabló, C.E.; Moehler, S.; Neeser, M.J. Automated data reduction workflows for astronomy. The ESO Reflex environment. *Astron. Astrophys.* **2013**, *559*, A96.
36. Shokry, A.; Nouh, M.I.; Saad, S.M.; Helmy, I. Fundamental parameters of some B-type stars using NOAO Indo-U.S. Library. *New Astron.* **2022**, *93*, 101780.
37. Valdes, F.; Gupta, R.; Rose, J.A.; Singh, H.P.; Bell, D.J. The Indo-US Library of Coudé Feed Stellar Spectra. *Astrophys. J. Suppl. Ser.* **2004**, *152*, 251–259.
38. Jaschek, M.; Slettebak, A.; Jaschek, C. Be star terminology. *Be Star Newsl.* **1981**, *4*, 9–11.
39. Collins, G.W., I. The Use of Terms and Definitions in the Study of be Stars (review Paper). In Proceedings of the 92nd Colloquium of the International Astronomical Union, Boulder, CO, USA, 18–22 August 1986; Cambridge University Press: Cambridge, UK, 1987; p. 3.
40. Rivinius, T.; Carciofi, A.C.; Martayan, C. Classical Be stars. Rapidly rotating B stars with viscous Keplerian decretion disks. *Astron. Astrophys. Rev.* **2013**, *21*, 69.
41. Barbier, D.; Chalonge, D. Remarques préliminaires sur quelques propriétés de la discontinuité de Balmer dans les spectres stellaires. *Annales d'Astrophysique* **1939**, *2*, 254.
42. Jaschek, M.; Groth, H.G. (Eds.) *IAU Symp. 89: Be Stars*; 1982; Volume 89.
43. Divan, L.; Zorec, J. Behavior of the energy distribution of 59 CYG in the far-ultraviolet and in the visible. In Proceedings of the Third European IUE Conference, Madrid, Spain, 10–13 May 1982; European Space Agency: Paris, France, 1982; Volume 176, pp. 291–293.
44. Hubert-Delplace, A.M.; Hubert, H. The recent peculiar Behaviour of the be star, HD 200120, 59 Cyg. *Astron. Astrophys.* **1981**, *44*, 109–113.
45. Divan, L.; Zorec, J. BCD spectrophotometry of the Be-shell star 88 Her. In *Proceedings of the Be Stars*; Jaschek, M., Groth, H.G., Eds.; Cambridge University Press: Cambridge, UK, 1982; Volume 98, pp. 61–63.
46. Doazan, V.; Marlborough, J.M.; Morossi, C.; Peters, G.J.; Rusconi, L.; Sedmak, G.; Stalio, R.; Thomas, R.N.; Willis, A. Ultraviolet and visual variability of theta CrB during a normal B-phase following a shell phase (1980–1985). *Astron. Astrophys.* **1986**, *158*, 1–13.
47. Doazan, V.; Franco, M.; Rusconi, L.; Sedmak, G.; Stalio, R. The long-term variations of γ Cas in the visual. *Astron. Astrophys.* **1983**, *128*, 171–180.
48. Hirata, R.; Kogure, T. The spectral variation of Pleione in 1969–1975. *Publ. Astron. Soc. Jpn.* **1976**, *28*, 509–515.
49. Kogure, T.; Hirata, R. The Be star phenomena. I. General properties. *Bull. Astron. Soc. India* **1982**, *10*, 281–309.
50. de Loore, C.; Altamore, A.; Baratta, G.B.; Bunner, A.N.; Divan, L.; Doazan, V.; Hensberge, H.; Sterken, C.; Viotti, R. First coordinated campaign of X-ray and ground-based observations of X-Persei = 3U 0352+30. *Astron. Astrophys.* **1979**, *78*, 287–291.
51. Divan, L.; Zorec, J.; Andrillat, Y. A Be type variation in an O star. *Astron. Astrophys.* **1983**, *126*, L8–L10.
52. Barbier, D. Le spectre continu de Gamma Cassiopeiae. *Annales d'Astrophysique* **1948**, *11*, 13.
53. Barbier, D.; Chalonge, D.; Vassy, E. In Proceedings of the Journal de Physique. Société Française de Physique, 1935; Volume 6, p. 137.
54. Chalonge, D.; Safir, H. *Proceedings of the C.R. Acad. Sci. Paris*; Elsevier of behalf of the French Academy of Sciences: Issy-les-Moulineaux, France, 1936; Volume 203, p. 1329.
55. Barbier, D.; Chalonge, D. Sur le spectre continu de Nova Herculis 1934 dans la région de courtes longueurs d'onde. *Annales d'Astrophysique* **1940**, *3*, 26.
56. Barbier, D.; Chalonge, D.; Canavaggia, R. Étude du rayonnement continu de quelques étoiles entre 3 100 et 4 600 Å-V: Nouvelles mesures de D et λ_1. *Annales d'Astrophysique* **1947**, *10*, 195.
57. Zorec, J.; Divan, L.; Hoeflich, P. Phase variations of 88 Herculis: Do the UV observations confirm a connection between these variations and the changes of the phospheric parameters of the underlying star? *Astron. Astrophys.* **1989**, *210*, 279–283.
58. Divan, L. Quantitative Spectral Classification of STARS Stars on Low Dispersion Spectra. In Proceedings of the IAU Colloq. 47: Spectral Classification of the Future, Vatican City, 11–15 July 1978; Vatican Observatory: Albano laziale, Italy, 1979; Volume 9, p. 247.
59. Mermilliod, J.C. Stellar content of young open clusters. II. Be stars. *Astron. Astrophys.* **1982**, *109*, 48–65.
60. Schmidt-Kaler, T. Die galaktischen Emissions-B-Sterne: (Spectralklassifikation, Photometrie, Entwicklung und Verteilung in der Milchstraszenebene). *Veroeffentlichungen Des Astron. Inst. Der Univ. Bonn* **1964**, *70*, 1.
61. Zorec, J.; Frémat, Y.; Cidale, L. On the evolutionary status of Be stars. I. Field Be stars near the Sun. *Astron. Astrophys.* **2005**, *441*, 235–248. [CrossRef]
62. Hubert, A.M. Variability in the Circumstellar Envelope of Be Stars. In *Proceedings of the Pulsation; Rotation; and Mass Loss in Early-Type Stars*; Balona, L.A., Henrichs, H.F., Le Contel, J.M., Eds.; Kluwer Academic Publishers: New York, NY, USA, 1994; Volume 162, p. 341.

63. Hubert, A.M.; Floquet, M.; Gomez, A.E.; Aletti, V. Photometric Variability of B and Be Stars. In Proceedings of the ESA Symposium 'Hipparcos-Venice 97', Venice, Italy, 13–16 May 1997; Volume 402, pp. 315–318.
64. Moujtahid, A.; Hubert, A.M.; Zorec, J.; Ballereau, D.; Chauville, J.; Floquet, M.; Mon, M. Correlated Spectroscopic and Spectrophotometric Behaviours of Be Stars. In Proceedings of the IAU Colloq. 175: The Be Phenomenon in Early-Type Stars, Alicantem, Spain, 28 June–2 July 1999; The University of Chicago Press: Chicago, NY, USA, 2000; Volume 214, p. 514.
65. Moujtahid, A.; Zorec, J.; Hubert, A.M. Long-term visual spectrophotometric behaviour of Be stars. II. Correlations with fundamental stellar parameters and interpretation. *Astron. Astrophys.* **1999**, *349*, 151–168.
66. Aubourg, E.; Bareyre, P.; Brehin, S.; Gros, M.; de Kat, J.; Lachieze-Rey, M.; Laurent, B.; Lesquoy, E.; Magneville, C.; Milsztajn, A.; et al. Search for very low-mass objects in the Galactic Halo. *Astron. Astrophys.* **1995**, *301*, 1.
67. de Wit, W.J.; Lamers, H.J.G.L.M.; Marquette, J.B.; Beaulieu, J.P. The remarkable light and colour variability of Small Magellanic Cloud Be stars. *Astron. Astrophys.* **2006**, *456*, 1027–1035. [CrossRef]
68. Hubert, A.M.; Floquet, M. Investigation of the variability of bright Be stars using HIPPARCOS photometry. *Astron. Astrophys.* **1998**, *335*, 565–572.
69. Mennickent, R.E.; Pietrzyński, G.; Gieren, W.; Szewczyk, O. On Be star candidates and possible blue pre-main sequence objects in the Small Magellanic Cloud. *Astron. Astrophys.* **2002**, *393*, 887–896. [CrossRef]
70. Alcock, C.; Akerlof, C.W.; Allsman, R.A.; Axelrod, T.S.; Bennett, D.P.; Chan, S.; Cook, K.H.; Freeman, K.C.; Griest, K.; Marshall, S.L.; et al. Possible gravitational microlensing of a star in the Large Magellanic Cloud. *Nature* **1993**, *365*, 621–623.
71. Keller, S.C.; Bessell, M.S.; Cook, K.H.; Geha, M.; Syphers, D. Blue Variable Stars from the MACHO Database. I. Photometry and Spectroscopy of the Large Magellanic Cloud Sample. *Astron. J.* **2002**, *124*, 2039–2044.
72. Udalski, A.; Kubiak, M.; Szymanski, M. Optical Gravitational Lensing Experiment. OGLE-2—The Second Phase of the OGLE Project. *Acta Astron.* **1997**, *47*, 319–344.
73. Limber, D.N.; Marlborough, J.M. The Support of the Envelopes of be Stars. *Astrophys. J.* **1968**, *152*, 181. [CrossRef]
74. Limber, D.N. Circumstellar Envelopes Formed Through Rotationally Forced Ejection. II. *Astrophys. J.* **1967**, *148*, 141. [CrossRef]
75. Marlborough, J.M. Models for the Envelopes of be Stars. *Astrophys. J.* **1969**, *156*, 135. [CrossRef]
76. Poeckert, R.; Marlborough, J.M. Intrinsic linear polarization of Be stars as a function of V sini. *Astrophys. J.* **1976**, *206*, 182–195. [CrossRef]
77. Poeckert, R.; Marlborough, J.M. Linear polarization of Halpha in the Be star Gamma Cassiopeiae. *Astrophys. J.* **1977**, *218*, 220–226. [CrossRef]
78. Poeckert, R.; Marlborough, J.M. A model for gamma Cassiopeiae. *Astrophys. J.* **1978**, *220*, 940–961. [CrossRef]
79. Poeckert, R.; Marlborough, J.M. Be star models: Observable effects of model parameters. *Astrophys. J. Suppl. Ser.* **1978**, *38*, 229–252. [CrossRef]
80. Marlborough, J.M. Models for the Circumstellar Envelopes of be Stars (review Paper). In *Proceedings of the Be and Shell Stars*; Slettebak, A., Ed.; Reidel Pub. Co.: Dordrecht, The Netherlands; Boston, MA, USA, 1976; Volume 70, p. 335.
81. Poeckert, R. Model atmospheres of Be stars. In *Proceedings of the Be Stars*; Jaschek, M., Groth, H.G., Eds.; Dordrecht, D. Reidel Publishing Co.: Dordrecht, The Netherlands; Boston, MA, USA, 1982; Volume 98, pp. 453–477.
82. Hirata, R.; Kogure, T. The Be star phenomena. II—Spectral formation and structure of envelopes. *Bull. Astron. Soc. India* **1984**, *12*, 109–151.
83. Slettebak, A.; Snow, T.P. Physics of Be stars. In Proceedings of the 92nd Colloquium of the International Astronomical Union, Boulder, CO, USA, 18–22 August 1986; University Press: Cambridge, UK, 1987.
84. Waters, L.B.F.M.; Marlborough, J.M. The Structure of the Circumstellar Material in Be Stars. In *Proceedings of the Pulsation; Rotation; and Mass Loss in Early-Type Stars*; Balona, L.A., Henrichs, H.F., Le Contel, J.M., Eds.; Kluwer Academic Publishers: New York, NY, USA, 1994; Volume 162, p. 399.
85. Doazan, V.; Stalio, R.; Thomas, R.N. Empirical atmospheric velocity patterns from combined IUE and visual observations: The Be-similar stars. In Proceedings of the Four Years of IUE Research, Conference Held in NASA, Goddard Space Flight Center, Greenbelt, MD, USA, 30 March–1 April 1982; NASA: Washington, DC, USA, 1982; Volume 2238, pp. 584–588.
86. Underhill, A.B.; Doazan, V.; Lesh, J.R.; Aizenman, M.L.; Thomas, R.N. *B Stars with and without Emission Lines. Monograph Series on Nonthermal Phenomena in Stellar Atmospheres*; NASA: Washington, DC, USA, 1982; Volume 456.
87. Thomas, R.N. *Stellar Atmospheric Structural Patterns*; NASA: Washington, DC, USA, 1983; Volume 471.
88. Catala, C.; Kunasz, P.B.; Praderie, F. Line formation in the wind of AB Aur. *Astron. Astrophys.* **1984**, *134*, 402–413.
89. Catala, C. Line formation in the winds of Herbig Ae/Be stars. The C IV resonancelines. *Astron. Astrophys.* **1988**, *193*, 222–228.
90. Cidale, L.S.; Ringuelet, A.E. Rigorous Treatment of the Radiative Transfer Problem in Stellar Winds: Significance of the Velocity Law and the Chromosphere in the H alpha Profile. *Astrophys. J.* **1993**, *411*, 874. [CrossRef]
91. Vazquez, A.C.; Cidale, L.S.; Ringuelet, A.E. Expanding Atmospheric Model Including a Chromosphere. I. Study of the Infrared Excess in Be Stars. *Astrophys. J.* **1993**, *419*, 286. [CrossRef]
92. Cidale, L.S.; Vazquez, A.C. Expanding Atmospheric Model Including a Chromosphere. II. Be Stars: Center-to-Limb Variation of the Emergent Intensity. *Astrophys. J.* **1995**, *453*, 393. [CrossRef]
93. Boehm, T.; Catala, C.; Donati, J.F.; Welty, A.; Baudrand, J.; Butler, C.J.; Carter, B.; Collier-Cameron, A.; Czarny, J.; Foing, B.; et al. Azimuthal structures in the wind and chromosphere of the Herbig AE star AB Aurigae. Results from the MUSICOS 1992 campaign. *Astron. Astrophys.* **1996**, *120*, 431–450. [CrossRef]

94. Cidale, L.S. Diagnosis of Stellar Winds and Temperature Structures in Be Stars through the Analysis of Mg II Lines. *Astrophys. J.* **1998**, *502*, 824–832. [CrossRef]
95. Bouret, J.C.; Catala, C. NLTE calculations of neutral helium lines in the wind of the Herbig Ae star AB Aurigae. *Astron. Astrophys.* **2000**, *359*, 1011–1024.
96. Zorec, J.; Arias, M.L.; Cidale, L.; Ringuelet, A.E. Be star disc characteristics near the central object. *Astron. Astrophys.* **2007**, *470*, 239–247. [CrossRef]
97. Hanuschik, R.W. Stellar V/sin/i and Optical Emission Linewidths in Be-Stars. *Astrophys. Space Sci.* **1989**, *161*, 61–73. [CrossRef]
98. Ballereau, D.; Chauville, J.; Zorec, J. Some Fe II emission-line profiles of nine southern Be stars. *Astron. Astrophys.* **1995**, *111*, 457.
99. Lee, U.; Osaki, Y.; Saio, H. Viscous excretion discs around Be stars. *Mon. Not. R. Astron. Soc.* **1991**, *250*, 432–437. [CrossRef]
100. Shakura, N.I.; Sunyaev, R.A. Black holes in binary systems. Observational appearance. *Astron. Astrophys.* **1973**, *24*, 337–355.
101. Pringle, J.E. Accretion discs in astrophysics. *Annu. Rev. Astron. Astrophys.* **1981**, *19*, 137–162. [CrossRef]
102. Pringle, J.E. The properties of external accretion discs. *Mon. Not. R. Astron. Soc.* **1991**, *248*, 754. [CrossRef]
103. Narita, S.; Kiguchi, M.; Hayashi, C. The Structure and Evolution of Thin Viscous Disks. I. Non-steady Accretion and Excretion. *Publ. Astron. Soc. Jpn.* **1994**, *46*, 575–587.
104. Okazaki, A.T. Viscous Transonic Outflow in Equatorial Discs of Be Stars. *Commun. Konkoly Obs. Hung.* **1997**, *100*, 407–412.
105. Okazaki, A.T. Viscous Transonic Decretion in Disks of Be Stars. *Publ. Astron. Soc. Jpn.* **2001**, *53*, 119–125.
106. Haubois, X.; Carciofi, A.C.; Rivinius, T.; Okazaki, A.T.; Bjorkman, J.E. Dynamical Evolution of Viscous Disks around Be Stars. I. Photometry. *Astrophys. J.* **2012**, *756*, 156.
107. Ghoreyshi, M.R.; Carciofi, A.C.; Jones, C.E.; Faes, D.M.; Baade, D.; Rivinius, T. A Multi-Observing Technique Study of the Dynamical Evolution of the Viscous Disk around the Be Star ω CMa. *Astrophys. J.* **2021**, *909*, 149.
108. Kurfürst, P.; Feldmeier, A.; Krtička, J. Two-dimensional modeling of density and thermal structure of dense circumstellar outflowing disks. *Astron. Astrophys.* **2018**, *613*, A75.
109. Carciofi, A.C.; Bjorkman, J.E. Non-LTE Monte Carlo Radiative Transfer. II. Nonisothermal Solutions for Viscous Keplerian Disks. *Astrophys. J.* **2008**, *684*, 1374–1383.
110. Millar, C.E.; Marlborough, J.M. Rates of Energy Gain and Loss in the Circumstellar Envelopes of Be Stars: The Poeckert-Marlborough Model. *Astrophys. J.* **1998**, *494*, 715–723. [CrossRef]
111. Millar, C.E.; Marlborough, J.M. Rates of Energy Gain and Loss in the Circumstellar Envelopes of BE Stars: Diffuse Radiation. *Astrophys. J.* **1999**, *516*, 276–279. [CrossRef]
112. Carciofi, A.C.; Bjorkman, J.E. Non-LTE Monte Carlo Radiative Transfer. I. The Thermal Properties of Keplerian Disks around Classical Be Stars. *Astrophys. J.* **2006**, *639*, 1081–1094.
113. Sigut, T.A.A.; Jones, C.E. The Thermal Structure of the Circumstellar Disk Surrounding the Classical Be Star γ Cassiopeiae. *Astrophys. J.* **2007**, *668*, 481–491.
114. McGill, M.A.; Sigut, T.A.A.; Jones, C.E. The Effect of Density on the Thermal Structure of Gravitationally Darkened Be Star Disks. *Astrophys. J. Suppl. Ser.* **2013**, *204*, 2.
115. Thomas, R.N. The Source Function in a Non-Equilibrium Atmosphere. I. The Resonance Lines. *Astrophys. J.* **1957**, *125*, 260. [CrossRef]
116. Thomas, R.N. *Some Aspects of Non-Equilibrium Thermodynamics in the Presence of a Radiation Field*; University of Colorado Press: Boulder, CO, USA, 1965.
117. Jefferies, J.T. *Spectral Line Formation*; Blaisdell: Waltham, MA, USA, 1968.
118. Mihalas, D. *Stellar Atmospheres*; W.H. Freeman: San Francisco, CA, USA, 1978.
119. Hubeny, I.; Mihalas, D. *Theory of Stellar Atmospheres*; Princeton University Press: Princeton, NJ, USA, 2014.
120. Viotti, R. Forbidden and permitted emission lines of singly ionized iron as a diagnostic in the investigation of stellar emission-line spectra. *Astrophys. J.* **1976**, *204*, 293–300. [CrossRef]
121. Curé, M. The Influence of Rotation in Radiation-driven Wind from Hot Stars: New Solutions and Disk Formation in Be Stars. *Astrophys. J.* **2004**, *614*, 929–941.
122. Hartquist, T.W.; Dyson, J.E.; Pettini, M.; Smith, L.J. Mass-loaded astronomical flows—I. General principles and their application to RCW 58. *Mon. Not. R. Astron. Soc.* **1986**, *221*, 715–726. [CrossRef]
123. Dyson, J.E.; Hartquist, T.W. Astronomical Bubbles with Clumpy Cores and Accelerating Haloes. *Astrophys. Lett. Commun.* **1992**, *28*, 301.
124. Arthur, S.J.; Dyson, J.E.; Hartquist, T.W. Mass-loaded flows—VII. Transonic flows from the cores of planetary nebulae. *Mon. Not. R. Astron. Soc.* **1994**, *269*, 1117–1122. [CrossRef]
125. Kroll, P.; Hanuschik, R.W. Dynamics of Self-Accreting Disks in Be Stars. In Proceedings of the IAU Colloq. 163: Accretion Phenomena and Related Outflows, Port Douglas, Queensland, Australia, 15–19 July 1996; Astronomical Society of the Pacific: San Francisco, CA, USA, 1997; Volume 121, p. 494.
126. Kee, N.D.; Owocki, S.; Townsend, R.; Müller, H.R. Pulsational Mass Ejection in Be Star Disks. In *Proceedings of the Bright Emissaries: Be Stars as Messengers of Star-Disk Physics*; Sigut, T.A.A., Jones, C.E., Eds.; Astronomical Society of the Pacific: San Francisco, CA, USA, 2016; Volume 506, p. 47.

127. Zorec, J.; Frémat, Y.; Domiciano de Souza, A.; Royer, F.; Cidale, L.; Hubert, A.M.; Semaan, T.; Martayan, C.; Cochetti, Y.R.; Arias, M.L.; et al. Critical study of the distribution of rotational velocities of Be stars. I. Deconvolution methods, effects due to gravity darkening, macroturbulence, and binarity. *Astron. Astrophys.* **2016**, *595*, A132. [CrossRef]
128. Krtička, J.; Owocki, S.P.; Meynet, G. Mass and angular momentum loss via decretion disks. *Astron. Astrophys.* **2011**, *527*, A84.
129. Granada, A.; Ekström, S.; Georgy, C.; Krtička, J.; Owocki, S.; Meynet, G.; Maeder, A. Populations of rotating stars. II. Rapid rotators and their link to Be-type stars. *Astron. Astrophys.* **2013**, *553*, A25.
130. Wade, G.A.; Neiner, C.; Alecian, E.; Grunhut, J.H.; Petit, V.; Batz, B.d.; Bohlender, D.A.; Cohen, D.H.; Henrichs, H.F.; Kochukhov, O.; et al. The MiMeS survey of magnetism in massive stars: Introduction and overview. *Mon. Not. R. Astron. Soc.* **2016**, *456*, 2–22.
131. Wade, G.A.; Petit, V.; Grunhut, J.H.; Neiner, C.; MiMeS Collaboration. Magnetic Fields of Be Stars: Preliminary Results from a Hybrid Analysis of the MiMeS Sample. In *Proceedings of the Bright Emissaries: Be Stars as Messengers of Star-Disk Physics, Astronomical Society of the Pacific: Proceedings of a Meeting held at The University of Western Ontario, London, ON, Canada, 11–13 August 2014*; Astronomical Society of the Pacific Conference Series; Sigut, T.A.A., Jones, C.E., Eds.; 2016; Volume 506, p. 207.
132. Kochukhov, O.; Sudnik, N. Detectability of small-scale magnetic fields in early-type stars. *Astron. Astrophys.* **2013**, *554*, A93.
133. Cantiello, M.; Langer, N.; Brott, I.; de Koter, A.; Shore, S.N.; Vink, J.S.; Voegler, A.; Lennon, D.J.; Yoon, S.C. Sub-surface convection zones in hot massive stars and their observable consequences. *Astron. Astrophys.* **2009**, *499*, 279–290.
134. Cantiello, M.; Braithwaite, J. Magnetic spots on hot massive stars. *Astron. Astrophys.* **2011**, *534*, A140.
135. Cantiello, M.; Braithwaite, J. Envelope Convection, Surface Magnetism, and Spots in A and Late B-type Stars. *Astrophys. J.* **2019**, *883*, 106.
136. Maeder, A.; Georgy, C.; Meynet, G. Convective envelopes in rotating OB stars. *Astron. Astrophys.* **2008**, *479*, L37–L40. [CrossRef]
137. Clement, M.J. On the equilibrium and secular instability of rapidly rotating stars. *Astrophys. J.* **1979**, *230*, 230–242. [CrossRef]
138. Smith, M.A.; Lopes de Oliveira, R.; Motch, C. A Census of the Class of X-ray Active γ Cas Stars. In *Proceedings of the Bright Emissaries: Be Stars as Messengers of Star-Disk Physics, Astronomical Society of the Pacific: Proceedings of a Meeting held at The University of Western Ontario, London, ON, Canada, 11–13 August 2014*; Astronomical Society of the Pacific Conference Series; Sigut, T.A.A., Jones, C.E., Eds.; 2016; Volume 506, p. 299.
139. Smith, M.A.; Lopes de Oliveira, R.; Motch, C. The X-ray emission of the γ Cassiopeiae stars. *Adv. Space Res.* **2016**, *58*, 782–808.
140. Owocki, S.P.; Castor, J.I.; Rybicki, G.B. Time-dependent Models of Radiatively Driven Stellar Winds. I. Nonlinear Evolution of Instabilities for a Pure Absorption Model. *Astrophys. J.* **1988**, *335*, 914. [CrossRef]
141. Zorec, J.; Frémat, Y.; Domiciano de Souza, A.; Delaa, O.; Stee, P.; Mourard, D.; Cidale, L.; Martayan, C.; Georgy, C.; Ekström, S. Differential rotation in rapidly rotating early-type stars. I. Motivations for combined spectroscopic and interferometric studies. *Astron. Astrophys.* **2011**, *526*, A87.
142. Apparao, K.M.V.; Antia, H.M.; Chitre, S.M. Rapidly rotating stars and the Be star phenomenon. *Astron. Astrophys.* **1987**, *177*, 198–200.
143. Aidelman, Y.; Cidale, L.S.; Zorec, J.; Panei, J.A. Open clusters. II. Fundamental parameters of B stars in Collinder 223, Hogg 16, NGC 2645, NGC 3114, and NGC 6025. *Astron. Astrophys.* **2015**, *577*, A45. [CrossRef]
144. Aidelman, Y.; Cidale, L.S.; Zorec, J.; Panei, J.A. Open clusters. III. Fundamental parameters of B stars in NGC 6087, NGC 6250, NGC 6383, and NGC 6530 B-type stars with circumstellar envelopes. *Astron. Astrophys.* **2018**, *610*, A30.
145. Ekström, S.; Georgy, C.; Eggenberger, P.; Meynet, G.; Mowlavi, N.; Wyttenbach, A.; Granada, A.; Decressin, T.; Hirschi, R.; Frischknecht, U.; et al. Grids of stellar models with rotation. I. Models from 0.8 to 120 M$_\odot$ at solar metallicity (Z = 0.014). *Astron. Astrophys.* **2012**, *537*, A146.
146. Cochetti, Y.R.; Zorec, J.; Cidale, L.S.; Arias, M.L.; Aidelman, Y.; Torres, A.F.; Frémat, Y.; Granada, A. Be and Bn stars: Balmer discontinuity and stellar-class relationship. *Astron. Astrophys.* **2020**, *634*, A18.
147. Hoffleit, D.; Jaschek, C. *The Bright Star Catalogue. Fourth Revised Edition. (Containing Data Compiled through 1979)*; Yale University Observatory: New Haven, CT, USA, 1982.
148. Hoffleit, D.; Saladyga, M.; Wlasuk, P. *A Supplement to the Bright Star Catalogue. Containing Data Compiled through 1981 for Stars 7.10 V and Brighter That Are Not in the Bright Star Catalogue*; Yale University Observatory: New Haven, CT, USA, 1983.
149. Zorec, J. On the Initial Mass Function of Be Stars and the Missing Be Stars of Late Spectral Types. In Proceedings of the IAU Colloq. 175: The Be Phenomenon in Early-Type Stars, Alicante, Spain, 28 June–2 July 1999; The University of Chicago Press: Chicago, NY, USA, 2000; Volume 214, p. 51.
150. Scalo, J.M. The Stellar Initial Mass Function. *Fund. Cosmic Phys.* **1986**, *11*, 1–278.
151. Rana, N.C. Mass function of stars in the solar neighbourhood. *Astron. Astrophys.* **1987**, *184*, 104–118.
152. Zorec, J.; Briot, D. Critical study of the frequency of Be stars taking into account their outstanding characteristics. *Astron. Astrophys.* **1997**, *318*, 443–460.
153. Zorec, J. Rotation and Properties of Be Stars (Invited Review). In *Proceedings of the Stellar Rotation*; Maeder, A., Eenens, P., Eds.; Cambridge University Press: Cambridge, UK, 2004; Volume 215, p. 73.
154. Maeder, A.; Meynet, G. The Evolution of Rotating Stars. *Annu. Rev. Astron. Astrophys.* **2000**, *38*, 143–190.
155. Cote, J.; van Kerkwijk, M.H. New bright Be stars and the Be star frequency. *Astron. Astrophys.* **1993**, *274*, 870–876.
156. Ghosh, K.K.; Apparao, K.M.V.; Pukalenthi, S. Observations of BN and AN stars: New Be stars. *Astron. Astrophys.* **1999**, *134*, 359–364. [CrossRef]

157. De Marco, O.; Lanz, T.; Ouellette, J.A.; Zurek, D.; Shara, M.M. First Evidence of Circumstellar Disks around Blue Straggler Stars. *Astrophys. J. Lett.* **2004**, *606*, L151–L154.
158. Porter, J.M.; Townsend, R.H.D. On the Evidence of Disks around Blue Straggler Stars. *Astrophys. J. Lett.* **2005**, *623*, L129–L132.
159. Jaschek, C.; Jaschek, M. *The Classification of Stars*; Cambridge University Press: Cambridge, UK, 1987.
160. Osmer, P.S.; Peterson, D.M. The composition and evolutionary status of the helium-rich stars. *Astrophys. J.* **1974**, *187*, 117–129. [CrossRef]
161. Cidale, L.S.; Arias, M.L.; Torres, A.F.; Zorec, J.; Frémat, Y.; Cruzado, A. Fundamental parameters of He-weak and He-strong stars. *Astron. Astrophys.* **2007**, *468*, 263–272. [CrossRef]
162. Hubeny, I.; Lanz, T. Non-LTE Line-blanketed Model Atmospheres of Hot Stars. I. Hybrid Complete Linearization/Accelerated Lambda Iteration Method. *Astrophys. J.* **1995**, *439*, 875. [CrossRef]
163. Dimitrijevic, M.S.; Sahal-Brechot, S. Stark broadening of neutral helium lines. *J. Quant. Spectrosc. Radiat. Transf.* **1984**, *31*, 301–313. [CrossRef]
164. Dimitrijevic, M.S.; Sahal-Brechot, S. Stark broadening of He I lines. *Astron. Astrophys.* **1990**, *82*, 519–529.
165. Freudenstein, S.A.; Cooper, J. A simple formula for estimating Stark widths of neutral lines. *Astrophys. J.* **1978**, *224*, 1079–1084. [CrossRef]
166. Allen, D.A.; Swings, J.P. Infrared Excesses and Forbidden Emission Lines in Early-Type Stars. *Astrophys. Lett.* **1972**, *10*, 83.
167. Allen, D.A.; Swings, J.P. The spectra of peculiar Be star with infrared excesses. *Astron. Astrophys.* **1976**, *47*, 293–302.
168. Zickgraf, F.J. Current Definition of B[e] Stars. In *Proceedings of the B[e] Stars*; Hubert, A.M., Jaschek, C., Eds.; Kluwer Academic Publishers: Dordrecht, The Netherlands; Boston, MA, USA, 1998; Volume 233, p. 1. [CrossRef]
169. Lamers, H.J.G.L.M.; Zickgraf, F.J.; de Winter, D.; Houziaux, L.; Zorec, J. An improved classification of B[e]-type stars. *Astron. Astrophys.* **1998**, *340*, 117–128.
170. Zickgraf, F.J.; Wolf, B.; Stahl, O.; Leitherer, C.; Klare, G. The hybrid spectrum of the LMC hypergiant R 126. *Astron. Astrophys.* **1985**, *143*, 421–430.
171. Zickgraf, F.J.; Wolf, B.; Stahl, O.; Leitherer, C.; Appenzeller, I. B(e)-supergiants of the Magellanic Clouds. *Astron. Astrophys.* **1986**, *163*, 119–134.
172. Zickgraf, F.J.; Wolf, B.; Stahl, O.; Humphreys, R.M. S 18: A new B(e) supergiant in the Small Magellanic Cloud with evidence for an excretion disk. *Astron. Astrophys.* **1989**, *220*, 206–214.
173. Zickgraf, F.J.; Stahl, O.; Wolf, B. IR survey of OB emission-line stars in the SMC: Detection of a new B E supergiant, AV 172. *Astron. Astrophys.* **1992**, *260*, 205–212.
174. Zickgraf, F.J.; Kovacs, J.; Wolf, B.; Stahl, O.; Kaufer, A.; Appenzeller, I. R4 in the Small Magellanic Cloud: A spectroscopic binary with a B[e]/LBV-type component. *Astron. Astrophys.* **1996**, *309*, 505–514.
175. Zickgraf, F.J.; Humphreys, R.M.; Lamers, H.J.G.L.M.; Smolinski, J.; Wolf, B.; Stahl, O. Spectroscopic study of the outflowing disk winds of B[e] supergiants in the Magellanic Clouds. *Astron. Astrophys.* **1996**, *315*, 510–520.
176. Gummersbach, C.A.; Zickgraf, F.J.; Wolf, B. B[e] phenomenon extending to lower luminosities in the Magellanic Clouds. *Astron. Astrophys.* **1995**, *302*, 409.
177. Wolf, B.; Stahl, O. The absorption spectrum of the Be star MWC 300. *Astron. Astrophys.* **1985**, *148*, 412–416.
178. McGregor, P.J.; Hyland, A.R.; Hillier, D.J. Atomic and Molecular Line Emission from Early-Type High-Luminosity Stars. *Astrophys. J.* **1988**, *324*, 1071. [CrossRef]
179. Winkler, H.; Wolf, B. An analysis of high resolution spectra of the Be -stars CPD -52 9243and MWC 300. *Astron. Astrophys.* **1989**, *219*, 151–157.
180. Kraus, M. A Census of B[e] Supergiants. *Galaxies* **2019**, *7*, 83,
181. Palla, F.; Stahler, S.W. The Pre-Main-Sequence Evolution of Intermediate-Mass Stars. *Astrophys. J.* **1993**, *418*, 414. [CrossRef]
182. Bibo, E.A.; The, P.S. The type of variability of Herbig Ae/Be stars. *Astron. Astrophys.* **1991**, *89*, 319.
183. Ciatti, F.; D'Odorico, S.; Mammano, A. Properties and evolution of BQ[] stars. *Astron. Astrophys.* **1974**, *34*, 181–186.
184. Kenyon, S.J. *The Symbiotic Stars*; Cambridge University Press: Cambridge, UK, 1986.
185. Miroshnichenko, A.S. Toward Understanding the B[e] Phenomenon. I. Definition of the Galactic FS CMa Stars. *Astrophys. J.* **2007**, *667*, 497–504. [CrossRef]
186. Miroshnichenko, A.S.; Manset, N.; Kusakin, A.V.; Chentsov, E.L.; Klochkova, V.G.; Zharikov, S.V.; Gray, R.O.; Grankin, K.N.; Gandet, T.L.; Bjorkman, K.S.; et al. Toward Understanding the B[e] Phenomenon. II. New Galactic FS CMa Stars. *Astrophys. J.* **2007**, *671*, 828–841. [CrossRef]
187. Kraus, M. The pre- versus post-main sequence evolutionary phase of B[e] stars. Constraints from ^{13}CO band emission. *Astron. Astrophys.* **2009**, *494*, 253–262. [CrossRef]
188. Kraus, M.; Liimets, T.; Moiseev, A.; Sánchez Arias, J.P.; Nickeler, D.H.; Cidale, L.S.; Jones, D. Resolving the Circumstellar Environment of the Galactic B[e] Supergiant Star MWC 137.II. Nebular Kinematics and Stellar Variability. *Astron. J.* **2021**, *162*, 150.
189. Barsukova, E.A.; Burenkov, A.N.; Goranskij, V.P. Sudden strengthening of He II emission line in the spectrum of B[e] star CI Cam. *Astron. Telegr.* **2021**, *14362*, 1.
190. Smith, D.; Remillard, R.; Swank, J.; Takeshima, T.; Smith, E. XTE J0421+560. *IAU Circ.* **1998**, *6855*, 1.

191. Korčáková, D.; Sestito, F.; Manset, N.; Kroupa, P.; Votruba, V.; Šlechta, M.; Danford, S.; Dvořáková, N.; Raj, A.; Chojnowski, S.D.; et al. First detection of a magnetic field in low-luminosity B[e] stars. New scenarios for the nature and evolutionary stages of FS CMa stars. *Astron. Astrophys.* **2022**, *659*, A35.
192. Aidelman, Y.; Cidale, L.S.; Kraus, M. Aias, M.L.; Zorec, J. Fundamental parameters of B[e] stars. *Astron. Astrophys.* **2022**, *submitted*.
193. Arias, M.L.; Cidale, L.S.; Kraus, M.; Torres, A.F.; Aidelman, Y.; Zorec, J.; Granada, A. Near-infrared Spectra of a Sample of Galactic Unclassified B[e] Stars. *Publ. Astron. Soc. Pac.* **2018**, *130*, 114201. [CrossRef]
194. Burnichon, M.L.; Chalonge, D.; Divan, L.; Swings, L. Etude de l'etoile Be HD 45677. *J. Obs.* **1967**, *50*, 391.
195. Merrill, P.W. Bright iron lines in the spectrum of HD 45677. *Astrophys. J.* **1928**, *67*, 405–408. [CrossRef]
196. Swings, P.; Struve, O. Spectrographic Observations of Peculiar Stars. *Astrophys. J.* **1940**, *91*, 546. [CrossRef]
197. Swings, P.; Struve, O. Spectrographic Observations of Peculiar Stars.VI. *Astrophys. J.* **1943**, *98*, 91. [CrossRef]
198. Zorec, J.; Moujtahid, A.; Ballereau, D.; Chauville, J. Fundamental Parameters of Two B[e] Stars: HD 45677 and HD 50138. In *Proceedings of the B[e] Stars*; Hubert, A.M., Jaschek, C., Eds.; Kluwer Academic Publishers: Dordrecht, The Netherlands; Boston, MA, USA, 1998; Volume 233, p. 55. [CrossRef]
199. Zorec, J. Distances, Kinematics and Distribution of B[e] Stars in Our Galaxy. In *Proceedings of the B[e] Stars*; Hubert, A.M., Jaschek, C., Eds.; Kluwer Academic Publishers: Dordrecht, The Netherlands; Boston, MA, USA, 1998; Volume 233, p. 27. [CrossRef]
200. Moujtahid, A.; Zorec, J.; Hubert, A.M. Ultraviolet Extinction of Circumstellar Dust in the B[e] Star HD 45677. In *Proceedings of the Ultraviolet Astrophysics Beyond the IUE Final Archive*; Wamsteker, W., Gonzalez Riestra, R., Harris, B., Eds.; ESA Publications Division: Noordwijk, The Netherlands, 1998; Volume 413, p. 261.
201. Grinin, V.P.; Kiselev, N.N.; Minikulov, N.K. Observations of "zodiacal light" of the isolated Herbig Ae-star BF Ori. *Pisma v Astronomicheskii Zhurnal* **1989**, *15*, 1028–1038.
202. Grinin, V.P.; Kiselev, N.N.; Minikulov, N.K.; Chernova, G.P.; Voshchinnikov, N.V. The investigations of 'zodiacal light' of isolated AE-Herbig stars with non-periodic Algol-type minima. *Astrophys. Space Sci.* **1991**, *186*, 283–298. [CrossRef]
203. Jeřábková, T.; Korčáková, D.; Miroshnichenko, A.; Danford, S.; Zharikov, S.V.; Kříček, R.; Zasche, P.; Votruba, V.; Šlechta, M.; Škoda, P.; et al. Time-dependent spectral-feature variations of stars displaying the B[e] phenomenon. III. HD 50138. *Astron. Astrophys.* **2016**, *586*, A116. [CrossRef]
204. Kluska, J.; Benisty, M.; Soulez, F.; Berger, J.P.; Le Bouquin, J.B.; Malbet, F.; Lazareff, B.; Thiébaut, E. A disk asymmetry in motion around the B[e] star MWC158. *Astron. Astrophys.* **2016**, *591*, A82. [CrossRef]
205. Miroshnichenko, A.S.; Zharikov, S.V.; Danford, S.; Manset, N.; Korčáková, D.; Kříček, R.; Šlechta, M.; Omarov, C.T.; Kusakin, A.V.; Kuratov, K.S.; et al. Toward Understanding the B[e] Phenomenon. V. Nature and Spectral Variations of the MWC 728 Binary System. *Astrophys. J.* **2015**, *809*, 129. [CrossRef]
206. Miroshnichenko, A.S.; Bernabei, S.; Polcaro, V.F.; Viotti, R.F.; Norci, L.; Manset, N.; Klochkova, V.G.; Rudy, R.J.; Lynch, D.K.; Venturini, C.C.; et al. Optical and Near-IR Observations of the B[e] Star AS 119. In *Proceedings of the Stars with the B[e] Phenomenon*; Kraus, M., Miroshnichenko, A.S., Eds.; University of Chicago Press: Vlieland, The Netherlands, 2006; Volume 355, p. 347.
207. Polster, J.; Korčáková, D.; Manset, N. Time-dependent spectral-feature variations of stars displaying the B[e] phenomenon. IV. V2028 Cygni: Modelling of Hα bisector variability. *Astron. Astrophys.* **2018**, *617*, A79. [CrossRef]
208. Zickgraf, F.J.; Schulte-Ladbeck, R.E. Polarization characteristics of galactic Be stars. *Astron. Astrophys.* **1989**, *214*, 274–284.
209. Condori, C.A.H.; Borges Fernandes, M.; Kraus, M.; Panoglou, D.; Guerrero, C.A. The study of unclassified B[e] stars and candidates in the Galaxy and Magellanic Clouds†. *Mon. Not. R. Astron. Soc.* **2019**, *488*, 1090–1110.
210. Langer, N.; Kudritzki, R.P. The spectroscopic Hertzsprung-Russell diagram. *Astron. Astrophys.* **2014**, *564*, A52.
211. Chentsov, E.L.; Klochkova, V.G.; Miroshnichenko, A.S. Spectral variability of the peculiar A-type supergiant 3Pup. *Astrophys. Bull.* **2010**, *65*, 150–163.
212. Miroshnichenko, A.S.; Danford, S.; Zharikov, S.V.; Klochkova, V.G.; Chentsov, E.L.; Vanbeveren, D.; Zakhozhay, O.V.; Manset, N.; Pogodin, M.A.; Omarov, C.T.; et al. Properties of Galactic B[e] Supergiants. V. 3 Pup-Constraining the Orbital Parameters and Modeling the Circumstellar Environments. *Astrophys. J.* **2020**, *897*, 48.
213. Langer, N.; Heger, A. B[e] Supergiants: What is Their Evolutionary Status? In *Proceedings of the B[e] Stars*; Hubert, A.M., Jaschek, C., Eds.; Kluwer Academic Publishers: Dordrecht, The Netherlands; Boston, MA, USA, 1998; Volume 233, p. 235.
214. Podsiadlowski, P.; Morris, T.S.; Ivanova, N. Massive Binary Mergers: A Unique Scenario for the sgB[e] Phenomenon? In *Proceedings of the Stars with the B[e] Phenomenon*; Kraus, M., Miroshnichenko, A.S., Eds.; University of Chicago Press: Vlieland, The Netherlands, 2006; Volume 355, p. 259.
215. de Mink, S.E.; Langer, N.; Izzard, R.G.; Sana, H.; de Koter, A. The Rotation Rates of Massive Stars: The Role of Binary Interaction through Tides, Mass Transfer, and Mergers. *Astrophys. J.* **2013**, *764*, 166.
216. de Mink, S.E.; Sana, H.; Langer, N.; Izzard, R.G.; Schneider, F.R.N. The Incidence of Stellar Mergers and Mass Gainers among Massive Stars. *Astrophys. J.* **2014**, *782*, 7.
217. Zickgraf, F.J. Discussion Session: Is there an Evolutionary Link between B[e] Supergiants and LBVs? In *Proceedings of the Stars with the B[e] Phenomenon*; Kraus, M., Miroshnichenko, A.S., Eds.; University of Chicago Press: Vlieland, The Netherlands, 2006; Volume 355, p. 211.
218. Meynet, G.; Maeder, A. Single Massive Stars at the Critical Rotational Velocity: Possible Links with Be and B[e] Stars. In *Proceedings of the Stars with the B[e] Phenomenon*; Kraus, M., Miroshnichenko, A.S., Eds.; University of Chicago Press: Vlieland, The Netherlands, 2006; Volume 355, p. 27.

219. Georgy, C.; Saio, H.; Ekström, S.; Meynet, G. The Advanced Stages of Stellar Evolution: Impact of Mass Loss, Rotation, and Link With B[e] Stars. In *Proceedings of the The B[e] Phenomenon: Forty Years of Studies*; Miroshnichenko, A., Zharikov, S., Korčáková, D., Wolf, M., Eds.; Insititut of Physics Publishing: Prague, Czech Republic, 2017; Volume 508, p. 99.
220. Cox, J.P.; Giuli, R.T. *Principles of Stellar Structure*; Gordon and Breach: New York, NY, USA, 1968.
221. Sweet, P.A. The importance of rotation in stellar evolution. *Mon. Not. R. Astron. Soc.* **1950**, *110*, 548. [CrossRef]
222. Zahn, J.P. Instability and Mixing Processes in Upper Main Sequence Stars. In *Proceedings of the Saas-Fee Advanced Course 13: Astrophysical Processes in Upper Main Sequence Stars*; Cox, A.N., Vauclair, S., Zahn, J.P., Eds.; Geneva Observatory: Sauverny, France, 1983; p. 253.
223. Zahn, J.P. Circulation and turbulence in rotating stars. *Astron. Astrophys.* **1992**, *265*, 115–132.
224. Ferraro, V.C.A. The non-uniform rotation of the Sun and its magnetic field. *Mon. Not. R. Astron. Soc.* **1937**, *97*, 458. [CrossRef]
225. Ferraro, V.C.A.; Plumpton, C. An Introduction to Magneto-Fluid Mechanics. *J. Plasma Phys.* **1967**, *1*, 499. [CrossRef]
226. Spruit, H.C. Differential rotation and magnetic fields in stellar interiors. *Astron. Astrophys.* **1999**, *349*, 189–202.
227. Spruit, H.C. Dynamo action by differential rotation in a stably stratified stellar interior. *Astron. Astrophys.* **2002**, *381*, 923–932. [CrossRef]
228. Spruit, H.C. Angular Momentum Transport and Mixing by Magnetic Fields (Invited Review). In *Proceedings of the Stellar Rotation*; Maeder, A., Eenens, P., Eds.; 2004; Volume 215, p. 356.
229. Maeder, A. *Physics, Formation and Evolution of Rotating Stars*; Springer: Berlin/Heidelberg, Germany, 2009. [CrossRef]
230. Fujimoto, M.Y. Dynamical stability of differentially rotating bodies to non-axisymmetric perturbations. *Astron. Astrophys.* **1987**, *176*, 53–58.
231. Tassoul, J.L. *Theory of Rotating Stars*; University Press: Princeton, NJ, USA, 1978.
232. Chandrasekhar, S. *Ellipsoidal Figures of Equilibrium*; New Haven Yale University Press: New Haven, CT, USA, 1969.
233. Kippenhahn, R.; Meyer-Hofmeister, E.; Thomas, H.C. Rotation in Evolving Stars. *Astron. Astrophys.* **1970**, *5*, 155.
234. Mark, J.W.K. Rapidly Rotating Stars. III. Massive Main-Sequence Stars. *Astrophys. J.* **1968**, *154*, 627. [CrossRef]
235. Bodenheimer, P. Rapidly Rotating Stars. VII. Effects of Angular Momentum on Upper-Main Models. *Astrophys. J.* **1971**, *167*, 153. [CrossRef]
236. Smith, R.C.; Collins, George W., I. Differential rotation and polar hollows. *Mon. Not. R. Astron. Soc.* **1992**, *257*, 340–352. [CrossRef]
237. Ostriker, J.P.; Mark, J.W.K. Rapidly rotating stars. I. The self-consistent-field method. *Astrophys. J.* **1968**, *151*, 1075–1088. [CrossRef]
238. Ostriker, J.P.; Bodenheimer, P. Rapidly Rotating Stars. II. Massive White Dwarfs. *Astrophys. J.* **1968**, *151*, 1089. [CrossRef]
239. Bodenheimer, P.; Ostriker, J.P. Rapidly Rotating Stars.VI. Pre-Main—Evolution of Massive Stars. *Astrophys. J.* **1970**, *161*, 1101. [CrossRef]
240. Jackson, S. Rapidly Rotating Stars. V. The Coupling of the Henyey and the Self-Consistent Methods. *Astrophys. J.* **1970**, *161*, 579. [CrossRef]
241. Bodenheimer, P.; Ostriker, J.P. Rapidly Rotating Stars. VIII. Zero-Viscosity Polytropic Sequences. *Astrophys. J.* **1973**, *180*, 159–170. [CrossRef]
242. Bisnovatyi-Kogan, G.S.; Blinnikov, S.I. Static Criteria for Stability of Arbitrarily Rotating Stars. *Astron. Astrophys.* **1974**, *31*, 391.
243. Blinnikov, S.I. Self-consistent field method in the theory of rotating stars. *Soviet Ast.* **1975**, *19*, 151–156.
244. Jackson, S.; MacGregor, K.B.; Skumanich, A. On the Use of the Self-consistent-Field Method in the Construction of Models for Rapidly Rotating Main-Sequence Stars. *Astrophys. J. Suppl. Ser.* **2005**, *156*, 245–264. [CrossRef]
245. Endal, A.S.; Sofia, S. The evolution of rotating stars. I. Method and exploratory calculations for a 7 M sun star. *Astrophys. J.* **1976**, *210*, 184–198. [CrossRef]
246. Eriguchi, Y.; Mueller, E. A general computational method for obtaining equilibria of self-gravitating and rotating gases. *Astron. Astrophys.* **1985**, *146*, 260–268.
247. Hachisu, I. A Versatile Method for Obtaining Structures of Rapidly Rotating Stars. *Astrophys. J. Suppl. Ser.* **1986**, *61*, 479. [CrossRef]
248. Eriguchi, Y.; Mueller, E. Structure of rapidly rotating axisymmetric stars. I—A numerical method for stellar structure and meridional circulation. *Astron. Astrophys.* **1991**, *248*, 435–447.
249. Clement, M.J. On the solution of Poisson's equation for rapidly rotating stars. *Astrophys. J.* **1974**, *194*, 709–714. [CrossRef]
250. Uryu, K.; Eriguchi, Y. Structures of Rapidly Rotating Baroclinic Stars-Part One—A Numerical Method for the Angular Velocity Distribution. *Mon. Not. R. Astron. Soc.* **1994**, *269*, 24. [CrossRef]
251. Uryu, K.; Eriguchi, Y. Structures of rapidly rotating baroclinic stars—II. an extended numerical method for realistic stellar models. *Mon. Not. R. Astron. Soc.* **1995**, *277*, 1411–1429. [CrossRef]
252. Fujisawa, K. A versatile numerical method for obtaining structures of rapidly rotating baroclinic stars: Self-consistent and systematic solutions with shellular-type rotation. *Mon. Not. R. Astron. Soc.* **2015**, *454*, 3060–3072.
253. Meynet, G.; Maeder, A. Stellar evolution with rotation. I. The computational method and the inhibiting effect of the μ-gradient. *Astron. Astrophys.* **1997**, *321*, 465–476.
254. Meynet, G.; Maeder, A. Stellar evolution with rotation. V. Changes in all the outputs of massive star models. *Astron. Astrophys.* **2000**, *361*, 101–120.
255. Heger, A.; Langer, N.; Woosley, S.E. Presupernova Evolution of Rotating Massive Stars. I. Numerical Method and Evolution of the Internal Stellar Structure. *Astrophys. J.* **2000**, *528*, 368–396.

256. Hubbard, W.B.; Slattery, W.L.; Devito, C.L. High zonal harmonics of rapidly rotating planets. *Astrophys. J.* **1975**, *199*, 504–516. [CrossRef]
257. Collins, George W., I. Continuum Emission from Rotating Non-Gray Stellar Atmospheres. II. *Astrophys. J.* **1966**, *146*, 914. [CrossRef]
258. Collins, II, G.W.; Smith, R.C. The photometric effect of rotation in the A stars. *Mon. Not. R. Astron. Soc.* **1985**, *213*, 519–552. [CrossRef]
259. Frémat, Y.; Zorec, J.; Hubert, A.M.; Floquet, M. Effects of gravitational darkening on the determination of fundamental parameters in fast-rotating B-type stars. *Astron. Astrophys.* **2005**, *440*, 305–320. [CrossRef]
260. Zorec, J.; Rieutord, M.; Espinosa Lara, F.; Frémat, Y.; Domiciano de Souza, A.; Royer, F. Gravity darkening in stars with surface differential rotation. *Astron. Astrophys.* **2017**, *606*, A32.
261. Milne, E.A. The equilibrium of a rotating star. *Mon. Not. R. Astron. Soc.* **1923**, *83*, 118–147. [CrossRef]
262. Sackmann, I.J. Rapid Uniform Rotation Along the Main Sequence II. *Astron. Astrophys.* **1970**, *8*, 76.
263. Deupree, R.G. Stellar Evolution with Arbitrary Rotation Laws. IV. Survey of Zero-Age Main-Sequence Models. *Astrophys. J.* **2001**, *552*, 268–277. [CrossRef]
264. Gillich, A.; Deupree, R.G.; Lovekin, C.C.; Short, C.I.; Toqué, N. Determination of Effective Temperatures and Luminosities for Rotating Stars. *Astrophys. J.* **2008**, *683*, 441–448. [CrossRef]
265. Higgins, E.R.; Vink, J.S. Massive star evolution: Rotation, winds, and overshooting vectors in the mass-luminosity plane. I. A calibrated grid of rotating single star models. *Astron. Astrophys.* **2019**, *622*, A50.
266. von Zeipel, H. The radiative equilibrium of a rotating system of gaseous masses. *Mon. Not. R. Astron. Soc.* **1924**, *84*, 665–683. [CrossRef]
267. Roxburgh, I.W. On stellar rotation, III. Thermally generated magnetic fields. *Mon. Not. R. Astron. Soc.* **1966**, *132*, 201. [CrossRef]
268. Lucy, L.B. Gravity-Darkening for Stars with Convective Envelopes. *Zeitschrift für Astrophysik* **1967**, *65*, 89.
269. Rieutord, M. Physical Processes Leading to Surface Inhomogeneities: The Case of Rotation. In *Lecture Notes in Physics: Cartograpy of the Sun and the Stars*; Rozelot, J.P., Neiner, C., Eds.; Springer International Publishing Switzerland: Cham, Switzerland, 2016; Volume 914, p. 101. [CrossRef]
270. Smith, R.C.; Worley, R. Gravity-darkening in stars for general rotation laws. *Mon. Not. R. Astron. Soc.* **1974**, *167*, 199–214. [CrossRef]
271. Monnier, J.D.; Zhao, M.; Pedretti, E.; Thureau, N.; Ireland, M.; Muirhead, P.; Berger, J.P.; Millan-Gabet, R.; Van Belle, G.; ten Brummelaar, T.; et al. Imaging the Surface of Altair. *Science* **2007**, *317*, 342.
272. Monnier, J.D.; Che, X.; Zhao, M.; Ekström, S.; Maestro, V.; Aufdenberg, J.; Baron, F.; Georgy, C.; Kraus, S.; McAlister, H.; et al. Resolving Vega and the Inclination Controversy with CHARA/MIRC. *Astrophys. J. Lett.* **2012**, *761*, L3.
273. Zhao, M.; Monnier, J.D.; Pedretti, E.; Thureau, N.; Mérand, A.; ten Brummelaar, T.; McAlister, H.; Ridgway, S.T.; Turner, N.; Sturmann, J.; et al. Imaging and Modeling Rapidly Rotating Stars: α Cephei and α Ophiuchi. *Astrophys. J.* **2009**, *701*, 209–224.
274. Che, X.; Monnier, J.D.; Zhao, M.; Pedretti, E.; Thureau, N.; Mérand, A.; ten Brummelaar, T.; McAlister, H.; Ridgway, S.T.; Turner, N.; et al. Colder and Hotter: Interferometric Imaging of β Cassiopeiae and α Leonis. *Astrophys. J.* **2011**, *732*, 68.
275. Domiciano de Souza, A.; Kervella, P.; Moser Faes, D.; Dalla Vedova, G.; Mérand, A.; Le Bouquin, J.B.; Espinosa Lara, F.; Rieutord, M.; Bendjoya, P.; Carciofi, A.C.; et al. The environment of the fast rotating star Achernar. III. Photospheric parameters revealed by the VLTI. *Astron. Astrophys.* **2014**, *569*, A10. [CrossRef]
276. Domiciano de Souza, A.; Bouchaud, K.; Rieutord, M.; Espinosa Lara, F.; Putigny, B. The evolved fast rotator Sargas. Stellar parameters and evolutionary status from VLTI/PIONIER and VLT/UVES. *Astron. Astrophys.* **2018**, *619*, A167.
277. Kippenhahn, R. Rotational darkening—Rotational brightening. *Astron. Astrophys.* **1977**, *58*, 267–271.
278. Maeder, A. Stellar evolution with rotation IV: Von Zeipel's theorem and anisotropic losses of mass and angular momentum. *Astron. Astrophys.* **1999**, *347*, 185–193.
279. Maunder, E.W.; Maunder, A.S.D. Sun, rotation period of the, from Greenwich sun-spot measures, 1879–1901. *Mon. Not. R. Astron. Soc.* **1905**, *65*, 813–825. [CrossRef]
280. Espinosa Lara, F.; Rieutord, M. Gravity darkening in rotating stars. *Astron. Astrophys.* **2011**, *533*, A43.
281. Maeder, A.; Peytremann, E. Stellar Rotation. *Astron. Astrophys.* **1970**, *7*, 120.
282. Collins, G. W., I. Further Note on Terminology—Specific Luminosity. *Astron. Astrophys.* **1973**, *26*, 315.
283. Moss, D.; Smith, R.C. REVIEW ARTICLE: Stellar rotation and magnetic stars. *Reports on Progress in Physics* **1981**, *44*, 831–891. [CrossRef]
284. Zorec, J. L'effet de la rotation sur la magnitude absolue des étoiles. In Proceedings of the Ecole de Printemps d'Astrophysique de Goutelas: HIPPARCOS, 1993; pp. 407–425.
285. Maeder, A.; Peytremann, E. Uniformly Rotating Stars with Hydrogen- and Metallic-Line Blanketed Model Atmospheres. *Astron. Astrophys.* **1972**, *21*, 279.
286. Collins, G. W., I.; Sonneborn, G.H. Some effects of rotation on the spectra of upper-main-sequence stars. *Astrophys. J. Suppl. Ser.* **1977**, *34*, 41–94. [CrossRef]
287. Slettebak, A.; Kuzma, T.J.; Collins, G. W., I. Effects of stellar rotation on spectral classification. *Astrophys. J.* **1980**, *242*, 171–187. [CrossRef]
288. Collins, Geroge W., I. Continuum Emission from a Rotating Non-Gray Stellar Atmosphere. *Astrophys. J.* **1965**, *142*, 265. [CrossRef]

289. Zorec, J.; Royer, F. Rotational velocities of A-type stars. IV. Evolution of rotational velocities. *Astron. Astrophys.* **2012**, *537*, A120.
290. Georgy, C.; Ekström, S.; Eggenberger, P.; Meynet, G.; Haemmerlé, L.; Maeder, A.; Granada, A.; Groh, J.H.; Hirschi, R.; Mowlavi, N.; et al. Grids of stellar models with rotation. III. Models from 0.8 to 120 $M_\odot$ at a metallicity Z = 0.002. *Astron. Astrophys.* **2013**, *558*, A103.
291. Sigut, T.A.A.; Ghafourian, N.R. Comparing Be Star Inclination Angles Determined from Hα Fitting and Gravitational Darkening. *arXiv* **2022**, arXiv:2209.06885.
292. Morel, T.; Blazère, A.; Semaan, T.; Gosset, E.; Zorec, J.; Frémat, Y.; Blomme, R.; Daflon, S.; Lobel, A.; Nieva, M.F.; et al. The Gaia-ESO survey: A spectroscopic study of the young open cluster NGC 3293. *Astron. Astrophys.* **2022**, *665*, A108.
293. Grant, L.P.; Peraiah, A. Spectral line formation in extended stellar atmo-spheres. *Mon. Not. R. Astron. Soc.* **1972**, *160*, 239. [CrossRef]
294. Simonneau, E. Radiative transfer in atmospheres with spherical symmetry. *J. Quant. Spectrosc. Radiat. Transf.* **1976**, *16*, 741–753. [CrossRef]
295. Simonneau, E. Radiative transfer in atmospheres with spherical symmetry—IV. The non-conservative problem. *J. Quant. Spectrosc. Radiat. Transf.* **1980**, *23*, 73–81. [CrossRef]
296. Lopez, R.; Simonneau, E.; Isern, J. Model atmospheres for type I supernovae—Curvature effects. *Astron. Astrophys.* **1987**, *184*, 249–255.
297. Korčáková, D.; Kubát, J. Emergent line profiles from rapidly rotating stars. *Memorie della Societa Astronomica Italiana Supplementi* **2005**, *7*, 130.
298. Frémat, Y.; Zorec, J.; Levenhagen, R.; Leister, N.; Hubert, A.M.; Floquet, M.; Neiner, C. Chemical Composition of Early Type Be Stars. In *Proceedings of the CNO in the Universe*; Charbonnel, C., Schaerer, D., Meynet, G., Eds.; University of Chicago Press: Chicago, NY, USA, 2003; Volume 304, p. 57.
299. Frémat, Y.; Zorec, J.; Hubert, A.M. Chemical Composition of the Pole-On Be Star HD 120991. In *Proceedings of the Stellar Rotation*; Maeder, A., Eenens, P., Eds.; Cambridge University Press: Cambridge, UK, 2004; Volume 215, p. 224.
300. Frémat, Y.; Zorec, J.; Levenhagen, R.; Leister, N.; Hubert, A.M.; Floquet, M.; Neiner, C. CNO abundances in Early Type Be Stars. In *Proceedings of the Stellar Rotation*; Maeder, A., Eenens, P., Eds.; Cambridge University Press: Cambridge, UK, 2004; Volume 215, p. 222.
301. Lennon, D.J.; Lee, J.K.; Dufton, P.L.; Ryans, R.S.I. A Be star with a low nitrogen abundance in the SMC cluster NGC 330. *Astron. Astrophys.* **2005**, *438*, 265–271. [CrossRef]
302. Hunter, I.; Dufton, P.L.; Smartt, S.J.; Ryans, R.S.I.; Evans, C.J.; Lennon, D.J.; Trundle, C.; Hubeny, I.; Lanz, T. The VLT-FLAMES survey of massive stars: Surface chemical compositions of B-type stars in the Magellanic Clouds. *Astron. Astrophys.* **2007**, *466*, 277–300. [CrossRef]
303. Hunter, I.; Brott, I.; Lennon, D.J.; Langer, N.; Dufton, P.L.; Trundle, C.; Smartt, S.J.; de Koter, A.; Evans, C.J.; Ryans, R.S.I. The VLT FLAMES Survey of Massive Stars: Rotation and Nitrogen Enrichment as the Key to Understanding Massive Star Evolution. *Astrophys. J. Lett.* **2008**, *676*, L29.
304. Hunter, I.; Brott, I.; Langer, N.; Lennon, D.J.; Dufton, P.L.; Howarth, I.D.; Ryans, R.S.I.; Trundle, C.; Evans, C.J.; de Koter, A.; et al. The VLT-FLAMES survey of massive stars: Constraints on stellar evolution from the chemical compositions of rapidly rotating Galactic and Magellanic Cloud B-type stars. *Astron. Astrophys.* **2009**, *496*, 841–853.
305. Trundle, C.; Dufton, P.L.; Hunter, I.; Evans, C.J.; Lennon, D.J.; Smartt, S.J.; Ryans, R.S.I. The VLT-FLAMES survey of massive stars: Evolution of surface N abundances and effective temperature scales in the Galaxy and Magellanic Clouds. *Astron. Astrophys.* **2007**, *471*, 625–643. [CrossRef]
306. Dunstall, P.R.; Brott, I.; Dufton, P.L.; Lennon, D.J.; Evans, C.J.; Smartt, S.J.; Hunter, I. The VLT-FLAMES survey of massive stars: Nitrogen abundances for Be-type stars in the Magellanic Clouds. *Astron. Astrophys.* **2011**, *536*, A65.
307. Potter, A.T.; Tout, C.A.; Brott, I. Towards a unified model of stellar rotation—II. Model-dependent characteristics of stellar populations. *Mon. Not. R. Astron. Soc.* **2012**, *423*, 1221–1233.
308. Brott, I.; Evans, C.J.; Hunter, I.; de Koter, A.; Langer, N.; Dufton, P.L.; Cantiello, M.; Trundle, C.; Lennon, D.J.; de Mink, S.E.; et al. Rotating massive main-sequence stars. II. Simulating a population of LMC early B-type stars as a test of rotational mixing. *Astron. Astrophys.* **2011**, *530*, A116.
309. Porter, J.M. On the possibility that rotation causes latitudinal abundance variations in stars. *Astron. Astrophys.* **1999**, *341*, 560–566.
310. Langer, N.; Cantiello, M.; Yoon, S.C.; Hunter, I.; Brott, I.; Lennon, D.; de Mink, S.; Verheijdt, M. Rotation and Massive Close Binary Evolution. In *Proceedings of the Massive Stars as Cosmic Engines*; Bresolin, F., Crowther, P.A., Puls, J., Eds.; 2008; Volume 250, pp. 167–178.
311. Shajn, G.; Struve, O. On the rotation of the stars. *Mon. Not. R. Astron. Soc.* **1929**, *89*, 222–239. [CrossRef]
312. Struve, O. Axial rotation as a major factor in stellar spectroscopy. *Observatory* **1931**, *54*, 80–84.
313. Elvey, C.T. The rotation of stars and the contours of Mg+ 4481. *Astrophys. J.* **1930**, *71*, 221–230. [CrossRef]
314. Carroll, J.A. The spectroscopic determination of stellar rotation and its effect on line profiles. *Mon. Not. R. Astron. Soc.* **1933**, *93*, 478–507. [CrossRef]
315. Slettebak, A. On the Axial Rotation of the Brighter O and B Stars. *Astrophys. J.* **1949**, *110*, 498. [CrossRef]
316. Howarth, I.D. Rotation and Line Broadening in OBA Stars (Invited Review). In *Proceedings of the Stellar Rotation*; Maeder, A., Eenens, P., Eds.; Cambridge University Press: Cambridge, UK, 2004; Volume 215, p. 33.

317. Slettebak, A.; Collins, G. W., I.; Boyce, P.B.; White, N.M.; Parkinson, T.D. A system of standard stars for rotational velocity determinations. *Astrophys. J. Suppl. Ser.* **1975**, *29*, 137–159. [CrossRef]
318. Dufton, P.L.; Smartt, S.J.; Lee, J.K.; Ryans, R.S.I.; Hunter, I.; Evans, C.J.; Herrero, A.; Trundle, C.; Lennon, D.J.; Irwin, M.J.; et al. The VLT-FLAMES survey of massive stars: Stellar parameters and rotational velocities in NGC 3293, NGC 4755 and NGC 6611. *Astron. Astrophys.* **2006**, *457*, 265–280. [CrossRef]
319. Díaz, C.G.; González, J.F.; Levato, H.; Grosso, M. Accurate stellar rotational velocities using the Fourier transform of the cross correlation maximum. *Astron. Astrophys.* **2011**, *531*, A143.
320. Howarth, I.D.; Siebert, K.W.; Hussain, G.A.J.; Prinja, R.K. Cross-correlation characteristics of OB stars from IUE spectroscopy. *Mon. Not. R. Astron. Soc.* **1997**, *284*, 265–285. [CrossRef]
321. Daflon, S.; Cunha, K.; de Araújo, F.X.; Wolff, S.; Przybilla, N. The Projected Rotational Velocity Distribution of a Sample of OB stars from a Calibration Based on Synthetic He I Lines. *Astron. J.* **2007**, *134*, 1570–1578.
322. Stoeckley, T.; Mihalas, D. *Limb Darkening and Rotation Broadening of Neutral Helium and Ionized Magnesium Line Profiles in Early-Type Stars*; NCAR: Boulder, CO, USA, 1973; Volume NCAR-TN/STR 84.
323. Huang, W.; Gies, D.R. Stellar Rotation in Young Clusters. I. Evolution of Projected Rotational Velocity Distributions. *Astrophys. J.* **2006**, *648*, 580–590.
324. Balona, L.A. Equivalent widths and rotational velocities of southern early-type stars. *Mem. R. Astron. Soc.* **1975**, *78*, 51.
325. Conti, P.S.; Ebbets, D. Spectroscopic studies of O-type stars. VII. Rotational velocities V sin i and evidence for macroturbulent motions. *Astrophys. J.* **1977**, *213*, 438–447. [CrossRef]
326. Slettebak, A. Spectral types and rotational velocities of the brighter Be stars and A-F type shell stars. *Astrophys. J. Suppl. Ser.* **1982**, *50*, 55–83. [CrossRef]
327. Wolff, S.C.; Edwards, S.; Preston, G.W. The origin of stellar angular momentum. *Astrophys. J.* **1982**, *252*, 322–336. [CrossRef]
328. Abt, H.A.; Morrell, N.I. The Relation between Rotational Velocities and Spectral Peculiarities among A-Type Stars. *Astrophys. J. Suppl. Ser.* **1995**, *99*, 135. [CrossRef]
329. Halbedel, E.M. Rotational Velocity Determinations for 164 Be and B Stars. *Publ. Astron. Soc. Pac.* **1996**, *108*, 833. [CrossRef]
330. Penny, L.R. Projected Rotational Velocities of O-Type Stars. *Astrophys. J.* **1996**, *463*, 737. [CrossRef]
331. Brown, A.G.A.; Verschueren, W. High S/N Echelle spectroscopy in young stellar groups. II. Rotational velocities of early-type stars in SCO OB2. *Astron. Astrophys.* **1997**, *319*, 811–838.
332. Steele, I.A.; Negueruela, I.; Clark, J.S. A representative sample of Be stars. I. Sample selection, spectral classification and rotational velocities. *Astron. Astrophys.* **1999**, *137*, 147–156. [CrossRef]
333. Chauville, J.; Zorec, J.; Ballereau, D.; Morrell, N.; Cidale, L.; Garcia, A. High and intermediate-resolution spectroscopy of Be stars 4481 lines. *Astron. Astrophys.* **2001**, *378*, 861–882. [CrossRef]
334. Abt, H.A.; Levato, H.; Grosso, M. Rotational Velocities of B Stars. *Astrophys. J.* **2002**, *573*, 359–365. [CrossRef]
335. Royer, F.; Gerbaldi, M.; Faraggiana, R.; Gómez, A.E. Rotational velocities of A-type stars. I. Measurement of v sin i in the southern hemisphere. *Astron. Astrophys.* **2002**, *381*, 105–121. [CrossRef]
336. Royer, F.; Grenier, S.; Baylac, M.O.; Gómez, A.E.; Zorec, J. Rotational velocities of A-type stars in the northern hemisphere. II. Measurement of v sin i. *Astron. Astrophys.* **2002**, *393*, 897–911. [CrossRef]
337. Keller, S.C. Rotation of Early B-type Stars in the Large Magellanic Cloud: The Role of Evolution and Metallicity. *Publ. Astron. Soc. Aust.* **2004**, *21*, 310–317.
338. Penny, L.R.; Sprague, A.J.; Seago, G.; Gies, D.R. Effects of Metallicity on the Rotational Velocities of Massive Stars. *Astrophys. J.* **2004**, *617*, 1316–1322.
339. Glebocki, R.; Gnacinski, P. *VizieR Online Data Catalog: Catalog of Stellar Rotational Velocities (Glebocki+ 2005)*; VizieR Online Data Catalog, 2005; p. III/244.
340. Strom, S.E.; Wolff, S.C.; Dror, D.H.A. B Star Rotational Velocities in h and χ Persei: A Probe of Initial Conditions during the Star Formation Epoch? *Astron. J.* **2005**, *129*, 809–828.
341. Wolff, S.C.; Strom, S.E.; Dror, D.; Lanz, L.; Venn, K. Stellar Rotation: A Clue to the Origin of High-Mass Stars? *Astron. J.* **2006**, *132*, 749–755.
342. Frémat, Y.; Neiner, C.; Hubert, A.M.; Floquet, M.; Zorec, J.; Janot-Pacheco, E.; Renan de Medeiros, J. Fundamental parameters of Be stars located in the seismology fields of COROT. *Astron. Astrophys.* **2006**, *451*, 1053–1063. [CrossRef]
343. Mokiem, M.R.; de Koter, A.; Evans, C.J.; Puls, J.; Smartt, S.J.; Crowther, P.A.; Herrero, A.; Langer, N.; Lennon, D.J.; Najarro, F.; et al. The VLT-FLAMES survey of massive stars: Mass loss and rotation of early-type stars in the SMC. *Astron. Astrophys.* **2006**, *456*, 1131–1151. [CrossRef]
344. Levenhagen, R.S.; Leister, N.V. Spectroscopic analysis of southern B and Be stars. *Mon. Not. R. Astron. Soc.* **2006**, *371*, 252–262.
345. Martayan, C.; Frémat, Y.; Hubert, A.M.; Floquet, M.; Zorec, J.; Neiner, C. Effects of metallicity, star-formation conditions, and evolution in B and Be stars. I. Large Magellanic Cloud, field of NGC 2004. *Astron. Astrophys.* **2006**, *452*, 273–284. [CrossRef]
346. Martayan, C.; Frémat, Y.; Hubert, A.M.; Floquet, M.; Zorec, J.; Neiner, C. Effects of metallicity, star-formation conditions, and evolution in B and Be stars. II. Small Magellanic Cloud, field of NGC 330. *Astron. Astrophys.* **2007**, *462*, 683–694. [CrossRef]
347. Wolff, S.C.; Strom, S.E.; Dror, D.; Venn, K. Rotational Velocities for B0-B3 Stars in Seven Young Clusters: Further Study of the Relationship between Rotation Speed and Density in Star-Forming Regions. *Astron. J.* **2007**, *133*, 1092–1103.

348. Wolff, S.C.; Strom, S.E.; Cunha, K.; Daflon, S.; Olsen, K.; Dror, D. Rotational Velocities for Early-Type Stars in the Young Large Magellanic Cloud Cluster R136: Further Study of the Relationship Between Rotation Speed and Density in Star-Forming Regions. *Astron. J.* **2008**, *136*, 1049–1060. [CrossRef]
349. Huang, W.; Gies, D.R. Stellar Rotation in Field and Cluster B Stars. *Astrophys. J.* **2008**, *683*, 1045–1051.
350. Hunter, I.; Lennon, D.J.; Dufton, P.L.; Trundle, C.; Simón-Díaz, S.; Smartt, S.J.; Ryans, R.S.I.; Evans, C.J. The VLT-FLAMES survey of massive stars: Atmospheric parameters and rotational velocity distributions for B-type stars in the Magellanic Clouds. *Astron. Astrophys.* **2008**, *479*, 541–555. [CrossRef]
351. Penny, L.R.; Gies, D.R. A FUSE Survey of the Rotation Rates of Very Massive Stars in the Small and Large Magellanic Clouds. *Astrophys. J.* **2009**, *700*, 844–858.
352. Fraser, M.; Dufton, P.L.; Hunter, I.; Ryans, R.S.I. Atmospheric parameters and rotational velocities for a sample of Galactic B-type supergiants. *Mon. Not. R. Astron. Soc.* **2010**, *404*, 1306–1320.
353. Huang, W.; Gies, D.R.; McSwain, M.V. A Stellar Rotation Census of B Stars: From ZAMS to TAMS. *Astrophys. J.* **2010**, *722*, 605–619.
354. Marsh Boyer, A.N.; McSwain, M.V.; Aragona, C.; Ou-Yang, B. Physical Properties of the B and Be Star Populations of h and χ Persei. *Astron. J.* **2012**, *144*, 158.
355. Bragança, G.A.; Daflon, S.; Cunha, K.; Bensby, T.; Oey, M.S.; Walth, G. Projected Rotational Velocities and Stellar Characterization of 350 B Stars in the Nearby Galactic Disk. *Astron. J.* **2012**, *144*, 130.
356. Dufton, P.L.; Langer, N.; Dunstall, P.R.; Evans, C.J.; Brott, I.; de Mink, S.E.; Howarth, I.D.; Kennedy, M.; McEvoy, C.; Potter, A.T.; et al. The VLT-FLAMES Tarantula Survey. X. Evidence for a bimodal distribution of rotational velocities for the single early B-type stars. *Astron. Astrophys.* **2013**, *550*, A109.
357. Ramírez-Agudelo, O.H.; Simón-Díaz, S.; Sana, H.; de Koter, A.; Sabín-Sanjulían, C.; de Mink, S.E.; Dufton, P.L.; Gräfener, G.; Evans, C.J.; Herrero, A.; et al. The VLT-FLAMES Tarantula Survey. XII. Rotational velocities of the single O-type stars. *Astron. Astrophys.* **2013**, *560*, A29.
358. Simón-Díaz, S.; Herrero, A. The IACOB project. I. Rotational velocities in northern Galactic O- and early B-type stars revisited. The impact of other sources of line-broadening. *Astron. Astrophys.* **2014**, *562*, A135.
359. Garmany, C.D.; Glaspey, J.W.; Bragança, G.A.; Daflon, S.; Borges Fernandes, M.; Oey, M.S.; Bensby, T.; Cunha, K. Projected Rotational Velocities of 136 Early B-type Stars in the Outer Galactic Disk. *Astron. J.* **2015**, *150*, 41. [CrossRef]
360. Holgado, G.; Simón-Díaz, S.; Herrero, A.; Barbá, R.H. The IACOB project. VII. The rotational properties of Galactic massive O-type stars revisited. *Astron. Astrophys.* **2022**, *665*, A150.
361. Solar, M.; Arcos, C.; Curé, M.; Levenhagen, R.S.; Araya, I. Automatic algorithm to obtain v sin i values via Fourier transform in the BeSOS database. *Mon. Not. R. Astron. Soc.* **2022**, *511*, 4404–4416.
362. Xiang, M.; Rix, H.W.; Ting, Y.S.; Kudritzki, R.P.; Conroy, C.; Zari, E.; Shi, J.R.; Przybilla, N.; Ramirez-Tannus, M.; Tkachenko, A.; et al. Stellar labels for hot stars from low-resolution spectra. I. The HotPayne method and results for 330 000 stars from LAMOST DR6. *Astron. Astrophys.* **2022**, *662*, A66.
363. Gaia Collaboration; Vallenari, A.; Brown, A.G.A.; Prusti, T.; de Bruijne, J.H.J.; Arenou, F.; Babusiaux, C.; Biermann, M.; Creevey, O.L.; Ducourant, C.; et al. Gaia Data Release 3: Summary of the content and survey properties. *arXiv* **2022**, arXiv:2208.00211.
364. Sartoretti, P.; Blomme, R.; David, M.; Seabroke, G. Gaia DR3 documentation Chapter 6: Spectroscopy. Gaia DR3 documentation, European Space Agency; Gaia Data Processing and Analysis Consortium. 2022. Available online: https://gea.esac.esa.int/archive/documentation/GDR3/index.html (accessed on 1 February 2022).
365. Creevey, O.L.; Sordo, R.; Pailler, F.; Frémat, Y.; Heiter, U.; Thévenin, F.; Andrae, R.; Fouesneau, M.; Lobel, A.; Bailer-Jones, C.A.L.; et al. Gaia Data Release 3: Astrophysical parameters inference system (Apsis) I—Methods and content overview. *arXiv* **2022**, arXiv:2206.05864.
366. Frémat, Y.; Royer, F.; Marchal, O.; Blomme, R.; Sartoretti, P.; Guerrier, A.; Panuzzo, P.; Katz, D.; Seabroke, G.M.; Thévenin, F.; et al. Gaia Data Release 3: Properties of the line broadening parameter derived with the Radial Velocity Spectrometer (RVS). *arXiv* **2022**, arXiv:2206.10986.
367. Blomme, R.; Frémat, Y.; Sartoretti, P.; Guerrier, A.; Panuzzo, P.; Katz, D.; Seabroke, G.M.; Thevenin, F.; Cropper, M.; Benson, K.; et al. Gaia Data Release 3: Hot-star radial velocities. *arXiv* **2022**, arXiv:2206.05486.
368. Katz, D.; Sartoretti, P.; Guerrier, A.; Panuzzo, P.; Seabroke, G.M.; Thévenin, F.; Cropper, M.; Benson, K.; Blomme, R.; Haigron, R.; et al. Gaia Data Release 3 Properties and validation of the radial velocities. *arXiv* **2022**, arXiv:2206.05902.
369. Gaia Collaboration.; Creevey, O.L.; Sarro, L.M.; Lobel, A.; Pancino, E.; Andrae, R.; Smart, R.L.; Clementini, G.; Heiter, U.; Korn, A.J.; et al. Gaia Data Release 3: A Golden Sample of Astrophysical Parameters. *arXiv* **2022**, arXiv:2206.05870.
370. Ulla, A.; Creevey, O.L.; Álvarez, M.A.; Andrae, R.; Bailer-Jones, C.A.L.; Bellas-Velidis, I.; Brugaletta, E.; Carballo, R.; Dafonte, C.; Delchambre, L.; et al. Gaia DR3 Documentation Chapter 11: Astrophysical Parameters. Gaia DR3 Documentation, European Space Agency; Gaia Data Processing and Analysis Consortium. 2022. Available online: https://gea.esac.esa.int/archive/documentation/GDR3/index.html (accessed on 1 February 2022).
371. Fouesneau, M.; Frémat, Y.; Andrae, R.; Korn, A.J.; Soubiran, C.; Kordopatis, G.; Vallenari, A.; Heiter, U.; Creevey, O.L.; Sarro, L.M.; et al. Gaia Data Release 3: Apsis II—Stellar Parameters. *arXiv* **2022**, arXiv:2206.05992.
372. Shridharan, B.; Mathew, B.; Bhattacharyya, S.; Robin, T.; Arun, R.; Kartha, S.S.; Manoj, P.; Nidhi, S.; Maheshwar, G.; Paul, K.T.; et al. Emission line star catalogues post-Gaia DR3: A validation of Gaia DR3 data using LAMOST OBA emission catalogue. *arXiv* **2022**, arXiv:2209.13221.

373. Plaskett, H.H. The formation of the magnesium b lines in the solar atmosphere. *Mon. Not. R. Astron. Soc.* **1931**, *91*, 870. [CrossRef]
374. Gray, D.F. *The Observation and Analysis of Stellar Photospheres*; Cambridge University Press: Cambridge, UK, 2008.
375. Levenhagen, R.S. A Fourier Transform Method for Vsin i Estimations under Nonlinear Limb-darkening Laws. *Astrophys. J.* **2014**, *797*, 29. [CrossRef]
376. Collins, George W., I.; Truax, R.J. Classical Rotational Broadening of Spectral Lines. *Astrophys. J.* **1995**, *439*, 860. [CrossRef]
377. Korčáková, D.; Kubát, J. Radiative transfer in moving media. I. Discontinuous finite element method for one-dimensional atmospheres. *Astron. Astrophys.* **2003**, *401*, 419–428. [CrossRef]
378. Korčáková, D.; Kubát, J. Radiative transfer in moving media. II. Solution of the radiative transfer equation in axial symmetry. *Astron. Astrophys.* **2005**, *440*, 715–725. [CrossRef]
379. Hillier, D.J.; Lanz, T. CMFGEN: A non-LTE Line-Blanketed Radiative Transfer Code for Modeling Hot Stars with Stellar Winds. In *Proceedings of the Spectroscopic Challenges of Photoionized Plasmas*; Ferland, G., Savin, D.W., Eds.; University of Chicago Press: Chicago, NY, USA, 2001; Volume 247, p. 343.
380. Hillier, D.J. The atomic physics underlying the spectroscopic analysis of massive stars and supernovae. *Astrophys. Space Sci.* **2011**, *336*, 87–93. [CrossRef]
381. Hillier, D.J. Hot Stars with Winds: The CMFGEN Code. In *Proceedings of the From Interacting Binaries to Exoplanets: Essential Modeling Tools*; Richards, M.T., Hubeny, I., Eds.; Cambridge Unversity Press: Cambridge, UK, 2012; Volume 282, pp. 229–234. [CrossRef]
382. Stoeckley, T.R. Distribution of rotational velocities in Be stars. *Mon. Not. R. Astron. Soc.* **1968**, *140*, 141. [CrossRef]
383. Stoeckley, T.R. Absorption line strengths in rotating stars. *Mon. Not. R. Astron. Soc.* **1968**, *140*, 149. [CrossRef]
384. Townsend, R.H.D.; Owocki, S.P.; Howarth, I.D. Be-star rotation: How close to critical? *Mon. Not. R. Astron. Soc.* **2004**, *350*, 189–195.
385. Cranmer, S.R. A Statistical Study of Threshold Rotation Rates for the Formation of Disks around Be Stars. *Astrophys. J.* **2005**, *634*, 585–601.
386. Heap, S.R. Ultraviolet Observations of be Stars (review Paper). In Proceedings of the Be and Shell Stars; Slettebak, A., Ed., 1976, Volume 70, p. 165.
387. Heap, S.R. Apparent wavelength dependence of v sin i for Zeta Tauri. *Astrophys. J. Lett.* **1977**, *218*, L17–L19. [CrossRef]
388. Hutchings, J.B. V sin i Values in the Far Ultraviolet. *Publ. Astron. Soc. Pac.* **1976**, *88*, 5. [CrossRef]
389. Hutchings, J.B.; Stoeckley, T.R. V and i in rotating stars from Copernicus UV data. *Publ. Astron. Soc. Pac.* **1977**, *89*, 19–22. [CrossRef]
390. Sonneborn, G.H.; Collins, G. W., I. On the wavelength dependence of rotational line broadening. *Astrophys. J.* **1977**, *213*, 787–790. [CrossRef]
391. Carpenter, K.G.; Slettebak, A.; Sonneborn, G. Rotational velocities of later B type and A type stars as determined from ultraviolet versus visual line profiles. *Astrophys. J.* **1984**, *286*, 741–746. [CrossRef]
392. Osaki, Y. On the Atmosphere of a Rotating Star. *Publ. Astron. Soc. Jpn.* **1966**, *18*, 7.
393. Pustyl'Nik, I. Radiative transfer in the atmospheres of rotating stars. *Izv. Akad. Nauk Ehstonskoj SSR* **1970**, *19*, 428–435. [CrossRef]
394. Hadrava, P. Radiative transfer in rotating stars. *Astron. Astrophys.* **1992**, *256*, 519–524.
395. Unsöld, A.; Struve, O. Curves of Growth and Line Contours. *Astrophys. J.* **1949**, *110*, 455. [CrossRef]
396. Huang, S.S.; Struve, O. A Study of Line Profiles: The Spectrum of Rho Leonis. *Astrophys. J.* **1953**, *118*, 463. [CrossRef]
397. Underhill, A.B. On the Effect of Radiation Pressure in the Atmospheres of Early-Type Stars. *Mon. Not. R. Astron. Soc.* **1949**, *109*, 562. [CrossRef]
398. Underhill, A.B. Numerical experiments concerning the rotational broadening of spectral lines. *Bull. Astron. Institutes Neth.* **1968**, *19*, 526.
399. Rosendhal, J.D. Evolutionary Effects in the Rotation of Supergiants. *Astrophys. J.* **1970**, *159*, 107. [CrossRef]
400. Grassitelli, L.; Fossati, L.; Simón-Díaz, S.; Langer, N.; Castro, N.; Sanyal, D. Observational Consequences of Turbulent Pressure in the Envelopes of Massive Stars. *Astrophys. J. Lett.* **2015**, *808*, L31.
401. Aerts, C.; Puls, J.; Godart, M.; Dupret, M.A. Collective pulsational velocity broadening due to gravity modes as a physical explanation for macroturbulence in hot massive stars. *Astron. Astrophys.* **2009**, *508*, 409–419.
402. Aerts, C.; Simón-Díaz, S.; Groot, P.J.; Degroote, P. On the use of the Fourier transform to determine the projected rotational velocity of line-profile variable B stars. *Astron. Astrophys.* **2014**, *569*, A118. [CrossRef]
403. Simón-Díaz, S. Asteroseismology of OB stars with hundreds of single snapshot spectra (and a few time-series of selected targets). In *Proceedings of the IAU Symposium*; Meynet, G., Georgy, C., Groh, J., Stee, P., Eds.; 2015; Volume 307, pp. 194–199.
404. Ryans, R.S.I.; Dufton, P.L.; Rolleston, W.R.J.; Lennon, D.J.; Keenan, F.P.; Smoker, J.V.; Lambert, D.L. Macroturbulent and rotational broadening in the spectra of B-type supergiants. *Mon. Not. R. Astron. Soc.* **2002**, *336*, 577–586. [CrossRef]
405. Gray, D.F. Atmospheric turbulence measured in stars above the main sequence. *Astrophys. J.* **1975**, *202*, 148–164. [CrossRef]
406. Gray, D.F. *The Observation and Analysis of Stellar Photospheres*; Cambridge Astrophysics, Series; Cambridge University Press: Cambridge, UK, 1992.
407. Simón-Díaz, S.; Herrero, A. Fourier method of determining the rotational velocities in OB stars. *Astron. Astrophys.* **2007**, *468*, 1063–1073. [CrossRef]
408. Dufton, P.L.; Ryans, R.S.I.; Simón-Díaz, S.; Trundle, C.; Lennon, D.J. B-type supergiants in the Small Magellanic Cloud: Rotational velocities and implications for evolutionary models. *Astron. Astrophys.* **2006**, *451*, 603–611. [CrossRef]

409. Sundqvist, J.O.; Simón-Díaz, S.; Puls, J.; Markova, N. The rotation rates of massive stars. How slow are the slow ones? *Astron. Astrophys.* **2013**, *559*, L10.
410. Stoeckley, T.R. Determination of aspect and degree of differential rotation, from line profiles in rapidly rotating stars. *Mon. Not. R. Astron. Soc.* **1968**, *140*, 121. [CrossRef]
411. Zorec, J.; Divan, L.; Mochkovitch, R.; Garcia, A. Differential rotation in B and Be stars. In *Proceedings of the IAU Colloq. 92: Physics of Be Stars*; Slettebak, A., Snow, T.P., Eds.; Cambridge University Press: Cambridge, UK, 1987; pp. 68–70.
412. Stoeckley, T.R.; Buscombe, W. Axial inclination and differential rotation for 19 rapidly rotating stars. *Mon. Not. R. Astron. Soc.* **1987**, *227*, 801–813. [CrossRef]
413. Zorec, J.; Mochkovitch, R.A.; Garcia, A. The Angular Momentum Loss and the Differential Rotation in B-Stars and Be-Stars. In *Proceedings of the NATO ASIC Proc. 316: Angular Momentum and Mass Loss for Hot Stars*; Willson, L.A., Stalio, R., Eds.; Springer: Berlin/Heidelberg, Germany, 1990; p. 239.
414. Cranmer, S.R.; Collins, II, G.W. The effects of zonal atmospheric currents on the spectra of rotating early-type stars. *Astrophys. J.* **1993**, *412*, 720–730. [CrossRef]
415. Rieutord, M. The dynamics of the radiative envelope of rapidly rotating stars. I. A spherical Boussinesq model. *Astron. Astrophys.* **2006**, *451*, 1025–1036. [CrossRef]
416. Espinosa Lara, F.; Rieutord, M. The dynamics of a fully radiative rapidly rotating star enclosed within a spherical box. *Astron. Astrophys.* **2007**, *470*, 1013–1022. [CrossRef]
417. Schou, J.; Antia, H.M.; Basu, S.; Bogart, R.S.; Bush, R.I.; Chitre, S.M.; Christensen-Dalsgaard, J.; Di Mauro, M.P.; Dziembowski, W.A.; Eff-Darwich, A.; et al. Helioseismic Studies of Differential Rotation in the Solar Envelope by the Solar Oscillations Investigation Using the Michelson Doppler Imager. *Astrophys. J.* **1998**, *505*, 390–417. [CrossRef]
418. Espinosa Lara, F.; Rieutord, M. Self-consistent 2D models of fast-rotating early-type stars. *Astron. Astrophys.* **2013**, *552*, A35,
419. Rieutord, M.; Espinosa Lara, F.; Putigny, B. An algorithm for computing the 2D structure of fast rotating stars. *J. Comput. Phys.* **2016**, *318*, 277–304.
420. Gagnier, D.; Rieutord, M.; Charbonnel, C.; Putigny, B.; Espinosa Lara, F. Evolution of rotation in rapidly rotating early-type stars during the main sequence with 2D models. *Astron. Astrophys.* **2019**, *625*, A69.
421. Jeffery, C.S. Quasi-emission lines in rotating B stars. *Mon. Not. R. Astron. Soc.* **1991**, *249*, 327. [CrossRef]
422. Takeda, Y.; Kawanomoto, S.; Ohishi, N. High-Resolution and High-S/N Spectrum Atlas of Vega. *Publ. Astron. Soc. Jpn.* **2007**, *59*, 245–261. [CrossRef]
423. Takeda, Y.; Kawanomoto, S.; Ohishi, N. Rotational Feature of Vega Revealed from Spectral Line Profiles. *Astrophys. J.* **2008**, *678*, 446–462. [CrossRef]
424. Zorec, J. Emission-like feature due to a latitudinal differential rotation. In *Proceedings of the Pulsation; Rotation; and Mass Loss in Early-Type Stars*; Balona, L.A., Henrichs, H.F., Le Contel, J.M., Eds.; Kluwer Academic Publishers: New York, NY, USA, 1994; Volume 162, pp. 257–258.
425. Reiners, A.; Schmitt, J.H.M.M. Rotation and differential rotation in field F- and G-type stars. *Astron. Astrophys.* **2003**, *398*, 647–661. [CrossRef]
426. Reiners, A.; Schmitt, J.H.M.M. Differential rotation in rapidly rotating F-stars. *Astron. Astrophys.* **2003**, *412*, 813–819. [CrossRef]
427. Sobolev, V.V. *Moving Envelopes of Stars*; Harvard University Press: Cambridge, UK, 1960. [CrossRef]
428. Kalkofen, W. *Methods in Radiative Transfer*; Cambridge University Press: Cambridge, UK, 1984.
429. Mihalas, D.; Mihalas, B.W. *Foundations of radiation hydrodynamics*; New York: Oxford University Press, 1984.
430. Kunasz, P.B. The theory of line transfer in expanding atmospheres. In *Progress in Stellar Spectral Line Formation Theory*; Beckman, J.E., Crivellari, L., Eds.; Springer Science & Business Media: Berlin, Germany, 1985; Volume 152, pp. 319–333.
431. Kalkofen, W. *Numerical Radiative Transfer*; Cambridge University Press: Cambridge, UK, 1987.
432. Conti, P.S.; Underhill, A.B.; Jordan, S.; Thomas, R.N.; Goldberg, L.; Pecker, J.C.; Baade, D.; Divan, L.; Garmany, C.D.; Henrichs, H.F.; et al. *O Stars and Wolf-Rayet Stars*; NASA: Washington, DC, USA, 1988; Volume 497.
433. Sen, K.K.; Wilson, S.J. *Radiative Transfer in Moving Media: Basic Mathematical Methods for Radiative Transfer In Spherically Symmetrical Moving Media*; Springer: Berlin/Heidelberg, Germany, 1998.
434. Stee, P. (Ed.) *Radiative Transfer and Hydrodynamics in Astrophysics*; EAS Publications Series; EDP Sciences: Les Ulis, France, 2002; Volume 5.
435. Ivan Hubeny, D.M.; Werner, K. (Eds.) *Stellar Atmosphere Modeling*; Astronomical Society of the Pacific Conference Series; University Chicago Press: Chicago, NY, USA, 2003; Volume 288.
436. Cannon, C.J. *The Transfer of Spectral Line Radiation*; Cambridge University Press: Cambridge, UK, 2012.
437. Furenlid, I.; Young, A. Mass loss and rotation in early-main-sequence B stars. *Astrophys. J. Lett.* **1980**, *240*, L59–L61. [CrossRef]
438. Cook, K.H.; Alcock, C.; Allsman, H.A.; Axelrod, T.S.; Freeman, K.C.; Peterson, B.A.; Quinn, P.J.; Rodgers, A.W.; Bennett, D.P.; Reimann, J.; et al. Variable Stars in the MACHO Collaboration Database. In *Proceedings of the IAU Colloq. 155: Astrophysical Applications of Stellar Pulsation*; Stobie, R.S., Whitelock, P.A., Eds.; Cambridge University Press: Cambridge, UK, 1995; Volume 83, p. 221.
439. Duval, P.; Karp, A.H. The combined effects of expansion and rotation on spectral line shapes. *Astrophys. J.* **1978**, *222*, 220–225. [CrossRef]

440. Chelli, A.; Petrov, R.G. Model fitting and error analysis for differential interferometry. I. General formalism. *Astron. Astrophys.* **1995**, *109*, 389–399.
441. Chelli, A.; Petrov, R.G. Model fitting and error analysis for differential interferometry. II. Application to rotating stars and binary systems. *Astron. Astrophys.* **1995**, *109*, 401–415.
442. Vakili, F.; Mourard, D.; Bonneau, D.; Morand, F.; Stee, P. Subtle structures in the wind of P Cygni. *Astron. Astrophys.* **1997**, *323*, 183–188.
443. Jankov, S.; Vakili, F.; Domiciano de Souza, A., J.; Janot-Pacheco, E. Interferometric-Doppler imaging of stellar surface structure. *Astron. Astrophys.* **2001**, *377*, 721–734. [CrossRef]
444. ten Brummelaar, T.A.; McAlister, H.A.; Ridgway, S.T.; Bagnuolo, W. G., J.; Turner, N.H.; Sturmann, L.; Sturmann, J.; Berger, D.H.; Ogden, C.E.; Cadman, R.; et al. First Results from the CHARA Array. II. A Description of the Instrument. *Astrophys. J.* **2005**, *628*, 453–465.
445. Armstrong, J.T.; Mozurkewich, D.; Rickard, L.J.; Hutter, D.J.; Benson, J.A.; Bowers, P.F.; Elias, N. M., I.; Hummel, C.A.; Johnston, K.J.; Buscher, D.F.; et al. The Navy Prototype Optical Interferometer. *Astrophys. J.* **1998**, *496*, 550–571. [CrossRef]
446. Colavita, M.M.; Wallace, J.K.; Hines, B.E.; Gursel, Y.; Malbet, F.; Palmer, D.L.; Pan, X.P.; Shao, M.; Yu, J.W.; Boden, A.F.; et al. The Palomar Testbed Interferometer. *Astrophys. J.* **1999**, *510*, 505–521.
447. Glindemann, A.; Albertsen, M.; Andolfato, L.; Avila, G.; Ballester, P.; Bauvir, B.; Delplancke, F.; Derie, F.; Dimmler, M.; Duhoux, P.; et al. VLTI technical advances: Present and future. In *Proceedings of the New Frontiers in Stellar Interferometry*; Traub, W.A., Ed.; SPIE: Washington, DC, USA, 2004; Volume 5491, p. 447. [CrossRef]
448. Petrov, R.G.; Malbet, F.; Weigelt, G.; Antonelli, P.; Beckmann, U.; Bresson, Y.; Chelli, A.; Dugué, M.; Duvert, G.; Gennari, S.; et al. AMBER, the near-infrared spectro-interferometric three-telescope VLTI instrument. *Astron. Astrophys.* **2007**, *464*, 1–12. [CrossRef]
449. Domiciano de Souza, A. Long Baseline Interferometry of Rotating Stars across the HR Diagram: Flattening, Gravity Darkening, Differential Rotation. In *Lecture Notes on Physics: The Rotation of Sun and Stars*; Rozelot, J.P., Neiner, C., Eds.; Springer: Berlin/Heidelberg, Germany, 2009; Volume 765, pp. 171–194. [CrossRef]
450. Domiciano de Souza, A. Interferometric Surface Mapping of Rapidly Rotating Stars: Application to the Be star Achernar. In *Lecture Notes in Physics*; Rozelot, J.P., Neiner, C., Eds.; Springer: Berlin/Heidelberg, Germany, 2016; Volume 914, p. 159. [CrossRef]
451. Brown, R.H. *The Intensity Interferometer; Its Application to Astronomy*; Taylor and Francis LTD London and Halsted Press: London, UK, 1974.
452. Domiciano de Souza, A.; Kervella, P.; Jankov, S.; Abe, L.; Vakili, F.; di Folco, E.; Paresce, F. The spinning-top Be star Achernar from VLTI-VINCI. *Astron. Astrophys.* **2003**, *407*, L47–L50. [CrossRef]
453. Vinicius, M.M.F.; Zorec, J.; Leister, N.V.; Levenhagen, R.S. α Eridani: Rotational distortion, stellar and circumstellar activity. *Astron. Astrophys.* **2006**, *446*, 643–660. [CrossRef]
454. Kanaan, S.; Meilland, A.; Stee, P.; Zorec, J.; Domiciano de Souza, A.; Frémat, Y.; Briot, D. Disk and wind evolution of Achernar: The breaking of the fellowship. *Astron. Astrophys.* **2008**, *486*, 785–798. [CrossRef]
455. Kervella, P.; Domiciano de Souza, A. The polar wind of the fast rotating Be star Achernar. VINCI/VLTI interferometric observations of an elongated polar envelope. *Astron. Astrophys.* **2006**, *453*, 1059–1066. [CrossRef]
456. van Belle, G.T. Interferometric observations of rapidly rotating stars. *Astron. Astrophys. Rev.* **2012**, *20*, 51.
457. Hadjara, M.; Petrov, R.G.; Jankov, S.; Cruzalèbes, P.; Spang, A.; Lagarde, S. Differential interferometry of the rapid rotator Regulus. *Mon. Not. R. Astron. Soc.* **2018**, *480*, 1263–1277.
458. Gray, R.O. The spectroscopic and photometric effects of rotation in the A-type stars. *J. R. Astron. Soc. Can.* **1988**, *82*, 336–348.
459. McAlister, H.A.; ten Brummelaar, T.A.; Gies, D.R.; Huang, W.; Bagnuolo, W. G., J.; Shure, M.A.; Sturmann, J.; Sturmann, L.; Turner, N.H.; Taylor, S.F.; et al. First Results from the CHARA Array. I. An Interferometric and Spectroscopic Study of the Fast Rotator α Leonis (Regulus). *Astrophys. J.* **2005**, *628*, 439–452.
460. Aufdenberg, J.P.; Mérand, A.; Coudé du Foresto, V.; Absil, O.; Di Folco, E.; Kervella, P.; Ridgway, S.T.; Berger, D.H.; ten Brummelaar, T.A.; McAlister, H.A.; et al. First Results from the CHARA Array. VII. Long-Baseline Interferometric Measurements of Vega Consistent with a Pole-On, Rapidly Rotating Star. *Astrophys. J.* **2006**, *645*, 664–675.
461. Peterson, D.M.; Hummel, C.A.; Pauls, T.A.; Armstrong, J.T.; Benson, J.A.; Gilbreath, G.C.; Hindsley, R.B.; Hutter, D.J.; Johnston, K.J.; Mozurkewich, D.; et al. Vega is a rapidly rotating star. *Nature* **2006**, *440*, 896–899.
462. Peterson, D.M.; Hummel, C.A.; Pauls, T.A.; Armstrong, J.T.; Benson, J.A.; Gilbreath, G.C.; Hindsley, R.B.; Hutter, D.J.; Johnston, K.J.; Mozurkewich, D.; et al. Resolving the Effects of Rotation in Altair with Long-Baseline Interferometry. *Astrophys. J.* **2006**, *636*, 1087–1097.
463. Challouf, M.; Nardetto, N.; Domiciano de Souza, A.; Mourard, D.; Tallon-Bosc, I.; Aroui, H.; Farrington, C.; Ligi, R.; Meilland, A.; Mouelhi, M. Flattening and surface-brightness of the fast-rotating star δ Persei with the visible VEGA/CHARA interferometer. *Astron. Astrophys.* **2017**, *604*, A51. [CrossRef]
464. Domiciano de Souza, A.; Vakili, F.; Jankov, S.; Janot-Pacheco, E.; Abe, L. Modelling rapid rotators for stellar interferometry. *Astron. Astrophys.* **2002**, *393*, 345–357. [CrossRef]
465. Domiciano de Souza, A.; Zorec, J.; Vakili, F. CHARRON: Code for High Angular Resolution of Rotating Objects in Nature. In *Proceedings of the SF2A-2012: Proceedings of the Annual meeting of the French Society of Astronomy and Astrophysics*; Boissier, S., de Laverny, P., Nardetto, N., Samadi, R., Valls-Gabaud, D., Wozniak, H., Eds.; EDP Sciences: Les Ulis, France, 2012; pp. 321–324.

466. Domiciano de Souza, A.; Zorec, J.; Jankov, S.; Vakili, F.; Abe, L.; Janot-Pacheco, E. Stellar differential rotation and inclination angle from spectro-interferometry. *Astron. Astrophys.* **2004**, *418*, 781–794. [CrossRef]
467. Delaa, O.; Zorec, J.; Domiciano de Souza, A.; Mourard, D.; Perraut, K.; Stee, P.; Frémat, Y.; Monnier, J.; Kraus, S.; Che, X.; et al. Spectrally resolved interferometric observations of α Cephei and physical modeling of fast rotating stars. *Astron. Astrophys.* **2013**, *555*, A100. [CrossRef]
468. Stee, P.; Allard, F.; Benisty, M.; Bigot, L.; Blind, N.; Boffin, H.; Borges Fernandes, M.; Carciofi, A.; Chiavassa, A.; Creevey, O.; et al. Science cases for a visible interferometer. *arXiv* **2017**, arXiv:1703.02395.
469. Meilland, A.; Millour, F.; Stee, P.; Domiciano de Souza, A.; Petrov, R.G.; Mourard, D.; Jankov, S.; Robbe-Dubois, S.; Spang, A.; Aristidi, E.; et al. An asymmetry detected in the disk of κ Canis Majoris with AMBER/VLTI. *Astron. Astrophys.* **2007**, *464*, 73–79. [CrossRef]
470. Meilland, A.; Millour, F.; Stee, P.; Spang, A.; Petrov, R.; Bonneau, D.; Perraut, K.; Massi, F. δ Centauri: A new binary Be star detected by VLTI/AMBER spectro-interferometry. *Astron. Astrophys.* **2008**, *488*, L67–L70. [CrossRef]
471. Meilland, A.; Stee, P.; Chesneau, O.; Jones, C. VLTI/MIDI observations of 7 classical Be stars. *Astron. Astrophys.* **2009**, *505*, 687–693.
472. Meilland, A.; Delaa, O.; Stee, P.; Kanaan, S.; Millour, F.; Mourard, D.; Bonneau, D.; Petrov, R.; Nardetto, N.; Marcotto, A.; et al. The binary Be star δ Scorpii at high spectral and spatial resolution. I. Disk geometry and kinematics before the 2011 periastron. *Astron. Astrophys.* **2011**, *532*, A80.
473. Meilland, A.; Millour, F.; Kanaan, S.; Stee, P.; Petrov, R.; Hofmann, K.H.; Natta, A.; Perraut, K. First spectro-interferometric survey of Be stars. I. Observations and constraints on the disk geometry and kinematics. *Astron. Astrophys.* **2012**, *538*, A110.
474. Meilland, A.; Stee, P.; Spang, A.; Malbet, F.; Massi, F.; Schertl, D. The binary Be star δ Scorpii at high spectral and spatial resolution. II. The circumstellar disk evolution after the periastron. *Astron. Astrophys.* **2013**, *550*, L5.
475. Delaa, O.; Stee, P.; Meilland, A.; Zorec, J.; Mourard, D.; Bério, P.; Bonneau, D.; Chesneau, O.; Clausse, J.M.; Cruzalebes, P.; et al. Kinematics and geometrical study of the Be stars 48 Persei and ψ Persei with the VEGA/CHARA interferometer. *Astron. Astrophys.* **2011**, *529*, A87. [CrossRef]
476. Stee, P.; Delaa, O.; Monnier, J.D.; Meilland, A.; Perraut, K.; Mourard, D.; Che, X.; Schaefer, G.H.; Pedretti, E.; Smith, M.A.; et al. The relationship between γ Cassiopeiae's X-ray emission and its circumstellar environment. II. Geometry and kinematics of the disk from MIRC and VEGA instruments on the CHARA Array. *Astron. Astrophys.* **2012**, *545*, A59. [CrossRef]
477. Stee, P.; Meilland, A. Stee, P.; Meilland, A. VLTI, CHARA and NPOI Observations of Be Stars. In *Proceedings of the Circumstellar Dynamics at High Resolution, Proceedings on the ESO/IAG/USP Workshop, Foz do Iguaçu, Brazil, 27 February–2 March 2012*; Astronomical Society of the Pacific Conference Series; Carciofi, A.C., Rivinius, T., Eds.; ASP, 2012; Volume 464, p. 167.
478. Cochetti, Y.R.; Arcos, C.; Kanaan, S.; Meilland, A.; Cidale, L.S.; Curé, M. Spectro-interferometric observations of a sample of Be stars. Setting limits to the geometry and kinematics of stable Be disks. *Astron. Astrophys.* **2019**, *621*, A123.
479. van Belle, G.T.; Ciardi, D.R.; ten Brummelaar, T.; McAlister, H.A.; Ridgway, S.T.; Berger, D.H.; Goldfinger, P.J.; Sturmann, J.; Sturmann, L.; Turner, N.; et al. First Results from the CHARA Array. III. Oblateness, Rotational Velocity, and Gravity Darkening of Alderamin. *Astrophys. J.* **2006**, *637*, 494–505.
480. Pelletier, G.; Pudritz, R.E. Hydromagnetic Disk Winds in Young Stellar Objects and Active Galactic Nuclei. *Astrophys. J.* **1992**, *394*, 117. [CrossRef]
481. Bodenheimer, P. Angular Momentum Evolution of Young Stars and Disks. *Annu. Rev. Astron. Astrophys.* **1995**, *33*, 199–238. [CrossRef]
482. Ray, T. Losing spin: The angular momentum problem. *Astron. Geophys.* **2012**, *53*, 5.19–5.22. [CrossRef]
483. Larson, R.B. Processes in Collapsing Interstellar Clouds. *Annu. Rev. Astron. Astrophys.* **1973**, *11*, 219. [CrossRef]
484. Burki, G.; Maeder, A. Observational tests on star formation I: Size variation of very young clusters through the Galaxy. *Astron. Astrophys.* **1976**, *51*, 247–254.
485. Guthrie, B.N.G. The rotation of early-type stars and the problem of star formation. *Mon. Not. R. Astron. Soc.* **1984**, *210*, 159–171. [CrossRef]
486. Bodenheimer, P. Evolution of rotating interstellar clouds. III. On the formation of multiple star systems. *Astrophys. J.* **1978**, *224*, 488–496. [CrossRef]
487. Deutsch, A.J. Maxwellian Distributions for Stellar Rotations. In *Proceedings of the IAU Colloq. 4: Stellar Rotation*; Slettebak, A., Ed.; Gordon and Breach Science Publishers: Philadelphia, PA, USA, 1970; p. 207.
488. Maeder, A.; Grebel, E.K.; Mermilliod, J.C. Differences in the fractions of Be stars in galaxies. *Astron. Astrophys.* **1999**, *346*, 459–464.
489. Huang, W.; Gies, D.R. Stellar Rotation in Young Clusters. II. Evolution of Stellar Rotation and Surface Helium Abundance. *Astrophys. J.* **2006**, *648*, 591–606.
490. Fabregat, J.; Torrejón, J.M. On the evolutionary status of Be stars. *Astron. Astrophys.* **2000**, *357*, 451–459.
491. Keller, S.C.; Grebel, E.K.; Miller, G.J.; Yoss, K.M. UBVI and Hα Photometry of the h and χ Persei Cluster. *Astron. J.* **2001**, *122*, 248–256.
492. McSwain, M.V.; Gies, D.R. The Evolutionary Status of Be Stars: Results from a Photometric Study of Southern Open Clusters. *Astrophys. J. Suppl. Ser.* **2005**, *161*, 118–146.
493. Maeder, A. Influence of axial stellar rotation of age estimates of open star clusters. *Astron. Astrophys.* **1971**, *10*, 354–361.
494. Packet, W. On the spin-up of the mass accreting component in a close binary system. *Astron. Astrophys.* **1981**, *102*, 17–19.

495. Pols, O.R.; Cote, J.; Waters, L.B.F.M.; Heise, J. The formation of Be stars through close binary evolution. *Astron. Astrophys.* **1991**, *241*, 419.
496. Klement, R.; Carciofi, A.C.; Rivinius, T.; Ignace, R.; Matthews, L.D.; Torstensson, K.; Gies, D.; Vieira, R.G.; Richardson, N.D.; Domiciano de Souza, A.; et al. Prevalence of SED Turndown among Classical Be Stars: Are All Be Stars Close Binaries? *Astrophys. J.* **2019**, *885*, 147.
497. Huang, S.S. A Statistical Study of the Rotation of the Stars. *Astrophys. J.* **1953**, *118*, 285. [CrossRef]
498. van Dien, E. Axial Rotation of the Brighter Stars in the Pleiades Cluster. *J. R. Astron. Soc. Can.* **1948**, *42*, 249.
499. Slettebak, A. The Spectra and Rotational Velocities of the Bright Stars of Draper Types B8-A2. *Astrophys. J.* **1954**, *119*, 146. [CrossRef]
500. Slettebak, A.; Howard, R.F. Axial Rotation in the Brighter Stars of Draper Types B2-B5. *Astrophys. J.* **1955**, *121*, 102. [CrossRef]
501. Kuiper, G.P. Problems of Double-Star Astronomy. I. *Publ. Astron. Soc. Pac.* **1935**, *47*, 15. [CrossRef]
502. Chandrasekhar, S.; Münch, G. On the Integral Equation Governing the Distribution of the True and the Apparent Rotational Velocities of Stars. *Astrophys. J.* **1950**, *111*, 142. [CrossRef]
503. Eddington, A.S. On a formula for correcting statistics for the effects of a known error of observation. *Mon. Not. R. Astron. Soc.* **1913**, *73*, 359–360. [CrossRef]
504. Lucy, L.B. An iterative technique for the rectification of observed distributions. *Astron. J.* **1974**, *79*, 745. [CrossRef]
505. Richardson, W.H. Bayesian-Based Iterative Method of Image Restoration. *J. Opt. Soc. Am.* **1972**, *62*, 55. [CrossRef]
506. Yudin, R.V. Statistical analysis of intrinsic polarization, IR excess and projected rotational velocity distributions of classical Be stars. *Astron. Astrophys.* **2001**, *368*, 912–931. [CrossRef]
507. Curé, M.; Rial, D.F.; Christen, A.; Cassetti, J. A method to deconvolve stellar rotational velocities. *Astron. Astrophys.* **2014**, *565*, A85.
508. Christen, A.; Escarate, P.; Curé, M.; Rial, D.F.; Cassetti, J. A method to deconvolve stellar rotational velocities II. The probability distribution function via Tikhonov regularization. *Astron. Astrophys.* **2016**, *595*, A50.
509. Orellana, R.; Escárate, P.; Curé, M.; Christen, A.; Carvajal, R.; Agüero, J.C. A method to deconvolve stellar rotational velocities. III. The probability distribution function via maximum likelihood utilizing finite distribution mixtures. *Astron. Astrophys.* **2019**, *623*, A138.
510. Yoon, S.C.; Langer, N.; Norman, C. Single star progenitors of long gamma-ray bursts. I. Model grids and redshift dependent GRB rate. *Astron. Astrophys.* **2006**, *460*, 199–208. [CrossRef]
511. Martayan, C.; Zorec, J.; Frémat, Y.; Ekström, S. Can massive Be/Oe stars be progenitors of long gamma ray bursts? *Astron. Astrophys.* **2010**, *516*, A103.
512. Ramachandran, V.; Hamann, W.R.; Oskinova, L.M.; Gallagher, J.S.; Hainich, R.; Shenar, T.; Sander, A.A.C.; Todt, H.; Fulmer, L. Testing massive star evolution, star formation history, and feedback at low metallicity. Spectroscopic analysis of OB stars in the SMC Wing. *Astron. Astrophys.* **2019**, *625*, A104.
513. Ramachandran, V.; Hamann, W.R.; Hainich, R.; Oskinova, L.M.; Shenar, T.; Sander, A.A.C.; Todt, H.; Gallagher, J.S. Stellar population of the superbubble N 206 in the LMC. II. Parameters of the OB and WR stars, and the total massive star feedback. *Astron. Astrophys.* **2018**, *615*, A40.
514. Puls, J.; Vink, J.S.; Najarro, F. Mass loss from hot massive stars. *Astron. Astrophys. Rev.* **2008**, *16*, 209–325.
515. Krumholz, M.R. Massive Star Formation: A Tale of Two Theories. In Proceedings of the New Horizons in Astronomy: Frank N. Bash Symposium, Proceedings, Frank N Bash Symp. No 2005: New Horizons in Astronomy : Austin, TX, USA, 16–18 October 2005; Kannappan, S.J., Redfield, S., Kessler-Silacci, J.E., Landriau, M., Drory, N., Eds.; Astronomical Society of the Pacific Conference Series; University of Chicago Press: Chicago, NY, USA, 2006; Volume 352, p. 31.
516. Bonnell, I.A.; Vine, S.G.; Bate, M.R. Massive star formation: Nurture, not nature. *Mon. Not. R. Astron. Soc.* **2004**, *349*, 735–741.
517. Bally, J.; Zinnecker, H. The Birth of High-Mass Stars: Accretion and/or Mergers? *Astron. J.* **2005**, *129*, 2281–2293.
518. Endal, A.S.; Sofia, S. Rotation in solar-type stars. I—Evolutionary models for the spin-down of the sun. *Astrophys. J.* **1981**, *243*, 625–640. [CrossRef]
519. Potter, A.T.; Tout, C.A.; Eldridge, J.J. Towards a unified model of stellar rotation. *Mon. Not. R. Astron. Soc.* **2012**, *419*, 748–759.
520. Potter, A.T.; Chitre, S.M.; Tout, C.A. Stellar evolution of massive stars with a radiative α-Ω dynamo. *Mon. Not. R. Astron. Soc.* **2012**, *424*, 2358–2370.
521. Denissenkov, P.A.; Ivanova, N.S.; Weiss, A. Main-sequence stars of 10 and 30 M_sun: Approaching the steady-state rotation. *Astron. Astrophys.* **1999**, *341*, 181–189.
522. Wolff, S.C.; Strom, S.E.; Hillenbrand, L.A. The Angular Momentum Evolution of 0.1-10 M_{solar} Stars from the Birth Line to the Main Sequence. *Astrophys. J.* **2004**, *601*, 979–999.
523. Tayler, R.J. Convection in rotating stars. *Mon. Not. R. Astron. Soc.* **1973**, *165*, 39. [CrossRef]
524. Deupree, R.G. Stellar Evolution with Arbitrary Rotation Laws. III. Convective Core Overshoot and Angular Momentum Distribution. *Astrophys. J.* **1998**, *499*, 340–347. [CrossRef]
525. Deupree, R.G. Two-dimensional Hydrodynamic Simulations of Zero-Age Main-Sequence Convective Cores. *Astrophys. J.* **2000**, *543*, 395–405. [CrossRef]
526. Kichatinov, L.L.; Rüdiger, G. Differential rotation in stellar convective envelopes. *Astron. Lett.* **1997**, *23*, 731–734.

527. Browning, M.K.; Brun, A.S.; Toomre, J. Simulations of Core Convection in Rotating A-Type Stars: Differential Rotation and Overshooting. *Astrophys. J.* **2004**, *601*, 512–529.
528. Arnett, W.D.; Meakin, C. Turbulent Mixing in Stars: Theoretical Hurdles. In *Proceedings of the Chemical Abundances in the Universe: Connecting First Stars to Planets*; Cunha, K., Spite, M., Barbuy, B., Eds.; 2010; Volume 265, pp. 106–110.
529. Augustson, K.C.; Brun, A.S.; Toomre, J. The Magnetic Furnace: Intense Core Dynamos in B Stars. *Astrophys. J.* **2016**, *829*, 92.
530. Rosen, A.L.; Krumholz, M.R.; Ramirez-Ruiz, E. What Sets the Initial Rotation Rates of Massive Stars? *Astrophys. J.* **2012**, *748*, 97.
531. Haemmerlé, L.; Eggenberger, P.; Meynet, G.; Maeder, A.; Charbonnel, C. Star formation with disc accretion and rotation. I. Stars between 2 and 22 $M_\odot$ at solar metallicity. *Astron. Astrophys.* **2013**, *557*, A112.
532. Haemmerlé, L.; Eggenberger, P.; Meynet, G.; Maeder, A.; Charbonnel, C.; Klessen, R.S. Massive star formation by accretion. II. Rotation: How to circumvent the angular momentum barrier? *Astron. Astrophys.* **2017**, *602*, A17.
533. Strittmatter, P.A. Stellar Rotation and Stellar Luminosity in Praesepe. *Astrophys. J.* **1966**, *144*, 430. [CrossRef]
534. Strittmatter, P.A.; Sargent, W.L.W. Stellar Rotation and the Position of the Metallic-Line Stars in the Color-Magnitude Diagram. *Astrophys. J.* **1966**, *145*, 130. [CrossRef]
535. Maeder, A. Stellar Rotation. *Publ. Obs. Geneva* **1968**, *75*, 125.
536. Smith, R.C. Effects of rotation in the colour-magnitude diagrams of Praesepe and the Hyades. *Mon. Not. R. Astron. Soc.* **1971**, *151*, 463. [CrossRef]
537. Cotton, A.; Smith, R.C. The theoretical spread of the main sequence due to stellar rotation. *Observatory* **1983**, *103*, 8–12.
538. Oke, J.B.; Greenstein, J.L. The Rotational Velocities of - - and G-Type Giant Stars. *Astrophys. J.* **1954**, *120*, 384. [CrossRef]
539. Sandage, A.R. Axial Rotation and Stellar Evolution. *Astrophys. J.* **1955**, *122*, 263. [CrossRef]
540. Danziger, I.J.; Faber, S.M. Rotation of evolving A and F stars. *Astron. Astrophys.* **1972**, *18*, 428.
541. Ekström, S.; Meynet, G.; Maeder, A.; Barblan, F. Evolution towards the critical limit and the origin of Be stars. *Astron. Astrophys.* **2008**, *478*, 467–485. [CrossRef]
542. Yang, W.; Bi, S.; Meng, X.; Tian, Z. Evolution of Rotational Velocities of A-type Stars. *Astrophys. J. Lett.* **2013**, *765*, L36.
543. Sun, W.; Duan, X.W.; Deng, L.; de Grijs, R. Exploring the Stellar Rotation of Early-type Stars in the LAMOST Medium-resolution Survey. II. Statistics. *Astrophys. J.* **2021**, *921*, 145.
544. Sun, W.; Duan, X.W.; Deng, L.; de Grijs, R.; Zhang, B.; Liu, C. Exploring the Stellar Rotation of Early-type Stars in the LAMOST Medium-resolution Survey. I. Catalog. *Astrophys. J. Suppl. Ser.* **2021**, *257*, 22.
545. Rhodes, E. J., J.; Deubner, F.L.; Ulrich, R.K. A new technique for measuring solar rotation. *Astrophys. J.* **1979**, *227*, 629–637. [CrossRef]
546. Deubner, F.L.; Ulrich, R.K.; Rhodes, E. J., J. Solar p-mode oscillations as a tracer of radial differential rotation. *Astron. Astrophys.* **1979**, *72*, 177–185.
547. Ando, H. A New Method for Determining the Internal Rotational Angular Velocity of the Stars. *Astrophys. Space Sci.* **1980**, *73*, 159–174. [CrossRef]
548. Reese, D.R. Internal rapid rotation and its implications for stellar structure and pulsations. *Eur. Phys. J. Conf.* **2015**, *101*, 05007. [CrossRef]
549. Aerts, C.; Thoul, A.; Daszyńska, J.; Scuflaire, R.; Waelkens, C.; Dupret, M.A.; Niemczura, E.; Noels, A. Asteroseismology of HD 129929: Core Overshooting and Nonrigid Rotation. *Science* **2003**, *300*, 1926–1928. [CrossRef]
550. Pamyatnykh, A.A.; Handler, G.; Dziembowski, W.A. Asteroseismology of the β Cephei star ν Eridani: Interpretation and applications of the oscillation spectrum. *Mon. Not. R. Astron. Soc.* **2004**, *350*, 1022–1028.
551. Briquet, M.; Morel, T.; Thoul, A.; Scuflaire, R.; Miglio, A.; Montalbán, J.; Dupret, M.A.; Aerts, C. An asteroseismic study of the β Cephei star θ Ophiuchi: Constraints on global stellar parameters and core overshooting. *Mon. Not. R. Astron. Soc.* **2007**, *381*, 1482–1488.
552. Dziembowski, W.A.; Pamyatnykh, A.A. The two hybrid B-type pulsators: ν Eridani and 12 Lacertae. *Mon. Not. R. Astron. Soc.* **2008**, *385*, 2061–2068.
553. Aerts, C.; Mathis, S.; Rogers, T.M. Angular Momentum Transport in Stellar Interiors. *Annu. Rev. Astron. Astrophys.* **2019**, *57*, 35–78.
554. Pedersen, M.G. Internal rotation and inclinations of slowly pulsating B stars: Evidence of interior angular momentum transport. *arXiv* **2022**, arXiv:2208.14497.
555. Hatta, Y.; Sekii, T.; Takata, M.; Kurtz, D.W. The Two-dimensional Internal Rotation of KIC 11145123. *Astrophys. J.* **2019**, *871*, 135.
556. Burssens, S.; Bowman, D.M.; Michielsen, M.; Simón-Díaz, S.; Aerts, C. Internal rotation and mixing in the massive star HD192575. In Proceedings of the Posters from the TESS Science Conference II (TSC2), Virtually, 2–6 August 2021; p. 75. [CrossRef]
557. Salmon, S.J.A.J.; Moyano, F.D.; Eggenberger, P.; Haemmerlé, L.; Buldgen, G. Backtracing the internal rotation history of the β Cep star HD 129929. *Astron. Astrophys.* **2022**, *664*, L1.
558. Salmon, S.J.A.J.; Montalbán, J.; Reese, D.R.; Dupret, M.A.; Eggenberger, P. The puzzling new class of variable stars in NGC 3766: Old friend pulsators? *Astron. Astrophys.* **2014**, *569*, A18.
559. Pedersen, M.G.; Aerts, C.; Pápics, P.I.; Michielsen, M.; Gebruers, S.; Rogers, T.M.; Molenberghs, G.; Burssens, S.; Garcia, S.; Bowman, D.M. Internal mixing of rotating stars inferred from dipole gravity modes. *Nat. Astron.* **2021**, *5*, 715–722.
560. Moravveji, E.; Townsend, R.H.D.; Aerts, C.; Mathis, S. Sub-inertial Gravity Modes in the B8V Star KIC 7760680 Reveal Moderate Core Overshooting and Low Vertical Diffusive Mixing. *Astrophys. J.* **2016**, *823*, 130.

561. Stoeckley, T.R.; Morris, C.S. Rotational distortion of stellar absorption lines. I. Parameters from photographic spectra. *Astrophys. J.* **1974**, *188*, 579–594. [CrossRef]
562. Buscombe, W.; Stoeckley, T.R. Absorption Line Profiles and Rotational Velocities for 59 Stars. *Astrophys. Space Sci.* **1975**, *37*, 197–220. [CrossRef]
563. Stoeckley, T.R.; Carroll, R.W.; Miller, R.D. Absorption line profiles for 39 rapidly rotating stars. *Mon. Not. R. Astron. Soc.* **1984**, *208*, 459. [CrossRef]
564. Elste, G.H. Line Shapes of Rotating Stars with Application to Alpha Lyrae. *Astrophys. J.* **1992**, *384*, 284. [CrossRef]
565. de Jager, C.; Nieuwenhuijzen, H.; van der Hucht, K.A. Mass loss rates in the Hertzsprung-Russell diagram. *Astron. Astrophys.* **1988**, *72*, 259–289.
566. Kudritzki, R.P.; Puls, J. Winds from Hot Stars. *Annu. Rev. Astron. Astrophys.* **2000**, *38*, 613–666. [CrossRef]
567. Vink, J.S.; de Koter, A.; Lamers, H.J.G.L.M. New theoretical mass-loss rates of O and B stars. *Astron. Astrophys.* **2000**, *362*, 295–309.
568. Keszthelyi, Z.; de Koter, A.; Götberg, Y.; Meynet, G.; Brands, S.A.; Petit, V.; Carrington, M.; David-Uraz, A.; Geen, S.T.; Georgy, C.; et al. The effects of surface fossil magnetic fields on massive star evolution: IV. Grids of models at Solar, LMC, and SMC metallicities. *Mon. Not. R. Astron. Soc.* **2022**, *517*, 2028–2055.
569. Donati, J.F.; Semel, M.; Carter, B.D.; Rees, D.E.; Collier Cameron, A. Spectropolarimetric observations of active stars. *Mon. Not. R. Astron. Soc.* **1997**, *291*, 658–682. [CrossRef]
570. Kochukhov, O.; Makaganiuk, V.; Piskunov, N. Least-squares deconvolution of the stellar intensity and polarization spectra. *Astron. Astrophys.* **2010**, *524*, A5.
571. Oksala, M.E.; Wade, G.A.; Townsend, R.H.D.; Owocki, S.P.; Kochukhov, O.; Neiner, C.; Alecian, E.; Grunhut, J. Revisiting the Rigidly Rotating Magnetosphere model for σ Ori E—I. Observations and data analysis. *Mon. Not. R. Astron. Soc.* **2012**, *419*, 959–970.
572. Grunhut, J.H.; Wade, G.A.; MiMeS Collaboration. The Incidence of Magnetic Fields in Massive Stars: An Overview of the MiMeS Survey Component. *AIP Conf. Proc.* **2012**, *1429*, 67.
573. Babel, J.; Montmerle, T. On the Periodic X-Ray Emission from the O7 V Star theta 1 Orionis C. *Astrophys. J. Lett.* **1997**, *485*, L29–L32. [CrossRef]
574. ud-Doula, A.; Owocki, S.P. Dynamical Simulations of Magnetically Channeled Line-driven Stellar Winds. I. Isothermal, Nonrotating, Radially Driven Flow. *Astrophys. J.* **2002**, *576*, 413–428.
575. Petit, V.; Owocki, S.P.; Wade, G.A.; Cohen, D.H.; Sundqvist, J.O.; Gagné, M.; Maíz Apellániz, J.; Oksala, M.E.; Bohlender, D.A.; Rivinius, T.; et al. A magnetic confinement versus rotation classification of massive-star magnetospheres. *Mon. Not. R. Astron. Soc.* **2013**, *429*, 398–422.
576. Townsend, R.H.D.; Owocki, S.P. A rigidly rotating magnetosphere model for circumstellar emission from magnetic OB stars. *Mon. Not. R. Astron. Soc.* **2005**, *357*, 251–264.
577. Ud Doula, A.; Owocki, S.P.; Townsend, R.H.D. Dynamical simulations of magnetically channelled line-driven stellar winds—II. The effects of field-aligned rotation. *Mon. Not. R. Astron. Soc.* **2008**, *385*, 97–108.
578. Shultz, M.; Wade, G.; Rivinius, T.; Neiner, C.; Alecian, E.; Petit, V.; Grunhut, J.; MiMeS Collaboration; BinaMIcS Collaboration. What can magnetic early B-type stars tell us about early B-type stars in general? In *Proceedings of the Lives and Death-Throes of Massive Stars*; Eldridge, J.J.; Bray, J.C.; McClelland, L.A.S..; Xiao, L., Eds.; Cambridge University Press: Cambridge, UK, 2017; Volume 329, pp. 126–130. [CrossRef]
579. Shultz, M.E.; Owocki, S.; Rivinius, T.; Wade, G.A.; Neiner, C.; Alecian, E.; Kochukhov, O.; Bohlender, D.; ud-Doula, A.; Landstreet, J.D.; et al. The magnetic early B-type stars—IV. Breakout or leakage? H α emission as a diagnostic of plasma transport in centrifugal magnetospheres. *Mon. Not. R. Astron. Soc.* **2020**, *499*, 5379–5395.
580. Shultz, M.E.; Owocki, S.P.; ud-Doula, A.; Biswas, A.; Bohlender, D.; Chandra, P.; Das, B.; David-Uraz, A.; Khalack, V.; Kochukhov, O.; et al. MOBSTER—VI. The crucial influence of rotation on the radio magnetospheres of hot stars. *Mon. Not. R. Astron. Soc.* **2022**, *513*, 1429–1448.
581. Shultz, M.E.; Wade, G.A.; Rivinius, T.; Alecian, E.; Neiner, C.; Petit, V.; Owocki, S.; ud-Doula, A.; Kochukhov, O.; Bohlender, D.; et al. The magnetic early B-type stars—III. A main-sequence magnetic, rotational, and magnetospheric biography. *Mon. Not. R. Astron. Soc.* **2019**, *490*, 274–295.
582. Weber, E.J.; Davis, Leverett, J. The Angular Momentum of the Solar Wind. *Astrophys. J.* **1967**, *148*, 217–227. [CrossRef]
583. Ud-Doula, A.; Owocki, S.P.; Townsend, R.H.D. Dynamical simulations of magnetically channelled line-driven stellar winds—III. Angular momentum loss and rotational spin-down. *Mon. Not. R. Astron. Soc.* **2009**, *392*, 1022–1033.
584. Shultz, M.E.; Wade, G.A.; Rivinius, T.; Neiner, C.; Alecian, E.; Bohlender, D.; Monin, D.; Sikora, J.; MiMeS Collaboration.; BinaMIcS Collaboration. The magnetic early B-type stars I: Magnetometry and rotation. *Mon. Not. R. Astron. Soc.* **2018**, *475*, 5144–5178.
585. Shultz, M.E.; Wade, G.A.; Rivinius, T.; Alecian, E.; Neiner, C.; Petit, V.; Wisniewski, J.P.; MiMeS Collaboration.; BinaMIcS Collaboration. The magnetic early B-type Stars II: Stellar atmospheric parameters in the era of Gaia. *Mon. Not. R. Astron. Soc.* **2019**, *485*, 1508–1527.
586. Shultz, M.; Rivinius, T.; Das, B.; Wade, G.A.; Chandra, P. The accelerating rotation of the magnetic He-weak star HD 142990. *Mon. Not. R. Astron. Soc.* **2019**, *486*, 5558–5566.

587. Grunhut, J.H.; Wade, G.A.; Neiner, C.; Oksala, M.E.; Petit, V.; Alecian, E.; Bohlender, D.A.; Bouret, J.C.; Henrichs, H.F.; Hussain, G.A.J.; et al. The MiMeS survey of Magnetism in Massive Stars: Magnetic analysis of the O-type stars. *Mon. Not. R. Astron. Soc.* **2017**, *465*, 2432–2470.
588. Fossati, L.; Castro, N.; Schöller, M.; Hubrig, S.; Langer, N.; Morel, T.; Briquet, M.; Herrero, A.; Przybilla, N.; Sana, H.; et al. B fields in OB stars (BOB): Low-resolution FORS2 spectropolarimetry of the first sample of 50 massive stars. *Astron. Astrophys.* **2015**, *582*, A45.
589. Alecian, E.; Neiner, C.; Wade, G.A.; Mathis, S.; Bohlender, D.; Cébron, D.; Folsom, C.; Grunhut, J.; Le Bouquin, J.B.; Petit, V.; et al. The BinaMIcS project: Understanding the origin of magnetic fields in massive stars through close binary systems. In *Proceedings of the New Windows on Massive Stars*; Meynet, G., Georgy, C., Groh, J., Stee, P., Eds.; Cambridge University Press: Cambridge, UK, 2015; Volume 307, pp. 330–335.
590. Castro, N.; Fossati, L.; Hubrig, S.; Simón-Díaz, S.; Schöller, M.; Ilyin, I.; Carrol, T.A.; Langer, N.; Morel, T.; Schneider, F.R.N.; et al. B fields in OB stars (BOB). Detection of a strong magnetic field in the O9.7 V star HD 54879. *Astron. Astrophys.* **2015**, *581*, A81.
591. Fossati, L.; Castro, N.; Morel, T.; Langer, N.; Briquet, M.; Carroll, T.A.; Hubrig, S.; Nieva, M.F.; Oskinova, L.M.; Przybilla, N.; et al. B fields in OB stars (BOB): On the detection of weak magnetic fields in the two early B-type stars beta CMa and epsilon CMa. Possible lack of a "magnetic desert" in massive stars. *Astron. Astrophys.* **2015**, *574*, A20.
592. Hubrig, S.; Schöller, M.; Fossati, L.; Morel, T.; Castro, N.; Oskinova, L.M.; Przybilla, N.; Eikenberry, S.S.; Nieva, M.F.; Langer, N. B fields in OB stars (BOB): FORS 2 spectropolarimetric follow-up of the two rare rigidly rotating magnetosphere stars HD 23478 and HD 345439. *Astron. Astrophys.* **2015**, *578*, L3.
593. Morel, T.; Castro, N.; Fossati, L.; Hubrig, S.; Langer, N.; Przybilla, N.; Schöller, M.; Carroll, T.; Ilyin, I.; Irrgang, A.; et al. The B Fields in OB Stars (BOB) Survey. In *Proceedings of the New Windows on Massive Stars*; Meynet, G., Georgy, C., Groh, J., Stee, P., Eds.; Cambridge University Press: Cambridge, UK, 2015; Volume 307, pp. 342–347.
594. Przybilla, N.; Fossati, L.; Hubrig, S.; Nieva, M.F.; Järvinen, S.P.; Castro, N.; Schöller, M.; Ilyin, I.; Butler, K.; Schneider, F.R.N.; et al. B fields in OB stars (BOB): Detection of a magnetic field in the He-strong star CPD-57° 3509. *Astron. Astrophys.* **2016**, *587*, A7.
595. Schöller, M.; Hubrig, S.; Fossati, L.; Carroll, T.A.; Briquet, M.; Oskinova, L.M.; Järvinen, S.; Ilyin, I.; Castro, N.; Morel, T.; et al. B fields in OB stars (BOB): Concluding the FORS 2 observing campaign. *Astron. Astrophys.* **2017**, *599*, A66.
596. Oksala, M.E.; Neiner, C.; Georgy, C.; Przybilla, N.; Keszthelyi, Z.; Wade, G.; Mathis, S.; Blazère, A.; Buysschaert, B. The evolution of magnetic fields in hot stars. In *Proceedings of the The Lives and Death-Throes of Massive Stars*; Eldridge, J.J., Bray, J.C., McClelland, L.A.S., Xiao, L., Eds.; 2017; Volume 329, pp. 141–145.
597. Neiner, C.; Oksala, M.E.; Georgy, C.; Przybilla, N.; Mathis, S.; Wade, G.; Kondrak, M.; Fossati, L.; Blazère, A.; Buysschaert, B.; et al. Discovery of magnetic A supergiants: The descendants of magnetic main-sequence B stars. *Mon. Not. R. Astron. Soc.* **2017**, *471*, 1926–1935.
598. Kobzar, O.; Khalack, V.; Bohlender, D.; Mathys, G.; Shultz, M.E.; Bowman, D.M.; Paunzen, E.; Lovekin, C.; David-Uraz, A.; Sikora, J.; et al. Analysis of eight magnetic chemically peculiar stars with rotational modulation. *Mon. Not. R. Astron. Soc.* **2022**.
599. Shultz, M.; Wade, G.A.; Neiner, C.; Kochukhov, O. Magnetic Stars Observed by BRITE. In *Proceedings of the 3rd BRITE Science Conference*; Wade, G.A., Baade, D., Guzik, J.A., Smolec, R., Eds.; Polish Astronomical Society: Warsawm, Poland, 2018; Volume 8, pp. 146–153.
600. Schneider, F.R.N.; Ohlmann, S.T.; Podsiadlowski, P.; Röpke, F.K.; Balbus, S.A.; Pakmor, R.; Springel, V. Stellar mergers as the origin of magnetic massive stars. *Nature* **2019**, *574*, 211–214.

Disclaimer/Publisher's Note: The statements, opinions and data contained in all publications are solely those of the individual author(s) and contributor(s) and not of MDPI and/or the editor(s). MDPI and/or the editor(s) disclaim responsibility for any injury to people or property resulting from any ideas, methods, instructions or products referred to in the content.

Article

Hidden Spectra Treasures in the Foster Archive: A Pilot Study of the Be Stars α Eri, α Col, ω Car and η Cen

Catalina Arcos [1,*], Leonardo Vanzi [2,3], Nikolaus Vogt [1], Stefano Garcia [2], Virginia Ortiz [2] and Ester Acuña [4]

1. Instituto de Física y Astronomía, Facultad de Ciencias, Universidad de Valparaíso, Valparaíso 2360102, Chile
2. Center of Astro Engineering, Pontificia Universidad Católica de Chile, Santiago 7820436, Chile
3. Department of Electrical Engineering, Pontificia Universidad Católica de Chile, Santiago 7820436, Chile
4. Observatorio UC Manuel Foster, Santiago 8420541, Chile
* Correspondence: catalina.arcos@uv.cl

Abstract: We present the archive of spectroscopic photographic plates of the Universidad Católica historic observatory Manuel Foster. The archive includes more than 4800 plates covering the period of time from 1928 to 1991. The spectra present in the archive are mostly those of bright variable or binary stars observed at different epochs. We developed a method of digitalization and data processing for the plates and verified it through the analysis of a selected sample of spectra. As an example of the potential relevance of this Foster archive we studied the variation of helium, Hβ and Hγ spectral lines over time (1980–1991), complementing with high resolution spectroscopic data from the "Be Star Observation Survey" (2012–2015), of four Be stars mainly, α Eri, α Col, ω Car and η Cen. The spectra of these stars show evidence of a circumstellar gas disk present in both periods of time. From the spectroscopic analysis, we found these stars are variable in helium and this variability presents an opposite behavior with the variability observed in the EW of the Hβ line profile. This archive represents a unique source of data from past that is available for the use of the community.

Keywords: stars: individual: α Eri, α Col, ω Car, η Cen; archive; stars: emission-line, Be

Citation: Arcos, C.; Vanzi, L.; Vogt, N.; Garcia, S.; Ortiz, V.; Acuña, E. Hidden Spectra Treasures in the Foster Archive: A Pilot Study of the Be Stars α Eri, α Col, ω Car and η Cen. *Galaxies* **2022**, *10*, 106. https://doi.org/10.3390/galaxies10060106

Academic Editors: Jorick Vink and Margo Aller

Received: 6 October 2022
Accepted: 19 November 2022
Published: 22 November 2022

Publisher's Note: MDPI stays neutral with regard to jurisdictional claims in published maps and institutional affiliations.

Copyright: © 2022 by the authors. Licensee MDPI, Basel, Switzerland. This article is an open access article distributed under the terms and conditions of the Creative Commons Attribution (CC BY) license (https://creativecommons.org/licenses/by/4.0/).

1. Introduction

The historical observatory Manuel Foster of Pontificia Universidad Católica de Chile (PUC) counts with an archive of spectroscopic photographic plates. The archive includes more than 4800 plates covering the period of time from 1928 to 1991 with wavelength range between 3900 and 5150 Å. All plates were obtained with the Cassegrain telescope of 93 cm in diameter and 16.9 m focal length (F/18). The spectrograph (a copy of the Mills spectrograph [1]) includes three prisms as dispersing elements, allowing the choice of three different spectral dispersion, using either 1, or 2 or all 3 prisms. The collimator of the spectrograph with 724 mm focal length provides a beam of 37.4 mm, the objective has an equivalent focal length of 16 inches [1]. The low resolution (∼700) optical spectra (λ 3900–5150 Å) present in the archive (see Section 2.1 for details) are mostly those of bright variable or binary stars observed at different epochs (see [2]). A full list of observed targets is available in an excel online (https://docs.google.com/spreadsheets/d/1h2o_MItVYwjV8F2V3rpG0eLDCQiuq-43mYGuExAUFAY/edit#gid=0, accessed on 5 October 2022 document belonging to PUC. The vast majority of these spectroscopic plates have to be scanned, and represent a unique database that can be of significant use to the community. Digitalization of photographic plates is an on going effort at PUC because of the historic and scientific relevance of this archive. The aim of this paper is to introduce the Foster archive to the community and highlight the importance of the hidden spectra treasures of variable stars. In particular Foster spectra can be used to:

- Identify hydrogen emission lines and quantify the strength through equivalent width (EW) measurements;

- Measure the violet to red peak intensity ratio (V/R) and the double-peak separation (DPS) in the case of emission lines in the spectra;
- Measure EW variations in photospheric spectral lines, e.g., HeI/MgII;
- Look for asymmetry in spectral lines.

In this work we analyze the spectral variability of four Southern Classical Be stars (CBes) observed with Foster. CBes are non-supergiant massive stars with spectral type spanning from late-O to early-A stars that show at least once in their life hydrogen lines and, sometimes, metal lines in emission in their spectra (e.g., Hα, Hβ, Brγ). This emission arises from a gas disk placed at the stellar equator in a thin geometric structure and rotating in a quasi-Keplerian motion [3–5]. The different disk stages, quiescence, active and disk-feeding (definitions proposed by Rivinius et al. [6] under which the disk goes through, are studied following temporal variations. The mid-term variations (order of months) are associated to V/R variations which are well explained by the theory of one-armed over-density wave [7]. The long-term line intensity variations (months to years) are directly associated to the formation or dissipation of the circumstellar disk.

Several works have been done in the last years to constrain Be stars' disk and stellar parameters. The main question converges in understanding the mechanism(s) triggering a higher mass-loss rate at the stellar equator capable of driving a rotating equatorial gas disk. The proposed mechanisms include fast rotation [8,9], non-radial pulsations [10,11], stellar winds [12,13] and binarity [14,15] (see [16] for a complete review and references therein). To investigate these mechanisms it is necessary to study a large number of Be stars in different disk phases and monitor them over time to follow the evolution of their disks. Furthermore, looking for a correlation between the behavior of the spectral lines formed in the photosphere and the emission lines from the disk can give insight into the physics of the disk's formation and dissipation. In this context, spectroscopic databases of massive stars covering extended period of time become very important to search for variability. Then, historical archives can reveal unexpected hidden treasures (see [17] as an example of the value of ancient spectra).

Spectroscopic plates from the Foster Observatory were used earlier in some research on Southern Be stars [18] including the study of optical spectra (3900–5100 Å) of 36 Be stars with measurements of hydrogen, HeI 4471 Å and MgII 4481 Å line profiles. They mostly focused on the variability of the HeI and MgII line profiles over time, finding an anti-correlation of the EW ratios that could be related to non-radial pulsations. A similar work was made by [19] for 10 Be stars and 4 "normal" B-type stars. They obtained similar values of HeI and MgII EWs when comparing a B star and a Be star of the same spectral type. However, they observed dependence of the HeI/MgII EW ratio with stellar rotation, i.e., HeI EWs become larger for faster rotators. Finally, in the case of [20] the authors found similar conclusions but using the measurement of Hα EWs in the optical spectra of 42 Be stars observed at CTIO. They found a dependence between Hα EWs and spectral types, being lower for later-type Be stars.

Other publications describing variability of Southern Be stars observed with Foster, as well as estimations of the disk cycle, disk size and rotational velocity of the star are reported in the literature (see [21–23] as examples), probing the potential information contained in these spectroscopic plates.

2. Extracting the Spectra from Foster Plates

2.1. Observations and Archive Properties

The 2-prism version of the Mills spectrograph, used during the latest activity period of the Manuel Foster observatory, gives a dispersion of 20 Å/mm at Hγ and a useful wavelength range from 3900 to 5150 Å. The spectrograms were secured on Kodak II-aO plates (after baking with Forming gas). He-Ar lamps served as wavelength comparison line sources acquired before and after the science observations. The slit configuration of the spectrograph and the position of the He-Ar lamps can be seen in Figure 1 of [2] which also describes the method for intensity calibration. Calibration plates were taken during most observing nights; however, their posterior analysis showed that the final calibration curves (density to intensity relation) did not vary significantly from night to night, because during the entire campaign always the same plate type and the same baking and developing procedures had been applied. Therefore, we finally decided to use in our present analysis a mean calibration curve, as already presented in Figure 4 of [2].

The archive contains more than 4800 stellar spectrograms of 276 different target stars (see online excel document (https://docs.google.com/spreadsheets/d/1h2o_MItVYwjV8F2V3rpG0eLDCQiuq-43mYGuExAUFAY/edit#gid=0, accessed on 5 October 2022). For 88 of them ≥ 10 spectral plates at different epochs are available, the maximum number is 219 plates for the Wolf-Rayet star γ Velorum. 99 of these targets refer to B III-Ve spectral types with an average coverage of 45 spectrograms per star mostly observed between 1984 and 1990. The line profile variability was the main aim of this campaign. However, the number of plates should be taken as an upper limit, because some of the plates are probably not very useful, affected by bad weather conditions or technical problems, resulting in saturation or underexposure. The remaining targets refer to OB supergiants, A-K supergiants, A-M giants, A-F main sequence stars and Wolf-Rayet stars. The faintest star is V = 5.7 mag, but most of them are in the range $0.5^m \leq V \leq 4.5^m$. We recompiled names and observation years of all Be stars in the Foster archive which are list in Table A1.

2.2. Digitalization and Data Processing

We developed a procedure of digitalization of the plates which include scanning, processing and data analysis. In this way we were able to extract scientific information at a much deeper level than it could be done with previously employed techniques. For the scanning of the plates we used a digital scanner EPSON Perfection V600 which allows to scan negative plates with resolution up to 12,800 dpi and 16 bits. The plates have sizes 100 × 27 mm, while the stellar spectra cover an area of approximately 50 × 0.8 mm. This allows to scan up to 6 plates at a time. The scanning time per plate is about 180 s at 3200 dpi. In the upper image of the Figure 1 we present a typical example of a digitalized plate. A zoomed region is shown in the lower image, where the photographic grains are just resolved at 3200 dpi, which is equivalent to 7.9 µm per pixel.

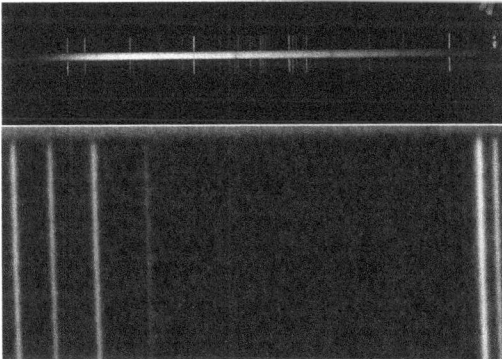

Figure 1. (**Upper**) Example of a digitalized plate from the Foster archive scanned with 3200 dpi. (**Lower**) Zoom of the previous plate where emission lines of the calibrated spectrum and the photographic grains are visible.

Data processing is necessary to transform the digitalized images delivered by the scanner in images that can be used to measure quantities of scientific interest. In particular the processing of the digitalized plates include:

- conversion from scanner units to photographic density,
- conversion from photographic density to intensity,
- extraction of the target and calibration spectra and

- wavelength calibration.

The scanner was set to work linearly and provide an output image which, in principle must be proportional to the amount of light transmitted by the plate, i.e., to the transmission T. All color or grey-level correction algorithms included in the software were set to OFF.

The photographic density is given by $D = -\log(T)$. To calibrate the relation between the 16-bit output of the scanner and the photographic density we used two neutral density filters spanning from density 0.04 to 2. One of them is a continuous filter with the density varying linearly with the position, in the other the density varies by known steps. We measured the transmission of the neutral density filters in the Lab and verified that their behavior was according to specs. The result obtained scanning the continuous filter is shown in Figure 2. The density to intensity conversion was made following the recipe of [2], in particular we fit the function (8) of [24] to the 130 points of Vogt & Barrera and inverted the function to calculate the relative illuminating 131 intensity.

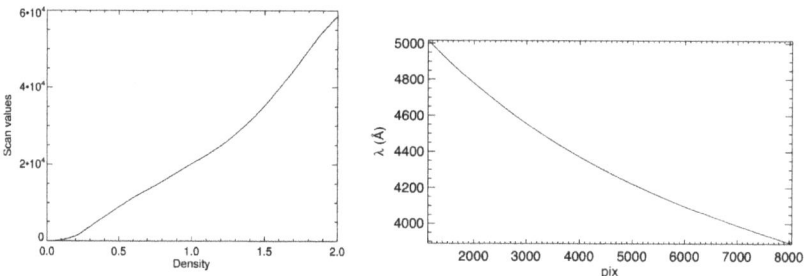

Figure 2. (**Left**) Signal generated by the scanner for the neutral density filter. The 2D image has been averaged in the spatial direction. The curve shows the conversion from scanner output to photographic density assuming linear behavior for the filter according to the manufacturer specifications and also verified by measurements in the Lab. (**Right**) Example of typical wavelength versus position solution for the 2 prism configuration.

A constant value of background was measured in the non exposed part of each plate and subtracted from each 2D spectrum once converted to intensity. The spectrum of the source was extracted with an aperture which traces the peak of emission in the direction of the spectral dispersion and averaged to produce a 1D spectrum. A linear trace proved to be good enough for this purpose, the optimization of the position and tilt of the extraction aperture is weighted for the brightness of the source. In this way possible misalignment of the plate or the scanning are compensated. Two calibration spectra are extracted following the same trace from the two sides of the spectrum of the scientific source. The extraction process was executed with an optimized procedure in IDL.

The wavelength calibration was performed with the task "identify" in IRAF using the comparison Argon spectra recorded in the plates on both sides of the scientific spectrum. Figure 2 shows a plot of lambda versus pixel position for the two-prism configuration of the spectrograph. Based on these results we found the average of the spectral dispersion to be 20 Å/mm and for the spectral resolution an average value $\lambda/\delta\lambda = 700$, for the two-prism configuration. A typical 1D spectrum wavelength calibrated and normalized is shown in Figure 3.

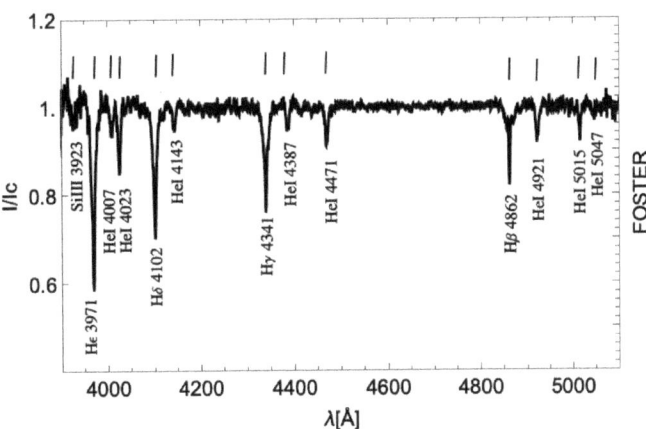

Figure 3. View of a typical 1D spectrum extracted from a scanned plate. The flux normalized spectrum corresponds to the star η Cen (HR 5540).

3. The Sample

To conduct a pilot study of the variability of CBes and highlight the importance of the archives containing spectra observed several years ago, we started with four CBes which count on a large number of observations available: η Cen, α Eri, ω Car, and α Col. These stars were observed from 1980 to 1991 with Foster. In addition recent information about the disk phases is available in the literature making them perfect targets for this work. We complemented the data with spectra from BeSOS database [1] obtained with the "PUC High Echelle Resolution Optical Spectrograph", Pucheros, [25] installed at the ESO 50 cm telescope of the Observatory Universidad Católica, Santa Martina, Santiago, Chile. These spectra cover the visible spectral range with a spectral resolution of $\sim$17,000. Observations were taken between 2012 and 2015, covering an effective wavelength range 4250–7000 Å. For details on the observations plan and reduction steps, the reader is referred to [26].

4. Method

Since the gas disk of Be stars is optically thin to the continuous radiation in the optical range, observing emission signatures in Balmer lines is usual, especially Hα and Hβ. We selected the most relevant spectral lines in common between the first (Foster) and second period (BeSOS) from Table 1 to perform the measurements (e.g., EW, DPS, and V/R). Spectral lines in Be stars are usually asymmetric due to non-radial pulsations. Here, the Foster spectra do not have enough spectral resolution to study asymmetries and in several cases the spectral lines are distinguishable from the spectra but very noisy. To avoid adding extra flux in the EWs calculation, we fit a Voigt function to the observed absorption line profiles (HeI and MgII lines) and then we integrated this function over the wavelength range. Because Balmer lines are stronger and emission features are presented in the spectral lines, we performed a piecewise (order-3) polynomial fit to the data points and then integrated this function over the spectral range to obtain the line EW (see [27]). The errors were calculated by following the procedure described in [27] for weak and strong lines. An example of the method is shown in Figure 4 for a Foster spectrum line (solid black line). The EW area is indicated as the shaded region in pink. In the left and right panel, the fitted function is represented by a dashed red line and the continuum was set to 1 in all cases (dashed black line). As this emission feature grows inside the absorption line, the EW value will be smaller. Then, small values of Balmer lines, in this work, will mean emission features appearing in the line.

Table 1. Lines present in at least one observed epoch in each star.

Lines (Å)	Hζ 3889	Hε 3970	HeI 4009	HeI 4026	Hδ 4101	HeI 4143	Hγ 4340	HeI 4387	HeI 4471	MgII 4481	HeI 4713	Hβ 4861	HeI 4921	HeI 5015	HeI 5047	FeII 5316	HeI 5876	Hα 6562	HeI 6678
Name/Sp.T																			
η Cen B2Ve	F	F	F	F	F	F	F/B	F/B	F/B	F/B	B	F/B	F/B	F/B	F/B	-	B	B	B
α Eri B6Vpe	F	F	F	F	F	F	F/B	F/B	F/B	F/B	B	F/B	F/B	B	B	-	B	B	B
ω Car B8IIIe	F	F	-	F	F	-	F/B	B	B	F/B	-	F/B	B	B	-	B	-	B	-
α Col B9Ve	F	F	-	F	F	-	F/B	F/B	F/B	F/B	-	F/B	F/B	B	B	-	-	B	-

Note: F and B symbols indicate if the line is present in Foster and BeSOS spectra, respectively. For cases without a line (or difficult to distinguish), the symbol–appears. In the case of FeII lines $\lambda\lambda$ 4549, 4556, 4584, 4629, 4667 and HeI $\lambda\lambda$ 4437, 4120, there are no signatures of these lines in the spectra.

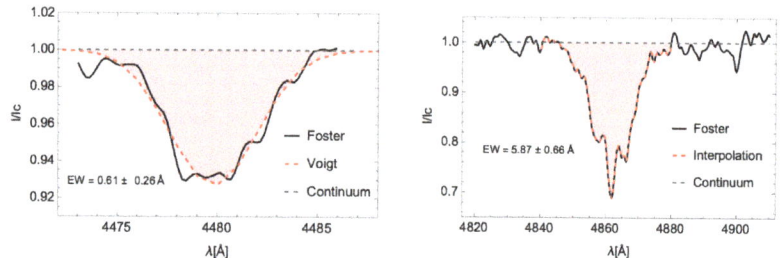

Figure 4. Example of the method used to measure the EW in Helium/MgII lines (**Left**) and Balmer lines (**Right**). The shaded region in pink indicates the EW.

The DPS and V/R measurements were performed over the Hβ line profile. The only star with no evidence of emission (in the first period) was α Eri and in this case we set the DPS as the width of the wings. We select the violet and red peaks by visual inspection. An example of the method is shown in the left panel of Figure 5 for one epoch of the star ω Car in the Foster archive. To quantify the difference between values obtained with different spectral resolutions, we selected a BeSOS spectrum of the same star and convoluted it with several Gaussian profiles of different widths (i.e., spectral resolutions) and then we re-binned the degraded spectra. We did it up to fit an observed Foster spectrum (see right panel of Figure 5). We obtained an effective spectral resolution of R∼1500 measured on HeI 4471 Å and MgII 4481 Å line profiles (dashed red line). After this step, we measured the EW in the HR and LR spectrum. We obtained a difference of 38% between both lines. We used this effective spectral resolution to degrade a Hβ line profile to quantify the V/R and DPS measured difference. We obtained a large difference of 61.7% for the DPS value and 1.6% for the V/R value. Also, we note that for a spectral resolution R∼1500 the minimum DPS resolved in terms of velocity is ∼200 km/s.

In the following, We present a detailed description of the EWs variation for Hγ and Hβ Balmer lines; HeI $\lambda\lambda$ 4387, 4921, and 4471; and MgII line. In addition, we calculate the DPS and V/R variation for the Hβ emission line for the four CBes studied in this paper (see Table 1). Information obtained from these measurements is used to study the connection between the variability of the photospheric lines with the disk phase. EW values for each star are displayed in the same figure from the latest (bottom) to the earliest spectral type (top). Each subplot is separated into two periods scaled by ten years (the period between 48,100 to 55,000 JD does not contain data) due to the lack of observations in both archives. Grid lines in the vertical axes of the plots indicate ranges of 500 Julian days. Then, the EWs for Balmer and helium/MgII lines are displayed in Appendix B in Figures A1 and A2, respectively. On the other hand, DPS and V/R variations are displayed together with the spectral lines for both periods and for each star.

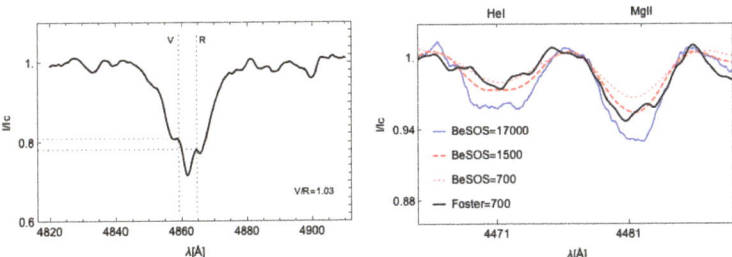

Figure 5. Example of the method used. (**left**) visual inspection to select V and R peaks on the Hβ line profile in the Foster archive. (**right**) degraded BeSOS spectrum (solid blue line) to lower resolutions of R = 1500 (dashed red line) and R = 700 (dotted pink line). In both panels the Foster spectrum is represented by a solid black line.

We discarded low exposure plates by selecting continuum regions around the analyzed lines and obtaining an average value, and its standard deviation, σ_s, for each normalized spectrum of each star[2]. The implemented selection criterion rejects all spectra with $\sigma_s \geq 0.03$. Because some σ_s values were more significant than the standard deviation selection criterion for the Foster spectra, the number of EW values is not the same for each line, and the EW ratio between HeI 4471 Å and MgII 4481 Å is not presented.

5. Results

5.1. η Cen-HR 5440-HD 127972

η Cen is a B2 Ve type star [28,29] showing a shell signature in Hα line profile since the '60s [30]. By modeling the spectral energy distribution, [26] estimated an earlier spectral type, B1 Ve, where the best Kurucz model is represented for the following stellar parameters: T_{eff} = 21,000 ± 210 K, log g – 3.95 + 0.04, $R_\star$ = 6.10 ± 0.12 $R_\odot$ and $v \sin i$ = 240 km/s. The truncation radius estimated by modeling the Hα emission line is 12.5 $R_\star$ with a low density disk [31]. η Cen is the third occurrence of Be stars showing the particular case of Stêfl frequencies [10,32,33]. These frequencies are usually temporary and about 10–20% slower than those associated to typical non-radial pulsations in Be stars. It is strongly suggested that they correlate with an outburst (mass-loss episodes that increase the emission line). From our observations archive, we have 40 spectra between 1984 and 1990 (first period) and 3 spectra spanning 2014 to 2015 (second period). As we can see in the EW plot for Balmer lines (see Figure A1), between 1984 to 1985, the values are between 3 and 4 Å and from 1987 a recurrent variation is observed, overall for the Hβ EW, with values ranging between 1 and 4 Å being on average around 2 Å and representing an increment of the emission features in the line. In the second period these values are similar to those of early 1984. Although, as mentioned above, this star presents non-radial pulsations that seem connected with mass-loss episodes, a further study comparing photometric variations simultaneously with ground-based observations could help to determine the connection between these episodes and the changes observed in the disk. Balmer lines for this star have an EW value between ∼1.0 and 5.0 Å. At the top of Figure A2 the EW variation for helium lines and MgII is shown. The three lines of HeI maintain a similar behavior and constant range variation over time. The lines λ4471 and 4921 Å are the strongest. MgII is weak during the first period, and it was distinguished only in a few plates. During the second period was very weak and not measurements could be performed. Hβ line profiles (see Figure 6) present emissions in the wings during a significant period and vary in time. This emission is present from 1984 to 1991 and it can still be seen from 2014 to 2016. V/R plot shows periodical variation and a small amplitude of ∼0.06. This variation correlates with the one observed in the DPS plot: V/R reaches the maximum value of ∼1.06 when also does the DPS, with a value of ∼750 km/s. The DPS ranges from 450 to 750 km/s. There are only three observations available in the second period and both quantities, V/R and DPS, also change in a correlated way. Since DPS is related to the rotational movement of the emitting region, e.g, Huang's relation [34], these changes are also related with the EW of the Hβ emission line. There is no evidence in the spectra, or information in the literature for a companion. In general terms, the disk of η Cen maintains almost constant in size and mass over the analyzed range of time presenting small changes most likely associated with recurrent episodes of mass-loss rates.

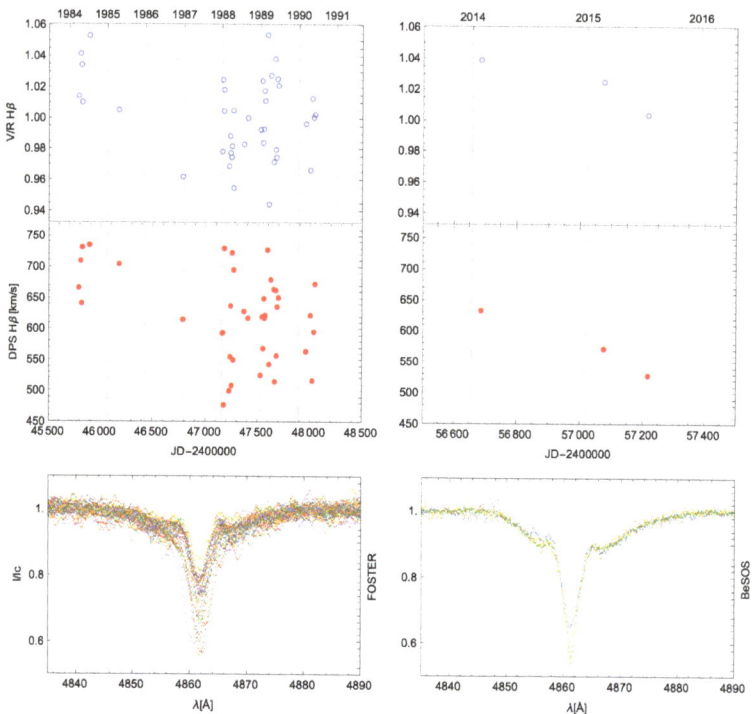

Figure 6. η Cen: V/R (**Top**) and DPS (**Middle**) variation measured on the Hβ emission line. Results are separated into two plots for the Foster (first column) and BeSOS (second column) periods. Spectral lines (**Bottom**) are over-plotted only to show variability purposes. Grid lines in the vertical axes are displayed every one year. DPS values measured in the first period can present until ∼62% of difference respect to the second period.

5.2. α Eri-HR 0472-HD 10144

α Eridani or HD 10144 is the brightest Be star with a visual magnitude of 0.4^{mag} and is located at a distance of 44 pc (Hipparcos). It is classified as a B6 Vpe spectral type, with a T_{eff} = 15,000 ± 600 K, log g = 3.60 ± 0.10 and a $v \sin i$ = 223 ± 15 km/s [28]. In a previous work, [31] studied the disk parameter for observations in BeSOS between 2013 and Jan 2014 (just a few months before a maximum intensity occurs in Hα). They found a base density disk of the order of $\sim 10^{-11}$ g/cm^3 and a truncation radius of 6.0 $R_\star$. From the literature [35], we know the binary nature of this star. Using IR images, Refs. [35,36] detected a companion at a distance of ∼12.3 AU, with a projected angular separation of ∼6.7 AU and a spectral type A1/3V. They do not have enough data in their work but proposed an orbital period of about 15 years. There is no evidence that the companion influences the disk formation. Ref. [37] used spectroscopic observations of Hα from 1991 to 2002, as well as data available in the literature to collect EWs values before 1990, to construct the line's variation history. The disk cycle proposed in their work is around 14–15 years. In addition, they studied the stellar surface activity by doing measurements on the line HeI 6678 Å, suggesting that the variation is due to non-radial pulsations. The observation dates studied in this work cover the years 1984, 1985, 1987, 1988, 1989, 1990, 2013, 2014, and 2015. We have 31 scanned plates from Foster and 12 spectra from BeSOS. The EWs variations for Balmer lines are shown in the second subplot from top to bottom in Figure A1. The EWs of both lines are very similar in their values and behavior in the first period of time, ranging between 2 and 4.5 Å. We know from the emission in the spectral lines that a disk formation phase is present for this star during the second period. However, the EW of Hβ keep around 4.0 Å, which could mean that the emission is also present during the first period, also in a low-emission phase (e.g., years 1985, 1987 and 1990). This is also observed in intensity changes in the Hβ line in Figure 7. The strongest Helium lines is 4471 Å with an almost constant EW value of 1 Å in both periods, and always stronger than MgII line. The EW variation for this star in the second period (active phase) is very low compared to the first one (see Figure A2). Figure 7 shows the V/R and DPS variations for this star. As we mentioned before, we cannot distinguish the emission from the Hβ line profiles in the first period, but thanks to the EW values we can infer a low-emission phase; therefore, for this star we have selected the width of the wings to represent the DPS. From the V/R plot, a small asymmetry remains constant with an amplitude of ~0.1, but with an opposite behavior compared to the second period. DPS values present an average value of ~700 km/s during the first period corresponding to a small emitting size (similar to η Cen in its maximum DPS). During the visibly active phase, a smaller value of DPS ~500 km/s is obtained. We remainder that a direct comparison between both periods, at least for the DPS, can not be done, and a difference of around 62% must be considered.

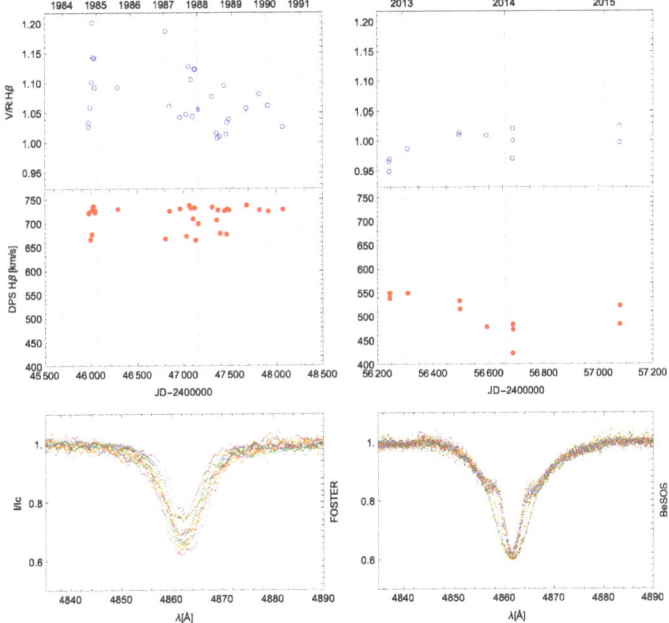

Figure 7. Same than Figure 6 but for α Eri. Changes in the spectral lines during the Foster period (first column) indicate a low-emission phase. In the Foster period we measured the DPS as the width of the wings.

5.3. ω Car-HR 4037-HD 89080

ω Car is a B8 IIIe spectral type star, the stellar parameters have been derived by several authors, e.g., Ref. [8] estimated a T_{eff} = 13,275 ± 251 K, log g = 3.581 ± 0.043 and a projected rotational velocity $v \sin i$ = 245 ± 13 km/s. Using stellar atmospheres model to reproduce the spectral energy distribution [26] found a lower temperature value of T_{eff} = 11,600 ± 116 K, a log g = 3.50 ± 0.04 and a stellar radius $R_\star$ = 7.20 ± 0.14 $R_\odot$. Optical and Near-IR spectroscopic and interferometric observations from 2012 show evidence of a Keplerian disk around ω Car. The emitting region has been estimated to be 3.1 ± 0.9 $R_\star$ at Brγ ([5], they considered a $R_\star$ = 6.2 $R_\odot$) and 25 $R_\star$ at Hα ([31], they considered a $R_\star$ = 7.20 $R_\odot$).

We used Foster spectra from 1985 to 1990 (52 observation dates) and the BeSOS database from 2013 to 2015 (2 spectra). The EWs of Balmer lines (see Figure A1) present a similar variability than η Cen but with deeper lines, where values go around 3.0 to 4.0 Å. In 1988/89 a higher dispersion is observed in both lines, and the emission increases (see Figure 8). The emission observed in the Hβ line profiles over time indicates that the disk has been present since the '85s. Only two points are available in the second period, and maintain similar to the variation range of the first period. We note the strong separation in EW between Hγ and Hβ in the second period. The helium lines for this star are weaker, with values between 0.5 and 0.6 Å, probably because of the low temperature. For the same reason a few EW values are available for this star because they do not fulfilled the criterion of σ. MgII is the strongest line present in this star with EW values between 0.3 and 0.6 Å. This is the only star of our sample showing FeII 5316 Å in its spectrum. The Hβ line profile shows a prominent DPS (see Figure 8) in both periods, but with the deepest central absorption during the second period. The variation of the V/R ratio is very similar to the one from η Cen, with a small amplitude of ∼0.06. The DPS ranges between ∼260 and 360 km/s.

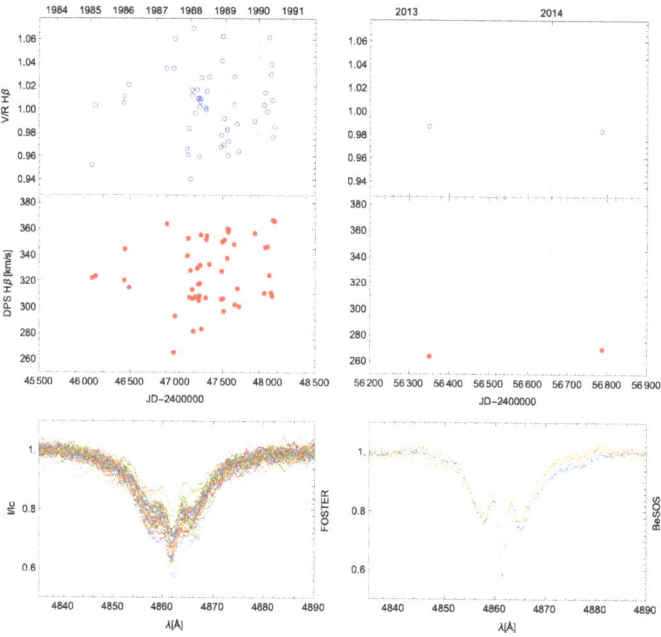

Figure 8. Same than Figure 6 but for ω Car. The disk has been present since the '85s.

5.4. α Col-HR 1956-HD 37795

α Col is classified as a B7 IIIe or B9 Ve spectral type [26,28]. First signatures of emission in Balmer lines was reported by [38] on Dec 1973. [8] derived the following stellar parameters: T_{eff} = 13,695 ± 437 K, log g = 3.559 ± 0.069 and $v \sin i$ = 192 ± 12 km/s. Similar values were found by [28] with a higher T_{eff} = 14,200 ± 400 K, a log g = 3.50 ± 0.10 and $v \sin i$ = 180 ± 15 km/s. Ref. [26] used rotational convoluted stellar atmosphere models and they found a T_{eff} = 12,200 ± 122 K, a log g = 3.50 ± 0.04 and a $R_\star$ = 7.00 ± 0.14 $R_\odot$. On the other hand, Ref. [5] obtained for the first time simultaneous Brγ spectroscopic observations and high angular resolution in the K-band of α Col using the ESO VLTI/AMBER instrument. To constrain the disk parameters, they used the stellar parameters derived by [8] in a simple two-dimensional kinematic model. A disk extension of 1.5 ± 0.3 stellar radii was estimated in the formation region of Brγ line. Their observations were carried out between 2007 and 2011, indicating the existence of a disk at that time. [31] analyzed observations between 2012 and 2015 and found a truncation radius disk measured in Hα of 50.0 $R_\star$ and a disk base density constant of 2.5×10^{-10} g/cm^3; they considered a B9 V spectral type for the central star. Our observations dates range from 1984 to 2015, corresponding to 50 scanned plates in Foster and 5 spectra in BeSOS. Figure A1 shows the EWs variation for Balmer lines. As α Col, Balmer lines are deeper compared to the other stars and the same EW separation from both lines, Hβ and Hγ, is observed. From the plot an increment of the emission is observed in 1988 and 1990. EWs values for HeI and MgII, are displayed in Figure A2. The HeI 4471 Å is the strongest Helium line for this star, but MgII has similar EW values. A high dispersion is presented during the fist period, with values ranging between 0.2 and 0.8 Å. In the second period, values are almost constant. On the other hand, the V/R ratio (see Figure 9) remains slightly below 1.0 until 1987 and above 1.0 up to 1990. Then in the second period, the V/R is symmetric and constant. The emission observed in the Hβ line profile increases in the second half of this period. The star presents the highest variation range in DPS with values between ~70 and 200 km/s and an almost constant value during the second period of 150 km/s. However, due to the low spectral resolution of Foster, we know that values lower than 200 km/s are not confident. Based on the DPS during the second period, we can trust values higher than 150 km/s in the first period of the plot, but having the consideration of 62% of difference that could exist between Foster and BeSOS data. From literature [17], a particular signature of a third peak is presented in Hα line-emission, varying in the order of days, shifting towards the central absorption, and sometimes disappearing. There is no evidence nor information in the literature of a companion for this star.

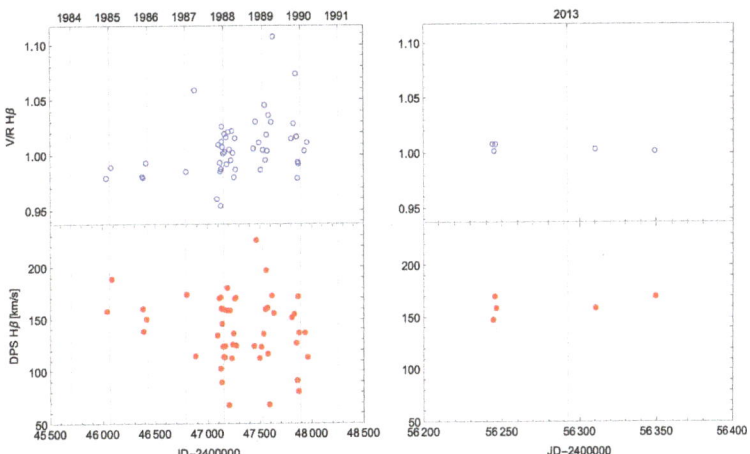

Figure 9. Cont.

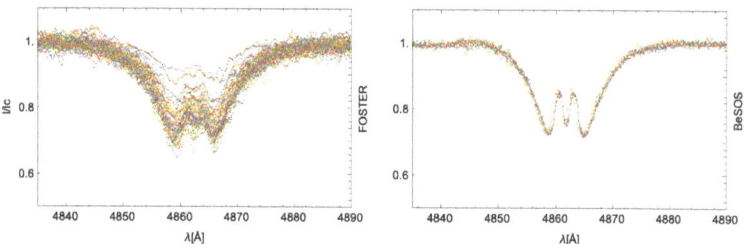

Figure 9. Same than Figure 6 but for α Col. The star present the lowest variation range in DPS. Only values higher than 150 km/s are reliable in the first period of data.

6. Summary and Discussion

We have compiled spectroscopic observations from Foster and BeSOS databases for the southern Be stars: η Cen, α Eri, ω Car, and α Col. The archives cover different optical wavelength ranges, 3900 to 5150 Å for Foster and 4250 to 7000 Å for BeSOS. Furthermore, the spectral resolution differs between databases; Foster has an effective spectral resolution of R~1500, while BeSOS spectra have a R~17,000. We look at the spectra of these stars for Balmer, helium, iron, and Mg lines. A compilation of the lines present in each star and instrument is shown in Table 1. For the four stars, a DPS is observed in Hβ for both periods, and only α Eri appears to be in a low-emission phase during the Foster period. The only star confirmed as binary is α Eri with an orbital period of ~15 years [37]. We note from Figure A1 that the coolest star, α Col, show a more pronounced difference in the EWs values of Hγ and Hβ line profiles, compared to the rest of the stars. In general B-types stars at different luminosity classes show a difference of around ~1 Å in Balmer lines EWs values [39]. In this case, the variation is in the order of ~2 Å. Therefore, this observed difference may be due to the star's disk or changes in the effective temperature of the star. As a comment and possible relation, [31] found that late spectral types host less massive disks with smaller emission region compared to early spectral types and thus, lower density disks.

We note that no star shows iron lines, except for the FeII 5313 Å observed in ω Car. This is the only star of our sample classified as a giant luminosity class. The hottest star in our work, η Cen, has the strongest helium lines with high values of EWs and very low, or no presence of MgII 4481 Å in both periods, agreeing with early spectra types classification. α Eri (mid spectral type) presents many HeI lines in the spectra with EWs values to η Cen but with the presence of MgII 4481 Å in low intensity during both periods. For the late spectral types, ω Car and α Col, both stars show weak helium lines, but the EW(HeI 4471 Å) ~ EW(MgII 4481 Å) in α Col, while for ω Car, EW(MgII 4481 Å) > EW(HeI 4471 Å). It is possible that this star corresponds to a later spectral type, but other studies must be done in comparison with line variability to re-classify it. Also, this star present a high Magnesium variability. This kind of variability has been reported before for magnetic chemically peculiar stars, which are mainly characterized by a dipolar magnetic field and their symmetry axis is inclined with respect to the rotational axis [40]. However, in the literature there is no evidence of magnetic fields for this star. Among the lines studied in this work, the four CBes are variable in helium. Also, the usual anti-correlation between MgII 4481 Å and HeI 4471 Å is also observed, i.e., as the EWs of MgII 4481 Å increase, the EWs of HeI 4471 Å decrease. Most likely, these changes are associated with the stellar photosphere rather than the disk.

From the EW variations plots, similar behavior is observed from Balmer and Helium lines. We remind the opposite meaning of EW values between absorption and emission features in the spectral lines. Then, when the EW value of Hβ increases, we can say that the emission coming from the disk is decreasing. By looking at Figure A1 and comparing the same observations years, when this happens, in general, EW values of Helium lines in Figure A2 increase. This is seen in η Cen, between 1989 and 1990, when the disk decreases

Helium lines, $\lambda 4387$ increases from an average of 0.6 to 0.9 Å, same for $\lambda 4921$, increases for an average of 1 to 1.2 Å. This is also observed in α Eri around 1988 and between 1989 to 1990. The other two targets present a larger error bar (because of the weakness of the lines and the low S/N in the Foster archive) and therefore it is difficult to attempt the same conclusion. A possible explanation could be the changes in the effective temperature of the star, which are also associated to non-radial pulsations. A next step will be to use stellar atmospheres models that consider the oblateness of the star due to fast rotation and try to measure the changes of the temperature. We test FT analysis of the V/R, DPS and EW data of the four targets, with the program PERIOD04 [41]. However, the data presented here include several gaps, leading spectral windows of the FT with very low signal to noise of the order of $\sim$3. Therefore, in order to estimate reliable periods is necessary to complement this set of data with information available in the literature or even in other spectroscopic plates.

Stellar parameters from the literature, together with DPS and V/R measurement on Hβ line profiles are summarized in Table 2. η Cen and α Eri have the largest DPS and, therefore, should have the smallest emitting region sizes. On the other hand, ω Car and α Col have the smallest DPS and the largest sizes. We note that the size of the emitting region increases toward late spectral types in our sample (based on DPS values).

Table 2. Summary of the **measurements** on the emission lines for the 4 Be Stars.

Name	Sp.T	Literature			N° Foster	N° BeSOS	V/R	DPS(Hβ) (km/s)
		T_{eff} (K)	$v \sin i$ (km/s)	$R_{H\alpha}$ (R_*)				
η Cen	B1Ve	21,000 [a]	240 [a]	12.5 [d]	40	3	0.95–1.05	450–750
α Eri	B6Ve	15,000 [b]	223 [b]	6 [d]	31	12	0.95–1.20	500–700
ω Car	B8IIIe	11,600 [a]	245 [c]	25 [d]	52	2	0.94–1.07	260–360
α Col	B9Ve	12,200 [a]	180 [b]	50 [d]	50	5	0.95–1.11	70–200

References: [a] [26], [b] [28], [c] [8] and [d] [31]. $R_{H\alpha}$ is the disk size calculated by comparing theoretical Hα line profiles with observations. DPS(Hβ) is the double-peak separation measured on the Hβ emission line. Comments: α Eri is in a binary system and is the only target with a low-emission phase during the first period. For the other targets there is not evidence of binary systems and they are in an active phase during both periods.

7. Conclusions

From 1928 to 1991, about 4800 spectroscopic plates were collected at the Observatory Universidad Católica Manuel Foster in Santiago, Chile. The plates contain spectra of various stellar sources in the southern hemisphere. We recovered the plates from the archive and developed a method of digitalization, processing, and analysis to extract valuable scientific information, which is useful to study long-term variability in different systems in the Foster archive. So the Foster archive stands in the tradition of those archive from the past that can contribute information about bright stars for several decades. In the same line, for example, the 2-meter telescope in Ondřejov, Czech Republic has been operating since 1967. The archive contains several thousands of photographic (up to 1992) and electronic (from 1992 to 2000) spectra. The spectra have been used to study Be stars, multiple spectroscopic systems, chemically peculiar stars, stellar pulsations, and symbiotic stars. A list of publications is available on the website https://stelweb.asu.cas.cz/en/telescope/date-archives/ (accessed on 5 October 2022). Recently, a digitalization project of spectroscopic plates taken in La Plata (Argentina) and Cerro Tololo (Chile) observatories between the '20s and '80s started at Universidad Nacional de La Plata, Argentina [42]. In total, there are around 15,000 photographic plates covering variable stars. Information on the ongoing project can be found on the website https://retroh.fcaglp.unlp.edu.ar/ (accessed on 5 October 2022).

As a final remark, the International Astronomical Union has a commission Working Group (B2) on the Preservation and Digitization of Photographic Plates. This archive is in the context of the group aims and the analyzed and published spectra will be available in a section in the BeSOS database for the use of the community.

Besides their historical value, these plates and their ongoing efforts to recover the spectra represent a rich database that we believe can still be of scientific interest to modern researchers.

Author Contributions: Conceptualization, C.A., L.V. and N.V.; Observations and data provider of Foster plates: S.G., V.O., N.V., E.A. and L.V.; Data extraction of Foster plates, S.G.; Data analysis, C.A. and S.G.; writing—original draft preparation, C.A., L.V. and N.V.; writing—review and editing, C.A., L.V. and N.V.; funding acquisition, C.A. and L.V. All authors have read and agreed to the published version of the manuscript.

Funding: C.A thanks to Fondecyt iniciación n. 11190945. This work has been possible thanks to the support of the fund comite mixto ESO Chile ETC. We also acknowledge support from CONICYT project Fondecyt n. 1130849. LV acknowledges support from ANID Fondecyt n. 1211162, Quimal ASTRO20-0025 and ANID BASAL CATA, FB210003.

Institutional Review Board Statement: Not applicable.

Informed Consent Statement: Not applicable.

Data Availability Statement: The data underlying this article will be shared on reasonable request to the corresponding author.

Acknowledgments: We are grateful for suggestions and comments from reviewers to improve this manuscript. We thanks to Cerro Tololo Interamerican Observatory plate donation and mirror aluminization and the collaboration of Ester Acuña, Abel Barrera and Gaston LeCerf. This work has made use of the BeSS database, operated at LESIA, Observatoire de Meudon, France: http://basebe.obspm.fr (accessed on 5 October 2022) and BeSOS archive, operated by the Instituto de Física y Astronomía, Universidad de Valparaíso, Chile: http://besos.ifa.uv.cl, (accessed on 5 October 2022) and funded by Fondecyt iniciación N 11130702.

Conflicts of Interest: The authors declare no conflict of interest.

Abbreviations

The following abbreviations are used in this manuscript:

Pucheros	Pontificia Universidad Católica High Echelle Resolution Optical Spectrograph
PUC	Pontificia Universidad Católica
BeSOS	Be Star Observation Survey
EW	Equivalent Width
DPS	Double Peak Separation
HR	High Resolution
LR	Low Resolution

Appendix A. List of Be Stars Observed in the Foster Archive

Table A1. Be stars in the Foster archive.

ID	HR	HD	HIP	Spectral Type	1981	1983	1984	1985	1986	1987	1988	1989	1990	1991
alf Eri	472	10,144	7588	B6Vpe			x	x	x	x	x	x	x	
228 Eri	1423	28,497	20,922	B2(V)ne				x	x	x	x			
56 Eri	1508	30,076	22,024	B2(V)nne				x	x		x			
3 Ori	1552	30,836	22,549	B2III			x							
69 Eri	1679	33,328	23,972	B2III(e)p				x	x	x	x	x		
HD 35165	1772	35,165	25,007	B2Vnpe					x	x	x			
25 Ori	1789	35,439	25,302	B1Vn					x	x	x	x		
123 Tau	1910	37,202	26,451	B1IVe$_{shell}$						x				
47 Ori	1934	37,490	26,594	B3Ve			x	x		x	x	x	x	

Table A1. Cont.

ID	HR	HD	HIP	Spectral Type	1981	1983	1984	1985	1986	1987	1988	1989	1990	1991
alf col	1956	37,795	26,634	B9Ve			x	x	x	x	x	x	x	
HD 41335	2142	41,335	28,744	B3/5Vnne				x	x	x	x			
HD 42054	2170	42,054	28,992	B5Ve					x	x				
HD 43285	2231	43,285	29,728	B6Ve					x	x				
HD 43544	2249	43,544	29,771	B2/B3Ve					x					
HD 44458	2284	44,458	30,214	B1.5IVe					x					
HD 44506	2288	44,506	30,143	B3V					x	x				
bet01 Mon	2356	45,725	-	B4Veshell				x	x	x	x	x	x	
HD 45871	2364	45,871	30,840	B3Ve					x					
15 Mon	2456	47,839	31,978	O7V+B1.5/2V				x	x					
10 Cma	2492	48,917	32,292	B2V					x	x		x	x	
13 Cma	2538	50,013	32,759	B1.5Ve			x	x	x	x	x	x	x	
HD 50123	2545	50,123	32,810	B6IVe+A					x					
19 Mon	2648	52,918	33,971	B2Vn(e)				x	x		x			
HD 54309	2690	54,309	34,360	B3Vne						x				
27 Cma	2745	56,014	34,981	B3IIIe					x	x	x	x	x	
28 Cma	2749	56,139	35,037	B2.5Ve			x	x	x	x	x	x	x	
ups01 Pup	2787	57,150	35,363	B2V+B3IVne			x	x	x	x	x			
ups02 Pup	2790	57,219	35,406	B3Vne			x	x	x	x	x			
HD 58155	2819	58,155	35,795	B3V					x	x	x			
HD 58343	2825	58,343	35,951	B2Vne					x	x	x			
bet Cmi	2845	58,715	36,188	B8Ve					x	x	x	x	x	
HD 58978	2855	58,978	36,168	B0.5IVe					x	x	x			
z Pup	2911	60,606	36,778	B2Vne					x		x			
HD 60855	2921	60,855	36,981	B2Ve					x	x	x			
omi Pup	3034	63,462	38,070	B1IVe			x	x		x	x	x		x
HD 66194	3147	66,194	38,994	B3Vn						x				
r Pup	3237	68,980	40,274	B2ne			x	x	x	x	x	x		
HD 71510	3330	71,510	41,296	B3IV					x	x	x			
HD 72067	3356	72,067	41,621	B2/3V						x				
t Car	3498	75,311	43,105	B3V(n)			x	x	x	x	x	x		
HD 77320	3593	77,320	44,213	B2Vnn(e)						x		x		
E Car	3642	78,764	44,626	B2(IV)n			x		x	x	x	x		
HD 81753	3745	81,753	46,329	B5V(e)						x				
I Hya	3858	83,953	47,522	B5V				x	x	x	x	x		
HD 86612	3946	86,612	48,943	B5Ve						x				
HD 88661	4009	88,661	49,934	B5Vne					x	x	x	x		
HD 88825	4018	88,825	50,044	B5(III)e						x				
ome Car	4037	89,080	50,099	B8IIIe			x		x	x	x	x	x	
J Vel	4074	89,890	50,676	B5II					x	x	x	x	x	
HD 91120	4123	91,120	51,491	B8/9IV/V					x	x	x	x		
p Car	4140	91,465	51,576	B4Vne		x	x	x	x	x	x	x		
HD 93237	4206	93,237	52,340	B5III						x				
HD 93563	4221	93,563	52,742	B8/9III					x	x				
A Cen	4460	100,673	56,480	B9V					x	x	x	x		
j Cen	4540	102,870	57,757	B3V	x									
HD 105382	4618	105,382	59,173	B3/5III			x	x	x	x				
del Cen	4621	105,435	59,196	B2Vne			x	x	x	x	x	x		
HD 105521	4625	105,521	59,232	B3IVe					x	x	x			
zet Crv	4696	107,348	60,189	B8V						x				
39 cru	4823	110,335	61,966	B5III					x	x	x	x		
HD 110432	4830	110,432	62,027	B0.5IVpe						x				
lam Cru	4897	112,078	63,007	B3Vne			x	x	x	x	x	x		
mu.02 Cru	4899	112,091	63,005	B5Vne			x	x		x	x	x		
mu. Cen	5193	120,324	67,472	B2Vnpe				x	x	x	x	x	x	

Table A1. Cont.

ID	HR	HD	HIP	Spectral Type	1981	1983	1984	1985	1986	1987	1988	1989	1990	1991
HD 120991	5223	120,991	67,861	B2Ve						x	x			
HD 124367	5316	124,367	69,618	B4Vne			x	x	x	x	x	x	x	
eta Cen	5440	127,972	71,352	B2Ve			x	x	x	x	x	x	x	
HD 129954	5500	129,954	72,438	B2V						x	x			
tet Cir	5551	131,492	73,129	B2IV/V					x	x	x	x		
kap Lup	5646	134,481	74,376	B9.5Vne			x		x	x	x	x	x	
mu. Lup	5683	135,734	74,911	B8Ve			x		x	x	x	x	x	
kap01 Aps	5730	137,387	76,013	B2Vnpe				x		x	x			
d Lup	5781	138,769	76,371	B3V			x					x	x	
HD 142184	5907	142,184	77,859	B2V						x				
48 Lib	5941	142,983	78,207	B5IIIp$_s$h			x		x	x	x	x		
7 Oph	6118	148,184	80,569	B2Vne			x	x	x	x	x	x		
eta01 TrA	6172	149,671	81,710	B7V						x				
13 Oph	6175	149,757	81,377	O9.2IVnn			x	x	x		x	x	x	
HD 155806	6397	155,806	84,401	O7.5V((f))z(e)						x	x			
iot Ara	6451	157,042	85,079	B2(V)nne				x	x	x	x		x	
alf Ara	6510	158,427	85,792	B2Vne			x	x	x	x	x	x	x	
51 Oph	6519	158,643	85,755	A0V				x		x				
66 Oph	6712	164,284	88,149	B2Ve			x	x	x	x	x	x		
HD 167128	6819	167,128	89,605	B3II/III				x		x	x			
lam Pav	7074	173,948	92,609	B2Ve			x	x	x	x	x	x	x	
HD 178175	7249	178,175	93,996	B2Ve				x		x	x			
46 Sgr	7342	181,615	95,176	B2Vp$_s$h				x	x	x	x			
39 Cap	8260	205,637	106,723	B3V				x	x	x	x	x	x	
12 PsA	8386	209,014	108,661	B8/9V+B8/9				x		x				
31 Aqr	8402	209,409	108,874	B7IVe				x	x	x	x	x		
HD 209522	8408	209,522	108,975	B4IVe				x		x				
31 Peg	8520	212,076	110,386	B2IV-Ve				x	x	x				
52 Aqr	8539	212,571	110,672	B1III-Ive				x	x	x	x	x		
18 PsA	8628	214,748	111,954	B8Ve				x	x	x	x	x		
4 Psc	8773	217,891	113,889	B6Ve				x	x	x	x	x		
eps Tuc	9076	224,686	118,322	B8V						x	x	x	x	

Appendix B. EWs Variation for the Four CBes Studied in This Work

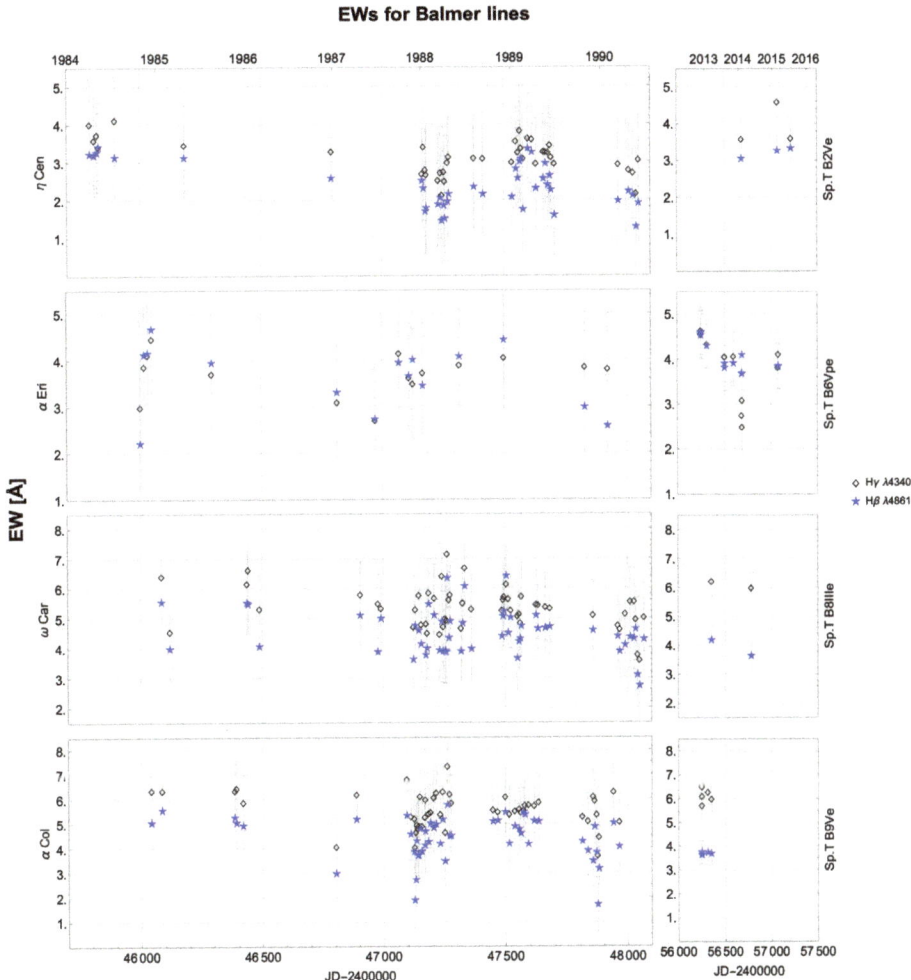

Figure A1. EWs variation for Hγ and Hβ for each star. Plots are sorted from the latest stellar type (**Bottom**), to the earliest type (**Top**). In each plot the name and spectral type of the star is indicated in the vertical axes. Vertical grid lines are disposed every 500 Julian days.

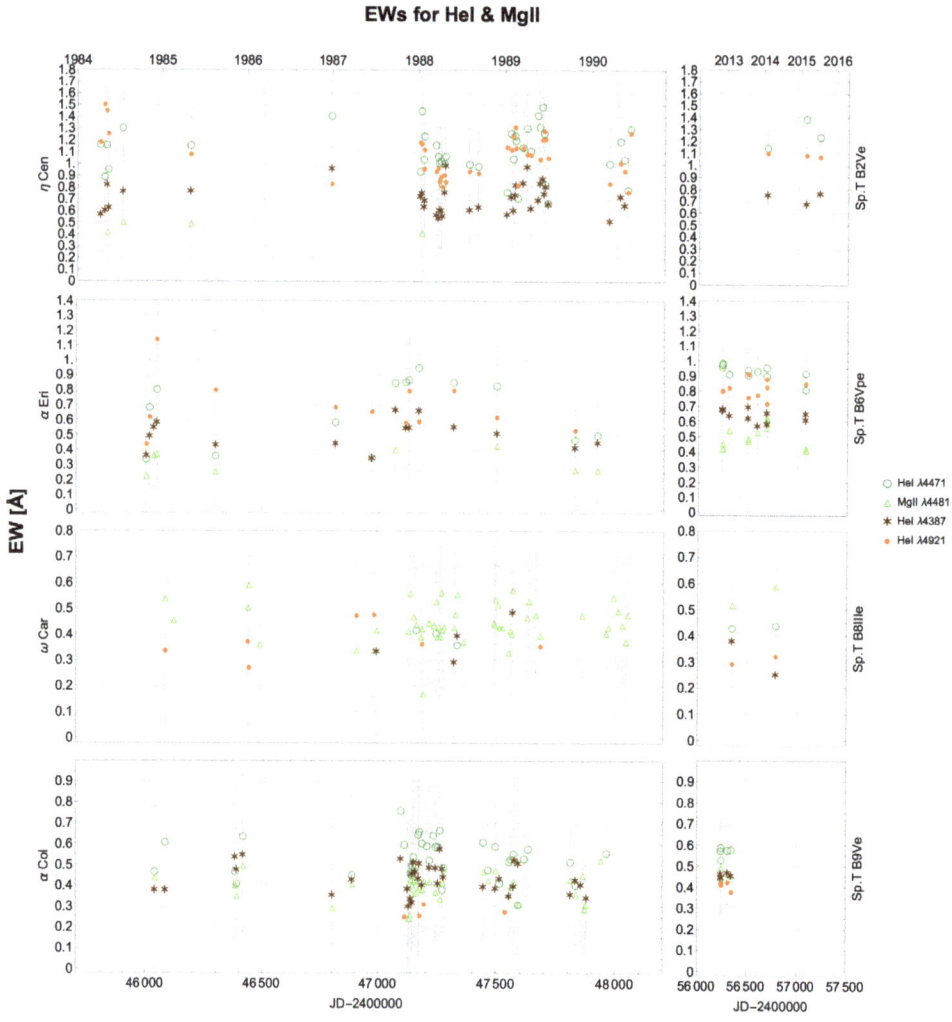

Figure A2. EWs variation for HeI lines $\lambda\lambda$ 4387, 4471, 4921 and MgII 4481 lines for each star.

Notes

1. http://besos.ifa.uv.cl/, (accessed on 5 October 2022)
2. Spectra are normalized by intensity leading the continuum to 1.

References

1. Campbell, W.W. The Mills spectrograph of the Lick Observatory. *Astrophys. J.* **1898**, *8*, 123-156. [CrossRef]
2. Vogt, N.; Barrera, L.H. A simple method to calibrate intensities of photographic slit spectrograms. *Astron. Astrophys.* **1985**, *148*, 237–239.
3. Quirrenbach, A.; Bjorkman, K.S.; Bjorkman, J.E.; Hummel, C.A.; Buscher, D.F.; Armstrong, J.T.; Mozurkewich, D.; Elias, N.M., II; Babler, B.L. Constraints on the Geometry of Circumstellar Envelopes: Optical Interferometric and Spectropolarimetric Observations of Seven Be Stars. *Astrophys. J.* **1997**, *479*, 477–496. [CrossRef]
4. Meilland, A.; Stee, P.; Vannier, M.; Millour, F.; Domiciano de Souza, A.; Malbet, F.; Martayan, C.; Paresce, F.; Petrov, R.G.; Richichi, A.; et al. First direct detection of a Keplerian rotating disk around the Be star α Arae using AMBER/VLTI. *Astron. Astrophys.* **2007**, *464*, 59–71. [CrossRef]

5. Meilland, A.; Millour, F.; Kanaan, S.; Stee, P.; Petrov, R.; Hofmann, K.H.; Natta, A.; Perraut, K. First spectro-interferometric survey of Be stars. I. Observations and constraints on the disk geometry and kinematics. *Astron. Astrophys.* **2012**, *538*, A110. [CrossRef]
6. Rivinius, T.; Baade, D.; Carciofi, A.C. Short-term variability and mass loss in Be stars. II. Physical taxonomy of photometric variability observed by the Kepler spacecraft. *Astron. Astrophys.* **2016**, *593*, A106. [CrossRef]
7. Okazaki, A.T. Long-term V/R variations of Be stars due to global one-armed oscillations of equatorial disks. *PASJ* **1991**, *43*, 75–94.
8. Frémat, Y.; Zorec, J.; Hubert, A.M.; Floquet, M. Effects of gravitational darkening on the determination of fundamental parameters in fast-rotating B-type stars. *Astron. Astrophys.* **2005**, *440*, 305–320. [CrossRef]
9. Granada, A.; Ekström, S.; Georgy, C.; Krtička, J.; Owocki, S.; Meynet, G.; Maeder, A. Populations of rotating stars. II. Rapid rotators and their link to Be-type stars. *Astron. Astrophys.* **2013**, *553*, A25. [CrossRef]
10. Rivinius, T.; Baade, D.; Štefl, S. Non-radially pulsating Be stars. *Astron. Astrophys.* **2003**, *411*, 229–247. [CrossRef]
11. Labadie-Bartz, J.; Carciofi, A.C.; Henrique de Amorim, T.; Rubio, A.; Luiz Figueiredo, A.; Ticiani dos Santos, P.; Thomson-Paressant, K. Classifying Be Star Variability with TESS. I. The Southern Ecliptic. *Astron. J.* **2022**, *163*, 226. [CrossRef]
12. Porter, J.M. On outflowing viscous disc models for Be stars. *Astron. Astrophys.* **1999**, *348*, 512–518.
13. Curé, M.; Meneses, R.; Araya, I.; Arcos, C.; Peña, G.; Machuca, N.; Rodriguez, A. Revisiting Viscous Transonic Decretion Disks of Be Stars. *arXiv* **2022**, arXiv:2206.09031.
14. Klement, R.; Carciofi, A.C.; Rivinius, T.; Ignace, R.; Matthews, L.D.; Torstensson, K.; Gies, D.; Vieira, R.G.; Richardson, N.D.; Domiciano de Souza, A.; et al. Prevalence of SED Turndown among Classical Be Stars: Are All Be Stars Close Binaries? *Astrophys. J.* **2019**, *885*, 147. [CrossRef]
15. Wang, L.; Gies, D.R.; Peters, G.J.; Götberg, Y.; Chojnowski, S.D.; Lester, K.V.; Howell, S.B. The Detection and Characterization of Be+sdO Binaries from HST/STIS FUV Spectroscopy. *Astron. J.* **2021**, *161*, 248. [CrossRef]
16. Rivinius, T.; Carciofi, A.C.; Martayan, C. Classical Be stars. Rapidly rotating B stars with viscous Keplerian decretion disks. *Astron. Astrophys. Rev.* **2013**, *21*, 69. [CrossRef]
17. Neiner, C.; de Batz, B.; Cochard, F.; Floquet, M.; Mekkas, A.; Desnoux, V. The Be Star Spectra (BeSS) Database. *Astron. J.* **2011**, *142*, 149. [CrossRef]
18. Barrera, L.H.; Vogt, N. Spectroscopic monitoring of southern Be stars. *Rev. Mex. Astron. Astrofis.* **1987**, *14*, 323–329.
19. Vogt, N.; Barrera, L.H.; Navarro, M. Correlation and variability of the He I 4471 and MG II 4481 absorption lines in Be stars—A possible diagnositc tool for nonradial pulsations. *Astrophys. Space Sci.* **1990**, *173*, 145–156. [CrossRef]
20. Mennickent, R.E.; Vogt, N.; Barrera, L.H.; Covarrubias, R.; Ramirez, A. On the rotation properties of Be stars and their envelopes. *Astron. Astrophys. Suppl. Ser.* **1994**, *106*, 427–439.
21. Mennickent, R.E.; Vogt, N. Spectroscopy of southern Be stars 1984–1987. *Astron. Astrophys. Suppl. Ser.* **1988**, *74*, 497–506.
22. Mennickent, R.E.; Vogt, N. V/R variations in H-beta emission profiles of Be stars. *Astron. Astrophys.* **1991**, *241*, 159.
23. Mennickent, R.E. H-beta line profile variability of seven southern Be stars. *Astron. Astrophys. Suppl. Ser.* **1991**, *88*, 1.
24. Moffat, A.F.J. A Theoretical Investigation of Focal Stellar Images in the Photographic Emulsion and Application to Photographic Photometry. *Astron. Astrophys.* **1969**, *3*, 455.
25. Vanzi, L.; Chacon, J.; Helminiak, K.G.; Baffico, M.; Rivinius, T.; Štefl, S.; Baade, D.; Avila, G.; Guirao, C. PUCHEROS: a cost-effective solution for high-resolution spectroscopy with small telescopes. *Mon. Not. R. Astron. Soc.* **2012**, *424*, 2770–2777. [CrossRef]
26. Arcos, C.; Kanaan, S.; Chávez, J.; Vanzi, L.; Araya, I.; Curé, M. Stellar parameters and H α line profile variability of Be stars in the BeSOS survey. *Mon. Not. R. Astron. Soc.* **2018**, *474*, 5287–5299. [CrossRef]
27. Vollmann, K.; Eversberg, T. Remarks on statistical errors in equivalent widths. *Astron. Nachrichten* **2006**, *327*, 862. [CrossRef]
28. Levenhagen, R.S.; Leister, N.V. Spectroscopic analysis of southern B and Be stars. *Mon. Not. R. Astron. Soc.* **2006**, *371*, 252–262. [CrossRef]
29. Vieira, R.G.; Carciofi, A.C.; Bjorkman, J.E.; Rivinius, T.; Baade, D.; Rímulo, L.R. The life cycles of Be viscous decretion discs: time-dependent modelling of infrared continuum observations. *Mon. Not. R. Astron. Soc.* **2017**, *464*, 3071–3089. [CrossRef]
30. Burton, W.M.; Evans, R.G. Spectroscopic Observations of the be Stars η CEN, γ CAS, and φ Per. Be and Shell Stars; Slettebak, A., Ed.; International Astronomical Union: Bass River, MA, USA, 1976; Volume 70, p. 199.
31. Arcos, C.; Jones, C.E.; Sigut, T.A.A.; Kanaan, S.; Curé, M. Evidence for Different Disk Mass Distributions between Early- and Late-type Be Stars in the BeSOS Survey. *Astrophys. J.* **2017**, *842*, 48. [CrossRef]
32. Stefl, S.; Baade, D.; Rivinius, T.; Stahl, O.; Wolf, B.; Kaufer, A. Circumstellar Quasi-periods Accompanying Stellar Periods of Be Stars. In Proceedings of the A Half Century of Stellar Pulsation Interpretation, Los Alamos, NM, USA, 16–20 June 1997; ASP Conference, Astronomical Society of the Pacific Conference Series; Bradley, P.A., Guzik, J.A., Eds.; 1998; Volume 135, p. 348.
33. Baade, D.; Rivinius, T.; Pigulski, A.; Carciofi, A.C.; Martayan, C.; Moffat, A.F.J.; Wade, G.A.; Weiss, W.W.; Grunhut, J.; Handler, G.; et al. Short-term variability and mass loss in Be stars. I. BRITE satellite photometry of η and μ Centauri. *Astron. Astrophys.* **2016**, *588*, A56. [CrossRef]
34. Huang, S.S. Profiles of Emission Lines in be Stars. *Astrophys. J.* **1972**, *171*, 549. [CrossRef]
35. Kervella, P.; Domiciano de Souza, A. The environment of the fast rotating star Achernar. High-resolution thermal infrared imaging with VISIR in BURST mode. *Astron. Astrophys.* **2007**, *474*, L49–L52. [CrossRef]
36. Kervella, P.; Domiciano de Souza, A.; Bendjoya, P. The close-in companion of the fast rotating Be star Achernar. *Astron. Astrophys.* **2008**, *484*, L13–L16. [CrossRef]

37. Vinicius, M.M.F.; Zorec, J.; Leister, N.V.; Levenhagen, R.S. α Eridani: rotational distortion, stellar and circumstellar activity. *Astron. Astrophys.* **2006**, *446*, 643–660. [CrossRef]
38. Mineva, V.A.; Kovachev, B.Z.; Radoslavova, T.B. Spectrophotometric exploration of the Alpha Col star. *Bolg. Akad. Nauk. Dokl.* **1981**, *34*, 1625–1628.
39. Didelon, P. Largeurs equivalentes de raies spectrales dans les etoiles B. *Astron. Astrophys. Suppl. Ser.* **1982**, *50*, 199–207.
40. Leone, F.; Catalano, F.A.; Malaroda, S. Magnesium abundance in main sequence B-type and magnetic chemically peculiar stars. *Astron. Astrophys.* **1997**, *325*, 1125–1131.
41. Breger, M. Amplitude variations of the multimode nonradial pulsator 4 Canum Venaticorum. *Astron. Astrophys.* **1990**, *240*, 308.
42. Meilán, N.; Collazo, S.; Alessandroni, M.R.; López Durso, M.; Peralta, R.A.; Aidelman, Y.; Cidale, L.S.; Gamen, R. Proyecto de digitalización de placas espectrográficas del Observatorio de La Plata. *Bol. Asoc. Argent. Astron. Plata Argentina* **2020**, *61B*, 251–253.

Article

Infrared Spectroscopy of Be Stars: Influence of the Envelope Parameters on Brackett-Series Behaviour

Yanina Roxana Cochetti [1,2,*], Anahi Granada [3], María Laura Arias [1,2], Andrea Fabiana Torres [1,2] and Catalina Arcos [4]

1. Instituto de Astrofísica de La Plata (CCT La Plata—CONICET, UNLP), Paseo del Bosque S/N, La Plata B1900FWA, Argentina; mlaura@fcaglp.unlp.edu.ar (M.L.A.); atorres@fcaglp.unlp.edu.ar (A.F.T.)
2. Departamento de Espectroscopía, Facultad de Ciencias Astronómicas y Geofísicas, Universidad Nacional de La Plata, La Plata B1900FWA, Argentina
3. Centro Interdisciplinario de Telecomunicaciones, Electrónica, Computación y Ciencia Aplicada (CITECCA), Sede Andina, Universidad Nacional de Río Negro, San Carlos de Bariloche R8400AHN, Argentina; agranada@unrn.edu.ar
4. Instituto de Física y Astronomía, Facultad de Ciencias, Universidad de Valparaíso, Valparaíso 2360102, Chile; catalina.arcos@uv.cl
* Correspondence: cochetti@fcaglp.unlp.edu.ar

Abstract: The IR spectra of Be stars display numerous hydrogen recombination lines, constituting a great resource for obtaining information on the physical and dynamic structures of different regions within the circumstellar envelope. Nevertheless, this spectral region has not been analysed in depth, and there is a lack of synthetic spectra with which to compare observations. Therefore, we computed synthetic spectra with the HDUST code for different disc parameters. Here, we present our results on the spectral region that includes lines of the Brackett series. We discuss the dependence of the line series strengths on several parameters that describe the structure of the disc. We also compared model line profiles, fluxes, and EWs with observational data for two Be stars (MX Pup and π Aqr). Even though the synthetic spectra adequately fit our observations of both stars and allow us to constrain the parameters of the disc, there is a discrepancy with the observed data in the EW and flux measurements, especially in the case of MX Pup. It is possible that by including Brackett lines of higher terms or adding the analysis of other series, we may be able to better constrain the parameters of the observed disc.

Keywords: stars: early-type; stars: emission-line, Be; stars: individual: MX Pup, π Aqr; circumstellar matter

Citation: Cochetti, Y.R.; Granada, A.; Arias, M.L.; Torres, A.F.; Arcos, C. Infrared Spectroscopy of Be Stars: Influence of the Envelope Parameters on Brackett-Series Behaviour. *Galaxies* **2023**, *11*, 90. https://doi.org/10.3390/galaxies11040090

Academic Editor: Jorick Sandor Vink

Received: 14 March 2023
Revised: 9 August 2023
Accepted: 11 August 2023
Published: 17 August 2023

Copyright: © 2023 by the authors. Licensee MDPI, Basel, Switzerland. This article is an open access article distributed under the terms and conditions of the Creative Commons Attribution (CC BY) license (https://creativecommons.org/licenses/by/4.0/).

1. Introduction

Classical Be stars are non-supergiant B-type stars whose spectra show or have shown the Hα line in emission [1,2]. These stars are rapidly rotating, and it is accepted that this emission originates in a circumstellar envelope, mostly compatible with a disc geometry in Keplerian rotation [3]. The model that best explains the observations of Be stars is the viscous decretion disc [4], which allows us not only to explain static discs but also to study their dynamical evolution [5].

In the infrared (IR) spectral region, Be spectra present a moderate flux excess and numerous hydrogen emission lines of the Paschen, Brackett, Pfund, and Humphreys series. Since these IR lines are formed in a region closer to the star than those observed in the optical spectral range, and the photospheric absorption is almost negligible for most of them, the study of these IR emission lines is a great tool to obtain information about the physical structure and dynamics within the innermost part of the disc [6–10].

This spectral region has gained importance in recent decades. An extensive atlas of the K- and L-bands of early-type stars, including Be stars, was published by Lenorzer et al. [11]. With those data, Lenorzer et al. [7] made a diagram with the flux ratios Hu14/Brα versus

Hu14/Pfγ. In this diagram, the location of the objects depends on the density of the emitting gas. Mennickent et al. [8] proposed a classification criterion based on the emission lines observed in the L-band: those stars on which the Humphreys lines present intensities similar to those of Brα and Pfδ lines constitute Group I; stars with Brα and Pfδ lines more intense than Humphreys lines are part of Group II; and stars without emission lines are part of Group III. Groups I and II fall in different regions of Lenorzer's diagram; thus, the group membership is probably connected to the density of the disc. Classifying a sample of Be stars using Mennickent's criterion, Granada et al. [9] found that the equivalent widths (EWs) of Brα and Brγ lines are similar for Group I objects, while for stars in Group II, EW(Brγ) is much larger (more than five times) than EW(Brα).

Up to the recent publication of Cochetti et al. [12]'s atlas, most of the works that focused on the IR data of Be stars showed low-resolution data, analysed a small sample, or covered a small spectral range. The already-mentioned atlas comprises medium-resolution spectra in the near-IR region of a sample of Galactic Be stars. By measuring different parameters of the observed hydrogen recombination lines, these authors diagnosed the physical conditions in the circumstellar environment. They defined new complementary criteria to classify Be stars according to their disc opacity. This atlas provides valuable observational material available to be analysed and modelled. In particular, the behaviour of the fluxes of the hydrogen lines for different stars may give us clues about the properties of the discs. To perform better analysis, modelling synthetic spectra and comparing them with observations is essential.

In the last years, different observables from Be star discs have been successfully modelled in the framework of the viscous decretion disc model (VDD; [4,13]). In this model, the rapidly rotating central star continuously loses mass and angular momentum and puts material in a Keplerian orbit at the base of the disc. Beyond this inner edge, the material is expected to be driven outwards through some turbulent viscous process. The rotationally distorted star irradiates the disc, which is assumed to extend out to a radius of R_d equatorial radii, and the star-plus-disc system is viewed at an inclination angle (angle between the star's rotation axis and the line of sight) of i. Then, in state-of-the-art disc models such as HDUST [14] or Bedisc/BeRay [15,16], each computed observable is a function of four model parameters: ρ_0, n, R_d, and i, where ρ_0 is the axisymmetric volume density at the innermost part of the disc, and n is the exponent of the radial power law usually used to describe the Be star disc's density.

The modelling of different observables has allowed constraints on Be disc parameters to be derived. In particular, modelling of the iconic Hα-line emission has shown that this line typically forms within the innermost tens of stellar radii [3,17,18]. The modelling of the Brγ line, in particular, from spectro-interferometric data [19] has allowed for constraining stellar parameters and also shown that the formation region of this line is, at most, 15 or 20 stellar radii. However, a smaller Brackett-series-forming region of a few stellar radii has been derived from near-IR spectra [20]. Unfortunately, the lack of simultaneous observations prevents a direct comparison between Hα- and Brγ-line-forming regions. The modelling of the near-IR hydrogen recombination lines allows us to explore the properties of their line-forming region, which is expected to be closer to the star than that of Hα and, thus, contributes to better constraints on the parameters of the disc.

With this aim, we started to compute a grid of synthetic spectra covering the near-IR spectral range with the HDUST code [14,21]. As a first step, we computed the first lines of the Brackett series for a set of disc parameters and one central B-type star. Even though we do not expect to be able to describe the whole complexity of the star–disc interphase, we seek to understand the impact on the emerging spectra of the different parameters involved in the modelling of Be stars in the viscous decretion disc framework, with the final goal of deducing disc properties from the IR spectra. From the synthetic spectra, we measured the equivalent widths and fluxes of the hydrogen lines and analysed the behaviour of the line series when varying the parameters that describe the disc density law. Finally, we compared the spectra and the line flux behaviour from the models with the observational data of two known Be stars: MX Pup and π Aqr.

2. Methods

We used the HDUST code, developed by Carciofi and Bjorkman [14,21]. This three-dimensional non-LTE Monte Carlo code solves the radiative transfer, radiative equilibrium, and statistical equilibrium equations in a 3D geometry for different gas density and velocity distributions. This code has been used to discuss the properties of Be stars via the analysis of the spectral energy distribution (SED) based on different wavelengths, as well as some optical hydrogen lines [22–25].

The code includes a 25-level model for the hydrogen atom, of which the first 12 are explicit non-LTE levels, and the upper 13 are implicit levels with the LTE population. This default configuration of the code allowed us to compute the Brackett series from the Brα line to Br12. We also modelled the continuum in the range 1.4–4.3 µm.

The disc is considered to start at the stellar radius $R_\star$, and the density throughout it is modelled with the expression

$$\rho(r,z) = \rho_0 \left(\frac{r}{R_\star}\right)^{-n} exp\left(-\frac{z^2}{2H^2}\right) \quad (1)$$

which includes a power law and a Gaussian distribution in the radial and vertical directions, respectively [14]. In this expression, ρ is the density at each pair of (r,z) coordinates, ρ_0 is the density at the base of the disc, and H is the scale height of the disc. For a vertically isothermal disc, H is obtained with the expression

$$H = H0 \left(\frac{r}{R_\star}\right)^{1.5}, H0 = \sqrt{\frac{kT}{\mu \cdot m_H}} \frac{1}{V_{orb}} R_\star \quad (2)$$

where k is the Boltzmann constant, μ is the mean molecular weight, m_H is the mass of the hydrogen atom, $V_{orb} = \sqrt{GM/R_{eq}}$ is the disc orbital speed at the stellar equator R_{eq}, and the temperature T has been set as 72% of the effective temperature T_{eff} of the star.

Despite the evidence of differential rotation reported in the literature (i.e., [26]), to date, there is not a comprehensive theory that covers all the observed physical phenomena occurring in the interior and photosphere of rapidly rotating stars. Then, the adoption of a solid and rigid approach is prevalent in their modelling, and particularly in this code.

We calculated models with ρ_0 values between 5×10^{-12} and 1×10^{-10} g cm^{-3}[1] and values of the exponent n in the range 2.5–4 [23]. These ρ_0 values correspond to a steady-state decretion rate $\dot{M}$ over the viscosity parameter α [27] in the range $2.13 \times 10^{-13} M_\odot$ yr$^{-1} \lesssim \dot{M}/\alpha \lesssim 2.13 \times 10^{-11} M_\odot$ yr^{-1}[23]. Following a procedure similar to the one described by Silaj et al. [28] for determining the convenient disc size for each n–ρ_0 pair, we analysed the change in the equivalent width of the modelled lines as a function of the disc size by increasing it by 10 $R_\star$ each time. Then, we determined the minimum disc size at which the EW reached a stable value for each model. The required radius is larger for lower n values or higher central densities. By using this radius as the disc's outer radius R_d, we ensure that we include the entire emitting region. To check the convergence of the simulation, we plotted the disc temperature for the last iterations, ensuring a decreasing profile in the inner part of the disc and an increasing temperature in the outer disc (forming a typical U shape, [14]).

In Table 1, we show the envelope's outer radius R_d chosen for each n–ρ_0 combination employed in the modelling[2]. We computed models that have been found adequate to fit Be-star Hα profiles and SEDs and are related to different dynamical stages of the disc: dissipating discs present $n \lesssim 3$, while the range $3 \lesssim n \lesssim 3.5$ is for stable discs, and $3.5 \lesssim n$ is related to discs in formation [23]. Models with a small exponent and high density ρ_0 trace the "forbidden zone", defined by Vieira et al. [23] as a region in the n–ρ_0 diagram with unobserved combinations of these parameters. Reaching this region would require very massive circumstellar discs.

As the first step for this work, we ran simulations for only one set of fundamental parameters for the central stellar model. We used a self-consistent rigid rotator with the following parameters: a mass $M = 10\,M_\odot$, a polar radius $R_{pole} = 5.5\,R_\odot$, a luminosity $L = 7500\,L_\odot$, and a rotational rate $W = 0.7$. We recall that $W = V_{rot}/V_{orb}$, where V_{rot} is the rotational velocity at the equator of the star, and V_{orb} is the equatorial circular orbital (Keplerian) velocity, as defined by Rivinius et al. [3]. This set of parameters corresponds to an early B star with $T_{eff} \simeq 24{,}250$ K and $V_{rot} \simeq 370\,\mathrm{km\,s^{-1}}$. For producing synthetic profiles with HDUST, the gravity-darkening effect is accounted for, considering a parameter $\beta = 0.188$ suitable for rapidly rotating stars [29]. As it has been proposed that late-type Be stars could have more tenuous discs than early-type Be stars due to the smaller variability observed [3,10], in a forthcoming article, we plan to include central stars with different spectral types to analyse the dependence on the disc parameters.

To obtain the synthetic spectra and measure the line parameters, the results were processed with PyHdust[3] and specutils[4] Python packages.

Table 1. Envelope's outer radius R_d for each n and ρ_0 combination. According to Vieira et al. [23], the n value is related to different disc stages: dissipating discs present $n \lesssim 3$, while the range $3 \lesssim n \lesssim 3.5$ is for stable discs, and $3.5 \lesssim n$ is related to discs in formation. For each model, we simulated the spectra for inclination angles in the range 0–60° with a 15° step.

n	$\rho_0 = 1 \times 10^{-12}$ (g cm^{-3})	$\rho_0 = 5 \times 10^{-12}$ (g cm^{-3})	$\rho_0 = 1 \times 10^{-11}$ (g cm^{-3})	$\rho_0 = 5 \times 10^{-11}$ (g cm^{-3})	$\rho_0 = 1 \times 10^{-10}$ (g cm^{-3})
2.5	30 $R_\star$	×	×	×	×
3.0	20 $R_\star$	20 $R_\star$	30 $R_\star$	50 $R_\star$	×
3.5	20 $R_\star$	20 $R_\star$	20 $R_\star$	30 $R_\star$	30 $R_\star$
4.0	20 $R_\star$	20 $R_\star$	20 $R_\star$	20 $R_\star$	20 $R_\star$

3. Results

For each simulation, we generated spectra for inclination angles in the range of 0–60°, with a 15° step. In Figure 1, we show as an example the normalised spectrum obtained for the model with $n = 3.5$ and $\rho_0 = 10^{-11}$ g cm^{-3}, viewed from a pole-on ($i = 0°$) orientation. In the following subsections, we will show the analysis performed for the 0°, 30°, and 60° orientations.

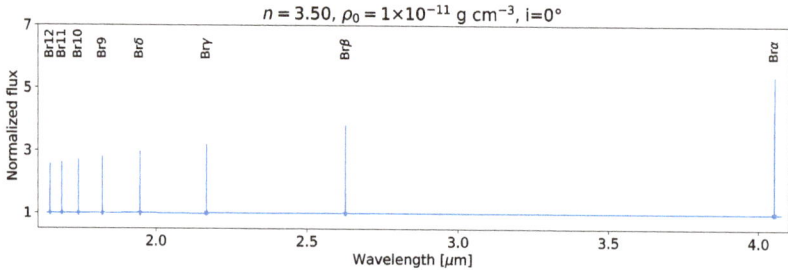

Figure 1. Spectrum from the simulation with $n = 3.5$ and $\rho_0 = 10^{-11}$ g cm^{-3}, with a pole-on ($i = 0°$) orientation. Brackett lines are labelled.

3.1. Brackett-Series Behaviour According to n and ρ_0

Figures 2–5 show the synthetic line profiles for the Brα to Br12 lines for the different simulations for $i = 0°$. Each figure corresponds to a different value of n, and in each plot, the different colours are for different ρ_0 values. Synthetic line profiles for $i = 30°$ and $i = 60°$ are in Appendix A.1.

For the behaviour across the series, we observe the following:

- For $n = 2.5$ (Figures 2 and A1), which is the lowest n value and can only be combined with the lowest density ($\rho_0 = 10^{-12}$ g cm^{-3}), there is a strong intensification for the lines from Br12 to Brα for all the inclinations. The lines are more intense for lower inclinations.
- For $n = 3.0$ (Figures 3 and A2), the height of each line relative to its adjacent continuum also increases from Br12 to Brα. The slope of the increase is different for each density: the lower the density, the stronger the increase. For $i = 0°$ and $i = 30°$, the increase is higher than for $i = 60°$.
- For $n = 3.5$ (Figures 4 and A3), the slope of the increase is also steeper for the lowest densities, but not as remarkable as for $n = 3.0$. The higher increase for lower densities means that, even though the higher-order lines are more intense for the highest densities, the first members of the series present similar intensities for intermediate densities.
- For $n = 4.0$ (Figures 5 and A4), the intensities are the smallest, with the increase not so different for each density from Br12 to Brα.

Since the n value is related to the dynamical state of the disc (forming, steady, or dissipating [23]), it was expected that different behaviour would be found for the line intensity across the series.

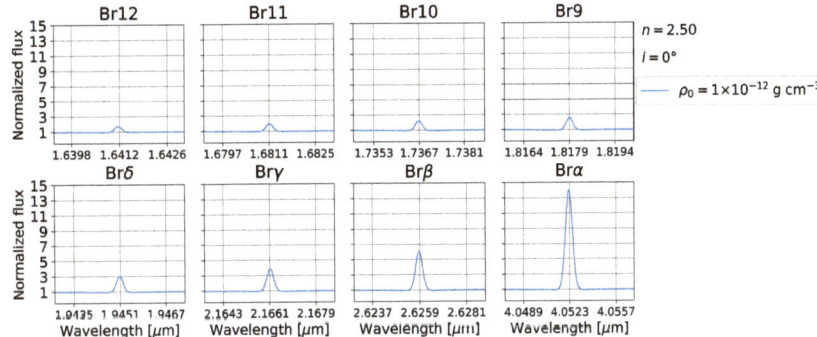

Figure 2. Synthetic Brackett line profiles obtained for $i = 0°$, an exponent of the density law $n = 2.5$, and $\rho_0 = 1 \times 10^{-12}$ g cm^{-3}.

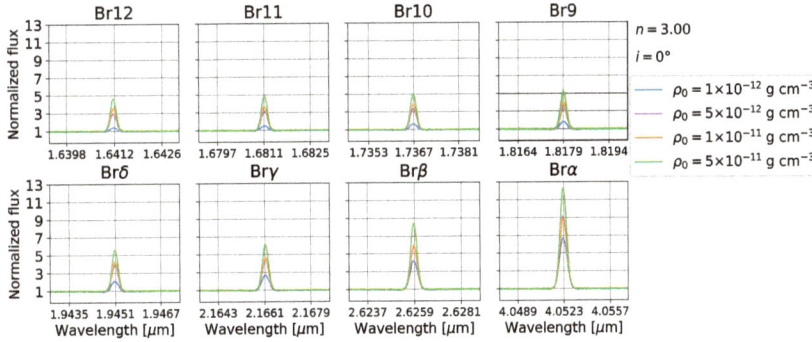

Figure 3. Synthetic Brackett line profiles obtained for $i = 0°$, an exponent of the density law $n = 3.0$, and different values of the central density.

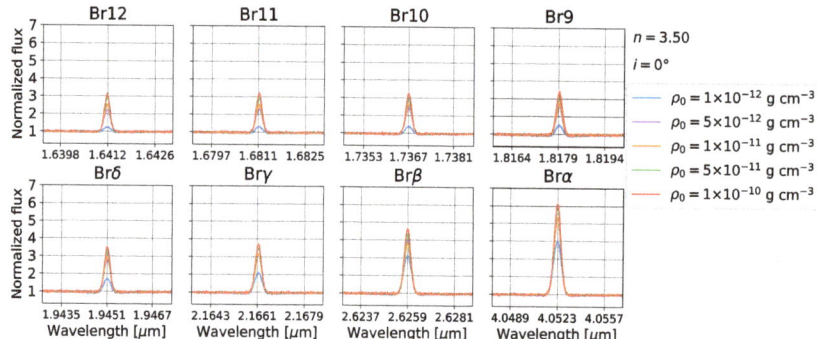

Figure 4. Synthetic Brackett line profiles obtained for $i = 0°$, an exponent of the density law $n = 3.5$, and different values of the central density.

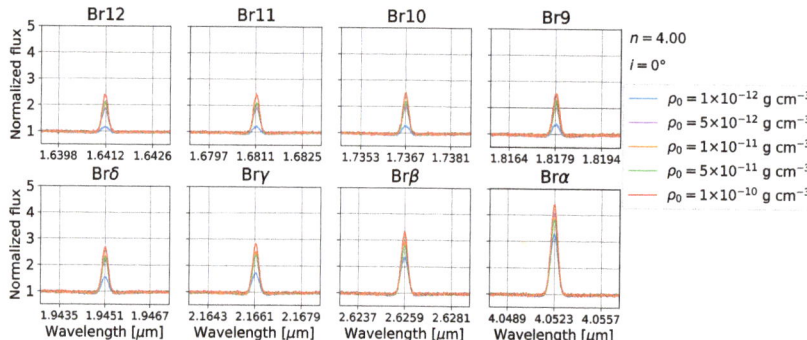

Figure 5. Synthetic Brackett line profiles obtained for $i = 0°$, an exponent of the density law $n = 4.0$, and different values of the central density.

3.2. EW and Flux Ratios Relative to Br12

To deduce the contribution of the disc, we corrected the measured emission line EWs for photospheric absorption [12]. The corrected EW values for $i = 0°$ for each n and ρ_0 normalised to EW(Br12) are plotted in Figure 6. Plots for $i = 30°$ and $i = 60°$ are in Appendix A.2. As we described before for the line profiles, for all the curves, there is an increase in the EW from Br12 to Brα, and in each panel (i.e., each density and inclination), the higher the n value, the smaller the slope of the curve.

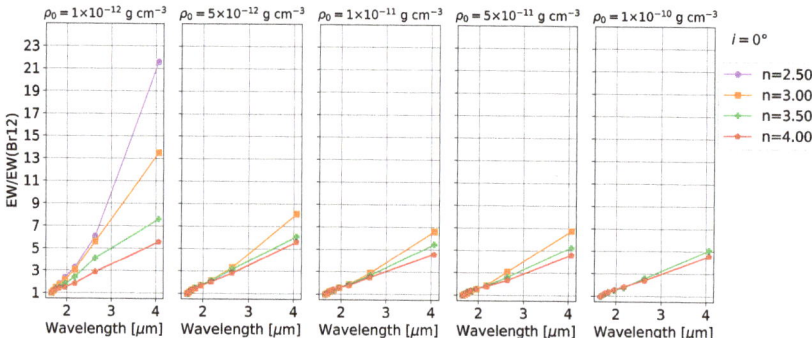

Figure 6. EW ratio for the Brackett lines relative to EW(Br12) for $i = 0°$ (for other inclinations see Figure A5). Each panel corresponds to a central density ρ_0, and each curve represents a given n value.

We obtained the flux for each line using the corrected line EW and the flux of the continuum at each wavelength. The fluxes normalised to flux(Br12) are plotted in Figure 7 for $i = 0°$ and in Appendix A.2 for $i = 30°$ and $i = 60°$. All the curves present non-monotonic behaviour, with fluxes increasing from Br12 to Br10 and then decreasing up to Brα. The lowest curve corresponds to the highest n value in each panel. For the lowest density, the maximum of the curve of line flux ratios is higher for lower n, but for the other densities, the maximum value is very similar for all the exponents.

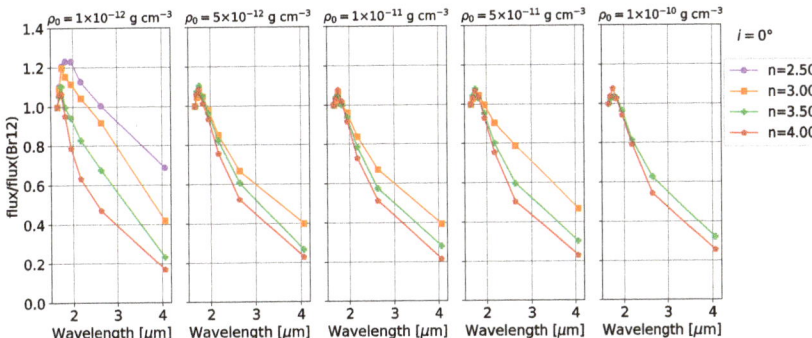

Figure 7. Flux ratio for the Brackett lines relative to flux(Br12) for $i = 0°$ (for other inclinations see Figure A6). Each panel corresponds to a central density ρ_0, and each curve represents a given n value.

In Figure 8, we show the plot of EW(Brα) versus EW(Brγ). We can see that the models with a higher exponent value present a smaller EW(Brα)/EW(Brγ) ratio and are close to the relation expected for Group I objects [9,12]. The Group II locus seems to only contain objects with envelopes with the smallest density. For each n, the higher the density, the higher the EWs.

Finally, Figure 9 shows EW(Br12) versus EW(Br11). This plot shows that the different models we calculated follow a linear relation close to the expected ratio for case B recombination, similarly to that obtained by Steele and Clark [31] (see their Figure 8, top). As for our Figure 8, larger values of EWs are obtained for smaller n values, and for a fixed n value, the higher the density, the higher the EWs. The behaviour is similar for the different inclination angles computed.

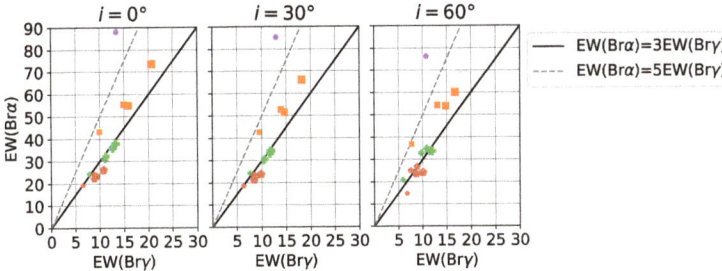

Figure 8. EW(Brα) versus EW(Brγ). Colours and symbols are the same as in Figure 7. The symbol size is proportional to the central density of the model.

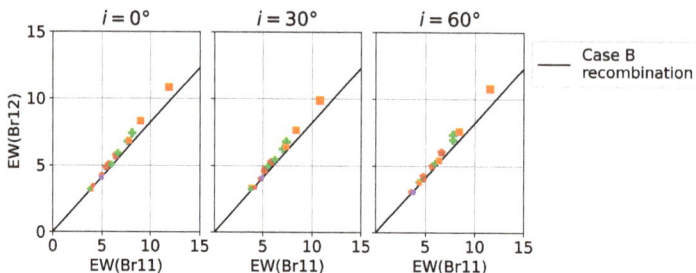

Figure 9. EW(Br12) versus EW(Br11). Colours and symbols are the same as in Figure 7. The symbol size is proportional to the central density of the model.

4. Discussion

We seek to explore whether the computed models successfully describe the observed data published by Cochetti et al. [12]. With this aim, we first perform Brackett line fitting and then compare the synthetic line ratios with those of the observed spectra. This way, we can analyse the possibility of determining the disc parameters ρ_0 and n. Using the published sample in Cochetti et al. [12], we have chosen two stars with stellar parameters close to those used in our modelling: MX Pup and π Aqr. For these stars, spectra were obtained in June 2017 with the FIRE spectrograph installed at the Magellan Baade Telescope at Las Campanas Observatory.

For the comparison with observational data, we allow the inclinations to span the whole range of our computations, 0°, 15°, 30°, 45°, and 60°. Theoretical spectra have been convolved with a Gaussian profile with FWHM = 60 km s^{-1} to match the observed data's spectral resolution. The uncertainties for the measurements are around 10% for the EW and flux due to possible errors in the continuum determination.

4.1. MX Pup (HD 68980)

For MX Pup, Frémat et al. [32] determined an effective temperature $T_{\text{eff}} = 25{,}125 \pm 642$ K and a projected rotational velocity $V \sin i = 152 \pm 8$ km s^{-1}. From fitting on the Hα profile, Silaj et al. [33] found an inclination angle $i = 20°$, while Arcos et al. [18] determined $i = 50°$ from data taken in 2013 and 2015.

We made a visual comparison between each model and the spectrum of MX Pup, and Figure 10 shows the best-fitting model. On the x-axis, in velocity units, we can see that the peak separation for higher-order members is larger than in Brγ, indicating that their forming region is closer to the central star. Since HDUST does not include non-coherent electron scattering, which can affect the line profiles with extended wings [34], a difference in the wings still remains, meaning that the non-coherent electron scattering is significant. This could be improved in the future by convolving the model with a Gaussian to simulate the electron velocity [35]. It is also possible that a single power law is not enough to fit the complexity of the disc if the wings and the core are formed in very different regions of the disc [36]. Because of this, as the best-fitting model, we chose a compromise solution for both the central part of the line and the wings. The parameters of the model are $n = 3.0$, $\rho_0 = 5 \times 10^{-11}$ g cm^{-3}, and $i = 60°$. Such parameters put this object in the stable/dissipating limit according to Vieira et al. [23]. With this inclination, the rotational velocity would be smaller than the one used for our modelling but inside the range of the true rotational velocity ratios expected for Be stars [37]. In the EW/EW(Br12) and flux/flux(Br12) plots (Figure 11), the observed data agree with the curve for the values determined from the spectrum. Nevertheless, in the EW(Br11) vs. EW(Br12) plot in Figure 12, the MX Pup position lies over the diagonal but far from the models. This difference may be caused by the poor fitting of the line wings.

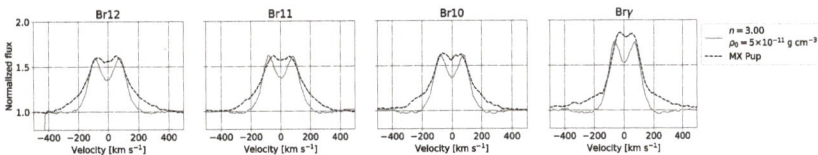

Figure 10. Best-fitting model for MX Pup.

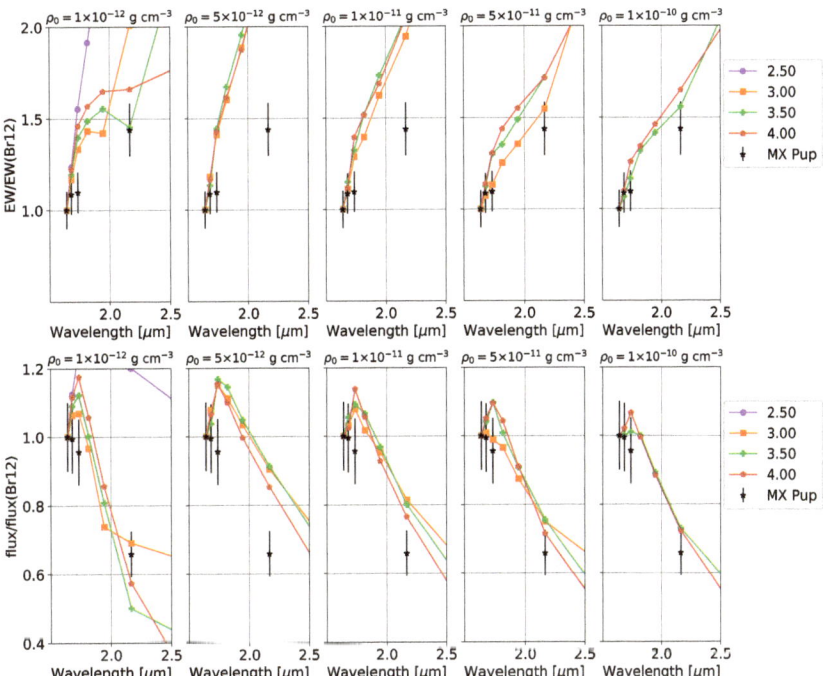

Figure 11. Comparison of the curves of EW and flux relative to those of Br12 for $i = 60°$ and different central densities ρ_0 and n values with the data obtained for the Be star MX Pup.

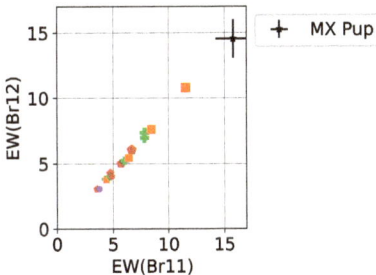

Figure 12. Location of the measurements for the star MX Pup in the EW(Br12) versus EW(Br11) plot for $i = 60°$. Colours and symbols are the same as in Figure 7. The symbol size is proportional to the central density of the model.

4.2. π Aqr (HD 212571)

For this object, Frémat et al. [32] determined an effective temperature $T_{\text{eff}} = 26{,}061 \pm 736$ K and a projected rotational velocity $V \sin i = 233 \pm 15\,\text{km}\,\text{s}^{-1}$. Both Silaj et al. [33]

and Arcos et al. [18] reported inclination angles for this object, with $i = 45°$ and $i = 60°$, respectively.

The best-fitting model found through a visual comparison is shown in Figure 13. The peak separation for the higher-order lines is not as different from the one for Brγ as in MX Pup. In this case, it does not seem to be necessary to add non-coherent scattering to fit the wings. The parameters are the following: $n = 3.5$, $\rho_0 = 10^{-11}$ g cm^{-3}, and $i = 45°$. This corresponds to a forming/stable disc [23]. Although in the EW/EW(Br12) plot, π Aqr data follow the tendency of the model from the spectral fitting, in the flux/flux(Br12) plot, the behaviour of the observed lines does not follow the model's tendency (Figure 14). Figure 15 shows the EW(Br12) vs. EW(Br11) plot, where π Aqr data agree with the parameters found previously.

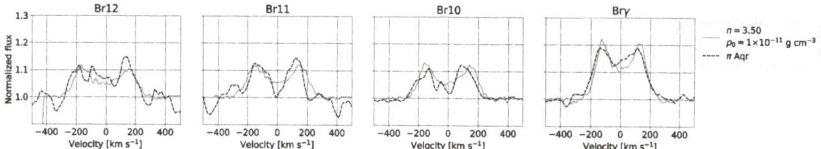

Figure 13. Best-fitting model for π Aqr.

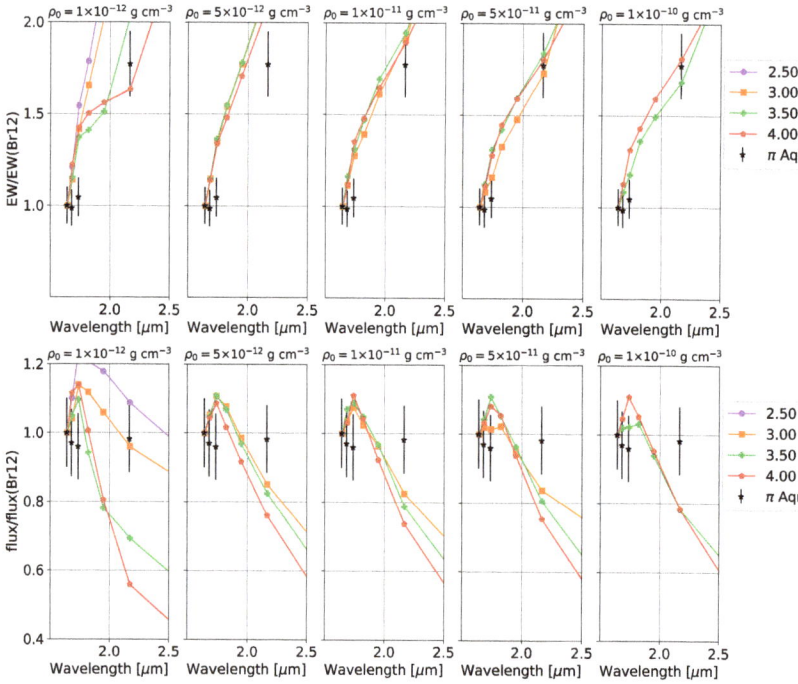

Figure 14. Comparison of the curves of EW and flux relative to those of Br12 for $i = 45°$ and different central densities ρ_0 and n values with the data obtained for the Be star π Aqr.

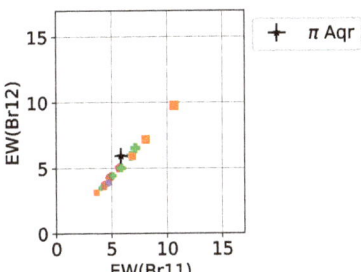

Figure 15. Location of the measurements for the star π Aqr in the EW(Br12) versus EW(Br11) plot for $i = 45°$. Colours and symbols are the same as in Figure 7. The symbol size is proportional to the central density of the model.

5. Conclusions

We used the HDUST code to perform simulations that allowed us to obtain synthetic Brackett-series line profiles for one set of parameters for the central star corresponding to an early-type object, with different parameters for the density law throughout the disc. The calculations of the synthetic models were carried out for different inclination angles of the system (star-plus-disc). The parameter space of the models that we used has been found suitable to fit Be-star Hα profiles (e.g., [23]).

After measuring the EW and flux of the Brackett-series line profiles and applying a correction for photospheric absorption, we analysed the behaviour of the series according to the disc parameters. We found that the line intensity increment through the series depends on those parameters. Regarding the EW(Brα) versus EW(Brγ) behaviour, the calculated models lie in different loci according to the central density. In the EW(Br12) versus EW(Br11) plot, the models follow a linear relation close to the expected ratio for case B recombination [31].

By comparing our simulations with the data obtained for MX Pup and π Aqr, we were allowed to set some constraints on the disc parameter values. Even though the synthetic spectra adequately fit our observations of both stars, the derived values show a discrepancy with the observed data in the EW and flux plots. The discrepancy is more remarkable in the case of MX Pup, where non-coherent electron scattering and a more complex radial dependence for the density seem to be needed.

It is possible that by including Brackett lines of higher terms or adding a similar analysis to the one performed throughout this article to other hydrogen series, we may be able to improve the determination of the parameters of the observed disc. We defer such analysis to a forthcoming article.

Author Contributions: Conceptualisation, Y.R.C., A.G., M.L.A. and A.F.T.; methodology, Y.R.C., A.G., M.L.A., A.F.T. and C.A.; formal analysis, Y.R.C., A.G., M.L.A. and A.F.T.; writing—original draft preparation, Y.R.C.; writing—review and editing, Y.R.C., A.G., M.L.A., A.F.T. and C.A.; visualisation, Y.R.C.; funding acquisition, A.G., M.L.A. and A.F.T. All authors have read and agreed to the published version of the manuscript.

Funding: The authors would like to thank the reviewers for all their useful comments, which improved our manuscript. Y.R.C. acknowledges support from a CONICET fellowship. M.L.A. and A.F.T. acknowledge financial support from the Universidad Nacional de La Plata (Programa de Incentivos 11/G160). A.G. acknowledges the research project from CONICET, PIBAA 28720210100879CO. C.A. thanks Fondecyt Regular N. 1230131 for the support. This project has received funding from the European Union's Framework Programme for Research and Innovation Horizon 2020 (2014–2020) under the Marie Skłodowska-Curie Grant Agreement No. 823734.

Data Availability Statement: The data presented in this study are available on request from the corresponding author.

Acknowledgments: This work made use of Astropy:[5] a community-developed core Python package and an ecosystem of tools and resources for astronomy [38–40]. The model calculations were performed with the cluster of CITECCA of Universidad Nacional de Río Negro, Argentina, which has four processors with eight threads, 2.2 GHz speed, and 180 GB of RAM. This paper includes data gathered with the 6.5 m Magellan Telescopes located at Las Campanas Observatory, Chile.

Conflicts of Interest: The authors declare no conflict of interest.

Abbreviations

The following abbreviations are used in this manuscript:

EW	Equivalent width
IR	Infrared
LTE	Local thermodynamic equilibrium
SED	Spectral energy distribution

Appendix A

Appendix A.1. Synthetic Brackett Line Profiles Obtained for $i = 30°$ and $i = 60°$

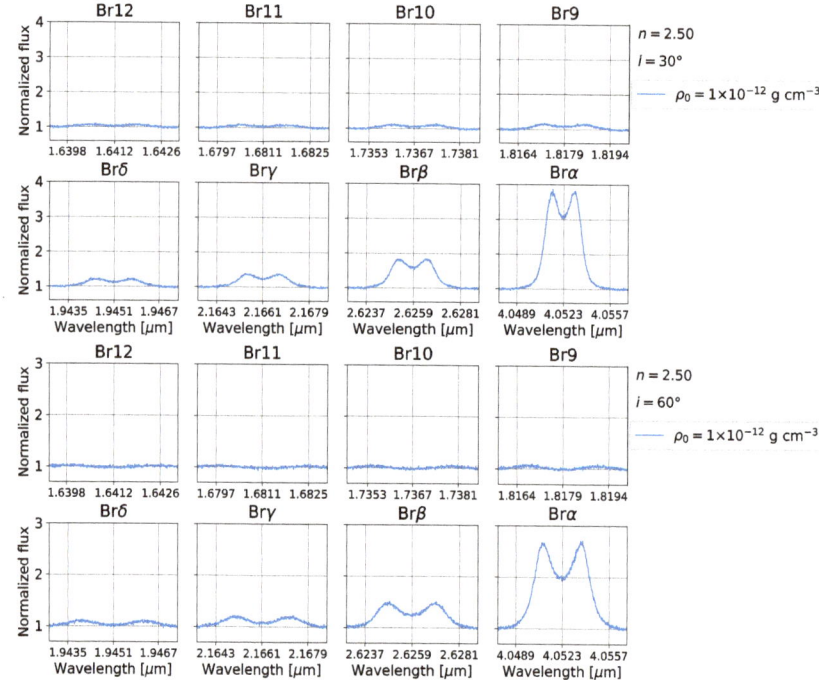

Figure A1. Synthetic Brackett line profiles obtained for an exponent of the density law $n = 2.5$, $\rho_0 = 1 \times 10^{-12}$ g cm^{-3}, and inclinations $i = 30°$ and $i = 60°$.

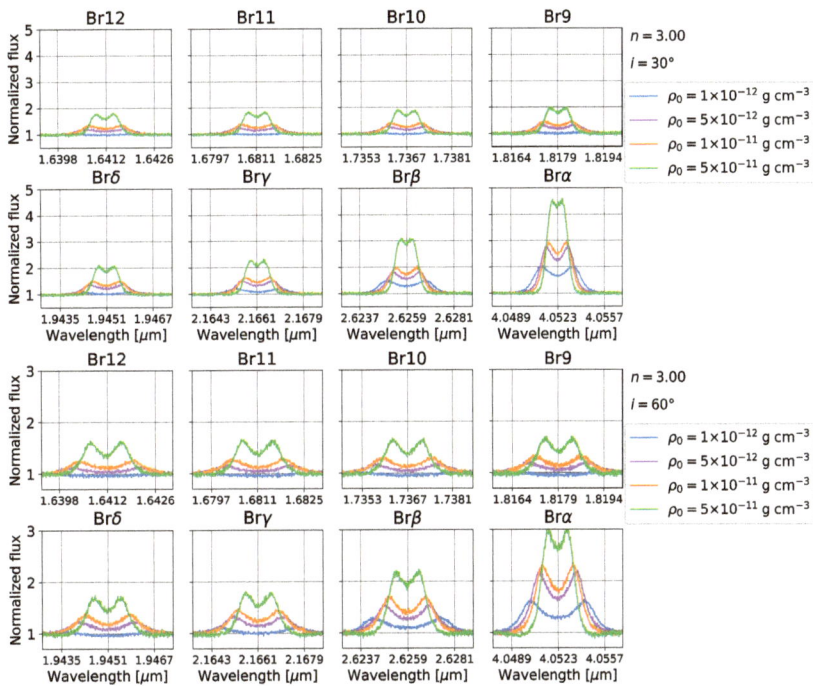

Figure A2. Synthetic Brackett line profiles obtained for an exponent of the density law $n = 3.0$, different values of the central density, and inclinations $i = 30°$ and $i = 60°$.

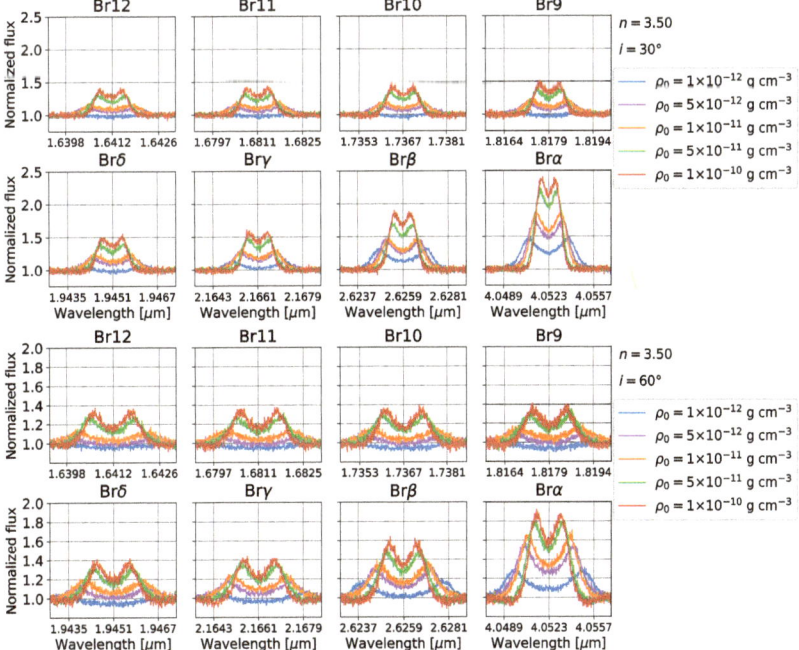

Figure A3. Synthetic Brackett line profiles obtained for an exponent of the density law $n = 3.5$, different values of the central density, and inclinations $i = 30°$ and $i = 60°$.

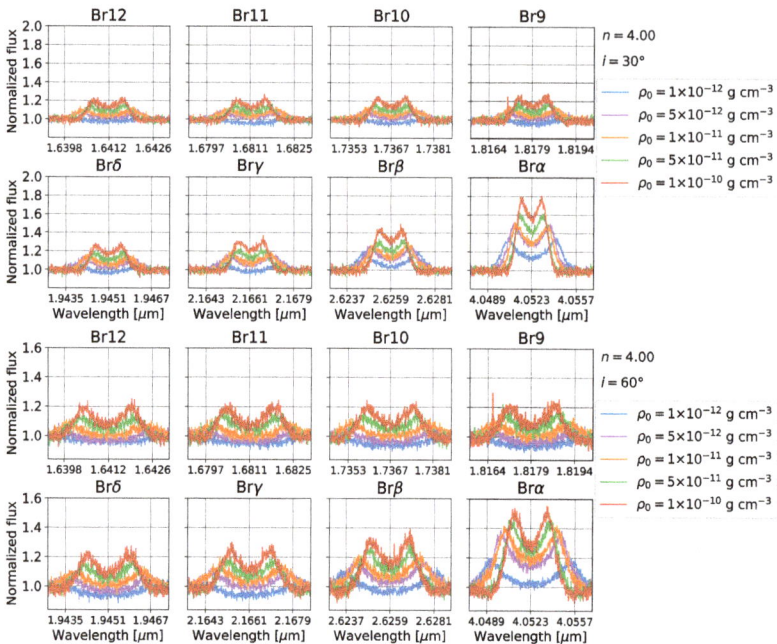

Figure A4. Synthetic Brackett line profiles obtained for an exponent of the density law $n = 4.0$, different values of the central density, and inclinations $i = 30°$ and $i = 60°$.

Appendix A.2. EW and Flux Ratios for $i = 30°$ and $i = 60°$

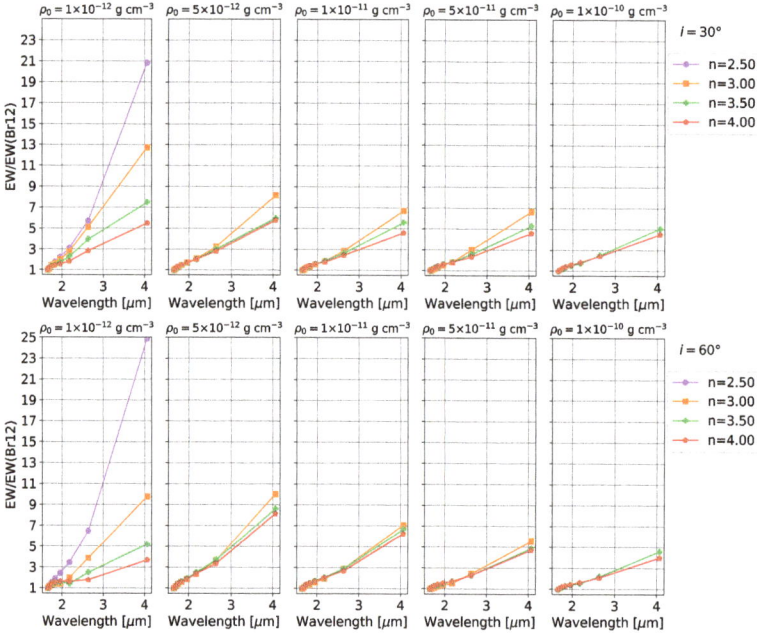

Figure A5. EW ratio for the Brackett lines relative to EW(Br12) for $i = 30°$ and $i = 60°$. Each panel corresponds to a given central density ρ_0, and each curve represents a given n value.

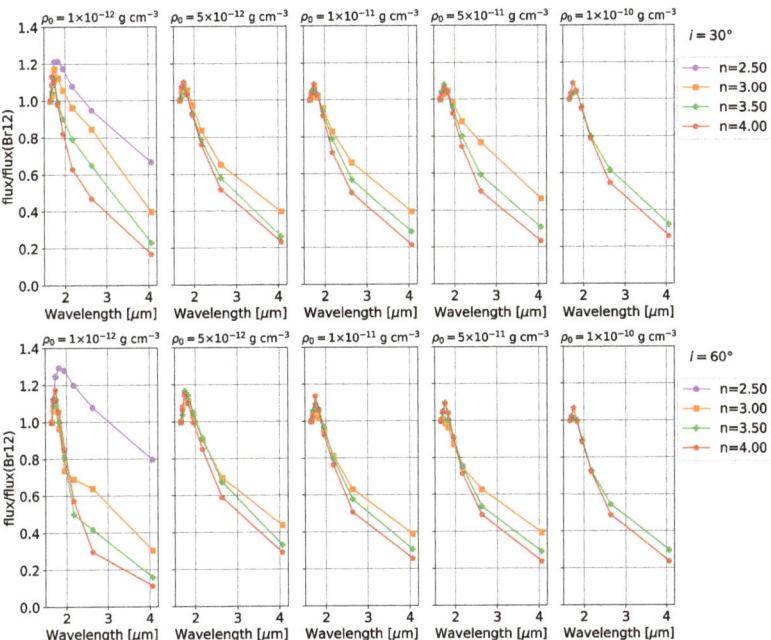

Figure A6. Flux ratio for the Brackett lines relative to flux(Br12) for $i = 30°$ and $i = 60°$. Each panel corresponds to a given central density ρ_0, and each curve represents a given n value.

Notes

1. The code actually uses n_0, the density in numerical units per cm^{-3}. Both parameters are related via the expression $\rho_0 = n_0 \cdot \mu \cdot m_H$, where $\mu = 0.6$ is the mean molecular weight, and $m_H = 1.67 \times 10^{-24}$ g is the mass of the hydrogen atom. Then, $\rho_0 \simeq n_o \times 10^{-24}$ g.
2. Apart from the models that we present in Table 1, we computed additional models for each to account for the disc's outer radius. We computed around 50 models, which took approximately 150 hours of calculation in CITECCA's cluster (see Acknowledgments for more information).
3. PyHdust provides analysis tools for multi-technique astronomical data and HDUST models.
4. Specutils is a Python package for representing, loading, manipulating, and analysing astronomical spectroscopic data [30].
5. http://www.astropy.org, accessed on 15 August 2023.

References

1. Jaschek, M.; Slettebak, A.; Jaschek, C. Be star terminology. *Be Star Newsl.* **1981**, *4*, 9–11.
2. Collins, G.W. The Use of Terms and Definitions in the Study of Be Stars (Review Paper). In *Proceedings of the 92nd Colloquium of the International Astronomical Union*, Boulder, CO, USA, 18–22 August 1986; Cambridge University Press: Cambridge, UK, 1987; p. 3. [CrossRef]
3. Rivinius, T.; Carciofi, A.C.; Martayan, C. Classical Be stars. Rapidly rotating B stars with viscous Keplerian decretion disks. *Astron. Astrophys. Rev.* **2013**, *21*, 69.
4. Lee, U.; Osaki, Y.; Saio, H. Viscous excretion discs around Be stars. *Mon. Not. R. Astron. Soc.* **1991**, *250*, 432–437. [CrossRef]
5. Carciofi, A.C. The circumstellar discs of Be stars. In *Proceedings of the Active OB Stars: Structure, Evolution, Mass Loss, and Critical Limits*; Neiner, C., Wade, G., Meynet, G., Peters, G., Eds.; Cambridge University Press: Cambridge, UK, 2011; Volume 272, pp. 325–336.
6. Hony, S.; Waters, L.B.F.M.; Zaal, P.A.; de Koter, A.; Marlborough, J.M.; Millar, C.E.; Trams, N.R.; Morris, P.W.; de Graauw, T. The infrared spectrum of the Be star gamma Cassiopeiae. *Astron. Astrophys.* **2000**, *355*, 187–193.
7. Lenorzer, A.; de Koter, A.; Waters, L.B.F.M. Hydrogen infrared recombination lines as a diagnostic tool for the geometry of the circumstellar material of hot stars. *Astron. Astrophys.* **2002**, *386*, L5–L8. [CrossRef]
8. Mennickent, R.E.; Sabogal, B.; Granada, A.; Cidale, L. L-Band Spectra of 13 Outbursting Be Stars. *Publ. Astron. Soc. Pac.* **2009**, *121*, 125.
9. Granada, A.; Arias, M.L.; Cidale, L.S. Simultaneous K- and L-band Spectroscopy of Be Stars: Circumstellar Envelope Properties from Hydrogen Emission Lines. *Astron. J.* **2010**, *139*, 1983–1992. [CrossRef]

10. Sabogal, B.E.; Ubaque, K.Y.; García-Varela, A.; Álvarez, M.; Salas, L. Evidence of Dissipation of Circumstellar Disks from L-band Spectra of Bright Galactic Be Stars. *Publ. Astron. Soc. Pac.* **2017**, *129*, 014203. [CrossRef]
11. Lenorzer, A.; Vandenbussche, B.; Morris, P.; de Koter, A.; Geballe, T.R.; Waters, L.B.F.M.; Hony, S.; Kaper, L. An atlas of 2.4 to 4.1 μm ISO/SWS spectra of early-type stars. *Astron. Astrophys.* **2002**, *384*, 473–490. [CrossRef]
12. Cochetti, Y.R.; Arias, M.L.; Cidale, L.S.; Granada, A.; Torres, A.F. Simultaneous J-, H-, K- and L-band spectroscopic observations of galactic Be stars. I. IR atlas. *Astron. Astrophys.* **2022**, *665*, A115.
13. Okazaki, A.T.; Bate, M.R.; Ogilvie, G.I.; Pringle, J.E. Viscous effects on the interaction between the coplanar decretion disc and the neutron star in Be/X-ray binaries. *Mon. Not. R. Astron. Soc.* **2002**, *337*, 967–980. [CrossRef]
14. Carciofi, A.C.; Bjorkman, J.E. Non-LTE Monte Carlo Radiative Transfer. I. The Thermal Properties of Keplerian Disks around Classical Be Stars. *Astrophys. J.* **2006**, *639*, 1081–1094.
15. Sigut, T.A.A.; Jones, C.E. The Thermal Structure of the Circumstellar Disk Surrounding the Classical Be Star γ Cassiopeiae. *Astrophys. J.* **2007**, *668*, 481–491.
16. Sigut, T.A.A. Spectral synthesis for Be stars. In *Proceedings of the Active OB Stars: Structure, Evolution, Mass Loss, and Critical Limits*; Neiner, C., Wade, G., Meynet, G., Peters, G., Eds.; Cambridge University Press: Cambridge, UK, 2011; Volume 6, pp. 426–427. [CrossRef]
17. Miroshnichenko, A.S.; Bjorkman, K.S.; Morrison, N.D.; Wisniewski, J.P.; Manset, N.; Levato, H.; Grosso, M.; Pollmann, E.; Buil, C.; Knauth, D.C. Spectroscopy of the growing circumstellar disk in the delta Scorpii Be binary. *Astron. Astrophys.* **2003**, *408*, 305–311. [CrossRef]
18. Arcos, C.; Jones, C.E.; Sigut, T.A.A.; Kanaan, S.; Curé, M. Evidence for Different Disk Mass Distributions between Early- and Late-type Be Stars in the BeSOS Survey. *Astrophys. J.* **2017**, *842*, 48.
19. Cochetti, Y.R.; Arcos, C.; Kanaan, S.; Meilland, A.; Cidale, L.S.; Curé, M. Spectro-interferometric observations of a sample of Be stars. Setting limits to the geometry and kinematics of stable Be disks. *Astron. Astrophys.* **2019**, *621*, A123.
20. Chojnowski, S.D.; Whelan, D.G.; Wisniewski, J.P.; Majewski, S.R.; Hall, M.; Shetrone, M.; Beaton, R.; Burton, A.; Damke, G.; Eikenberry, S.; et al. High-Resolution H-Band Spectroscopy of Be Stars With SDSS-III/Apogee: I. New Be Stars, Line Identifications, and Line Profiles. *Astron. J.* **2015**, *149*, 7,
21. Carciofi, A.C.; Bjorkman, J.E. Non-LTE Monte Carlo Radiative Transfer. II. Nonisothermal Solutions for Viscous Keplerian Disks. *Astrophys. J.* **2008**, *684*, 1374–1383.
22. Klement, R.; Carciofi, A.C.; Rivinius, T.; Matthews, L.D.; Vieira, R.G.; Ignace, R.; Bjorkman, J.E.; Mota, B.C.; Faes, D.M.; Bratcher, A.D.; et al. Revealing the structure of the outer disks of Be stars. *Astron. Astrophys.* **2017**, *601*, A74.
23. Vieira, R.G.; Carciofi, A.C.; Bjorkman, J.E.; Rivinius, T.; Baade, D.; Rímulo, L.R. The life cycles of Be viscous decretion discs: Time-dependent modelling of infrared continuum observations. *Mon. Not. R. Astron. Soc.* **2017**, *464*, 3071–3089.
24. Richardson, N.D.; Thizy, O.; Bjorkman, J.E.; Carciofi, A.; Rubio, A.C.; Thomas, J.D.; Bjorkman, K.S.; Labadie-Bartz, J.; Genaro, M.; Wisniewski, J.P.; et al. Outbursts and stellar properties of the classical Be star HD 6226. *Mon. Not. R. Astron. Soc.* **2021**, *508*, 2002–2018.
25. Marr, K.C.; Jones, C.E.; Carciofi, A.C.; Rubio, A.C.; Mota, B.C.; Ghoreyshi, M.R.; Hatfield, D.W.; Rímulo, L.R. The Be Star 66 Ophiuchi: 60 Years of Disk Evolution. *Astrophys. J.* **2021**, *912*, 76.
26. Zorec, J.; Rieutord, M.; Espinosa Lara, F.; Frémat, Y.; Domiciano de Souza, A.; Royer, F. Gravity darkening in stars with surface differential rotation. *Astron. Astrophys.* **2017**, *606*, A32.
27. Shakura, N.I.; Sunyaev, R.A. Black holes in binary systems. Observational appearance. *Astron. Astrophys.* **1973**, *24*, 337–355.
28. Silaj, J.; Jones, C.E.; Sigut, T.A.A.; Tycner, C. The Hα Profiles of Be Shell Stars. *Astrophys. J.* **2014**, *795*, 82. [CrossRef]
29. Espinosa Lara, F.; Rieutord, M. Gravity darkening in rotating stars. *Astron. Astrophys.* **2011**, *533*, A43.
30. Earl, N.; Tollerud, E.; O'Steen, R.; Brechmos.; Kerzendorf, W.; Busko, I.; Shailesahuja.; D'Avella, D.; Robitaille, T.; Ginsburg, A.; et al. astropy/specutils: v1.10.0. *Zenodo* **2023**. [CrossRef]
31. Steele, I.A.; Clark, J.S. A representative sample of Be stars III: H band spectroscopy. *Astron. Astrophys.* **2001**, *371*, 643–651. [CrossRef]
32. Frémat, Y.; Zorec, J.; Hubert, A.M.; Floquet, M. Effects of gravitational darkening on the determination of fundamental parameters in fast-rotating B-type stars. *Astron. Astrophys.* **2005**, *440*, 305–320. [CrossRef]
33. Silaj, J.; Jones, C.E.; Tycner, C.; Sigut, T.A.A.; Smith, A.D. A Systematic Study of Hα Profiles of Be Stars. *Astrophys. J. Suppl.* **2010**, *187*, 228–250. [CrossRef]
34. Hummel, W.; Dachs, J. Non-coherent scattering in vertically extended Be star disks : winebottle-type emission-line profiles. *Astron. Astrophys.* **1992**, *262*, L17.
35. Marr, K.C.; Jones, C.E.; Tycner, C.; Carciofi, A.C.; Silva, A.C.F. The Role of Disk Tearing and Precession in the Observed Variability of Pleione. *Astrophys. J.* **2022**, *928*, 145.
36. Sigut, T.A.A.; Tycner, C.; Jansen, B.; Zavala, R.T. The Circumstellar Disk of the Be Star o Aquarii as Constrained by Simultaneous Spectroscopy and Optical Interferometry. *Astrophys. J.* **2015**, *814*, 159.
37. Zorec, J.; Frémat, Y.; Domiciano de Souza, A.; Royer, F.; Cidale, L.; Hubert, A.M.; Semaan, T.; Martayan, C.; Cochetti, Y.R.; Arias, M.L.; et al. Critical study of the distribution of rotational velocities of Be stars. I. Deconvolution methods, effects due to gravity darkening, macroturbulence, and binarity. *Astron. Astrophys.* **2016**, *595*, A132. [CrossRef]

38. Astropy Collaboration.; Robitaille, T.P.; Tollerud, E.J.; Greenfield, P.; Droettboom, M.; Bray, E.; Aldcroft, T.; Davis, M.; Ginsburg, A.; Price-Whelan, A.M.; et al. Astropy: A community Python package for astronomy. *Astron. Astrophys.* **2013**, *558*, A33.
39. Astropy Collaboration.; Price-Whelan, A.M.; Sipőcz, B.M.; Günther, H.M.; Lim, P.L.; Crawford, S.M.; Conseil, S.; Shupe, D.L.; Craig, M.W.; Dencheva, N.; et al. The Astropy Project: Building an Open-science Project and Status of the v2.0 Core Package. *Astron. J.* **2018**, *156*, 123.
40. Astropy Collaboration.; Price-Whelan, A.M.; Lim, P.L.; Earl, N.; Starkman, N.; Bradley, L.; Shupe, D.L.; Patil, A.A.; Corrales, L.; Brasseur, C.E.; et al. The Astropy Project: Sustaining and Growing a Community-oriented Open-source Project and the Latest Major Release (v5.0) of the Core Package. *Astrophys. J.* **2022**, *935*, 167.

Disclaimer/Publisher's Note: The statements, opinions and data contained in all publications are solely those of the individual author(s) and contributor(s) and not of MDPI and/or the editor(s). MDPI and/or the editor(s) disclaim responsibility for any injury to people or property resulting from any ideas, methods, instructions or products referred to in the content.

Article

New Method to Detect and Characterize Active Be Star Candidates in Open Clusters

Anahí Granada [1,2,*], Maziar R. Ghoreyshi [3], Carol E. Jones [3] and Tõnis Eenmäe [4]

[1] Centro Interdisciplinario de Telecomunicaciones, Electrónica, Computación y Ciencia Aplicada (CITECCA), Sede Andina, Universidad Nacional de Río Negro, Anasagasti 1463, San Carlos de Bariloche R8400AHN, Río Negro, Argentina
[2] Consejo Nacional de Investigaciones Científicas y Técnicas (CONICET), Godoy Cruz 2290, Buenos Aires 1461, Argentina
[3] Physics and Astronomy Department, The University of Western Ontario, London, ON N6A 3K7, Canada
[4] Faculty of Science and Technology, University of Tartu, Observatooriumi 1, 61602 Tõravere, Estonia
* Correspondence: agranada@unrn.edu.ar

Abstract: With the aim of better understanding the physical conditions under which Be stars form and evolve, it is imperative to further investigate whether poorly studied young open clusters host Be stars. In this work, we explain how data from Gaia DR2 and DR3 can be combined to recover and characterize active Be stars in open clusters. We test our methodology in four open clusters broadly studied in the literature, known for hosting numerous Be stars. In addition, we show that the disk formation and dissipation approach that is typically used to model long term Be star variability, can explain the observed trends for Be stars in a (G_{DR3}-G_{DR2}) versus G_{DR3} plot. We propose that extending this methodology to other open clusters, and, in particular, those that are poorly studied, will help to increase the number of Be candidates. Eventually, Be stars may eclipse binary systems in open clusters.

Keywords: Be stars; early-type emission stars; circumstellar disks; early-type variable stars; open star clusters; Gaia

Citation: Granada, A.; Ghoreyshi, M.R.; Jones, C.E.; Eenmäe, T. New Method to Detect and Characterize Active Be Star Candidates in Open Clusters. *Galaxies* **2023**, *11*, 37. https://doi.org/10.3390/galaxies11010037

Academic Editor: Jorick Sandor Vink

Received: 23 January 2023
Revised: 13 February 2023
Accepted: 16 February 2023
Published: 19 February 2023

Copyright: © 2023 by the authors. Licensee MDPI, Basel, Switzerland. This article is an open access article distributed under the terms and conditions of the Creative Commons Attribution (CC BY) license (https://creativecommons.org/licenses/by/4.0/).

1. Introduction

Be stars are rapidly rotating main sequence B-type stars that have exhibited hydrogen emission lines at least once, a signature of the presence of a circumstellar envelope, usually described in the viscous decretion disk framework [1]. Even though Be stars constitute about 30% of early B-type stars, or even more in some young open clusters [2–4], the mechanisms involved in the development of the disk are still under study, and neither the origin of the rapid rotation of these stars, nor how close to critical they rotate, is well understood [1]. In the single star scenario, stellar evolution allows stars with a sufficiently large initial angular momentum content to evolve towards the critical limit [5–7]. Episodes of mass transfer in binary systems [8,9] could also lead to the formation of a rapidly rotating star that could potentially become a Be star.

Following the single rotating star scenario, it is expected that clusters with log(age[Myr]) around 7.1 to 7.4 are likely host a number of Be stars, and, indeed, this is observed (e.g., [10]). In this framework, these authors proposed that the Be phenomenon is an evolutionary effect, appearing at the end of the main-sequence lifetime of a rapidly rotating B star.

Yet, why do some clusters of this age range host a very large fraction of Be stars while others have only a handful of them? Is this just an effect of small number statistics or are there *real* differences in the clusters where these stars form and evolve?

With the aim of better understanding the physical conditions under which Be stars form and evolve, it is imperative to further investigate whether poorly studied young open clusters host Be stars. This is not an easy task, because characterizing Be stars relies on obtaining spectroscopic data of individual stars, which is usually expensive in terms of telescope time. The results from spectroscopic surveys, such as the Apache Point Observatory Galactic Evolution Experiment (APOGEE) [11] or the Large sky Area Multi-Object fiber Spectroscopic Telescope (LAMOST) [12], have successfully increased the number of Be stars (e.g., Chojnowski et al. [13], Lin et al. [14], Vioque et al. [15], Wang et al. [16]). However, due to the transient nature of these objects, developing new methods of detecting Be candidates is called for. In particular, a new method, utilizing photometric archival data from Gaia Data Release 2 (DR2) [17] and Data Release 3 (DR3) [18], seems very promising, and devising such a method was the aim of the present article.

In Section 2, we explain how data from Gaia DR2 and DR3 can be combined to recover and characterize active Be stars in four broadly studied open clusters, two of which constitute the Double Cluster NGC869/NGC884. Then, in Section 3 we present a toy model of disc formation and dissipation around a B type star which can help explain the observations. Finally, we present our results and conclusions.

2. Materials and Methods
2.1. Gaia DR2 and DR3 Photometry

One of the main goals of the Gaia mission is to deliver multi-band photometry from the spectral energy distribution of stars in order to derive stellar fundamental parameters and identify peculiar objects [19]. Up to now, Gaia has provided three data releases (DRs). In particular, Gaia's second data release (DR2) was published during April, 2018, and Gaia Data Release 3 was split into the early release, called Gaia Early Data Release 3 (Gaia EDR3) and the full Gaia Data Release 3 (Gaia DR3), which was finally released in June, 2022. While DR2 spanned 22 months of data, (E)DR3 data spanned 34 months, including those of DR2.

At each Gaia release, a fundamental step in data processing, referred to as photometric external calibration, was performed. As a consequence, each release has its own definition of the set of passbands G, G_{BP} and G_{RP}. Basically, each of these three passbands changes between different releases. For this reason, a direct comparison of individual stars from different releases is usually discouraged [20]. Instead, for a comparison between releases it is recommended to use carefully selected datasets. For further details on Gaia photometry the reader is referred to Riello et al. [20], as well as the Gaia documentation pages.

In the present article, we proposed comparing Gaia DR2 and DR3 photometry for four well studied open clusters hosting numerous Be stars: the double cluster NGC869/NGC884, NGC663 and NGC7419.

2.2. Taking Advantage of the Variable Nature of Be Stars

First of all, it is important to recall that Be stars usually undergo photometric and spectroscopic variability on different timescales [1]. The characteristics of their mid- and long-term photometric variability, typically lasting from months to years, can be mostly explained in the Viscous Decretion Disk (VDD) framework [21] in terms of disk formation or dissipation processes, or due to disk perturbations (e.g., Rivinius et al. [1], Labadie-Bartz et al. [22], and references therein).

For the clusters under study, we claimed that stars which had not exhibited variability during the Gaia mission, small amplitude variable stars and, even, unresolved binaries, would not only have smaller error bars in Gaia photometric data than active Be stars, but would also define a narrow sequence in the (G_{DR3}-G_{DR2}) versus G_{DR3}. We refer to stars in this tight sequence as *stable stars*. Due to the typical amplitude of their long-term variability, of the order of one magnitude [23], active Be stars that exhibited variability during the Gaia mission would depart from this sequence. In this case, we describe stars with a significant disk variability within the epoch of each release as 'active'. Stars with a stable disk, or even those having variability much smaller than the duration of the mission, would also remain close to the narrow band of *stable stars*. In the next subsection we detail our findings for each cluster.

If the Gaia filters were identical between releases, the narrow band of *stable stars* would cluster around zero. As described above, this is not the case, as filters were redefined at each release [17,18]. In this work, we center our analysis on the Gaia G filter variability between DR2 and DR3, as in both releases the errors in the G band were significantly smaller than those of the other two bands, G_B and G_R [20].

2.3. Open Cluster Data

As mentioned previously, we focused on four galactic open clusters gathered in three different samples, with ages between 14 and 40 Myr, notorious for their large number of Be stars, which have been broadly studied in the literature: the double cluster NGC869/NGC884 [24–26], NGC663 [3,27,28] and NGC7419 [2]. As these are rich in Be stars, it was expected that a fraction of them would either be forming or dissipating a disk in the epoch of observation.

For each cluster, we considered as cluster members those stars having membership probabilities larger than 0.5, according to [29]. All the data analyzed in this work are provided in the Appendix A.

3. Results

3.1. NGC869/NGC884

The pair consisting of NGC869 and NGC884, centered at right ascension, RA, and declination, dec, 34.741°, +57.134° and 35.584°, +57.149°, respectively, is a physically bound system. The distance to NGC869 is 2246 pc, its age is 12.9 Myr and its mean absorption in the V band is 1.749, while for NGC 884 the distance is 2150 pc, with an age of 15.4 Myr and a mean extinction in the V band of 1.709, according to [30], and, in agreement with [29,31,32], within the errors.

For this double cluster, we selected objects with $G_{DR3} < 14.5$ because all their known B stars are more than one magnitude brighter than this value. To start with, we investigated errors in the G band for these bright cluster members. In Figure 1a we plotted the errors in G_{DR3} (errG_{DR3}) versus G_{DR3}. Small violet symbols indicate cluster members, and red squares indicate known Be stars from the literature [25,33]. Cyan symbols indicate stars classified as eclipsing binaries (EBs) in SIMBAD, and green triangles are non-Be pulsating variable stars in NGC 884 by [34]. Blue pentagons belong to the RS Canum Venaticorum class, a type of active eclipsing binary star, according to [35].

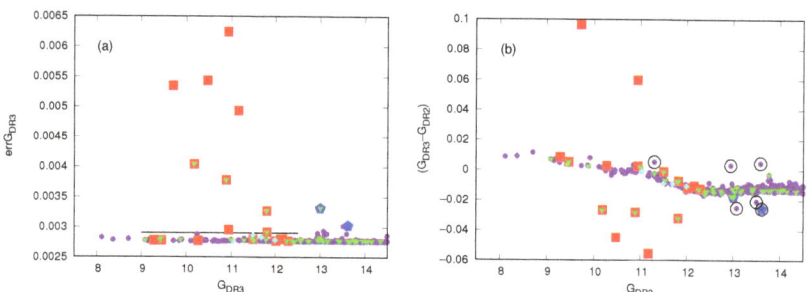

Figure 1. Data for NGC869/NGC884. (**a**) Error in the G_{DR3} band versus G_{DR3} magnitude. (**b**) (G_{DR3}-G_{DR2}) versus G_{DR3}. Violet small symbols indicate cluster members, and red squares indicate known Be stars, cyan symbols indicate stars classified as eclipsing binaries, green triangles are pulsating variable stars. Blue pentagons belong to the RS Canum Venaticorum class. The open black circles enclose objects that depart significantly from the tight relation for stable stars.

The values of errG_{DR3} centered around a median value of 0.00277 for most stars in the range of G plotted. Performing a detailed analysis of these photometric errors would be a complex task [20] and was beyond the scope of the present article. However, and very interestingly, all stars with $G_{DR3} < 12.5$ with $\text{err}G_{DR3} > 0.0029$ are known Be stars. Values of $G_{DR3} > 12.5$ delimited the transition from late B to early A stars at around G = 14 mag, and, also, intriguingly, two of the stars with the largest departure from the median value were stars classified as rotational variables by [35], and one of them had shown pulsations [34].

In Figure 1b we plotted (G_{DR3}-G_{DR2}) versus G_{DR3}. Again, and not surprisingly, we can clearly see that the seven Be stars with the largest $\text{err}G_{DR3}$ also departed from the violet trend, which indicated that these objects were changing significantly between the two different releases. In addition, we can see that while two of the Be stars faded (above the violet trend), another five Be stars brightened (below the violet trend). In Section 4, we interpret the behavior of these objects in the context of a disk formation/dissipation scheme, as seen from different inclination angles.

An inspection of Figure 1b led us to propose that six stars departed from the violet trend that gathered most stars, or *stable stars*. These are indicated with open black circles and were considered to be Be candidates. Interestingly, one of them was an EB. Together with the 16 known Be stars and other interesting variable stars, we listed the Be candidates, as shown in Table A1.

3.2. NGC663

The open cluster NGC663 is located at RA of 26.586° and dec of +61.212°, at a distance of 2950 pc, having an age of 30 Myr and an average extinction in the V band of 2.18 [32].

Similar to the double cluster, we considered objects with $G_{DR3} < 14.5$, which easily included B-type stars.

The color coding in Figure 2 is identical to Figure 1: violet symbols indicate cluster members in the quoted range, while red squares are known Be stars. Figure 2a shows that stars with $\text{err}G_{DR3} > 0.0029$ and $G_{DR3} < 13.5$ were all known Be stars. There were two stars beyond this limit, with $\text{err}G_{DR3} > 0.0029$, that were not known Be stars. Figure 2b shows that most of the Be stars were at, or above, the violet trend of *stable* stars. Only three Be stars were below. An eye inspection of Figure 2 led us to propose 5 Be candidates, which are indicated as open circles, in addition to the 30 known Be stars. We list them in Table A2.

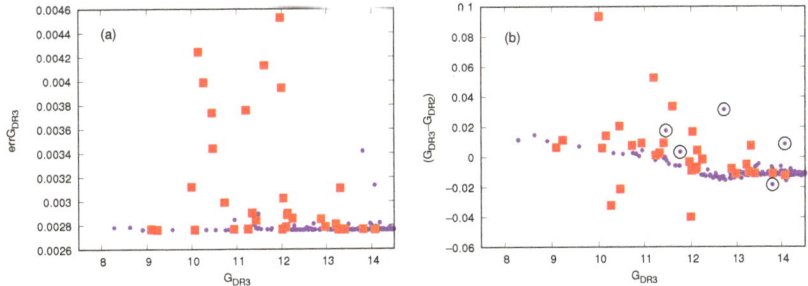

Figure 2. Data for the cluster NGC 663. (a) Error in the G_{DR3} band versus G_{DR3} magnitude. (b) (G_{DR3}-G_{DR2}) versus G_{DR3}. Violet small symbols indicate cluster members and red squares indicate known Be stars. The open black circles enclose objects that depart significantly from the tight relation for stable stars.

3.3. NGC7419

The cluster NGC 7419 is located at RA 343.579°, dec +60.814°, and, according to the literature, at a distance between 3105 pc [30] and 3236 pc [36]. Different values for its age are found in the literature, between 5 Myr [30] and 30 Myr [31]. The number of red supergiant and Be stars observed in this cluster may favor an intermediate age of 14 Myr [37]. The average extinction in the V band of this cluster is large, with a value of 4.291 [30].

NGC 7419 is more distant than the other three clusters mentioned above, and suffers frpm a heavy intra-cluster reddening, as can be deduced from its broad Color-Magnitude Diagram [2]. Objects close to a magnitude of G = 17 have been classified as Be stars by these authors, so this is why we included all objects brighter than G_{EDR3} = 18 in this analysis.

Figure 3a shows that all but one star with $G_{DR3} < 15$ and $errG_{DR3} > 0.003$ were known Be stars. As in the previous cases, Be stars were characterized by their large errors when compared to the non-Be cluster stars (*stable stars*). In Figure 3b, the values of G_{DR3}-G_{DR2} had a significantly larger dispersion than those in Figures 1b and 2b. This is why we color-coded according to the (B_G-R_G) color of each star. Red open squares indicate Be stars and the empty red square corresponds to a Be star for which no B_G or R_G was available. Black open circles in Figure 3b enclose stars that significantly departed from the main *stable star* distribution. The dispersion was much larger in this highly reddened cluster, so we arbitrarily considered stars with G_{DR3}-$G_{DR2} > -0.015$ or G_{DR3}-$G_{DR2} < -0.034$ as candidates. Using this criteria, we obtained 19 new Be candidates, two of which were actually EBs, in addition to the 37 known Be stars that were cluster members with probability higher than 0.5, according to [29]. We list the Be candidates and EBs in Table A3.

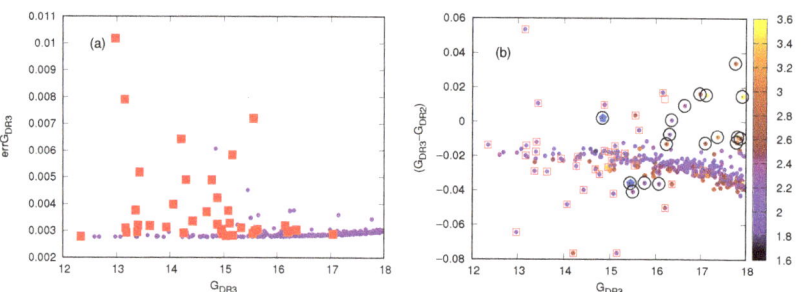

Figure 3. (**a**) The same as Figure 1a for the cluster NGC 7419. (**b**) The same plot as Figure 1b, but the color coding corresponds to the color ($B_G - R_G$). The blue symbols indicate EBs.

3.4. A Model for Be Disk Formation and Dissipation

It was beyond the scope of the present article to model each Be star developing, or not developing, variability between the DR2 and DR3 releases, in detail. However, we could gain insight as to whether a typical model for disk formation and dissipation could explain the observed trends for Be stars in the (G_{DR3}-G_{DR2}) versus GDR3 plot.

With this aim, we used the SINGLEBE code [38,39] in order to compute the dynamical 1D surface density of an isothermal viscous decretion disk. SINGLEBE is a hydrodynamic code, which solves the time-dependent fluid equations [40] in the thin disk approximation. The vertical hydrostatic equilibrium solution, with a power-law scale height $H = H_0 (r/R)^{1.5}$, and H_0 being the scale height at the base of the disk, r being the distance from the central star and R the stellar radius, was used to convert the output of SINGLEBE to a volume density. Then, this density structure was used by the 3D non-LTE Monte Carlo radiative transfer code HDUST [41,42], which calculated the synthetic observables from the star plus disk system, including the spectral energy distribution, SED.

In the models presented here, we considered only two different stellar models, corresponding to B2 and B7 stars. Their stellar parameters were identical to those presented by Ghoreyshi et al. [43] and the stars were considered to rotate at 75% of their critical speeds. We assumed that the disks were built up for 50 years with a steady mass injection rate of 7.7×10^{-9} M$\odot$/year and 1.14×10^{-10} M$\odot$/year for the B2 and B7 stellar model, respectively, which were typical mass loss rates for these spectral types [43]. After build-up, the disks were allowed to dissipate for 50 years.

For the B2 star, a base surface density of 0.8 g cm^{-2} (volume density, $\rho_0 = 2.13 \times 10^{-11}$ g cm^{-3}) was adopted. Similarly, for the B7 star a disk base surface density of 0.1 g cm^{-2} (volume density, $\rho_0 = 4.4 \times 10^{-12}$ g cm^{-3}) was used. We considered the inclination angles of 0° (pole-on), 30°, 70° and 90° (equator-on), as seen by an observer.

Along the synthetic build-up/dissipation sequence, for each SED, we computed Gaia magnitudes G, G_B and G_R using the passbands and zero point magnitudes available in the Gaia DR2[1] and DR3[2] web pages, which enabled us to make lightcurves. Of course, this type of synthetic lightcurve represents average long term variability of Be stars, and does not capture other short term variability often observed in Be stars.

We note that, for most stars, Gaia DR3 provides time-averaged magnitudes in the different bands, and only for some stars is a time-series also available [20].

It was the goal of this article to interpret these average values delivered by Gaia DR2 and DR3. To do so, at each time of the build-up/dissipation sequence, we computed the simple average values of the G, G_B and G_R within the previous 22 months for DR2 and 34 months for DR3, and assumed that, prior to the disk development, there was no disk at all.

Then, at each time of the build-up/dissipation sequence, we computed the difference between the average G magnitude DR3 (G_{DR3}) and the average G magnitude DR2 (G_{DR2}), *twelve months before*. This was to account for the fact that DR3 included 12 more months of data than DR2 data.

We denoted (G_{DR3}-G_{DR2})$_0$ as the value of (G_{DR3}-G_{DR2}) for the diskless star. In Figure 4a we plotted the predicted (G_{DR3}-G_{DR2}), normalized relative to (G_{DR3}-G_{DR2})$_0$, versus G_{DR3} for the B2 stellar model with different colors indicating different inclination angles. The dotted part of the curve corresponds to buildup, while the continuous line indicates the dissipation phase. Full circles mark the start of the cycle. Coloured squares show the average value within a full disk formation–dissipation cycle (100 years in the case presented here), but a similar result would be obtained for a shorter cycle, when formation and dissipation have the same length. This suggested that stars that undergo full cycles within the Gaia releases might tend to cluster above the zero level.

Figure 4b is the same Figure 4a for the B7 model. While the global behavior was similar for both stellar models, a larger departure from the zero level was obtained for the earlier spectral type.

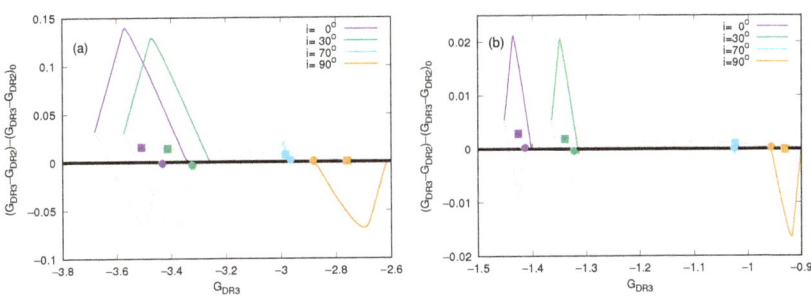

Figure 4. (**a**) Modeled (G_{DR3}-G_{DR2}), normalized relative to the value at time zero, with subindex 0, versus G_{DR3} for the B2 model. Different colors indicate different inclination angles. The dotted part of the curve corresponds to build-up and the continuous line to dissipation. The black line indicates the zero level (no variability). The full circles indicate the beginning of the cycle. The squares indicate the average value within a complete cycle and the gray line joins these squares. (**b**) Same as (**a**) for the B7 model.

We can see that for small inclination angles the values of (G_{DR3}-G_{DR2})-(G_{DR3}-G_{DR2})$_0$ were positive during the disc build-up phase and negative during dissipation, while for large inclinations the opposite was observed. Moreover, the predicted differences in brightness were in good agreement with what was observed for early and late Be stars, as seen in the previous subsections.

Interestingly, if the dissipation phases were longer than the formation phases, as suggested in the literature [44,45], then observation of stars during their disk dissipation stage would be more likely. Furthermore, observation of active Be stars with a small inclination angle, with positive values of G_{DR3}-G_{DR2}, and those with large inclination angles with negative values, relative to the stable stars would be more likely.

We investigated this point for the Double Cluster NGC869/NGC884, for which we had extensive data for cluster B-type stars. Figure 5 is similar to Figure 1b, but normalized to the stable star sequence, through a linear piecewise fit. This way, we could directly compare the data with the models of Figure 4. The color coding indicates the projected rotational velocity (in km s^{-1}) for the stars with available data. In Table A1 we indicate the reference for each velocity measurement.

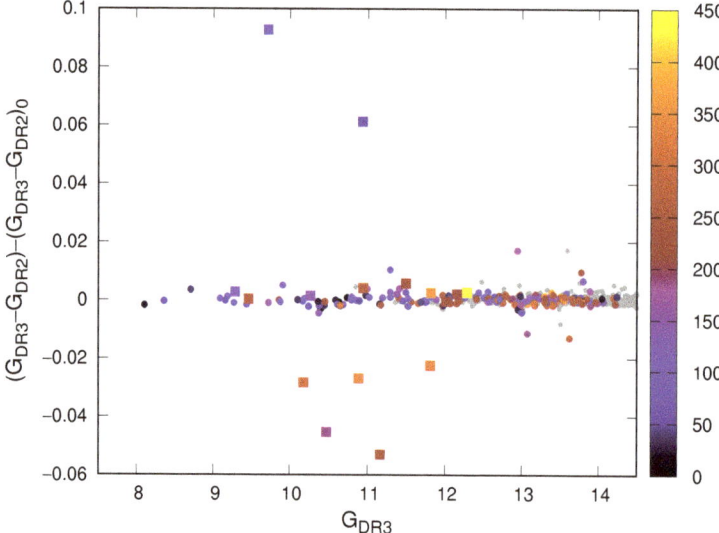

Figure 5. Similar to Figure 1b, but normalized to the *stable star* sequence, color coded according their projected rotational velocity (in units of km s^{-1}).

We can see that the stars with a positive difference in G magnitude between the two releases had small projected rotational velocities, while among the five stars with negative differences, three had values larger than 300 km s^{-1} and only one had a projected rotational velocity smaller than 200 km s^{-1}.

4. Discussion

We showed that the error in the Gaia G band and G_{DR3}-G_{DR2} were both excellent quantities to detect active Be stars in open clusters. Additionally, the latter quantity could be useful for insight into the inclination angle of the star as well. This could be particularly useful for stars for which no Gaia spectroscopic data, nor lightcurve, is yet available.

Increasing the number of Be candidates in poorly studied open clusters certainly helps in better understanding the environments where these stars form and evolve, so this paper opens a new opportunity in the Be star research community.

Among the few objects without a Be classification, that departed from the tight stellar sequence in the G_{DR3}-G_{DR2} versus G_{DR3} diagram, we found one RS Canum Venaticorum star in the Double Cluster and two Eclipsing Binary stars in NGC 7419. These types of binaries with short period variability, could exhibit large differences within different releases, which could certainly influence the average magnitude of the object if a different proportion of light minima were caught between DR2 and DR3.

Among the new Be candidates, we could potentially have a few eclipsing binaries, which are themselves interesting objects to study in clusters (e.g., [46] and references therein). The study of these objects allows us to accurately measure stellar parameters of the binary stars, having the same age and chemical composition, and, together with the cluster isochrone studies, put solid constraints on stellar evolution models.

Author Contributions: Conceptualization, A.G.; Formal analysis, A.G. and M.R.G.; Investigation, A.G., M.R.G., C.E.J. and T.E.; Methodology, A.G., M.R.G. and C.E.J.; Project administration, A.G.; Resources, A.G.; Writing—original draft, A.G., M.R.G., C.E.J. and T.E. All authors have read and agreed to the published version of the manuscript.

Funding: A.G. acknowledges the research project from Universidad Nacional de Río Negro PI-UNRN2020 40-B-890. M.R.G. acknowledges the grant awarded by the Western University Postdoctoral Fellowship Program (WPFP). C.E.J. wishes to acknowledge support through the Natural Sciences and Engineering Research Council of Canada. T.E. gratefully acknowledges financial support from the Estonian Ministry of Education and Research through the Estonian Research Council institutional research funding IUT40-1, from the European Union European Regional Development Fund project KOMEET 2014-2020.4.01.16-0029 and from Marie Skłodowska-Curie Research and Innovation Staff Exchanges RISE grant "Physics of Extreme Massive Stars", agreement no 823734. The authors thank the APC Discount Voucher from Western University b7f7871016d98506.

Institutional Review Board Statement: Not applicable.

Informed Consent Statement: Not applicable.

Data Availability Statement: This work made use of data from the European Space Agency (ESA) mission *Gaia* https://www.cosmos.esa.int/gaia (accessed on 15 February 2023), processed by the *Gaia* Data Processing and Analysis Consortium , DPAC, https://www.cosmos.esa.int/web/gaia/dpac/consortium (accessed on 15 February 2023). Funding for the DPAC was provided by national institutions; in particular, the institutions participating in the *Gaia* Multilateral Agreement. This research made use of the SIMBAD database, operated at CDS, Strasbourg, France.

Acknowledgments: The authors thank the reviewers of this manuscript for their constructive feedback.

Conflicts of Interest: The authors declare no conflict of interest.

Appendix A

Appendix A.1. NGC869/NGC884

We present Table A1 that contains relevant data for the Double Cluster. The first columns indicate position (RA and Dec), then Gaia DR2 brightness in G,GB and GR passbands, Gaia DR3 brightness in G, its error , GB and GR passbands. The column "Group" indicates different types of interesting objects: **1** corresponds to known Be stars either from SIMBAD or [47], **2** Eclipsing binaries from SIMBAD, **3** β Ceph stars from SIMBAD, **4** Pulsational variable according to [34], **5** Candidate Be stars according to the present work and **6** RS Canum Venaticorum stars according to [35]. The asterisk in the last column indicates the two-peak separation of an intense near IR hydrogen line from APOGEE [48], for a Be star without V sin(i) available in the literature.

Table A1. Relevant data for variable stars of the Double Cluster. See text for a description.

RA	DEC	G_{DR2}	GB_{DR2}	GR_{DR2}	G_{DR3}	$errG_{DR3}$	GB_{DR3}	GR_{DR3}	Group	$V\sin(i)$	Ref.
34.9493	57.1110	10.8742	11.1487	10.3406	10.9341	0.0062	11.2242	10.4313	1	51	[24]
34.5823	57.1462	12.1688	12.3107	11.8610	12.1583	0.0028	12.3166	11.8579	1	207	[24]
34.8701	57.1179	10.9403	11.2315	10.4876	10.9430	0.0030	11.2193	10.4826	1	264	[24]
34.7368	57.1286	12.0243	12.2062	11.6728	12.0125	0.0028	12.2074	11.6691	1	219	[24]
34.7263	57.1582	10.2641	10.4417	9.9624	10.2670	0.0028	10.4313	9.9551	1	151	[24]
34.8644	57.1382	9.2773	9.4955	8.9362	9.2856	0.0028	9.4795	8.9269	1	127	[24]
34.7244	57.1395	9.4608	9.6454	9.1537	9.4659	0.0028	9.6308	9.1476	1	258	[24]
35.8538	57.3176	10.5171	10.9499	9.9197	10.4723	0.0054	10.8947	9.8582	1	187	[33]
35.4289	57.0918	11.2214	11.4029	10.9038	11.1660	0.0049	11.3458	10.7962	1	242	[24]
35.7402	57.3940	12.3000	12.5393	11.8834	12.2877	0.0028	12.5352	11.8769	1	426	*
35.4706	57.1664	9.6194	9.9104	9.1552	9.7160	0.0053	9.9660	9.2597	1	79	[24]
35.7503	57.2039	11.8232	12.0240	11.4525	11.8161	0.0029	12.0239	11.4485	1–4	360	[24]
35.7095	57.1474	11.8411	12.0159	11.5078	11.8091	0.0033	12.0022	11.4536	1–4	338	[24]
35.7674	57.1274	10.2063	10.4515	9.8175	10.1797	0.0040	10.4128	9.7811	1–4	300	[49]
35.4353	57.1812	11.4977	11.7296	11.1105	11.4967	0.0028	11.7209	11.1072	1–4	229	[33]
35.5103	57.1557	10.9163	11.1057	10.5768	10.8882	0.0038	11.0594	10.5437	1–4	345	[26]
35.5127	57.1348	11.5075	11.6759	11.1950	11.4989	0.0029	11.6788	11.1861	2		
35.7670	57.1691	12.0040	12.2191	11.6201	11.9934	0.0028	12.2141	11.6150	2		
35.5034	57.1255	11.0243	11.1382	10.8016	11.0235	0.0028	11.1329	10.7970	2	108	[24]
35.5651	57.2247	11.6959	11.8949	11.3398	11.6868	0.0028	11.8861	11.3327	2	42	[24]
34.7403	57.1383	10.9792	11.1288	10.6958	10.9792	0.0028	11.1266	10.6901	3	29	[24]
35.5116	57.1403	9.8773	10.0418	9.6027	9.8797	0.0028	10.0249	9.5944	3–4	101	[24]
35.4569	57.0266	13.0194	13.2094	12.6567	13.0016	0.0033	13.2051	12.6650	6–4	67	[24]
35.4656	57.2516	13.0872	13.3849	12.6004	13.0761	0.0028	13.3813	12.5933	4		
35.5632	57.1613	12.2744	12.4501	11.9421	12.2618	0.0028	12.4466	11.9218	4	74	[24]
35.3907	57.0655	12.9961	13.2280	12.5881	12.9839	0.0028	13.2265	12.5862	4	33	[24]
35.3685	57.1429	12.7480	12.9406	12.3892	12.7346	0.0028	12.9385	12.3857	4		
35.6083	57.1148	13.6839	13.8988	13.2909	13.6709	0.0028	13.8914	13.2902	4	261	[24]
35.5357	57.1245	9.0813	9.2409	8.8145	9.0882	0.0028	9.2231	8.8073	4	106	[24]
35.6736	57.1745	13.7828	14.0480	13.3374	13.7699	0.0028	14.0369	13.3322	4		
35.7984	57.1721	9.4555	9.6140	9.1837	9.4608	0.0028	9.5966	9.1803	4		
35.6791	57.2053	14.0299	14.2640	13.5991	14.0169	0.0028	14.2585	13.5901	4		
35.5610	57.1038	12.7677	12.9296	12.4428	12.7528	0.0028	12.9300	12.4399	4	261	[24]
35.5740	57.1038	13.9658	14.1573	13.6065	13.9531	0.0028	14.1542	13.6056	4		
35.3572	57.0727	13.4464	13.6912	13.0258	13.4346	0.0028	13.6834	13.0205	4	170	[24]
35.6499	57.2089	14.3629	14.6035	13.9438	14.3493	0.0028	14.5967	13.9436	4		
35.3535	57.0403	13.1761	13.3697	12.8174	13.1619	0.0028	13.3636	12.8114	4	318.4	[50]
35.5546	57.0228	13.0086	13.2169	12.6250	12.9914	0.0028	13.2075	12.6215	4	203	[24]
35.8416	57.0071	13.6374	13.8506	13.2549	13.6243	0.0028	13.8428	13.2505	4		
35.7392	57.0089	12.4026	12.5392	12.1126	12.3868	0.0028	12.5385	12.1088	4		
35.5774	57.1972	11.5349	11.7150	11.1979	11.5283	0.0028	11.7130	11.1917	4	112	[24]
35.8269	57.2589	13.4186	13.6603	13.0026	13.4065	0.0028	13.6522	12.9944	4		
35.4569	57.1393	11.2951	11.4272	11.0250	11.2929	0.0028	11.4336	11.0207	4	136	[33]
35.4455	57.1649	13.4276	13.6494	13.0349	13.4139	0.0028	13.6431	13.0301	4	116	[24]
35.4415	57.0836	12.5769	12.7887	12.1972	12.5632	0.0028	12.7839	12.1909	4	248	[24]
35.4350	57.1483	11.8044	11.9496	11.5095	11.7970	0.0028	11.9506	11.5091	4	67	[26]
35.5934	57.1841	13.7854	14.0243	13.3578	13.7725	0.0028	14.0223	13.3545	4	371	[24]
35.5205	57.1005	13.1930	13.3577	12.8670	13.1802	0.0028	13.3576	12.8634	4	154	[24]
35.5464	57.1067	14.1656	14.3810	13.7706	14.1522	0.0028	14.3773	13.7698	4		
35.6210	57.0288	12.9666	13.1444	12.6263	12.9498	0.0028	13.1403	12.6211	4	30	[24]
35.5146	57.1190	12.5907	12.7674	12.2621	12.5758	0.0028	12.7574	12.2585	4	141	[24]
35.5571	57.0881	13.0428	13.1915	12.7294	13.0287	0.0028	13.1966	12.7270	4	108	[24]
35.5302	57.2033	12.5708	12.7529	12.2199	12.5593	0.0028	12.7542	12.2175	4	81	[24]
35.5116	57.1479	12.4469	12.6046	12.1357	12.4332	0.0028	12.6024	12.1317	4		
35.5968	57.1682	14.0780	14.3092	13.6478	14.0658	0.0028	14.3022	13.6473	4		
35.4447	57.1241	10.9932	11.1498	10.7138	10.9954	0.0028	11.1461	10.7073	4	169	[24]
35.6619	57.1955	12.4247	12.6093	12.0751	12.4104	0.0028	12.6058	12.0690	4	241	[24]
35.5470	57.1093	13.8381	14.0395	13.4629	13.8253	0.0028	14.0377	13.4634	4		

Table A1. Cont.

RA	DEC	G_{DR2}	GB_{DR2}	GR_{DR2}	G_{DR3}	$errG_{DR3}$	GB_{DR3}	GR_{DR3}	Group	$V \sin(i)$	Ref.
35.6244	57.2080	9.4355	9.6155	9.1426	9.4389	0.0028	9.5958	9.1318	4	116	[24]
34.4696	57.1242	12.9408	13.1021	12.6158	12.9441	0.0029	13.1196	12.6285	5	180	[50]
34.8142	57.0939	13.0988	13.3639	12.6514	13.0739	0.0029	13.3422	12.6408	5	185	[24]
34.8834	57.1218	13.6437	13.9132	13.1905	13.6183	0.0030	13.9044	13.1716	5–6	272	[24]
35.4618	57.2643	13.5838			13.5885	0.0029			5		
35.5035	57.1597	11.2881	11.4393	11.0116	11.2935	0.0028	11.4408	11.0079	5	79	[24]
35.7434	57.0802	13.7731	14.0176	13.3364	13.7703	0.0028	14.0095	13.3302	5–4	228	[24]

EBs are identified with the symbol * next to the RA value.

Appendix A.2. NGC663

Be candidates of NGC663 proposed in this article as described in Section 3.2. All the data for the star cluster members [29] are available online in Gaia DR2 and DR3.

Table A2. Be candidates of NGC663. See text for a description.

RA	DEC	G_{DR2}	GB_{DR2}	GR_{DR2}	G_{DR3}	$errG_{DR3}$	GB_{DR3}	GR_{DR3}
26.1345	61.0357	11.7643	11.9475	11.4064	11.7675	0.0025	11.9430	11.3990
26.5405	61.2060	12.6804	12.9845	12.1887	12.7118	0.0028	13.0160	12.2311
26.5817	61.2636	11.4527	11.8009	10.9543	11.4701	0.0029	11.7979	10.9528
26.6511	61.2011	14.0593	14.4175	13.5280	14.0679	0.0031	14.4219	13.5397
26.8232	61.3506	13.8222	14.1819	13.2778	13.8036	0.0034	14.1629	13.2602

Appendix A.3. NGC7419

17 Be candidates of NGC7419 proposed in this article as described in Section 3.3, and 2 EBs. All the data for the star cluster members [29] are available online in Gaia DR2 and DR3.

Table A3. Be candidates of NGC7419. See text for a description.

RA	DEC	G_{DR2}	GB_{DR2}	GR_{DR2}	G_{DR3}	$errG_{DR3}$	GB_{DR3}	GR_{DR3}
343.3217	60.8266	16.2447	17.9178	14.9481	16.2317	0.0038	17.8946	14.9813
343.3976	60.7236	17.8185	19.6804	16.4386	17.8098	0.0030	19.7270	16.4657
343.5181	60.7933	17.1032	18.5614	15.8439	17.0907	0.0029	18.5888	15.8691
343.5192	60.7611	16.3070	17.4165	15.0216	16.2996	0.0029	17.4446	15.0234
343.5481	60.8442	15.7882	16.9885	14.6624	15.7526	0.0030	16.9658	14.6565
343.5646	60.8256	16.6244	17.6384	15.2435	16.6339	0.0031	17.6608	15.2752
343.5754	60.8065	16.3449			16.3460	0.0030	17.3938	15.1945
343.5891 *	60.8316	15.4774	16.6915	14.3236	15.4416	0.0045	16.6786	14.3248
343.5972	60.8085	16.9569	17.8607	15.2027	16.9731	0.0031	17.9529	15.2617
343.6272	60.7987	17.7249	18.8056	15.8962	17.7590	0.0035	18.8473	15.9640
343.6694	60.7861	15.5368	16.7586	14.4025	15.4961	0.0035	16.7250	14.3865
343.6737	60.8330	17.3693	19.2339	16.0482	17.3605	0.0030	19.2504	16.0646
343.6776	60.7436	17.7969	19.5841	16.4758	17.7850	0.0030	19.5888	16.4931
343.6802	60.8398	17.0955			17.1111	0.0030	19.3139	15.7285
343.6902	60.8920	16.3127	17.3359	15.2738	16.3054	0.0028	17.3156	15.2716
343.7105	60.7716	16.1045	17.3278	14.9738	16.0684	0.0030	17.2937	14.9631
343.7106	60.8031	17.8798	19.3287	16.6495	17.8704	0.0030	19.3064	16.6636
343.7752 *	60.7680	14.8305	15.8472	13.7992	14.8327	0.0061	15.8423	13.8122
343.8161	60.8345	17.8965			17.9114	0.0031	20.0666	16.5431

EBs are identified with the symbol * next to the RA value.

Notes

1 https://www.cosmos.esa.int/web/gaia/iow_20180316 (accessed on 15 February 2023).

2 https://www.cosmos.esa.int/web/gaia/edr3-passbands (accessed on 15 February 2023).

References

1. Rivinius, T.; Carciofi, A.C.; Martayan, C. Classical Be stars. Rapidly rotating B stars with viscous Keplerian decretion disks. *Astron. Astrophys. Rev.* **2013**, *21*, 69. [CrossRef]
2. Pigulski, A.; Kopacki, G. NGC 7419: An open cluster rich in Be stars. *Astron. Astrophys. Suppl. Ser.* **2000**, *146*, 465–469. [CrossRef]
3. Pigulski, A.; Kopacki, G.; Kołaczkowski, Z. The young open cluster NGC 663 and its Be stars. *Astron. Astrophys.* **2001**, *376*, 144–153. [CrossRef]
4. Mathew, B.; Subramaniam, A.; Bhatt, B.C. Be phenomenon in open clusters: Results from a survey of emission-line stars in young open clusters. *Mon. Not. R. Astron. Soc.* **2008**, *388*, 1879–1888. [CrossRef]
5. Ekström, S.; Meynet, G.; Maeder, A.; Barblan, F. Evolution towards the critical limit and the origin of Be stars. *Astron. Astrophys.* **2008**, *478*, 467–485. [CrossRef]
6. Georgy, C.; Ekström, S.; Granada, A.; Meynet, G.; Mowlavi, N.; Eggenberger, P.; Maeder, A. Populations of rotating stars. I. Models from 1.7 to 15 M$_\odot$ at Z = 0.014, 0.006, and 0.002 with Ω/Ω_{crit} between 0 and 1. *Astron. Astrophys.* **2013**, *553*, A24. [CrossRef]
7. Granada, A.; Ekström, S.; Georgy, C.; Krtička, J.; Owocki, S.; Meynet, G.; Maeder, A. Populations of rotating stars. II. Rapid rotators and their link to Be-type stars. *Astron. Astrophys.* **2013**, *553*, A25. [CrossRef]
8. Bodensteiner, J.; Shenar, T.; Sana, H. Investigating the lack of main-sequence companions to massive Be stars. *Astron. Astrophys.* **2020**, *641*, A42. [CrossRef]
9. El-Badry, K.; Conroy, C.; Quataert, E.; Rix, H.W.; Labadie-Bartz, J.; Jayasinghe, T.; Thompson, T.; Cargile, P.; Stassun, K.G.; Ilyin, I. Birth of a Be star: An APOGEE search for Be stars forming through binary mass transfer. *Mon. Not. R. Astron. Soc.* **2022**, *516*, 3602–3630. [CrossRef]
10. Fabregat, J.; Torrejón, J.M. On the evolutionary status of Be stars. *Astron. Astrophys.* **2000**, *357*, 451–459.
11. Majewski, S.R. APOGEE—SDSS-III's Other Milky Way Experiment. In *Proceedings of the American Astronomical Society Meeting Abstracts #219*; 2012; Volume 219, p. 205.06.
12. Luo, A.L.; Zhao, Y.H.; Zhao, G.; Deng, L.C.; Liu, X.W.; Jing, Y.P.; Wang, G.; Zhang, H.T.; Shi, J.R.; Cui, X.Q.; et al. The first data release (DR1) of the LAMOST regular survey. *Res. Astron. Astrophys.* **2015**, *15*, 1095. [CrossRef]
13. Chojnowski, S.D.; Whelan, D.G.; Wisniewski, J.P.; Majewski, S.R.; Hall, M.; Shetrone, M.; Beaton, R.; Burton, A.; Damke, G.; Eikenberry, S.; et al. High-Resolution H-Band Spectroscopy of Be Stars With SDSS-III/Apogee: I. New Be Stars, Line Identifications, and Line Profiles. *Astron. J.* **2015**, *149*, 7. [CrossRef]
14. Lin, C.C.; Hou, J.L.; Chen, L.; Shao, Z.Y.; Zhong, J.; Yu, P.C. Searching for classical Be stars in LAMOST DR1. *Res. Astron. Astrophys.* **2015**, *15*, 1325. [CrossRef]
15. Vioque, M.; Oudmaijer, R.D.; Schreiner, M.; Mendigutía, I.; Baines, D.; Mowlavi, N.; Pérez-Martínez, R. Catalogue of new Herbig Ae/Be and classical Be stars. A machine learning approach to Gaia DR2. *Astron. Astrophys.* **2020**, *638*, A21. [CrossRef]
16. Wang, L.; Li, J.; Wu, Y.; Gies, D.R.; Liu, J.Z.; Liu, C.; Guo, Y.; Chen, X.; Han, Z. Identification of New Classical Be Stars from the LAMOST Medium Resolution Survey. *Astrophys. J. Suppl. Ser.* **2022**, *260*, 35. [CrossRef]
17. Gaia Collaboration; Brown, A.G.A.; Vallenari, A.; Prusti, T.; de Bruijne, J.H.J.; Babusiaux, C.; Bailer-Jones, C.A.L.; Biermann, M.; Evans, D.W.; Eyer, L.; et al. Gaia Data Release 2. Summary of the contents and survey properties. *Astron. Astrophys.* **2018**, *616*, A1. [CrossRef]
18. Gaia Collaboration; Vallenari, A.; Brown, A.G.A.; Prusti, T.; de Bruijne, J.H.J.; Arenou, F.; Babusiaux, C.; Biermann, M.; Creevey, O.L.; Ducourant, C.; et al. Gaia Data Release 3: Summary of the content and survey properties. *arXiv* **2022**, arXiv:2208.00211.
19. Gaia Collaboration; Prusti, T.; de Bruijne, J.H.J.; Brown, A.G.A.; Vallenari, A.; Babusiaux, C.; Bailer-Jones, C.A.L.; Bastian, U.; Biermann, M.; Evans, D.W.; et al. The Gaia mission. *Astron. Astrophys.* **2016**, *595*, A1. [CrossRef]
20. Riello, M.; De Angeli, F.; Evans, D.W.; Montegriffo, P.; Carrasco, J.M.; Busso, G.; Palaversa, L.; Burgess, P.W.; Diener, C.; Davidson, M.; et al. Gaia Early Data Release 3. Photometric content and validation. *Astron. Astrophys.* **2021**, *649*, A3. [CrossRef]
21. Lee, U.; Osaki, Y.; Saio, H. Viscous excretion discs around Be stars. *Mon. Not. R. Astron. Soc.* **1991**, *250*, 432–437. [CrossRef]
22. Labadie-Bartz, J.; Pepper, J.; McSwain, M.V.; Bjorkman, J.E.; Bjorkman, K.S.; Lund, M.B.; Rodriguez, J.E.; Stassun, K.G.; Stevens, D.J.; James, D.J.; et al. Photometric Variability of the Be Star Population. *Astron. J.* **2017**, *153*, 252. [CrossRef]
23. Haubois, X.; Mota, B.C.; Carciofi, A.C.; Draper, Z.H.; Wisniewski, J.P.; Bednarski, D.; Rivinius, T. Dynamical Evolution of Viscous Disks around Be Stars. II. Polarimetry. *Astrophys. J.* **2014**, *785*, 12. [CrossRef]
24. Strom, S.E.; Wolff, S.C.; Dror, D.H.A. B Star Rotational Velocities in h and χ Persei: A Probe of Initial Conditions during the Star Formation Epoch? *Astron. J.* **2005**, *129*, 809–828. [CrossRef]
25. McSwain, M.V.; Gies, D.R. The Evolutionary Status of Be Stars: Results from a Photometric Study of Southern Open Clusters. *Astrophys. J. Suppl. Ser.* **2005**, *161*, 118. [CrossRef]
26. Huang, W.; Gies, D.R.; McSwain, M.V. A Stellar Rotation Census of B Stars: From ZAMS to TAMS. *Astrophys. J.* **2010**, *722*, 605. [CrossRef]
27. Pandey, A.K.; Upadhyay, K.; Ogura, K.; Sagar, R.; Mohan, V.; Mito, H.; Bhatt, H.C.; Bhatt, B.C. Stellar contents of two young open clusters: NGC 663 and 654. *Mon. Not. R. Astron. Soc.* **2005**, *358*, 1290–1308. [CrossRef]
28. Torrejon, J.M.; Fabregat, J.; Bernabeu, G.; Alba, S. Be stars in open clusters. II. Balmer line spectroscopy. *Astron. Astrophys. Suppl. Ser.* **1997**, *124*, 329–347. [CrossRef]

29. Cantat-Gaudin, T.; Jordi, C.; Vallenari, A.; Bragaglia, A.; Balaguer-Núñez, L.; Soubiran, C.; Bossini, D.; Moitinho, A.; Castro-Ginard, A.; Krone-Martins, A.; et al. A Gaia DR2 view of the open cluster population in the Milky Way. *Astron. Astrophys.* **2018**, *618*, A93. [CrossRef]
30. Dias, W.S.; Monteiro, H.; Moitinho, A.; Lépine, J.R.D.; Carraro, G.; Paunzen, E.; Alessi, B.; Villela, L. Updated parameters of 1743 open clusters based on Gaia DR2. *Mon. Not. R. Astron. Soc.* **2021**, *504*, 356–371. [CrossRef]
31. Kharchenko, N.V.; Piskunov, A.E.; Roeser, S.; Schilbach, E.; Scholz, R.D. VizieR Online Data Catalog: Milky Way global survey of star clusters. II. (Kharchenko+, 2013). *Vizier Online Data Cat.* **2013**, *355*, 80053.
32. Cantat-Gaudin, T.; Anders, F.; Castro-Ginard, A.; Jordi, C.; Romero-Gómez, M.; Soubiran, C.; Casamiquela, L.; Tarricq, Y.; Moitinho, A.; Vallenari, A.; et al. Painting a portrait of the Galactic disc with its stellar clusters. *Astron. Astrophys.* **2020**, *640*, A1. [CrossRef]
33. Huang, W.; Gies, D.R. Stellar Rotation in Young Clusters. I. Evolution of Projected Rotational Velocity Distributions. *Astrophys. J.* **2006**, *648*, 580–590. [CrossRef]
34. Saesen, S.; Briquet, M.; Aerts, C.; Miglio, A.; Carrier, F. Pulsating B-type Stars in the Open Cluster NGC 884: Frequencies, Mode Identification, and Asteroseismology. *Astron. J.* **2013**, *146*, 102. [CrossRef]
35. Chen, X.; Wang, S.; Deng, L.; de Grijs, R.; Yang, M.; Tian, H. The Zwicky Transient Facility Catalog of Periodic Variable Stars. *Astrophys. J. Suppl. Ser.* **2020**, *249*, 18. [CrossRef]
36. Cantat-Gaudin, T.; Anders, F. Clusters and mirages: Cataloguing stellar aggregates in the Milky Way. *Astron. Astrophys.* **2020**, *633*, A99. [CrossRef]
37. Marco, A.; Negueruela, I. NGC 7419 as a template for red supergiant clusters. *Astron. Astrophys.* **2013**, *552*, A92. [CrossRef]
38. Okazaki, A.T.; Bate, M.R.; Ogilvie, G.I.; Pringle, J.E. Viscous effects on the interaction between the coplanar decretion disc and the neutron star in Be/X-ray binaries. *Mon. Not. R. Astron. Soc.* **2002**, *337*, 967–980. [CrossRef]
39. Okazaki, A.T. Theory vs. Observation of Circumstellar Disks and Their Formation. In *Proceedings of the Active OB-Stars: Laboratories for Stellar and Circumstellar Physics*; Okazaki, A.T., Owocki, S.P., Stefl, S., Eds.; Astronomical Society of the Pacific Conference Series, ASP: San Francisco, CA, USA, 2007; Volume 361, p. 230.
40. Lynden-Bell, D.; Pringle, J.E. The evolution of viscous discs and the origin of the nebular variables. *Mon. Not. R. Astron. Soc.* **1974**, *168*, 603–637. [CrossRef]
41. Carciofi, A.C.; Bjorkman, J.E. Non-LTE Monte Carlo Radiative Transfer. I. The Thermal Properties of Keplerian Disks around Classical Be Stars. *Astrophys. J.* **2006**, *639*, 1081–1094. [CrossRef]
42. Carciofi, A.C.; Bjorkman, J.E. Non-LTE Monte Carlo Radiative Transfer. II. Nonisothermal Solutions for Viscous Keplerian Disks. *Astrophys. J.* **2008**, *684*, 1374–1383. [CrossRef]
43. Ghoreyshi, M.R.; Jones, C.E.; Granada, A. Angular momentum loss rates in Be stars determined by the viscous decretion disc model. *Mon. Not. R. Astron. Soc.* **2023**, *518*, 30–38. [CrossRef]
44. Vieira, R.G.; Carciofi, A.C.; Bjorkman, J.E.; Rivinius, T.; Baade, D.; Rímulo, L.R. The life cycles of Be viscous decretion discs: Time-dependent modelling of infrared continuum observations. *Mon. Not. R. Astron. Soc.* **2017**, *464*, 3071–3089. [CrossRef]
45. Rímulo, L.R.; Carciofi, A.C.; Vieira, R.G.; Rivinius, T.; Faes, D.M.; Figueiredo, A.L.; Bjorkman, J.E.; Georgy, C.; Ghoreyshi, M.R.; Soszyński, I. The life cycles of Be viscous decretion discs: Fundamental disc parameters of 54 SMC Be stars. *Mon. Not. R. Astron. Soc.* **2018**, *476*, 3555–3579. [CrossRef]
46. Southworth, J.; Clausen, J.V. Eclipsing Binaries in Open Clusters. *Astrophys. Space Sci.* **2006**, *304*, 199–202. [CrossRef]
47. Laur, J.; Kolka, I.; Eenmäe, T.; Tuvikene, T.; Leedjärv, L. Variability survey of brightest stars in selected OB associations. *Astron. Astrophys.* **2017**, *598*, A108. [CrossRef]
48. Jönsson, H.; Holtzman, J.A.; Allende Prieto, C.; Cunha, K.; García-Hernández, D.A.; Hasselquist, S.; Masseron, T.; Osorio, Y.; Shetrone, M.; Smith, V.; et al. APOGEE Data and Spectral Analysis from SDSS Data Release 16: Seven Years of Observations Including First Results from APOGEE-South. *Astron. J.* **2020**, *160*, 120. [CrossRef]
49. Glebocki, R.; Gnacinski, P. VizieR Online Data Catalog: Catalog of Stellar Rotational Velocities (Glebocki+ 2005). *Vizier Online Data Cat.* **2005**, *III*, 244.
50. Sun, W.; Duan, X.W.; Deng, L.; de Grijs, R.; Zhang, B.; Liu, C. Exploring the Stellar Rotation of Early-type Stars in the LAMOST Medium-resolution Survey. I. Catalog. *Astrophys. J. Suppl. Ser.* **2021**, *257*, 22. [CrossRef]

Disclaimer/Publisher's Note: The statements, opinions and data contained in all publications are solely those of the individual author(s) and contributor(s) and not of MDPI and/or the editor(s). MDPI and/or the editor(s) disclaim responsibility for any injury to people or property resulting from any ideas, methods, instructions or products referred to in the content.

Article

Discovering New B[e] Supergiants and Candidate Luminous Blue Variables in Nearby Galaxies

Grigoris Maravelias [1,2,*], Stephan de Wit [1,3], Alceste Z. Bonanos [1], Frank Tramper [4], Gonzalo Munoz-Sanchez [1,3] and Evangelia Christodoulou [1,3]

[1] IAASARS, National Observatory of Athens, GR-15326 Penteli, Greece; sdewit@noa.gr (S.d.W.); bonanos@noa.gr (A.Z.B.); gonzalom@noa.gr (G.M.-S.); evachris@noa.gr (E.C.)
[2] Institute of Astrophysics FORTH, GR-71110 Heraklion, Greece
[3] Department of Physics, National and Kapodistrian University of Athens, Panepistimiopolis, GR-15784 Zografos, Greece
[4] Institute of Astronomy, KU Leuven, Celestijnlaan 200D, 3001 Leuven, Belgium; frank.tramper@kuleuven.be
* Correspondence: maravelias@noa.gr

Abstract: Mass loss is one of the key parameters that determine stellar evolution. Despite the progress we have achieved over the last decades we still cannot match the observational derived values with theoretical predictions. Even worse, there are certain phases, such as the B[e] supergiants (B[e]SGs) and the Luminous Blue Variables (LBVs), where significant mass is lost through episodic or outburst activity. This leads to various structures forming around them that permit dust formation, making these objects bright IR sources. The ASSESS project aims to determine the role of episodic mass in the evolution of massive stars, by examining large numbers of cool and hot objects (such as B[e]SGs/LBVs). For this purpose, we initiated a large observation campaign to obtain spectroscopic data for ∼1000 IR-selected sources in 27 nearby galaxies. Within this project we successfully identified seven B[e] supergiants (one candidate) and four Luminous Blue Variables of which six and two, respectively, are new discoveries. We used spectroscopic, photometric, and light curve information to better constrain the nature of the reported objects. We particularly noted the presence of B[e]SGs at metallicity environments as low as 0.14 $Z_\odot$.

Keywords: massive stars; mass-loss stars; star evolution; emission line; Be circumstellar matter; supergiant stars; star variables; S Doradus infrared; galaxies: individual; WLM; NGC 55; NGC 247; NGC 253; NGC 300; NGC 3109; NGC 7793

Citation: Maravelias, G.; de Wit, S.; Bonanos, A.Z.; Tramper, F.; Munoz-Sanchez, G.; Christodoulou, E. Discovering New B[e] Supergiants and Candidate Luminous Blue Variables in Nearby Galaxies. *Galaxies* 2023, 11, 79. https://doi.org/10.3390/galaxies11030079

Academic Editor: Roberta M. Humphreys

Received: 15 March 2023
Revised: 6 June 2023
Accepted: 8 June 2023
Published: 19 June 2023

Copyright: © 2023 by the authors. Licensee MDPI, Basel, Switzerland. This article is an open access article distributed under the terms and conditions of the Creative Commons Attribution (CC BY) license (https://creativecommons.org/licenses/by/4.0/).

1. Introduction

How *exactly* single massive stars, born as O/B-type main-sequence stars, progress to more evolved phases and eventually die remains an open question. Binarity, which has an important implication in the evolution, even further complicates the quest for an answer. Observational data has revealed a number of transitional phases in which massive stars can be found, also known as the massive star "zoo". Whether they pass through certain phases or not depends on the following: initial mass (≥ 8 $M_\odot$), metallicity (Z), rotational velocity ($v_{\rm rot}$), mass loss properties and binarity [1–4]. Although some of them are quite distinct (e.g., Wolf–Rayet stars, as opposed to Red Supergiants, RSGs), there are phases which display common observables, such as B[e] supergiants (B[e]SGs) and Luminous Blue Variables (LBVs).

The B[e] phenomenon is characterized by numerous emission lines in the optical spectra [5]. In particular, there is strong Balmer emission, low excitation permitted (e.g., Fe II), and forbidden lines (of [Fe II], and [O I]), as well as strong near- or mid-IR excess due to hot circumstellar dust. However, this can be observed in sources at different evolutionary stages (such as in Herbig AeBe stars, symbiotic systems, and compact planetary nebulae, see [5] for detailed classification criteria). The B[e]SGs form a distinct subgroup based on a number of secondary criteria. They are luminous stars (log $L/L_\odot \gtrsim 4.0$), showing broad

Balmer emission lines with P Cygni or double-peaked profiles. They may also display evidence of chemically processed material (e.g., ^{13}CO enrichment, TiO) which points to an evolved nature, although it is not yet certain if they are in pre- or post-RSG phases [6,7]. The presence of the hot circumstellar dust is due to a complex circumstellar environment (CSE) formed by two components, a stellar wind radiating from the poles and a denser equatorial ring-like structure [8–12]. However, the formation mechanism of this structure remains elusive. A variety of mechanisms have been proposed, such as the following: fast rotation [13], the bi-stability mechanism [14], slow-wind solutions [15], magneto-rotational instability [16], mass transfer in binaries [17], mergers [18], non-radial pulsations or the presence of objects that clear their paths [19]. Although poorly constrained, their initial masses range from roughly 10 $M_\odot$ to less than 40 $M_\odot$ (Mehner 2023, IAU S361, subm.).

The LBVs are another rare subgroup of massive evolved stars, considered to represent a transitional phase from massive O-type main-sequence to Wolf–Rayet stars (e.g., [3,20–22]). They experience instabilities that lead to photometric variability, typically referred to as S Dor cycles [22], as well as outbursts and episodic mass loss, similar to the giant eruption of η Carina that resulted in large amounts of mass lost through ejecta (e.g., [23]). It is not yet fully understood whether these two types of variability are related (e.g., [24]). Apart from the evident photometric variability, their spectral appearance changes significantly during their outburst activities (S Dor cycle). It is typical to experience loops from hot (spectra of O/B type) to cool states (A/F spectral types while in outbursts). Depending on the luminosity, the brightest LBVs ($\log L/L_\odot > 5.8$) seem to directly originate from main-sequence stars (with mass > 50 $M_\odot$), while the less luminous ones are possibly post-RSG objects that have lost almost half of their initial masses (within the range of $\sim$25–40 $M_\odot$) during the RSG phase (Mehner 2023, IAU S361, subm.). Currently, various mechanisms have been suggested, such as radiation and pressure instabilities, stellar rotation, and binarity (see the reviews on the theory and observational evidence in [22,24], Mehner 2023, IAU S361, subm., and the references therein) and, as such, no comprehensive theory exists to explain them.

Therefore, if and how these two phases are linked remains an open question. B[e]SGs tend to have initial masses with a wide range below the most luminous LBVs, and in accordance with the less luminous ones. The presence of similar lines in their spectra points to similarities in their CSEs, with shells and bipolar nebulae observed in both cases [22,25,26].

Due to their photometric variability, LBVs are more commonly detected in other galaxies compared to B[e]SGs, which generally display less variability[1]. Therefore, B[e]SGs need to be searched for to be discovered. This has only been successful for 56 (candidate) sources in the Galaxy and for the Magellanic Clouds (MCs), M31 and M33, and M81 [7], and only recently in NGC 247 [27]. On the other hand, LBVs have been found in more galaxies (additional to the aforementioned), such as IC 10, IC 1613, NGC 2366, NGC 6822, NGC 1156, DDO 68, and PHL 293B [22,28–31], summing up to about 150 sources (including candidates).

This paper presents the discovery of new B[e]SGs and LBV candidates found with a systematic survey to identify massive, evolved, dusty sources in nearby galaxies ($\leq$5 Mpc), as part of the ASSESS project[2] (Bonanos 2023, IAU S361, subm.). In Section 2 we provide a short summary of the observations and of our approach, in Section 3 we present the new sources, and in Sections 4 and 5 we discuss and conclude our work.

2. Materials and Methods

2.1. Galaxy Sample

For the ASSESS project, a list of 27 nearby galaxies ($\leq$5 Mpc) was compiled (see Bonanos 2023, IAU S361, subm.). In this paper, we present our results from a sub-sample of these galaxies (Table 1) for which the spectral classification is final, while for another set we have scheduled observations in queue and have submitted proposals. For some galaxies (e.g., MCs) data have been collected through other catalogs/surveys and are presented separately (e.g., [32–35]).

Table 1. Properties of galaxies examined in this work: galaxy ID (column 1), sky coordinates (columns 2 and 3), galaxy type (column 4), distance (column 5), metallicity (column 6), and radial velocity (RV, column 7).

ID (1)	R.A. (J2000) (2)	Dec. (J2000) (3)	Gal. Type (4)	Distance (Mpc) (5)	Metal.[1] ($Z_\odot$) (6)	RV [1,2] (km s^{-1}) (7)
WLM	00:01:58	−15:27:39	SB(s)m: sp	0.98 ± 0.04 [36]	0.14 [37]	−130 ± 1 [38]
NGC 55	00:14:53	−39:11:48	SB(s)m: sp	1.87 ± 0.02 [36]	0.27 [39]	129 ± 2 [38]
IC 10	00:20:17	+59:18:14	dIrr IV/BCD	0.80 ± 0.03 [40]	0.45 [41]	−348 ± 1 [38]
NGC 247	00:47:09	−20:45:37	SAB(s)d	3.03 ± 0.03 [36]	0.40 [42]	156 [43]
NGC 253	00:47:33	−25:17:18	SAB(s)c	3.40 ± 0.06 [44]	0.72 [45]	259 [46]
NGC 300	00:54:53	−37:41:04	SA(s)d	1.97 ± 0.06 [36]	0.41 [47]	146 ± 2 [38]
NGC 1313	03:18:16	−66:29:54	SB(s)d	4.61 ± 0.17 [48]	0.57 [49]	470 [50]
NGC 3109	10:03:07	−26:09:35	SB(s)m edge-on	1.27 ± 0.03 [36]	0.21 [51]	403 ± 2 [38]
Sextans A	10:11:01	−04:41:34	IBm	1.34 ± 0.02 [52]	0.06 [53]	324 ± 2 [38]
M83	13:37:01	−29:51:56	SAB(s)c	4.90 ± 0.20 [54]	1.58 [55]	519 [46]
NGC 6822	19:44:58	+14:48:12	IB(s)m	0.45 ± 0.01 [36]	0.32 [56]	−57 ± 2 [38]
NGC 7793	23:57:50	−32:35:28	SA(s)d	3.47 ± 0.04 [36]	0.42 [57]	227 [46]

[1] The numbers presented here reflect the mean value per galaxy. [2] The RV errors correspond to the statistical error and not the systemic one, which is (typically) larger.

2.2. Target Selection

The aim of the ASSESS project is to determine the role of episodic mass loss by detecting and analyzing dusty evolved stars that are primary candidates to exhibit episodic mass loss events (Bonanos 2023, IAU S361, subm.). This mass loss results in the formation of complex structures, such as shells and bipolar nebulae in Wolf–Rayet stars and LBVs (e.g., [25,58]), detached shells in AGBs and RSGs (e.g., [59]), disks and rings around B[e]SGs (e.g., [7,11], or even the dust-enshrouded shells within which the progenitors of Super-Luminous Supernovae lay (e.g., [60–62]). The presence of these dusty CSEs makes these sources bright in mid-IR imaging. Therefore, we based our catalog construction on published point-source *Spitzer* catalogs [63]. Since IR data alone cannot distinguish between these sources, the base catalogs were supplemented with other optical and near-IR surveys (Pan-STARRS1; [64], VISTA Hemisphere Survey—VHS; [65], *Gaia* DR2; [66,67]). *Gaia* information was also used to remove foreground sources when possible (see [68], and Tramper et al., in prep., for more details).

Given this data collection, we performed a selection process to minimize contamination by AGB stars and background IR galaxies/quasars. An absolute magnitude cut of $M_{[3.6]} \leq -9.0$ [34] and an apparent magnitude cut at $m_{[4.5]} \leq 15.5$ [69] were applied to avoid AGB stars and background galaxies, respectively. In order to select the dusty targets we considered all sources with an IR excess, defined by the color term $m_{[3.6]} - m_{[4.5]} > 0.1$ mag (to exclude the majority of foreground stars, for which this is approximately 0, and to select the most dusty IR sources). The three aforementioned criteria served as a minimum to consider a source as a priority target. Consequently, the reddest and brightest point-sources in the *Spitzer* catalogs were given the highest priority. An extensive priority list/system was constructed by imposing certain limits for the color term, $M_{[3.6]}$, and the presence of an optical counterpart (for more details, see Tramper et al., in prep.). Depending on the galaxy size we ended up with a few tens to hundreds of targets per galaxy.

To obtain spectroscopic data for such a large number of targets we required instruments with multi-object spectroscopic modes. With these we could allocate up to a few tens of objects per pointing. Multiple pointings (with dithering and/or overlap) were applied to cover more extended galaxies and when the density of the target was high. Therefore, when we were creating the necessary multi-object masks we were forced to select sources based on the spatial limitations (e.g., located out of the field-of-view or at the sensor's gap) and spectral overlaps. Consequently, some priority targets were dropped and, additionally, non-priority targets ("fillers", i.e., sources dropped through the target selection approach described previously) were added to fill the space.

2.3. Observations

To verify the nature of our selected targets we needed spectroscopic information. Since this is not available for the majority of the ASSESS galaxies, we initiated an observation campaign to obtain low resolution spectra. Given the large number of targets, along with the sizes of the galaxies, we used the multi-object spectroscopic modes of the Optical System for Imaging and low-Intermediate-Resolution Integrated Spectroscopy (OSIRIS; [70]), on the 10.4 m GTC ([71], for the galaxies visible from the Northern hemisphere, i.e., IC 10 and NGC 6822). We used the FOcal Reducer/low dispersion Spectrograph 2 (FORS; [72]), at 8.2 m ESO/VLT (for the Southern galaxies, i.e., the rest of Table 1). The resolving power and wavelength coverage was similar for both instruments, at $\sim$500–700 over the range $R \sim$5300–9800 Å for GTC/OSIRIS and $R \sim$1000 over the range $\sim$5200–8700 Å for VLT/FORS2. Details for the observations and data reduction can be found at Munoz-Sanchez et al., in prep., for the GTC/OSIRIS campaign and Tramper et al., in prep., for the VLT/FORS2 campaign. Here we provide only a short overview of the data reduction followed.

For the OSIRIS data we used the *GTCMOS* package[3] which is an IRAF[4]. This pipeline for spectroscopic data combines (for each raw exposure) the two CCD images from the detector (correcting for geometric distortions) and performs bias subtraction. Although it can perform the wavelength calibration and can correct the curvature across the spatial direction in 2D images, we noticed that it was not perfect. For this reason we opted to perform a manual approach and extracted a small cut in the image around each slit. We performed the wavelength calibration individually for each of these images (slits) and tilt was corrected when necessary. The science and sky spectra were extracted (in 1D), and followed by flux calibration. We used IRAF to extract the long-slit spectra for standard stars, and then the routine `standard` and `sensfunc` to obtain the sensitivity curve. This was applied through the `calibrated` routine to the science spectrum.

For the FORS2 data, we used the FORS2 pipeline v5.5.7 under the EsoReflex environment [74]. This resulted in flux-calibrated, sky-subtracted 1D spectra for each slit on the mask. However, for some slits the pipeline did not produce suitable spectra, due to multiple objects in the slit, strongly variable nebular emission, slit overlap, and/or strong vignetting at the top of the CCD. For this reason, we also performed the reduction without sky subtraction and manually selected the object and sky extraction regions from the 2D spectrum. For each slit, the automatically and manually extracted spectra were visually inspected, and the best reduction was chosen.

2.4. Spectral Classification

The resolution and wavelength range (as described in the previous section) provide access to a number of spectral features, such as Hα (a mass loss tracer for high $\dot{M}$ stars), the TiO bands (present in cool stars), He I and He II lines (indicative of hot stars), various metal lines (notably Fe lines), and the Ca triplet (luminosity indicator). Therefore, we were able to effectively classify the vast majority of our targets.

Both B[e]SGs and LBVs are characterized by strong emission lines, indicative of their complex CSEs. Hα is usually found in very strong emissions and is significantly broadened in the presence of strong stellar winds and/or the presence of a (detached) disk (e.g., [7,12]). There were a number of He I lines (at $\lambda\lambda$5876.6, 6678.2, 7065.2, 7281.4) within our observed range, which manifest in the hottest sources. In the quiescence state of LBVs, the presence of He lines indicates hotter sources (of B/A spectral type, which can be observed even with P-Cygni profiles when stellar winds are strong, such as, for example, in [21]). However, when an outburst is triggered and evolves outwards, the temperature temporarily decreases until the ejecta become optically thin. As a result of this temperature shift, the spectral lines typical for the quiescent LBV weaken and metal emission lines strengthen (e.g., [75]). During this phase, and depending on the temperature and density conditions of the circumstellar material, they may also display some forbidden Fe lines. B[e]SGs display additional forbidden emission lines, due to their more complex CSEs, with typical examples being [O I] $\lambda\lambda 5577, 6300, 6364$ and [Ca II] $\lambda 7291, 7324$. The latter is more evident in the more luminous sources (e.g., [10,21]).

Therefore, among all sources identified with strong Hα emissions, we classified as being B[e]SGs those with evident [O I] $\lambda 6300$ [5,21], and as being LBVs those without. Both classes may display forbidden emission lines from Fe and Ca (e.g., [21,76,77]), while all of them display Fe emission lines. We note here that these LBVs are candidate sources, since there is no absolute way to characterize an LBV from a single-epoch spectrum (in contrast to B[e]SGs). It has to be supplemented with more spectroscopic or photometric observations that reveal variability (and possibly the return to a hotter state). We also note that our sample contained more interesting sources that displayed Hα in emission (i.e., main sequence O-stars and blue supergiants), but these were left for future papers (e.g., Munoz-Sanchez et al. 2023, IAU S361, submission).

3. Results

3.1. Statistics

From our large observational campaign, we were able to robustly classify (after careful visual inspection) 465 objects in the 12 targeted galaxies (see Table 1). Only 11 out of all of these ($\sim$3%) contained features in their optical spectra that indicated a B[e] SG/LBV nature (which was the subject of the current work, with the rest being left for future papers). Other stellar sources related to massive stars included mainly RSGs ($\sim$37%), other Blue Supergiants ($\sim$7%), and Yellow Supergiants ($\sim$5%). There was a small number of emission objects ($\sim$2%), carbon stars ($\sim$6%), and AGN/QSO and other background galaxies ($\sim$4%), while another bulk of sources were classified as H II regions ($\sim$22%) and foreground sources ($\sim$14%). In Table 2 we present the identified objects. We note that, although the same approach was followed for all 12 galaxies, we obtained null results for five of them: IC 10, NGC 1313, Sextans A, M83, NGC 6822. In addition, there were only four objects ($\sim$36%) with previous spectral information, for which we confirmed or updated classification. It is also interesting to note that $\sim$64% of these sources were considered priority targets in our survey (Table 2, col. 4), while the rest failed to pass our selection criteria (see Section 2.2). We further discuss these facts in Section 4.

Table 2. Properties of the sources identified in this work: source ID in this work (column 1), sky coordinates (columns 2 and 3), priority target (column 4), source ID in *Spitzer* (base) catalog (column 5), SNR (column 6), spectral type from this work and literature (columns 7 and 8), and radial velocity from this work (RV, column 9).

Name	RA (J2000)	Dec (J2000)	Prio.	ID [1]	SNR [2]	SpT	Prev. SpT	RV (km s^{-1})
(1)	(2)	(3)	(4)	(5)	(6)	(7)	(8)	(9)
WLM-1	00:02:02.32	−15:27:43.81	Y	95	30	B[e]SG	Fe star [78]	−48 ± 10
NGC55-1	00:15:09.31	−39:12:41.62	Y	178	18	B[e]SG	–	156 ± 31
NGC55-2	00:15:18.54	−39:13:12.32	N	736	46	LBVc	LBVc/WN11 [79]	105 ± 38
NGC55-3	00:15:37.66	−39:13:48.68	N	2924	50	LBVc	LBVc/WN11 [79]	202 ± 30
NGC247-1	00:47:02.17	−20:47:40.13	Y	246	26	B[e]SG	B[e]SG [27]	217 ± 12
NGC247-2	00:47:03.91	−20:43:17.22	N	1192	44	LBVc	–	114 ± 41
NGC253-1	00:47:04.90	−25:20:44.12	Y	739	3	B[e]SG	–	283 ± 56
NGC300-1	00:55:27.93	−37:44:19.61	Y	67	44	B[e]SG	–	58 ± 27
NGC300-2	00:55:19.17	−37:40:56.53	Y	389	9	B[e]SG	–	121 ± 34
NGC3109-1	10:03:02.11	−26:08:58.06	Y	188	70	LBVc	–	371 ± 29
NGC7793-1	23:57:43.28	−32:34:01.81	N	111	19	B[e]SG c	–	317 ± 32

[1] This ID corresponds to the *Spitzer* source numbering, as used throughout the ASSESS project (see Tramper et al., in prep., and Munoz-Sanchez et al., in prep., for the use with full catalogs). [2] Estimated by averaging the SNR over the ranges 6000–6150Å and 6950–7100Å.

3.2. Spectra

All spectra showed a strong, broadened Hα component, accompanied by several other characteristic emission lines. We present their spectra in Figures 1 and 2, where the strength of the Hα emission for all objects is highlighted in the right panel. The order of the spectra (from top to bottom) was one of decreasing Hα strength.

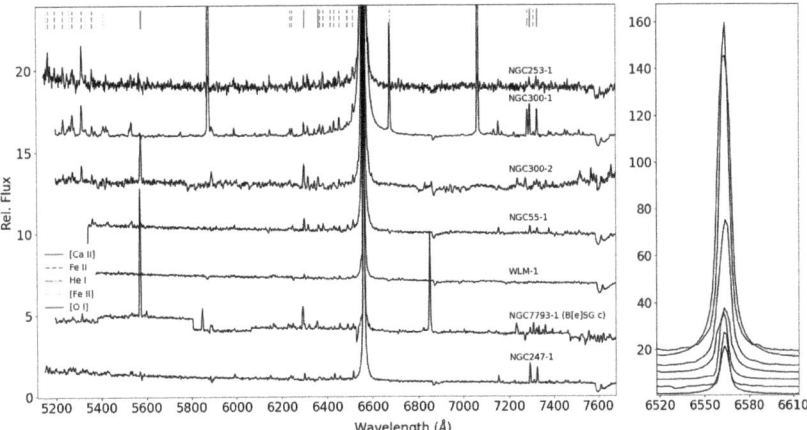

Figure 1. Spectra of objects classified as B[e]SGs (including the B[e]SG candidate NGC7793-1). (Left) The full spectra for all stars with small offsets for better illustration purposes. The most prominent emission features are indicated. (Right) The region around Hα is highlighted to emphasize the relative strength of the emission compared to the continuum.

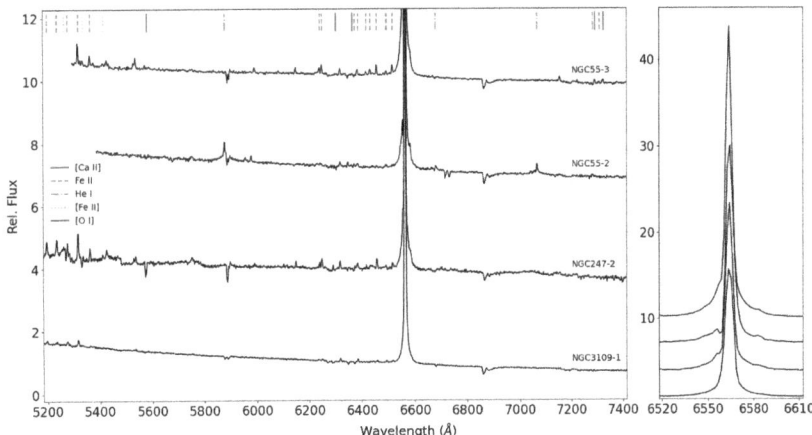

Figure 2. Similar to Figure 1, but for LBVc. We note the lack of forbidden emission lines.

We identified a series of Fe II emission lines in the left wing of Hα (∼6200–6500 Å), and, when the spectrum extended far enough to bluer wavelengths, we identified another series ranging from roughly ∼5100–5400 Å. Figure 3 showcases these lines in a zoom-in on the ∼6200–6500 Å region. We used the Fe II emission lines in this region to correct for the radial velocity (RV) shift. The obtained RV values are shown in column 9 of Table 2. Therefore, we verified that the RVs were in agreement with the motion of their host galaxies, confirming that these stars were, indeed, of extragalactic origin.

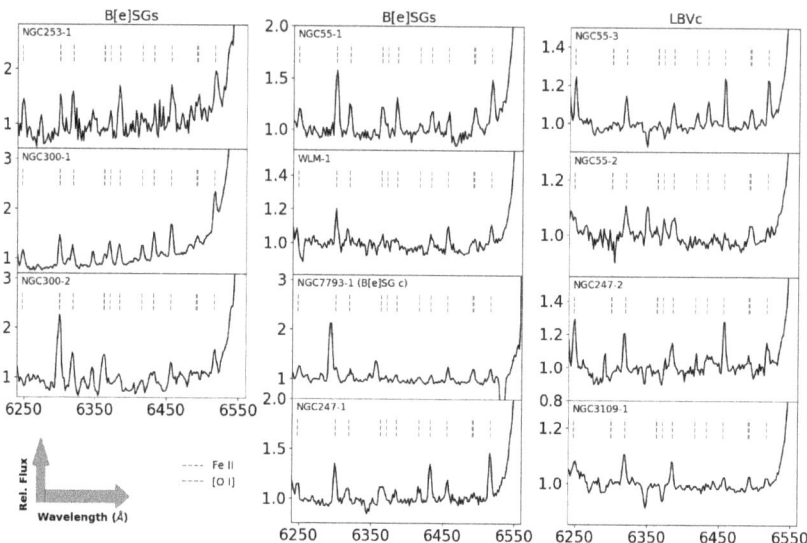

Figure 3. The region between the [O I] and Hα line, that showcases multiple Fe II emission lines. We note the clear presence of [O I] λ6300 line for the B[e]SGs (left and middle panels, with the exception of the candidate NGC7793-1, due to the problematic spectrum; see text for more) and its absence from the LBVc spectra (right panel).

According to the classification criteria presented in Section 2.4, we robustly identified 6 sources as being B[e]SGs: WLM-1, NGC55-1, NGC247-1, NGC253-1, NGC300-1, and NGC300-2. Figure 1 presents the full spectra for the B[e]SGs, while Figure 3 shows the characteristic [O I] λ6300 line. It is particularly interesting to note the very strong He I lines of NGC300-1. These emission lines require a hotter formation region, such as a spherical or a bipolar shell formed by a strong stellar wind, in addition to the structures that give rise to the forbidden emission features. We also note the absence of [Fe II] lines for the WLM-1, NGC253-1, and NGC300-2 sources. Half of the sources (NGC55-1, NGC247-1, NGC300-1) displayed strong [Ca II] emission lines, while for one source (NGC253-1) they were very faint (limited by the noise), and were totally absent for two of the sources (WLM-1, NGC300-2; see Figure 4). These lines were stronger in luminous sources (e.g., [76,77]). The very low SNR for the NGC253-1 and NGC300-2 (see Table 2, column 6) justified the lack of Fe and Ca lines. In the case of WLM-1, the SNR was sufficiently good that the lack of forbidden Fe lines should be considered a real non-detection (similar to source WLM 23 from [78]. We further discuss this in Section 4.2). Unfortunately, due to overlapping slits in the mask design, some of these spectra suffered from artifacts from the reduction processing (in particular, NGC300-2). Although the B[e] phenomenon can also characterize other types of objects, we noticed a lack of dominant emission lines, such as nebular lines ([N II] λλ6548, 6583, [S II] λλ6717, 6731, [Ar III] λ7135), present in planetary nebulae (e.g., [80,81]), O VI Raman-scattered lines (λλ6830, 7088) of symbiotic systems (e.g., [81,82]), or even the absorption lines of Li I 6708 present in young stellar objects (e.g., [83]). Moreover, during the visual screening of all these spectra, objects with such characteristic lines would be classified differently, as all possible objects were considered. Additionally, at the distances we were looking at, we were mainly probing the upper part of the Hertzsprung–Russell diagram, while their RVs were relatively compatible (within their error margins) with those of their host galaxies. Our *Gaia* cleaning approach removed the majority of the foreground sources (naturally, a small fraction remained hidden in our target lists). Therefore, we consider these objects to be strong supergiant candidates.

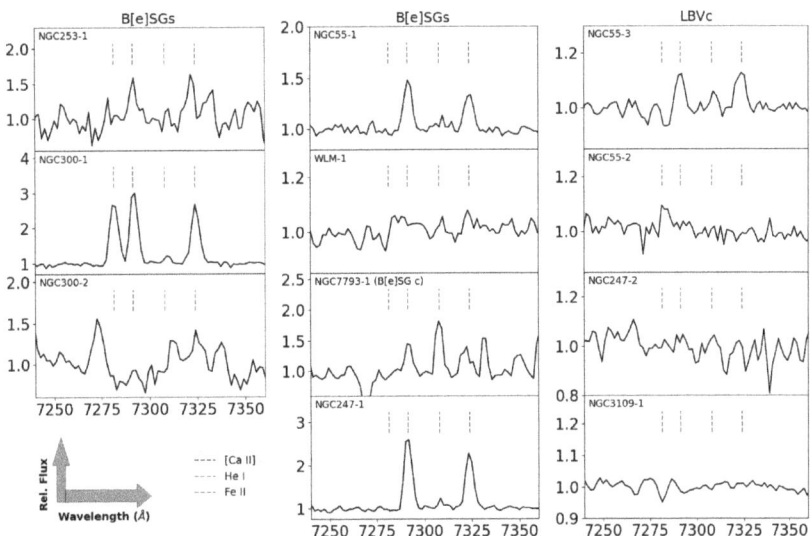

Figure 4. The region around the [Ca II] emission doublet. Its presence is evident in some B[e]SGs (left and middle panels, including NGC7793-1 candidate source, that suffers from data reduction artifacts due to slit overlaps), while LBVc (right panel) do not typically exhibit these lines (except for NGC55-3).

We characterized as LBVc the following 4 sources: NGC55-2, NGC55-3, NGC247-2, and NGC3109-1 (see Figure 2). NGC55-2 was the hotter of all these sources as it was the only LBVc with all He I lines in emission. NGC3109-1 displayed He I lines in absorption, while the rest did not show any of these lines. During the outbursts the He I lines decrease and vanish, as the temperature and the density (due to the expanding pseudo-photosphere) drop significantly to allow for other lines to form. It is during these cooler states that Fe lines become evident in LBVs. Depending on the conditions, forbidden emission lines may form. This was the case with NGC55-3, which displayed the [Ca II] lines in emission, along with a few [Fe II] lines. The other sources did not show any forbidden lines. Similar to the B[e]SG spectra, there were unavoidable residuals and artifacts, due to the slit overlap and reduction issues.

Of these cases, NGC7793-1 was the most extreme example[5]. The region at [O I] $\lambda6300$ was highly contaminated with a sky residual line from another source in the slit. Therefore, we could not conclude whether this line existed or not. We noticed the presence of some [Fe II] and the [Ca II] lines, but a B[e]SG or LBV classification solely from this spectrum was not possible. However, additional information could be retrieved from photometry (see Section 4.3), so that we could propose a B[e]SG candidate (B[e]SG c) classification for NGC7793-1.

The final classification for each star is provided in column 7 of Table 2.

3.3. Light Curves and Variability

We collected variability information for all targets from both Pan-STARRS DR2[6] and the VizieR[7] services. We found four sources (WLM-1, NGC247-1, NGC274-2, and NGC3109-1) with data in the Pan-STARRS DR2 release[8] (with an approximate coverage between 2010 and 2014). We considered only values with `psfQfPerfect`>0.9 to select the best data. For three sources (NGC55-2, NGC55-3, and NGC247-1) we found additional data in the catalog of large-amplitude variables from *Gaia* DR2 (covering 2014 to 2016; [84]), and NGC3109-1 had already been reported as a variable [85]. In Table 3 we summarize the collected information for all sources and their corresponding magnitude differences (peak-to-peak) for all (5) Pan-STARRS filters, the two *Gaia* filters (for which we doubled the quoted values in the catalog to match the Pan-STARRS definition of magnitude difference, and some additional variability studies).

Table 3. Variability information for our sample (source IDs, column 1), as provided by Pan-STARRS DR2 data (columns 2-6), *Gaia* (columns 7 and 8), and literature (column 9).

Name	Pan-STARRS DR2					Gaia DR2 [84]		Other
	Δg (mag)	Δr (mag)	Δi (mag)	Δz (mag)	Δy (mag)	ΔBP (mag)	ΔRP (mag)	
(1)	(2)	(3)	(4)	(5)	(6)	(7)	(8)	(9)
WLM-1	0.13	0.20	0.25	0.22	0.50	-	-	-
NGC55-1	-	-	-	-	-	-	-	-
NGC55-2	-	-	-	-	-	0.24	0.18	-
NGC55-3	-	-	-	-	-	0.24	0.18	-
NGC247-1	0.37	0.27	0.20	0.28	0.35	0.36	0.18	$\Delta V = 0.29 \pm 0.09$ [27] $\Delta g' \sim 0.1$ [86]
NGC247-2	0.37	0.27	0.31	0.23	1.00	-	-	-
NGC253-1 *	-	-	-	-	-	-	-	-
NGC300-1	-	-	-	-	-	-	-	-
NGC300-2	-	-	-	-	-	-	-	-
NGC3109-1	0.12	0.26	0.43	0.37	0.49	-	-	$\Delta J = 0.08$ [85]
NGC7793-1	-	-	-	-	-	-	-	-

* Only three epochs of observations, so not considered.

In total, we found light curves for two B[e]SGs (WLM-1 and NGC247-1) and four LBVc (NGC55-2, NGC55-3, NGC247-1, and NGC3109-1). We show the Pan-STARRS light curves in Figure 5 and 6, where we plot the magnitude difference at each epoch with the mean for the particular filter (indicated on the y-axis label). For the B[e]SGs we noticed a (mean) variability of 0.25-0.3 mag, while for the LBVs it was slightly larger, at 0.3–0.44 mag. There were no obvious trends in the B[e]SG light curves, while, in the case of NGC3109-1, a dimming across all filters was observed. Menzies et al. [85] also detected such a trend, although smaller, for this target, due to the different filters used. Limited by the photometric data, they argued that a background galaxy or AGN was not excluded, but, given our spectrum and its consistent RV value with its host galaxy, we could actually verify its stellar nature. For NGC247-2, the light curves were generally flatter. There was a noticeable peak present in the y light curve (at MJD$\sim$56,300 days), which was not evident in the other filters (although we note that there were no observations around the same epoch). The quality flags corresponding to these particular points did not show any issue. However, we should be cautious with this, as further mining of the data is needed to reveal if this is a real event or an artifact.

NGC247-1 was the only source for which we had multiple sources of variability information. Very good agreement between the Pan-STARRS and *Gaia* data is evident , and consistent with the value quoted by Solovyeva et al. [27] ($\Delta V = 0.29 \pm 0.09$ mag). Although Davidge [86] quoted a smaller value ($\Delta g' \sim 0.1$ mag), their time coverage was limited to about 6 months, a time frame that definitely does not cover the whole variability cycles for these sources.

Traditionally, LBVs are considered variable at many scales (e.g., [22,24,87,88]). The (optical) S Dor variability is of the order of 0.1 mag to about 2.5 mag with cycles ranging from years to decades. The giant eruptions, although much more energetic ($\sim$5 mag) are less frequent events (a time frame in the order of centuries), and, therefore, a smaller subgroup of LBVs have been observed to display such events. On the other hand, the B[e]SGs are considered more stable, with variability that does not exceed $\sim$0.2 mag (optical; [5]). However, this is changing and significant variability is observed, due to binary interactions and possible pulsations (e.g., [12,19,89]). Therefore, it is not surprising to observe similar magnitude differences between the two classes.

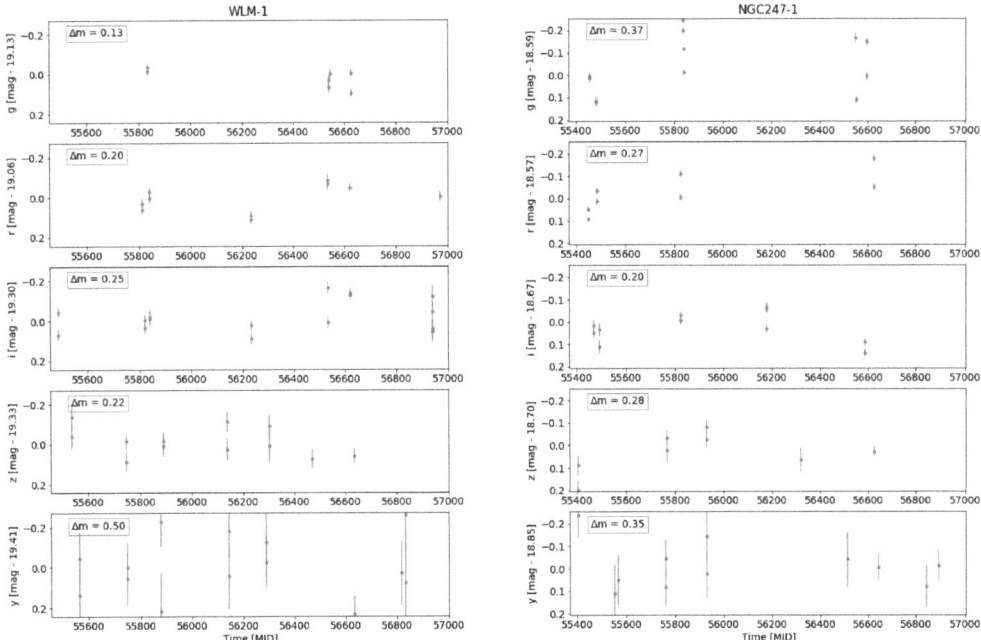

Figure 5. The light curves from the Pan-STARRS survey for the B[e]SGs WLM-1 and NGC247-1. Each panel (per filter) shows the difference of each epoch from the mean value (noted on the y-axis label). See text for more.

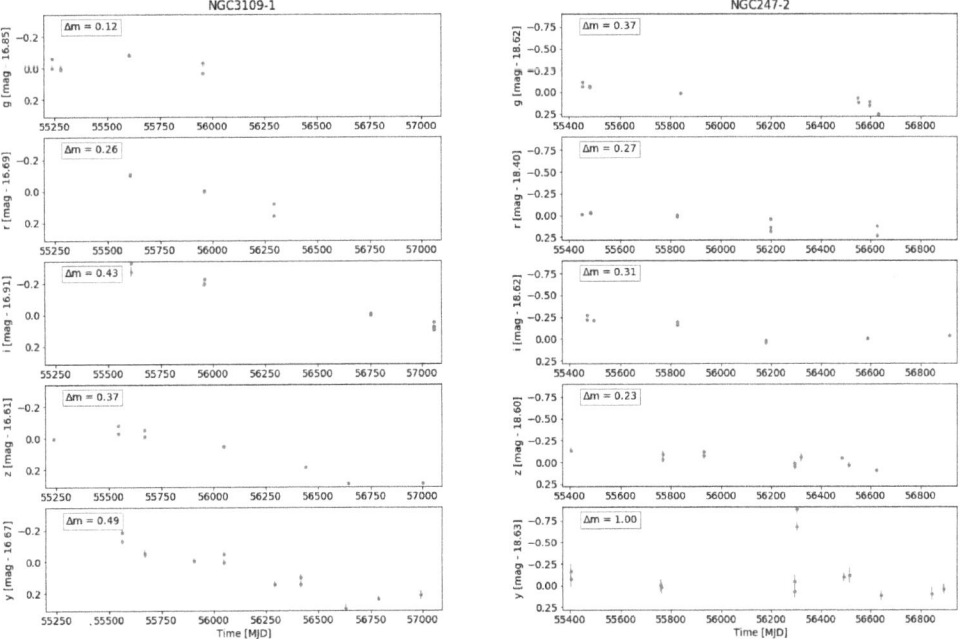

Figure 6. Same as Figure 5, but for the candidate LBVs NGC3109-1 and NGC247-2.

4. Discussion

4.1. Demographics

As mentioned already in Section 3.1, we did not detect B[e]SGs or LBVs in the following five (out of 12) galaxies: IC 10, NGC 1313 Sextans A, M83, NGC 6822.

M83 and NGC 1313 are the most distant galaxies (at 4.9 and 4.6 Mpc, respectively) and confusion becomes an important issue (unsurprisingly, M83 is the galaxy for which we detected the most H II regions; see Tramper et al., in prep.). Due to the spatial resolution of *Spitzer* and the increasing distance of some of our target galaxies, H II regions or other point-like objects (e.g., clusters) were included in the point-source catalogs and, therefore, considered to be viable targets in our priority system. The farthest galaxies, for which the majority of observed targets were, indeed, resolved point sources and at least one was either an LBVc or a B[e]SG, were NGC 7793 and NGC 253 (at ~3.4 Mpc). Therefore, the null detections for IC 10 and NGC 6822 (less than 1 Mpc) and for Sextans A (at 1.34 Mpc) were not due to distance and confusion.

Massey et al. [90] detected one LBV in NGC 6822 (J194503.77-145619.1) and three in IC 10 (J002012.13+591848.0, J002016.48+591906.9, J002020.35+591837.6). Our inability to recover these targets was due to two reasons. Firstly, we imposed strict criteria to prioritize our target selection (see Section 2.2) based on relative strong IR luminosity and color. Almost all of these targets (except for IC 10 J002020.35+591837.6) had $m_{[4.5]} > 15.5$ mags, which directly excluded them from further consideration. This was further supported by the fact that four out of or our 11 discoveries initially did not pass as a priority target (see Table 2), but were observed as "filler" stars (see Section 2.2). This was particularly important for galaxies with smaller sizes, where only one (IC 10, Sextans A) or two pointings (NGC 6822) were performed. Therefore, the second reason was the limitations that arose from the particular pointing(s) to the galaxy, as targets might have been located? out of the field-of-view or at a sensor's gap (which was the case for IC 10 J002020.35 + 591837.6), and therefore not be observable. Other reasons (not corresponding to the aforementioned targets) that could impact the selection of a target or render its spectrum useless include overlapping slits, a poor wavelength calibration and/or SNR, or other reduction issues.

In the case of NGC 55, four LBVs (including two candidates) are known [79]. Two of them (the candidates) were recovered from our survey (NGC55-2 and NGC55-3 as B_13 and B_34, respectively) as LBVc (see Section 3.2). The other two (C1_30,00:14:59.91,-39:12:11.88 and A_42,0:16:09.69,-39:16:13.44) were sources outside the region investigated by Williams and Bonanos [91], so without any *Spitzer* data to be included in our base catalogs.

In total, our approach was successful in detecting these populations, and it was mainly limited by technical issues.

4.2. Comparison with Previous Classifications

Four of our sources had previous classifications (see Table 2). WLM-1 had been identified as an Hα source previously [90], through a photometric survey, and identified as an Fe line star through spectroscopic observations (WLM 23 in [78]). Even though the presence of the [O I] $\lambda 6300$ line was noted, the source was not classified as a B[e]SG, due to the lack of forbidden Fe lines (see e.g., [20,21,92] on Fe stars). Therefore, we updated its classification to a B[e]SG from an Fe star. We also noted that our spectrum (obtained on November 2020) was very similar to theirs (obtained on December 2012), which might indicate that the star was rather stable over this eight-year period (however, this should be treated with caution due to the lack of systematic observations).

NGC55-2 and NGC55-3 had been identified as candidate LBV/WN11 (ids B_34 and B_13, respectively), with both Balmer and He I lines in emission and with P-Cygni profiles [79]. Their spectra were within the 3800–5000 Å range and outside ours. However, given that the diagnostic [O I] line was not present, we classified both of these sources as LBVc, consistent with the previous results[9].

For NGC247-1 we provided a classification of B[e]SG, similar to what was suggested by Solovyeva et al. [27]. We note here that their spectral coverage was ~4400–7400 Å which overlapped with our observed range. Hence, we can also comment that no significant differences existed between the two observations (October 2018 and December 2020 by Solovyeva et al. [27] and our observations, respectively), although this time difference is rather small with respect to the variability timescales for these sources [5,12,19].

Therefore, we confirmed the previous classifications for three out of four sources, leaving us with 6 new B[e]SGs (including the reclassified Fe star and the candidate NGC7793-1) and 2 LBVc. The majority (~72%) of our findings are genuine discoveries and, as such, contribute greatly to the pool of extragalactic B[e]SGs, in particular.

4.3. Separating the Two Classes with Photometry

The total numbers of B[e]SGs and LBVs (even including candidates) are definitely small. Combined with the uncertainty pertaining to their roles in stellar evolution theory (e.g., B[e]SGs are not predicted by any code) it is easy to grasp why we really need larger samples and from different galactic environments, to fully understand these sources. Photometric data are typically used to pinpoint interesting candidates. These kinds of diagnostics exist mainly for IR, due to the presence of dust around these objects.

Bonanos et al. [93] found the B[e]SG, LBVs and RSGs to be among the most luminous sources in the mid-IR, using a color-magnitude diagram (CMD) with a combination of near-IR (2MASS) and mid-IR (*Spitzer*) J-[3.6] and [3.6]-[4.5] for the massive stars in the Large Magellanic Cloud (with a similar work for the Small Magellanic Cloud presented in [94]). In the most recent census of B[e]SGs, Kraus [7] presented color–color diagrams (CCD) to highlight the separation between B[e]SGs and LBVs (see their Figure 5). Indeed, by using the 2MASS near-IR colors $H - K$ and $J - H$ and mid-IR *WISE* W2-W4 and W1-W2 the two classes are distinct. This is the result of the hot dust component in the B[e]SGs, (formed in the denser disk/ring-like CSE closer to the star) which intensifies the near- and mid-IR excesses, compared to the LBVs (which form dust further away as the wind mass-loss and/or outburst material dissipates). Therefore, the location of a source in these diagrams may be used to verify its nature. We attempted to replicate these aforementioned diagrams by adding the new sources. However, one strong limitation was the lack of data for our sample. For the mid IR *WISE* [95] we found data for 5 (out of 11) sources (see Table 4) Using the data for 21 stars (excepting LHA 120-S 111) provided in [7] we plot, in Figure 7, the *WISE* colors for the MC sources and our 5 objects. We notice that, in general, the newly discovered sources are almost consistent with the loci of the MC sources, with the exception of NGC55-1. The new B[e]SG extend the W2-W4 color further to the red, while the LBVc NGC55-3 extended the W1-W2 color to the blue. Errors were plotted in the cases where they were available[10]. The errors provided for NGC55-1 were (numerically) small and placed it within the locus of LBV. However, caution should be taken with *WISE* photometry, as the resolution from W1 to W4 worsens significantly, and, combined with the distance of our galaxies, the photometric measurements could be strongly affected by confusion due to crowding (e.g., for both NGC 55 and NGC300 at ~2 Mpc). Combined with the position (and the uncertainty) of the LBVc NGC55-3, we might also be looking at a systematic offset of these populations. Unfortunately, the points in this plot are too scarce to make a robust examination of how the different galactic environments (e.g., metallicity, extinction effects) affect the positions of these populations.

We were unable to construct the *J-H* vs. *H-K* CCD because of the lack of 2MASS data for our sources (only for NGC3109-1 did data exist; 2MASS point source catalog; [96]), due to the shallowness of the survey and the distances of our target galaxies. However, we were able to acquire *J* photometry from the VHS DR5 for 5 of our sources (including NGC3109-1; [97]). Equipped with both *J* and [3.6] photometry we plot, in Figure 7, the equivalent CMD plot presented in [93], where the underlying MC objects were the same as in [7]. We notice excellent agreement of all new sources to their corresponding classes.

Once again we were hampered by a lack of data for our sample. We could remedy this using the complete data from *Spitzer* and *Gaia* surveys (missing NGC253-1 from our sample without *Gaia* data). In order to consider the MC sources, we used the *Gaia* DR3 [66,98] and *Spitzer* data from the SAGE survey [93,94]. This time, we only lost two targets (CPD-69 463 and LHA 120-S 83 without *Spitzer* data), but were still left with 19 sources.

Table 4. Photometry in *Gaia* DR3 (columns 2–7), VHS DR5 (columns 8-13), *Spitzer* (columns 14–23), and WISE (columns 24–31).

ID	BP [mag]	σ_{BP} [mag]	RP [mag]	σ_{RP} [mag]	G [mag]	...	W4 [mag]	σ_{W4} [mag]
(1)	(2)	(3)	(4)	(5)	(6)	...	(30)	(31)
WLM-1	19.132	0.029	18.919	0.025	19.252	...	8.278	−999
NGC55-1	19.797	0.053	19.199	0.045	19.552	...	8.115	0.233
NGC55-2	18.620	0.018	18.332	0.024	18.691	...	−999	−999
NGC55-3	18.266	0.023	17.772	0.026	18.148	...	9.091	−999
NGC247-1	18.602	0.036	18.341	0.028	18.723	...	−999	−999
NGC247-2	18.798	0.026	18.344	0.029	18.573	...	−999	−999
NGC300-1	18.557	0.013	17.784	0.013	18.323	...	8.897	−999
NGC300-2	20.866	0.093	20.556	0.113	20.742	...	9.013	−999
NGC3109-1	17.173	0.014	16.775	0.018	17.060	...	−999	−999
NGC7793-1	19.635	0.043	19.460	0.053	19.520	...	−999	−999

Note: The table is available in its entirety at the CDS.

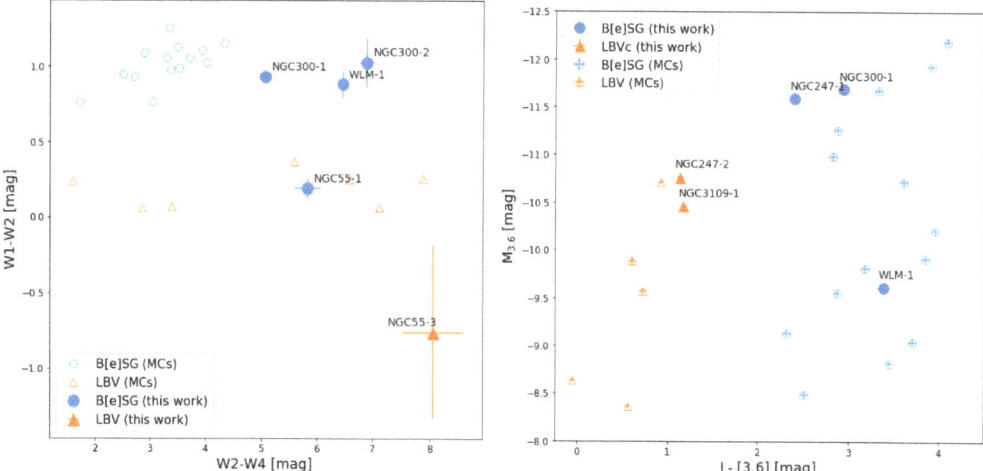

Figure 7. (Left) The mid-IR *WISE* CCD for B[e]SGs and LBVs, including sources from the MCs (after [7]) and our sample (for 5 out of 11 sources with *WISE* data). In general, the separation also holds for the new sources, with the exception of NGC55-1 (see text for more). (Right) IR CMD combining near-IR *J*-band (available for only five of our sources) with *Spitzer* [3.6]. We notice that, in this case, the newly found sources are consistent with the positions of the MC sources.

In Figure 8 we present the optical (*Gaia*) CMD, plotting BP–RP vs. M_G band. We notice the lack of any correlation in the optical.

In Figure 8 we also present the mid-IR (*Spitzer*) CMD, plotting [3.6]–[4.5] vs. $M_{[3.6]}$ band. The separation between the two classes becomes more evident in this case. The presence of hotter dusty environments becomes more significant for B[e]SGs, as they looked redder than LBVs (with a [3.6]–[4.5] range between 0.5 to 0.65 mag). They also tend to be much more luminous in the [3.6] than the LBVs. We highlighted the position of NGC7793-1 in this plot. Although, from its spectrum alone, we could not determine a secure classification (due to issues with the obtained spectrum) it is located among the B[e]SGs of our sample and of the MCs. Therefore, we considered it a candidate B[e]SG. A future spectrum is needed to verify the existence of the [O I] λ6300 line, similar to the rest of the secure B[e]SGs in our sample.

We also tried to combine the optical and IR data in a CMD where we plot the [3.6]–[4.5] vs. M_G magnitude (Figure 9). The result was actually similar to the previous IR CMD (as the x-axis did not change). In this case, the plot can be more helpful, as the LBVs are populating the upper left part of the plot. Therefore, very bright optical sources with IR color up to ∼0.5 mag were most probably LBVs, while sources with color > 0.5 mag would be B[e]SG (at almost any G magnitude).

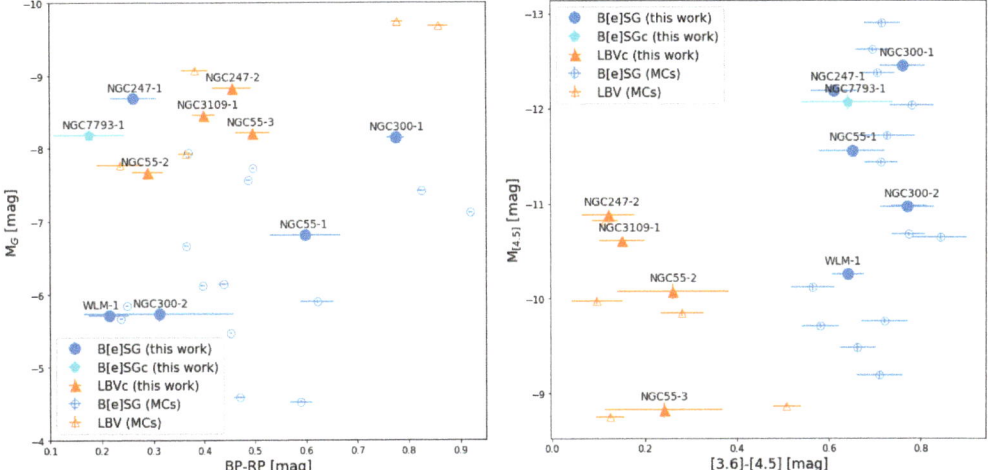

Figure 8. (Left) The optical (*Gaia*) CMD, plotting BP–RP vs. M_G magnitude. We included all our sample and the MC sources from [7] (except for two sources without a complete dataset in both *Gaia* and *Spitzer* surveys). (Right) The mid-IR (*Spitzer*) CMD using the IR color [3.6]–[4.5] vs. $M_{[4.5]}$. In this case, there is a significant improvement in the separation between the two classes. The position of NGC7793-1 favors a B[e]SG nature (see text for more).

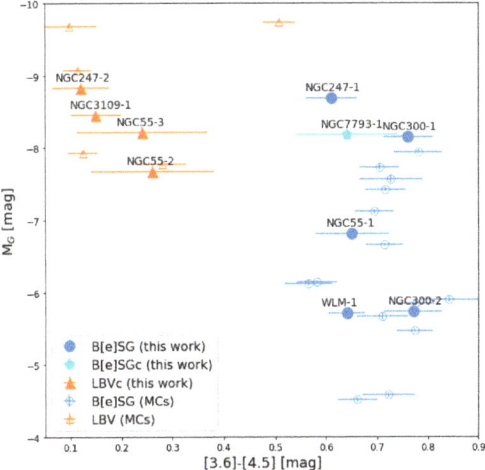

Figure 9. Similar to Figure 8 but plotting the IR color [3.6]–[4.5] vs. the optical M_G magnitude. Similar to the IR CMD we saw relatively good separation between the two classes, with LBVs being brighter in the optical and less dusty compared to the B[e]SGs.

4.4. Metallicity Dependence of Populations

In this section, we examine the populations of the two classes as a function of metallicity. For this, we plot the cumulative distribution function with metallicity (Figure 10), considering all detected and known objects in our sample of galaxies. Namely, the numbers presented in Table 2, as well as the two LBVs in NGC55 [79], one in NGC 6822 and three in IC 10 [90], resulting in 7 B[e]SGs (including the NGC7793-1 candidate) and 10 LBVs in our sample of 12 galaxies.

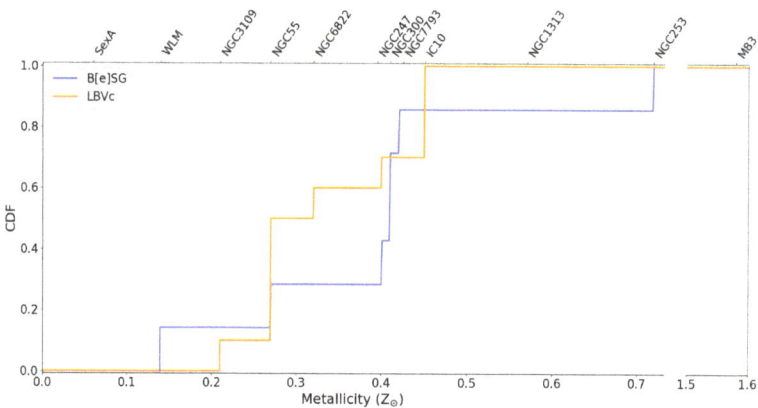

Figure 10. The cumulative distribution function of the B[e]SGs and LBVs (including candidates) from this work and the literature. We notice (for the first time) the presence of B[e]SGs in lower metallicity environments and the fact that the two populations are not totally different (see text for more).

We notice the presence of B[e]SGs at metallicity as low as $\sim$0.14 $Z_\odot$ (WLM). The current work is the first to detect these sources at such low metallicities. The population of LBVs begins at $\sim$0.21 $Z_\odot$ (NGC 3109), and then increases steadily as we move towards higher metallicities. B[e]SGs presents an important step (increase) around $\sim$0.4 $Z_\odot$. In total, the two populations do not look significantly different. We have to be cautious interpreting this figure, however, due to the low number of statistics and completeness issues, as, for example, depending on the angle under which we observe a galaxy, we may not be able to fully observe its stellar content (e.g., NGC 253).

5. Conclusions

In this work, we report the detection of 6 secure B[e]SGs, 1 candidate B[e]SG, and 4 LBV candidates sources, of which 6 B[e]SGs and 2 LBVs are new discoveries. They are based on spectroscopic and photometric diagnostics, supplemented with RVs that are consistent with their host galaxies. By inspecting the available IR (2MASS, *WISE*, *Spitzer*) and optical (*Gaia*) CMDs we find that the new sources are totally consistent with the loci of these populations from MCs. This adds further support regarding their natures. Building the cumulative distribution function of both populations with metallicity we notice the presence of B[e]SGs at environments with Z$\sim$ 0.14 $Z_\odot$, which increases the pool of extragalactic B[e]SGs and, especially, at lower metallicities. This is particularly important in order to investigate (with increased samples) these phases of massive stars. Since B[e]SGs and LBVs are among the classes with the most important episodic and outburst activities they provide valuable information on the role of episodic mass loss and insights into stellar evolution in general.

Author Contributions: Conceptualization, G.M. and A.Z.B.; Funding acquisition, A.Z.B.; Investigation, G.M., S.d.W., A.Z.B., G.M.-S. and E.C.; Methodology, G.M., S.d.W. and F.T.; Software, F.T. and G.M.-S.; Supervision, A.Z.B.; Visualization, G.M. and S.d.W.; Writing—original draft, G.M., S.d.W.

and A.Z.B.; Writing—review & editing, G.M., S.d.W., A.Z.B., F.T., G.M.-S. and E.C. All authors have read and agreed to the published version of the manuscript.

Funding: This research was funded by the European Research Council (ERC) under the European Union's Horizon 2020 research and innovation programme (Grant agreement No. 772086).

Data Availability Statement: Photometry and 1D extracted spectra will become available through the VizieR/CDS catalog tool.

Acknowledgments: GM acknowledges feedback from Francisco Najarro and Michaela Kraus. Based on observations collected at the European Southern Observatory under the ESO programme 105.20HJ and 109.22W2. Based on observations made with the Gran Telescopio Canarias (GTC), installed at the Spanish Observatorio del Roque de los Muchachos of the Instituto de Astrofísica de Canarias, on the island of La Palma (programme GTC83/20A). This work was (partly) based on data obtained with the instrument OSIRIS, built by a Consortium led by the Instituto de Astrofísica de Canarias in collaboration with the Instituto de Astronomía of the Universidad Autónoma de México. OSIRIS was funded by GRANTECAN and the National Plan of Astronomy and Astrophysics of the Spanish Government. This work was based, in part, on observations made with the *Spitzer* Space Telescope, which is operated by the Jet Propulsion Laboratory, California Institute of Technology, under a contract with NASA. This work made use of data from the European Space Agency (ESA) mission *Gaia* (https://www.cosmos.esa.int/gaia), processed by the *Gaia* Data Processing and Analysis Consortium (DPAC, https://www.cosmos.esa.int/web/gaia/dpac/consortium). Funding for the DPAC was provided by national institutions, in particular, the institutions participating in the *Gaia* Multilateral Agreement. Based on observations made with ESO Telescopes at the La Silla or Paranal Observatories under programme ID(s) 179.A-2010(A), 179.A-2010(B), 179.A-2010(C), 179.A-2010(D), 179.A-2010(E), 179.A-2010(F), 179.A-2010(G), 179.A-2010(H), 179.A-2010(I), 179.A-2010(J), 179.A-2010(K), 179.A-2010(L), 179.A-2010(M), 179.A-2010(N), 179.A-2010(O) (regarding VISTA Hemisphere Survey). This publication made use of data products from the Two Micron All Sky Survey, which is a joint project of the University of Massachusetts and the Infrared Processing and Analysis Center/California Institute of Technology, funded by the National Aeronautics and Space Administration and the National Science Foundation. This publication used data products from the Wide-field Infrared Survey Explorer, which is a joint project of the University of California, Los Angeles, and the Jet Propulsion Laboratory/California Institute of Technology, funded by the National Aeronautics and Space Administration. This work used Astropy http://www.astropy.org: a community-developed core Python package and an ecosystem of tools and resources for astronomy [99–101], NumPy (https://numpy.org/; [102]), and matplotlib (https://matplotlib.org/; [103])

Conflicts of Interest: The authors declare no conflict of interest.

Abbreviations

The following abbreviations are used in this manuscript:

B[e]SG	B[e] Supergiant
CCD	Color-Color Diagram
CMD	Color-Magnitude Diagram
CSE	circumstellar environment
LBV	Luminous Blue Variable
MC	Magellanic Cloud
SNR	Signal to Noise Ratio
RSG	Red Supergiant
RV	Radial Velocity

Notes

1 Lamers et al. [5] commented on a relative low variation of up to 0.2 mag, which was not the case in more recent studies, see Section 3.3 for more details.
2 https://assess.astro.noa.gr
3 https://www.inaoep.mx/~ydm/gtcmos/gtcmos.html accessed 1 September 2022—see also [73].
4 IRAF is distributed by the National Optical Astronomy Observatory, which is operated by the Association of Universities for Research in Astronomy (AURA) under cooperative agreement with the National Science Foundation.

5 Features at $\lambda\lambda$ ~5577, 5811 (step), 5846, 6855, and the region around the [Ca II] lines.
6 https://catalogs.mast.stsci.edu/panstarrs/.
7 http://vizier.cds.unistra.fr/.
8 There were only a couple of detections (epochs) for NGC253-1, which did not provide any meaningful information, and, therefore, we did not consider them. A declination of about $-25°$ was very close to the limit of the survey. All other galaxies with southern declination than $-30°$, i.e., NGC 55, NGC 300, and NGC 7793, were not visible.
9 As our observations were obtained from different epochs (October–December 2020) than those by Castro et al. [79] (November 2004) the spectra appearance might have changed, but there was no wavelength overlap to confirm this.
10 Only NGC55-1 had an error estimate in the W4 band, while the rest of the sources did not. For all other sources we could only plot W1–W2 errors.

References

1. Ekström, S.; Georgy, C.; Eggenberger, P.; Meynet, G.; Mowlavi, N.; Wyttenbach, A.; Granada, A.; Decressin, T.; Hirschi, R.; Frischknecht, U.; et al. Grids of stellar models with rotation. I. Models from 0.8 to 120 $M_\odot$ at solar metallicity (Z = 0.014). *Astron. Astrophys.* **2012**, *537*, A146. [CrossRef]
2. Georgy, C.; Ekström, S.; Granada, A.; Meynet, G.; Mowlavi, N.; Eggenberger, P.; Maeder, A. Populations of rotating stars. I. Models from 1.7 to 15 $M_\odot$ at Z = 0.014, 0.006, and 0.002 with Ω/Ω_{crit} between 0 and 1. *Astron. Astrophys.* **2013**, *553*, A24. [CrossRef]
3. Smith, N. Mass Loss: Its Effect on the Evolution and Fate of High-Mass Stars. *Annu. Rev. Astron. Astrophys.* **2014**, *52*, 487–528. [CrossRef]
4. Eldridge, J.J.; Stanway, E.R. New Insights into the Evolution of Massive Stars and Their Effects on Our Understanding of Early Galaxies. *Annu. Rev. Astron. Astrophys.* **2022**, *60*, 455–494. [CrossRef]
5. Lamers, H.J.G.L.M.; Zickgraf, F.J.; de Winter, D.; Houziaux, L.; Zorec, J. An improved classification of B[e]-type stars. *Astron. Astrophys.* **1998**, *340*, 117–128.
6. Kraus, M. The pre- versus post-main sequence evolutionary phase of B[e] stars. Constraints from ^{13}CO band emission. *Astron. Astrophys.* **2009**, *494*, 253–262. [CrossRef]
7. Kraus, M. A Census of B[e] Supergiants. *Galaxies* **2019**, *7*, 83. [CrossRef]
8. Zickgraf, F.J.; Wolf, B.; Stahl, O.; Leitherer, C.; Klare, G. The hybrid spectrum of the LMC hypergiant R 126. *Astron. Astrophys.* **1985**, *143*, 421–430.
9. Zickgraf, F.J.; Schulte-Ladbeck, R.E. Polarization characteristics of galactic Be stars. *Astron. Astrophys.* **1989**, *214*, 274–284.
10. Aret, A.; Kraus, M.; Muratore, M.F.; Borges Fernandes, M. A new observational tracer for high-density disc-like structures around B[e] supergiants. *Mon. Not. R. Astron. Soc.* **2012**, *423*, 284–293. [CrossRef]
11. de Wit, W.J.; Oudmaijer, R.D.; Vink, J.S. Dusty Blue Supergiants: News from High-Angular Resolution Observations. *Adv. Astron.* **2014**, *2014*, 270848. [CrossRef]
12. Maravelias, G.; Kraus, M.; Cidale, L.S.; Borges Fernandes, M.; Arias, M.L.; Curé, M.; Vasilopoulos, G. Resolving the kinematics of the discs around Galactic B[e] supergiants. *Mon. Not. R. Astron. Soc.* **2018**, *480*, 320–344. [CrossRef]
13. Kraus, M. Ionization structure in the winds of B[e] supergiants. II. Influence of rotation on the formation of equatorial hydrogen neutral zones. *Astron. Astrophys.* **2006**, *456*, 151–159. [CrossRef]
14. Petrov, B.; Vink, J.S.; Gräfener, G. Two bi-stability jumps in theoretical wind models for massive stars and the implications for luminous blue variable supernovae. *Mon. Not. R. Astron. Soc.* **2016**, *458*, 1999–2011. [CrossRef]
15. Curé, M.; Rial, D.F.; Cidale, L. Outflowing disk formation in B[e] supergiants due to rotation and bi-stability in radiation driven winds. *Astron. Astrophys.* **2005**, *437*, 929–933. .:20052686. [CrossRef]
16. Krtička, J.; Kurfürst, P.; Krtičková, I. Magnetorotational instability in decretion disks of critically rotating stars and the outer structure of Be and Be/X-ray disks. *Astron. Astrophys.* **2015**, *573*, A20. [CrossRef]
17. Wheelwright, H.E.; de Wit, W.J.; Weigelt, G.; Oudmaijer, R.D.; Ilee, J.D. AMBER and CRIRES observations of the binary sgB[e] star HD 327083: Evidence of a gaseous disc traced by CO bandhead emission. *Astron. Astrophys.* **2012**, *543*, A77. [CrossRef]
18. Podsiadlowski, P.; Morris, T.S.; Ivanova, N. Massive Binary Mergers: A Unique Scenario for the sgB[e] Phenomenon? In Proceedings of the Stars with the B[e] Phenomenon, Vlieland, The Netherlands, 10–16 July 2005; Astronomical Society of the Pacific Conference Series; Kraus, M., Miroshnichenko, A.S., Eds.; Astronomical Society of the Pacific: San Francisco, CA, USA, 2006; Volume 355, p. 259.
19. Kraus, M.; Cidale, L.S.; Arias, M.L.; Maravelias, G.; Nickeler, D.H.; Torres, A.F.; Borges Fernandes, M.; Aret, A.; Curé, M.; Vallverdú, R.; et al. Inhomogeneous molecular ring around the B[e] supergiant LHA 120-S 73. *Astron. Astrophys.* **2016**, *593*, A112. [CrossRef]
20. Humphreys, R.M.; Weis, K.; Davidson, K.; Bomans, D.J.; Burggraf, B. Luminous and Variable Stars in M31 and M33. II. Luminous Blue Variables, Candidate LBVs, Fe II Emission Line Stars, and Other Supergiants. *Astrophys. J.* **2014**, *790*, 48. [CrossRef]
21. Humphreys, R.M.; Gordon, M.S.; Martin, J.C.; Weis, K.; Hahn, D. Luminous and Variable Stars in M31 and M33. IV. Luminous Blue Variables, Candidate LBVs, B[e] Supergiants, and the Warm Hypergiants: How to Tell Them Apart. *Astrophys. J.* **2017**, *836*, 64. [CrossRef]

22. Weis, K.; Bomans, D.J. Luminous Blue Variables. *Galaxies* **2020**, *8*, 20. [CrossRef]
23. Vink, J.S. Eta Carinae and the Luminous Blue Variables. In *Eta Carinae and the Supernova Impostors*; Astrophysics and Space Science Library; Davidson, K., Humphreys, R.M., Eds.; Springer: Boston, MA, USA, 2012, Volume 384, p. 221. [CrossRef]
24. Davidson, K. Radiation-Driven Stellar Eruptions. *Galaxies* **2020**, *8*, 10. [CrossRef]
25. Wachter, S.; Mauerhan, J.C.; Van Dyk, S.D.; Hoard, D.W.; Kafka, S.; Morris, P.W. A Hidden Population of Massive Stars with Circumstellar Shells Discovered with the Spitzer Space Telescope. *Astron. J.* **2010**, *139*, 2330–2346. [CrossRef]
26. Liimets, T.; Kraus, M.; Moiseev, A.; Duronea, N.; Cidale, L.S.; Fariña, C. Follow-Up of Extended Shells around B[e] Stars. *Galaxies* **2022**, *10*, 41. [CrossRef]
27. Solovyeva, Y.; Vinokurov, A.; Sarkisyan, A.; Atapin, K.; Fabrika, S.; Valeev, A.F.; Kniazev, A.; Sholukhova, O.; Maslennikova, O. New luminous blue variable candidates in the NGC 247 galaxy. *Mon. Not. R. Astron. Soc.* **2020**, *497*, 4834–4842. [CrossRef]
28. Richardson, N.D.; Mehner, A. The 2018 Census of Luminous Blue Variables in the Local Group. *Res. Notes Am. Astron. Soc.* **2018**, *2*, 121. [CrossRef]
29. Wofford, A.; Ramírez, V.; Lee, J.C.; Thilker, D.A.; Della Bruna, L.; Adamo, A.; Van Dyk, S.D.; Herrero, A.; Kim, H.; Aloisi, A.; et al. Candidate LBV stars in galaxy NGC 7793 found via HST photometry + MUSE spectroscopy. *Mon. Not. R. Astron. Soc.* **2020**, *493*, 2410–2428. [CrossRef]
30. Guseva, N.G.; Thuan, T.X.; Izotov, Y.I. Decade-long time-monitoring of candidate luminous blue variable stars in the two very metal-deficient star-forming galaxies DDO 68 and PHL 293B. *Mon. Not. R. Astron. Soc.* **2022**, *512*, 4298–4307. [CrossRef]
31. Solovyeva, Y.; Vinokurov, A.; Tikhonov, N.; Kostenkov, A.; Atapin, K.; Sarkisyan, A.; Moiseev, A.; Fabrika, S.; Oparin, D.; Valeev, A. Search for LBVs in the Local Volume galaxies: Study of two stars in NGC 1156. *Mon. Not. R. Astron. Soc.* **2023**, *518*, 4345–4356. [CrossRef]
32. de Wit, S.; Bonanos, A.Z.; Tramper, F.; Yang, M.; Maravelias, G.; Boutsia, K.; Britavskiy, N.; Zapartas, E. Properties of luminous red supergiant stars in the Magellanic Clouds. *arXiv* **2022**, arXiv:2209.11239.
33. Yang, M.; Bonanos, A.Z.; Jiang, B.W.; Gao, J.; Gavras, P.; Maravelias, G.; Ren, Y.; Wang, S.; Xue, M.Y.; Tramper, F.; et al. Evolved massive stars at low-metallicity. I. A source catalog for the Small Magellanic Cloud. *Astron. Astrophys.* **2019**, *629*, A91. [CrossRef]
34. Yang, M.; Bonanos, A.Z.; Jiang, B.W.; Gao, J.; Gavras, P.; Maravelias, G.; Wang, S.; Chen, X.D.; Tramper, F.; Ren, Y.; et al. Evolved massive stars at low metallicity. II. Red supergiant stars in the Small Magellanic Cloud. *Astron. Astrophys.* **2020**, *639*, A116. [CrossRef]
35. Yang, M.; Bonanos, A.Z.; Jiang, B.; Gao, J.; Gavras, P.; Maravelias, G.; Wang, S.; Chen, X.D.; Lam, M.I.; Ren, Y.; et al. Evolved massive stars at low-metallicity. III. A source catalog for the Large Magellanic Cloud. *Astron. Astrophys.* **2021**, *646*, A141. [CrossRef]
36. Zgirski, B.; Pietrzyński, G.; Gieren, W.; Górski, M.; Wielgórski, P.; Karczmarek, P.; Bresolin, F.; Kervella, P.; Kudritzki, R.P.; Storm, J.; et al. The Araucaria Project. Distances to Nine Galaxies Based on a Statistical Analysis of their Carbon Stars (JAGB Method). *Astrophys. J.* **2021**, *916*, 19. [CrossRef]
37. Urbaneja, M.A.; Kudritzki, R.P.; Bresolin, F.; Przybilla, N.; Gieren, W.; Pietrzyński, G. The Araucaria Project: The Local Group Galaxy WLM—Distance and Metallicity from Quantitative Spectroscopy of Blue Supergiants. *Astrophys. J.* **2008**, *684*, 118–135. [CrossRef]
38. McConnachie, A.W. The Observed Properties of Dwarf Galaxies in and around the Local Group. *Astron. J.* **2012**, *144*, 4. [CrossRef]
39. Hartoog, O.E.; Sana, H.; de Koter, A.; Kaper, L. First Very Large Telescope/X-shooter spectroscopy of early-type stars outside the Local Group. *Mon. Not. R. Astron. Soc.* **2012**, *422*, 367–378. [CrossRef]
40. Sanna, N.; Bono, G.; Stetson, P.B.; Monelli, M.; Pietrinferni, A.; Drozdovsky, I.; Caputo, F.; Cassisi, S.; Gennaro, M.; Prada Moroni, P.G.; et al. On the Distance and Reddening of the Starburst Galaxy IC 10. *apjl* **2008**, *688*, L69. [CrossRef]
41. Tehrani, K.; Crowther, P.A.; Archer, I. Revealing the nebular properties and Wolf–Rayet population of IC10 with Gemini/GMOS. *Mon. Not. R. Astron. Soc.* **2017**, *472*, 4618–4633. [CrossRef]
42. Kacharov, N.; Neumayer, N.; Seth, A.C.; Cappellari, M.; McDermid, R.; Walcher, C.J.; Böker, T. Stellar populations and star formation histories of the nuclear star clusters in six nearby galaxies. *Mon. Not. R. Astron. Soc.* **2018**, *480*, 1973–1998. [CrossRef]
43. Tully, R.B.; Courtois, H.M.; Sorce, J.G. Cosmicflows-3. *Astron. J.* **2016**, *152*, 50. [CrossRef]
44. Madore, B.F.; Freedman, W.L. Astrophysical Distance Scale: The AGB J-band Method. I. Calibration and a First Application. *Astrophys. J.* **2020**, *899*, 66. [CrossRef]
45. Spinoglio, L.; Fernández-Ontiveros, J.A.; Malkan, M.A.; Kumar, S.; Pereira-Santaella, M.; Pérez-Díaz, B.; Pérez-Montero, E.; Krabbe, A.; Vacca, W.; Colditz, S.; et al. SOFIA Observations of Far-IR Fine-structure Lines in Galaxies to Measure Metallicity. *Astrophys. J.* **2022**, *926*, 55. [CrossRef]
46. Meyer, M.J.; Zwaan, M.A.; Webster, R.L.; Staveley-Smith, L.; Ryan-Weber, E.; Drinkwater, M.J.; Barnes, D.G.; Howlett, M.; Kilborn, V.A.; Stevens, J.; et al. The HIPASS catalogue—I. Data presentation. *Mon. Not. R. Astron. Soc.* **2004**, *350*, 1195–1209. [CrossRef]
47. Kudritzki, R.P.; Urbaneja, M.A.; Bresolin, F.; Przybilla, N.; Gieren, W.; Pietrzyński, G. Quantitative Spectroscopy of 24 A Supergiants in the Sculptor Galaxy NGC 300: Flux-weighted Gravity-Luminosity Relationship, Metallicity, and Metallicity Gradient. *Astrophys. J.* **2008**, *681*, 269–289. [CrossRef]
48. Qing, G.; Wang, W.; Liu, J.F.; Yoachim, P. The Distance Measurement of NGC 1313 with Cepheids. *Astrophys. J.* **2015**, *799*, 19. [CrossRef]

49. Hernandez, S.; Winch, A.; Larsen, S.; James, B.L.; Jones, L. Chemical Abundances of Young Massive Clusters in NGC 1313. *Astron. J.* **2022**, *164*, 89. [CrossRef]
50. Koribalski, B.S.; Staveley-Smith, L.; Kilborn, V.A.; Ryder, S.D.; Kraan-Korteweg, R.C.; Ryan-Weber, E.V.; Ekers, R.D.; Jerjen, H.; Henning, P.A.; Putman, M.E.; et al. The 1000 Brightest HIPASS Galaxies: H I Properties. *Astron. J.* **2004**, *128*, 16–46. [CrossRef]
51. Hosek, M.W.; Kudritzki, R.P.; Bresolin, F.; Urbaneja, M.A.; Evans, C.J.; Pietrzyński, G.; Gieren, W.; Przybilla, N.; Carraro, G. Quantitative Spectroscopy of Blue Supergiants in Metal-poor Dwarf Galaxy NGC 3109. *Astrophys. J.* **2014**, *785*, 151. [CrossRef]
52. Tammann, G.A.; Reindl, B.; Sandage, A. New period-luminosity and period-color relations of classical Cepheids. IV. The low-metallicity galaxies IC 1613, WLM, Pegasus, Sextans A and B, and Leo A in comparison to SMC. *Astron. Astrophys.* **2011**, *531*, A134. [CrossRef]
53. Kniazev, A.Y.; Grebel, E.K.; Pustilnik, S.A.; Pramskij, A.G.; Zucker, D.B. Spectrophotometry of Sextans A and B: Chemical Abundances of H II Regions and Planetary Nebulae. *Astron. J.* **2005**, *130*, 1558–1573. [CrossRef]
54. Bresolin, F.; Kudritzki, R.P.; Urbaneja, M.A.; Gieren, W.; Ho, I.T.; Pietrzyński, G. Young Stars and Ionized Nebulae in M83: Comparing Chemical Abundances at High Metallicity. *Astrophys. J.* **2016**, *830*, 64. [CrossRef]
55. Hernandez, S.; Larsen, S.; Aloisi, A.; Berg, D.A.; Blair, W.P.; Fox, A.J.; Heckman, T.M.; James, B.L.; Long, K.S.; Skillman, E.D.; et al. The First Metallicity Study of M83 Using the Integrated UV Light of Star Clusters. *Astrophys. J.* **2019**, *872*, 116. [CrossRef]
56. Dopita, M.A.; Seitenzahl, I.R.; Sutherland, R.S.; Nicholls, D.C.; Vogt, F.P.A.; Ghavamian, P.; Ruiter, A.J. Calibrating Interstellar Abundances Using Supernova Remnant Radiative Shocks. *Astron. J.* **2019**, *157*, 50. [CrossRef]
57. Della Bruna, L.; Adamo, A.; Lee, J.C.; Smith, L.J.; Krumholz, M.; Bik, A.; Calzetti, D.; Fox, A.; Fumagalli, M.; Grasha, K.; et al. Studying the ISM at ~10 pc scale in NGC 7793 with MUSE. II. Constraints on the oxygen abundance and ionising radiation escape. *Astron. Astrophys.* **2021**, *650*, A103. [CrossRef]
58. Gvaramadze, V.V.; Kniazev, A.Y.; Fabrika, S. Revealing evolved massive stars with Spitzer. *Mon. Not. R. Astron. Soc.* **2010**, *405*, 1047–1060. [CrossRef]
59. Cox, N.L.J.; Kerschbaum, F.; van Marle, A.J.; Decin, L.; Ladjal, D.; Mayer, A.; Groenewegen, M.A.T.; van Eck, S.; Royer, P.; Ottensamer, R.; et al. A far-infrared survey of bow shocks and detached shells around AGB stars and red supergiants. *Astron. Astrophys.* **2012**, *537*, A35. [CrossRef]
60. Smith, N.; Li, W.; Miller, A.A.; Silverman, J.M.; Filippenko, A.V.; Cuillandre, J.C.; Cooper, M.C.; Matheson, T.; Van Dyk, S.D. A Massive Progenitor of the Luminous Type IIn Supernova 2010jl. *Astrophys. J.* **2011**, *732*, 63. [CrossRef]
61. Zhang, T.; Wang, X.; Wu, C.; Chen, J.; Chen, J.; Liu, Q.; Huang, F.; Liang, J.; Zhao, X.; Lin, L.; et al. Type IIn Supernova SN 2010jl: Optical Observations for over 500 Days after Explosion. *Astron. J.* **2012**, *144*, 131. [CrossRef]
62. Gal-Yam, A. The Most Luminous Supernovae. *Annu. Rev. Astron. Astrophys.* **2019**, *57*, 305–333. [CrossRef]
63. Werner, M.W.; Roellig, T.L.; Low, F.J.; Rieke, G.H.; Rieke, M.; Hoffmann, W.F.; Young, E.; Houck, J.R.; Brandl, B.; Fazio, G.G.; et al. The Spitzer Space Telescope Mission. *Astrophys. J. Suppl. Ser.* **2004**, *154*, 1–9. [CrossRef]
64. Chambers, K.C.; Magnier, E.A.; Metcalfe, N.; Flewelling, H.A.; Huber, M.E.; Waters, C.Z.; Denneau, L.; Draper, P.W.; Farrow, D.; Finkbeiner, D.P.; et al. The Pan-STARRS1 Surveys. *arXiv* **2016**, arXiv:1612.05560.
65. McMahon, R.G.; Banerji, M.; Gonzalez, E.; Koposov, S.E.; Bejar, V.J.; Lodieu, N.; Rebolo, R.; VHS Collaboration. First Scientific Results from the VISTA Hemisphere Survey (VHS). *Messenger* **2013**, *154*, 35–37.
66. Gaia Collaboration; Prusti, T.; de Bruijne, J.H.J.; Brown, A.G.A.; Vallenari, A.; Babusiaux, C.; Bailer-Jones, C.A.L.; Bastian, U.; Biermann, M.; Evans, D.W.; et al. The Gaia mission. *Astron. Astrophys.* **2016**, *595*, A1. [CrossRef]
67. Gaia Collaboration; Brown, A.G.A.; Vallenari, A.; Prusti, T.; de Bruijne, J.H.J.; Babusiaux, C.; Bailer-Jones, C.A.L.; Biermann, M.; Evans, D.W.; Eyer, L.; et al. Gaia Data Release 2. Summary of the contents and survey properties. *Astron. Astrophys.* **2018**, *616*, A1. [CrossRef]
68. Maravelias, G.; Bonanos, A.Z.; Tramper, F.; de Wit, S.; Yang, M.; Bonfini, P. A machine-learning photometric classifier for massive stars in nearby galaxies. I. The method. *Astron. Astrophys.* **2022**, *666*, A122. [CrossRef]
69. Williams, S.J.; Bonanos, A.Z.; Whitmore, B.C.; Prieto, J.L.; Blair, W.P. The infrared massive stellar content of M 83. *Astron. Astrophys.* **2015**, *578*, A100. [CrossRef]
70. Cepa, J.; Aguiar-Gonzalez, M.; Bland-Hawthorn, J.; Castaneda, H.; Cobos, F.J.; Correa, S.; Espejo, C.; Fragoso-Lopez, A.B.; Fuentes, F.J.; Gigante, J.V.; et al. OSIRIS tunable imager and spectrograph for the GTC. Instrument status. In Proceedings of the Instrument Design and Performance for Optical/Infrared Ground-based Telescopes, Waikoloa, HI, USA, 25–28 August 2002; Society of Photo-Optical Instrumentation Engineers (SPIE) Conference Series; Iye, M., Moorwood, A.F.M., Eds.; SPIE: Bellingham, WA, USA, 2003; Volume 4841, pp. 1739–1749. [CrossRef]
71. Rodríguez Espinosa, J.M.; Alvarez, P.; Sánchez, F. The GTC: An Advanced 10m Telescope for the ORM. *Astrophys. Space Sci.* **1998**, *263*, 355–360. [CrossRef]
72. Appenzeller, I.; Fricke, K.; Fürtig, W.; Gässler, W.; Häfner, R.; Harke, R.; Hess, H.J.; Hummel, W.; Jürgens, P.; Kudritzki, R.P.; et al. Successful commissioning of FORS1 - the first optical instrument on the VLT. *Messenger* **1998**, *94*, 1–6.
73. Gómez-González, V.M.A.; Mayya, Y.D.; Rosa-González, D. Wolf-Rayet stars in M81: Detection and characterization using GTC/OSIRIS spectra and HST/ACS images. *Mon. Not. R. Astron. Soc.* **2016**, *460*, 1555–1566. [CrossRef]
74. Freudling, W.; Romaniello, M.; Bramich, D.M.; Ballester, P.; Forchi, V.; García-Dabló, C.E.; Moehler, S.; Neeser, M.J. Automated data reduction workflows for astronomy. The ESO Reflex environment. *Astron. Astrophys.* **2013**, *559*, A96. [CrossRef]

75. Ritchie, B.W.; Clark, J.S.; Negueruela, I.; Najarro, F. Spectroscopic monitoring of the luminous blue variable Westerlund1-243 from 2002 to 2009. *Astron. Astrophys.* **2009**, *507*, 1597–1611. [CrossRef]
76. Aret, A.; Kraus, M.; Šlechta, M. Spectroscopic survey of emission-line stars—I. B[e] stars. *Mon. Not. R. Astron. Soc.* **2016**, *456*, 1424–1437. [CrossRef]
77. Condori, C.A.H.; Borges Fernandes, M.; Kraus, M.; Panoglou, D.; Guerrero, C.A. The study of unclassified B[e] stars and candidates in the Galaxy and Magellanic Clouds†. *Mon. Not. R. Astron. Soc.* **2019**, *488*, 1090–1110. [CrossRef]
78. Britavskiy, N.E.; Bonanos, A.Z.; Mehner, A.; Boyer, M.L.; McQuinn, K.B.W. Identification of dusty massive stars in star-forming dwarf irregular galaxies in the Local Group with mid-IR photometry. *Astron. Astrophys.* **2015**, *584*, A33. [CrossRef]
79. Castro, N.; Herrero, A.; Garcia, M.; Trundle, C.; Bresolin, F.; Gieren, W.; Pietrzyński, G.; Kudritzki, R.P.; Demarco, R. The Araucaria Project: VLT-spectroscopy of blue massive stars in NGC 55. *Astron. Astrophys.* **2008**, *485*, 41–50. [CrossRef]
80. Stasińska, G.; Peña, M.; Bresolin, F.; Tsamis, Y.G. Planetary nebulae and H ii regions in the spiral galaxy NGC 300. Clues on the evolution of abundance gradients and on AGB nucleosynthesis. *Astron. Astrophys.* **2013**, *552*, A12. [CrossRef]
81. Iłkiewicz, K.; Mikołajewska, J. Distinguishing between symbiotic stars and planetary nebulae. *Astron. Astrophys.* **2017**, *606*, A110. [CrossRef]
82. Akras, S.; Gonçalves, D.R.; Alvarez-Candal, A.; Pereira, C.B. Discovery of five new Galactic symbiotic stars in the VPHAS+ survey. *Mon. Not. R. Astron. Soc.* **2021**, *502*, 2513–2517. [CrossRef]
83. Megeath, S.T.; Gutermuth, R.A.; Kounkel, M.A. Low Mass Stars as Tracers of Star and Cluster Formation. *Publ. Astron. Soc. Pac.* **2022**, *134*, 042001. [CrossRef]
84. Mowlavi, N.; Rimoldini, L.; Evans, D.W.; Riello, M.; De Angeli, F.; Palaversa, L.; Audard, M.; Eyer, L.; Garcia-Lario, P.; Gavras, P.; et al. Large-amplitude variables in Gaia Data Release 2. Multi-band variability characterization. *Astron. Astrophys.* **2021**, *648*, A44. [CrossRef]
85. Menzies, J.W.; Whitelock, P.A.; Feast, M.W.; Matsunaga, N. Luminous AGB variables in the dwarf irregular galaxy, NGC 3109. *Mon. Not. R. Astron. Soc.* **2019**, *483*, 5150–5165. [CrossRef]
86. Davidge, T.J. New Blue and Red Variable Stars in NGC 247. *Astron. J.* **2021**, *162*, 152. [CrossRef]
87. van Genderen, A.M. S Doradus variables in the Galaxy and the Magellanic Clouds. *Astron. Astrophys.* **2001**, *366*, 508–531. [CrossRef]
88. Martin, J.C.; Humphreys, R.M. Multi-epoch BVRI Photometry of Luminous Stars in M31 and M33. *Astron. J.* **2017**, *154*, 81. [CrossRef]
89. Porter, A.; Blundell, K.; Podsiadlowski, P.; Lee, S. GG Carinae: Discovery of orbital-phase-dependent 1.583-day periodicities in the B[e] supergiant binary. *Mon. Not. R. Astron. Soc.* **2021**, *503*, 4802–4814. [CrossRef]
90. Massey, P.; McNeill, R.T.; Olsen, K.A.G.; Hodge, P.W.; Blaha, C.; Jacoby, G.H.; Smith, R.C.; Strong, S.B. A Survey of Local Group Galaxies Currently Forming Stars. III. A Search for Luminous Blue Variables and Other Hα Emission-Line Stars. *Astron. J.* **2007**, *134*, 2474–2503. [CrossRef]
91. Williams, S.J.; Bonanos, A.Z. Spitzer mid-infrared point sources in the fields of nearby galaxies. *Astron. Astrophys.* **2016**, *587*, A121. [CrossRef]
92. Clark, J.S.; Castro, N.; Garcia, M.; Herrero, A.; Najarro, F.; Negueruela, I.; Ritchie, B.W.; Smith, K.T. On the nature of candidate luminous blue variables in M 33. *Astron. Astrophys.* **2012**, *541*, A146. [CrossRef]
93. Bonanos, A.Z.; Massa, D.L.; Sewiło, M.; Lennon, D.J.; Panagia, N.; Smith, L.J.; Meixner, M.; Babler, B.L.; Bracker, S.; Meade, M.R.; et al. Spitzer SAGE Infrared Photometry of Massive Stars in the Large Magellanic Cloud. *Astron. J.* **2009**, *138*, 1003–1021. [CrossRef]
94. Bonanos, A.Z.; Lennon, D.J.; Köhlinger, F.; van Loon, J.T.; Massa, D.L.; Sewiło, M.; Evans, C.J.; Panagia, N.; Babler, B.L.; Block, M.; et al. Spitzer SAGE-SMC Infrared Photometry of Massive Stars in the Small Magellanic Cloud. *Astron. J.* **2010**, *140*, 416–429. [CrossRef]
95. Cutri, R.M. VizieR Online Data Catalog: WISE All-Sky Data Release (Cutri+ 2012). *VizieR Online Data Cat.* **2012**, II/311.
96. Cutri, R.M.; Skrutskie, M.F.; van Dyk, S.; Beichman, C.A.; Carpenter, J.M.; Chester, T.; Cambresy, L.; Evans, T.; Fowler, J.; Gizis, J.; et al. VizieR Online Data Catalog: 2MASS All-Sky Catalog of Point Sources (Cutri+ 2003). *VizieR Online Data Cat.* **2003**, II/246.
97. McMahon, R.G.; Banerji, M.; Gonzalez, E.; Koposov, S.E.; Bejar, V.J.; Lodieu, N.; Rebolo, R.; VHS Collaboration. VizieR Online Data Catalog: The VISTA Hemisphere Survey (VHS) catalog DR5 (McMahon+, 2020). *VizieR Online Data Cat.* **2021**, II/367.
98. Gaia Collaboration.; Vallenari, A.; Brown, A.G.A.; Prusti, T.; de Bruijne, J.H.J.; Arenou, F.; Babusiaux, C.; Biermann, M.; Creevey, O.L.; Ducourant, C.; et al. Gaia Data Release 3: Summary of the content and survey properties. *arXiv* **2022**, arXiv:2208.00211.
99. Astropy Collaboration.; Robitaille, T.P.; Tollerud, E.J.; Greenfield, P.; Droettboom, M.; Bray, E.; Aldcroft, T.; Davis, M.; Ginsburg, A.; Price-Whelan, A.M.; et al. Astropy: A community Python package for astronomy. *Astron. Astrophys.* **2013**, *558*, A33. [CrossRef]
100. Astropy Collaboration.; Price-Whelan, A.M.; Sipőcz, B.M.; Günther, H.M.; Lim, P.L.; Crawford, S.M.; Conseil, S.; Shupe, D.L.; Craig, M.W.; Dencheva, N.; et al. The Astropy Project: Building an Open-science Project and Status of the v2.0 Core Package. *Astron. J.* **2018**, *156*, 123. [CrossRef]
101. Astropy Collaboration.; Price-Whelan, A.M.; Lim, P.L.; Earl, N.; Starkman, N.; Bradley, L.; Shupe, D.L.; Patil, A.A.; Corrales, L.; Brasseur, C.E.; et al. The Astropy Project: Sustaining and Growing a Community-oriented Open-source Project and the Latest Major Release (v5.0) of the Core Package. *Astrophys. J.* **2022**, *935*, 167. [CrossRef]

102. Harris, C.R.; Millman, K.J.; van der Walt, S.J.; Gommers, R.; Virtanen, P.; Cournapeau, D.; Wieser, E.; Taylor, J.; Berg, S.; Smith, N.J.; et al. Array programming with NumPy. *Nature* **2020**, *585*, 357–362. [CrossRef] [PubMed]
103. Hunter, J.D. Matplotlib: A 2D graphics environment. *Comput. Sci. Eng.* **2007**, *9*, 90–95. [CrossRef]

Disclaimer/Publisher's Note: The statements, opinions and data contained in all publications are solely those of the individual author(s) and contributor(s) and not of MDPI and/or the editor(s). MDPI and/or the editor(s) disclaim responsibility for any injury to people or property resulting from any ideas, methods, instructions or products referred to in the content.

Article

Dense Molecular Environments of B[e] Supergiants and Yellow Hypergiants

Michaela Kraus [1,*], Michalis Kourniotis [1], María Laura Arias [2,3], Andrea F. Torres [2,3] and Dieter H. Nickeler [1]

1 Astronomical Institute, Czech Academy of Sciences, Fričova 298, 251 65 Ondřejov, Czech Republic
2 Departamento de Espectroscopía, Facultad de Ciencias Astronómicas y Geofísicas, Universidad Nacional de La Plata, Paseo del Bosque S/N, La Plata B1900FWA, Argentina
3 Instituto de Astrofísica de La Plata (CCT La Plata-CONICET, UNLP) Paseo del Bosque S/N, La Plata B1900FWA, Argentina
* Correspondence: michaela.kraus@asu.cas.cz

Abstract: Massive stars expel large amounts of mass during their late evolutionary phases. We aim to unveil the physical conditions within the warm molecular environments of B[e] supergiants (B[e]SGs) and yellow hypergiants (YHGs), which are known to be embedded in circumstellar shells and disks. We present K-band spectra of two B[e]SGs from the Large Magellanic Cloud and four Galactic YHGs. The CO band emission detected from the B[e]SGs LHA 120-S 12 and LHA 120-S 134 suggests that these stars are surrounded by stable rotating molecular rings. The spectra of the YHGs display a rather diverse appearance. The objects 6 Cas and V509 Cas lack any molecular features. The star [FMR2006] 15 displays blue-shifted CO bands in emission, which might be explained by a possible close to pole-on oriented bipolar outflow. In contrast, HD 179821 shows blue-shifted CO bands in absorption. While the star itself is too hot to form molecules in its outer atmosphere, we propose that it might have experienced a recent outburst. We speculate that we currently can only see the approaching part of the expelled matter because the star itself might still block the receding parts of a (possibly) expanding gas shell.

Keywords: stars: massive; stars: supergiants; stars: winds; outflows; circumstellar matter

Citation: Kraus, M.; Kourniotis, M.; Arias, M.L.; Torres, A.F.; Nickeler, D.H. Dense Molecular Environments of B[e] Supergiants and Yellow Hypergiants. *Galaxies* **2023**, *11*, 76. https://doi.org/10.3390/galaxies11030076

Academic Editor: Oleg Malkov

Received: 14 March 2023
Revised: 5 June 2023
Accepted: 12 June 2023
Published: 16 June 2023

Copyright: © 2023 by the authors. Licensee MDPI, Basel, Switzerland. This article is an open access article distributed under the terms and conditions of the Creative Commons Attribution (CC BY) license (https:// creativecommons.org/licenses/by/ 4.0/).

1. Introduction

The evolution of massive stars ($M_{\rm ini} \gtrsim 8\,M_\odot$) bears many uncertainties, which render it difficult to trace such objects from the cradle up to their spectacular explosion as supernova. One major hindrance is the poorly constrained mass loss due to stellar winds that the stars experience along the course of their evolution. Furthermore, the post-main sequence evolution of massive stars encounters phases in which the stars lose a significant amount of mass due to episodically enhanced mass loss or occasional mass eruptions, both of poorly understood origin. The ejected mass can accumulate around the star in rings, shells, or bipolar lobes, as seen in some B- or B[e]-type supergiants (e.g., Sher 25 [1,2], MWC 137 [3,4], SBW1 [5]), yellow hypergiants (IRC+10 420 [6] and Hen 3-1379 [7]), many luminous blue variables [8,9], and Wolf-Rayet stars [10–14].

Two groups of evolved massive stars are particularly interesting. These are the B[e] supergiants (B[e]SGs) and the yellow hypergiants (YHGs). Both types of objects have dense and warm circumstellar environments, and representatives of both classes of objects show (at least occasionally) emission from hot molecular gas.

1.1. B[e] Supergiants

The group of B[e]SGs consists of luminous ($\log(L_*/L_\odot) \geq 4.0$) post-main sequence B-type emission line stars. Besides large-scale ejecta (with sizes of several pc) detected in some B[e]SGs [3,9], all objects have intense winds and are surrounded by massive disks on small scales (up to ∼100 AU) [15–17], giving rise to the specific emission features characterizing stars with the B[e] phenomenon [18]. These disks give shelter to a diversity of molecular and dust species, and the near-infrared (NIR) is an ideal wavelength regime to detect molecular emission features that, when resolved with high spectral resolution, provide insight into the physical properties of the disks and reveal the disk dynamics.

The most commonly observed molecule is CO. The emission from its first-overtone bands arises in the K-band around 2.3 μm and has been detected in about 50% of the B[e]SGs [19]. Besides the main isotope ^{12}CO, emission from ^{13}CO is seen in considerable amounts [20,21], confirming that the matter from which the disks have formed contains processed material that must have been released from the stellar surface [22].

Emission from the first-overtone bands of SiO, arising in the L-band around 4 μm, has been reported for some Galactic B[e]SGs [23]. SiO has a lower binding energy than CO. It thus forms at distances farther away from the star than CO. The individual ro-vibrational lines in both CO and SiO are kinematically broadened with a double-peaked profile, and (quasi-)Keplerian rotation of the molecular gas around the central object has been suggested as the most likely explanation to interpret the spectral appearance [17,23–28].

Observations of B[e]SGs in the NIR are sparse. But persistent CO band emission over years and decades has been detected in numerous objects [21,29,30] and has been used as one of the criteria to identify and classify stars as B[e]SGs in Local Group galaxies [31,32]. However, in a few cases, considerable variability in these emission features has been reported as well. The most striking object is certainly the B[e]SG star LHA 115-S 65 in the Small Magellanic Cloud (SMC), for which a sudden appearance of CO band emission has been recorded [33]. The disk around this object is seen edge-on, and in addition to its rotation around the central object, it also drifts outwards, very slowly and with a velocity decreasing with distance from the star and reaching about zero [34]. This slowdown might have resulted in a build-up of density in regions favorable for molecule condensation and for excitation of the CO bands.

Furthermore, LHA 115-S 18 in the SMC showed no CO band emission back in 1987/1989 [35], whereas follow-up observations in November 1995 taken with a more than three times higher resolution displayed intense CO bands [36], which were also seen in the observations acquired in October 2009 [21].

In the spectrum of the Galactic B[e]SG MWC 349, intense CO band emission appeared in the early 1980s [37]. It was still observable in 2013, but by then the CO gas had clearly cooled and the emission intensity had significantly decreased, which has been interpreted as due to expansion and dilution of the circumstellar disk [38]. Two more objects in the Large Magellanic Cloud (LMC) displayed indications of CO band variability, most likely related to inhomogeneities within the distribution of the molecular gas around the central star. These are LHA 120-S 73 [24] and LHA 120-S 35 [27].

In the optical range, indications for emission from TiO molecular bands have been found in six B[e]SGs [24,27,39,40]. All six objects reside in the Magellanic Clouds, and five of them also have CO band emission[1]. No Galactic B[e]SG has been reported to date to display TiO band emission [19].

1.2. Yellow Hypergiants

With temperatures in the range $T_{\rm eff} \simeq 4000$–8000 K and luminosities $\log(L/L_\odot)$ spreading from 5.2 to 5.8, the YHGs populate a rather narrow domain in the Hertzsprung-Russel (HR) diagram. The stars are in their post-red supergiant (post-RSG) evolutionary phase [41], and their luminosities place them on evolutionary tracks of stars with initial masses in the range $M_{\rm ini} \simeq 20$–$40\,M_\odot$. Evolutionary calculations of (rotating) stars in this mass range have shown that these objects can indeed evolve back to the blue, hot side of the HR diagram [42], whereas stars with lower initial mass just reach the RSG stage before they explode as SNe of type II-P. Support for this theoretical scenario is provided by the lack of high-mass ($M_{\rm ini} \geq 18\,M_\odot$, i.e., with luminosities $\log L/L_\odot > 5.1$) RSG progenitors for this type of supernovae [43,44].

As post-RSGs, the YHGs might be expected to be embedded in envelopes, remnants of the previous mass-losing activities during the RSG stage. However, surprisingly, so far, only about half of the YHGs have been reported to have a dusty and/or cold molecular envelope. These are the Galactic objects IRC +10420 [45–47], HR 5171A [48], HD 179821 [49] and Hen 3-1379 [7,50], three YHGs in the LMC (HD 269953, HD 269723, HD 268757, [51]), as well as Var A [52] and three more YHG candidates in the galaxy M33 [53].

A typical classification characteristic of YHGs is the occurrence of outbursts that can be clearly discriminated from the more regular (cyclic) brightness variability due to stellar pulsations. During such an outburst event, the star inflates, its brightness drastically decreases, and the object seems to undergo a red loop evolution in the HR diagram. Molecules such as TiO and CO can form in the cool, outer atmospheric layers, leading to intense absorption structures in the optical and NIR, respectively, and the object's entire spectral appearance resembles that of a much later spectral type. The outbursts are most likely connected with enhanced mass loss or mass eruptions from the star, which might be connected to non-linear instabilities such as finite-time singularities or blow-ups typically occurring in fluid dynamics [54] or to strange-mode instabilities [55,56] as recent computations propose [57]. The duration of the outbursts can range from months to decades before the star appears back at its real location in the HR diagram.

The bona-fide YHG, ρ Cas, experienced four documented outbursts during the past 80 years with variable duration (from weeks up to three years) and amplitude (0.29 to 1.69 mag), connected with significant changes in its spectral appearance [58–61], whereas Var A in M33 presumably underwent an eruption around 1950 that lasted $\sim$45 years [52]. The object V509 Cas experienced mass-loss events in the seventies, during which the star's apparent temperature decreased significantly. Since then, the star has displayed a steady increase in its effective temperature from $\sim$5000 K in 1973 to $\sim$8000 K in 2001 [62], and since then it has stabilized at that temperature [63]. Furthermore, IRC +10420 changed its spectral type from F8 to mid- to early A, connected with an increase in temperature over a period of about 30 years with an average rate of $\sim$120 K per year [64,65]. The light curves of other YHG candidates also display outburst activity in connection with variable mass loss [51,53]. However, in many cases, the mass-loss episodes appear to be short, so that the released material expands and dilutes without creating detectable large-scale circumstellar envelopes [66].

Nevertheless, many YHGs are surrounded (or were for some period in the past) by hot molecular gas traced by first-overtone CO band emission, suggesting that the objects are embedded in a dense and hot environment. Whether the molecular gas is arranged in a ring revolving around the object, as in the case of the B[e]SGs, is currently not known. However, the CO band spectral features in YHGs seem to be much more variable than in their hotter B[e]SG counterparts, especially because they often appear superimposed on photospheric CO band absorption that forms during the expansion and cooling periods of the long-term pulsation cycles, especially of the cooler YHGs, or during outburst events. One such candidate with cyclic CO band variability is ρ Cas. During its pulsation cycles, CO band emission appears when the star is hottest (maximum brightness) and most compact, whereas CO bands are seen in absorption when the star is coolest (minimum brightness) and most inflated [67]. It has been speculated that the appearance of CO bands in emission might be related to propagating pulsation-driven shock waves in the outer atmosphere of the star [67]. An alternative scenario would also be conceivable, in which the CO emission could be permanent, arising from a circumstellar ring or shell and being detectable only during phases in which no photospheric absorption compensates the emission [60]. Support for the latter scenario is provided by the fact that the CO emission features remain at constant radial velocities (along with other emission lines formed in the circumstellar environment such as [Ca II] and Fe I, [60]) whereas the absorption components change from red- to blue-shifted, in phase with the pulsation cycle.

The object ρ Cas is, to date, the best monitored YHG in the NIR. For many other YHGs observations in the K-band have been taken only sporadically. Hence, not much can be said or concluded about the variability rhythm of their CO band features. Occasional CO band emission has been reported for the galactic objects HD 179821 [68,69], V509 Cas [70], [FMR2006] 15 [71], and the two LMC objects HD 269723 and HD 269953 [21,29]. The latter even displays emission from hot water vapor and is, so far, the only evolved massive star with such emission features from its environment [72].

1.3. Motivation and Aims

The appearance of CO band emission in the NIR spectra of B[e]SGs and YHGs suggests that similar physical conditions may prevail in the circumstellar environments of both groups of objects. In depth studies of these conditions are rare though.

For the B[e]SGs in the Magellanic Clouds CO, column densities, temperatures, and ^{12}CO/^{13}CO isotopic ratios were determined based on medium-resolution K-band spectra [21]. The resolution of these spectra was, however, too low to derive the gas kinematics with high confidence. On the other hand, high-resolution K-band spectra of the Galactic B[e]SG sample allowed to obtain the CO dynamics [17,73], whereas in most cases the spectral coverage was too short to infer about the density and temperature of their hot molecular environment.

The situation for the YHGs is even worse. Besides for ρ Cas [67] and for the LMC object HD 269953 [21,72] no attempts have been undertaken so far to study their warm molecular environments in more detail and to derive the parameters of the CO band emitting regions.

Therefore, we started to systematically observe both the B[e]SGs and the YHGs in the Milky Way and Magellanic Clouds to fill this knowledge gap. A further motivation for our research is provided by the location of the B[e]SGs with CO band emission in the HR diagram. As has been mentioned previously for the Magellanic Clouds sample [19], these objects cluster around luminosity values $\log L/L_\odot \simeq$ 5.0–5.9, whereas B[e]SGs that are more luminous than $\sim$5.9 or less luminous than $\sim$5.0 do not show CO band emission. The same holds when inspecting the Galactic B[e]SGs. In the left panel of Figure 1 we depict the LMC objects, whereas in the right panel we display the Galactic sample[2] from [17,38]. The YHGs in the LMC [51] and the Milky Way [75] have been added to the plots[3]. Their positions with maximum and minimum effective temperature values, as reported in the literature, have been connected with dashed lines. Errors in temperature and luminosity are indicated when provided by the corresponding studies.

Interestingly, the YHGs and the B[e]SGs with CO band emission share similar evolutionary tracks, which is particularly evident for the Galactic objects. This raises the question of whether evolved stars in this particular mass range suffer from specific instabilities in the blue and yellow temperature regimes independent of their evolutionary state. Such instabilities need to have the potential to drive mass ejections or eruptions, and the released mass would have to be dense and cool enough to create the required conditions for the formation of significant amounts of molecules generating intense band emission.

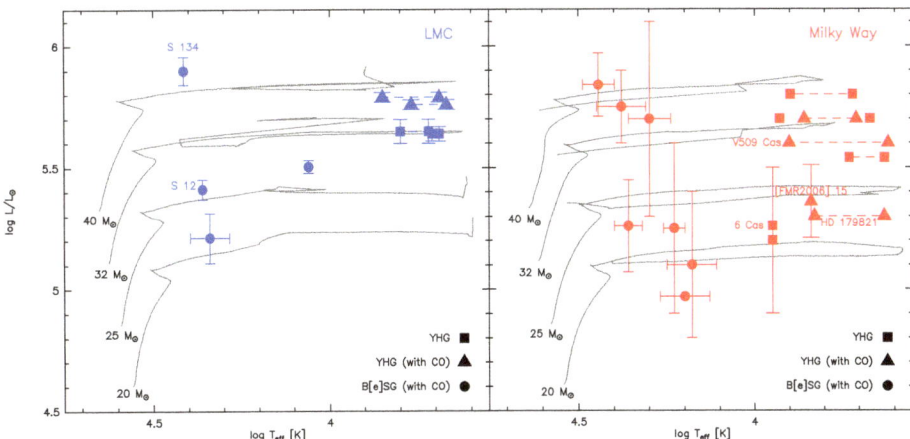

Figure 1. HR diagram with evolutionary tracks for LMC ($Z = 0.006$, [76], left panel) and solar ($Z = 0.014$, [42], right panel) metallicity for models of rotating stars ($v/v_{\mathrm{crit}} = 0.4$) with initial masses from 20–40 $M_\odot$. Shown are the positions of LMC (blue symbols) and Galactic (red symbols) objects. Only B[e]SGs with hot circumstellar molecular CO gas are shown. These populate similar evolutionary tracks as the YHGs. The minimum and maximum temperature values (where known) of the YHGs are connected by dashed lines. YHGs with reported (at least once) CO band emission are shown with triangles. The stars of the current study are labeled.

1.4. Selection of Targets

For our current study, we have selected two B[e]SGs from the LMC. These are the objects LHA 120-S 12 and LHA 120-S 134. Both are known to display CO band emission (see, e.g., [21]), but both of them lack high-resolution K-band spectra to derive the CO kinematics. Furthermore, we have selected four Galactic YHGs, [FMR2006] 15, HD 179821, V509 Cas, and 6 Cas. Of these, only the former three objects were reported in the literature to display (at some epochs) CO band emission. The basic stellar parameters of all objects, as obtained from the literature, are listed in Table 1.

Table 1. Stellar parameters. Errors are given where available.

Object	G [mag]	K_s [mag]	log T_{eff} [K]	log $L/L_\odot$	Ref.	d [kpc]	Ref.
LHA 120-S 12	12.4	10.2	4.36	5.34 ± 0.04	[15]	49.6 ± 0.5	[77]
LHA 120-S 134	11.4	8.6	4.41	5.90 ± 0.06	[15]	49.6 ± 0.5	[77]
[FMR2006] 15	19.5	6.7	3.84	5.36 ± 0.15	[71]	6.6 ± 0.9	[71]
6 Cas	—	3.4	3.93	5.13	[78]	2.8 ± 0.3	[79]
V509 Cas	5.0	1.7	3.90	5.60	[63,80]	1.4 ± 0.5	[62] [a]
HD 179821	7.5	4.7	3.83	5.30	[75,81]	5.3 ± 0.3	[81,82]

Note: The G and K_s-band magnitudes are from *GAIA* Early Data Release 3 [82] and from the 2MASS point source catalog [83], respectively. The YHG effective temperatures refer to the hot state. No *GAIA* G-band measurement is available for 6 Cas. [a] Listed distance is based on the *HIPPARCOS* parallax of 0.73 ± 0.25. The new *GAIA* Early Data Release 3 parallax of 0.2507 ± 0.0633 [82] places the object at a distance of ~4 kpc.

2. Observations and Data Reduction

High-resolution spectra ($R \sim 50{,}000$) of the two B[e]SGs were acquired with the visitor spectrograph Phoenix [84] mounted at GEMINI-South. The spectra were taken on 20 December 2004 and 30 November 2017 under program IDs GS-2004B-Q-54 and GS-2017B-Q-32. The observations were carried out in the K-band with two different filters,

K4396 and K4308. The central wavelength was chosen such that the wavelength ranges cover the first and second band heads of the first-overtone CO band emission.

Medium-resolution K-band spectra of the YHGs have been acquired with the Gemini Near-InfraRed Spectrograph (GNIRS, [85,86]) at GEMINI-North under Program IDs GN-2019A-Q-204, GN-2019B-Q-418, and GN-2021A-Q-315.

The spectrum of [FMR2006] 15 was observed on 12 May 2019 centered on $\lambda = 2.35$ µm. The instrument configuration was a short camera (0.15″ per pixel) with the 0.3″ slit and the 111 l mm^{-1} grating, resulting in a resolving power of $R \sim 5900$.

V509 Cas and 6 Cas were observed on 21 December 2019 with the instrumental configuration: Long camera (0.05″ per pixel) with the 0.10″ slit and the 321 mm^{-1} grating which provides a resolving power of $R \sim 5100$. The observations were centered on $\lambda = 2.35$ µm.

HD 179821 was observed on 7 April 2021 with two different central wavelengths $\lambda = 2.14$ µm and 2.33 µm and with the following instrument configuration: a short camera (0.15″ per pixel), the 0.3″ slit, and the 111 l mm^{-1} grating, resulting in a resolving power of $R \sim 5900$.

For all objects, a telluric standard star (usually a late B-type main sequence star) was observed close in time and airmass. For optimal sky subtraction, the star was positioned at two different locations along the slit (A and B), and the observations were carried out in ABBA cycles. Data reduction and telluric correction were performed using standard IRAF[4] tasks. The reduction steps consist of subtraction of AB pairs, flat-fielding, wavelength calibration (using the telluric lines), and telluric correction. The observing log is given in Table 2, where we list the star name, object class, observing date (UT), used instrument, covered wavelength range, spectral resolution R, and resulting signal-to-noise ratio (SNR).

Table 2. Observation log.

Object	Class	Obs Date [yyyy-mm-dd]	Instrument	$\lambda_{\min}$–$\lambda_{\max}$ [µm]	R	SNR
LHA 120-S 12	B[e]SG	2004-12-20	Phoenix	2.291–2.300	50,000	35
		2017-11-30	Phoenix	2.319–2.330	50,000	20
LHA 120-S 134	B[e]SG	2017-11-30	Phoenix	2.290–2.299	50,000	40
		2017-11-30	Phoenix	2.319–2.330	50,000	100
[FMR2006] 15	YHG	2019-05-12	GNIRS	2.257–2.440	5900	300
6 Cas	YHG	2019-12-21	GNIRS	2.232–2.453	5100	100
V509 Cas	YHG	2019-12-21	GNIRS	2.233–2.453	5100	140
HD 179821	YHG	2021-04-07	GNIRS	2.046–2.424	5900	250

3. Results

3.1. Description of the Spectra

CO band head emission is detected in both B[e]SGs (black lines in Figure 2) despite the low quality of some of the spectral pieces and some telluric remnants in the red parts of the short wavelength portions (left lower panels). The spectrum of LHA 120-S 134 contains additional emission lines from the hydrogen Pfund series. Intense emission in these recombination lines has already been reported from that star based on medium-resolution spectra [21]. No contribution from the Pfund lines is seen in the spectra of LHA 120-S 12.

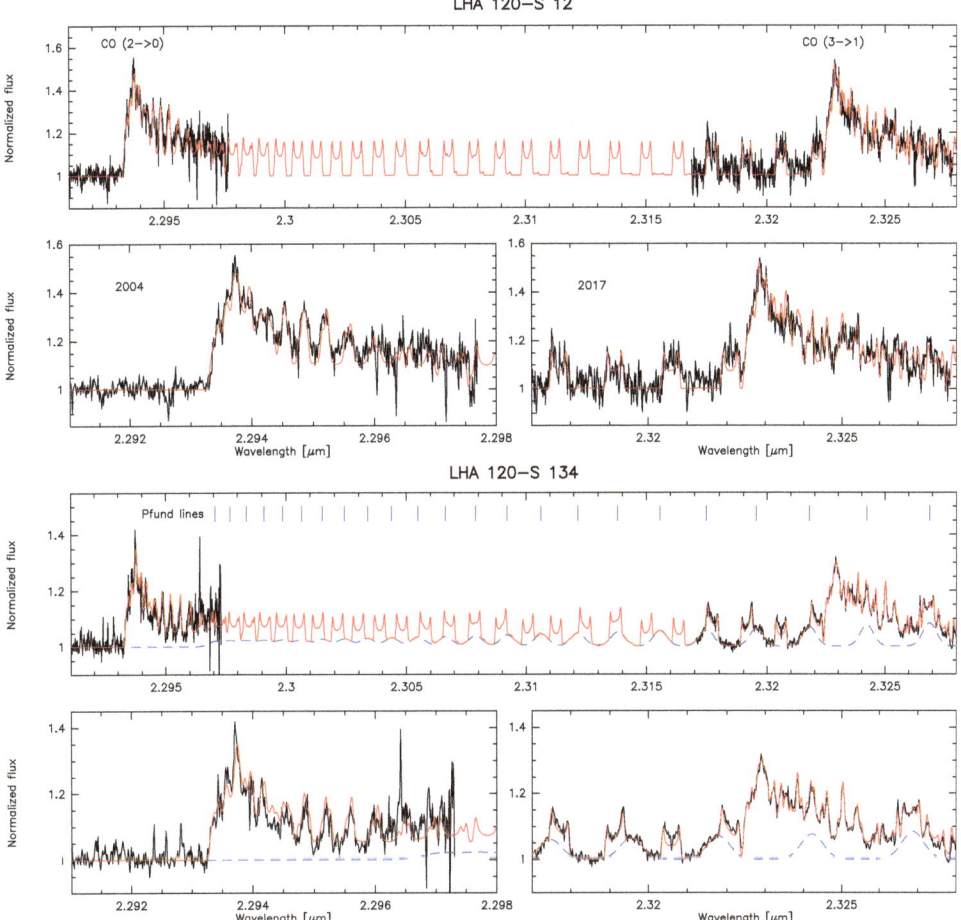

Figure 2. Best fitting model (red) to the normalized Phoenix spectra (black) of LHA 120-S 12 (**top**) and LHA 120-S 134 (**bottom**). For each star we display the entire fit to the total spectrum (top panels) and the zoom to the band heads (bottom panels). Emission from the Pfund series (blue dashed line), detected in the spectrum of LHA 120-S 134, is included in the total fit.

The K-band spectra of the YHGs display a diverse appearance, as depicted in Figure 3. Two stars possess just Pfund lines in absorption (V509 Cas and 6 Cas) and an otherwise featureless spectrum, in agreement with their high effective temperature ($T_{\rm eff} \geq 8000$ K, see Table 1). The star [FMR2006] 15 shows CO band emission overimposed on the atmospheric spectrum of a presumably late-type star. In HD 179821 the CO bands and the Br γ line are in absorption along with numerous other photospheric lines, whereas the Na I $\lambda\lambda$2.206,2.209 doublet shows prominent emission.

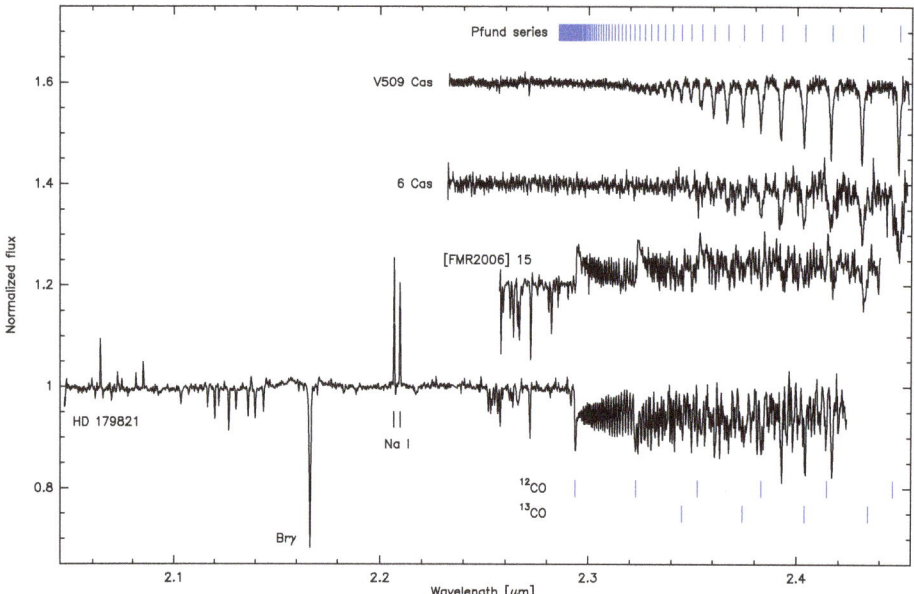

Figure 3. Normalized medium-resolution *K*-band spectra of the YHGs taken with GNIRS. For better visualization, the spectra have been offset along the flux axis. Positions of the CO band heads and of the lines from the Pfund series are marked by ticks. The lines of Brγ and of the Na I doublet are labeled as well.

3.2. Modeling of the CO Band and Pfund Line Emission

We model the CO emission using our molecular disk code [87] that has been developed to compute the ro-vibrational bands from a rotating ring (or disk) of circumstellar gas under local thermodynamic equilibrium conditions. The calculations are carried out for the two main isotopes, ^{12}CO and ^{13}CO [21,22].

The high-resolution spectra of the two B[e]SGs display kinematic broadening of the individual ro-vibrational lines in the form of a double-peaked profile (Figure 2). Such a profile can be interpreted either as rotation around the central object or as an equatorial outflow. Since B[e]SGs are known to be surrounded by (quasi-)Keplerian rotating disks, the assumption of rotation as the most likely broadening mechanism seems to be justified. The situation for the YHG star [FMR2006] 15 is less clear. The medium-resolution spectrum provides no clear hint about a possible double-peaked shape of the individual lines. Therefore, we can only derive an upper limit for a possible rotational (or outflow) contribution to the total dynamics.

Due to the high sensitivity of the CO band intensity to the gas temperature and column density, the observed emission traces the hottest and densest molecular regions. Therefore, it is usually sufficient to consider a single ring of gas with constant density and temperature, reducing the number of free parameters to the column density N_{CO}, temperature T_{CO}, the isotope ratio ^{12}CO/^{13}CO, and the gas kinematics split into contributions of the rotation velocity projected to the line of sight $v_{\rm rot,los}$, and of a combined thermal and turbulent velocity in the form of a Gaussian component $v_{\rm Gauss}$.

The best-fitting CO parameters obtained for the three objects are listed in Table 3, and the total emission spectra are included (in red) in Figure 2 for the two B[e]SGs. It is noteworthy that the two CO band heads of LHA 120-S 12 can be reproduced fairly well with the same model parameters, despite the fact that the spectral pieces have been observed 13 years apart. This implies that the ring of CO gas around LHA 120-S 12 is stable on longer timescales.

Table 3. Best fitting CO parameters.

Object	T_{CO} [K]	N_{CO} [$\times 10^{21}$ cm^{-2}]	$v_{rot,los}$ [km s^{-1}]	v_{Gauss} [km s^{-1}]	$^{12}CO/^{13}CO$
LHA 120-S 12	2800 ± 200	1.5 ± 0.2	27 ± 2	3 ± 1	20 ± 2
LHA 120-S 134	2300 ± 100	2.0 ± 0.1	30 ± 1	1.5 ± 0.5	15 ± 2
[FMR2006] 15	3000 ± 200	2.0 ± 0.2	20 ± 5	15 ± 5	4 ± 2
[FMR2006] 15	3000 ± 200	2.0 ± 0.2	0	20 ± 5	4 ± 2

Note: The $^{12}CO/^{13}CO$ values for LHA 120-S 12 and LHA 120-S 134 have been derived by [20,21], respectively. Our Phoenix spectra do not reach the wavelength region of the ^{13}CO bands.

The spectrum of LHA 120-S 134 displays emission from the hydrogen Pfund line series superimposed on the CO band spectrum[5]. These Pfund lines appear to be broad with no indication of a double-peaked profile shape. To include the contribution of these lines to the total emission spectrum, we apply our code developed for the computation of the hydrogen series according to Menzel case B recombination, assuming that the lines are optically thin [87]. We fix the electron temperature at 10,000 K, which is a reasonable value for ionized gas around an OB supergiant star and, using a Gaussian profile, we obtain a velocity of 53 ± 3 km s^{-1}. Similar velocity values for the lines from the Pfund series have been found for various B[e]SGs (see [20,21,26]). These rather low values compared with the wind velocities of classical B supergiants might suggest that the Pfund lines form in a wind emanating from the surface of the ionized part of the circumstellar disk. The electron density can be derived from the maximum number of the Pfund series visible in the spectrum. For LHA 120-S 134, this number corresponds to the line Pf(57), resulting in an electron density of $(5.8 \pm 0.5) \times 10^{12}$ cm^{-3} within the Pfund line forming region. Having the parameters for the Pfund emission fixed, we compute the contribution of the Pfund series to the total emission spectrum of LHA 120-S 134. This contribution is shown in blue in Figure 2.

The best-fitting CO model for the YHG [FMR2006] 15 is depicted in Figure 4 (top). It should be noted that the contribution of the rotation (respectively outflow) component should be considered an upper limit. A double-peaked profile corresponding to such a velocity might be hidden within the CO band structure. Only high-resolution observations will tell us whether such a profile component is really included. We found that a model omitting this velocity component and using instead just a single Gaussian profile (added to Table 3) results in a similar but slightly less satisfactory fit because the intensity of many individual ro-vibrational lines is overestimated in the short wavelength domain (Figure 4, bottom).

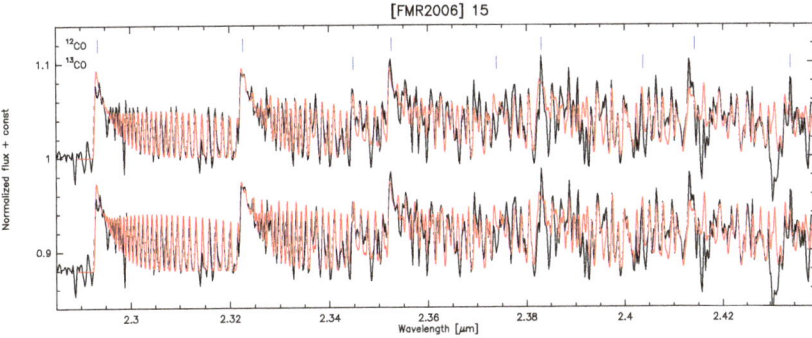

Figure 4. Best fitting model (red) to the observed (black) K-band spectrum of [FMR2006] 15 for the model including a rotational component (**top**) and a pure Gaussian broadening (**bottom**).

The rotation velocity of the CO gas allows us to estimate the distances of the CO-emitting rings from the central object. For this, we need to know the current stellar masses. Considering the stellar effective temperatures, luminosities (from Table 1), and the carbon isotope ratios (from Table 3), we search for the best matching evolutionary models for each of our targets utilizing the stellar evolution track interpolation tool SYCLIST[6]. To reproduce the observed carbon isotope ratios, we vary the initial stellar rotation rate between zero and the maximum offered value of 0.4 and select those evolutionary tracks that fit both the stellar location in the HRD and the carbon isotope ratio. The metallicities used are 0.006 for the LMC objects and 0.014 for the Galactic star. Table 4 lists the stellar radii of our targets along with the most likely initial masses, initial rotation rate, and the resulting current masses. We note that according to these evolutionary tracks, the B[e]SGs seem to have evolved just off the main sequence, whereas the low carbon isotope ratio measured in [FMR2006] 15 clearly places this object on the post-RSG path. For the reason that no proper values for the disk inclination angles are known, we provide the distances of the CO-emitting rings as a function of the inclination angle. These are lower limits to the real distances. We refrain from adding a distance of the CO emitting region for [FMR2006] 15 because of the speculative nature of its rotation component.

Table 4. Estimated current stellar masses and distances of the CO emitting rings from the star.

Object	R_* [$R_\odot$]	M_{in} [$M_\odot$]	v/v_{crit} [km s^{-1}]	M_* [$M_\odot$]	$r_{CO}/(\sin i)^2$ [cm]	$r_{CO}/(\sin i)^2$ [R_*]
LHA 120-S 12	30	26	0.25	25	4.6×10^{14}	218.0
LHA 120-S 134	44	45	0.36	40	5.9×10^{14}	192.7
[FMR2006] 15	333	25	0.40	13	—	—

4. Discussion

We have detected CO band emission in three objects and CO band absorption in one object in our sample. To assess a possible CO variability, we summarize in Table 5 information about previous detections of CO band features in the K-band from the literature, including our results.

The number of observations in the K-band over the last 3–4 decades is sparse for all objects, but the two B[e]SGs continuously display emission over a time interval of 32 years. The diversity in spectral resolution used for the observations makes it difficult to compare the shape and intensity of the emission bands and, hence, to judge variability. Only for LHA 120-S 12, we can say that no significant variability seems to have taken place between our observations acquired in 2004 and in 2017.

For the YHGs, the situation is different. Three of them clearly show variability in their CO bands, ranging from emission over complete disappearance to absorption (or a combination of emission and absorption). The changes are on time scales ranging from months to years. The exception is 6 Cas, which has not been reported previously to have CO bands, and our own observations taken 25 years later also lack any CO band features.

In the following, we briefly describe the known characteristics of each of our targets with respect to stellar variability and the properties of the circumstellar environments in order to incorporate our new observational results.

Table 5. CO band detection in the K-band and CO variability.

Object	Obs Date	R	CO Bands	Ref.
LHA 120-S 12	1985-12-28	450	em	[29]
	2004-12-17	50,000	em	TW
	2009-10-14	4500	em	[20,21]
	2017-11-30	50,000	em	TW
LHA 120-S 134	1985-12-29	450	em	[29]
	2009-11-10	4500	em	[21]
	2017-11-30	50,000	em	TW
[FMR2006] 15	2005-09-15	1000	weak abs	[88]
	2006-05-05	17,000	em	[71]
	2006-08-12	17,000	none	[71]
	2019-05-12	5900	em	TW
6 Cas	1996-08-31	1800	none	[89]
	2019-12-21	5100	none	TW
V509 Cas	1979-1980	32,000	em + abs (variable)	[70]
	1988	N.A.	none	[67]
	2003-11-20	300	none	[67]
	2004-10-30	300	none	[67]
	2019-12-21	5100	none	TW
HD 179821	1989-07-14	1600	em	[69]
	1990-09-26	760	em	[68]
	1991-11-04	330	em	[68]
	1992 August	N.A.	none	[69]
	1997-04-19	1800	none	[89]
	2000-10-18	N.A.	none	[67]
	2021-04-07	5900	abs	TW

Note: em = emission; abs = absorption; TW = this work; N.A. = no information available.

LHA 120-S 12 (=SK −67 23)

The object LHA 120-S 12 was first mentioned in the catalog of Hα-emission stars in the Magellanic Clouds [90]. Follow-up observations recorded an intense IR excess due to hot dust [91]. The two-component wind associated with the star led to its classification as B[e]SG [15], and the high degree of intrinsic polarization suggested a high, but not fully edge-on, viewing angle towards the object and its dusty disk [92]. Shell-line profiles were seen in the NIR, consistent with the high inclination angle of the system [21]. Observations with the *Spitzer Space Telescope* revealed silicate dust within the circumstellar disk, based on weak features in the star's mid-IR spectrum, but no IR nebulosity has been detected in association with the object [93].

The first detection of CO band emission from LHA 120-S 12 was in 1985 based on a low-resolution K-band spectrum [29]. This spectrum clearly depicted four first-overtone band heads, similar to the observations collected in 2009. Modeling of the latter [20] resulted in similar parameters for the CO temperature and column density to those we found from the high-resolution spectra taken before (2004) and afterwards (2017). Moreover, we see no changes in the rotation velocity, projected to the line of sight, within the time span of 13 years between our two observations. Therefore, we believe that the CO-emitting ring revolving around LHA 120-S 12 is rather stable, with no detectable outflow or inflow.

A similar (projected) rotation velocity to that for CO can be inferred from the double-peaked profiles of the [CaII] lines resolved in high-resolution optical spectra of LHA 120-S 12 [94], although a Gaussian component with a higher value might be necessary to smooth out the sharp synthetic double-peaked rotation profile in order to reproduce the shape of the observed forbidden lines.

LHA 120-S 134 (=HD 38489, SK −69 259, MWC 126)

The object LHA 120-S 134 was listed in the catalog of B and A-type stars with bright hydrogen lines published in 1933 [95] under the Mount Wilson Catalogue (MWC) number 126 with classification as Beq. Its hybrid spectra with narrow (equivalent widths of 30–50 km s^{-1}) emission lines of neutral and low-ionized metals in the optical spectral range [15] and very broad P Cygni profiles (implying a wind terminal velocity of ∼2300 km s^{-1}) of high-ionized metals in the ultraviolet [96], along with the detected intense IR excess emission characteristic of hot circumstellar dust [91], resulted in the classification of the star as B[e]SG. A small inclination angle has been proposed for LHA 120-S 134 [15], and the relatively low measured degree of polarization supports a close to pole-on orientation of the star plus disk system [92].

LHA 120-S 134 is one of only two B[e]SGs showing broad emission from He II λ4686 [15]. The other object is LHA 115-S 18 in the SMC, which displays this (time variable) He II line in emission in concert with Raman-scattered O VI emission and TiO molecular emission features [40]. While for LHA 115-S 18, it was proposed that the peculiar spectral characteristics might point towards an LBV-like status of the star [40], a possible Wolf-Rayet companion was postulated to explain the spectral features of LHA 120-S 134 [97]. Although a solid proof for a Wolf-Rayet companion is still missing, LHA 120-S 134 has appeared since then in the catalog of LMC Wolf-Rayet stars (WR 146).

The mid-IR spectrum of LHA 120-S 134, obtained with the *Spitzer Space Telescope*, shows intense 10 μm and weak 20 μm emission features of amorphous silicate dust, and a faint and wispy nebulosity around the IR bright star was found with the telescope's imaging facilities [93]. Follow-up optical imaging revealed that LHA 120-S 134 is located on the northeast rim of the superbubble of DEM L269 and on the western rim of the H II region SGS LMC-2 [98]. Therefore, it is unclear if and how much of the optical and IR nebulosity might be related to LHA 120-S 134 itself.

In the NIR regime, the first mention of CO first-overtone emission dates back to 1985, when the star was observed with low-resolution [29]. The next *K*-band spectrum was taken only 24 years later (see Table 5). The new spectrum had higher spectral resolution, but the CO bands appeared to be similar to the previous detection. Our new, high-resolution spectrum was acquired after another time gap of 8 years. Our modeling of the band spectrum revealed basically the same CO parameters (temperature and column density) as in 2009, with one addition, the projected rotational velocity. As for LHA 120-S 12, this velocity is comparable to the one that might be inferred from the [Ca II] lines [94], and we may conclude that also LHA 120-S 134 is surrounded by a stable, rotating ring of atomic and molecular gas.

[FMR2006] 15 (=2MASS J18375778-0652320)

The star [FMR2006] 15 was recorded as object number 15 in a survey of the cool supergiant population of the massive young star cluster RSGC1 [88]. Based on the weak CO absorption detected in its *K*-band spectrum, a spectral type of G6 I was allocated. In a follow-up investigation, it was proposed that [FMR2006] 15 is most likely a YHG, based on the star's luminosity and spectral similarity to ρ Cas [71]. The star was assigned an effective temperature of 6850 ± 350 K and a luminosity of $\log L/L_\odot = 5.36^{+0.14}_{-0.16}$. With this temperature, the spectral type of [FMR2006] 15 is more likely G0 (± 2 subtypes), and the luminosity implies that the star lies with about $M_\mathrm{ini} \sim 25\,M_\odot$ in the lower mass range of stars developing into YHGs (see Figure 1). This relatively low initial mass was considered to be the reason for the lack of detectable intense IR excess emission and of maser emission in contrast to the high-mass YHGs such as, e.g., IRC +10420 [71].

The first low-resolution *K*-band spectrum of [FMR2006] 15 was taken in September 2005. At that time, very weak CO band absorption was seen [88]. Soon thereafter, the object was re-observed twice with considerably higher spectral resolution. In May 2006, the spectrum displayed CO band emission and in August 2006, the emission had disappeared [71]. Our detection of intense CO band emission 13 years later is a clear indication that [FMR2006] 15 is embedded in circumstellar matter, even if the molecular gas is with 3000 K much too hot for the condensation of dust grains. When modeling the CO bands from [FMR2006] 15, we noticed that the CO emission displays a blue-shift of $-77\,\mathrm{km\,s^{-1}}$ with respect to the atmospheric absorption line spectrum. A similar behavior has been seen in IRC +10420. This star displays blue-shifted emission in its IR hydrogen recombination lines [99] and in its CO rotational transitions [45], which has been interpreted as emission formed in a close to pole-on seen bipolar outflow with the central star eclipsing the receding part of the emission. The outflow velocity from IRC +10420 has been measured to be on the order of $\sim 40\,\mathrm{km\,s^{-1}}$, which is about half the value we measure for [FMR2006] 15. To date, nothing is known about a possible orientation of [FMR2006] 15, but we might speculate that a similar scenario as for IRC +10420 might hold for that object as well. In such a case, we would expect to have a Gaussian-like distribution of the velocity (see bottom model in Figure 4).

On the other hand, if we consider the contribution from the rotation as real, the blue-shifted CO gas might revolve around a hidden companion, as in the case of the YHG HD 269953 in the LMC [51]. In this object, the companion was proposed to be surrounded by a gaseous disk traced by numerous emission lines that display a time-variable radial velocity offset with respect to the photospheric lines of the YHG star. If [FMR2006] 15 is indeed a binary system, time-resolved K-band observations would be essential to derive the orbital parameters and to characterize the hidden companion.

The best-fitting CO model has been subtracted from the K-band spectrum of [FMR2006] 15, and the residual is shown in Figure 5. Also included in this plot is a synthetic spectrum of a cool supergiant star with $T_\mathrm{eff} = 6800\,\mathrm{K}$ and $\log g = 1.5$, similar to the values derived for [FMR2006] 15 [71]. This spectrum has been computed with the spectrum synthesis code Turbospectrum using MARCS atmospheric models [100]. We note that the main photospheric features are decently well represented by this model. The short wavelength coverage of our spectrum and a possible alteration of the intrinsic stellar spectrum caused by the absorbing circumstellar gas impede a more decent classification of the object.

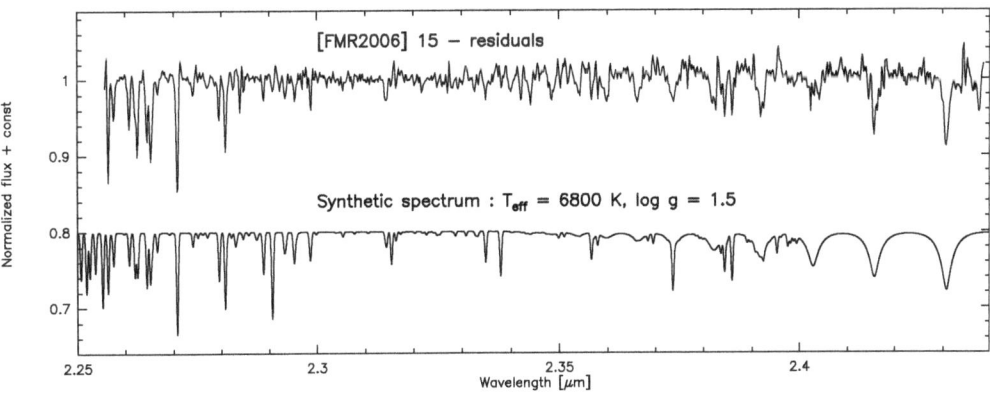

Figure 5. Residual spectrum of [FMR2006] 15 after subtraction of the blue-shifted CO band emission. For illustration purposes, a synthetic model spectrum of a cool supergiant is shown as well (shifted down along the flux axis for better visualization) with parameters similar to those determined for [FMR2006] 15.

6 Cas (=V566 Cas, HR 9018, HD 223385)

The object 6 Cas was reported to be a spectroscopic binary based on its composite spectrum in the ultraviolet [101]. While the main component is traced by the resonance lines of low-ionized metals with terminal wind velocities of about $330\,\mathrm{km\,s^{-1}}$, typical for A-type supergiants, the high-ionized resonance lines with P Cygni profiles and terminal wind velocities of about $2400\,\mathrm{km\,s^{-1}}$ can be assigned to a significantly hotter O-type companion.

Disentangling the spectra in the optical, to which the O star only very weakly contributes, resulted in the classification of the system as a A3 Ia (A) and a O9.5 II (B) component [79]. The positions of the two stars are separated by about 1″.5 at a position angle of ∼195°. Whether these two stars form indeed a bound binary system or whether they are just close in projection could not be solved yet. The small radial velocity variations detected for the A supergiant seem to be due to pulsations rather than orbital motion in a binary system [79].

The Hα profile of 6 Cas resembles those seen in other hypergiants, such as HD 33579 and Schulte 12. It is superimposed on broad electron scattering wings [102]. Discrete absorption components (DACs) traveling through the broad absorption components of the wind profiles of H and Fe II lines were detected in 6 Cas [103]. Based on the atypical characteristics of a regular A-type supergiant, the star was assigned the status of a hypergiant.

We are aware of only one former K-band spectrum of 6 Cas, which has been taken with the ISO-SWS instrument [89]. This spectrum shows no indication of CO bands, just the lines of the Pfund series in absorption, similar to our spectrum (Figure 3). Such an otherwise featureless spectrum is consistent with a hot ($T_{\text{eff}} > 8000$ K) star, in agreement with its previous classification (see Table 1).

V509 Cas (=HR 8752, HD 217476)

The star V509 Cas is one of the YHGs for which an outburst was recorded based on combined photometric and spectroscopic monitoring of the star [62]. This outburst must have taken place around 1973, after a preceding 16-year-long period of reddening and cooling of the object from about 5000 K down to about 4000 K [104]. Thereafter, the effective temperature of the star gradually increased until it reached a value of about 8000 K around 2001 [62], where it has stabilized since then [63]. During the "heating" period, several short-term drops in temperature were recorded, which were associated with phases of enhanced mass loss [62].

Despite the outburst and the multiple mass-loss events, no extended nebulosity was detected at optical wavelengths so far [66]. Nevertheless, the star must be embedded in an ionized circumstellar envelope traced by thermal emission at radio wavelengths [105,106]. The ionization of this envelope is most likely performed by the radiation field of a distant, hot main-sequence B1-type companion [107]. The envelope is also the place where other emission lines are formed. Most prominent are the nebular lines of [N II] $\lambda\lambda$ 6548,6583 [108], but also those of [O I] $\lambda\lambda$ 6300,6364 and [Ca II] $\lambda\lambda$ 7291,7324 were identified [63], which were proposed to trace a possible Keplerian disk or ring based on the double-peaked profiles, particularly of the [Ca II] lines, for which a rotation velocity, projected to the line of sight, of about 40 km s^{-1} was derived [109]. Support for such an interpretation comes from an optical spectroscopic monitoring of V509 Cas between 2015 and 2022. The observations revealed that the [Ca II] lines are stable in position and shape over the observing period of ∼7 years, in contrast to the photospheric absorption lines, whose shape and radial velocity are strongly influenced by the pulsation activity of the star [110].

In the NIR, observations dating back to 1979–1980 detected CO features displaying both emission and absorption components with variable strength (see Table 5, [70]). Comparison with the brightness curve revealed that the CO emission was strongest when the star was around maximum brightness [67]. Furthermore, the CO emission appeared at the stellar systemic velocity, whereas the absorption was either blue- or red-shifted. This behavior is very similar to what has been observed for the YHG star ρ Cas [67]. Since about 1988, the CO features have disappeared from the NIR spectra of V509 Cas, and our spectrum also shows no traces of molecular emission. The absorption lines of the hydrogen Pfund series seen in our otherwise featureless K-band spectrum are in agreement with a stellar temperature of about 8000 K.

HD 179821 (=RAFGL 2343, IRAS 19114+0002, V1427 Aql)

The star is embedded within a detached, almost spherical shell of cold (120–140 K) dust [49] responsible for an intense far-IR excess emission [111]. In the radio regime, CO ($J = 1 \longrightarrow 0$) emission has been detected, which traces a cold molecular outflow with a velocity of $\sim$33–35 km s^{-1} [112]. The object is oxygen-rich, as is inferred from its OH maser emission [113].

The evolutionary state of the star is highly debated in the literature due to its uncertain distance value. Distance estimates range from about 1.5 kpc to about 6 kpc, which would classify the star as either a post-asymptotic giant branch star or a YHG. Considering the newest parallax measurements of 0.1893 ± 0.0206 provided by the GAIA Early Data Release 3 [82], a distance closer to the upper value seems to be more likely. Such a high value is also in agreement with the star's kinematic distance derived from its large heliocentric systemic velocity of 84–88 km s^{-1} [81,112,113].

The spectral classification of HD 179821 has been rather controversial as well in the past decades ranging from F3–5 [81,114,115], over G5 [116] to K4 [117]. The value for $\log g$, derived from high-resolution spectroscopy, ranges around 0.5 ± 0.5. Such a low $\log g$ value assigns the star a luminosity class I.

Extinction values obtained for HD 179821 range from $A_V = 2.0$ [116] over 3.1 [118] to $\sim$4 [115], and it has been suggested that the total extinction might be variable due to possible changes in the circumstellar contribution from the dust shell [119].

In a recent work, data from long-term photometric and spectroscopic monitoring have been presented [120]. The colors imply that the star first became bluer between 1990 and 1995 and has displayed a systematic reddening since 2002. The fastest change in color took place between 2013 and 2017, with a simultaneous brightening of the star. The spectra confirm this trend and record a more or less stable temperature of $T_{\text{eff}} = 6800 \pm 100$ K between 1994 and 2008, in agreement with previous temperature determinations [7] [81,114,115,120]. Since then, it decreased and reached a value of about 5900 K in 2017 [120]. The reddening and change in temperature indicate the onset of a possible new red loop evolution of HD 179821, i.e., an excursion to the cool edge of the HR diagram, related to a significant increase in the stellar radius and an increase in mass loss.

In the NIR, HD 179821 displayed CO-band emission, during observations taken between 1989 and 1991 (see Table 5), whereas no indication for CO band features (neither emission nor absorption) was seen between 1992 and 2000. The presence and disappearance of CO band emission might indicate a prior phase of higher mass loss or some mass ejection episode followed by the subsequent expansion and dilution of the released circumstellar material. Such a scenario might be supported by the redder color of the star around 1990 [120]. The absence of CO bands in the spectra between 1992 and 2000 supports the classification of the star as early to mid F-type, because stars in this temperature range are too hot for the formation of molecules in their atmospheres.

In contrast to all previous NIR observations, our data from 2021 clearly display CO band absorption. One possible explanation could be that the trend of cooling has continued since 2017. When comparing our observed spectra with synthetic spectra, we found that the intensity of the first band head of CO can be achieved for a stellar effective temperature of $\sim$5400 K (see Figure 6), although the entire CO band structure signals a considerably cooler temperature for the molecular gas due to the only weakly pronounced higher band heads. Hotter stars display less intense and cooler stars more intense CO bands. However, a star with a temperature of 5400 K should show significantly stronger absorption in all other atomic photospheric lines, which is not the case. Instead, the intensity and specific line ratios in the K-band spectrum are more in line with an effective temperature of about 6600 K. But such a hot stellar photosphere contains no CO molecular absorption features (Figure 6). Based on this discrepancy, we believe that our K-band spectrum is composite. It shows a hotter stellar photosphere along with CO absorption formed in a presumably cooler gas shell or outflow. Support for such a scenario is provided by the fact that the CO absorption bands display a blue-shift of about $-43 \pm 1\,\mathrm{km\,s^{-1}}$ with respect to all other photospheric atomic lines and the circumstellar emission lines, such as the Na I doublet. If interpreted as outflow velocity, then this value is comparable to the outflow seen in IRC +10420 [45,99] but is slightly higher than the isotropic expanding cold molecular gas and dust shell around HD 179821 ($\sim$35 km s^{-1}, [112,122]). We exclude a cool binary component as an explanation for the velocity-shifted CO absorption bands because, in this case, the cool companion would imprint (besides CO) significantly stronger blue-shifted absorption lines onto the K-band spectrum, which are not observed. If the proposed scenario of a new outflowing shell or gas layer is correct, possibly initiated during the reddening of the star and the increase in stellar radius recorded in the years 2014–2017 [120], or a possible outburst event that might have followed this reddening as in the case of V509 Cas (as we mentioned before, see [62,104]), then we may speculate that with further expansion of this matter, future K-band observations will display CO bands in emission before the material dilutes and the CO features might disappear again.

Besides CO, our spectrum of HD 179821 also shows Br γ in absorption and intense emission of the Na I doublet. The line of Br γ was in absorption in previous observations taken in 1989 [69] and 1990 [68]. The latter spectrum also displays intense emission of the (blended) Na I doublet, as well as the spectrum taken in 2000 [67]. The remaining previous spectra either do not cover the spectral region of Br γ and/or the Na I doublet, or these lines have not been mentioned by the corresponding authors.

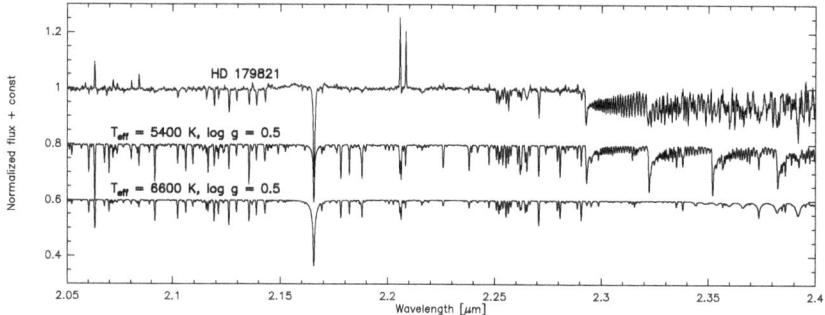

Figure 6. Comparison of the K-band spectrum of HD 179821 (**top**) with synthetic spectra for effective temperatures of 5400 K (**middle**) and 6600 K (**bottom**). For illustration purposes, the synthetic model spectra are included in this figure and shifted down along the flux axis for better visualization.

Na I emission is a clear indicator for circumstellar material. It has been reported from the NIR spectra of numerous evolved massive stars: (i) the YHGs ρ Cas, V509 Cas, Hen 3-1379 (the Fried Egg Nebula), IRC +10420 [70,123], and the IRC +10420 analog IRAS 18357-0604 [124], (ii) most of the B[e]SGs [21,29,125], and (iii) also many luminous blue variables [21,53]. The equivalent widths of the intense Na I lines from the YHG IRC +10420 [6] are about three times higher than for HD 179821, for which we measured values of -1.0670 ± 0.015 Å and -0.9231 ± 0.018 Å. Recent spatially resolved observations revealed that the Na I emission in IRC +10420 and Hen 3-1379 is confined within a compact spherical envelope around the star [50,123]. The emission lines in our spectrum are symmetric, and their wavelengths coincide with the systemic velocity of the star. While their formation region can be a compact spherical shell as well, they cannot be related to the possible new blue-shifted outflow traced by the CO band absorption.

5. Conclusions

We present new medium- and high-resolution K-band spectra for two B[e]SGs and four YHGs. The spectra of both B[e]SGs show rotationally broadened CO band emission, from which we could derive, for the first time, the projected rotation velocity of the CO gas for both stars. On the other hand, our model parameters for the CO temperature and column density are very similar to those reported in previous studies based on spectra with significantly lower resolution [20,21]. The similarities of the detected CO band features over more than 30 years suggest that the CO emitting gas rings around these two B[e]SGs are stable structures, neatly fitting to the findings of most of the B[e]SGs.

With respect to the YHGs, we detect CO band emission from only one star, the highly reddened cluster member [FMR2006] 15, which previously showed time-variable CO features (see Table 5), and which has an effective temperature that is clearly too high to form molecules within its atmosphere. Consequently, the CO emission must be of circumstellar origin. A second object, HD 179821, shows CO bands in absorption, while it had CO emission during 1989–1991 but lacked any CO features in its spectrum since 1992 (Table 5). The latter is consistent with the star's high effective temperature, which prevents the formation of molecules. For both YHGs, our detected CO features are clearly blue-shifted with respect to the photospheric absorption lines, suggesting that both stars most likely had recent mass ejection events and the CO emission/absorption forms within the expelled matter. For [FMR2006] 15, we propose that the blue-shifted emission arises in a possible pole-on seen bipolar outflow as in the case of the YHG star IRC +10420 [6], but nothing can be said yet about the geometry of the outflow seen from HD 179821 because the (highly inflated) star itself might still block large portions of the possibly receding parts of the ejecta.

Author Contributions: Conceptualization, M.K. (Michaela Kraus), M.K. (Michalis Kourniotis) and D.H.N.; methodology, M.K. (Michaela Kraus), M.K. (Michalis Kourniotis), M.L.A. and A.F.T.; formal analysis, M.K. (Michaela Kraus) and M.K. (Michalis Kourniotis); investigation, M.K. (Michaela Kraus), M.K. (Michalis Kourniotis) and D.H.N.; resources, M.K. (Michaela Kraus), M.K. (Michalis Kourniotis), M.L.A. and A.F.T.; writing—original draft preparation, review and editing, M.K. (Michaela Kraus), M.K. (Michalis Kourniotis), M.L.A., A.F.T. and D.H.N.; visualization, M.K. (Michaela Kraus) and M.K. (Michalis Kourniotis); funding acquisition, M.K. (Michaela Kraus), M.L.A. and A.F.T. All authors have read and agreed to the published version of the manuscript.

Funding: This research was funded by the Czech Science foundation (GA ČR, grant number 20-00150S), by CONICET (PIP 1337), by the Universidad Nacional de La Plata (Programa de Incentivos 11/G160), Argentina, and by the European Union's Framework Programme for Research and Innovation Horizon 2020 (2014-2020) under the Marie Skłodowska-Curie Grant Agreement No. 823734. The Astronomical Institute of the Czech Academy of Sciences is supported by the project RVO:67985815.

Institutional Review Board Statement: Not applicable.

Informed Consent Statement: Not applicable.

Data Availability Statement: The data underlying this article will be shared on reasonable request to the corresponding author.

Acknowledgments: We thank the anonymous referees for their valuable comments and suggestions. This research made use of the NASA Astrophysics Data System (ADS) and of the SIMBAD database, operated at CDS, Strasbourg, France. This paper is based on observations obtained with the Phoenix infrared spectrograph, developed and operated by the National Optical Astronomy Observatory and based on observations obtained at the international Gemini Observatory, a program of NSF's NOIRLab, which is managed by the Association of Universities for Research in Astronomy (AURA) under a cooperative agreement with the National Science Foundation on behalf of the Gemini Observatory partnership: the National Science Foundation (United States), National Research Council (Canada), Agencia Nacional de Investigación y Desarrollo (Chile), Ministerio de Ciencia, Tecnología e Innovación (Argentina), Ministério da Ciência, Tecnologia Inovações e Comunicações (Brazil), and Korea Astronomy and Space Science Institute (Republic of Korea) under program IDs GS-2004B-Q-54, GS-2017B-Q-32, GN-2019A-Q-204, GN-2019B-Q-418, and GN-2021A-Q-315.

Conflicts of Interest: The authors declare no conflict of interest.

Abbreviations

The following abbreviations are used in this manuscript:

B[e]SG   B[e] supergiant
NIR   Near infrared
LMC   Large Magellanic Cloud
SMC   Small Magellanic Cloud
YHG   Yellow hypergiant

Notes

1. The sixth object is LHA 120-S 111. To our knowledge, it has been observed in the K-band only once, in January 1987 [29]. At that time, no CO band emission was detected.
2. We excluded two Galactic B[e]SGs from this plot. With a literature luminosity value of $\log L/L_\odot = 4.33 \pm 0.09$ [74], the luminosity of HD 62623 is considerably lower than for the other B[e]SGs with CO bands. However, its distance with a parallax value of 0.59 ± 0.17 is not well constrained. The object HD 327083 turned out to be misclassified and has been removed from the B[e]SGs list (Cidale et al., in preparation).
3. We omit the SMC objects, because we do not have new data for any of the SMC B[e]SGs, and there are currently no confirmed YHGs in the SMC.
4. IRAF is distributed by the National Optical Astronomy Observatory, which is operated by the Association of Universities for Research in Astronomy (AURA) under cooperative agreement with the National Science Foundation.
5. Pfund line emission is also reported from LHA 120-S 12 [20], but in that star the maximum detected Pfund transition is with Pf(31) arising at 2.34 µm clearly outside our spectral coverage.
6. https://www.unige.ch/sciences/astro/evolution/en/database/syclist/ (accessed on 1 April 2023).
7. We note that a higher effective temperature of $\sim 7350\,\mathrm{K}$ was proposed in the same period [121].

References

1. Brandner, W.; Grebel, E.K.; Chu, Y.H.; Weis, K. Ring Nebula and Bipolar Outflows Associated with the B1.5 Supergiant Sher 25 in NGC 3603. *Astrophys. J.* **1997**, *475*, L45–L48.
2. Hendry, M.A.; Smartt, S.J.; Skillman, E.D.; Evans, C.J.; Trundle, C.; Lennon, D.J.; Crowther, P.A.; Hunter, I. The blue supergiant Sher 25 and its intriguing hourglass nebula. *Mon. Not. R. Astron. Soc.* **2008**, *388*, 1127–1142.
3. Marston, A.P.; McCollum, B. Extended shells around B[e] stars. Implications for B[e] star evolution. *Astron. Astrophys.* **2008**, *477*, 193–202. [CrossRef]
4. Kraus, M.; Liimets, T.; Cappa, C.E.; Cidale, L.S.; Nickeler, D.H.; Duronea, N.U.; Arias, M.L.; Gunawan, D.S.; Oksala, M.E.; Fernandes, M.B.; et al. Resolving the Circumstellar Environment of the Galactic B[e] Supergiant Star MWC 137 from Large to Small Scales. *Astron. J.* **2017**, *154*, 186.
5. Smith, N.; Arnett, W.D.; Bally, J.; Ginsburg, A.; Filippenko, A.V. The ring nebula around the blue supergiant SBW1: Pre-explosion snapshot of an SN 1987A twin. *Mon. Not. R. Astron. Soc.* **2013**, *429*, 1324–1341.
6. Oudmaijer, R.D.; de Wit, W.J. Neutral and ionised gas around the post-red supergiant IRC +10 420 at AU size scales. *Astron. Astrophys.* **2013**, *551*, A69.
7. Lagadec, E.; Zijlstra, A.A.; Oudmaijer, R.D.; Verhoelst, T.; Cox, N.L.J.; Szczerba, R.; Mékarnia, D.; van Winckel, H. A double detached shell around a post-red supergiant: IRAS 17163-3907, the Fried Egg nebula. *Astron. Astrophys.* **2011**, *534*, L10.
8. Weis, K. Nebulae around Luminous Blue Variables-large bipolar variety. In *Active OB Stars: Structure, Evolution, Mass Loss, and Critical Limits*; Proceedings of the International Astronomical Union; Neiner, C., Wade, G., Meynet, G., Peters, G., Eds.; Cambridge University Press: Cambridge, UK, 2011; IAU Symposium No. 272, pp. 372–377. [CrossRef]
9. Liimets, T.; Kraus, M.; Moiseev, A.; Duronea, N.; Cidale, L.S.; Fariña, C. Follow-Up of Extended Shells around B[e] Stars. *Galaxies* **2022**, *10*, 41.
10. Chu, Y.H. Galactic ring nebulae associated with Wolf-rayet stars. I. Introduction and classification. *Astrophys. J.* **1981**, *249*, 195–200. [CrossRef]
11. Marston, A.P.; Chu, Y.H.; Garcia-Segura, G. A Survey of Nebulae around Galactic Wolf-Rayet Stars in the Southern Sky. I. *Astrophys. J. Suppl.* **1994**, *93*, 229. [CrossRef]
12. Marston, A.P.; Yocum, D.R.; Garcia-Segura, G.; Chu, Y.H. A Survey of Nebulae around Galactic Wolf-Rayet Stars in the Southern Sky. II. *Astrophys. J. Suppl.* **1994**, *95*, 151. [CrossRef]
13. Maryeva, O.V.; Koenigsberger, G.; Karpov, S.V.; Lozinskaya, T.A.; Egorov, O.V.; Rossi, C.; Calabresi, M.; Viotti, R.F. Asymmetrical nebula of the M33 variable GR290 (WR/LBV). *Astron. Astrophys.* **2020**, *635*, A201.
14. Sévigny, M.; St-Louis, N.; Drissen, L.; Martin, T. New insights into the WR nebula M1-67 with SITELLE. *Mon. Not. R. Astron. Soc.* **2021**, *501*, 5350–5361.

15. Zickgraf, F.J.; Wolf, B.; Stahl, O.; Leitherer, C.; Appenzeller, I. B(e)-supergiants of the Magellanic Clouds. *Astron. Astrophys.* **1986**, *163*, 119–134.
16. Domiciano de Souza, A.; Driebe, T.; Chesneau, O.; Hofmann, K.H.; Kraus, S.; Miroshnichenko, A.S.; Ohnaka, K.; Petrov, R.G.; Preisbisch, T.; Stee, P.; et al. AMBER/VLTI and MIDI/VLTI spectro-interferometric observations of the B[e] supergiant CPD-57°2874. Size and geometry of the circumstellar envelope in the near- and mid-IR. *Astron. Astrophys.* **2007**, *464*, 81–86. [CrossRef]
17. Maravelias, G.; Kraus, M.; Cidale, L.S.; Borges Fernandes, M.; Arias, M.L.; Curé, M.; Vasilopoulos, G. Resolving the kinematics of the discs around Galactic B[e] supergiants. *Mon. Not. R. Astron. Soc.* **2018**, *480*, 320–344.
18. Lamers, H.J.G.L.M.; Zickgraf, F.J.; de Winter, D.; Houziaux, L.; Zorec, J. An improved classification of B[e]-type stars. *Astron. Astrophys.* **1998**, *340*, 117–128.
19. Kraus, M. A Census of B[e] Supergiants. *Galaxies* **2019**, *7*, 83.
20. Liermann, A.; Kraus, M.; Schnurr, O.; Fernandes, M.B. The ^{13}Carbon footprint of B[e] supergiants. *Mon. Not. R. Astron. Soc.* **2010**, *408*, L6–L10.
21. Oksala, M.E.; Kraus, M.; Cidale, L.S.; Muratore, M.F.; Borges Fernandes, M. Probing the ejecta of evolved massive stars in transition. A VLT/SINFONI K-band survey. *Astron. Astrophys.* **2013**, *558*, A17.
22. Kraus, M. The pre-versus post-main sequence evolutionary phase of B[e] stars. Constraints from ^{13}CO band emission. *Astron. Astrophys.* **2009**, *494*, 253–262. .:200811020. [CrossRef]
23. Kraus, M.; Oksala, M.E.; Cidale, L.S.; Arias, M.L.; Torres, A.F.; Borges Fernandes, M. Discovery of SiO Band Emission from Galactic B[e] Supergiants. *Astrophys. J.* **2015**, *800*, L20.
24. Kraus, M.; Cidale, L.S.; Arias, M.L.; Maravelias, G.; Nickeler, D.H.; Torres, A.F.; Borges Fernandes, M.; Aret, A.; Curé, M.; Vallverdú, R.; et al. Inhomogeneous molecular ring around the B[e] supergiant LHA 120-S 73. *Astron. Astrophys.* **2016**, *593*, A112.
25. Kraus, M.; Oksala, M.E.; Nickeler, D.H.; Muratore, M.F.; Borges Fernandes, M.; Aret, A.; Cidale, L.S.; de Wit, W.J. Molecular emission from GG Carinae's circumbinary disk. *Astron. Astrophys.* **2013**, *549*, A28.
26. Muratore, M.F.; Kraus, M.; Oksala, M.E.; Arias, M.L.; Cidale, L.; Borges Fernandes, M.; Liermann, A. Evidence of the Evolved Nature of the B[e] Star MWC 137. *Astron. J.* **2015**, *149*, 13.
27. Torres, A.F.; Cidale, L.S.; Kraus, M.; Arias, M.L.; Barbá, R.H.; Maravelias, G.; Borges Fernandes, M. Resolving the clumpy circumstellar environment of the B[e] supergiant LHA 120-S 35. *Astron. Astrophys.* **2018**, *612*, A113.
28. Cidale, L.S.; Borges Fernandes, M.; Andruchow, I.; Arias, M.L.; Kraus, M.; Chesneau, O.; Kanaan, S.; Curé, M.; de Wit, W.J.; Muratore, M.F. Observational constraints for the circumstellar disk of the B[e] star CPD-52 9243. *Astron. Astrophys.* **2012**, *548*, A72. [CrossRef]
29. McGregor, P.J.; Hillier, D.J.; Hyland, A.R. CO Overtone Emission from Magellanic Cloud Supergiants. *Astrophys. J.* **1988**, *334*, 639. [CrossRef]
30. McGregor, P.J.; Hyland, A.R.; Hillier, D.J. Atomic and Molecular Line Emission from Early-Type High-Luminosity Stars. *Astrophys. J.* **1988**, *324*, 1071. [CrossRef]
31. Kraus, M.; Cidale, L.S.; Arias, M.L.; Oksala, M.F.; Borges Fernandes, M. Discovery of the First B[e] Supergiants in M 31. *Astrophys. J.* **2014**, *780*, L10.
32. Sholukhova, O.; Bizyaev, D.; Fabrika, S.; Sarkisyan, A.; Malanushenko, V.; Valeev, A. New luminous blue variables in the Andromeda galaxy. *Mon. Not. R. Astron. Soc.* **2015**, *447*, 2459–2467.
33. Oksala, M.E.; Kraus, M.; Arias, M.L.; Borges Fernandes, M.; Cidale, L.; Muratore, M.F.; Curé, M. The sudden appearance of CO emission in LHA 115-S 65. *Mon. Not. R. Astron. Soc.* **2012**, *426*, L56–L60.
34. Kraus, M.; Borges Fernandes, M.; de Araújo, F.X. Neutral material around the B[e] supergiant star LHA 115-S 65. An outflowing disk or a detached Keplerian rotating disk? *Astron. Astrophys.* **2010**, *517*, A30.
35. McGregor, P.J.; Hyland, A.R.; McGinn, M.T. Emission-line stars in the Magellanic Clouds: Infrared spectroscopy of Be and Ofpe/WN9 stars. *Astron. Astrophys.* **1989**, *223*, 237–240.
36. Morris, P.W.; Eenens, P.R.J.; Hanson, M.M.; Conti, P.S.; Blum, R.D. Infrared Spectra of Massive Stars in Transition: WNL, Of, Of/WN, Be, B[e], and Luminous Blue Variable Stars. *Astrophys. J.* **1996**, *470*, 597. [CrossRef]
37. Hamann, F.; Simon, M. Velocity-resolved Infrared Spectroscopy of MWC 349. *Astrophys. J.* **1986**, *311*, 909. [CrossRef]
38. Kraus, M.; Arias, M.L.; Cidale, L.S.; Torres, A.F. Evidence of an evolved nature of MWC 349A. *Mon. Not. R. Astron. Soc.* **2020**, *493*, 4308–4314.
39. Zickgraf, F.J.; Wolf, B.; Stahl, O.; Humphreys, R.M. S 18: A new B(e) supergiant in the Small Magellanic Cloud with evidence for an excretion disk. *Astron. Astrophys.* **1989**, *220*, 206–214.
40. Torres, A.F.; Kraus, M.; Cidale, L.S.; Barbá, R.; Borges Fernandes, M.; Brandi, E. Discovery of Raman-scattered lines in the massive luminous emission-line star LHA 115-S 18. *Mon. Not. R. Astron. Soc.* **2012**, *427*, L80–L84.
41. Oudmaijer, R.D.; Davies, B.; de Wit, W.J.; Patel, M. Post-Red Supergiants. In *Proceedings of the Biggest, Baddest, Coolest Stars: Proceedings of a Workshop Held at the Millennium Centre, Johnson City, TN, USA, 16–18 July 2007*; Astronomical Society of the Pacific Conference Series; Luttermoser, D.G., Smith, B.J., Stencel, R.E., Eds.; Astronomical Society of the Pacific: San Francisco, CA, USA, 2009; Volume 412, p. 17,

42. Ekström, S.; Georgy, C.; Eggenberger, P.; Meynet, G.; Mowlavi, N.; Wyttenbach, A.; Granada, A.; Decressin, T.; Hirschi, R.; Frischknecht, U.; et al. Grids of stellar models with rotation. I. Models from 0.8 to 120 $M_\odot$ at solar metallicity (Z = 0.014). *Astron. Astrophys.* **2012**, *537*, A146.
43. Smartt, S.J.; Eldridge, J.J.; Crockett, R.M.; Maund, J.R. The death of massive stars-I. Observational constraints on the progenitors of Type II-P supernovae. *Mon. Not. R. Astron. Soc.* **2009**, *395*, 1409–1437.
44. Smartt, S.J. Observational Constraints on the Progenitors of Core-Collapse Supernovae: The Case for Missing High-Mass Stars. *Publ. Astron. Soc. Aust.* **2015**, *32*, e016.
45. Oudmaijer, R.D.; Groenewegen, M.A.T.; Matthews, H.E.; Blommaert, J.A.D.L.; Sahu, K.C. The spectral energy distribution and mass-loss history of IRC + 10420. *Mon. Not. R. Astron. Soc.* **1996**, *280*, 1062–1070. [CrossRef]
46. Castro-Carrizo, A.; Lucas, R.; Bujarrabal, V.; Colomer, F.; Alcolea, J. SiO emission from a huge, detached shell in IRC + 10420. *Astron. Astrophys.* **2001**, *368*, L34–L37. [CrossRef]
47. Tiffany, C.; Humphreys, R.M.; Jones, T.J.; Davidson, K. The Morphology of IRC + 10420's Circumstellar Ejecta. *Astron. J.* **2010**, *140*, 339–349.
48. Deguchi, S.; Nakada, Y.; Sahai, R. SiO and CO emission from carbon stars with silicate features and southern IRAS sources. *Astron. Astrophys.* **1990**, *230*, 339–354.
49. Jura, M.; Werner, M.W. The Detached Dust Shell around the Massive Star HD 179821. *Astrophys. J.* **1999**, *525*, 113–L116. [CrossRef]
50. Koumpia, E.; Oudmaijer, R.D.; Graham, V.; Banyard, G.; Black, J.H.; Wichittanakom, C.; Ababakr, K.M.; de Wit, W.J.; Millour, F.; Lagadec, E.; et al. Optical and near-infrared observations of the Fried Egg Nebula. Multiple shell ejections on a 100 yr timescale from a massive yellow hypergiant. *Astron. Astrophys.* **2020**, *635*, A183.
51. Kourniotis, M.; Kraus, M.; Maryeva, O.; Borges Fernandes, M.; Maravelias, G. Revisiting the evolved hypergiants in the Magellanic Clouds. *Mon. Not. R. Astron. Soc.* **2022**, *511*, 4360–4376.
52. Humphreys, R.M.; Jones, T.J.; Polomski, E.; Koppelman, M.; Helton, A.; McQuinn, K.; Gehrz, R.D.; Woodward, C.E.; Wagner, R.M.; Gordon, K.; et al. M33's Variable A: A Hypergiant Star More Than 35 YEARS in Eruption. *Astron. J.* **2006**, *131*, 2105–2113.
53. Kourniotis, M.; Bonanos, A.Z.; Yuan, W.; Macri, L.M.; Garcia-Alvarez, D.; Lee, C.H. Monitoring luminous yellow massive stars in M 33: New yellow hypergiant candidates. *Astron. Astrophys.* **2017**, *601*, A76.
54. Nickeler, D.H.; Karlický, M. On the validity of ideal MHD in the vicinity of stagnation points in the heliosphere and other astrospheres. *Astrophys. Space Sci. Trans.* **2008**, *4*, 7–12. [CrossRef]
55. Glatzel, W. On the origin of strange modes and the mechanism of related instabilities. *Mon. Not. R. Astron. Soc.* **1994**, *271*, 66. [CrossRef]
56. Glatzel, W.; Kiriakidis, M.; Chernigovskij, S.; Fricke, K.J. The non-linear evolution of strange-mode instabilities. *Mon. Not. R. Astron. Soc.* **1999**, *303*, 116–124. [CrossRef]
57. Kraus, M.; Cidale, L.S.; Arias, M.L.; Torres, A.F.; Kolka, I.; Maravelias, G.; Nickeler, D.H.; Glatzel, W.; Liimets, T. Environments of evolved massive stars: Evidence for episodic mass ejections. *IAU Symp.* **2022**, *366*, 51–56. [CrossRef]
58. Lobel, A.; Dupree, A.K.; Stefanik, R.P.; Torres, G.; Israelian, G.; Morrison, N.; de Jager, C.; Nieuwenhuijzen, H.; Ilyin, I.; Musaev, F. High-Resolution Spectroscopy of the Yellow Hypergiant ρ Cassiopeiae from 1993 through the Outburst of 2000–2001. *Astrophys. J.* **2003**, *583*, 923–954.
59. Klochkova, V.G.; Panchuk, V.E.; Tavolzhanskaya, N.S. Changes of the Optical Spectrum of the Hypergiant ρ Cas due to a Shell Ejection in 2013. *Astron. Rep.* **2018**, *62*, 623–635.
60. Kraus, M.; Kolka, I.; Aret, A.; Nickeler, D.H.; Maravelias, G.; Eenmäe, T.; Lobel, A.; Klochkova, V.G. A new outburst of the yellow hypergiant star ρ Cas. *Mon. Not. R. Astron. Soc.* **2019**, *483*, 3792–3809.
61. Maravelias, G.; Kraus, M. Bouncing against the Yellow Void-Exploring the Outbursts of rho Cassiopeiae from Visual Observations. *J. Am. Assoc. Variable Star Observers* **2022**, *50*, 49.
62. Nieuwenhuijzen, H.; De Jager, C.; Kolka, I.; Israelian, G.; Lobel, A.; Zsoldos, E.; Maeder, A.; Meynet, G. The hypergiant HR 8752 evolving through the yellow evolutionary void. *Astron. Astrophys.* **2012**, *546*, A105. [CrossRef]
63. Aret, A.; Kolka, I.; Kraus, M.; Maravelias, G. Similarities in the Structure of the Circumstellar Environments of B[e] Supergiants and Yellow Hypergiants. In *Proceedings of the The B[e] Phenomenon: Forty Years of Studies, Prague, Czech Republic, 27 June–1 July 2016*; Astronomical Society of the Pacific Conference Series; Miroshnichenko, A., Zharikov, S., Korčáková, D., Wolf, M., Eds.; Astronomical Society of the Pacific: San Francisco, CA, USA, 2017; Volume 508, p. 239,
64. Oudmaijer, R.D. High resolution spectroscopy of the post-red supergiant IRC+10420. I. The data. *Astron. Astrophys. Suppl. Ser.* **1998**, *129*, 541–552. .:1998404. [CrossRef]
65. Klochkova, V.G.; Yushkin, M.V.; Chentsov, E.L.; Panchuk, V.E. Evolutionary Changes in the Optical Spectrum of the Peculiar Supergiant IRC + 10420. *Astron. Rep.* **2002**, *46*, 139–151. [CrossRef]
66. Schuster, M.T.; Humphreys, R.M.; Marengo, M. The Circumstellar Environments of NML Cygni and the Cool Hypergiants. *Astron. J.* **2006**, *131*, 603–611.
67. Gorlova, N.; Lobel, A.; Burgasser, A.J.; Rieke, G.H.; Ilyin, I.; Stauffer, J.R. On the CO Near-Infrared Band and the Line-splitting Phenomenon in the Yellow Hypergiant ρ Cassiopeiae. *Astrophys. J.* **2006**, *651*, 1130–1150.

68. Hrivnak, B.J.; Kwok, S.; Geballe, T.R. Near-Infrared Spectroscopy of Proto–Planetary Nebulae. *Astrophys. J.* **1994**, *420*, 783. [CrossRef]
69. Oudmaijer, R.D.; Waters, L.B.F.M.; van der Veen, W.E.C.J.; Geballe, T.R. Near-infrared spectroscopy of post-AGB stars. *Astron. Astrophys.* **1995**, *299*, 69.
70. Lambert, D.L.; Hinkle, K.H.; Hall, D.N.B. Circumstellar shells of luminous supergiants. I. Carbon monoxide in rho CAS and HR 8752. *Astrophys. J.* **1981**, *248*, 638–650. [CrossRef]
71. Davies, B.; Figer, D.F.; Law, C.J.; Kudritzki, R.P.; Najarro, F.; Herrero, A.; MacKenty, J.W. The Cool Supergiant Population of the Massive Young Star Cluster RSGC1. *Astrophys. J.* **2008**, *676*, 1016–1028.
72. Kraus, M.; Arias, M.L.; Cidale, L.S.; Torres, A.F.; Kourniotis, M. Molecular environment of the yellow hypergiant star HD 269953. *Bol. Asoc. Argent. Astron. Plata Argent.* **2022**, *63*, 65–67.
73. Muratore, M.F.; Kraus, M.; de Wit, W.J. Near-infrared spectroscopic survey of galactic B[e] stars. *Bol. Asoc. Argent. Astron. Plata Argent.* **2012**, *55*, 123–127.
74. Millour, F.; Meilland, A.; Chesneau, O.; Stee, P.; Kanaan, S.; Petrov, R.; Mourard, D.; Kraus, S. Imaging the spinning gas and dust in the disc around the supergiant A[e] star HD 62623. *Astron. Astrophys.* **2011**, *526*, A107.
75. de Jager, C. The yellow hypergiants. *Astron. Astrophys. Rev.* **1998**, *8*, 145–180. [CrossRef]
76. Eggenberger, P.; Ekström, S.; Georgy, C.; Martinet, S.; Pezzotti, C.; Nandal, D.; Meynet, G.; Buldgen, G.; Salmon, S.; Haemmerlé, L.; et al. Grids of stellar models with rotation. VI. Models from 0.8 to 120 $M_\odot$ at a metallicity Z = 0.006. *Astron. Astrophys.* **2021**, *652*, A137.
77. Pietrzyński, G.; Graczyk, D.; Gallenne, A.; Gieren, W.; Thompson, I.B.; Pilecki, B.; Karczmarek, P.; Górski, M.; Suchomska, K.; Taormina, M.; et al. A distance to the Large Magellanic Cloud that is precise to one per cent. *Nature* **2019**, *567*, 200–203.
78. Kudritzki, R.P.; Puls, J.; Lennon, D.J.; Venn, K.A.; Reetz, J.; Najarro, F.; McCarthy, J.K.; Herrero, A. The wind momentum-luminosity relationship of galactic A- and B-supergiants. *Astron. Astrophys.* **1999**, *350*, 970–984.
79. Maíz Apellániz, J.; Barbá, R.H.; Fariña, C.; Sota, A.; Pantaleoni González, M.; Holgado, G.; Negueruela, I.; Simón-Díaz, S. Lucky spectroscopy, an equivalent technique to lucky imaging. II. Spatially resolved intermediate-resolution blue-violet spectroscopy of 19 close massive binaries using the William Herschel Telescope. *Astron. Astrophys.* **2021**, *646*, A11.
80. de Jager, C.; Nieuwenhuijzen, H. An obstacle to the late evolution of massive stars. *Mon. Not. R. Astron. Soc.* **1997**, *290*, L50–L54. [CrossRef]
81. Kipper, T. Optical Spectroscopy of a Post-Agb Star HD 179821 (V1427 Aql). *Balt. Astron.* **2008**, *17*, 87–102.
82. Gaia Collaboration. VizieR Online Data Catalog: Gaia EDR3 (Gaia Collaboration, 2020). *VizieR Online Data Catalog* **2020**, I/350. [CrossRef]
83. Cutri, R.M.; Skrutskie, M.F.; van Dyk, S.; Beichman, C.A.; Carpenter, J.M.; Chester, T.; Cambresy, L.; Evans, T.; Fowler, J.; Gizis, J.; et al. VizieR Online Data Catalog: 2MASS All-Sky Catalog of Point Sources (Cutri + 2003). *VizieR Online Data Catalog* **2003**, II/246.
84. Hinkle, K.H.; Blum, R.D.; Joyce, R.R.; Sharp, N.; Ridgway, S.T.; Bouchet, P.; van der Bliek, N.S.; Najita, J.; Winge, C. The Phoenix Spectrograph at Gemini South. In *Discoveries and Research Prospects from 6- to 10-Meter-Class Telescopes II*; Guhathakurta, P., Ed.; Society of Photo-Optical Instrumentation Engineers (SPIE) Conference Series; 2003; Volume 4834, pp. 353–363. [CrossRef]
85. Elias, J.H.; Rodgers, B.; Joyce, R.R.; Lazo, M.; Doppmann, G.; Winge, C.; Rodríguez-Ardila, A. Performance of the Gemini near-infrared spectrograph. In *Proceedings of the Society of Photo-Optical Instrumentation Engineers (SPIE) Conference Series*; McLean, I.S., Iye, M., Eds.; Society of Photo-Optical Instrumentation Engineers (SPIE): Bellingham, WA, USA, 2006; Volume 6269, p. 626914. [CrossRef]
86. Elias, J.H.; Joyce, R.R.; Liang, M.; Muller, G.P.; Hileman, E.A.; George, J.R. Design of the Gemini near-infrared spectrograph. In *Proceedings of the Society of Photo-Optical Instrumentation Engineers (SPIE) Conference Series*; McLean, I.S., Iye, M., Eds.; Society of Photo-Optical Instrumentation Engineers (SPIE): Bellingham, WA, USA, 2006; Volume 6269, p. 62694C. [CrossRef]
87. Kraus, M.; Krügel, E.; Thum, C.; Geballe, T.R. CO band emission from MWC 349. I. First overtone bands from a disk or from a wind? *Astron. Astrophys.* **2000**, *362*, 158–168.
88. Figer, D.F.; MacKenty, J.W.; Robberto, M.; Smith, K.; Najarro, F.; Kudritzki, R.P.; Herrero, A. Discovery of an Extraordinarily Massive Cluster of Red Supergiants. *Astrophys. J.* **2006**, *643*, 1166–1179.
89. Vandenbussche, B.; Beintema, D.; de Graauw, T.; Decin, L.; Feuchtgruber, H.; Heras, A.; Kester, D.; Lahuis, F.; Lenorzer, A.; Lorente, R.; et al. The ISO-SWS post-helium atlas of near-infrared stellar spectra. *Astron. Astrophys.* **2002**, *390*, 1033–1048. [CrossRef]
90. Henize, K.G. Catalogues of Hα-emission Stars and Nebulae in the Magellanic Clouds. *Astrophys. J. Suppl.* **1956**, *2*, 315. [CrossRef]
91. Stahl, O.; Leitherer, C.; Wolf, B.; Zickgraf, F.J. Three new hot stars with dust shells in the Magellanic clouds. *Astron. Astrophys.* **1984**, *131*, L5–L6.
92. Magalhaes, A.M. Polarization and the Envelopes of B[e] Supergiants in the Magellanic Clouds. *Astrophys. J.* **1992**, *398*, 286. [CrossRef]
93. Kastner, J.H.; Buchanan, C.; Sahai, R.; Forrest, W.J.; Sargent, B.A. The Dusty Circumstellar Disks of B[e] Supergiants in the Magellanic Clouds. *Astron. J.* **2010**, *139*, 1993–2002. [CrossRef]

94. Aret, A.; Kraus, M.; Muratore, M.F.; Borges Fernandes, M. A new observational tracer for high-density disc-like structures around B[e] supergiants. *Mon. Not. R. Astron. Soc.* **2012**, *423*, 284–293.
95. Merrill, P.W.; Burwell, C.G. Catalogue and Bibliography of Stars of Classes B and A whose Spectra have Bright Hydrogen Lines. *Astrophys. J.* **1933**, *78*, 87. [CrossRef]
96. Shore, S.N.; Sanduleak, N. The extreme LMC supergiant HD 38489: An optical and ultraviolet study. *Astrophys. J.* **1983**, *273*, 177–186. [CrossRef]
97. Massey, P.; Neugent, K.F.; Morrell, N.; Hillier, D.J. A Modern Search for Wolf-Rayet Stars in the Magellanic Clouds: First Results. *Astrophys. J.* **2014**, *788*, 83.
98. Hung, C.S.; Ou, P.S.; Chu, Y.H.; Gruendl, R.A.; Li, C.J. A Multiwavelength Survey of Wolf-Rayet Nebulae in the Large Magellanic Cloud. *Astrophys. J. Suppl.* **2021**, *252*, 21.
99. Oudmaijer, R.D.; Geballe, T.R.; Waters, L.B.F.M.; Sahu, K.C. Discovery of near-infrared hydrogen line emission in the peculiar F8 hypergiant IRC + 10420. *Astron. Astrophys.* **1994**, *281*, L33–L36.
100. Plez, B. *Turbospectrum: Code for Spectral Synthesis*; record ascl:1205.004; ASCL.net, Michigan Technological University: Houghton, MI, USA, 2012.
101. Talavera, A.; Gomez de Castro, A.I. The UV high resolution spectrum of A-type supergiants. *Astron. Astrophys.* **1987**, *181*, 300–314.
102. Klochkova, V.G.; Chentsov, E.L. The Problem of Spectral Mimicry of Supergiants. *Astron. Rep.* **2018**, *62*, 19–30.
103. Chentsov, E.L. Unstable Wind of 6 Cassiopeiae. *Astrophys. Space Sci.* **1995**, *232*, 217–232. [CrossRef]
104. Luck, R.E. An analysis of the superluminous star HR 8752. *Astrophys. J.* **1975**, *202*, 743–754. [CrossRef]
105. Smolinski, J.; Feldman, P.A.; Higgs, L.A. A search for radio emission from late-type supergiant stars. *Astron. Astrophys.* **1977**, *60*, 277–280.
106. Higgs, L.A.; Feldman, P.A.; Smolinski, J. The radio source associated with the G-type supergiant HR 8752. *Astrophys. J.* **1978**, *220*, L109–L112. [CrossRef]
107. Stickland, D.J.; Harmer, D.L. The discovery of a hot companion to HR 8752. *Astron. Astrophys.* **1978**, *70*, L53–L56.
108. Sargent, W.L.W. Forbidden emission lines in the spectrum of a G-type supergiant. *Observatory* **1965**, *85*, 33–35.
109. Aret, A.; Kraus, M.; Kolka, I.; Maravelias, G. The Yellow Hypergiant-B[e] Supergiant Connection. In *Proceedings of the Stars: From Collapse to Collapse, Nizhny Arkhyz, Russia, 3–7 October 2016*; Astronomical Society of the Pacific Conference Series; Balega, Y.Y., Kudryavtsev, D.O., Romanyuk, I.I., Yakunin, I.A., Eds.; Astronomical Society of the Pacific: San Francisco, CA, USA, 2017; Volume 510, p. 162,
110. Kasikov, A.; Kolka, I.; Aret, A. Following V509 Cas into the void with FIES. In *Proceedings of the NOT—A Telescope for the Future, La Palma, Canary Islands, Spain, 7–10 June 2022*; CERN: Geneva, Switzerland, 2022; p. 14. [CrossRef]
111. van der Veen, W.E.C.J.; Habing, H.J.; Geballe, T.R. Objects in transition from the AGB to the planetary nebula stage: New visual and infrared observations. *Astron. Astrophys.* **1989**, *226*, 108–136.
112. Zuckerman, B.; Dyck, H.M. Dust Grains and Gas in the Circumstellar Envelopes around Luminous Red-Giant Stars. *Astrophys. J.* **1986**, *311*, 345. [CrossRef]
113. Likkel, L. OH and H_2O Observations of Cold IRAS Stars. *Astrophys. J.* **1989**, *344*, 350. [CrossRef]
114. Zacs, L.; Klochkova, V.G.; Panchuk, V.E.; Spelmanis, R. The chemical composition of the protoplanetary nebula candidate HD 179821. *Mon. Not. R. Astron. Soc.* **1996**, *282*, 1171–1180. [CrossRef]
115. Reddy, B.E.; Hrivnak, B.J. Spectroscopic Study of HD 179821 (IRAS 19114 + 0002): Proto-Planetary Nebula or Supergiant? *Astron. J.* **1999**, *117*, 1834–1844. [CrossRef]
116. Hrivnak, B.J.; Kwok, S.; Volk, K.M. A Study of Several F and G Supergiant-like Stars with Infrared Excesses as Candidates for Proto–Planetary Nebulae. *Astrophys. J.* **1989**, *346*, 265. [CrossRef]
117. Odenwald, S.F. An IRAS Survey of Infrared Excesses in G-Type Stars. *Astrophys. J.* **1986**, *307*, 711. [CrossRef]
118. Arkhipova, V.P.; Ikonnikova, N.P.; Noskova, R.I.; Sokol, G.V.; Shugarov, S.Y. Light Variations in the Candidate for Protoplanetary Objects HD 179821 = V1427 Aql in 1899–1999. *Astron. Lett.* **2001**, *27*, 156–162. [CrossRef]
119. Arkhipova, V.P.; Esipov, V.F.; Ikonnikova, N.P.; Komissarova, G.V.; Tatarnikov, A.M.; Yudin, B.F. Photometric variability and evolutionary status of the supergiant with an infrared excess HD 179821 = V1427 aquilae. *Astron. Lett.* **2009**, *35*, 764–779. [CrossRef]
120. Ikonnikova, N.P.; Taranova, O.G.; Arkhipova, V.P.; Komissarova, G.V.; Shenavrin, V.I.; Esipov, V.F.; Burlak, M.A.; Metlov, V.G. Multicolor Photometry and Spectroscopy of the Yellow Supergiant with Dust Envelope HD 179821 = V1427 Aquilae. *Astron. Lett.* **2018**, *44*, 457–473.
121. Şahin, T.; Lambert, D.L.; Klochkova, V.G.; Panchuk, V.E. HD 179821 (V1427 Aql, IRAS 19114 + 0002)—A massive post-red supergiant star? *Mon. Not. R. Astron. Soc.* **2016**, *461*, 4071–4087.
122. Castro-Carrizo, A.; Quintana-Lacaci, G.; Bujarrabal, V.; Neri, R.; Alcolea, J. Arcsecond-resolution ^{12}CO mapping of the yellow hypergiants IRC + 10420 and AFGL 2343. *Astron. Astrophys.* **2007**, *465*, 457–467. [CrossRef]
123. Koumpia, E.; Oudmaijer, R.D.; de Wit, W.J.; Mérand, A.; Black, J.H.; Ababakr, K.M. Tracing a decade of activity towards a yellow hypergiant. The spectral and spatial morphology of IRC + 10420 at au scales. *Mon. Not. R. Astron. Soc.* **2022**, *515*, 2766–2777.

124. Clark, J.S.; Negueruela, I.; González-Fernández, C. IRAS 18357-0604—An analogue of the galactic yellow hypergiant IRC + 10420? *Astron. Astrophys.* **2014**, *561*, A15.
125. Arias, M.L.; Vallverdú, R.; Torres, A.F.; Kraus, M. *High-Resolution, Near-Infrared Observations of B[e] Supergiants*; Boletin de la Asociacion Argentina de Astronomia: La Plata, Argentina, 2021; Volume 62, pp. 104–106.

Disclaimer/Publisher's Note: The statements, opinions and data contained in all publications are solely those of the individual author(s) and contributor(s) and not of MDPI and/or the editor(s). MDPI and/or the editor(s) disclaim responsibility for any injury to people or property resulting from any ideas, methods, instructions or products referred to in the content.

Article

New Insight into the FS CMa System MWC 645 from Near-Infrared and Optical Spectroscopy

Andrea Fabiana Torres [1,2,*], María Laura Arias [1,2], Michaela Kraus [3], Lorena Verónica Mercanti [1,2] and Tõnis Eenmäe [4]

1 Departamento de Espectroscopía, Facultad de Ciencias Astronómicas y Geofísicas, Universidad Nacional de La Plata, Paseo del Bosque S/N, La Plata B1900FWA, Argentina
2 Instituto de Astrofísica de La Plata (CCT La Plata-CONICET, UNLP), Paseo del Bosque S/N, La Plata B1900FWA, Argentina
3 Astronomical Institute, Czech Academy of Sciences, Fričova 298, 251 65 Ondřejov, Czech Republic
4 Tartu Observatory, University of Tartu, Observatooriumi 1, 61602 Tõravere, Estonia
* Correspondence: atorres@fcaglp.unlp.edu.ar

Citation: Torres, A.F.; Arias, M.L.; Kraus, M.; Mercanti, L.V.; Eenmäe, T. New Insight into the FS CMa System MWC 645 from Near-Infrared and Optical Spectroscopy. *Galaxies* **2023**, *11*, 72. https://doi.org/10.3390/galaxies11030072

Academic Editor: Jorick Sandor Vink

Received: 24 April 2023
Revised: 31 May 2023
Accepted: 1 June 2023
Published: 10 June 2023

Copyright: © 2023 by the authors. Licensee MDPI, Basel, Switzerland. This article is an open access article distributed under the terms and conditions of the Creative Commons Attribution (CC BY) license (https://creativecommons.org/licenses/by/4.0/).

Abstract: The B[e] phenomenon is manifested by a heterogeneous group of stars surrounded by gaseous and dusty circumstellar envelopes with similar physical conditions. Among these stars, the FS CMa-type objects are suspected to be binary systems, which could be experiencing or have undergone a mass-transfer process that could explain the large amount of material surrounding them. We aim to contribute to the knowledge of a recently confirmed binary, MWC 645, which could be undergoing an active mass-transfer process. We present near-infrared and optical spectra, identify atomic and molecular spectral features, and derive different quantitative properties of line profiles. Based on publicly available photometric data, we search for periodicity in the light curve and model the spectral energy distribution. We have detected molecular bands of CO in absorption at 1.62 μm and 2.3 μm for the first time. We derive an upper limit for the effective temperature of the cool binary component. We found a correlation between the enhancement of the Hα emission and the decrease in optical brightness that could be associated with mass-ejection events or an increase in mass loss. We outline the global properties of the envelope, possibly responsible for brightness variations due to a variable extinction, and briefly speculate on different possible scenarios.

Keywords: stars: emission-line, Be; stars: peculiar; stars: individual: MWC 645; circumstellar matter; binaries: general; techniques: spectroscopic

1. Introduction

In their evolution, some B-type stars undergo phases that are still puzzling for astrophysicists even after years of study since they develop certain peculiarities that are not yet well understood. The B[e] phenomenon displayed by several B-type stars in their optical spectra is an example of them. Its manifestation can be seen through the presence of permitted and forbidden low-excitation emission lines of neutral and low ionization metals arising from circumstellar (CS) gas and large infrared excess due to CS dust [1]. The phenomenon is associated with stars with different initial masses, isolated or in binary systems, transiting different evolutionary stages, such as supergiants, compact planetary nebulae, Herbig Ae/Be stars, and symbiotic systems [2]. Despite the cited differences among the stars, the physical conditions of their CS gaseous and dusty envelopes are similar, which is a crucial factor when seeking to understand the development of the phenomenon. Furthermore, since the CS envelopes veil the photospheric features of the central objects, it is difficult to determine their spectral types and evolutionary states. Therefore, stars without a proper classification comprise the category named "Unclassified B[e] stars" (UnclB[e]).

Nearly a decade after Lamers' classification, Miroshnichenko proposed a new group called FS CMa stars [3]. The observational defining criteria are (1) the presence of a hot star (between O9 and A2 spectral types) continuum with emission lines of H I, Fe II, O I, [Fe II], [O I], Ca II; (2) an infrared spectral energy distribution (SED) that shows a large excess with a maximum at 10–30 μm and a strong decrement beyond these wavelengths; and (3) a star location outside a region of star formation. According to these properties, almost all the UnclB[e] objects in Lamers et al.'s publication are in this group [4], which has approximately seventy members between confirmed and candidate ones [5]. They are suspected to be binaries at a post-mass-exchange evolutionary phase, with a secondary component fainter and cooler than the primary or degenerate [3,6,7]. Since the predictions of mass-loss rates of single-stars theory [8,9] cannot explain the existence of a large amount of CS matter, a mass-transfer process in a binary system could be a likely explanation. However, only a few objects of this class have been confirmed as binary systems, probably due to the scarcity of available observational data to discover them and the difficulties in detecting signs of binarity due to the presence of CS matter, the intrinsic stellar variability, and the low brightness of most members of the FS CMa group [10]. Recently, Miroshnichenko et al. [11] published a review of FS CMa objects, where they reported fifteen stars as binaries and six as binary system candidates.

MWC 645 (= V2211 Cyg, α = 21:53:27.49, δ = +52:59:58.01; V = 13.0, H-K = 1.53, J-H = 1.267) was originally included in the supplement of the Catalogue of Mount Wilson about A and B stars with bright H I spectral lines [12]. The presence of strong double emission lines of Fe II and [Fe II] (with a radial velocity difference between the red and blue peaks of 150 km s^{-1}) and triple-peaked profiles of the Hγ and Hδ transitions were reported by Swings and Allen [13]. They also remarked striking spectral similarities between MWC 645 and η Car. Also, permitted and forbidden transitions of low excitation and ionization potential belonging to Ti II, Cr II, [O I], and [N II] were observed [1]. Photometric variations were found by Gottlieb and Liller [14] with an amplitude of 0.3 mag and a possible period of 23.6 years. A deep spectroscopic study was done by Jaschek et al. [15] that revealed no stellar absorption features. These authors concluded that possibly MWC 645 is a late B-type object based on the absence of He II lines and the weakness of the He I lines at λ 6678 Å and λ 7065 Å possibly detected once, each one at different years. They did not find spectral transitions from C, Ne, and Mg atoms or ions, but they found lines of K I and Cu II (typically seen in stellar types later than F) and Zr II (usually seen in stars later than A0-type). They highlighted the extreme spectroscopic variability of MWC 645 over the years. Lamers et al. [2] included it in the UnclB[e] stars group. MWC 645 has IRAS flux ratios that locate it in the region occupied by OH/IR stars [16]. Zickgraf [17] detected a characteristic asymmetric profile for the emission metal lines, with a steep red flank and a blue wing. He reported the splitting in the central emission of [O I] and [Fe II] lines and a peculiar emission profile of the Hα line showing a broad blue and a narrow red component with a full width at half maximum (FWHM) of 5.0 Å and 1.3 Å, respectively. He proposed a latitude-dependent wind model with a large optical depth dust disk at an intermediate inclination to explain the asymmetric line profiles and their splitting. Marston and McCollum [18] obtained Hα narrow band imaging and found no visible extended emission associated with the star.

Recently, Nodyarov et al. [19,20] studied high-resolution optical spectra of MWC 645 taken in two different years. He found absorption lines of neutral metals, such as Li I, Ca I, Fe I, Ti I, V I, and Ni I, typically present in cool stellar spectra, with a different average radial velocity in each spectrum, that revealed the binary nature of the object. However, they did not find any absorption line typical of a B-type object in any of their spectra. They disentangled the contribution of each stellar component and estimated their surface temperatures and luminosities (T_{eff} = 18,000 ± 2000 K and 4250 ± 250 K, log (L/L$_\odot$) = 4.0 ± 0.5 and 3.1 ± 0.3 for the hot and cool components, respectively). Low-resolution near-IR spectra displayed emission lines of the H I Paschen and Brackett series, as well as of Fe II, O I, N I, and He I. Photometric monitoring in the optical and near-IR regions showed quasi-cyclic variations of both short and long periods (months and ∼4 years, respectively). The authors conclude that the star can be classified as an FS CMa-type object, where its intermediate-mass components (7 $M_\odot$ and 2.8 $M_\odot$) undergo an ongoing mass-transfer process. According to the shape displayed by the spectral energy distribution with weak emission peaks at about 10 µm and 18 µm, they inferred the presence of silicates in an optically thin dusty shell. MWC 645 is one of the eight FS CMa objects in which absorption lines of neutral metals typical of late-type secondaries have been detected. To contribute to the study of this intriguing object, we decided to observe it in the near-IR to search for signatures of both stars, mainly of the cool component, that help to characterize it. In addition, the acquisition of new optical spectra and the public availability of data (spectroscopic and photometric) that could shed some light on this complex system motivated us to analyze them. The paper is organized as follows: We present the infrared and optical observations used in this work in Section 2. In Sections 3 and 4, we analyze the data. In Section 5, we discuss the results. Finally, Section 6 contains the main conclusions.

2. Observations

2.1. Near-Infrared Spectra

Near-infrared spectra were taken using the Gemini Near-Infrared Spectrograph (GNIRS, [21]) attached to the 8 m telescope at GEMINI-North (Hawaii) under the programs GN-2017A-Q-62, GN-2018A-Q-406, and GN-2022B-Q-225. On 6 June 2017, we obtained K-band spectra in long-slit mode centered at 2.35 µm. The instrumental configuration used was a 110.5 l/mm grating, a 0.3 arcsec slit, and the short camera (0.15 arcsec/pix). We also acquired spectra with the same configuration but in cross-dispersed mode centered at 2.19 µm and 2.36 µm on 24 and 30 July 2018, respectively. The effective spectral coverage by these set-ups was 0.90 µm–2.27 µm and 0.85 µm–2.45 µm, respectively, with gaps between the orders (the interval 1.36 µm to 1.46 µm is unusable due to saturated telluric lines). The resulting mean spectral resolving power of the spectra was R ∼ 5500. On 24 August 2022, L-band spectra were obtained with a different long-slit configuration: a 31.7 l/mm grating, a 0.1 arcsec slit, and the long camera (0.05 arcsec/pix), with two different central wavelengths (3.48 and 4.00 µm). This configuration resulted in R ∼ 5100. The spectra were taken in two ABBA nodding sequences along the slit. To account for telluric absorption, a late-B- or an early-A-type star close to the target in both time and position was observed. Stars of these spectral types are featureless in the observed wavelength range, except for hydrogen absorption lines that can be successfully removed in the reduction process by fitting theoretical line profiles. Flats were also acquired. The data were reduced with the Image Reduction and Analysis Facility (IRAF)/Gemini tasks. The sky contribution was removed by subtracting the AB pairs. The spectra were flat-fielded and telluric corrected. The wavelength calibration was performed using the telluric lines. The data were normalized to unity.

A and B positions were added to increase the signal-to-noise ratio (S/N). The final S/N ratio varies for the different spectral ranges, as it is affected by the quality of the telluric correction. Some regions are very polluted with telluric lines and it was impossible to make a complete cancellation, thus some residuals remain. In addition, for some spectral regions heavily crowded by emission lines, it becomes difficult to make accurate S/N ratio estimates. Table 1 summarizes the mean values of the S/N ratio for all our GNIRS near-IR observations.

Table 1. Mean values of the S/N ratio for GNIRS near-IR observations in different spectral ranges.

Observations Program ID	Spectral Range [Å]	Mean S/N Ratio
GN-2017A-Q-62	22,570–24,400	200
GN-2018A-Q-406	21,000–24,500	200
GN-2018A-Q-406	15,800–18,200	100
GN-2018A-Q-406	11,000–13,600	90
GN-2018A-Q-406	8400–11,000	60
GN-2022B-Q-225	33,311–36,446	200
GN-2022B-Q-225	38,500–41,786	100

2.2. Complementary Data

Optical observations were carried out at Ondřejov Observatory, Czech Republic, using the Coudé spectrograph [22] attached to the Perek 2 m telescope. We obtained spectra with a resolving power of R $\sim$ 12,000 covering a spectral range from 6262 Å to 6735 Å on 12 and 13 September 2018. We also acquired a spectrum centered at 8600 Å. We used a grating of 830.77 l/mm with a SITe 2030 × 800 CCD and a slit width of 0.7 arcsec. Additional observations were done at Tartu Observatory, Estonia, with the 1.5 m Cassegrain reflector AZT-12 on 1 November 2021, using the long-slit spectrograph ASP-32 with a 600 l/mm. The wavelength coverage extended from 5450 Å to 7480 Å. Data were processed using standard IRAF tasks. Spectra were bias and flat-field corrected, wavelength calibrated, heliocentric velocity corrected, and flux normalized.

We also searched for available optical spectra in the BeSS database [23]. We downloaded sixteen spectra taken between 2019 and 2022, with a resolving power R $\sim$ 14,000/16,000 in the spectral range 6500–6600 Å. In addition, we collected a lower resolution spectrum (R $\sim$ 5000) acquired on 30 August 2019 that covers the range 6150–7000 Å. The spectra were corrected by heliocentric velocity and normalized to the continuum using the standard IRAF tasks. We chose the same sample of continuum points to normalize all spectra. The telluric correction has not been applied.

In addition, we extracted from the public database ASAS-SN (All-Sky Automated Survey for Supernovae, Shappee et al. [24], Kochanek et al. [25]) (https://www.astronomy.ohio-state.edu/asassn/ (accessed on 2 February 2023)) survey photometric data of this star obtained over eight years. The collection is composed of V-band magnitudes from 16 December 2014 to 29 November 2018 and data in the g-band from 12 April 2018 up to 17 January 2023. Furthermore, we collected ground- and spaced-based multicolor photometry from Vizier service from 0.3 µm to 140 µm.

3. Analysis of the IR Data

Figure 1 shows the near-IR spectrum of MWC 645 from 8400 to 13,600 Å. It displays numerous emission lines, particularly of the H I Paschen series. The strongest lines, except those of H I, correspond to O I, Fe II, and the Ca II triplet. Many transitions of N I can be identified in emission along this spectral range. Forbidden lines of [Fe II] and [S II] are also present. The moderate-resolution data reveal several absorption lines of Fe I that could be associated with the cool stellar companion.

We carefully searched for He I lines in our spectra. If the lines are present, they are incipient and hidden in the noise. The most intense transitions in the interval λ8400–24,500 Å (cited in the NIST database [26]) correspond to λ10,830 Å and λ20,587 Å. Nodyarov et al. [20] reported the presence of the He I λ10,830 Å transition in emission from a low-resolution spectrum (R $\sim$ 700). We identified a group of Fe II lines with the three emission peaks at the interval 10,826–10,862 µm. However, the bluest feature of this group (see Figure 1) is broad, and thus, the He I λ10,830 Å line could be blended with the Fe II lines. Our data have a higher resolution than the spectrum of Nodyarov et al., but not sufficient to separate the Fe II lines from the He I line. These authors also identified the He I λ20,587 Å line. Unfortunately, it lies outside our spectral coverage.

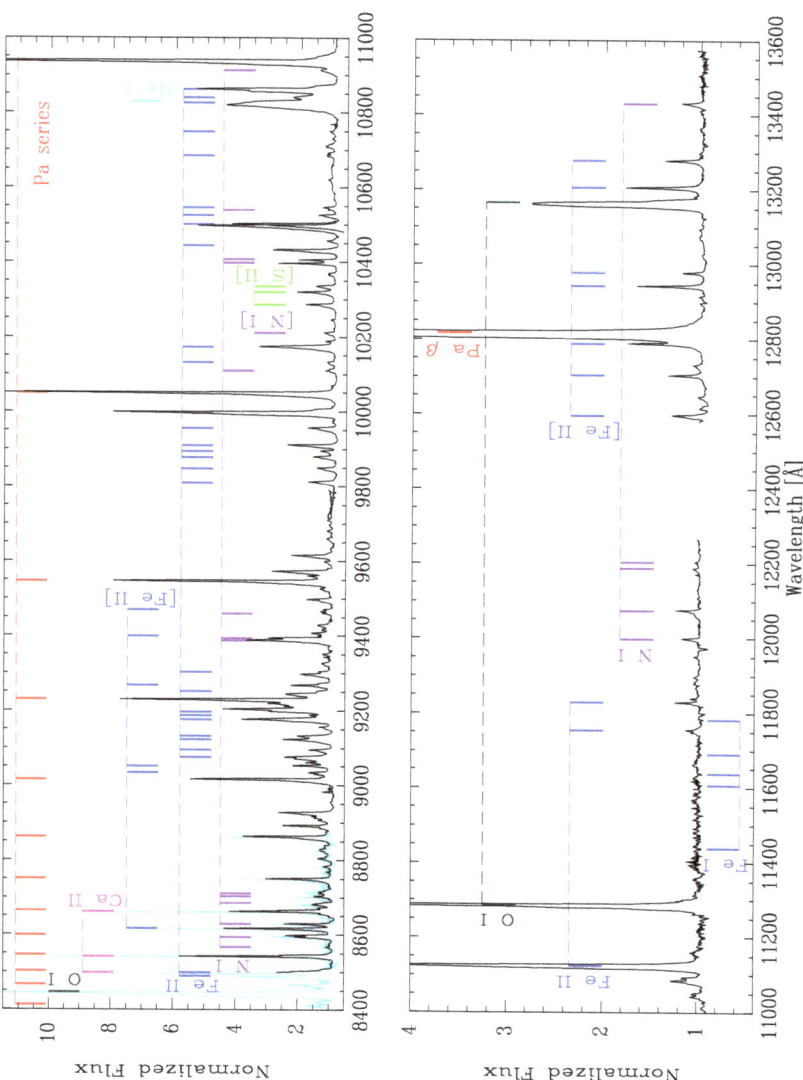

Figure 1. Normalized medium-resolution spectrum of MWC 645 taken with Gemini/GNIRS on July 2018 from 8500 Å to 13,600 Å. The normalized Ondřejov spectrum from 8400 Å to 8870 Å, acquired on September 2018, is shown in cyan. Main spectral lines are identified by colored markings. The spectral features of a given element (either permitted or forbidden and of different ionization states) are joined by a dashed line of the same color: hydrogen is indicated in red, oxygen in gray, iron in blue, calcium in pink, nitrogen in violet, sulfur in green, and helium in cyan. Wavelengths are given in angstroms.

The H-band spectrum of MWC 645 (upper panel of Figure 2) is dominated by the H I Brackett series and several permitted and forbidden lines of Fe II. In the K-band, the most intense feature is the Brγ line (see lower panel of Figure 2), which stands out among several emission lines corresponding to Fe II, [Fe II], and presumably [Ni II]. The Mg II doublet at $\lambda\lambda$ 21,374 Å and 21,437 Å is also in emission. The Pfund series extends from 2.3 microns longward. In addition, absorption features of neutral metals characteristic of late-type stars, such as Ca I, Mg I, and Na I, are present. For the first time, we have detected the presence of CO band heads in absorption around 2.3 µm and around 1.6 µm, which are typical photospheric features of late-type luminous stars.

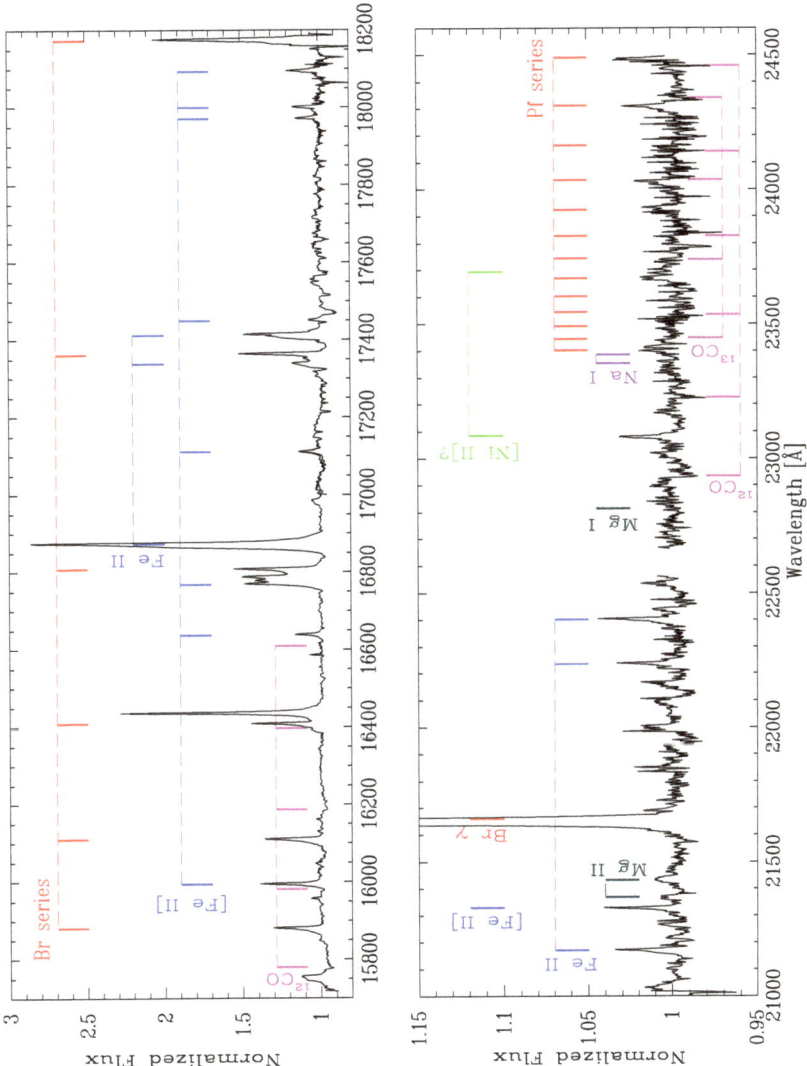

Figure 2. Normalized medium-resolution spectrum of MWC 645 taken with Gemini/GNIRS in 2018, covering the *H*-(upper panel) and *K*-bands (lower panel). Main spectral lines and molecular bands are identified by colored markings. The spectral features of a given element (either permitted or forbidden and of different ionization states) or molecule (of different isotopes) are joined by a dashed line of the same color: hydrogen is indicated in red, magnesium in gray, iron in blue, sodium in violet, nitrogen in green, and carbon monoxide in pink. Wavelengths are given in angstroms.

Figure 3 shows the first obtained *L*-band spectrum of MWC 645 in two different spectral regions. The first interval between 33,310 Å and 36,410 Å is relatively featureless (see left panel), except for the presence of the permitted emission line of Fe II λ 35,423 Å, which is clearly seen above the continuum level, and some H I lines of the Humphreys series in emission, where the strongest is the one corresponding to the 20-6 transition. Lower-order members of the Humphreys series can be seen in the second spectral interval (right panel) that ranges from 38,520 Å to 41,730 Å, where the strong emission of the Brα line can also be observed. We searched for absorption bands of the first-overtone of silicon monoxide (SiO) around 4 µm but found none.

Figure 4 plots three H I lines: Paβ, Brγ, and Brα. Their profiles are single-peaked but asymmetric. We measured the total equivalent width (EW) of each line using the 'e' function in the IRAF splot routine. The total measured EWs are 184 Å, 15 Å, and 48 Å, respectively. The percentage uncertainty of the EW measurements is 5%. Unfortunately, the wavelength calibration of our near-IR spectra is not accurate enough to determine reliable radial velocity measurements.

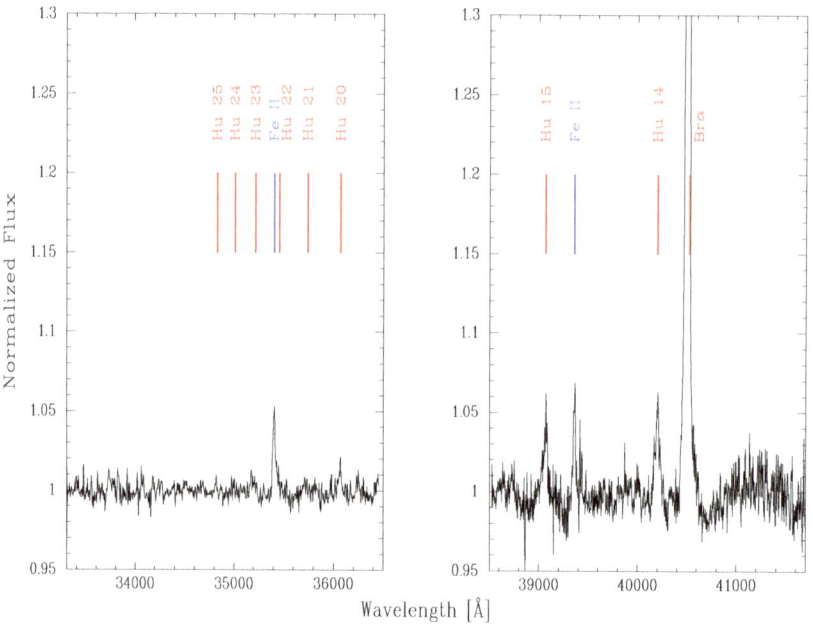

Figure 3. Normalized *L*-band spectrum of MWC 645 obtained in 2022. The emission lines of H I and Fe II are marked in red and blue, respectively. The "bump" longward of the Brα line is a remnant from telluric correction. Wavelengths are in angstroms.

Co Absorption Bands

Figure 5 (upper panel) displays the second-overtone band heads of ^{12}CO in absorption from the 2018 *H*-band spectrum. The lower panel compares the *K*-band spectra taken in 2017 (in red) and 2018 (in black), where the variation of the first-overtone band heads of ^{12}CO is clearly seen. The positions of the ^{13}CO band heads are also marked and clearly detected in the spectrum from 2017. The 2018 spectrum is too weak to see these faint features.

The strength of the CO absorption bands in the near-IR spectra of classical late-type stars depends on the stellar effective temperature, T_{eff}, and surface gravity, log g [27,28]. The CO absorption becomes deeper when the effective temperature decreases and the luminosity increases. Thus, hot star spectra display no trace of CO features ($T_{eff} \geq$ 5800 K–6000 K, Ali et al. [29]), and dwarf stars present weaker CO absorption bands than supergiants. To characterize the cool companion of MWC 645, responsible for the CO absorption features, and estimate its fundamental parameters, we used the IRTF (NASA Infrared Telescope Facility) Spectral Library [30,31], which collects stellar spectra observed with the spectrograph SpeX at a resolving power of R ∼ 2000 and an S/N ratio of about 100 at $\lambda < 4$ μm. We looked for late-type stars with spectral types between F and M and luminosity classes between I and V to compare their spectra with our spectrum from 2017 in the wavelength range from 2.26 to 2.44 μm. Figure 6 shows this comparison. The MWC 645 spectrum (solid black line) was degraded to the resolution of the template spectrum (dashed red line), which corresponds to a G0 Ib-II star (HD 185018). The intensity of the first ^{12}CO band head of MWC 645 coincides reasonably well with that of the early G-type star; however, the rest of the band heads are less intense. The blue edge of the CO(2-0) band head might present an incipient emission. The absorption of the first ^{13}CO band head is more intense than that displayed by the library star. According to Wallace and Hinkle [32], the ^{13}CO isotope is prominent in the supergiants and giants but is not apparent in the dwarfs, although its strength also depends on the initial rotation velocity of the star and the mixing processes that can cause a surface enrichment in ^{13}C. Otherwise, the absorption lines of neutral metals are less intense than those in the template spectrum, indicating an earlier spectral type (F8-F9 subtypes). Furthermore, the lack of SiO band heads at 4 μm, often observed in K0-type stars and later, also points towards an earlier type [33].

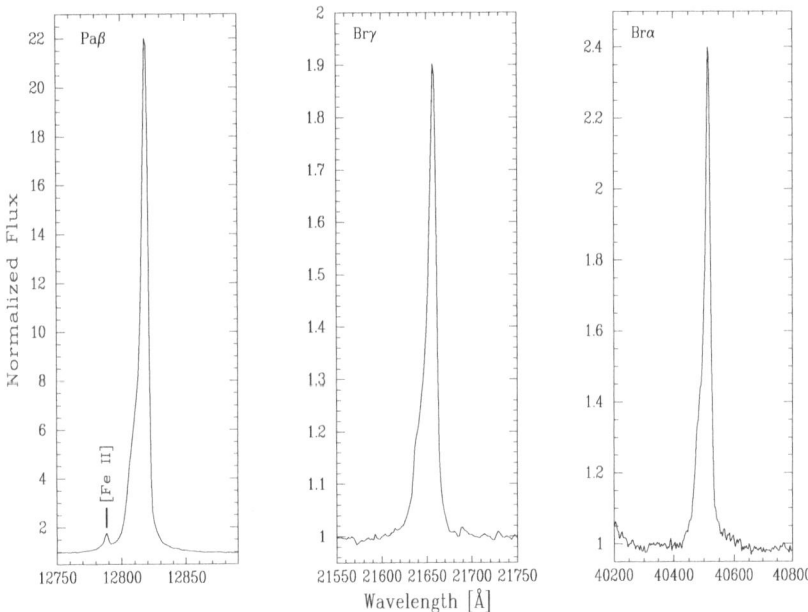

Figure 4. Strongest H I lines detected in our IR spectra of MWC 645. They display an asymmetric profile, where the red flank is steeper than the blue one. Wavelengths are in angstroms.

Winge et al. [34] presented a spectroscopic library of late spectral-type stellar templates in the K-band at a resolving power of R ∼ 5900. The authors plotted the equivalent width (EW) of the first CO overtone as a function of T_{eff} (see their Figure 3) for a stellar sample with T_{eff} in the range 3200–5200 K and different luminosity classes. They measured the EW from the blue edge of the (2-0) band head to the blue edge of the (3-1) band head, more precisely in the window 2.293–2.322 μm. We measured the EW of the CO(2,0) band head from the spectrum of 2017 and obtained 2.55 ± 0.5 Å. A visual extrapolation of the relation seen in the figure between EW and T_{eff} in the hottest edge of the plot gives an estimation of T_{eff} around 5200 ± 100 K. From the spectrum obtained in 2018, we measured an EW of the CO(2,0) band head equal to 1.22 ± 0.1 and estimated a $T_{eff} \sim 5300 \pm 100$ K.

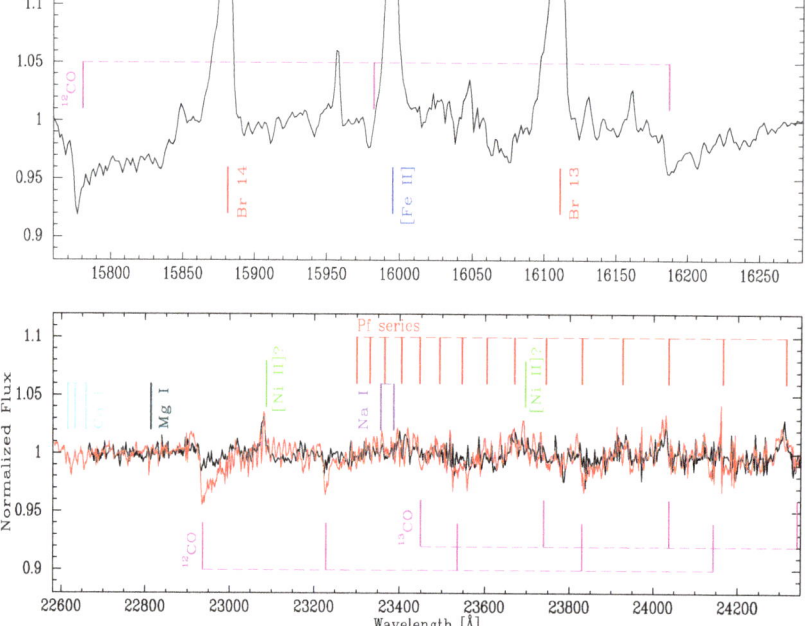

Figure 5. CO molecular bands of MWC 645. Some emission and absorption lines are also identified. The spectral features of a given element or molecule (of different isotopes) are indicated by colored markings and joined by a dashed line of the same color: hydrogen is indicated in red, magnesium in gray, iron in blue, sodium in violet, nitrogen in green, calcium in cyan, and carbon monoxide in pink. Upper panel: ^{12}CO second-overtone band heads seen in the *H*-band spectrum taken in 2018. Lower panel: ^{12}CO and ^{13}CO band heads in absorption detected in the *K*-band. The spectra obtained in 2017 (in red) and 2018 (in black) revealed the variability in the strength of the observed bands. Wavelengths are given in angstroms.

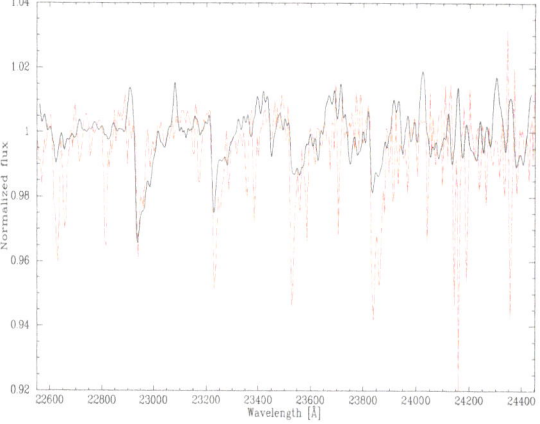

Figure 6. Comparison between the degraded spectrum of MWC 645 to R ∼ 2000 (solid black line) and a G0 Ib-II star, HD 185018 (dashed red line), where the CO(2-0) band head fits well. The other ^{12}CO absorption bands from MWC 645 are shallower than those of the template; perhaps they are filled by emission. Wavelengths are given in angstroms.

4. Analysis of the Optical Data

4.1. Photometric Light Curve

Figure 7 shows the light curve of MWC 645 taken from ASAS-SN. We applied the relationship derived by Nodyarov et al. [20] to convert the g-band magnitudes into V-band magnitudes. The optical photometry in the V-band acquired by the authors mentioned above is also included. The dates of the spectroscopic observations presented in this work are marked in the light curve as vertical lines.

The brightness fluctuations of the star up to July 2022 have been reported by Nodyarov et al. [20], who suggested that a new minimum in the light curve might take place in the second half of 2022. As can be seen in the plot, the star continued fading up to the end of October approximately, reaching a minimum of $\sim$0.1 mag brighter than the minimum that occurred in August/September 2018. Then, it began to strengthen in brightness again.

Nodyarov et al. [20] searched for periodicity, excluding visual magnitudes greater than 13.2 mag from their analysis. They derived 69, 145, and 295 days. They attributed the quasi-cyclic photometric variations to variable CS extinction. We applied the Lomb–Scargle method using the IRSA (https://irsa.ipac.caltech.edu/irsaviewer/timeseries (accessed on 30 November 2022)) time series tool to the V-band light curve shown in Figure 7. We discarded the magnitudes with errors greater than 0.03. The scan of periodic signals with values below ten days gave strong peaks at one day and harmonics of the sidereal day due to the observing cycle.

The periodogram for periods greater than one day is shown in Figure 8. The six peaks at or above a confidence level of 20 in the power spectrum correspond to periods of approximately 65, 112, 162, 298, 461, and 709 days. We dismissed the last period since the time coverage of the observations is not enough for its precise determination. The phase diagram for each period shows a large scatter of the magnitude points and a small amplitude in their modulation ($\sim$0.2–0.3 mag). We should note that the 65- and 298-day periods were also found by Nodyarov et al. [20]. As our data spread over a more extended baseline than the one used by the authors mentioned above, this might be a possible explanation for the differences in the other identified periods.

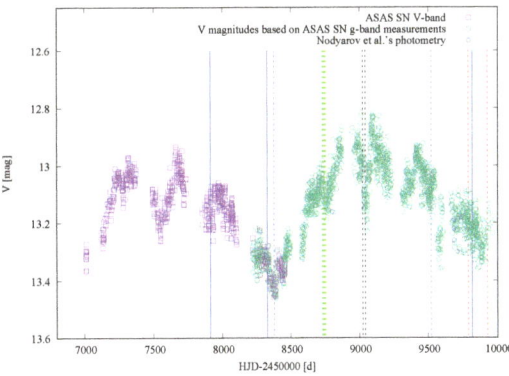

Figure 7. Light curve of MWC 645 from 16 December 2014 up to 11 November 2022 taken from ASAS-SN. The purple squares indicate the V-band magnitudes, and the green circles represent the g-band measurements converted to the V-band magnitudes. The conversion has been carried out with the relationship found by Nodyarov et al. [20]. Their optical photometric observations are also included (blue circles). Vertical solid blue lines mark the dates of our IR observations (2017, 2018, and 2022, respectively); dotted blue lines mark the dates of our optical spectra (2018 and 2021, respectively); and those dotted in green, gray, and red correspond to the spectra downloaded from the BeSS database taken in 2019, 2020, and 2022, respectively. Time is given in heliocentric Julian dates (HJD) minus 2.45×10^6 days.

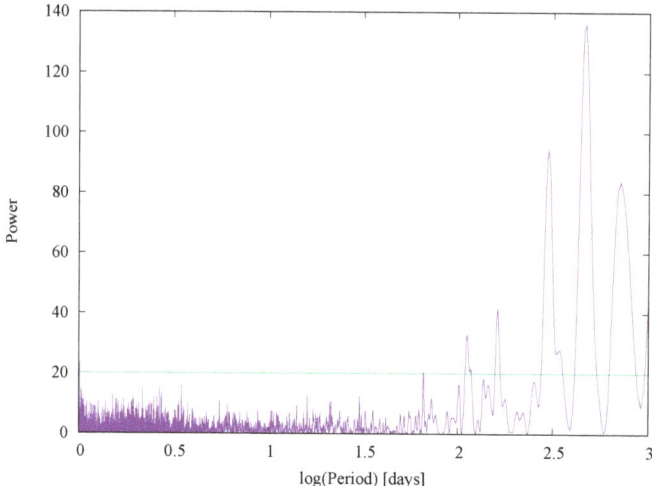

Figure 8. Fourier power spectrum of the ASAS-SN light curve of MWC 645. The periods in days are on a logarithmic scale. The green line shows the confidence level.

4.2. Spectroscopic Data

The peculiar profile of the Hα line of MWC 645, composed of a broad blue-shifted peak and a narrow red-shifted one, can be seen in Figure 9, where all the spectra are normalized to the continuum level. This plot shows not only the profile changes over different years (spectra from the same year are displayed in the same color) but also daily. We note that as the BeSS spectra are not corrected by telluric lines, some of the profiles present one or two absorption features superimposed on the blue-shifted emission peak corresponding to water vapor lines of the Earth's atmosphere, which affect the shape of the profile. A variation in the emission strength of both peaks is seen. Using the 'e' task in the IRAF splot routine, we measured the intensity of the blue (V) and red (R) emission peaks and the total equivalent width of the Hα profile, except for the lowest resolution spectrum. Table 2 presents these values and the calculated V/R ratios. We can see that the V/R ratio presents changes over four years, even doubling its value. We note that the ratio of V/R $\sim$ 0.3 corresponds to observations close in time (except for one). For this subset, the changes in EW might be mainly due to the continuum-level variations.

The average radial velocity of the blue and red emission components derived from the Ondřejov spectra are -225 ± 5 km s^{-1} and -31 ± 3 km s^{-1}, respectively, which are in agreement with the values reported by Zickgraf [17] of -218 km s^{-1} and -30 km s^{-1} and Nodyarov et al. [20] of -252 ± 9 km s^{-1} and -30 ± 2 km s^{-1}, respectively. Fitting a Gaussian profile to the narrow red component of the Hα line, we obtained an average FWHM of 90 ± 1 km s^{-1}. To fit the broad emission component, we built a profile with a red wing symmetrical to the observed blue one and obtained an average FWHM of 318 ± 4 km s^{-1}.

Even though the BeSS material is not accurate enough to measure radial velocities, we have estimated them from the different spectra for both emission peaks fitting the components with Gaussian profiles. The average value is -229 km s^{-1} and -26 km s^{-1} for the blue and red emission peaks, respectively. In Figure 9, a variation in the central wavelength of the Hα red emission peak (which is not distorted by telluric lines) can be observed from the different spectra; however, as the wavelength calibration of the BeSS spectra is not well suited for radial velocity determination, we cannot confirm if this change is real. The average FWHMs of the blue and red components are 256 ± 4 km s^{-1} and 80 ± 2 km s^{-1}, respectively. The Hα broad component line profile from the BeSS spectra has a smaller average FWHM than the Ondřejov spectra. In the latter, the broad component presents a blue wing with a gentler slope that is also outlined for the red wing, giving a greater width.

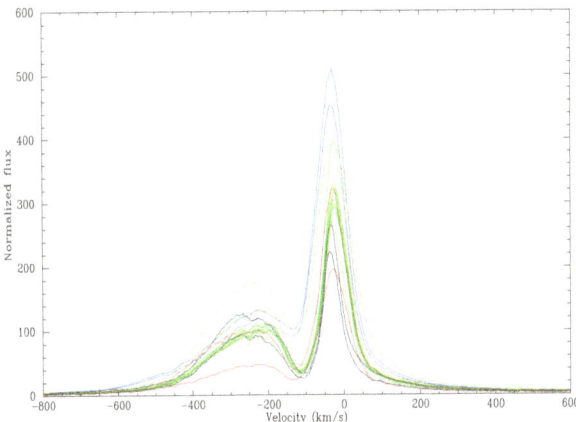

Figure 9. Hα line variation of MWC 645. Spectra taken in 2018, 2019, 2020, and 2022 are displayed in blue, green, gray, and red, respectively. The spectrum with the lowest resolution is plotted with a dashed-line. The heliocentric radial velocity scale is shown in km s^{-1}.

Table 2. Hα line parameters of MWC 645. Column 1 indicates the observing date; column 2 the heliocentric Julian date (minus 2.45×10^6 d); column 3 the observatory/database where the spectrum was obtained; columns 4 and 5 the emission intensities of the blue and red peaks (V and R) in continuum units, respectively; column 6 the emission intensity ratio of both peaks (V/R); and column 7 the total equivalent width (EW) in Å. The measurement errors are on the order of 1% for the intensities and 10% for the EW values.

Obs. Date (yyyy-mm-dd)	HJD-2450000	Observatory	V	R	V/R	EW [Å]
2018-09-11	8373.3290	Ondřejov	118.4	452.0	0.26	−1980.3
2018-09-12	8374.2823	Ondřejov	132.8	509.0	0.26	−2122.8
2019-08-28	8724.3870	BeSS [1]	90.2	268.9	0.33	−1132.2
2019-08-30	8726.4279	BeSS [2]	75.1	391.9	0.45	−2130.7
2019-09-01	8728.4429	BeSS	95.7	283.9	0.34	−1237.2
2019-09-02	8729.4419	BeSS	105.7	316.0	0.33	−1291.7
2019-09-03	8730.4365	BeSS	101.5	306.1	0.33	−1353.7
2019-09-05	8732.4433	BeSS	100.1	292.3	0.34	−1147.3
2019-09-07	8734.3942	BeSS	109.7	323.5	0.34	−1372.1
2019-09-10	8737.3883	BeSS	99.8	296.0	0.34	−1292.5
2019-09-11	8738.3925	BeSS	132.3	397.1	0.33	−1642.3
2019-09-14	8741.4241	BeSS	110.5	330.9	0.33	−1348.1
2019-09-19	8746.3989	BeSS	100.5	297.7	0.34	−1129.5
2019-09-20	8747.3862	BeSS	103.1	304.3	0.34	−1264.6
2019-09-24	8751.3722	BeSS	109.1	326.1	0.33	−1283.5
2020-06-25	9025.5535	BeSS	94.1	223.7	0.42	−933.8
2020-07-12	9043.5126	BeSS	127.4	264.8	0.48	−1216.8
2021-11-01	9520.3161	Tartu	—	—	—	—
2022-07-26	9787.5322	BeSS	47.3	195.6	0.24	−718.2
2022-12-09	9923.2775	BeSS	100.5	321.2	0.31	−1392.3

[1] For details about the instruments and observers, please visit the web page BeSS database (http://basebe.obspm.fr/basebe/) (accessed on 2 February 2023). [2] Spectrum with the lowest resolution.

Apart from the Hα line, our medium-resolution spectroscopic observations and the 2019 low-resolution BeSS spectrum also show the lines of [O I] λλ 6300, 6364 Å. These lines appear single-peaked, although asymmetric, as was previously mentioned by Zickgraf [17], which might suggest that the lines are composed of two blended components (see Figure 10). Several permitted Fe II lines and the forbidden lines of [N II] λ 6583 Å and [S II] λλ 6716, 6731 Å are apparent. We calculated the average heliocentric radial velocity of the emission lines of the Ondřejov spectra and the standard error of the mean, obtaining -43 ± 2 km s^{-1}. Jaschek et al. [15] derived -76 ± 5 km s^{-1}, and Nodyarov et al. [20] found -61 ± 4.3 km s^{-1}, which indicates a variation in radial velocity. The He I λ 6678 Å transition is absent, as in the spectra studied by the last-mentioned authors and Zickgraf [17].

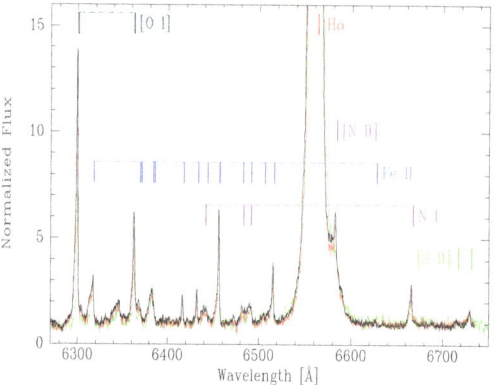

Figure 10. Example of the emission lines in the surroundings of the Hα line of MWC 645. Ondřejov normalized spectra taken in 2018 on 12 September (in red line) and 13 September (in black line) and the 2019 low-resolution spectrum (in green) are shown with the main lines identified by colored markings. The spectral features of a given element (either permitted or forbidden and of different ionization states) are joined by a dashed line of the same color: hydrogen is indicated in red, oxygen in gray, iron in blue, nitrogen in violet, and sulfur in green. Wavelengths are given in angstroms.

4.3. Global Properties of the Circumstellar Material

Figure 11 shows the spectral energy distribution (SED) of MWC 645, built from the photometry publicly available from the ultraviolet to the far-IR (0.3 μm–140 μm). The low-resolution spectrum taken in 2021 in the spectral region of the Hα line is also included.

To derive the global physical properties of the CS material, we considered a simple model presented by Marchiano et al. [35] and Arias et al. [36]. The numerical code allows for obtaining the SED assembled by different envelope components. The model assumes the presence of a spherical envelope composed of gas close to the star ($\leq 5\,R_*$) and (or) dust further away from it ($\geq 100\,R_*$) [37,38]. The emergent flux is computed from the central star and the envelope (considering that the latter can be reduced to an equivalent shell), applying a plane-parallel solution for the transfer equation. The optical depth τ_λ^G and the source function characterize the gaseous shell, which can be described by adding as free parameters the electron temperature T_G and the effective radius R_G. The dusty region is treated using an analogous scheme with similar parameters describing the shell: an optical depth τ_λ^D, a temperature T_D, and an effective radius R_D. The model allows several dust shell components to be added. The interstellar extinction is also included by an optical depth τ_λ^{ISM}. The absorption A(λ) is related to each optical depth through the expression $\tau = 0.4 \ln(10)\,A(\lambda)$. Using the law given by Cardelli et al. [39], it can be written as $A(\lambda) = [R_V\,a(1/\lambda) + b(1/\lambda)]\,E(B-V)$, where R_V is the total to selective extinction and E(B-V) is the color excess. We took $R_V^{ISM} = 3.1$ for the interstellar dust and tried different values of R_V^D greater than 3.1 for the CS dust shell components. The temperature of the dust grains depends on the stellar radiation and the distance from the star center [40]. That is, $T_D(r) = T_{eff} W(r)^{[1/(4+p)]}$, where W(r) is the geometrical dilution factor. The parameter p depends on the nature of the dust, but it is usually on the order of one. Furthermore, as the equilibrium temperature should be lower than the dust condensation temperature (typically around 1500 K) to allow the formation of grains, it constrains the distance where condensation can occur, e.g., for a T_{eff} of 18,000 K, the condensation distance is about 249 R_*.

The code gives the observed flux normalized to that at a reference wavelength λ_{ref}; we chose λ_{ref} = 0.55 μm. We assumed that the flux of the central object results from the contribution of both stellar fluxes, for which we considered the Kurucz [41] atmosphere models. Regarding what we know about the stars, we selected models between 17,000 and 20,000 K for the hot binary component and between 4500 and 6000 K for the cool one and explored the flux contribution of the hot and cool components to the total flux in the range of 70%–90% and 30%–10%, respectively (similar percentages were suggested by Nodyarov et al. [20]). Figure 11 displays our best fit of the theoretical SED (solid blue line) to the observed data. The best-fitting model was computed by taking the photospheric fluxes from a star with T_{eff} = 18,000 K, log g = 4.0, and R_* = 3.73 $R_\odot$ and a cool star with T_{eff} = 5000 K and log g = 1.5. The contribution of each stellar flux to the total flux is 80% and 20% for the hot and cold components, respectively. The resulting envelope has one gaseous shell at R_G = 1.15 $R_\odot$ with T_G = 16780 K and τ_V = 0.1. The dusty region comprises three different shells with the following parameters: R_D^1 = 348 R_*, T_D^1 = 1310 K, R_D^2 = 3750 R_*, T_D^2 = 507 K, R_D^3 = 0.02 pc, T_D^3 = 98 K. The computed total CS visual absorption is A_V^D = 0.097 mag. We obtained a color excess due to the interstellar medium of $E(B-V)^{ISM}$ = 0.98 ± 0.02 mag, which results in a total visual absorption A_V = 3.13 ± 0.11 mag. This value agrees with the one derived by Nodyarov et al. [20]. A disagreement between the theoretical and observed SEDs in the region of log λ = 1.0–1.5 is observed. We should note that the employed code can model the thermal emission of the dust, but it cannot address the computation of silicate bands. Thus, the presence of silicate particles might be responsible for the observed difference between the SEDs at 10 μm and 18 μm. In fact, Nodyarov et al. [20] have already reported weak emission bumps at these wavelengths.

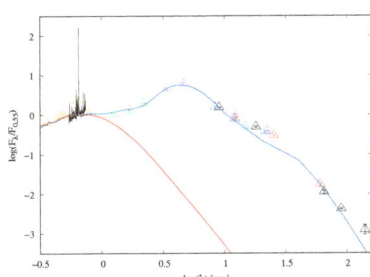

Figure 11. Spectral energy distribution of MWC 645. The open triangles represent the observed photometric data: optical bands (yellow), 2MASS (green) [42], WISE (violet) [43], MSX (light blue) [44], IRAS (red) [45], and AKARI (black) [46]. The error bars of the photometric data are included (in most photometric bands, they fall inside the symbols). The low-resolution spectrum acquired in 2021 over the Hα region is also displayed (with a dashed black line). The solid red line shows the SED modeled considering the contribution of the photospheric fluxes from both stars, the thermal emission from a gaseous shell close to the system and the effect of the interstellar medium extinction. The solid blue line shows our best-fitting theoretical SED, obtained by adding to the SED plotted in red the contribution of three dusty shells surrounding the stellar system. The flux is normalized to that at λ_{ref} = 0.55 μm and displayed on a logarithmic scale. The wavelengths are in microns.

5. Discussion

The forest of emission features in the IR spectral ranges presented in this work account for the presence of large amounts of CS gas embedding and veiling MWC 645, which makes it challenging to characterize the components of the binary system. In addition, the existence of dust revealed by the strong IR excess of radiation above the photospheric fluxes is another element to consider when building a probable scenario.

In previous studies, the hot component of the system was considered to be a B-type star [13,15], until Nodyarov et al. [20] assigned it as an early-B subtype. This assignment was mainly constrained by the absence of He II lines and the possible presence of He I absorption lines suggested by Jaschek et al. [15] (see Section 1) in the optical spectral range and the identification of the He I lines in emission in the near-IR. We were not able to detect neutral helium lines in either emission or absorption in our data, at least with a signal intensity above the noise level. However, we cannot discard a possible blend of the He I line at 1.083 µm (the transition of He I with the highest theoretical intensity that falls in our observed spectral ranges) with a group of Fe II emission lines precisely identified in the same spectral region. We detected Mg II lines in emission at 2.138 µm and 2.144 µm. According to the works of Clark and Steele [47] and Steele and Clark [48], who studied a representative sample of Be stars in the H- and K-bands, no evidence of He I features and the simultaneous presence of Brγ and Mg II lines in emission indicate a spectral type between B2 and B4. If He I lines are present, the spectral type is B3 or earlier. New high-resolution IR spectra could be valuable to clarify the presence of He I lines. Nevertheless, we cannot discard any variability in neutral helium lines.

The ^{12}CO absorption bands detected for the first time at 1.62 µm and 2.3 µm allow us to constrain the spectral type and effective temperature of the cool binary component. The spectral type derived by the best fit of the first ^{12}CO band head with that of a G0-type star does not agree with the strength of metallic lines, which are seen very weakly in the spectra of MWC 645. The effective temperature associated with this spectral type for supergiant/giant stars is around 5600 K [49,50]. Using the EW of the CO(2,0) band head, we determined a T_{eff} average value of about 5250 K. However, since the SED of the star in the near-IR has a significant contribution from the hot companion and from the ionized envelope (free-free emission), all spectral features from the cool companion are reduced in their intensity (due to the false continuum). Thus, the temperature determination should be interpreted as an upper limit to the effective temperature.

On the other hand, we might consider that instead of tracing a cool companion, the detected CO absorption might also uncover some mass-ejection episode, such as can be seen in the yellow hypergiant ρ Cas [51] or in the eruptive variable V838 Mon [52], when molecular absorption bands start to develop in the spectrum while the star's brightness fades. In the case of ρ Cas, the CO bands turn from absorption into emission when the star reaches the next maximum brightness phase in which it is too hot for molecule condensation in its extended outer layers [53]. This finding agrees with the scenario that mass was ejected into the environment, radiating while it is expanding and cooling. Since we have only two observations, we cannot trace the evolution of the molecular absorption along the light curve. However, it is interesting to mention that the most intense CO absorption (seen in 2017) occurs when the brightness begins to increase after a seeming local minimum. This observation presents an emission peak blueward of the CO(2,0) absorption band head that is blue-shifted at about 308 km s^{-1}. If it is indeed due to CO, it might suggest a high-velocity molecular outflow. Furthermore, the considerably weaker CO absorption spectrum observed in 2018 might be interpreted as partially filled with circumstellar emission. Moreover, the absorption-line spectrum ascribed to the cool companion might have originated in an optically thick disk. Polster et al. [54] suggested that the absorption features seen in the spectrum of the FS CMa star, MWC 623, are formed in an equatorial disk viewed nearly edge-on, which acts as a pseudo-photosphere.

When we look at the light curve, we see global qualitative similarities between the range mostly recorded in V-band magnitudes up to the deepest minimum (which occurred in August/September 2018) and that traced by the *g*-band magnitudes from this minimum up to January 2023. In both time lapses of about four years, we can distinguish a well-outlined dip (at HJD-2450000 $\sim$ 7500 d and $\sim$ 9000 d) less intense than the main minimum. Although they present slightly different shapes and depths, and the light curve afterwards reaches a different maximum magnitude level, they are alike. This similar pattern suggests that the dominant source of variability is the same. Variable CS (or circumbinary) extinction along the line of sight due to dust clumps might be responsible for these photometric variations [55–57]. Our simple model to fit the SED allowed us to derive the global properties of the dusty envelope. Despite its spherical geometry, it traces different components of optically thin dust, the first located at a distance of $\sim$6 AU. This distance agrees with the innermost dusty disk radius of $\sim$5 AU derived for the star FS CMa through aperture-synthesis imaging in the *L* and *N* bands [58]. Material orbiting the system at this distance at Keplerian velocity would have a period of $\sim$5 years. A warped inner edge of the disk can also produce variable extinction [59].

Previous studies of MWC 645 have not reported variations in the V/R ratio of the Hα line. Observations made over the last 30 years, although discontinuous in time, have not revealed an inversion in the intensity ratio of the peaks, always showing V/R < 1. We noticed that the V/R ratio derived from our spectra varies from 0.2 to 0.5 (see Table 2). We also calculated the V/R ratio from the intensities of both Hα peaks included in the work of Nodyarov et al. [20]. Their observations are from 2004 to 2021 and not continuous. We found a variable V/R ratio with values between 0.3 and 0.9. The only two spectra from dates included in the light curve range, October 2016 (HJD-2450000 $\sim$ 7680 d) and November 2021 (HJD-2450000 $\sim$ 9545 d), present a V/R ratio of 0.7 and 0.3, respectively. Spectroscopic monitoring of the Hα line could help scan and characterize this variability and search for any periodicity in the changes of the peak intensities that could be related to the rotation of a density perturbation in the disk [60] or an orbital motion [61]. Zickgraf computed line profiles assuming a latitude-dependent wind model with a dust disk and obtained similar line profiles to the observed Hα profile with V/R < 1 for an intermediate inclination angle.

The dimming in optical brightness occurs over a long time, and, along this phase, the Hα emission changes. The V/R ratio decreases, and the EW increases (with the smallest value of the V/R relation and the maximum EW at the minimum of the light curve). This fact seems to be associated with a change in the amount of the circumstellar material, and not only due to a natural increase in emission intensity during the light curve minimum. Also, the FWHM of the blue emission component is the largest at this point. This Hα line strengthening might be attributed to enhanced mass loss (or a mass ejection episode). Similar variations in the emission of the spectral lines and brightness have been observed in the FS CMa star, MWC 728. The light curve of MWC 645 suggests a (possible) periodic behavior. If this were true, the Hα enhancement might result from a mass transfer process during periastron passage in an eccentric binary system, with a period of the order of 4 years. A denser observational grid with high-quality spectra is needed to study the link between mass loss and brightness behavior.

A detailed calculation of the Hα line profile based on a physically consistent model is a difficult task. However, it would be valuable to explore models with simplified assumptions to gain insight into the system geometry and the structure of the CS matter [9,54]. More complex scenarios considering non-conservative mass transfer between the binary components should be considered to draw a picture of the structures involved in the emission processes [62,63].

6. Conclusions

In this paper, we have studied the FS CMa-type object, MWC 645, a recently confirmed binary system. We have presented IR medium-resolution spectra covering the *J*-, *H*-, *K*-, and *L*-bands and identified the main spectral features. We have reported the presence of CO bands in absorption for the first time. We have searched for periodicity in the light curve and a possible correlation between its behavior and the spectroscopic optical data. We found that the photometric variations could be explained by variable extinction along the line of sight. In addition, we noted that the stellar brightness fading is accompanied by the enhancement of the Hα line emission, which might be due to mass ejection events. Finally, a proper fitting to the observed SED was found, giving a global picture of the gaseous and dusty structures that could enshroud the binary.

Simultaneous optical and near-IR spectroscopy during the following brightness minimum would be very useful for tracing the onset and progress of the possible mass transfer. Such an understanding is utmost for deepening the comprehension of binary evolution in general and of the nature of this fascinating object in particular.

Author Contributions: Conceptualization, A.F.T., M.L.A. and M.K.; methodology, A.F.T., M.L.A. and M.K.; software, A.F.T., M.L.A. and L.V.M.; formal analysis, A.F.T. and M.L.A.; investigation, A.F.T., M.L.A., M.K., L.V.M. and T.E.; resources, A.F.T., M.L.A. and M.K.; data curation, A.F.T., M.L.A., M.K., L.V.M. and T.E.; writing—original draft preparation, review and editing, A.F.T., M.L.A., M.K., L.V.M. and T.E.; visualization, A.F.T., M.L.A. and M.K.; funding acquisition, A.F.T., M.L.A., M.K. and T.E. All authors have read and agreed to the published version of the manuscript.

Funding: A.F.T. and M.L.A. acknowledge financial support from the Universidad Nacional de La Plata (Programa de Incentivos 11/G160) and CONICET (PIP 1337), Argentina. M.K. acknowledges financial support from the Czech Science Foundation (GAČR, grant number 20-00150S). The Astronomical Institute of the Czech Academy of Sciences, Ondřejov, is supported by project RVO: 67985815. This project has received funding from the European Union's Framework Programme for Research and Innovation Horizon 2020 (2014-2020) under the Marie Skłodowska-Curie Grant Agreement No. 823734. T.E. gratefully acknowledges financial support from the Estonian Ministry of Education and Research through the Estonian Research Council institutional research funding IUT40-1, from the European Union European Regional Development Fund project KOMEET 2014-2020.4.01.16-0029.

Institutional Review Board Statement: Not applicable.

Informed Consent Statement: Not applicable.

Data Availability Statement: The data involved in this research are available on request from the authors.

Acknowledgments: The authors are grateful to the referees, whose comments and suggestions helped to improve the paper. This work has made use of IRAF, which is distributed by the National Optical Astronomy Observatory, operated by the Association of Universities for Research in Astronomy (AURA) under a cooperative agreement with the National Science Foundation; the BeSS database operated at LESIA, Observatoire de Meudon, France (http://basebe.obspm.fr (accessed on 2 February 2023)); the SIMBAD database and the VizieR catalog access tool, both operated at CDS, Strasbourg, France; the NASA Astrophysics Data System (ADS); the NASA IRTF (Infrared Telescope Facility) Spectral Library; and the NASA IRSA period search tool and the All-Sky Automated Survey for Supernovae (ASAS-SN). This paper is based on observations obtained at (i) Ondřejov Observatory (Czech Republic) with the Perek 2 m telescope; (ii) Tartu Observatory (Estonia); and (iii) the international Gemini Observatory, a program of NSF's NOIRLab, which is managed by the Association of Universities for Research in Astronomy (AURA) under a cooperative agreement with the National Science Foundation on behalf of the Gemini Observatory partnership: the National Science Foundation (United States), National Research Council (Canada), Agencia Nacional de Investigación y Desarrollo (Chile), Ministerio de Ciencia, Tecnología e Innovación (Argentina), Ministério da Ciência, Tecnologia, Inovações e Comunicações (Brazil), and Korea Astronomy and Space Science Institute (Republic of Korea) under program IDs GN-2017A-Q-62, GN-2018A-Q-406, and GN-2022B-Q-225. This work has made use of the ground-based research infrastructure of Tartu Observatory, funded through the projects TT8 (Estonian Research Council) and KosEST (EU Regional Development Fund). The authors thank L.S. Cidale for many fruitful discussions.

Conflicts of Interest: The authors declare no conflict of interest.

Abbreviations

The following abbreviations are used in this manuscript:

CS	Circumstellar
CCD	Charged-coupled device
EW	Equivalent width
V	Blue peak intensity
R	Red peak intensity
V/R	Blue-to-red emission peak ratio
FWHM	Full width at half maximum
HJD	Heliocentric Julian date
IR	Infrared
SED	Spectral energy distribution

References

1. Allen, D.A.; Swings, J.P. The spectra of peculiar Be star with infrared excesses. *Astron. Astrophys.* **1976**, *47*, 293–302.
2. Lamers, H.J.G.L.M.; Zickgraf, F.J.; de Winter, D.; Houziaux, L.; Zorec, J. An improved classification of B[e]-type stars. *Astron. Astrophys.* **1998**, *340*, 117–128.
3. Miroshnichenko, A.S. Toward Understanding the B[e] Phenomenon. I. Definition of the Galactic FS CMa Stars. *Astrophys. J.* **2007**, *667*, 497–504. [CrossRef]
4. Korčáková, D.; Sestito, F.; Manset, N.; Kroupa, P.; Votruba, V.; Šlechta, M.; Danford, S.; Dvořáková, N.; Raj, A.; Chojnowski, S.D.; et al. First detection of a magnetic field in low-luminosity B[e] stars. New scenarios for the nature and evolutionary stages of FS CMa stars. *Astron. Astrophys.* **2022**, *659*, A35. [CrossRef]
5. Kuratova, A.K.; Miroshnichenko, A.S.; Zharikov, S.V.; Manset, N.; Khokhlov, S.A.; Raj, A.; Kusakin, A.V.; Reva, I.V.; Kokumbaeva, R.I.; Usenko, I.A.; et al. Spectroscopic Monitoring of the B[E] Objects Fscma and MOCAM. *Odessa Astron. Publ.* **2019**, *32*, 63. [CrossRef]
6. Miroshnichenko, A.S.; Manset, N.; Kusakin, A.V.; Chentsov, E.L.; Klochkova, V.G.; Zharikov, S.V.; Gray, R.O.; Grankin, K.N.; Gandet, T.L.; Bjorkman, K.S.; et al. Toward Understanding the B[e] Phenomenon. II. New Galactic FS CMa Stars. *Astrophys. J.* **2007**, *671*, 828–841. [CrossRef]
7. Miroshnichenko, A.S.; Chentsov, E.L.; Klochkova, V.G.; Zharikov, S.V.; Grankin, K.N.; Kusakin, A.V.; Gandet, T.L.; Klingenberg, G.; Kildahl, S.; Rudy, R.J.; et al. Toward Understanding the B[e] Phenomenon. III. Properties of the Optical Counterpart of IRAS 00470+6429. *Astrophys. J.* **2009**, *700*, 209–220. [CrossRef]
8. Vink, J.S.; de Koter, A.; Lamers, H.J.G.L.M. Mass-loss predictions for O and B stars as a function of metallicity. *Astron. Astrophys.* **2001**, *369*, 574–588. [CrossRef]
9. Carciofi, A.C.; Miroshnichenko, A.S.; Bjorkman, J.E. Toward Understanding the B[e] Phenomenon. IV. Modeling of IRAS 00470+6429. *Astrophys. J.* **2010**, *721*, 1079–1089. [CrossRef]
10. Korčáková, D.; Miroshnichenko, A.S.; Zharikov, S.V.; Manset, N.; Danford, S.; Votruba, V.; Dvořáková, N. Mystery of lithium in FS CMa stars. *Mem. Soc. Astron. Ital.* **2020**, *91*, 118.
11. Miroshnichenko, A.S.; Zharikov, S.V.; Manset, N.; Khokhlov, S.A.; Nodyarov, A.S.; Klochkova, V.G.; Danford, S.; Kuratova, A.K.; Mennickent, R.; Chojnowski, S.D.; et al. Recent Progress in Finding Binary Systems with the B[e] Phenomenon. *Galaxies* **2023**, *11*, 36. [CrossRef]
12. Merrill, P.W.; Burwell, C.G. No. 682. Supplement to the Mount Wilson catalogue and bibliography of stars of classes B and A whose spectra have bright hydrogen lines. *Contrib. Mt. Wilson Obs. Carnegie Inst. Wash.* **1943**, *682*, 1–32.
13. Swings, J.P.; Allen, D.A. MWC 645 and MWC 819: Two Stars Resembling Eta Carinae. *Astrophys. Lett.* **1973**, *14*, 65.
14. Gottlieb, E.W.; Liller, W. Photometric histories of six infrared objects and three highly reddened blue supergiants. *Astrophys. J.* **1978**, *225*, 488–495. [CrossRef]
15. Jaschek, M.; Andrillat, Y.; Jaschek, C. B[e] stars. III. MWC 645. *Astron. Astrophys. Suppl.* **1996**, *120*, 99–105. [CrossRef]
16. Landaberry, S.J.C.; Pereira, C.B.; de Araújo, F.X. The eta Carinae spectrum of SS73 11 (=Ve 2–27). *Astron. Astrophys.* **2001**, *376*, 917–927. [CrossRef]
17. Zickgraf, F.J. Kinematical structure of the circumstellar environments of galactic B[e] -type stars. *Astron. Astrophys.* **2003**, *408*, 257–285. [CrossRef]
18. Marston, A.P.; McCollum, B. Extended shells around B[e] stars. Implications for B[e] star evolution. *Astron. Astrophys.* **2008**, *477*, 193–202. [CrossRef]
19. Nodyarov, A.S.; Miroshnichenko, A.S.; Khokhlov, S.A.; Zharikov, S.V.; Manset, N.; Klochkova, V.G.; Usenko, I.A. High-Resolution Spectroscopy of the B[e] Star MWC 645. *Odessa Astron. Publ.* **2021**, *34*, 59. [CrossRef]
20. Nodyarov, A.S.; Miroshnichenko, A.S.; Khokhlov, S.A.; Zharikov, S.V.; Manset, N.; Klochkova, V.G.; Grankin, K.N.; Arkharov, A.A.; Efimova, N.; Klimanov, S.; et al. Toward Understanding the B[e] Phenomenon. IX. Nature and Binarity of MWC645. *Astrophys. J.* **2022**, *936*, 129. [CrossRef]

21. Elias, J.H.; Joyce, R.R.; Liang, M.; Muller, G.P.; Hileman, E.A.; George, J.R. Design of the Gemini near-infrared spectrograph. In *Proceedings of the Ground-Based and Airborne Instrumentation for Astronomy, Orlando, FL, USA, 24–31 May 2006, Society of Photo-Optical Instrumentation Engineers (SPIE) Conference Series*; McLean, I.S., Iye, M., Eds.; SPIE, 2006; Volume 6269, p. 62694C. [CrossRef]
22. Slechta, M.; Skoda, P. 2-meter telescope devices: Coudé slit spectrograph and HEROS. *Publ. Astron. Inst. Czechoslov. Acad. Sci.* **2002**, *90*, 1–4.
23. Neiner, C.; de Batz, B.; Cochard, F.; Floquet, M.; Mekkas, A.; Desnoux, V. The Be Star Spectra (BeSS) Database. *Astron. J.* **2011**, *142*, 149. [CrossRef]
24. Shappee, B.; Prieto, J.; Stanek, K.Z.; Kochanek, C.S.; Holoien, T.; Jencson, J.; Basu, U.; Beacom, J.F.; Szczygiel, D.; Pojmanski, G.; et al. All Sky Automated Survey for SuperNovae (ASAS-SN or "Assassin"). In Proceedings of the American Astronomical Society Meeting, Washington, DC, USA, 5–9 January 2014; p. 236.03.
25. Kochanek, C.S.; Shappee, B.J.; Stanek, K.Z.; Holoien, T.W.S.; Thompson, T.A.; Prieto, J.L.; Dong, S.; Shields, J.V.; Will, D.; Britt, C.; et al. The All-Sky Automated Survey for Supernovae (ASAS-SN) Light Curve Server v1.0. *Publ. ASP* **2017**, *129*, 104502. [CrossRef]
26. Kramida, A.; Ralchenko, Y.; Reader, J.; NIST ASD Team. *NIST Atomic Spectra Database (Ver. 5.10)*; National Institute of Standards and Technology: Gaithersburg, MD, USA, 2022. Available online: https://physics.nist.gov/asd (accessed on 20 May 2023).
27. Kleinmann, S.G.; Hall, D.N.B. Spectra of Late-Type Standard Stars in the Region 2.0–2.5 Microns. *Astrophys. J. Suppl.* **1986**, *62*, 501. [CrossRef]
28. Mármol-Queraltó, E.; Cardiel, N.; Cenarro, A.J.; Vazdekis, A.; Gorgas, J.; Pedraz, S.; Peletier, R.F.; Sánchez-Blázquez, P. A new stellar library in the region of the CO index at 2.3 μm. New index definition and empirical fitting functions. *Astron. Astrophys.* **2008**, *489*, 885–909. [CrossRef]
29. Ali, B.; Carr, J.S.; Depoy, D.L.; Frogel, J.A.; Sellgren, K. Medium-Resolution Near-Infrared (2.15-2.35 micron) Spectroscopy of Late-Type Main-Sequence Stars. *Astron. J.* **1995**, *110*, 2415. [CrossRef]
30. Cushing, M.C.; Rayner, J.T.; Vacca, W.D. An Infrared Spectroscopic Sequence of M, L, and T Dwarfs. *Astrophys. J.* **2005**, *623*, 1115–1140. [CrossRef]
31. Rayner, J.T.; Cushing, M.C.; Vacca, W.D. The Infrared Telescope Facility (IRTF) Spectral Library: Cool Stars. *Astrophys. J. Suppl.* **2009**, *185*, 289–432. [CrossRef]
32. Wallace, L.; Hinkle, K. Medium-Resolution Spectra of Normal Stars in the K Band. *Astrophys. J. Suppl.* **1997**, *111*, 445–458. [CrossRef]
33. Heras, A.M.; Shipman, R.F.; Price, S.D.; de Graauw, T.; Walker, H.J.; Jourdain de Muizon, M.; Kessler, M.F.; Prusti, T.; Decin, L.; Vandenbussche, B.; et al. Infrared spectral classification of normal stars. *Astron. Astrophys.* **2002**, *394*, 539–552. [CrossRef]
34. Winge, C.; Riffel, R.A.; Storchi-Bergmann, T. The Gemini Spectral Library of Near-IR Late-Type Stellar Templates and Its Application for Velocity Dispersion Measurements. *Astrophys. J. Suppl.* **2009**, *185*, 186–197. [CrossRef]
35. Marchiano, P.E.; Cidale, L.S.; Brandi, E.; Muratore, M.F. Spectral Energy Distribution in the symbiotic system BI Cru. *Bol. Asoc. Argent. Astron. Plata Argent.* **2013**, *56*, 163–166.
36. Arias, M.L.; Cidale, L.S.; Kraus, M.; Torres, A.F.; Aidelman, Y.; Zorec, J.; Granada, A. Near-infrared Spectra of a Sample of Galactic Unclassified B[e] Stars. *Publ. ASP* **2018**, *130*, 114201. [CrossRef]
37. Zorec, J. Distances, Kinematics and Distribution of B[e] Stars in Our Galaxy. In *Proceedings of the B[e] Stars, Astrophysics and Space Science Library*; Hubert, A.M., Jaschek, C., Eds.; Springer: Dordrecht, The Netherlands, 1998; Volume 233, p. 27. [CrossRef]
38. Muratore, M.F.; Kraus, M.; Liermann, A.; Schnurr, O.; Cidale, L.S.; Arias, M.L. Unveiling the evolutionary phase of B[e] supergiants. *Bol. Asoc. Argent. Astron. Plata Argent.* **2010**, *53*, 123–126.
39. Cardelli, J.A.; Clayton, G.C.; Mathis, J.S. The Relationship between Infrared, Optical, and Ultraviolet Extinction. *Astrophys. J.* **1989**, *345*, 245. [CrossRef]
40. Lamers, H.J.G.L.M.; Cassinelli, J.P. *Introduction to Stellar Winds*; Cambridge University Press: Cambridge, UK, 1999.
41. Kurucz, R.L. Model atmospheres for G, F, A, B, and O stars. *Astrophys. J. Suppl.* **1979**, *40*, 1–340. [CrossRef]
42. Skrutskie, M.F.; Cutri, R.M.; Stiening, R.; Weinberg, M.D.; Schneider, S.; Carpenter, J.M.; Beichman, C.; Capps, R.; Chester, T.; Elias, J.; et al. The Two Micron All Sky Survey (2MASS). *Astron. J.* **2006**, *131*, 1163–1183. [CrossRef]
43. Wright, E.L.; Eisenhardt, P.R.M.; Mainzer, A.K.; Ressler, M.E.; Cutri, R.M.; Jarrett, T.; Kirkpatrick, J.D.; Padgett, D.; McMillan, R.S.; Skrutskie, M.; et al. The Wide-field Infrared Survey Explorer (WISE): Mission Description and Initial On-orbit Performance. *Astron. J.* **2010**, *140*, 1868–1881. [CrossRef]
44. Egan, M.P.; Price, S.D.; Kraemer, K.E. The Midcourse Space Experiment Point Source Catalog Version 2.3. *Proc. Am. Astron. Soc. Meet. Abstr.* **2003**, *203*, 1301.
45. Neugebauer, G.; Beichman, C.A.; Soifer, B.T.; Aumann, H.H.; Chester, T.J.; Gautier, T.N.; Gillett, F.C.; Hauser, M.G.; Houck, J.R.; Lonsdale, C.J.; et al. Early Results from the Infrared Astronomical Satellite. *Science* **1984**, *224*, 14–21. [CrossRef]
46. Yamamura, I. The AKARI Far-Infrared Bright Source Catalogue. In Proceedings of the 38th COSPAR Scientific Assembly, Bremen, Germany, 18–25 July 2010; Volume 38, p. 2.
47. Clark, J.S.; Steele, I.A. A representative sample of Be stars. II. K band spectroscopy. *Astron. Astrophys. Suppl.* **2000**, *141*, 65–77. [CrossRef]
48. Steele, I.A.; Clark, J.S. A representative sample of Be stars III: H band spectroscopy. *Astron. Astrophys.* **2001**, *371*, 643–651. [CrossRef]

49. Popper, D.M. Stellar masses. *Annu. Rev. Astron Astrophys.* **1980**, *18*, 115–164. [CrossRef]
50. Kovtyukh, V.V. High-precision effective temperatures of 161 FGK supergiants from line-depth ratios. *Mon. Not. RAS* **2007**, *378*, 617–624. [CrossRef]
51. Gorlova, N.; Lobel, A.; Burgasser, A.J.; Rieke, G.H.; Ilyin, I.; Stauffer, J.R. On the CO Near-Infrared Band and the Line-splitting Phenomenon in the Yellow Hypergiant ρ Cassiopeiae. *Astrophys. J.* **2006**, *651*, 1130–1150. [CrossRef]
52. Geballe, T.R.; Rushton, M.T.; Eyres, S.P.S.; Evans, A.; van Loon, J.T.; Smalley, B. Infrared spectroscopy of carbon monoxide in V838 Monocerotis during 2002–2006. *Astron. Astrophys.* **2007**, *467*, 269–275. [CrossRef]
53. Kraus, M.; Kolka, I.; Aret, A.; Nickeler, D.H.; Maravelias, G.; Eenmäe, T.; Lobel, A.; Klochkova, V.G. A new outburst of the yellow hypergiant star ρ Cas. *Mon. Not. RAS* **2019**, *483*, 3792–3809. [CrossRef]
54. Polster, J.; Korčáková, D.; Manset, N. Time-dependent spectral-feature variations of stars displaying the B[e] phenomenon. IV. V2028 Cygni: Modelling of Hα bisector variability. *Astron. Astrophys.* **2018**, *617*, A79. [CrossRef]
55. Miroshnichenko, A.; Ivezić, Ž.; Elitzur, M. On Protostellar Disks in Herbig Ae/Be Stars. *Astrophys. J. Lett.* **1997**, *475*, L41–L44. [CrossRef]
56. Herbst, W.; Shevchenko, V.S. A Photometric Catalog of Herbig AE/BE Stars and Discussion of the Nature and Cause of the Variations of UX Orionis Stars. *Astron. J.* **1999**, *118*, 1043–1060. [CrossRef]
57. Bouvier, J.; Grankin, K.N.; Alencar, S.H.P.; Dougados, C.; Fernández, M.; Basri, G.; Batalha, C.; Guenther, E.; Ibrahimov, M.A.; Magakian, T.Y.; et al. Eclipses by circumstellar material in the T Tauri star AA Tau. II. Evidence for non-stationary magnetospheric accretion. *Astron. Astrophys.* **2003**, *409*, 169–192. [CrossRef]
58. Hofmann, K.H.; Bensberg, A.; Schertl, D.; Weigelt, G.; Wolf, S.; Meilland, A.; Millour, F.; Waters, L.B.F.M.; Kraus, S.; Ohnaka, K.; et al. VLTI-MATISSE L- and N-band aperture-synthesis imaging of the unclassified B[e] star FS Canis Majoris. *Astron. Astrophys.* **2022**, *658*, A81. [CrossRef]
59. Bartlett, E.S.; Clark, J.S.; Coe, M.J.; Garcia, M.R.; Uttley, P. Timing and spectral analysis of the unusual X-ray transient XTE J0421+560/CI Camelopardalis. *Mon. Not. RAS* **2013**, *429*, 1213–1220. [CrossRef]
60. Okazaki, A.T. Long-Term V/R Variations of Be Stars Due to Global One-Armed Oscillations of Equatorial Disks. *Publ. ASJ* **1991**, *43*, 75–94.
61. Kriz, S.; Harmanec, P. A Hypothesis of the Binary Origin of Be Stars. *Bull. Astron. Institutes Czechoslov.* **1975**, *26*, 65.
62. Deschamps, R.; Braun, K.; Jorissen, A.; Siess, L.; Baes, M.; Camps, P. Non-conservative evolution in Algols: Where is the matter? *Astron. Astrophys.* **2015**, *577*, A55. [CrossRef]
63. Brož, M.; Mourard, D.; Budaj, J.; Harmanec, P.; Schmitt, H.; Tallon-Bosc, I.; Bonneau, D.; Božić, H.; Gies, D.; Šlechta, M. Optically thin circumstellar medium in the β Lyr A system. *Astron. Astrophys.* **2021**, *645*, A51. [CrossRef]

Disclaimer/Publisher's Note: The statements, opinions and data contained in all publications are solely those of the individual author(s) and contributor(s) and not of MDPI and/or the editor(s). MDPI and/or the editor(s) disclaim responsibility for any injury to people or property resulting from any ideas, methods, instructions or products referred to in the content.

Article

Large-Scale Ejecta of Z CMa—Proper Motion Study and New Features Discovered

Tiina Liimets [1,*], Michaela Kraus [1], Lydia Cidale [2,3], Sergey Karpov [4] and Anthony Marston [5]

1. Astronomical Institute, Czech Academy of Sciences, Fričova 298, 25165 Ondřejov, Czech Republic
2. Instituto de Astrofísica de La Plata (CCT La Plata-CONICET, UNLP), Paseo del Bosque S/N, La Plata B1900FWA, Buenos Aires, Argentina
3. Departamento de Espectroscopía, Facultad de Ciencias Astronómicas y Geofísicas, Universidad Nacional de La Plata, Paseo del Bosque S/N, La Plata B1900FWA, Buenos Aires, Argentina
4. Institute of Physics of the Czech Academy of Sciences (FZU AV ČR), Na Slovance 2, Praha 8, 18200 Prague, Czech Republic
5. European Space Agency, European Space Astronomy Centre, Camino Bajo del Castillo, s/n Urbanización Villafranca del Castillo Villañueva de la Cañada, E-28692 Madrid, Spain
* Correspondence: tiina.liimets@asu.cas.cz

Abstract: Z Canis Majoris is a fascinating early-type binary with a Herbig Be primary and a FU Orionis-type secondary. Both of the stars exhibit sub-arcsecond jet-like ejecta. In addition, the primary is associated with the extended jet as well as with the large-scale outflow. In this study, we investigate further the nature of the large-scale outflow, which has not been studied since its discovery almost three and a half decades ago. We present proper motion measurements of individual features of the large-scale outflow and determine their kinematical ages. Furthermore, with our newly acquired deep images, we have discovered additional faint arc-shaped features that can be associated with the central binary.

Keywords: circumstellar matter: jets and outflows; stars: individual Z CMa; stars: emission-line; Be

Citation: Liimets, T.; Kraus, M.; Cidale, L.; Karpov, S.; Marston, A. Large-Scale Ejecta of Z CMa—Proper Motion Study and New Features Discovered. *Galaxies* **2023**, *11*, 64. https://doi.org/10.3390/galaxies11030064

Academic Editor: Sun Kwok

Received: 15 March 2023
Revised: 20 April 2023
Accepted: 30 April 2023
Published: 4 May 2023

Copyright: © 2023 by the authors. Licensee MDPI, Basel, Switzerland. This article is an open access article distributed under the terms and conditions of the Creative Commons Attribution (CC BY) license (https:// creativecommons.org/licenses/by/ 4.0/).

1. Introduction

Z Canis Majoris (Z CMa) has intrigued astronomers for decades. It is an active early-type emission line binary consisting of a Herbig Be primary and a FU Orionis-type (FU Ori) secondary separated by $0''.1$ (e.g., Bonnefoy et al. [1]). Current high-spatial resolution observations show that the primary is located in the northwest (NW) direction from the secondary (e.g., Bonnefoy et al. [1]; Dong et al. [2]). The light curve of the system is rich in long-term variability, months to years, as well as in day-by-day variability with a non-periodic nature and varying amplitudes (Sicilia-Aguilar et al. [3] and references therein).

The system is surrounded with multiple circumstellar and large-scale outflows. The brightest feature around Z CMa is a reflection nebula extending up to about $35''$ toward the NW from the central binary. It was discovered from photographic plates from 1953 by Herbig [4]. On these plates, the reflection nebula has a bar-shaped morphology (see Figure 10 in Herbig [4]). However, on the newer, higher resolution and more sensitive CCD images, the morphology more closely resembles a comma (Figure 1 left and Figure 2). We refer to this feature as comma nebula.

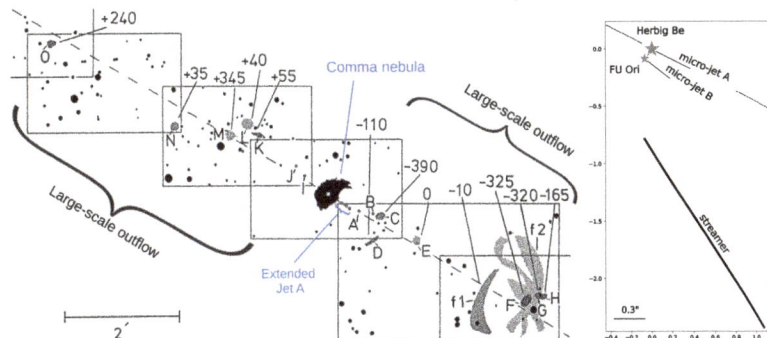

Figure 1. (**Left**): Schematic view of the various extended nebular features surrounding Z CMa. The stars in the field of view are drawn as black filled circles, large-scale outflow features are marked with grayish areas or circles with black contures. The dashed line is drawn at the 60° position angle. Individual numbers refer to radial velocities (Poetzel et al. [5]). Field of view is $9'.5 \times 5'.6$. The base of the figure is depicted from Poetzel et al. [5]. Reproduced with permission © ESO. (**Right**): Schematic view of the sub-arcsecond features around Z CMa presented in true proportions. The lengths of the micro-jets are taken from Whelan et al. [6] and those for the streamer are taken from Dong et al. [2]. The dashed line represents the position angle of the large-scale outflow. Field of view is $1''.6 \times 2''.7$. A white square representing the same size is drawn on the left panel at the position of the central binary inside the comma nebula. On both panels, north is up and east is to the left.

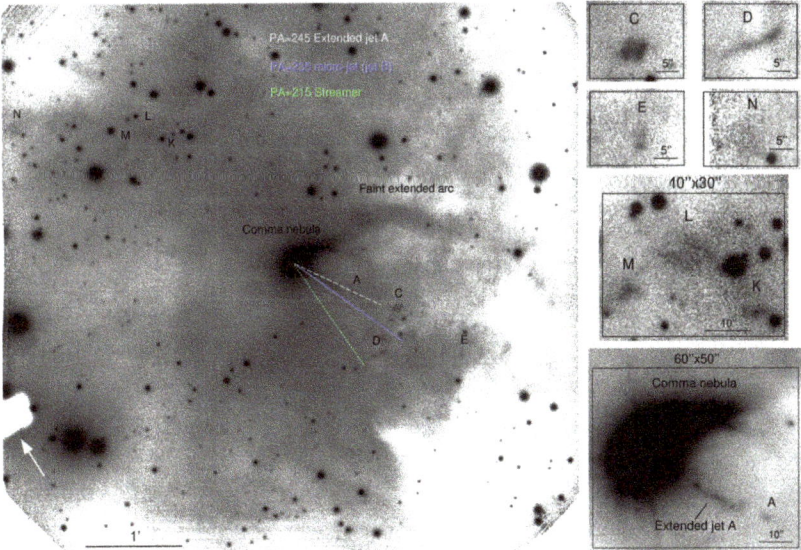

Figure 2. (**Left**): [S II] image of Z CMa acquired with GMOS attached to Gemini-South. The white square shape feature, indicated with a white arrow, is an artifact from vignetting of the guiding probe. (**Right**): Insets of the resolved features in the [S II] image. The FOVs of the smallest insets are $20'' \times 15''$ each. On all images, north is up, east is to the left, and the intensity is in log scale to improve the contrast. See text for more details.

In the sub-arcsecond scale, jets emerge from both components—micro-jet A from the primary and micro-jet B from the secondary at the position angle (PA, measured from north to east) 245° and 235°, respectively (Whelan et al. [6]; Figure 1 right). Both jets are slightly wiggly and show associated knots (Whelan et al. [6]; Antoniucci et al. [7]). The micro-jet A extends out to about $30''$ and is referred to as (extended) jet A (Figure 1 left) in the literature. This jet was discovered by Poetzel et al. [5], and it has also a wiggly nature (Whelan et al. [6]).

Millan-Gabet and Monnier [8] discovered another jet-like small-scale feature at PA 215°. This feature is designated in the literature as a streamer, and its length is about $2''$ (Figure 1 right). However, it does not emanate from either of the binary components. In fact, it appears to start $0''.7$ toward the south (S) from the central binary (Dong et al. [2]). The same authors find a point source a further $\sim 2''$ away from the streamer at the same PA and therefore confirm that the streamer is most likely created in a rare flyby event. Furthermore, these authors point out that the flyby event explains also the anomalous double-jet activity in this system, which considering the masses of the binary components could happen only with the probability of less than 1%.

Looking at greater distances, further out from the extended jet A at the same PA, a large-scale outflow was discovered by Poetzel et al. [5] from the narrow-band Hα and [S II] images acquired in the end of 1990s. While the micro and extended jets are primarily detected as one-sided objects in the southwest (SW) direction, the large-scale outflow has emission features also toward the northeast (NE) (Figure 1 left). The large-scale outflow consists of blobby and elongated features. The kinematics of the features refer to a bipolar nature.

The NE features are all red-shifted, while the SW ones appear blue-shifted. Eight features are identified by Poetzel et al. [5] in the SW side extending up to $4'.7$ and seven features are identified in the NE reaching up to $6'$ from the central object. This is the largest known outflow[1] for this type of stars extending across 3.5 pc when considering the distance of 1125 pc (Dong et al. [2]). At the discovery years, the average PA of the outflow features was 60° (equivalent to 240°). This PA aligns with that of the extended jet A, which is associated with the primary, and it is therefore widely accepted that the large-scale outflow is a result of the ejections from the Herbig Be component.

The large-scale outflow is what we concentrate on in this paper. In particular, we will measure proper motions of the individual features, and in combination with their respective radial velocities, we aim to reveal the true 3D nature of this huge nebulosity. For that, an accurate distance estimation is essential. Several distance estimates for Z CMa exist in the literature. They are all based on the fact that Z CMa is a member of the OB association CMa OB1. Published values are 1150 ± 50 pc by Clariá [12], 990 ± 50 pc by Whelan et al. [6], and more recently 1125 ± 30 pc by Dong et al. [2]. Throughout this paper, we use the latest estimate, 1125 pc, because it was calculated by using the largest number of members of the association (50) and is therefore the most accurate one. We note here that the estimated Gaia Data Release 3 (Gaia DR3) (Gaia Collaboration et al. [13]; Gaia Collaboration et al. [14]) distance of the Z CMa is not reliable due to the very large value of RUWE[2], as described in Dong et al. [2].

2. Observations and Data Reduction

Our first imaging data were obtained with the 60-inch telescope at Mt. Palomar on 2002 February 28. A single 20-min exposure in a narrow-band Hα filter ($\lambda = 6564.8$ Å, $\Delta\lambda = 20$ Å) was secured with a seeing of $1''.5$. The field of view (FOV) was $12'.5$, and a chosen binning of 2×2 provided a pixel scale of $0''.756$ pix^{-1}. This image was reduced using the standard routines in IRAF[3] (Tody [15,16]).

The second set of images of Z CMa was acquired on 2019 September 27 with the 8.1 m telescope. We used the Gemini Multi-Object Spectrographs (GMOS, Hook et al. [17]) mounted at Gemini-South as part of the observing proposal AR-2019B-020. The images were collected in the narrow band Hα G0336 ($\lambda = 6567.0$ Å, $\Delta\lambda = 70$ Å) and [S II] G0335 ($\lambda = 6717.2$ Å, $\Delta\lambda = 43$ Å) filters with the total exposure time of 145 and 435 s, respectively. The observations in both filters consisted of several shorter exposures that have been dithered to eliminate the gaps between the detectors and to minimize contamination (saturation effects) due to the bright central star. The bin 2×2 was used, yielding a pixel scale of $0''.16$ pix^{-1}. The FOV of the final reduced images is $6' \times 5'.5$. The observations were carried out with a seeing between $1''.3$ and $1''.4$. Data reduction was performed using the Gemini software DRAGONS (Labrie et al. [18]). Details of the observations are in Table 1.

Table 1. Log of the observations. The first column lists the start date of the observing night. Column 2 lists the telescope used. Column 3 contains the central wavelength (λ) and the width ($\Delta\lambda$) of the filter. The last column is the total exposure time in seconds.

Date	Telescope	Filter $\lambda/\Delta\lambda$ (Å)	Total Exp. Time (s)
28 February 2002	60-inch Mt. Palomar	Hα 6564.8/20	1200
27 September 2019	Gemini-South	Hα 6567.0/70	145
27 September 2019	Gemini-South	[S II] 6717.2/43	435

We also acquired a set of stacked images from the Pan-STARRS images archive (Waters et al. [19]) that are results of co-adding multiple exposures made between 2010 and 2015 during the 3π survey (Chambers et al. [20]). We downloaded stacked images covering the region around Z CMa in g, r, i, and z filters, re-scaled them to the common photometric zero point, and created mosaics in each individual filter with the original spatial resolution of the Pan-STARRS stacked images of $0''.25$ pix^{-1}. We then created a composite RGB image from z, i, and g mosaics with logarithmic intensity scaling applied. We excluded the r filter from the composite image as it shows the largest number of stacking artifacts in the background, and it is mostly unusable for studying the morphology of the nebular features. The FOV of the final image was $9' \times 9'$.

2.1. Pre-Analysis Processing of the Narrow-Band Images

To accurately analyze the possible morphological and/or kinematical changes between our two epochs, our narrow-band Hα images first had to be matched pixel by pixel. For this, we used 32 stars in the FOV whose proper motions were smaller or equal to ± 5 mas year^{-1} and RUWE < 1.4; all values were taken from the Gaia DR3. The 2019 frame was matched against the 2002 frame, because the latter had a larger pixel scale. The matching was completed in IRAF using the tasks *geomap* and *geotran*. The errors of the matching were $\sigma_{RA} = 0''.18$ and $\sigma_{DEC} = 0''.23$. With this procedure, both frames were given the same pixel scale, $0''.756$ pix^{-1}. The last step in matching the coordinates is to compensate for the possible proper motion of the central star. In our case, this effect is insignificant, considering the small proper motion of Z CMa (see Section 3.4) and that our two datasets are separated by 17.58 years. In this stage of the image processing, the frames were ready to be compared by blinking to find any obvious movement of the outflow features in the plane of the sky or to measure directly the coordinates of individual features to calculate proper motions.

For the features for which the blinking of the frames did not reveal any visual expansion and/or which, due to their elongated shape, are not suitable for direct coordinate measuring, further processing was needed in order to use the magnification method (see Section 3.2). These steps included seeing and flux matching. The first was not needed because the seeing of the original frames was already similar, and after pixel by pixel matching, it became equal. Flux matching was completed using the analyzed feature (in our case feature *D*; see Section 3.2) by summing up all the flux in a rectangle-shaped area equivalent to the size and shape of the feature and then arithmetically matching it with the same area flux on the second epoch image. Beforehand, the sky was removed. We estimate that the flux matching is accurate down to a few percent.

3. Results

In Figure 3, we present our 2002 Hα image which covers the whole large-scale outflow of Z CMa, extending $10'.7$ from NE to SW. Our GMOS images from 2019 have a smaller FOV as demonstrated with the black rectangle in Figure 3. The GMOS image taken in the lines of [S II] $\lambda\lambda 6716, 6731$ is considerably deeper and presents the individual features with a better S/N (Figure 2). For a meaningful analysis (see Sections 3.1 and 3.2), it is important to use data in the same filter/spectral lines, especially when the aim is to find any morphological and/or kinematical changes between two epochs. The reason is that the excitation of emission lines from diverse elements can occur under different physical conditions so that the lines do not necessarily trace the same gaseous regions. Therefore, we restrict our proper motion analysis to the Hα frames, because we do not have a [S II] frame from 2002. In addition, our two Hα frames have a similar S/N, hence presenting similar detectability of the features, further making them a suitable match for the analysis. However, we note here that all the features that are resolvable in the GMOS Hα frame have the same morphology and position in the GMOS [S II] frame.

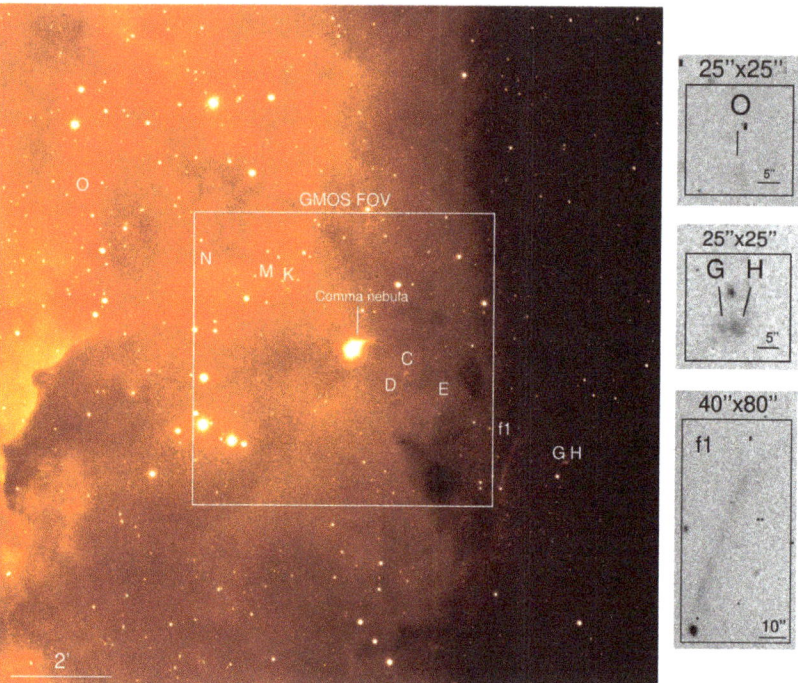

Figure 3. (**Left**): Hα image of Z CMa taken in 2002. GMOS FOV is shown for comparison. (**Right**): Insets of the features resolvable in the Hα image and which are outside GMOS FOV. On all images, north is up, east is to the left, and the intensity is in log scale to improve the contrast. See text for more details.

We refer to the individual features as they have been named by previous authors. The features of the large-scale outflow were named by Poetzel et al. [5] using capital letters from A to O (Figure 1). In addition, the designation of $f1$ and $f2$ was given to refer to the filaments in the SW side nearby the blobby features F, G, and H. On our figures, the labels of the features are always directly above the feature itself, apart from the label $f1$ which is to the left from the feature (Figure 3). We note here that not all the features presented by Poetzel et al. [5] are detectable and/or resolvable in our 2002 Hα frame due to the slightly lower S/N. From our GMOS [S II] frame, we could identify the features A, B, C, D, E, M, N, K, and L. Features O, G, H, and $f1$ are outside the GMOS FOV. We could not identify features I and J, situated between the central star and the feature K, on any of our images, which is probably due to a slightly lower S/N of our images compared to the images from Poetzel et al. [5].

The morphology of the large-scale outflow has not changed during the past 30 years when comparing our 2019 image with the 2002 one and the one from 1988–1989 from Poetzel et al. [5] (compare their Figure 1 with our Figures 2 and 3). The large-scale outflow has a bipolar nature and it consists of individual features (features A to O) with varying shapes—blobby, elongated, filamentary, arced. The PA of the outflow is ∼60° (or ∼240°), as measured from our images. The approximate value is due to the slightly different PAs of individual features. Nevertheless, this shows that the PA of the outflow has not changed during the past 30 years either (∼60° is measured also by Poetzel et al. [5]).

In our [S II] frame (Figure 2), we refer to a few other features related to the ejections from Z CMa: in particular, the extended jet A (see also Figure 3 in Whelan et al. [6]), the PA of the micro-jet B (see also Figure 1 in Antoniucci et al. [7]), and the PA of the streamer (see also Figure 2 in Canovas et al. [21]). Figure 2 shows also the previously known bright comma nebula which is almost perpendicular to the large-scale outflow. Furthermore, our image reveals another, fainter and previously not detected extended arc-shaped feature in the NW direction, which will be discussed further in Section 3.4 and 4.

As a first step in finding any expansion in the plane of the sky, we have used the simple blinking of the two Hα matched frames. It reveals that feature C has a visible expansion, while the rest of the features appear to stand still. Overall, a reliable analysis is only possible for the brightest features, which are those labeled with C and D. Therefore, we focus in the following on these two and determine their proper motions before we take a closer look at the faint arc structures.

3.1. Proper Motion Calculations of Feature C

Feature C is one of the brightest among the large-scale outflow. It has a roundish shape and is clearly detectable in both of our Hα frames taken in 2002 and 2019. The exact temporal separation of the two images is 17.58 years. The shape of feature C does not change during that period. As mentioned above, feature C presents the fastest motion from the central star compared to other features. In addition, feature C also has the largest radial velocity compared to other features as measured by Poetzel et al. [5]. Therefore, considering the inclination angle out of the plane of the sky (see below), it is not surprising that this feature would show a clear expansion in the plane of the sky while others do not. The movement in the plane of the sky of feature C is in accordance with the general direction of the features in SW direction, confirming that it must have been ejected from Z CMa.

Due to the roundish shape of feature C, it was possible to measure directly its central coordinates on both of our images. The total movement in the plane of the sky during the 17.58 years considered is $1''.4$, yielding a proper motion of $0''.08$ year^{-1} and a tangential velocity of ∼420 km s^{-1}. Considering the radial velocity of −390 km s^{-1} of feature C (Poetzel et al. [5]), its expansion velocity is about 580 km s^{-1} and the inclination out of the plane of the sky is 43° using the ordinary cosine relation between the velocity vectors ($i = arccos(v_{sky}/v_{exp})$). The found inclination angle agrees with the estimates made for the micro-jet B, which was proposed to have an inclination angle between 28° and 64° (Antoniucci et al. [7]) according to the tangential and radial velocity estimates by Whelan et al. [6].

The distance of feature C from the central star is $68''$ and $69''$ during the observations taken in 2002 and 2019, respectively. The position angle of feature C has not changed during the time duration between our 2002 and 2019 images, and it is 246°. Precise measurements with errors for all the calculated values are in Table 2.

Table 2. Results of the calculations from direct measuring of feature C and from using the magnification method for feature D. A distance of 1125 pc toward Z CMa was adopted. See text for more details.

Description	Feature C	Feature D
Total movement in the plane of the sky ($''$)	1.4 ± 0.3	-
Proper motion μ ($''$ year^{-1})	0.08 ± 0.02	0.013 ± 0.004
Radial velocity [a] v_{rad} (km s^{-1})	-390 ± 24	-110 ± 24
Tangential velocity v_{sky} (km s^{-1})	423 ± 88	69 ± 23
Expansion velocity v_{exp} (km s^{-1})	576 ± 67	130 ± 24
Inclination out of the plane of the sky i (°)	43 ± 15	58 ± 14
Distance from the central star in 2002 d_{2002} ($''$)	67.8 ± 0.3	75.3 ± 0.8
Distance from the central star in 2019 d_{2019} ($''$)	69.2 ± 0.3	75.5 [b] ± 0.8
PA at 2002 (°)	246.5 ± 0.2	222.9 ± 0.6
PA at 2019 (°)	246.6 ± 0.2	222.9 ± 0.6
Age at 2002 (years)	854 ± 177	5859 ± 1953
Magnification factor M	-	1.003 ± 0.001

[a] From Poetzel et al. [5]. [b] Calculated from d_{2002} and our derived proper motion.

Using the above calculated proper motion, the distance from the central star, and assuming constant expansion velocity since the ejection, it is possible to calculate the age of feature C at our first epoch, 2002. It is on the order of 850 years, which is in accordance with the estimates in Poetzel et al. [5].

3.2. Proper Motion Calculations of Feature D

The second feature for which we were able to calculate the expansion in the plane of the sky is the arc-shaped feature D. Due to its elongated shape, a direct measuring of the coordinates was not an appropriate method. For that reason the magnification method (see, e.g., Santander-García et al. [22]; Liimets et al. [23]) was used, which is suitable to find the proper motion of extended structures without a clear central point and/or when the total movement in the plane of the sky is as small as a tenth of a pixel. Both criteria are valid for feature D. The magnification method is based on finding a magnification factor M, which represents an image with minimum residuals of the magnified first epoch image, which is subtracted from the second epoch image. The method provides the proper motion, tangential velocity, and age. In order to use the magnification method, the frames being analyzed have to have coordinates, seeing, and flux matched. This was completed using the procedures described in Section 2.1. Further details about the magnification method and the derivation of the formulas used in the following can be found in Section 3.3 of the PhD thesis by Liimets [24]. The best magnification factor for feature D was determined to be $M = 1.003 \pm 0.001$. However, we note here that this result should be used with caution. We can say with confidence that M is not larger than 1.003. Consequently, all the following numerical values should be taken as upper limits. The proper motion can be calculated, in convenient units, in the following way

$$\mu ['' \text{year}^{-1}] = \frac{(M-1) \cdot d['']}{\Delta t [\text{year}]}, \quad (1)$$

where d is the distance of feature D from the central star on the first epoch, in our case the year 2002, and Δt is the time interval between the two epochs. In the case of the elongated feature D, the distance from the central star is somewhat challenging to estimate. However, we are confident that when considering a somewhat larger error of 1 pixel, it is accurate enough to serve the purpose of the simple calculations presented in this paper. Hence, the distance of feature D from the central star at our 2002 epoch is $75'' \pm 1''$, and considering Equation (1), the proper motion becomes $\mu = 0''.013$ year^{-1}. As for feature C, the precise values with their errors for all the calculations for feature D are in Table 2.

Using the proper motion and the distance to feature D, it is possible to calculate the tangential velocity in the following convenient units

$$v_{sky} \, [\text{km s}^{-1}] = 4.74 \cdot \mu \, [''\, \text{year}^{-1}] \cdot D \, [pc]. \qquad (2)$$

We consider the distance to Z CMa to be 1125 pc and therefore $v_{sky} = 69$ km s^{-1}. The radial velocity of feature D has been measured to be -110 km s^{-1} (Poetzel et al. [5]), which results in an expansion velocity of 130 km s^{-1}. The inclination angle would therefore be slightly larger than for feature C but with a value of 58° again matching with the estimates in Antoniucci et al. [7].

The magnification factor can also be used to calculate the age of the feature at the first epoch,

$$T \, [\text{year}] = \frac{\Delta t \, [\text{year}]}{(M-1)}, \qquad (3)$$

which for feature D gives a value of about 6000 years.

The PA of feature D is constant between our two observing epochs and has a value of 223°.

3.3. Proper Motion of Other Features

We tried to measure the expansion in the plane of the sky for the two other features, E and K, which were considerably fainter than the features C and D but still resolvable compared to the rather marginal detections of the features L, M, and N. Features E and K have an irregular shape, and therefore, the magnification method was used. We find no measurable movement in the plane of the sky for both features. For feature E, it is somewhat expected due to its RV being 0 km s^{-1} (Poetzel et al. [5]), while the RV of feature K is quoted as +55 km s^{-1} by the same authors. Considering the pixel scale of $0''.756$ of our matched images and the fact that the magnification method is able to measure an expansion of about one tenth of a pixel, the smallest tangential velocity that we should be able to detect is ~ 20 km s^{-1}. This, in return, would mean an inclination out of the plane of the sky 70°, which, within our error estimate of $\pm 10°$, agrees with the estimates by Antoniucci et al. [7].

3.4. Faint Extended Arc

From our deep GMOS [S II] image, we have discovered that the bright comma nebula has a fainter continuation. We designate this feature a faint extended arc (see Figure 2). Despite the lower S/N of our GMOS Hα image, the faint arc is also detectable on that frame, but we refrain from showing it because it does not provide new information. With an orientation of the faint arc toward the NW, it is perpendicular to the main large-scale outflow. The arc is more pronounced on the Pan-STARRS RGB image (Figure 4) on which additional related features become visible. We detect a repeating pattern of filaments, inside the main arc, which seem to mimic "feathers" (see white arrows in Figure 4). Interestingly, while most of the feathers do not have a direct connection to the star, the closest feather to the central binary, designated with number 1, seems to have connecting filaments. These filaments start from the bright comma nebula as a small arc resembling a "fishtail" (marked with solid red arrow) and then continue toward feather 1 with less homogeneous emission. From the image, it is visible that the diffuse emission in the direction of feather 2 is continuous further than that of our Pan-STARRS FOV. However, the faint extended arc which ends with feather 4 extends up to $3'$ toward the west.

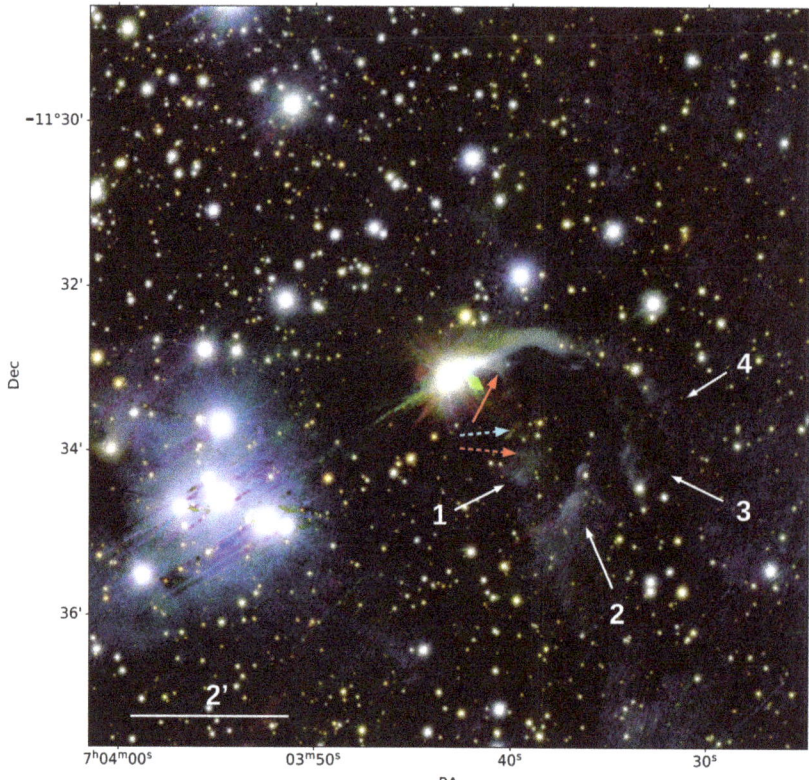

Figure 4. Pan-STARRS RGB image presenting the faint extended features around Z CMa. Red channel corresponds to z filter, green channel corresponds to i, and blue channel corresponds to g filter. The intensity is in log scale to improve the contrast. North is up and east is to the left. FOV $9' \times 9'$. See text for more details.

4. Discussion

Looking at the Figure 4, it is tempting to assume that the newly discovered faint extended features, as well as the bright comma-shaped feature, are created by matter expelled from the central star during a single or multiple independent mass ejection event(s) and now appear misplaced from the central position of the object due to the movement of Z CMa through the interstellar medium. However, the proper motion of the star, $\mu_{RA} = -4.4$ mas year^{-1} and $\mu_{DEC} = 2.0$ mas year^{-1} (Hipparcos-2 catalog, van Leeuwen [25]), does not support that. Z CMa is moving toward the faint extended arc and not away from it. Our choice of using the proper motion values from the Hipparcos-2 catalog is due to the mentioned large RUWE value of Z CMa in Gaia DR3 (see Section 1), referring to a problematic astrometric solution. The latter is most probably a result of the binary nature of the object and its generally rather bright photometric values. However, we note that Gaia DR3 proper motion values agree relatively well with the Hipparcos-2 catalog.

Another possibility, due to the roundish shape, is that the faint extended arc with the feathers is related to the orbital motion of the binary. Considering the recent high-contrast imaging polarimetry observations (Canovas et al. [21]), the orbital movement of the FU Ori companion around the primary is $0°.7$ year^{-1} when considering a circular orbit. This implies an orbital period of about 500 years. Currently, the FU Ori companion is located in the southeast (SE) direction from the primary (e.g., Figure 2 in Bonnefoy et al. [1]). Therefore, considering the hypothesis of a single ejection and the position of the faint extended arc and its feathers, they could have been ejected half a period ago. However, taking into account the distance of these features from the central binary, it would imply a peculiarly large tangential velocity of ∼3700 km s^{-1}, which has not been measured in any other features related to the small- or large-scale outflows of Z CMa. The obvious repeating pattern of the feathers could also suggest consecutive mass ejections occurring on every orbit at a specific location. We can then assume that the closest feather could have been ejected half a period ago and the next ones could have been ejected 1.5, 2.5, and 3.5 periods ago, respectively. The equivalent tangential velocities would then be about 1800, 900, 600, and 530 km s^{-1}, considering distance estimates for each feather from the central binary of 85, 130, 148, and 180″, respectively. While the radial velocities of −600 km s^{-1} have been measured to be associated with the micro-jets (Poetzel et al. [5]; Whelan et al. [6]), all these estimated values exceed our measured tangential velocities of the large-scale outflow features (see Table 2). If we consider our largest tangential velocity of 420 km s^{-1}, and the largest extent of the faint arc structures (3′), it would have had to be ejected ∼2300 years ago. This timescale is comparable with the ages we have found for the large-scale outflow features C and D, 850 and 5900 years, respectively. However, if these newly discovered features are related to the mass ejections occurring at the particular location during the orbit of the FU Ori-type companion around the primary, it is more likely that the orbit is not circular but elliptical. The latter would include periastron passage, which can enhance the mass-loss and possibly initiate outflows, as has been seen in other binary systems (e.g., in a case of symbiotic binary R Aquarii (Liimets et al. [26]) and proposed for the formation of the circumbinary molecular ring in the B[e] supergiant system GG Car (Kraus et al. [27])). At the same time, while feathers 2 and 3 do not seem to be physically connected with the central binary, the faint extended arc and its extension feather 4 are clearly a continuation from the star and its bright comma nebula, implying a constant flow of matter. At this point, we also mention that when inspecting the Figure 4 more thoroughly, it is possible that there is a dark cloud blocking the connections of the feathers with the central binary or with the faint arc, as there are no stars detected in the west from the binary below the faint arc. However, according to the study based on 2 Micron All Sky Survey by Dobashi [28], there are no dark clouds in the FOV of our Figure 4.

We can further calculate the possible tangential velocities related to the new discovered features when considering their maximum extent of 3′ and the ages found for the features C and D. The age of 850 years would result in a velocity of about 1100 km s^{-1}, while the 5900 years (feature D) would result in a velocity of about 160 km s^{-1}. The latter is more in line with the tangential velocities measured in this work and with the radial velocities measured by Poetzel et al. [5]. It is possible that the elongated feature D, which indeed has a larger deviation from the average PA of the large-scale outflow compared to other features, potentially referring to a different origin, is related to the feather 1. However, the seemingly similarly shaped feature marked with dashed cyan arrow in Figure 4 is not feature D. Feature D is 15″ further away from the central star and has a slightly smaller PA. A red dashed arrow is indicating its position in Figure 4, and it shows that this feature is situated on the edge of the feather 1, but there is no brightness enhancement in that position other than the boarder line of the feathery feature.

On the other hand, it cannot be ruled out that the faint features are accidentally aligned with Z CMa and that they are actually part of the huge nebula Sh 2-296 on which edge Z CMa is located (see Figure 6 in Fernandes et al. [29]). However, due to the obvious positional proximity to Z CMa, we are inclined to favor the idea that the discovered features are connected to Z CMa rather than being aligned by chance.

The new detected faint features as well as our measured different ages of different features (850 versus 5900 years for feature C and D, respectively) are in accordance with the nature of the Z CMa binary, which has had several eruptions in the past. The knotty nature of the large-scale outflow as well as the (micro) jets (e.g., Whelan et al. [6], Antoniucci et al. [7]) additionally refer to several discrete mass ejections. In addition, the RVs of the individual features vary a lot with values reaching up to ∼±400 km s^{-1}

(Poetzel et al. [5]), further supporting the scenario that the central object has experienced in the past several mass ejections with different initial velocities.

Another possible explanation could be that the measured tangential velocities of feature C and D have not been constant since their ejection from the central binary, which, in return, would affect the calculated ages. However, observationally, we cannot assess it at this point. We have not found suitable observational data prior to 2002 to check the potential velocity canges before our first dataset. In addition, considering the measured velocities, we will have to wait for at least another $\sim$20 years to obtain a new set of observations, which could potentially show a change in tangential velocities. However, independently of the two scenarios, we wish to emphasize that our measured ages are more precise than previous estimates (Poetzel et al. [5]) because we have accurately measured the tangential velocities, while formerly, those were approximated according to the radial velocities and the possible inclination angle of the large-scale outflow.

As measured from our 2002 and 2019 images, the PA of feature D is 223° and does not change during our observing period, 17.58 years. To expand the epoch of analsyis, we estimate from the schematic Figure 1 in Poetzel et al. [5] (they do not publish any numerical values of the PAs nor the distances from the central star) that the PA of feature D at their observing time between 1989 and 1990 was $\sim$225°. Therefore, we conclude that the PA of feature D has not changed during the past 30 years. Following this result, the claim made by Whelan et al. [6] that feature D is related with the micro-jet B (emanating from the FU Ori-type companion with a PA $\sim$235°, which is indicated with the blue solid line in Figure 2), according to their observations, is not supported by our precise PA measurements.

We investigate further whether feature D could be related to the additional sub-arcsecond component emerging from Z CMa, the jet-like structure identified as a streamer (see Figure 3 in Millan-Gabet and Monnier [8], Figure 2 in Canovas et al. [21], Figure 1 in Liu et al. [30], and Figure 1 in Dong et al. [2]). Unfortunately, these authors do not provide any PA measurements for the straight part of the streamer (the outer edge of this feature is slightly curved toward west), but from their figures, we can estimate it to be approximately 215°. We also note that the streamer does not start from the central binary but about $0''.7$ straight toward the south from the binary (see Figure 2 in Canovas et al. [21]). We indicate the PA of this feature with the green dotted line in our Figure 2. Even though the angle of the streamer is more similar to the PA of feature D than the angle of the micro-jet B is, it is still clearly evident that their PAs do not align. Dong et al. [2] explains the streamer as a result of the flyby event because they discovered a faint component whose PA matches with the one of the streamer. The fact that feature D is not aligned with the streamer provides an additional support for the flyby event because it shows that the streamer is not related to any small or large-scale ejecta.

5. Conclusions

We have presented the first proper motion study of features C and D within the large-scale outflow of Z CMa. The two very different proper motion values obtained for these two features confirm the previous suggestion that the large-scale outflow is a result of several active ejection phases with varying initial velocities in the life of Z CMa. Our precise position angle measurements of the same features reveal that they are not aligned with the feature streamer, providing further support for the occurrence of the flyby event in this complex system.

We have discovered new features most probably related to Z CMa—a faint extended arc with several features mimicking feathers. It is very likely that these features are connected to the central binary and are the result of previous mass ejection(s) possibly related to the orbital motion of the binary system.

Author Contributions: Conceptualization and Project administration, T.L., M.K. and L.C.; Resources and Data curation: T.L., L.C., S.K. and A.M.; Formal analysis, Investigation, Software, Methodology and Validation, T.L.; Supervision, M.K.; Visualization, T.L. and S.K.; Writing—original draft preparation, T.L. and S.K.; Writing—review and editing, T.L., M.K., S.K., L.C. and A.M.; Funding acquisition, M.K., L.C. and S.K. All authors have read and agreed to the published version of the manuscript.

Funding: This research was funded by the Czech Science Foundation (GA ČR, grant number 20-00150S), by CONICET (PIP 1337) and the Universidad Nacional de La Plata (Programa de Incentivos 11/G160). The Astronomical Institute of the Czech Academy of Sciences is supported by the project RVO:67985815. This project has also received funding from the European Union's Framework Programme for Research and Innovation Horizon 2020 (2014–2020) under the Marie Skłodowska-Curie Grant Agreement No. 823734. S.K. acknowledges support from the European Structural and Investment Fund and the Czech Ministry of Education, Youth and Sports (Project CoGraDS-CZ.02.1.01/0.0/0.0/15_003/0000437).

Data Availability Statement: Publicly available GMOS-S and Pan-STARRS datasets analyzed in this study can be retrieved from dedicated archives https://archive.gemini.edu/searchform (accessed on 27 February 2023), and https://ps1images.stsci.edu/cgi-bin/ps1cutouts (accessed on 27 February 2023), respectively. The Mt. Palomar image is available from the corresponding author on reasonable request.

Acknowledgments: This research made use of the NASA Astrophysics Data System (ADS) and of the SIMBAD database, which is operated at CDS, Strasbourg, France. This publication is based on observations obtained at the international Gemini Observatory, which is a program of NSF's NOIRLab (processed using DRAGONS (Data Reduction for Astronomy from Gemini Observatory North and South)), which is managed by the Association of Universities for Research in Astronomy (AURA) under a cooperative agreement with the National Science Foundation on behalf of the Gemini Observatory partnership: the National Science Foundation (United States), National Research Council (Canada), Agencia Nacional de Investigación y Desarrollo (Chile), Ministerio de Ciencia, Tecnología e Innovación (Argentina), Ministério da Ciência, Tecnologia, Inovações e Comunicações (Brazil), and Korea Astronomy and Space Science Institute (Republic of Korea) under program ID GS-2019B-Q-210. The Pan-STARRS1 Surveys (PS1) and the PS1 public science archive have been made possible through contributions by the Institute for Astronomy, the University of Hawaii, the Pan-STARRS Project Office, the Max–Planck Society and its participating institutes, the Max Planck Institute for Astronomy, Heidelberg and the Max Planck Institute for Extraterrestrial Physics, Garching, The Johns Hopkins University, Durham University, the University of Edinburgh, the Queen's University Belfast, the Harvard–Smithsonian Center for Astrophysics, the Las Cumbres Observatory Global Telescope Network Incorporated, the National Central University of Taiwan, the Space Telescope Science Institute, the National Aeronautics and Space Administration under Grant No. NNX08AR22G issued through the Planetary Science Division of the NASA Science Mission Directorate, the National Science Foundation Grant No. AST-1238877, the University of Maryland, Eotvos Lorand University (ELTE), the Los Alamos National Laboratory, and the Gordon and Betty Moore Foundation. This work has made use of data from the European Space Agency (ESA) mission *Gaia* (https://www.cosmos.esa.int/gaia, accessed on 27 February 2023), processed by the *Gaia* Data Processing and Analysis Consortium (DPAC, https://www.cosmos.esa.int/web/gaia/dpac/consortium) (accessed on 27 February 2023). Funding for the DPAC has been provided by national institutions, in particular the institutions participating in the *Gaia* Multilateral Agreement.

Conflicts of Interest: The authors declare no conflict of interest.

Abbreviations

The following abbreviations are used in this manuscript:

Z CMa	Z Canis Majors
FU Ori	FU Orionis
PA	position angle
FOV	field of view

RUWE	renormalized unit weight error
S	south
NW	northwest
NE	northeast
SW	southwest
SE	southeast

Notes

1 Although evolved massive stars can have nebula exceeding this size by far, such as the huge bipolar nebula of the B[e] supergiant MWC 314 extending across 13 pc (Marston and McCollum [9]; Liimets et al. [10]) as well as the 10 pc size elaborate filamentary structures around the Luminous Blue Variable P Cygni (see Boumis et al. [11] and references therein).

2 Renormalised Unit Weight Error (RUWE). RUWE is expected to be around 1.0 for sources where the single-star model provides a good fit to the astrometric observations. A value significantly greater than 1.0 (e.g., >1.4) could indicate that the source is non-single or otherwise problematic for the astrometric solution.

3 IRAF is distributed by the National Optical Astronomy Observatory, which is operated by the Association of Universities for Research in Astronomy (AURA) under cooperative agreement with the National Science Foundation.

References

1. Bonnefoy, M.; Chauvin, G.; Dougados, C.; Kóspál, Á.; Benisty, M.; Duchêne, G.; Bouvier, J.; Garcia, P.J.V.; Whelan, E.; Antoniucci, S.; et al. The 2008 outburst in the young stellar system Z CMa. III. Multi-epoch high-angular resolution images and spectra of the components in near-infrared. *Astron. Astrophys.* **2017**, *597*, A91.
2. Dong, R.; Liu, H.B.; Cuello, N.; Pinte, C.; Ábrahám, P.; Vorobyov, E.; Hashimoto, J.; Kóspál, Á.; Chiang, E.; Takami, M.; et al. A likely flyby of binary protostar Z CMa caught in action. *Nat. Astron.* **2022**, *6*, 331–338. [CrossRef]
3. Sicilia-Aguilar, A.; Bouvier, J.; Dougados, C.; Grankin, K.; Donati, J.F. Reading between the lines. Disk emission, wind, and accretion during the Z CMa NW outburst. *Astron. Astrophys.* **2020**, *643*, A29. [CrossRef]
4. Herbig, G.H. The Spectra of Be- and Ae-Type Stars Associated with Nebulosity. *Astrophys. J. Suppl.* **1960**, *4*, 337. [CrossRef]
5. Poetzel, R.; Mundt, R.; Ray, T.P. Z CMa: A large-scale high velocity bipolar outflow traced by Herbig-Haro objects and a jet. *Astron. Astrophys.* **1989**, *224*, L13–L16.
6. Whelan, E.T.; Dougados, C.; Perrin, M.D.; Bonnefoy, M.; Bains, I.; Redman, M.P.; Ray, T.P.; Bouy, H.; Benisty, M.; Bouvier, J.; et al. The 2008 Outburst in the Young Stellar System Z CMa: The First Detection of Twin Jets. *Astrophys. J.* **2010**, *720*, L119–L124.
7. Antoniucci, S.; Podio, L.; Nisini, B.; Bacciotti, F.; Lagadec, E.; Sissa, E.; La Camera, A.; Giannini, T.; Schmid, H.M.; Gratton, R.; et al. Sub-0.1″ optical imaging of the Z CMa jets with SPHERE/ZIMPOL. *Astron. Astrophys.* **2016**, *593*, L13.
8. Millan-Gabet, R.; Monnier, J.D. Discovery of a Near-Infrared Jetlike Feature in the Z Canis Majoris System. *Astrophys. J.* **2002**, *580*, L167–L170.
9. Marston, A.P.; McCollum, B. Extended shells around B[e] stars. Implications for B[e] star evolution. *Astron. Astrophys.* **2008**, *477*, 193–202. [CrossRef]
10. Liimets, T.; Kraus, M.; Moiseev, A.; Duronea, N.; Cidale, L.S.; Fariña, C. Follow-Up of Extended Shells around B[e] Stars. *Galaxies* **2022**, *10*, 41.
11. Boumis, P.; Meaburn, J.; Redman, M.P.; Mavromatakis, F. A deep mosaic of [O III]5007 Å CCD images of the environment of the LBV star P Cygni. *Astron. Astrophys.* **2006**, *457*, L13–L16. [CrossRef]
12. Clariá, J.J. A study of the stellar association Canis Major OB 1. *Astron. Astrophys.* **1974**, *37*, 229–236.
13. Prusti, T. et al. [Gaia Collaboration]. The Gaia mission. *Astron. Astrophys.* **2016**, *595*, A1.
14. Brown, A.G.A. et al. [Gaia Collaboration]. Gaia Early Data Release 3. Summary of the contents and survey properties. *Astron. Astrophys.* **2021**, *649*, A1.
15. Tody, D. The IRAF Data Reduction and Analysis System. In Proceedings of the 1986 Astronomy Conferences, Tucson, AZ, USA, 1–2 March 1986; Volume 627, p. 733. [CrossRef]
16. Tody, D. IRAF in the Nineties. *Astron. Soc. Pac. Conf. Ser.* **1993**, *52*, 173.
17. Hook, I.M.; Jørgensen, I.; Allington-Smith, J.R.; Davies, R.L.; Metcalfe, N.; Murowinski, R.G.; Crampton, D. The Gemini-North Multi-Object Spectrograph: Performance in Imaging, Long-Slit, and Multi-Object Spectroscopic Modes. *Publ. Astron. Soc. Pacific* **2004**, *116*, 425–440. [CrossRef]
18. Labrie, K.; Anderson, K.; Cárdenes, R.; Simpson, C.; Turner, J.E.H. DRAGONS—Data Reduction for Astronomy from Gemini Observatory North and South. *Astron. Soc. Pac. Conf. Ser.* **2019**, *523*, 321.
19. Waters, C.Z.; Magnier, E.A.; Price, P.A.; Chambers, K.C.; Burgett, W.S.; Draper, P.W.; Flewelling, H.A.; Hodapp, K.W.; Huber, M.E.; Jedicke, R.; et al. Pan-STARRS Pixel Processing: Detrending, Warping, Stacking. *Astrophys. J. Suppl.* **2020**, *251*, 4.
20. Chambers, K.C.; Magnier, E.A.; Metcalfe, N.; Flewelling, H.A.; Huber, M.E.; Waters, C.Z.; Denneau, L.; Draper, P.W.; Farrow, D.; Finkbeiner, D.P.; et al. The Pan-STARRS1 Surveys. *arXiv* **2016**. [CrossRef]
21. Canovas, H.; Perez, S.; Dougados, C.; de Boer, J.; Ménard, F.; Casassus, S.; Schreiber, M.R.; Cieza, L.A.; Caceres, C.; Girard, J.H. The inner environment of Z Canis Majoris: High-contrast imaging polarimetry with NaCo. *Astron. Astrophys.* **2015**, *578*, L1.

22. Santander-García, M.; Corradi, R.L.M.; Whitelock, P.A.; Munari, U.; Mampaso, A.; Marang, F.; Boffi, F.; Livio, M. HST and VLT observations of the symbiotic star Hen 2-147. Its nebular dynamics, its Mira variable and its distance. *Astron. Astrophys.* **2007**, *465*, 481–491. [CrossRef]
23. Liimets, T.; Corradi, R.L.M.; Jones, D.; Verro, K.; Santander-García, M.; Kolka, I.; Sidonio, M.; Kankare, E.; Kankare, J.; Pursimo, T.; et al. New insights into the outflows from R Aquarii. *Astron. Astrophys.* **2018**, *612*, A118.
24. Liimets, T. Nebulosities and Jets from Outbursting Evolved Stars. Ph.D. Thesis, University of Tartu, Tartu, Estonia, 2019. arXiv:1910.04157.
25. van Leeuwen, F. Validation of the new Hipparcos reduction. *Astron. Astrophys.* **2007**, *474*, 653–664. [CrossRef]
26. Liimets, T.; Corradi, R.M.L.; Jones, D.; Kolka, I.; Santander-Garcia, M.; Sidonio, M.; Verro, K. Nebulosities of the Symbiotic Binary R Aquarii—A Short Review. *Gold. Age Cataclysmic Var. Relat. Objects V* **2021**, *368*, 41.
27. Kraus, M.; Oksala, M.E.; Nickeler, D.H.; Muratore, M.F.; Borges Fernandes, M.; Aret, A.; Cidale, L.S.; de Wit, W.J. Molecular emission from GG Carinae's circumbinary disk. *Astron. Astrophys.* **2013**, *549*, A28.
28. Dobashi, K. Atlas and Catalog of Dark Clouds Based on the 2 Micron All Sky Survey. *Publ. Astron. Soc. Jpn.* **2011**, *63*, S1–S362. [CrossRef]
29. Fernandes, B.; Montmerle, T.; Santos-Silva, T.; Gregorio-Hetem, J. Runaways and shells around the CMa OB1 association. *Astron. Astrophys.* **2019**, *628*, A44.
30. Liu, H.B.; Takami, M.; Kudo, T.; Hashimoto, J.; Dong, R.; Vorobyov, E.I.; Pyo, T.S.; Fukagawa, M.; Tamura, M.; Henning, T.; et al. Circumstellar disks of the most vigorously accreting young stars. *Sci. Adv.* **2016**, *2*, e1500875.

Disclaimer/Publisher's Note: The statements, opinions and data contained in all publications are solely those of the individual author(s) and contributor(s) and not of MDPI and/or the editor(s). MDPI and/or the editor(s) disclaim responsibility for any injury to people or property resulting from any ideas, methods, instructions or products referred to in the content.

Article

Reddening-Free Q Parameters to Classify B-Type Stars with Emission Lines

Yael Aidelman [1,2,*,†] and Lydia Sonia Cidale [1,2,†]

1. Instituto de Astrofísica de La Plata, CCT La Plata, CONICET-UNLP, Paseo del Bosque S/N, La Plata B1900FWA, Argentina
2. Departamento de Espectroscopía, Facultad de Ciencias Astronómicas y Geofísicas, Universidad Nacional de La Plata (UNLP), Paseo del Bosque S/N, La Plata B1900FWA, Argentina
* Correspondence: aidelman@fcaglp.unlp.edu.ar or ialp@fcaglp.unlp.edu.ar
† Member of the Carrera del Investigador Científico y Tecnológico, CONICET, Argentina.

Abstract: The emission-line B-type stars constitute a heterogeneous group. Many of these stars show similar optical spectroscopic features and color indices, making it difficult to classify them adequately by means of photometric and spectroscopic techniques. Thus, it is relevant to deal with appropriate classification criteria to avoid as many selection effects as possible. For this purpose, we analyzed different reddening-free Q parameters, taking advantage of the Gaia and 2MASS photometric surveys, for both main sequence and emission-line B-type stars. Along with this work, we provided various criteria to search for normal and emission-line B-type stars, using different color–color, Q–color, and Q–Q diagrams. It was also possible to identify stars in different transition phases (i.e., $(Rp - J)$ vs. $(J - Ks)$ diagrams) and to classify them according to their NIR radiation excesses (i.e., the $(Bp - Rp)$ vs. $(H - Ks)$ diagram). Other diagrams, such as the Q_{JKHK} vs. $(H - Ks)$ or Q_{BpJHK} vs. $(Bp - Ks)$, were very useful to search for and classify different classes of B-type stars with emission lines. These diagrams highlighted the presence of several stars, classified as CBe, with large color excesses that seemed to be caused by the presence of dust in their envelopes. Therefore, these stars would be misclassified. Three groups of HAeBe stars with different intrinsic dust properties were also distinguished. The amount of intrinsic dust emission in the diverse groups of emission-line stars was well-recognized via the Q_{JHK} vs. Q_{BpRpHK} diagram. The different selection criteria are very important tools for automated designs of machine learning and optimal search algorithms.

Keywords: stars: early-type; stars: emission-line, Be; (stars:) circumstellar matter; stars: peculiar (except chemically peculiar)

Citation: Aidelman, Y.; Cidale, L.S. Reddening-Free Q Parameters to Classify B-Type Stars with Emission Lines. *Galaxies* **2023**, *11*, 31. https://doi.org/10.3390/galaxies11010031

Academic Editor: Dimitris M. Christodoulou

Received: 5 January 2023
Revised: 1 February 2023
Accepted: 2 February 2023
Published: 15 February 2023

Copyright: © 2023 by the authors. Licensee MDPI, Basel, Switzerland. This article is an open access article distributed under the terms and conditions of the Creative Commons Attribution (CC BY) license (https:// creativecommons.org/licenses/by/ 4.0/).

1. Introduction

In some particular evolutionary phases, and even during the main sequence stage, some B-type stars exhibit emission lines in their optical spectra, mainly the Hα line (cf. [1]). These are known as emission-line B-type stars. Usually, these stars are embedded in dense gaseous circumstellar envelopes (CEs), although many also contain a dusty environment. The presence of the CE not only gives rise to the emission lines but also to infrared (IR) excess over the photospheric radiation due to bound–free, free–free, and thermal emission from gaseous and dust components, respectively [2].

Within the emission-line B-type stars group, we can find objects with very different evolutionary stages, such as the following: (a) the Herbig Ae/Be (HAeBe) stars, pre-main sequence (PMS) objects of spectral types B, A, or F, with stellar masses between 2 M$_\odot$ and 10 M$_\odot$ [3,4]; (b) the classical Be (CBe) stars, B-type non-supergiant stars that show rapid rotation, which facilitates the formation of gaseous envelopes [5]; (c) the B supergiant (BSG) stars with strong wind outflows and rings [6]; (d) the stars with the B[e] phenomenon, a heterogeneous group identified by an emission spectrum with additional forbidden lines of single-ionized atoms and a large IR radiation excess [7]. This last group gathers together the following: (i) young stellar objects, like HAeB[e] stars; (ii) the Luminous Blue Variables (LBVs), post-main sequence massive objects characterized by intense eruptive mass-loss events [8–11]; (iii) the B[e] supergiants (B[e]SGs), massive stars found beyond the main-sequence, with a luminosity of $10^4 - 10^6$ L$_\odot$ (cf. [12]). Their CEs are represented by a two-component wind model [13] comprising; (iv) the interactive binaries with B-type companions [14]; (v) the post-AGB stars, a very short evolutionary phase, in which the objects increase their effective temperatures at constant luminosity and turn into a planetary nebula (cf. [15]). Table 1 summarizes the emission-line B-type stars analyzed in this work.

Table 1. Summary of emission-line B-type stars analyzed in this work.

Emission-Line B-Type Star	Evolutionary Stage
HAeB[e]/HAeBe	Pre-main sequence object with or without the B[e] phenomenon
CBe	Non-supergiant B-type star
LBV	Post-main sequence massive object
B[e]SG	B-supergiant star with the B[e] phenomenon

However, there is some inevitable contamination among the above-mentioned categories. The HAeBes share many properties with the CBe stars, as their gaseous discs generate very similar observable features [5,16]. As many as around half of the known HAeBes display the B[e] phenomenon [17]. Other objects have a doubtful nature, the so-called unclassified B[e] (unclB[e]) stars, so they are easily confused with members of other groups [7]. These objects simultaneously show features of either early or evolved stars with dust in their envelopes, yet lack cold dust components [13]. Sheikina et al. [18] and Miroshnichenko [14] proposed that the unclB[e] stars with little cold dust, called FS CMa-type stars, are binary systems undergoing mass transfer. The FS CMa-type stars have comparable near-infrared excesses as Group I Herbig stars (those with very large IR radiation excess) but with mid-infrared excess even weaker than Group III (stars with small IR radiation excess), see Chen et al. [19].

Thus, the emission-line B-type stars constitute a heterogeneous group. Usually, they are either associated with star-forming regions or ejecta-enriched material from an evolved star. As many stars show similar spectroscopic features and color indices, it has become difficult to classify them adequately through photometric and spectroscopic techniques. Moreover, since these kinds of objects are often very distant, they also have uncertain luminosity. Therefore, determinations of their photospheric parameters, spectral types, and evolutionary stages are very challenging (e.g., [12,20]).

In the last decade, there has been an exponential growth of data volume from space missions and synoptic sky surveys that have transformed how astronomy is conducted. To take full advantage of this wealth of information, mainly as a tool to optimize object selection, catalog stellar objects, and perform statistical analyses on them, it is very important to deal with appropriate criteria of classification that avoid as many selection effects as possible.

The near-infrared (NIR) nature of the surveys is particularly useful for the analysis of early-type emission-line stars, since photometry is less sensitive to dust obscuration than optical observations. The NIR intrinsic colors of early-type, and most late-type, stars cause them to lie along separated sequences in a $(H - Ks)$ vs. $(J - H)$ color–color diagram (CCD) when reddened by an arbitrary amount, as Comerón et al. [21] illustrated for the nearby Cygnus OB2 association. The NIR CCDs have also been used to classify emission-line B-type stars, such as HAeBes, B[e]SGs, and LBVs [22,23], but there is still a certain mix of objects due to interstellar reddening effects and, possibly, to similar physical properties that lie in their discs [12]. As a consequence, a selection method based only on excess emission in infrared radiation may easily generate a sample contaminated with Young Stellar Objects (YSOs) and post-main-sequence stars.

An alternative method of selection is to use the reddening-free quantity $Q_{JHK} = (J - H) - 1.70 (H - Ks)$ [24], equivalent to the Johnson's Q parameter [25]. This Q_{JHK} parameter allows us, for example, to separate early-type normal stars, which are characterized by $Q_{JHK} \simeq 0$, from background red giants, that are expected to have $Q_{JHK} \simeq 0.5$ [26]. Comerón and Pasquali [26] showed that by choosing stars with $Q_{JHK} < 0.30$, together with traditional selection criteria through magnitudes and colors, all the early-type stars would be recognized, including those displaying NIR excess produced by circumstellar environments, which would further decrease the value of Q_{JHK}. Notwithstanding, the sample might be potentially contaminated by oxygen-rich AGB stars and carbon-rich giants that display JHKs colors similar to those of normal reddened early-type stars [27]. Later, Garcia et al. [28] also found that the use of Q diagrams in the optical ranges, such as Q_{UBV} vs. V_{corr} (the Q parameter in the UBV bands versus the apparent visual magnitude corrected from extinction) partially broke the color degeneracy of massive blue stars. Therefore, this diagram was also helpful in estimating stellar parameters. They also showed that the different types of massive stars (O, B, Be stars, and the WO) are located in well-defined loci with little mixing. In the same direction, Aparicio Villegas et al. [29] developed a methodology for stellar classification and physical parameter estimation (T_{eff}, $\log g$, [Fe/H], and color excess $E(B - V)$), based on 18 independent reddening-free Q-values.

One of the purposes of this work was to design strategies to select specific groups of emission-line B-type stars from their photometric properties avoiding contamination with other objects to the greatest extent possible. To fulfill this goal, we carried out an exhaustive analysis of the behavior of the Q parameter, taking advantage of the photometric measurements provided by the Global Astrometric Interferometer for Astrophysics mission (Gaia, [30,31]) and the Two Micron All Sky Survey (2MASS, [32,33]). In this way, we would be able to perform a stellar classification based solely on apparent magnitudes, minimizing as much as possible the effect of interstellar reddening. This technique would also enable distinction between the various emission components: star, free–free emission, and thermal emission. There are different possibilities for using Q parameters for stellar classification purposes. For example, the simplest method is based on Q-Q diagrams (QQD) calibrated in terms of spectral classes (or temperatures) and absolute magnitudes (or gravities). Another more powerful method is to use more than two Q-parameters, forming multidimensional space cells [34].

In this work, we used G_{Bp} and G_{Rp} photometry from Gaia Early Data Release 3 (EDR3, [35]) and J, H, and Ks from 2MASS to construct reddening-free Q parameters. We also showed that Q-color index diagram (QCD) and QQD enabled us to separate different types of B-type stars in transition phases, namely, HAeBes, LBVs, and B[e]SG stars, from each other. The paper is organized as follows: in Section 2, we define the reddening-free Q parameter; in Section 3, we describe the criteria to calculate Q parameters using Gaia EDR3 and 2MASS photometry. The behavior of the Q parameters with effective temperatures is also addressed; Section 4 shows the loci of the selected groups of emission-line B-type stars in the different diagrams (CCD, QCD, and QQD). Finally, Sections 5 and 6 present a discussion of this work's results and the main conclusions.

2. Reddening-Free Q Parameter

In the $(U - B)$ vs. $(B - V)$ color–color diagram, Johnson and Morgan [25] noted that the lines connecting reddened and unreddened stars of the same spectral type had very nearly the same slope. Thus, they defined the quantity:

$$Q_{UBV} = (U - B) - \frac{E(U - B)}{E(B - V)} (B - V) \tag{1}$$

that turned out to be an interstellar reddening-free parameter and could be used to determine the spectral types and reddening of stars between B1 and B9, inclusive (as explained by Johnson and Morgan [25]).

The definition of the reddening-free Q parameter can be extended to any other photometric system:

$$Q_{m_1 m_2 m_3 m_4} = (m_1 - m_2) - \frac{E(m_1 - m_2)}{E(m_3 - m_4)}(m_3 - m_4) \qquad (2)$$

considering either four different magnitudes (m_1, m_2, m_3, and m_4) corresponding to different photometric filters or three different magnitudes (adopting $m_2 = m_3$).

The knowledge of the slope, $E(m_1 - m_2)/E(m_3 - m_4)$, in the $(m_1 - m_2)$ vs. $(m_3 - m_4)$ CCD enables elimination of the effect of interstellar reddening and derives the intrinsic color indices of stars. Thus, Q represents the ordinate to the origin of the line connecting the reddened and unreddened positions of each star; see green circles in Figure 1.

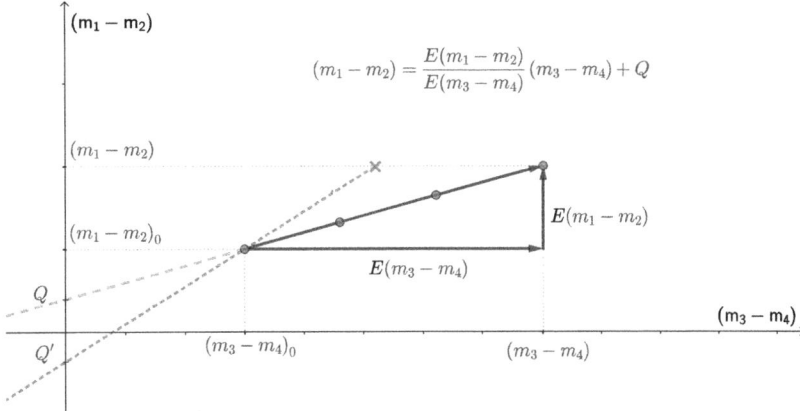

Figure 1. The reddening-free Q parameter definition. Stars with the same spectral types but different color excesses (green circle symbols) have the same Q. If the star has an anomalous color excess (red cross symbol), the Q parameter changes to Q' (see details in the text).

On the other hand, the interstellar absorption at a given magnitude, i.e., A_{m_2} for the m_2 magnitude, as well as the ratio $E(m_1 - m_2)/E(m_3 - m_4)$, both depend on the interstellar extinction law[1]. However, dust extinction has different properties in different regions of the sky, on both small and large scales. Therefore, variations in the interstellar medium (ISM) law would affect not only the color-excess ratios but also the Q parameters [36,37]. This change is represented by:

$$\Delta Q = Q' - Q = -\left[\left(\frac{E(m_1 - m_2)}{E(m_3 - m_4)}\right)_{anomalous} - \left(\frac{E(m_1 - m_2)}{E(m_3 - m_4)}\right)_{normal}\right](m_3 - m_4) \qquad (3)$$

as shown in Figure 1. Thus, the changes of Q due to different $E(m_1 - m_2)/E(m_3 - m_4)$ ratios may help to detect the peculiarity of the extinction law towards particular sky regions.

It is interesting to note that in a QCD, stars with the same spectral type, but different "normal" reddening, lie on a straight line parallel to the x-axis, as shown in Figure 2. By using an appropriate set of photometric filters, e.g., the automated two-dimensional classification from multicolor photometry in the Vilnius system [38–42], it would be possible to make a stellar classification with a set of reddening-free Q parameters.

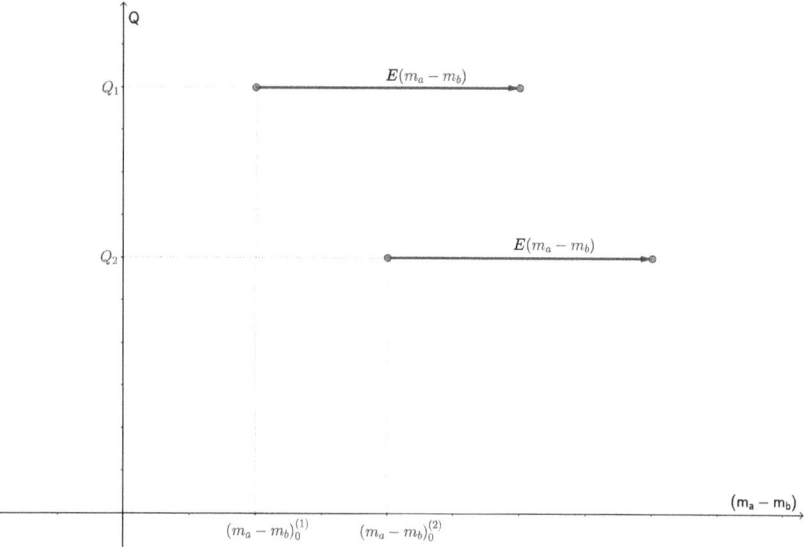

Figure 2. Free-reddening Q parameter vs. color-index diagram (QCD). Stars with the same spectral type but different "normal" reddening lie on a straight line parallel to the x-axis.

To calculate the standard error (σ_Q), we should consider that the Q parameter depends on the accuracy of the magnitudes and the color-excess ratio. Thus,

$$\sigma_Q^2 = \sigma_{m_1}^2 + \sigma_{m_2}^2 + \left(\frac{E(m_1 - m_2)}{E(m_3 - m_4)}\right)^2 \left(\sigma_{m_3}^2 + \sigma_{m_4}^2\right) + (m_3 - m_4)^2 \sigma_E^2 \quad (4)$$

where $E = E(m_1 - m_2)/E(m_3 - m_4)$. Then, considering that the accuracy of all magnitudes is the same, σ_m, and $\sigma_E = 0$, it results in:

$$\sigma_Q^2 = 2\sigma_m^2 \left[1 + \left(\frac{E(m_1 - m_2)}{E(m_3 - m_4)}\right)^2\right]. \quad (5)$$

3. Study of Gaia and 2MASS Reddening-Free Q Parameters

We chose to work with Gaia EDR3 [35] and 2MASS [32,33] catalogs because the corresponding missions surveyed the entire sky. Furthermore, B-type stars and nearby stars have unsaturated magnitudes. In this way, we had, in the visual and NIR regions, respectively, the G_{Bp} (blue) and G_{Rp} (red) passbands of the Gaia system and the 2MASS J, H, and Ks photometric bands. We denote, for simplicity, the Gaia Bp and Rp bands. We did not use the Gaia G-band because, in such a high bandwidth, the extinction coefficient varies strongly with temperature and the extinction itself but less with surface gravity and metallicity [43,44].

3.1. The Q Parameters and Color-Color Diagrams for Normal B-Type Stars

Since there is no previous study of Q parameters that combine the Gaia and 2MASS photometry, we needed, first, to analyze the position of the B-type stars without emission lines (from now on normal B-type stars) in the different diagrams, Q vs. effective temperature (T_{eff}), and their corresponding color–color relations.

The sample of "normal" B-type stars was selected from Zorec et al. [45,46], Aidelman et al. [47,48,49] and Cochetti et al. [50]. Our sample did not consider stars classified as Be (that at least once presented the Hα line in emission) or Bn (with hydrogen lines deformed by high rotation). This was a homogeneous sample with confident stellar parameters obtained with the spectrophotometric BCD classification system [46,51,52]. The BCD system has a relevant advantage; the photospheric Balmer discontinuity, D^*, is not affected by interstellar extinction or circumstellar effects (extinction or emission). The reason for this is that the D^* value is a linear combination of the Q_{UBV} parameter [53].

The cross-matches between the B-type star sample and the Gaia EDR3 and 2MASS catalogs were done using TOPCAT[2] (*Tool for OPerations on Catalogs And Tables* [54,55]). The search radius was set at 0.2 arcsec to avoid duplicated sources. We also selected stars with accurate photometry. This meant they satisfied the following criteria: (a) the magnitude $G < 16$ mag, (b) mean errors in Bp and Rp little than 0.01 mag for the Gaia EDR3, and (c) $JHKs$ photometry with quality flags ("Qfl" or "ph_qual") equal "AAA" for the 2MASS. In this way, we guaranteed that the σ_{m_i} contributions at σ_Q (in Equation (4)) were as small as possible.

To calculate the reddening-free Q values (Equation (2)) we used the combination of all color indices: $(Bp - Rp)$, $(Bp - J)$, $(Bp - H)$, $(Bp - Ks)$, $(Rp - J)$, $(Rp - H)$, $(Rp - Ks)$, $(J - H)$, $(J - Ks)$, and $(H - Ks)$, which led to a total of $C(10,2) = \frac{10!}{2!(10-2)!} = 45$ Q parameters. The color-excess ratios were obtained from the relative absorption coefficients, $r_{m_a} = A_{m_a}/A_V$, for each of the selected five bands, which were estimated using the "extinction" code Fast Interstellar Dust Extinction Laws in Python[3] [56]. With this tool, we estimated the r_{m_a} values for different wavelengths and different extinction laws for each of the three available models [57–59], which also permitted the use of different values for R_V. These values are given in Table 2. This code also calculated the extinction for Fitzpatrick and Massa [60]'s model but only for $R_V = 3.1$. Therefore, this last model was not adopted in this work. It is important to stress that the dependence of Bp and Rp relative absorption coefficients with the T_{eff} are still unknown. However, as the filters were narrower than the one of the G band, we would expect a lower effect.

Table 2. Values of relative absorption coefficients for the Gaia EDR3 and 2MASS filters

Filter	λ Central [Å]	$r_{m_a} = A_{m_a}/A_V$			
		$R_V = 2.1$	$R_V = 3.1$	$R_V = 4.1$	$R_V = 5.1$
		Cardelli et al. [57]			
Bp	5110	1.128	1.091	1.073	1.061
Rp	7770	0.550	0.631	0.673	0.698
J	12,350	0.230	0.288	0.317	0.335
H	16,620	0.143	0.178	0.197	0.208
Ks	21,590	0.094	0.117	0.129	0.136
		O'Donnell [58]			
Bp	5110	1.153	1.105	1.080	1.066
Rp	7770	0.573	0.643	0.678	0.700
J	12,350	0.230	0.288	0.317	0.335
H	16,620	0.143	0.178	0.197	0.208
Ks	21,590	0.094	0.117	0.129	0.136
		Fitzpatrick [59]			
Bp	5110	1.130	1.087	1.065	1.052
Rp	7770	0.503	0.581	0.621	0.645
J	12,350	0.263	0.262	0.262	0.261
H	16,620	0.170	0.163	0.160	0.158
Ks	21,590	0.115	0.112	0.111	0.110

In order to estimate typical Q errors for normal B-type stars, we calculated ΔQ and σ_Q (Equations (3) and (4), respectively). On the one hand, we estimated $|\Delta Q|$ where the Q' values were calculated using Fitzpatrick's model for $R_V = 2.1$, 4.1, and 5.1 while the Q values corresponded to $R_V = 3.1$. On the other hand, we calculated σ_Q for each reddening law (see Table 2) using $R_V = 3.1$ and $\sigma_E = 0$. Both $|\Delta Q|$ and σ_Q values were calculated for the 45 combinations of color-excess ratios corresponding to our entire sample of normal B-type stars. These values are shown in Figure 3. The results showed that the mean values were: $|\Delta Q| \sim 0.05$ and $\sigma_Q \sim 0.15$. Therefore, any star would show anomalous reddening if its $|Q|$ value was larger than $3|\Delta Q| \sim 0.15$ of the expected one for its spectral type. By analogy, we could define a Q range of $3\,\sigma_Q \sim 0.45$ that would contain all the normal B-type stars. This criterion could be used to identify peculiar stars since they would be located outside the defined Q range.

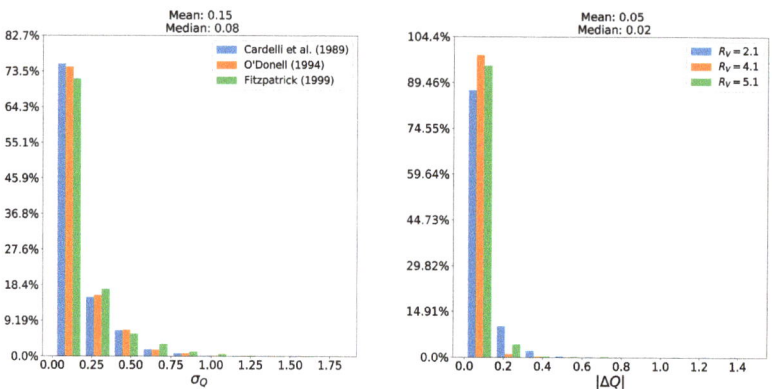

Figure 3. Frequency histogram of error distributions for different reddening laws. The σ_Q (left panel) and $|\Delta Q|$ (right panel) values were calculated for our entire sample of "normal" B-type stars using the 45 reddening-free Q parameters. To calculate σ_Q we used $R_V = 3.1$ for the different reddening laws, while to calculate $|\Delta Q|$ we used $R_V = 2.1$, 4.1, and 5.1. See details in the text.

Figure 4 shows some examples of the behavior of the parameter Q with T_{eff} and their respective color–color indices. These diagrams were selected considering the color indices frequently used in the literature. To study this behavior, we plotted our B-type star sample, together with the main sequence (MS) taken from the Modern Mean Dwarf Stellar Color and Effective Temperature Sequence[4] (much of the content of this table, but not all, was incorporated into the Table 5 of [61]). The Q values for both the B star sample (red points) and the MS (solid gray line) were calculated using the model created by Fitzpatrick [59] with $R_V = 3.1$. To show the effect of anomalous and normal reddening on color–color diagrams, the MS was reddened considering $A_V = 1$ mag, adopting the model done by Fitzpatrick [59] for the different R_V coefficients listed in Table 2, while in the Q-T_{eff} diagrams the Q values for MS were calculated also using the Fitzpatrick's model with $R_V = 2.1$, 4.1, and 5.1 (solid and dashed lines, see the caption to Figure 4). Then, considering that $\sigma_Q = 0.15$ (as set out above), we could define the following limits in the Q-T_{eff} diagrams and CCDs (shown in brown dashed lines):

$$-0.10 \leq Q_{JKRpJ} \leq 0.05 \qquad (6)$$

$$-0.07 \leq Q_{HKBpRp} \leq 0.08 \qquad (7)$$

$$-0.11 \leq Q_{JHK} \leq 0.04 \qquad (8)$$

that encloses almost all the B-type stars of the sample (around 86%, 93%, and 64% of a total of 163 stars for each of the limits defined above).

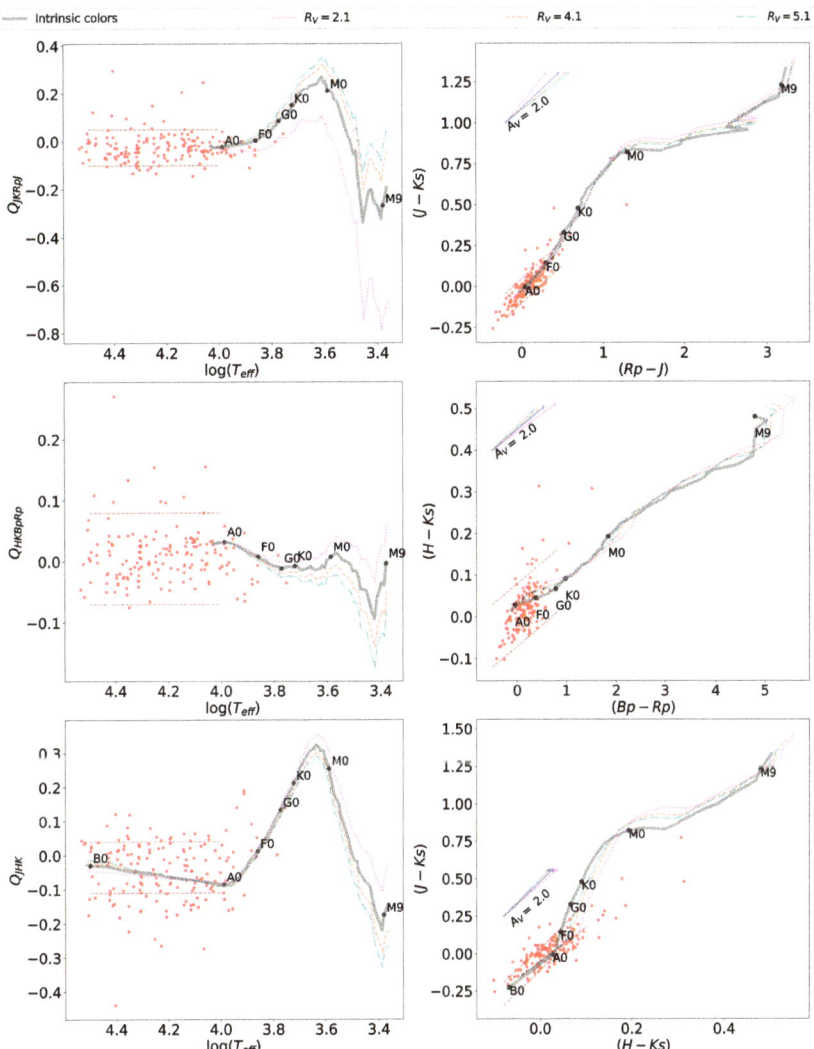

Figure 4. Behavior of Q_{JKRpJ}, Q_{HKBpRp} and Q_{JHK} parameters with T_{eff} (left panels) and their corresponding color–color diagrams (right panels). The red points represent the loci of the "normal" B-type star of the sample [45–50]. The main sequence band (solid gray line, [61]) was reddened considering $A_V = 1$ mag adopting the model done by Fitzpatrick [59] for different R_V coefficients: $R_V = 2.1$ (pink dotted line), $R_V = 3.1$ (solid blue line), $R_V = 4.1$ (orange dashed line), $R_V = 5.1$ (green dashed line). The effect of the reddening vector, i.e., $A_V = 2$ mag, for different ISM laws, is indicated in the top-left part of the plots. The brown dashed lines enclosed the cloud of points within $\sigma_Q < 0.15$.

3.2. The Q–Color and Q–Q Diagrams for Normal B-Type Stars

Generally, the effective temperature of the B-type stars is a priori an unknown quantity and is often derived from spectroscopic data. Therefore, an alternative way to describe the behavior of the Q parameters with T_{eff}, or the spectral types, is to use the color indices. Thus, the $(Bp - Rp)$ color index is useful, for example, to classify stars according to their T_{eff}. Jordi et al. [43] remarked that $(Bp - Rp)$ plays the same role as $(V - I_C)$, thus the relationship between $(Bp - Rp)$ and T_{eff} could be almost equivalent to the well-known relation of $T_{\text{eff}} = f(V - I_C)$. On the other hand, Mucciarelli et al. [62] also found that $(Bp - Ks)_0$ and $(G - Ks)_0$ were the best choice to derive precise and accurate T_{eff} values among late-type stars.

Then, to explore the behavior of Q parameters with the color indices, we present some QCD in Figure 5. The Q parameters for the sample of B-type stars and the MS were constructed as explained in Section 3.1. In these plots, the MS was also reddened. These diagrams allowed early-type stars (red cloud of points) to be separated from late types on the MS. In particular, in the Q_{BpJRpK} vs. $(J - Ks)$ and Q_{BpRpHK} vs. $(Bp - Rp)$ diagrams (left panels), the B-type stars were located in a well-delimited region. This group was located at

$$-0.11 < Q_{BpJRpK} < 0.20 \quad \text{and} \quad -0.20 < (J - Ks) < 0.14 \tag{9}$$
$$-0.60 < Q_{BpRpHK} < 0.50 \quad \text{and} \quad -0.30 < (Bp - Rp) < 0.40 \tag{10}$$

where the upper limits for the color indices correspond to an F0 V star. Another relevant characteristic of the diagrams of Figures 4 and 5 is that the differences between the MS and anomalous reddening ones seemed to be very small, at least for the selected B-type star group.

Our sample of B-type stars in the Q_{BpKJH} vs. $(Bp - Ks)$ diagram (top-right panel of Figure 5) was located in the range $-0.17 < Q_{BpKJH} < 1.0$, where the lower limit corresponded to the position of an F0 V-type star. This region could be compared with that defined by Poggio et al. [63] in the $(J - H)$ vs. $(G_{DR2} - Ks)$. These authors made a preliminary selection of OB candidates that satisfied both $(J - H) < 0.14 (G_{DR2} - Ks) + 0.02$ and $(J - Ks) < 0.23 (G_{DR2} - Ks)$. The first condition could be expressed in terms of Q as $Q_{JHG_{DR2}K} = (J - H) - 0.14(G_{DR2} - Ks) < 0.02$. Then, by multiplying both sides by $(-0.14)^{-1}$, we obtained $Q_{G_{DR2}KJH} = (G_{DR2} - Ks) - 0.07(J - H) > -0.14$ which was similar to the region defined for Q_{BpKJH}.

As was already mentioned in Section 2, the reddening vector in the QCD has a horizontal direction (Figure 2), for this reason, the reddened B-type stars were overplotted to A and F-type MS stars. This effect was more evident in the Q_{JHK} vs. $(H - Ks)$ diagram (bottom-right panel of Figure 5).

Finally, Figure 6 shows other examples of QQDs. Here, we find that in the Q_{JHK} vs. Q_{BpRpK} (top-left panel) and Q_{RpJK} vs. Q_{BpRpHK} (bottom-left panel) diagrams, the loci of the B-type stars were restricted to a small region around the coordinate point $(0,0)$:

$$-0.18 < Q_{JHK} < 0.01 \quad \text{and} \quad -0.12 < Q_{BpRpK} < 0.18 \tag{11}$$
$$-0.05 < Q_{RpJK} < 0.24 \quad \text{and} \quad -0.6 < Q_{BpRpHK} < 0.7 \tag{12}$$

These relations could be used as possible criteria for selecting B-type stars. In particular, the first diagram is very sensitive to the ISM reddening law, and it could also be helpful for detecting stars with anomalous reddening.

On the other hand, the combination of parameters Q_{JHK} vs. Q_{BpRpHK} and Q_{BpKHK} vs. Q_{BpRpHK} (right panels) were not convenient at all, since they mixed late- and early-type stars. Notwithstanding, it drew attention to the fact that the B-type stars were located in a very narrow line that was almost perpendicular to the reddening direction.

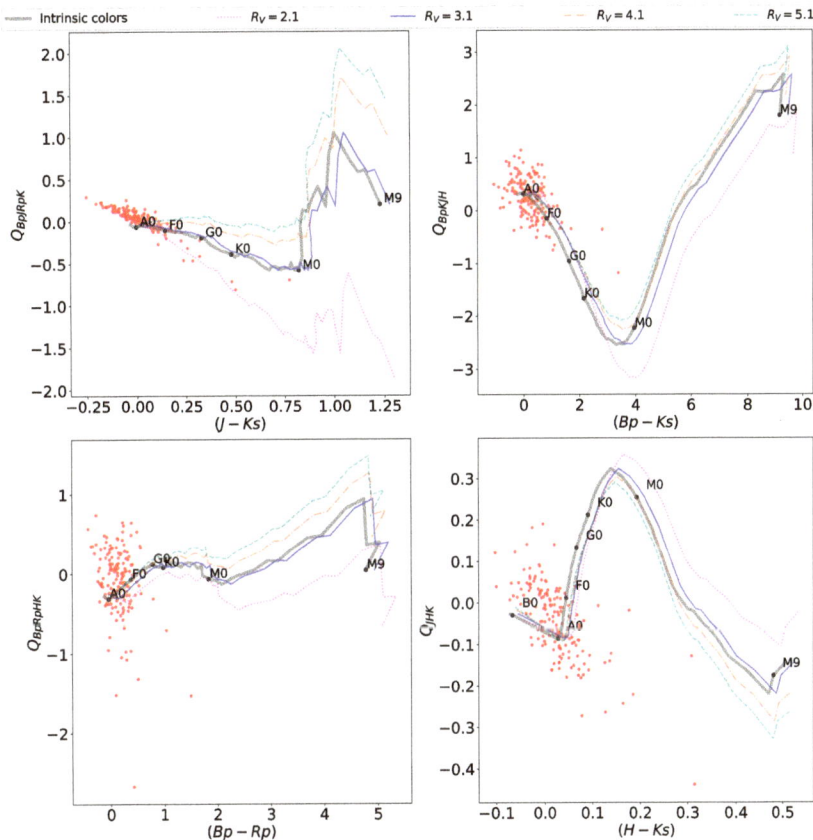

Figure 5. Examples of the behavior of the Q parameter with the color indices. The red points represent the loci of the "normal" B-type star of the sample [45–50]. The main sequence band (solid gray line) was reddened considering $A_V = 1$ mag for different R_V coefficients: $R_V = 2.1$ (pink dotted line), $R_V = 3.1$ (solid blue line), $R_V = 4.1$ (orange dashed line), and $R_V = 5.1$ (green dashed line). See the text for details.

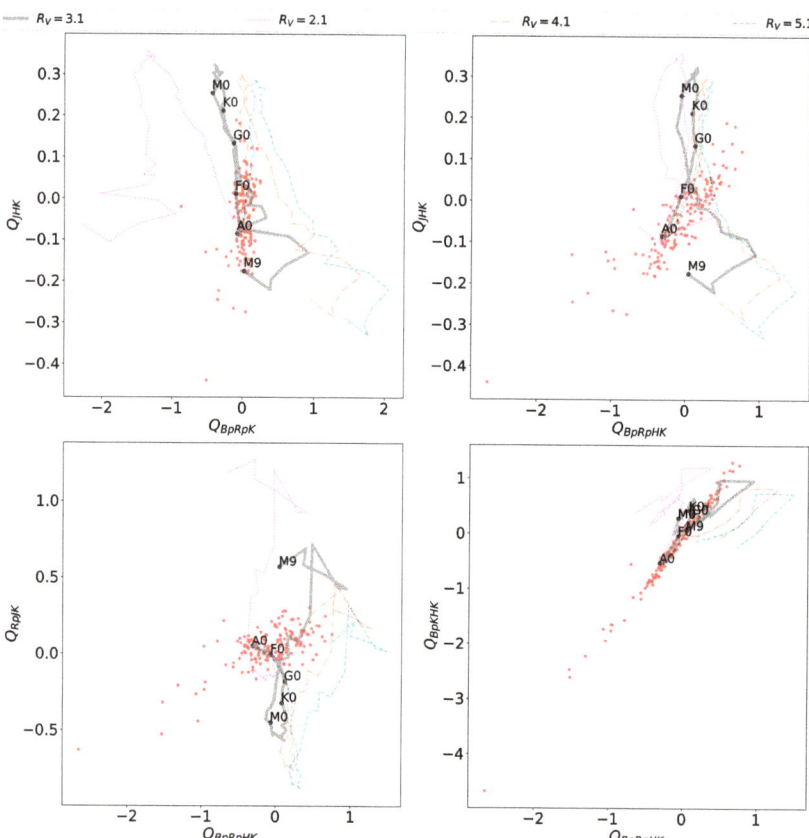

Figure 6. Examples of selected Q-Q diagrams using different R_V absorption coefficients. The red points represent the loci of the "normal" B-type stars of the sample [45–50]. The main sequence band was calculated using different R_V coefficients: $R_V = 2.1$ (pink dotted line), $R_V = 3.1$ (gray solid line), $R_V = 4.1$ (orange dashed line), and $R_V = 5.1$ (green dashed line). See the text for details.

4. B-Type Stars with Emission Lines

In the previous section, using several Gaia and 2MASS color–color, Q–color, and Q–Q diagrams, we delimited regions where normal B-type stars locate. In this section, we analyzed the location in the color–color, Q–color, and Q–Q diagrams of different kinds of emission-line B-type stars with respect to the normal ones. Our goal was to be able to find criteria that would allow us to separate among the diverse groups of stars with emission lines.

To analyze the position of the emission-line B-type stars in the different diagrams, we constructed our sample using stars classified by Lamers et al. [7], Kraus [12], Miroshnichenko [14], Cidale et al. [20], Oksala et al. [22], Zorec et al. [45,46], Aidelman et al. [47,48,49], Cochetti et al. [50], Clark et al. [64] and Guzmán-Díaz et al. [65]. In this way, in addition to the 163 normal B-type stars already studied, our sample was composed of 84 classical Be stars (CBe), 204 HAeBe and HAeB[e] stars, 33 LBVs, and 23 B[e]SGs. We searched for Gaia and 2MASS photometric data, and, finally, we obtained a sample of 156 emission-line B-type stars. It was striking that only 34 HAeBe stars, among the 204 selected stars, had photometric data in the Gaia EDR3 and 2MASS catalogs.

4.1. Color–Color Diagrams for Emission-Line B-Type Stars

The Gaia and 2MASS color–color diagrams are shown in Figure 7. Again, the solid gray line is the main sequence. The reddening vector ($A_v = 2$ mag) is indicated in the upper left. The cloud of red and cyan points are, respectively, the normal B and CBe stars. Blue symbols are for HAeBes (and a few HAeB[e]s), orange triangles for B[e]SGs, and green squares for LBVs. Symbols surrounded by black open circles correspond to stars with accurate photometry. These stars satisfied the stated criteria (a), (b), and (c) in Section 3.1. We distinguished two groups of CBe stars in all the plots. The first group behaved as normal B-type stars do. These stars seemed to have small or no infrared excess. The second group encompassed stars with a large color excess, sharing similar photometric characteristics to some LBV stars. These CBe stars presented interstellar reddening and intrinsic color excesses (because the reddening vector points toward a slightly different direction). The large color excess observed in the last group of stars is discussed in Section 5. Particularly interesting are the $(Rp - Ks)$ vs. $(Bp - Rp)$ and $(J - Ks)$ vs. $(Bp - Ks)$ diagrams (top-left and top-right panels in Figure 7, respectively). These are very useful for searching for CBe stars with large infrared excesses because the normal B stars distribute along a narrow and well-defined band.

The best studied NIR color–color diagram in the literature is the $(J - H)$ vs. $(H - Ks)$ (see Figure 7, middle-left panel). Even when the stars are affected by reddening, this NIR two-color diagram permits separate LBVs from B[e]SG stars [22,23]. However, there is still significant contamination between LBVs and HAeBe stars. This diagram also allows distinguishing, although some mix is observed; the HAeBe stars from both the CBe stars [66] and the B[e]SGs [23]. For illustrative purposes, two straight brown dashed lines ($(J - H) = 1.93(H - Ks) - 1.11$ and $(J - H) = 1.93(H - Ks) - 0.55$) were plotted to indicate the regions mostly occupied by each group of stars: LBVs, HAeBes and B[e]SGs. Similarly, a $(J - H)$ vs. $(Bp - Ks)$ diagram (Figure 7, bottom-right panel) shows some mixture between young and evolved emission-line stars.

A relevant result was the behavior of the two-color diagram $(Bp - Rp)$ vs. $(H - Ks)$ (Figure 7, bottom-left panel), which separated groups of stars (but without recognizing their evolutionary stages) in regions or bands of different color excesses. This behavior was not caused by the reddening vector, which pointed almost perpendicularly to the horizontal shift shown for the different groups of stars. Moreover, this color–color diagram also separated the LBVs from the MS late-type stars. Similarly, the $(Rp - J)$ vs. $(J - Ks)$ diagram discriminated, but to lesser degree, LBVs from the other groups. Although, it seemed to gather groups of stars with different circumstellar reddening (Figure 7, middle-right panel).

4.2. Q–Color and Q-Q Diagrams for Emission-Line B-Type Stars

As shown in Section 4.1, it was very difficult to delimit regions for HAeBe and B[e]SG stars using two-color diagrams. Particularly, in all these diagrams, these two types of stars were, to more or less degree, mixed. The same happened with the population of HAeBe and LBV stars. Part of this degeneracy could be solved using the Q-color and Q-Q diagrams.

Figure 8 presents several examples of QCD. The top-left panel, Q_{BpRpHK} vs. $(Bp - Rp)$ located the LBVs over the line cut $Q_{BpRpHK} = -4.5$ (brown dashed line). Particularly, the Q_{JHK} vs. $(J - Ks)$ diagram (top-right panel) separated the different groups according to the amount of IR emission. The greater the amount of hot dust, the higher the $(H - Ks)$ color index, which implied a more negative value for $Q_{JHKs} = (J - H) - \frac{E(J-H)}{E(H-Ks)}(H - Ks)$. In addition, the dashed brown lines at $Q_{JHK} = -1.11$ and $Q_{JHK} = -0.55$ represent, respectively, the straight lines plotted on the $(J - H)$ vs. $(H - Ks)$ diagram. According to this, there were three groups of HAeBe stars, one of them located between the main sequence and the line cut $Q_{JHK} = -0.55$. It shared its properties with the LBV group of stars. A second group was located within a region with $-1.11 < Q_{JHK} < -0.55$, while the last group had properties similar to the B[e]SGs.

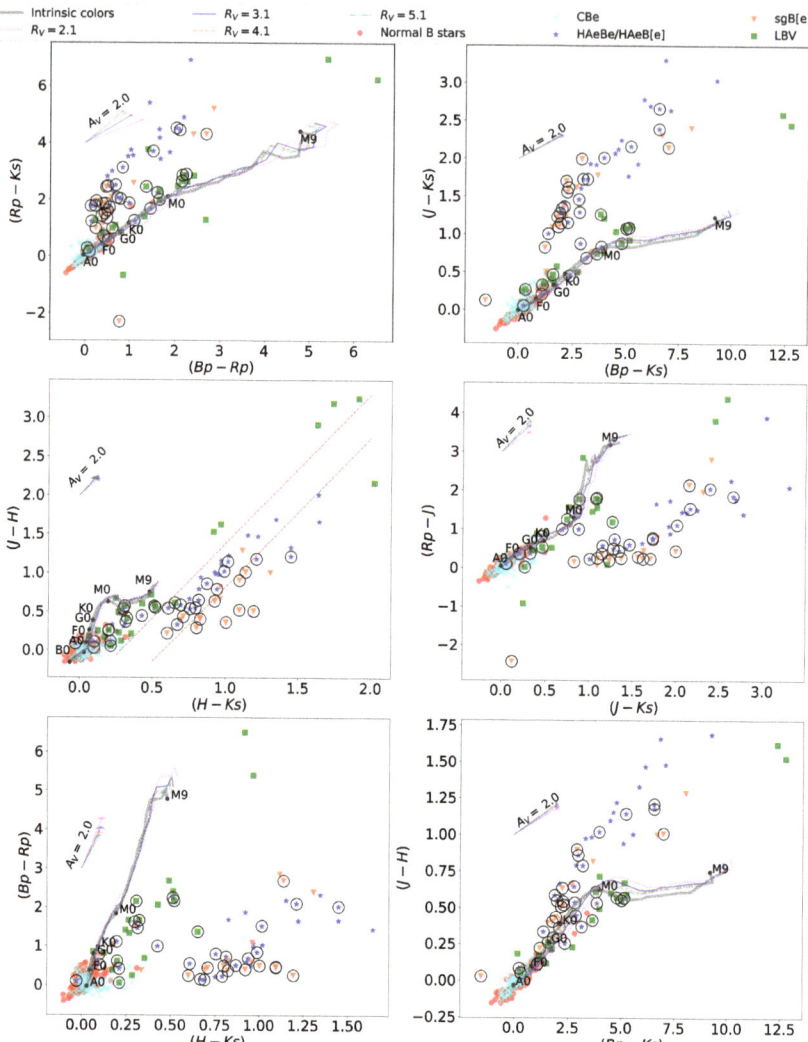

Figure 7. Selected Gaia and 2MASS color–color diagrams for B-type stars with emission lines. The main sequence was taken from Pecaut and Mamajek [61]. It was reddened using the extinction law calculated by Fitzpatrick [59], applying $A_V = 1$ mag and different R_V values. Symbols represent the normal B-type stars (in red), the CBe stars (in cyan), the HAeBes/HAeB[e]s (in blue), the LBVs (in green), and the B[e]SGs (in orange). Black open circles point to stars with accurate photometry (see Section 3.1). The reddening vector $A_V = 2$ mag is indicated in the upper-left side of the plots.

The middle-left panel, Q_{JKHK} vs. $(H - Ks)$, better separated HAeBes from B[e]SGs. There, the HAeBe stars were more reddened than the B[e]SGs. It is also interesting to note that in both Q_{JHK} vs. $(J - Ks)$ and Q_{JKHK} vs. $(H - Ks)$ diagrams, it was possible to distinguish a group of five strongly reddened LBV stars as they were shifted horizontally to the right.

The most interesting diagram was Q_{BpJHK} vs. $(Bp - Ks)$ (middle-right panel), which enabled identification of LBV, B[e]SG, and HAeBe stars from each other. Almost all LBVs had $Q_{BpJHK} > -7$. The B[e]SG stars were also well-separated from HAeBe by the straight brown dashed line ($Q_{BpJHK} = -4.33(Bp - Ks) - 0.57$) drawn in the figure. However, in a Q_{RpJK} vs. $(J - Ks)$ diagram (bottom-left panel), both CBe and B[e]SG stars seemed to lie on the same straight line ($Q_{RpJK} = -1.99(J - Ks) + 0.08$). Instead, the LBVs had values of $Q_{RpJK} > -1.7$. The Q_{BpKJK} vs. $(Rp - Ks)$ diagram showed similar behavior.

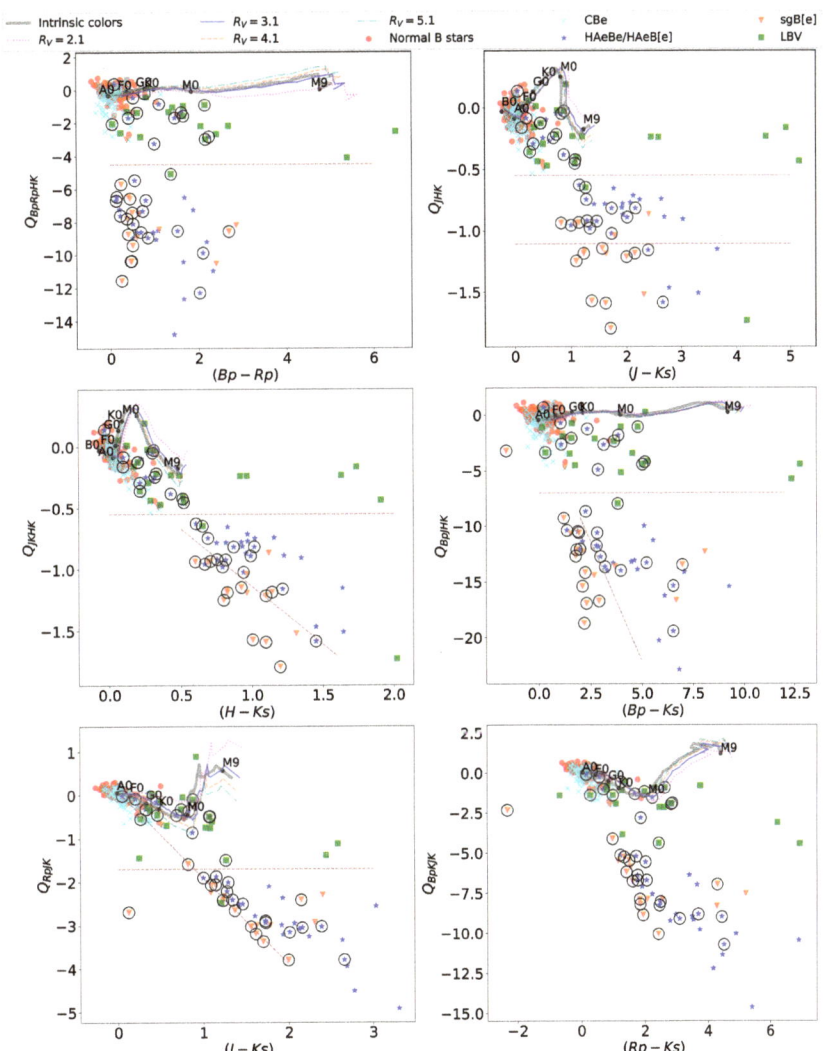

Figure 8. Q-color diagrams of emission-line B-type stars. The main sequence was reddened applying $A_V = 1$ mag and different R_V values (see Section 3 for details). Symbols represent the normal B-type stars (in red), the CBe stars (in cyan), the HAeBes/HAeB[e]s (in blue), the LBVs (in green), and the B[e]SGs (in orange). Black open circles point to stars that have accurate photometry (see Section 3.1). The brown dashed lines delimit regions of stars that share similar properties.

Figure 9 illustrates several examples of QQD. LBVs and B[e]SGs which are clearly apart from each other. In most of these diagrams, the HAeBe stars are also separate from the B[e]SGs. For instance, on a Q_{JHK} vs. Q_{BpRpHK} diagram (top-left panel), most of the LBVs lie on the right of the vertical line at $Q_{BpRpHK} = -4.5$ and above the horizontal at $Q_{JHK} = -0.55$, thereby complementing information given in Figure 8 (top panels).

In a diagram Q_{JHK} vs. Q_{BpKJK} (middle-left panel), the LBVs are inside a box delimited by $-4.8 < Q_{BpKJK} < -0.25$, and $-0.55 < Q_{JHK} < 0.11$. In addition, it is also possible to separate HAeBes from B[e]SGs through the line $Q_{JHK} = 0.1\, Q_{BpKJK} - 0.35$. Finally, the Q_{JHK} vs. Q_{BpRpK} diagram (bottom-left panel) shows similar behavior.

Concerning the panels on the right side, they show a greater mix between the group of stars than the ones on the left side, e.g., in the Q_{RpJHK} vs. Q_{BpKJK} diagram (middle-right panel), the LBVs have values of $Q_{RpJHK} > -2.4$, and the HAeBes and B[e]SGs can be separated with the line $Q_{RpJHK} = 0.48\, Q_{BpKJK} - 1.4$.

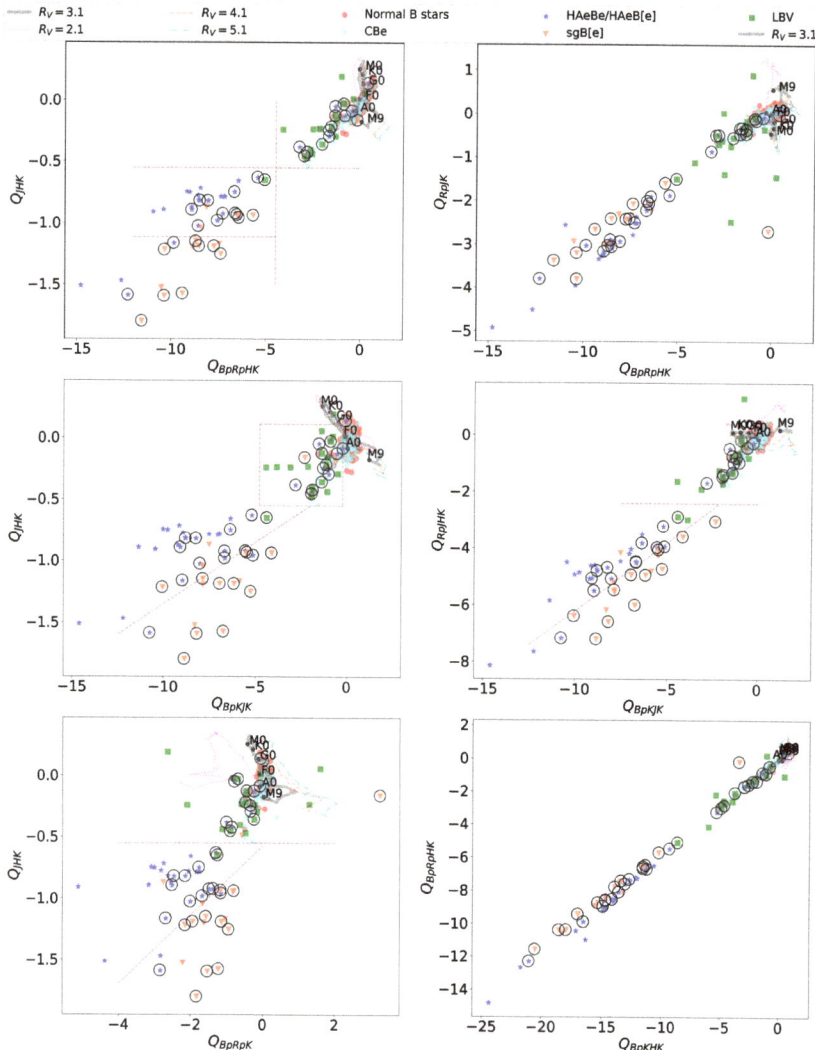

Figure 9. Example of Q–Q diagrams for B stars with emission lines. The main sequence was reddened applying $A_V = 1$ mag and different R_V values (see Section 3 for details). Symbols represent the normal B-type stars (in red), the CBe stars (in cyan), the HAeBes/HAeB[e]s (in blue), the LBVs (in green), and the B[e]SGs (in orange). Black open circles point to stars that have accurate photometry (see Section 3.1). The brown dashed lines delimit regions of stars that share similar properties.

5. Discussion

Traditionally, the positions of the stars in different photometric diagrams, or even in the HR diagram, have been adopted as selection criteria to classify stars according to their intrinsic properties, memberships in clusters or associations, variability types, etc. In this work, we focused on studying various color–color diagrams, taking advantage of the Gaia and 2MASS photometric surveys as powerful analysis tools to select and classify B-type stars with and without emission lines. Notably, one would expect that the NIR radiation excess emitted from gaseous and dust CEs of early-type stars would show different properties according to their chemical composition (gas-to-dust ratio), the absorption and heating mechanisms of particles, dust temperature, and the size of dust grains. As each diagram has its own limitations, mainly because the colors are affected by ISM, it was important to adopt multiple criteria. For this purpose, we also analyzed different reddening-free Q parameters, since the method presents high confidence. For instance, Mohr-Smith et al. [67] found that the extension of the Q method to the Sloan filter system had a confidence of 97%. The study was based on follow-up spectroscopy of a sample of OB stars selected in a previous study using the Q method [68].

In general, from the two-color diagrams of Figure 7, the HAeBe stars were very difficult to separate from B[e]SGs or LBVs. The best option was the $(J-H)$ vs. $(H-Ks)$ diagram that enabled distinguishing HAeBe from B[e]SGs [23], and LVBs from B[e]SGs [22], delimiting particular regions through straight lines (middle-left panel). However, there was still a substantial mixture between the HAeBe stars and the LBVs group (according to [23]). This preliminary classification could be complemented by the information derived from a $(Bp-Rp)$ vs. $(H-Ks)$ diagram that permitted, in addition, the identification of hot-HAeBe and B[e]SG stars with strong intrinsic IR emissions and no reddening interstellar excess. These stars were located within a region defined by $0 \lesssim (Bp-Rp) \lesssim 1$ and $0.5 \lesssim (H-Ks) \lesssim 1.25$. All this information could also be retrieved in the Q_{JHK} vs. Q_{BpRpHK} diagram, where the straight lines, plotted on the $JHKs$ two-color diagram, were transformed into horizontal lines at $Q_{JHK} = -0.55$ and $Q_{JHK} = -1.11$. The box in the $(Bp-Rp)$ vs. $(H-Ks)$ two-color diagram resulted in the range $-12.4 \lesssim Q_{BpRpHK} \lesssim -3.96$. Furthermore, in this QQD the three groups of stars, that is the HAeBes, LBVs, and B[e]SGs, were well separated from each other. According to Oksala et al. [22], the B[e]SGs were surrounded by larger amounts of hot dust than the LBVs, whereas the LBVs hosted huge amounts of cool dust at far distances.

We also recognized three kinds of HAeBe stars that presented different amounts of NIR radiation excess. In all CCDs, QCDs, and QQDs, we observed a group of HAeBes located in the same region as LBVs and some CBe stars. Figure 9 (top-left panel) shows that this group of stars had $Q_{JHK} > -0.55$. The other two groups had $-1.11 < Q_{JHK} < -0.55$, and $Q_{JHK} < -1.11$, respectively. The latter corresponded to HAeBe stars with significant IR excess. Our result agreed with that found by Chen et al. [19] in a $(J-H)$ vs. $(H-Ks)$ diagram. These authors noted that the majority of HAeBe stars were located in the region of $0 < (H-Ks) < 2$ and $-0.2 < (J-H) < 2.0$, which means $-4.08 < Q_{JHK} < 2$. Only a small group of sources presented less or no infrared excess showing a similar distribution as the majority of the CBe stars ($-0.2 < (H-Ks) < 0.2$ and $-0.2 < (J-H) < 0.2$), which implied that $-0.59 < Q_{JHK} < 0.59$. Finally, they found sources that had very large infrared excesses with $(H-K) > 1.5$. Sources with less excess IR radiation probably originated from free ionized gas emissions. Instead, those sources with large excesses of IR radiation were likely indicative of the influence of a bright nebulosity related to low temperatures or mainly thermal emissions from the surrounding dust [4].

A somewhat similar result was achieved for the CBe stars. Our sample of CBe stars encompassed bright Be field stars and Galactic cluster members. Among them, we recognized two types of CBe stars. One group, with small near-IR excess, resembled the normal B-type stars. The second group consisted of a few CBe stars (and a small group of stars classified as normal B-type stars) with a moderate IR excess, sharing properties with the LBVs. The latter could be better identified in the QCDs of Figure 8, particularly, in the

Q_{BpRpHK} vs. $(Bp - Rp)$ (top-left panel) and Q_{BpJHK} vs. $(Bp - Ks)$ (middle-right panel) diagrams. Zhang et al. [66] explained that a small radiation excess could be accounted for by hot circumstellar plasma, but a larger IR excess would require thermal emission from circumstellar dust. Lee and Chen [69] proposed that the dust grains were produced by condensation in the expanding stellar atmosphere, and, therefore, the dust particles should be very small in size to reprocess starlight efficiently. In contrast, the HAeBe stars showed a very large infrared excess which arose from relatively large grains as part of surplus star-forming materials. However, Rivinius et al. [5] asserted that dust had never been found in the envelopes of Be stars and that it was precisely the presence of dust that made it possible to distinguish a CBe from a HAeBe. As a consequence, stars with large color excesses are likely misclassified and would not be CBe.

6. Conclusions

This work discussed the properties of different color–color diagrams and equivalent Johnson's reddening-free Q parameters built using a combination of the Gaia and 2MASS photometry. Although the Q values were free from the ISM reddening effect, they still showed a dependence on the reddening law in the sky region. Nevertheless, using appropriate combinations of color indices, it was possible to minimize this effect among the O, B, and A stars. Furthermore, by analyzing the different diagrams, we were able to establish criteria to identify stars of spectral type B (Equations (8), (10) and (11)), as well as to identify early-type stars with emission lines of different nature. Main selection criteria are summarize in Table 3.

Table 3. Summary of classification criteria for normal B-type stars, LBVs, and B[e]SGs.

Normal B-Type Stars	
$QT_{\text{eff}}D$	$-0.1 \leq Q_{JKRpJ} \leq 0.05$
	$-0.07 \leq Q_{HKBpRp} \leq 0.08$
	$-0.11 \leq Q_{JHK} \leq 0.04$
QCD	$-0.11 < Q_{BpJRpK} < 0.20 \wedge -0.20 < (J - Ks) < 0.14$
	$-0.60 < Q_{BpRpHK} < 0.50 \wedge -0.30 < (Bp - Rp) < 0.40$
QQD	$-1.8 < Q_{JHK} < 0.01 \wedge -0.12 < Q_{BpRpK} < 0.18$
	$-0.6 < Q_{BpRpHK} < 0.7 \wedge -0.05 < Q_{RpJK} < 0.24$
LBVs	
QCD	$Q_{BpRpHK} > -4.5$
	$Q_{JHK} > -0.55$
	$Q_{BpJHK} > -7$
	$Q_{RpJK} > -1.7$
QQD	$-4.8 < Q_{BpKJK} < -0.25 \wedge -0.55 < Q_{JHK} < 0.11$
	$Q_{RpJHK} > -2.4$
B[e]SGs	
QCD	$Q_{JHK} < -1.11$
	$Q_{BpJHK} < -4.33\,(Bp - Ks) - 0.57$
QQD	$Q_{JHK} < 0.1\,Q_{BpKJK} - 0.35$
	$Q_{RpJHK} < 0.48\,Q_{BpKJK} - 1.4$

In addition to the well-known $(J - H)$ vs. $(H - Ks)$ color–color diagram, which is an excellent dust tracer of circumstellar envelopes, we also found that the $(Bp - Rp)$ vs. $(H - Ks)$ diagram could be used to identify emission-line B-type stars corrected by interstellar reddening, but with strong intrinsic IR emission. Particularly interesting were the $(Rp - Ks)$ vs. $(Bp - Rp)$ and $(J - Ks)$ vs. $(Bp - Ks)$ diagrams, as they might be helpful to search for CBe stars with large infrared excesses, because the normal B-type stars distribute along a narrow and well-defined band.

The main advantage of the QCDs over the color–color ones was that the B-type stars did not mix with the M-type stars, as shown, particularly in the Q_{BpRpHK} vs. $(Bp - Rp)$, and Q_{BpJHK} vs. $(Bp - Ks)$ diagrams. Moreover, early-type stars that have anomalous reddening, due to the presence of circumstellar envelopes, can be easily recognized. This is because the effect of interstellar extinction causes stars of the same spectral types to shift to the right of the diagram. At the same time, the presence of the circumstellar envelope makes the Q value more negative (when using an IR color index). We suggest using different selection criteria, such as those that arise from the Q_{JKHK} vs. $(H - Ks)$ and Q_{BpJHK} vs. $(Bp - Ks)$ diagrams, to search for, and classify, B-type stars with emission lines. These diagrams also allow the LBV stars to be well-recognized from HAeBes and B[e]SGs. In addition, the region of the HAeBe stars can be separated from that of the B[e]SG stars by a straight line. However, there was still significant contamination between HAeBes and LBVs, that it would be desirable to get rid of when making source selections.

As was already stressed, QQDs are free from interstellar extinction, although they are sensitive to anomalous reddening. However, the presence of a circumstellar envelope produces a shift of the Q-values in a direction almost perpendicular to the effect of the anomalous reddening. Then, the QQDs are powerful tools to identify stars with intrinsic IR emission.

Using the Q parameters allowed us to recognize two classes of CBe stars. One group resembled the normal B-type stars with small near-IR excess, and the other group had a few objects showing moderate to large IR excess. This second group might show dusty envelopes formed as a result of the evolution and would not be CBe stars. We also detected three kinds of HAeBe stars. The first group had small IR radiation excess. Stars in this group shared properties with CBe stars and LBVs. A second group showed a moderate contribution from dust, and a third group showed a significant IR excess comparable with the properties of B[e]SGs. All these groups need further study to understand their dust grain properties and formation processes.

Thus, based on the location of the stars in different CCDs, QCDs, and QQDs, we were able to state various selection criteria that allowed separating, as much as possible, the diverse groups of B-type stars with a moderate and large IR excess (e.g., CBes, HAeBes, LBVs, and B[e]SGs). Both the diagrams and stated criteria can be very useful tools for automated designs of machine learning and optimal search algorithms.

Author Contributions: Conceptualization, Y.A. and L.S.C.; methodology, Y.A.; formal analysis, Y.A.; investigation, Y.A.; writing—original draft preparation, Y.A. and L.S.C.; writing—review and editing, L.S.C.; visualization, Y.A.; supervision, L.S.C.; funding acquisition, Y.A. and L.S.C. All authors have read and agreed to the published version of the manuscript.

Funding: This research was funded by CONICET (PIP 1337) and the Universidad Nacional de La Plata (Programa de Incentivos 11/G160), Argentina. In addition, this project has received funding from the European Union's Framework Programme for Research and Innovation Horizon 2020 (2014-2020) under the Marie Skłodowska-Curie Grant Agreement No. 823734.

Data Availability Statement: Not applicable.

Acknowledgments: This publication made use of data products from the Two Micron All Sky Survey, which is a joint project of the University of Massachusetts and the Infrared Processing and Analysis Center/California Institute of Technology, funded by the National Aeronautics and Space Administration and the National Science Foundation. This work presented results from the European Space Agency (ESA) space mission Gaia. Gaia data are being processed by the Gaia Data Processing and Analysis Consortium (DPAC). Funding for the DPAC is provided by national institutions, in particular, the institutions participating in the Gaia MultiLateral Agreement (MLA). The Gaia mission website is https://www.cosmos.esa.int/gaia (the last accessed date was 1 February 2023). The Gaia archive website is https://archives.esac.esa.int/gaia (the last accessed date was 1 February 2023).

Conflicts of Interest: The authors declare no conflict of interest.

Abbreviations

The following abbreviations are used in this manuscript:

CE	Circumstellar envelope
IR	infrared
HAeBe	Herbig Ae/Be star
PMS	Pre-main sequence star
CBe	Classical Be star
BSG	B supergiant star
LBV	Luminous Blue Variable star
B[e]SG	B[e] supergiant star
unclB[e]	unclassified B[e] star
NIR	near infrared
CCD	Color-color diagram
YSO	Young Stellar Object
Gaia	Global Astrometric Interferometer for Astrophysics mission
2MASS	Two Micron All Sky Survey
QQD	Q-Q diagrams
EDR3	Early Data Release 3
QCD	Q-color index diagram
ISM	Interstellar medium
T_{eff}	Effective temperature
Topcat	Tool for OPerations on Catalogues And Tables
MS	Main sequence

Notes

1 The color-excess ratio is $\frac{E(m_1 - m_2)}{E(m_3 - m_4)} = \frac{A_{m_1} - A_{m_2}}{A_{m_3} - A_{m_4}} = \frac{R_{m_1} - R_{m_2}}{R_{m_3} - R_{m_4}} = \frac{r_{m_1} - r_{m_2}}{r_{m_3} - r_{m_4}}$ where R_{m_a} and r_{m_a} are the relative absorption coefficients referring to the color excess $E(B - V)$, $R_{m_a} = A_{m_a}/E(B - V)$, and to the extinction A_V, $r_{m_a} = A_{m_a}/A_V$, respectively.

2 http://www.starlink.ac.uk/topcat/. The last accessed date is 2 January 2023.

3 https://extinction.readthedocs.io/en/latest/. The last accessed date was 2 January 2023.

4 Version 2019.3.22, in http://www.pas.rochester.edu/ emamajek/EEM_dwarf_UBVIJHK_colors_Teff.txt. The last accessed date was 2 January 2023.

References

1. Porter, J.M.; Rivinius, T. Classical Be Stars. *Publ. Astron. Soc. Pac.* **2003**, *115*, 1153–1170. [CrossRef]
2. Allen, D.A. Near infra-red magnitudes of 248 early-type emission-line stars and related objects. *Mon. Not. R. Astron. Soc.* **1973**, *161*, 145–166. [CrossRef]
3. Herbig, G.H. The Spectra of Be- and Ae-Type Stars Associated with Nebulosity. *Astrophys. J. Suppl. Ser.* **1960**, *4*, 337. [CrossRef]
4. Hillenbrand, L.A.; Strom, S.E.; Vrba, F.J.; Keene, J. Herbig Ae/Be Stars: Intermediate-Mass Stars Surrounded by Massive Circumstellar Accretion Disks. *Astrophys. J.* **1992**, *397*, 613. [CrossRef]
5. Rivinius, T.; Carciofi, A.C.; Martayan, C. Classical Be stars. Rapidly rotating B stars with viscous Keplerian decretion disks. *Astron. Astrophys. Rev.* **2013**, *21*, 69. [CrossRef]
6. Chita, S.M.; Langer, N.; van Marle, A.J.; García-Segura, G.; Heger, A. Multiple ring nebulae around blue supergiants. *Astron. Astrophys.* **2008**, *488*, L37–L41. [CrossRef]
7. Lamers, H.J.G.L.M.; Zickgraf, F.J.; de Winter, D.; Houziaux, L.; Zorec, J. An improved classification of B[e]-type stars. *Astron. Astrophys.* **1998**, *340*, 117–128.
8. Humphreys, R.M.; Davidson, K. The Luminous Blue Variables: Astrophysical Geysers. *Publ. Astron. Soc. Pac.* **1994**, *106*, 1025. [CrossRef]
9. Mehner, A.; Baade, D.; Groh, J.H.; Rivinius, T.; Hambsch, F.J.; Bartlett, E.S.; Asmus, D.; Agliozzo, C.; Szeifert, T.; Stahl, O. Spectroscopic and photometric oscillatory envelope variability during the S Doradus outburst of the luminous blue variable R71. *Astron. Astrophys.* **2017**, *608*, A124. [CrossRef]
10. Campagnolo, J.C.N.; Borges Fernandes, M.; Drake, N.A.; Kraus, M.; Guerrero, C.A.; Pereira, C.B. Detection of new eruptions in the Magellanic Clouds luminous blue variables R 40 and R 110. *Astron. Astrophys.* **2018**, *613*, A33. [CrossRef]
11. Weis, K.; Bomans, D.J. Luminous Blue Variables. *Galaxies* **2020**, *8*, 20. [CrossRef]
12. Kraus, M. A Census of B[e] Supergiants. *Galaxies* **2019**, *7*, 83. [CrossRef]
13. Zickgraf, F.J.; Schulte-Ladbeck, R.E. Polarization characteristics of galactic Be stars. *Astron. Astrophys.* **1989**, *214*, 274–284.

14. Miroshnichenko, A.S. Toward Understanding the B[e] Phenomenon. I. Definition of the Galactic FS CMa Stars. *Astrophys. J.* **2007**, *667*, 497–504. [CrossRef]
15. Kwok, S. Proto-planetary nebulae. *Annu. Rev. Astron. Astrophys.* **1993**, *31*, 63–92. [CrossRef]
16. Klement, R.; Carciofi, A.C.; Rivinius, T.; Matthews, L.D.; Vieira, R.G.; Ignace, R.; Bjorkman, J.E.; Mota, B.C.; Faes, D.M.; Bratcher, A.D.; et al. Revealing the structure of the outer disks of Be stars. *Astron. Astrophys.* **2017**, *601*, A74. [CrossRef]
17. Oudmaijer, R.D. The B[e] Phenomenon in Pre-Main-Sequence Herbig Ae/Be Stars. In Proceedings of the The B[e] Phenomenon: Forty Years of Studies, Prague, Czech Republic, 27 June–1 July 2016; Miroshnichenko, A., Zharikov, S., Korčáková, D., Wolf, M., Eds.; Astronomical Society of the Pacific Conference Series; Astronomical Society of the Pacific: California, CA, USA, 2017; Volume 508, p. 175.
18. Sheikina, T.A.; Miroshnichenko, A.S.; Corporon, P. B-type Emission-line Stars with Warm Circumstellar Dust. In Proceedings of the IAU Colloq. 175: The Be Phenomenon in Early-Type Stars, Alicante, Spain, 28 June–2 July 1999; Smith, M.A., Henrichs, H.F., Fabregat, J., Eds.; Astronomical Society of the Pacific Conference Series; Astronomical Society of the Pacific: California, CA, USA, 2000; Volume 214, p. 494.
19. Chen, P.S.; Shan, H.G.; Zhang, P. A new photometric study of Herbig Ae/Be stars in the infrared. *New Astron.* **2016**, *44*, 1–11. [CrossRef]
20. Cidale, L.; Zorec, J.; Tringaniello, L. BCD spectrophotometry of stars with the B[e] phenomenon. I. Fundamental parameters. *Astron. Astrophys.* **2001**, *368*, 160–174. [CrossRef]
21. Comerón, F.; Pasquali, A.; Rodighiero, G.; Stanishev, V.; De Filippis, E.; López Martí, B.; Gálvez Ortiz, M.C.; Stankov, A.; Gredel, R. On the massive star contents of Cygnus OB2. *Astron. Astrophys.* **2002**, *389*, 874–888. [CrossRef]
22. Oksala, M.E.; Kraus, M.; Cidale, L.S.; Muratore, M.F.; Borges Fernandes, M. Probing the ejecta of evolved massive stars in transition. A VLT/SINFONI K-band survey. *Astron. Astrophys.* **2013**, *558*, A17. [CrossRef]
23. Cochetti, Y.R.; Kraus, M.; Arias, M.L.; Cidale, L.S.; Eenmäe, T.; Liimets, T.; Torres, A.F.; Djupvik, A.A. Near-infrared Characterization of Four Massive Stars in Transition Phases. *Astron. J.* **2020**, *160*, 166. [CrossRef]
24. Negueruela, I.; Schurch, M.P.E. A search for counterparts to massive X-ray binaries using photometric catalogues. *Astron. Astrophys.* **2007**, *461*, 631–639. [CrossRef]
25. Johnson, H.L.; Morgan, W.W. Fundamental stellar photometry for standards of spectral type on the Revised System of the Yerkes Spectral Atlas. *Astrophys. J.* **1953**, *117*, 313. [CrossRef]
26. Comerón, F.; Pasquali, A. The ionizing star of the North America and Pelican nebulae. *Astron. Astrophys.* **2005**, *430*, 541–548. [CrossRef]
27. Bessell, M.S.; Brett, J.M. JHKLM Photometry: Standard Systems, Passbands, and Intrinsic Colors. *Publ. Astron. Soc. Pac.* **1988**, *100*, 1134. [CrossRef]
28. Garcia, M.; Herrero, A.; Castro, N.; Corral, L.; Rosenberg, A. The young stellar population of IC 1613. II. Physical properties of OB associations. *Astron. Astrophys.* **2010**, *523*, A23. [CrossRef]
29. Aparicio Villegas, T.; Alfaro, E.J.; Cabrera-Caño, J.; Moles, M.; Benítez, N.; Perea, J.; del Olmo, A.; Fernández-Soto, A.; Cristóbal-Hornillos, D.; Aguerri, J.A.L.; et al. Stellar physics with the ALHAMBRA photometric system. *J. Phys. Conf. Ser.* **2011**, *328*, 012004. [CrossRef]
30. Gaia Collaboration; Prusti, T.; de Bruijne, J.H.J.; Brown, A.G.A.; Vallenari, A.; Babusiaux, C.; Bailer-Jones, C.A.L.; Bastian, U.; Biermann, M.; Evans, D.W.; et al. The Gaia mission. *Astron. Astrophys.* **2016**, *595*, A1. [CrossRef]
31. Gaia Collaboration; Brown, A.G.A.; Vallenari, A.; Prusti, T.; de Bruijne, J.H.J.; Mignard, F.; Drimmel, R.; Babusiaux, C.; Bailer-Jones, C.A.L.; Bastian, U.; et al. Gaia Data Release 1. Summary of the astrometric, photometric, and survey properties. *Astron. Astrophys.* **2016**, *595*, A2. [CrossRef]
32. Cutri, R.M.; Skrutskie, M.F.; van Dyk, S.; Beichman, C.A.; Carpenter, J.M.; Chester, T.; Cambresy, L.; Evans, T.; Fowler, J.; Gizis, J.; et al. 2MASS All Sky Catalog of Point Sources. 2003. Available online: https://vizier.cds.unistra.fr/viz-bin/VizieR?-source=II/246 (accessed on 2 January 2023).
33. Skrutskie, M.F.; Cutri, R.M.; Stiening, R.; Weinberg, M.D.; Schneider, S.; Carpenter, J.M.; Beichman, C.; Capps, R.; Chester, T.; Elias, J.; et al. The Two Micron All Sky Survey (2MASS). *Astron. J.* **2006**, *131*, 1163–1183. [CrossRef]
34. Straizys, V.; Lazauskaite, R.; Liubertas, R.; Azusienis, A. Star Classification Possibilities with Broad-Band Photometric Systems. I. The Sloan System. *Balt. Astron.* **1998**, *7*, 605–623. [CrossRef]
35. Gaia Collaboration; Brown, A.G.A.; Vallenari, A.; Prusti, T.; de Bruijne, J.H.J.; Babusiaux, C.; Biermann, M.; Creevey, O.L.; Evans, D.W.; Eyer, L.; et al. Gaia Early Data Release 3. Summary of the contents and survey properties. *Astron. Astrophys.* **2021**, *649*, A1. [CrossRef]
36. Johnson, H.L.; Morgan, W.W. Some Evidence for a Regional Variation in the Law of Interstellar Reddening. *Astrophys. J.* **1955**, *122*, 142. [CrossRef]
37. Straižys, V. *Multicolor Stellar Photometry*; Pachart Pub. House: Tucson, AZ, USA, 1992.
38. Bartkevicius, A.; Zdanavicius, K. Calculation of reddening-free parameters Q of the Vilnius photometric system by the iteration method. *Vilnius Astron. Obs. Biul.* **1975**, *41*, 30.
39. Pucinskas, A. Photographic photometry of stars in the region of open cluster IC 4996 in the Vilnius photometric system. *Vilnius Astron. Obs. Biul.* **1982**, *59*, 3.

40. Jasevicius, V. Photometric quantification of stars in the Vilnius system using the method of independent Q parameters. *Vilnius Astron. Obs. Biul.* **1986**, *74*, 40.
41. Smriglio, F.; Boyle, R.P.; Straizys, V.; Janulis, R.; Nandy, K.; MacGillivray, H.T.; McLachlan, A.; Coluzzi, R.; Segato, C. Automated two-dimensional classification from multicolour photometry in the Vilnius system. *Astron. Astrophys.* **1986**, *66*, 181–190.
42. Smriglio, F.; Dasgupta, A.K.; Nandy, K.; Boyle, R.P. A comparison of automated spectral classification using Vilnius photometry with MK classification. *Astron. Astrophys.* **1990**, *228*, 399–402.
43. Jordi, C.; Gebran, M.; Carrasco, J.M.; de Bruijne, J.; Voss, H.; Fabricius, C.; Knude, J.; Vallenari, A.; Kohley, R.; Mora, A. Gaia broad band photometry. *Astron. Astrophys.* **2010**, *523*, A48. [CrossRef]
44. Danielski, C.; Babusiaux, C.; Ruiz-Dern, L.; Sartoretti, P.; Arenou, F. The empirical Gaia G-band extinction coefficient. *Astron. Astrophys.* **2018**, *614*, A19. [CrossRef]
45. Zorec, J.; Frémat, Y.; Cidale, L. On the evolutionary status of Be stars. I. Field Be stars near the Sun. *Astron. Astrophys.* **2005**, *441*, 235–248. .:20053051. [CrossRef]
46. Zorec, J.; Cidale, L.; Arias, M.L.; Frémat, Y.; Muratore, M.F.; Torres, A.F.; Martayan, C. Fundamental parameters of B supergiants from the BCD system. I. Calibration of the (λ_1, D) parameters into T_{eff}. *Astron. Astrophys.* **2009**, *501*, 297–320. [CrossRef]
47. Aidelman, Y.; Cidale, L.S.; Zorec, J.; Arias, M.L. Open clusters. I. Fundamental parameters of B stars in NGC 3766 and NGC 4755. *Astron. Astrophys.* **2012**, *544*, A64. [CrossRef]
48. Aidelman, Y.; Cidale, L.S.; Zorec, J.; Panei, J.A., II. Fundamental parameters of B stars in Collinder 223, Hogg 16, NGC 2645, NGC 3114, and NGC 6025. *Astron. Astrophys.* **2015**, *577*, A45. [CrossRef]
49. Aidelman, Y.; Cidale, L.S.; Zorec, J.; Panei, J.A. Open clusters. III. Fundamental parameters of B stars in NGC 6087, NGC 6250, NGC 6383, and NGC 6530 B-type stars with circumstellar envelopes. *Astron. Astrophys.* **2018**, *610*, A30. [CrossRef]
50. Cochetti, Y.R.; Zorec, J.; Cidale, L.S.; Arias, M.L.; Aidelman, Y.; Torres, A.F.; Frémat, Y.; Granada, A. Be and Bn stars: Balmer discontinuity and stellar-class relationship. *Astron. Astrophys.* **2020**, *634*, A18. [CrossRef]
51. Barbier, D.; Chalonge, D. Étude du rayonnement continu de quelques étoiles entre 3 100 et 4 600 Å (4e Partie-discussion générale). *Ann. D'Astrophys.* **1941**, *4*, 30.
52. Chalonge, D.; Divan, L. Recherches sur les spectres continus stellaires. V. Etude du spectre continu de 150 etoiles entre 3150 et 4600 A. *Ann. D'Astrophys.* **1952**, *15*, 201.
53. Moujtahid, A.; Zorec, J.; Hubert, A.M.; Garcia, A.; Burki, G. Long-term visual spectrophotometric behaviour of Be stars. *Astron. Astrophys.* **1998**, *129*, 289–311. [CrossRef]
54. Taylor, M.B. TOPCAT & STIL: Starlink Table/VOTable Processing Software. In Proceedings of the Astronomical Data Analysis Software and Systems XIV, San Francisco, CA, USA, 30 November 2005; Shopbell, P., Britton, M., Ebert, R., Eds.; Astronomical Society of the Pacific Conference Series. Astronomical Society of the Pacific: California, CA, USA, 2005; Volume 347, p. 29.
55. Taylor, M.B. STILTS—A Package for Command-Line Processing of Tabular Data. In Proceedings of the Astronomical Data Analysis Software and Systems XV, San Francisco, CA, USA, 1 July 2006; Astronomical Society of the Pacific: California, CA, USA, 2006; Volume 351, p. 666.
56. Barbary, K. *Extinction v0.3.0*; Zenodo: Les Ulis, France, 2016. [CrossRef]
57. Cardelli, J.A.; Clayton, G.C.; Mathis, J.S. The Relationship between Infrared, Optical, and Ultraviolet Extinction. *Astrophys. J.* **1989**, *345*, 245. [CrossRef]
58. O'Donnell, J.E. R v-dependent Optical and Near-Ultraviolet Extinction. *Astrophys. J.* **1994**, *422*, 158. [CrossRef]
59. Fitzpatrick, E.L. Correcting for the Effects of Interstellar Extinction. *Publ. Astron. Soc. Pac.* **1999**, *111*, 63–75. [CrossRef]
60. Fitzpatrick, E.L.; Massa, D. An Analysis of the Shapes of Interstellar Extinction Curves. V. The IR-through-UV Curve Morphology. *Astrophys. J.* **2007**, *663*, 320–341. [CrossRef]
61. Pecaut, M.J.; Mamajek, E.E. Intrinsic Colors, Temperatures, and Bolometric Corrections of Pre-main-sequence Stars. *Astrophys. J. Suppl. Ser.* **2013**, *208*, 9. [CrossRef]
62. Mucciarelli, A.; Bellazzini, M.; Massari, D. Exploiting the Gaia EDR3 photometry to derive stellar temperatures. *Astron. Astrophys.* **2021**, *653*, A90. [CrossRef]
63. Poggio, E.; Drimmel, R.; Lattanzi, M.G.; Smart, R.L.; Spagna, A.; Andrae, R.; Bailer-Jones, C.A.L.; Fouesneau, M.; Antoja, T.; Babusiaux, C.; et al. The Galactic warp revealed by Gaia DR2 kinematics. *Mon. Not. R. Astron. Soc. Lett.* **2018**, *481*, L21–L25. [CrossRef]
64. Clark, J.S.; Larionov, V.M.; Arkharov, A. On the population of galactic Luminous Blue Variables. *Astron. Astrophys.* **2005**, *435*, 239–246. [CrossRef]
65. Guzmán-Díaz, J.; Mendigutía, I.; Montesinos, B.; Oudmaijer, R.D.; Vioque, M.; Rodrigo, C.; Solano, E.; Meeus, G.; Marcos-Arenal, P. Homogeneous study of Herbig Ae/Be stars from spectral energy distributions and Gaia EDR3. *Astron. Astrophys.* **2021**, *650*, A182. [CrossRef]
66. Zhang, P.; Chen, P.S.; Yang, H.T. 2MASS observations of Be stars. *New Astron.* **2005**, *10*, 325–352. [CrossRef]
67. Mohr-Smith, M.; Drew, J.E.; Napiwotzki, R.; Simón-Díaz, S.; Wright, N.J.; Barentsen, G.; Eislöffel, J.; Farnhill, H.J.; Greimel, R.; Monguió, M.; et al. The deep OB star population in Carina from the VST Photometric Hα Survey (VPHAS+). *Mon. Not. R. Astron. Soc. Lett.* **2017**, *465*, 1807–1830. [CrossRef]

68. Mohr-Smith, M.; Drew, J.E.; Barentsen, G.; Wright, N.J.; Napiwotzki, R.; Corradi, R.L.M.; Eislöffel, J.; Groot, P.; Kalari, V.; Parker, Q.A.; et al. New OB star candidates in the Carina Arm around Westerlund 2 from VPHAS+. *Mon. Not. R. Astron. Soc. Lett.* **2015**, *450*, 3855–3873. [CrossRef]
69. Lee, C.D.; Chen, W.P. Dust formation of Be stars with large infrared excess. In Proceedings of the Active OB Stars: Structure, Evolution, Mass Loss, and Critical Limits; Neiner, C., Wade, G., Meynet, G., Peters, G., Eds.; Cambridge University Press: Cambridge, UK, 2011; Volume 272, pp. 366–371. [CrossRef]

Disclaimer/Publisher's Note: The statements, opinions and data contained in all publications are solely those of the individual author(s) and contributor(s) and not of MDPI and/or the editor(s). MDPI and/or the editor(s) disclaim responsibility for any injury to people or property resulting from any ideas, methods, instructions or products referred to in the content.

Article

Synthetic Light Curve Design for Pulsating Binary Stars to Compare the Efficiency in the Detection of Periodicities

Aldana Alberici Adam [1,*,†], Gunther F. Avila Marín [2], Alejandra Christen [2] and Lydia Sonia Cidale [1,3]

1. Instituto de Astrofísica de La Plata, La Plata B1900FWA, Argentina; lydia@fcaglp.fcaglp.unlp.edu.ar
2. Instituto de Estadística, Universidad de Valparaíso, Valparaíso 2340000, Chile; gunther.avila@postgrado.uv.cl (G.F.A.M.); alejandra.christen@uv.cl (A.C.)
3. Facultad de Ciencias Astronómicas y Geofísicas, La Plata B1900FWA, Argentina
* Correspondence: aldialb@fcaglp.unlp.edu.ar
† Current address: Observatorio Astronómico, Paseo del Bosque, La Plata B1900FWA, Argentina.

Citation: Alberici Adam, A.; Avila Marín, G.F.; Christen, A.; Cidale, L.S. Synthetic Light Curve Design for Pulsating Binary Stars to Compare the Efficiency in the Detection of Periodicities. *Galaxies* **2023**, *11*, 69. https://doi.org/10.3390/galaxies11030069

Academic Editor: Oleg Malkov

Received: 25 March 2023
Revised: 19 May 2023
Accepted: 22 May 2023
Published: 31 May 2023

Copyright: © 2023 by the authors. Licensee MDPI, Basel, Switzerland. This article is an open access article distributed under the terms and conditions of the Creative Commons Attribution (CC BY) license (https://creativecommons.org/licenses/by/4.0/).

Abstract: B supergiant stars pulsate in regular and quasi-regular oscillations resulting in intricate light variations that might conceal their binary nature. To discuss possible observational bias in a light curve, we performed a simulation design of a binary star affected by sinusoidal functions emulating pulsation phenomena. The Period04 tool and the WaveletComp package of R were used for this purpose. Thirty-two models were analysed based on a combination of two values on each of the $k = 6$ variables, such as multiple pulsations, the amplitude of the pulsation, the pulsation frequency, the beating phenomenon, the light-time effect, and regular or quasi-regular periods. These synthetic models, unlike others, consider an ARMA (1, 1) statistical noise, irregular sampling, and a gap of about 4 days. Comparing Morlet wavelet with Fourier methods, we observed that the orbital period and its harmonics were well detected in most cases. Although the Fourier method provided more accurate period detection, the wavelet analysis found it more times. Periods seen with the wavelet method have a shift due to the slightly irregular time scale used. The pulsation period hitting rate depends on the wave amplitude and frequency with respect to eclipse depth and orbital period. None of the methods was able to distinguish accurate periods leading to a beating phenomenon when they were longer than the orbital period, resulting, in both cases, in an intermediate value. When the beating period was shorter, the Fourier analysis found it in all cases except for unsolved quasi-regular periods. Overall, the Morlet wavelet analysis performance was lower than the Fourier analysis. Considering the strengths and disadvantages found in these methods, we recommend using at least two diagnosis tools for a detailed time series data analysis to obtain confident results. Moreover, a fine-tuning of trial periods by applying phase diagrams would be helpful for recovering accurate values. The combined analysis could reduce observational bias in searching binaries using photometric techniques.

Keywords: methods: statistical methods: numerical; binaries: eclipsing; stars: oscillations (including pulsations)

1. Introduction

Observational surveys of massive stars of spectral type O show that most are forming binary systems [1–3]. Studies carried out in the region of 30 Dor also show that the binarity frequencies among dwarf and giant B-type stars agree with those of the O-type stars [4,5]. However, in contrast with the high binary incidence among dwarf massive stars, binary systems with B supergiant components are very rare. Barbá et al. [6] reported a noticeable sharp drop in the number of spectroscopic binaries between the supergiants of spectral types O9.7 and B0. These authors attributed this peculiarity to observational bias but suggested also discussing the action of possible evolutive scenarios such as binary mergers, binary disruptions, etc., to explain this effect.

On the other hand, massive stars pulsate in regular and quasi-regular (strange modes) oscillations (cf. [7]). They may show β Cephei type modes (low-order p and g modes) and Slowly Pulsating B (SPB) type variability (high-order g modes) [8,9]. Strange modes might also be excited with long periods, of the order of hours or days [10–13]. The pulsation mechanism is, in general, well understood (e.g., the κ mechanism, ϵ mechanism, or strange-mode instabilities) but for other groups, such as the γ Dor stars, roAp stars, and S Dor stars, no consensus has yet been reached [14]. These pulsations are revealed by photometric variations that can be identified by asteroseismological methods, while in the stellar spectrum, they generate variability in the broadening of the line profiles [15,16]. However, when pulsating stars are found in binary systems, their luminosity variations are modulated by the orbital motion [17]. This is the light-time effect, equivalent to a periodic Doppler shift of the pulsation frequency. Therefore, these effects generate complex patterns in the light curves that could make it difficult to detect or estimate periods or even conceal a binary detection.

The importance behind the detection of pulsating stars in massive binary systems lies in their impact on the study of the mass discrepancy problem, the theory of stellar interiors, and the verification of stellar evolution models [18], as they allow the masses of each component to be measured accurately and independently.

The search for periodicities is one of the most basic tasks in time series data analysis. It is a crucial feature for classifying variable stars and deriving stellar parameters. It is usually solved using an estimation function called a periodogram. Various types of periodograms are used in practice; for example, the Lomb–Scargle periodogram [19,20] is a known algorithm for detecting periodic signals in unevenly and regularly sampled time series. When periodic or quasi-periodic fluctuations arise in an interrupted or transient event [21], the wavelet transform is well suited for detecting changes in the parameters of signals due to its characteristic of focusing on a limited time interval of the data [22]. Nevertheless, the wavelet analysis is appropriate for time series with regular sampling. Today this problem has been partially addressed since satellite missions (Kepler, TESS, among others) provide photometric data of pulsating stars and binary systems with almost fixed cadences of a few seconds or minutes. Hence, using techniques for regularly sampled data is pertinent in this context [21,23–25].

On the other hand, it is well-known that power spectra can often be misleading. The performance of many different methods for period detection and estimation in light curves has been tested in the literature, for example by Graham et al. [26]. Those authors examined a huge number of sources from three datasets (CRTS, MACHO, ASAS) and ran the data through eleven different algorithms such as the Fourier transform, Lomb–Scargle, phase dispersion minimisation, among others. The study of different observed light curves showed that, at best, period-finding algorithms could retrieve the period of a regularly periodic object with a reasonable degree of accuracy in only about 50% of cases. However, the situation is much worse for stellar objects with semi-periodic, quasi-periodic or multi-periodic variability; typically, only around 10–20% of the cases are the periods successfully recovered, evidencing the low level of accuracy for most of the methods, as shown in Figure 19 given by Graham et al. [26]. Although the authors performed an excellent comparative study among many methods for period searching, it still has the disadvantage that the authentic periods of the signal could be miss-detected because, even when they are accurate, they might be inexact. Based on this weakness, it is necessary to explore simulation designs for which the true values of the periods are known but compare the performance of a method not analysed in the above work, such as wavelet analysis, with the revisited Fourier analysis. This is the primary goal of the work.

In addition, one of the usual drawbacks when analysing observed data is the noise level of the signal, which generates consequently noisy periodograms. Thus, specific peaks in a periodogram might be spurious and not due to the presence of any real periodic phenomena [27,28]. Other drawbacks can also arise, for example, when the sample data are not evenly spaced in time. This situation is very frequent in the case of astronomical observations, where the data are out of phase in time due to various natural phenomena. It is well known that these non-equispaced data produce the so-called alias [29], i.e., false periods, where it is difficult to discern which peak of the periodogram is the real one and which is its alias. Unfortunately, simple and rigorous methods are unavailable to solve the alias discrimination task. In practice, we need to have prior information about the phenomenon under investigation to discuss the number and reliability of the possible periods. For example, for binary systems, the half period is often the most significant peak in a periodogram [26].

In summary, for a thorough analysis of the light curves of massive stars, it is necessary to have (i) a valid mathematical model to describe the time series, (ii) a period analysis tool that supports the detection of variable frequencies, and (iii) a homogeneous time distribution of the data to ensure a stable solution. Although the Fourier transformation provides a confident signal frequency spectrum, we often lose information about when this frequency happens. An alternative method is the wavelet analysis [30–32] that allows searching for local periods in a given window of time to obtain both frequency and time resolution.

In this work, we proposed to use a simulation design developed to study how different periodic or quasi-periodic simulated phenomena are reproduced in a wavelet analysis scalogram and a Fourier periodogram. For comparison purposes, we generated multi-periodic synthetic light curves (with pre-fixed periods) for a binary system with a pulsating companion. These synthetic models, unlike others, consider a statistical noise, a gap, and a slightly irregular sampling. In addition, phenomena such as beating and light time effects were also emulated. We obtained surprising and interesting results that will be useful in future studies of the periodicities of light curves and which have possible implications for the low incidence rate of binary systems among B supergiants.

Section 2 introduces the simulation design by detailing the variables and their features. Section 3 briefly describes the time series analysis methods used. Section 4 shows the results obtained from the wavelet and Fourier analyses. Section 5 discusses our results in the context of the expected objectives and future work. Finally, Appendix A shows the complete frequency list for each model and all the periodograms resulting from the Fourier analysis. Appendix B shows the list of periods detected for the synthetic light curves with the wavelet analysis and the power spectrum for each one.

2. Simulation Design

In statistics, the area of experimental designs is concerned with finding the design that best helps to determine and study the factors that influence a given variable. There are many different types of designs: the factorial, fractional factorial, block, Latin square, and nested designs, among others [33,34]. Factorial designs are those in which experiments corresponding to all possible combinations of the "levels" or values of the factors or measurements of interest are carried out. The 2^k designs are factorial designs with k factors or variables for which only two levels or values are considered for each. These designs with few values for each factor are often used when it is desired to determine which factors are affecting a certain process or phenomenon. Extreme values of quantitative variables (such as temperature or concentration) or extreme categories of qualitative variables (such as the presence-absence of an attribute, among other possibilities) are considered for the levels. This type of design makes it possible to determine which factors or variables really influence the change or variation of the process or phenomenon. They are used in the first stage of the study when many variables are known about a process, and it is desired to determine its most influential factors. Once the relevant factors have been determined, it is possible to move on to a more exhaustive design, which considers more levels or values for each factor or even a more complex design. Each power-of-2 design considers 2^k different runs, which can be a considerable amount depending on the number k. Because it is not always feasible to perform 2^k experiments or runs, one can consider fractional factorial designs (e.g., 2^{k-p}), which consist of running only a fraction of the total amount of runs given by $2^k/2^p$. Therefore, it is important to optimally select this fraction. For this purpose, the fraction of the design is chosen in a way that allows obtaining a complete factorial design when p factors are discarded, for instance, because they are considered negligible after running the experiment. In this case, the number of factors p to be removed agrees with the power p in the denominator of $2^k/2^p$. This methodology is based on the belief that if one considers a large number of variables, only a smaller number is relevant, and it does not produce significant variations in the phenomenon. In a later step, the remaining executions could be carried out to obtain a complete 2^k design.

One of the most important uses of fractional factorial experiments is in factor-screening[1] [33,34]. Furthermore, one of the benefits of using a statistical methodology is to be able to have hypothesis tests that will give us ways to quantify the significance of the findings. In this work, a fractional factorial design was used to establish the models to be simulated and underlined through Fourier and wavelet analysis. The design is detailed in the following section.

Designs of Synthetic Light Curves

In this stage, instead of performing a real experiment, we considered using numerical simulations of light curves to know in advance the result that we should find and thus be able to make a diagnosis of the goodness of each of the methods, Fourier analysis and wavelet analysis, to be used to detect periodicity. In the context of this research, we performed the simulations of synthetic light curves of a pulsating star in an eclipsing binary system. To build them, we used the photometric data of a known binary, HD 19,356, obtained by the TESS (Transiting Exoplanet Survey Satellite) mission [35] with a temporal resolution of 120 s and a gap of about 4 days. Then, to obtain the synthetic light curve, we modelled the data with the software PHOEBE (PHysics Of Eclipsing BinariEs) [36,37], adopting the orbital period of 2.87 days corresponding to the selected binary system [38]. The resulting synthetic curve, referred to as model No. 0, is shown in Figure 1. This model will be the basis for the complete simulation design.

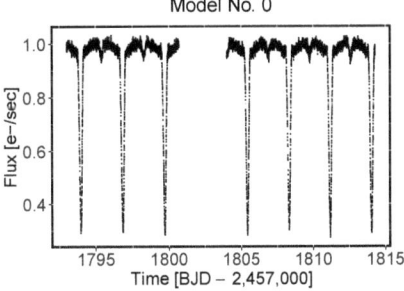

Figure 1. Synthetic light curve of an eclipsing binary affected by ARMA (1, 1) noise.

As the time series of many physical processes show stochastic and auto-correlated properties [39,40], we contaminated the synthetic curves with an ARMA (1, 1) noise [41]. Simulations of light curves with ARMA noise have already been used by authors such as Caceres et al. [40,42] to study different phenomena, such as exoplanetary transits. According to these authors, autoregressive modelling is more effective after minimising systematic variations due to instrumental or atmospheric conditions. Alberici [37] performed numerical simulations of light curves with three different types of stochastic processes to model noise: two stationary, a Gaussian process (white noise), an ARMA (1, 1) process (dependent of the past), and a non-stationary GARCH (1, 1) process (also dependent of the past). This author independently tested the Fourier and wavelet analysis and found no difference among the diverse types of noise models when they were analyzed with the same techniques.

Therefore, according to Alberici [37] and Caceres et al. [40,42], we opted for using an ARMA (1, 1) noise model, which has some dependence on the prior values, for the random fluctuation of the star luminosity [40]. The ARMA processes were used in this work to model the synthetic light curves, considered without systematic effects. The magnitude of the variability of the ARMA process was determined at a low value for simplicity, and it is left for future work to perform a study with higher variability of the noise process (i.e., where the variability of the luminosity could affect the detection of the binary star eclipse or pulsations, for example).

Therefore, a simulation design was planned with six variables and two categories (factors) each, defined as follows:

1. Beating phenomenon (X_1).

Beating is the periodic variation in amplitude at a certain wave point due to the superposition of two waves having slightly different frequencies f_i and f_j, producing a time disturbance between constructive and destructive interference. For simplicity, we considered two waves of equal amplitude, A, and wave number, k, but with slightly different angular frequencies:

$$y_i(x,t) = A\cos(kx - \omega_i t), \qquad y_j(x,t) = A\cos(kx - \omega_j t). \tag{1}$$

The resulting wave is the sum of $y_i + y_j$, which gives

$$y(x,t) = 2A\cos\left(\frac{\omega_i - \omega_j}{2}t\right)\cos\left(kx - \frac{\omega_i + \omega_j}{2}t\right), \tag{2}$$

where

$$\omega = 2\pi f. \tag{3}$$

Simulations with and without beating phenomena were considered.

2. Period of the pulsation (X_2).

In this variable, we explored the effects on the light curve considering regular pulsation cycles with a period longer or shorter than the binary's orbital period.

3. Amplitude of the pulsation with respect to the depth of the primary eclipse (X_3).

Two categories were considered for this variable. The first is that the pulsation amplitude is 20% of the depth of the primary eclipse, which we call a large amplitude. The second is that the pulsation amplitude is 4% of the depth of the primary eclipse, referred to as a small amplitude. The values of the amplitudes of the pulsations were chosen in this way because they agree with those observed in TESS light curves of some pulsating binary stars [43].

4. Number of pulsation periods (X_4).

Scenarios where the light curve is affected by simultaneously one or two pulsation periods were considered.

5. Light-time effect (X_5).

This scenario considers a star in a binary system that pulsates sinusoidally with a single frequency. Its luminosity varies with time as a consequence of this pulsation. However, the orbital motion of the star leads to a periodic variation of the distance between the observer and the star, so the phase of the observed luminosity variation also varies with the orbital period [17]. The light-time effect leads to frequency multiplets in the Fourier spectra for pulsating binary stars. To model this effect in our synthetic light curves, we relied on the simplest case of a pulsating binary star with a circular orbital motion. Furthermore, to emulate the splitting of the pulsation period, we proceeded to affect the curve with three sinusoidal pulsations: the intrinsic pulsation frequency of the star and two additional oscillation frequencies due to the light-time effect, where the spacing between each of them and the intrinsic frequency of the pulsation is equal to the orbital frequency (for more details see [17]).

Let P_o be the orbital period of the star and P_p the period of the pulsation, then the multiple components of the luminosity variations will have the following frequencies:

$$f_+ = \frac{1}{P_o} + \frac{1}{P_p}, \quad f_- = \left| \frac{1}{P_o} - \frac{1}{P_p} \right|. \quad (4)$$

Furthermore, the amplitudes and phases of the multiple components can be used to derive meaningful information about the orbital elements and the radial velocity curve. Scenarios with and without the light-time effect were considered in this work.

6. Quasi-regular pulsations (X_6).

Let us now consider a set of harmonic oscillators. Periodic motion means that there must be some time τ at which all the different oscillators have completed an integer number of cycles, i.e.:

$$\begin{aligned} \omega_0 \tau &= 2\pi n_0, \\ \omega_1 \tau &= 2\pi n_1, \\ \omega_2 \tau &= 2\pi n_2, \\ &\cdots \end{aligned} \quad (5)$$

This means that all n_k are integers; therefore, the quotient ω_i / ω_j must be a rational number. If, by contrast, it is irrational, any oscillator i or j might not complete an integer number of cycles in the time τ, giving rise to quasi-periodic pulsations (QPP). This kind of behaviour is a common feature in the solar atmosphere and flare stars where the flaring energy is released by magnetic reconnection (cf. [44]). Quasi-periodic pulsations are also observed in the α Cyg variables and are attributed to strange-mode oscillations [45,46]. A star oscillating with some periodic phenomenon would appear in a power spectrum as a peak at precisely that frequency (Dirac-δ function). Otherwise, it will appear as a broader (sometimes Lorentz-shaped) peak.

Simulations with and without quasi-regular periods were considered.

Since we have defined X_i variables (with $i = 1, \cdots, 6$) and each has two levels or categories, the total design is $2^6 = 64$ models. Since the design is optimal, we can consider only a fraction of the total design without losing information about the variables involved. Thus, in this work, we analysed $2^6/2^1 = 32$ models. The design parameter model is defined in Table 1.

Table 1. Simulation design features. The table lists the phenomena, properties, and periods fixed to simulate the pulsations in each model. The light-time effect and the beating phenomenon generate periods that are labeled with the indices "t" and "b", respectively. In all cases, these effects are present only in the lowest pulsation period.

Model	Beating Phenomena	Light Time Effect	Relation to the Orbital Period	N°. of Pul.	Quasi-Regular Periods	Amplitude [a] [%]	Fixed Period Values [c] [Days]
No. 1	Yes	No	Higher	2	Yes	20	3e–8.96^b–10
No. 2	Yes	No	Higher	2	No	20	7– 7.70^b–10
No. 3	Yes	Yes	Higher	1	Yes	20	3e–2.12^t–4.42^t–8.96^b
No. 4	Yes	Yes	Higher	1	No	20	7–2.03^t–4.86^t–7.70^b
No. 5	Yes	No	Higher	2	Yes	4	3e–8.96^b–10
No. 6	Yes	No	Higher	2	No	4	7– 7.70^b–10
No. 7	Yes	Yes	Higher	1	Yes	4	3e–2.12^t–4.42^t–8.96^b
No. 8	Yes	Yes	Higher	1	No	4	7–2.03^t–4.86^t–7.70^b

Table 1. Cont.

Model	Beating Phenomena	Light Time Effect	Relation to the Orbital Period	N°. of Pul.	Quasi-Regular Periods	Amplitude [a] [%]	Fixed Period Values [c] [Days]
No. 9	Yes	No	Lower	2	Yes	20	$\sqrt{2}$–1.27^b–1.6
No. 10	Yes	No	Lower	2	No	20	1.2–1.32^b–1.6
No. 11	Yes	Yes	Lower	1	Yes	20	$\sqrt{2}$–0.94^t–2.78^t–1.27^b
No. 12	Yes	Yes	Lower	1	No	20	1.2–0.84^t–2.06^t–1.32^b
No. 13	Yes	No	Lower	2	Yes	4	$\sqrt{2}$–1.27^b–1.6
No. 14	Yes	No	Lower	2	No	4	1.2–1.32^b–1.6
No. 15	Yes	Yes	Lower	1	Yes	4	$\sqrt{2}$–0.94^t–2.78^t–1.27^b
No. 16	Yes	Yes	Lower	1	No	4	1.2–0.84^t–2.06^t–1.32^b
No. 17	No	No	Higher	2	Yes	20	3e–10
No. 18	No	No	Higher	2	No	20	7–10
No. 19	No	Yes	Higher	1	Yes	20	3e–2.12^t–4.42^t
No. 20	No	Yes	Higher	1	No	20	7–2.03^t–4.86^t
No. 21	No	No	Higher	2	Yes	4	3e–10
No. 22	No	No	Higher	2	No	4	7–10
No. 23	No	Yes	Higher	1	Yes	4	3e–2.12^t–4.42^t
No. 24	No	Yes	Higher	1	No	4	7–2.03^t–4.86^t
No. 25	No	No	Lower	2	Yes	20	$\sqrt{2}$–1.6
No. 26	No	No	Lower	2	No	20	1.2–1.6
No. 27	No	Yes	Lower	1	Yes	20	$\sqrt{2}$–0.94^t–2.78^t
No. 28	No	Yes	Lower	1	No	20	1.2–0.84^t–2.06^t
No. 29	No	No	Lower	2	Yes	4	$\sqrt{2}$–1.6
No. 30	No	No	Lower	2	No	4	1.2–1.6
No. 31	No	Yes	Lower	1	Yes	4	$\sqrt{2}$–0.94^t–2.78^t
No. 32	No	Yes	Lower	1	No	4	1.2–0.84^t–2.06^t

[a] The pulsation amplitude is with respect to the depth of the primary eclipse. [c] e: Euler number.

3. Time Series Analysis Methods

Signals were differentiated into stationary and non-stationary. The first ones were localised in the time since their frequencies did not vary. The spectral analysis of this type of signal was carried out using the Fourier transform, which allows its decomposition into infinite sinusoidal terms, transforming the signal from the time base to the frequency base and back again. When switching from the frequency domain to the time domain, valuable information is lost, which, due to the stationary nature of the signal, is irrelevant. However, in non-stationary signals, i.e., those whose frequencies vary with time, the loss of information becomes very relevant, and it is difficult to determine when a frequency change occurs [47]. This is why wavelet analysis is presented as a tool to obtain a detailed decomposition and reconstruction of signals with abrupt changes, using a multi-resolution analysis with windows of a variable length adapted to the change in frequency of the signal. This is the main property of the wavelet, as it minimises the computational cost of the analysis by not needing infinite terms. Moreover, it improves the detail of the detection by allowing the use of long time intervals in those segments where greater precision is required at low frequencies and smaller intervals where information is required at high frequencies. Unlike Fourier, where the basis functions are sines and cosines of infinite length, in the wavelet analysis, the basis functions are localised functions in frequency and time [48]. Thus, they are a suitable method for studying quasi-regular, transient and discontinuous phenomena, such as light curves of binary stars.

The software used to carry out data analysis are detailed below.

3.1. The Period04 Tool

The Period04 tool [49] is a hybrid program, written in Java and C++, dedicated to the statistical analysis of large astronomical time series which contain gaps or missing data over a significant time interval. The program allows the extraction of individual frequencies from the multi-periodic content of the time series through an analysis based on the Discrete Fourier Transform (DFT). The light curves were fitted using the formula:

$$\chi(t) = Z + \sum A_i \sin[2\pi(f_i t + \phi_i)], \tag{6}$$

where Z is the zero-point, A_i is the amplitude, f_i is the frequency, and ϕ_i is the corresponding phase.

3.2. The WaveletComp Package

WaveletComp is an R package [50–52] for continuous time series analysis based on the Morlet wavelet family [32]. This family of wavelets leads to a wavelet transform of the time series. The transform can be separated into real and imaginary parts, thus providing information about the local amplitude and instantaneous phase of any periodic process over time.

The Morlet "mother" wavelet [32], in the version implemented by WaveletComp, is:

$$\phi(t) = \pi^{-1/4} e^{i\omega t} e^{-t^2/2}. \tag{7}$$

In this frame, the angular frequency ω is set to 6 [51], therefore the relation between the wavelet scale, a obtained with wavelet analysis, and the "equivalent Fourier period" [53] is given by:

$$\frac{1}{f} = \frac{4\pi a}{\omega + \sqrt{2 + \omega^2}}, \tag{8}$$

where

$$\frac{1}{f} \sim 1.033\, a \qquad \text{if} \qquad \omega = 6.$$

This correction is given directly by the WaveletComp package.

The Morlet wavelet transform of a time series $f(t)$ is defined as the convolution of the series with a set of "daughter wavelets" generated by the mother wavelet through translations in time, τ, and dilatations in frequency, a, being:

$$W(\tau, a) = \int_{-\infty}^{\infty} x_t \frac{1}{\sqrt{a}} \phi^*\left(\frac{t-\tau}{a}\right), \tag{9}$$

where ϕ^* denotes the complex conjugate function.

The local amplitude of any periodic component of the series can be expressed as:

$$\text{Ampl}(\tau, a) = \frac{1}{a} |W(\tau, a)|. \tag{10}$$

Then, the square of the amplitude can be interpreted as the energy density of the wave in the time-frequency domain and is called the *wavelet power spectrum*:

$$\text{Power}(\tau, a) = \frac{1}{a^2} |W(\tau, a)|^2. \tag{11}$$

This power spectrum is represented by a scalogram, see Figure 2, where a color code, from blue (low power) to red (high power), is used to indicate the power spectrum range. Maximum probability values are represented with black lines. White lines delimit probability regions with a 0.1 significance level against a white noise null. The WaveletComp package returns by default a scalogram whose axes are period vs index (ordinate and abscissa, respectively), where index refers to the amount of data in the time series and is directly related to the time axis of the series. More information can be obtained from the average wavelet power over time, which will depend on how many cycles a certain period completes in the series and not only on the magnitude of the wavelet power. The red curve represents values with a significance level less than or equal to 0.05, all of which differ from white noise. The values with the highest average powers indicate the most relevant periods for the signal. A practical example of wavelet analysis for multi-periodic time series with periods 30, 38, and 80 is shown in Figure 2. Periods 30 and 38 generate a beating effect in the series due to their proximity. All periods with higher power are detected with solid black lines in the scalogram, and the significance decreases towards nearby periods. The color code on the right-hand sidebar indicates a higher significance for the period 30 and decreases until the period 80 (see central panel). This is due to the number of cycles each period completes in the time range of the series. In the right panel, on the average wavelet power plot, periods 30 and 38 are more difficult to identify individually, unlike period 80, although they show a striking feature in the scalogram over the entire timeline identifying the presence of the beating. The results show the mixing effect produced in power obtained with the wavelet analysis for nearby periods. In agreement with the power spectrum, the maximum average wavelet power is around period 30 and decreases for longer periods.

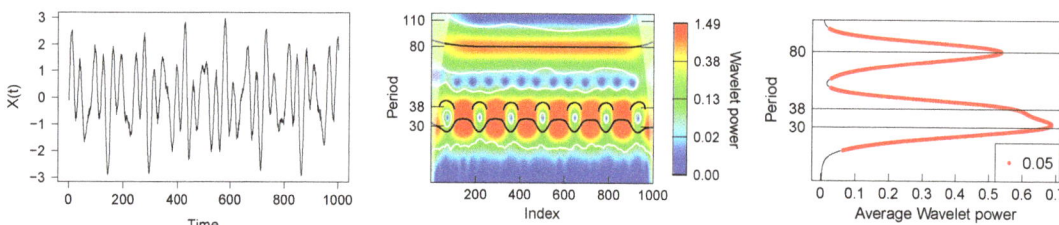

Figure 2. Example of applying wavelet analysis for a multi-periodic time series. Left panel: sinusoidal series with periods 30, 38, and 80. Central panel: Scalogram or wavelet power spectrum of the series. The top in the color code indicates a higher significance for periods. Periods with a higher power are detected with solid black lines. Right panel: average wavelet power of the series in logarithm scale. In this diagram, the beating periods 30 and 38 are more difficult to identify individually, unlike period 80, but they show a striking feature in the scalogram. Although all the periods represented in red have a significance level less than or equal to 0.05, the most relevant ones, about 30, 38, and 80, are the three highest average wavelet power. They were obtained using the WaveletComp Monte Carlo sampling-based test.

Estimating the spectral power using the continuous wavelet transform of a non-periodic time series can generate false spectral structures at the beginning and the end of the wavelet spectrum [54]. This is a known problem in the periodogram analysis caused by replacing the integrals over an infinite continuous signal by finite sums [55]. The "zero padding" method is used to solve this problem, which completes with zero values on both edges of the time series after a sample mean is removed. Then, the wavelet spectrum is estimated [53,54]. One of the drawbacks of applying this method is that it produces discontinuities at the edges of the wavelet spectrum. The region of the wavelet spectrum where padding effects become important is known as the cone of influence (COI) [53]. Therefore, the portions of the wavelet spectrum between the time axis and the COI should be considered unreliable because statistical significance tests in that region cannot be trusted [53,56]. Additionally, wavelet analysis performed with the WaveletComp package can only produce accurate results if the time series is equispaced.

3.3. Period Detection Criteria

To identify the periods detected with both tools in the synthetic light curves, the following selection criteria were established:

- For the wavelet analysis, each light-curve model was separated into two parts because they present a discontinuity (or a gap) in the time sequence (see Figure 3). For all periods that showed a maximum average wavelet power greater than 10, the mean value of each period across time and its standard deviation were calculated as suggested by Roesch and Schmidbauer [51]. The latter was used as a measurement uncertainty of the period. They were considered more or less significant according to the color code of the scalogram, even those outside the cone of influence, since there is a significant improvement in the reconstruction of the light curves when considering those periods as well [37]. The period's harmonics were identified for all independent periods. For the harmonics of the orbital period, we searched for the values included in the intervals with endpoints given by $P_h \pm \text{FWHM}/2$ (see Table A2), where P_h is an harmonic in the Model No. 0 and $\text{FWHM}/2 = \sqrt{2\log(2)} * \sigma_{P_h}$, being FWHM the full width at half maximum of a normal distribution with standard deviation σ_{P_h} at any expectation value.
- In the Fourier analysis, frequencies with a signal/noise ratio ≥ 5 were selected [57]. Then, the uncertainties of the frequencies were calculated following a nonlinear least squares fitting procedure available in the software [49]. According to Bognár, Zs. et al. [58] and Lenz and Breger [49], we accepted a peak as a combination of frequencies if the amplitudes of the main frequencies were greater than that of their presumed combination term and the difference between the observed and predicted frequency was not greater than the Rayleigh resolution criterion of the data sample. Once the independent frequencies were selected, we reported the corresponding periods.

4. Results

Each synthetic light curve generated with the design model of Table 1 is shown in Figure 3. These curves were analysed using the wavelet and Fourier analysis. To give a detailed description of the results obtained and how they are interpreted, we selected a group of four models with specific characteristics in common and compared them with Model No. 0 (shown in Figure 1). For the remaining 28 models, we will comment on the general results obtained, following the same analysis criteria. The results for the complete set of models are in Appendices A and B.

Table 2 lists the independent periods detected for Models No. 0, 1, 5, 17, and 21 with each method and their corresponding errors. Among these five models, the magnitude of the pulsation amplitude, the presence or absence of the beating phenomenon, and the values of the pulsation periods vary. The respective wavelet power spectrum, average wavelet power, and periodograms are illustrated in Figures 4 and 5.

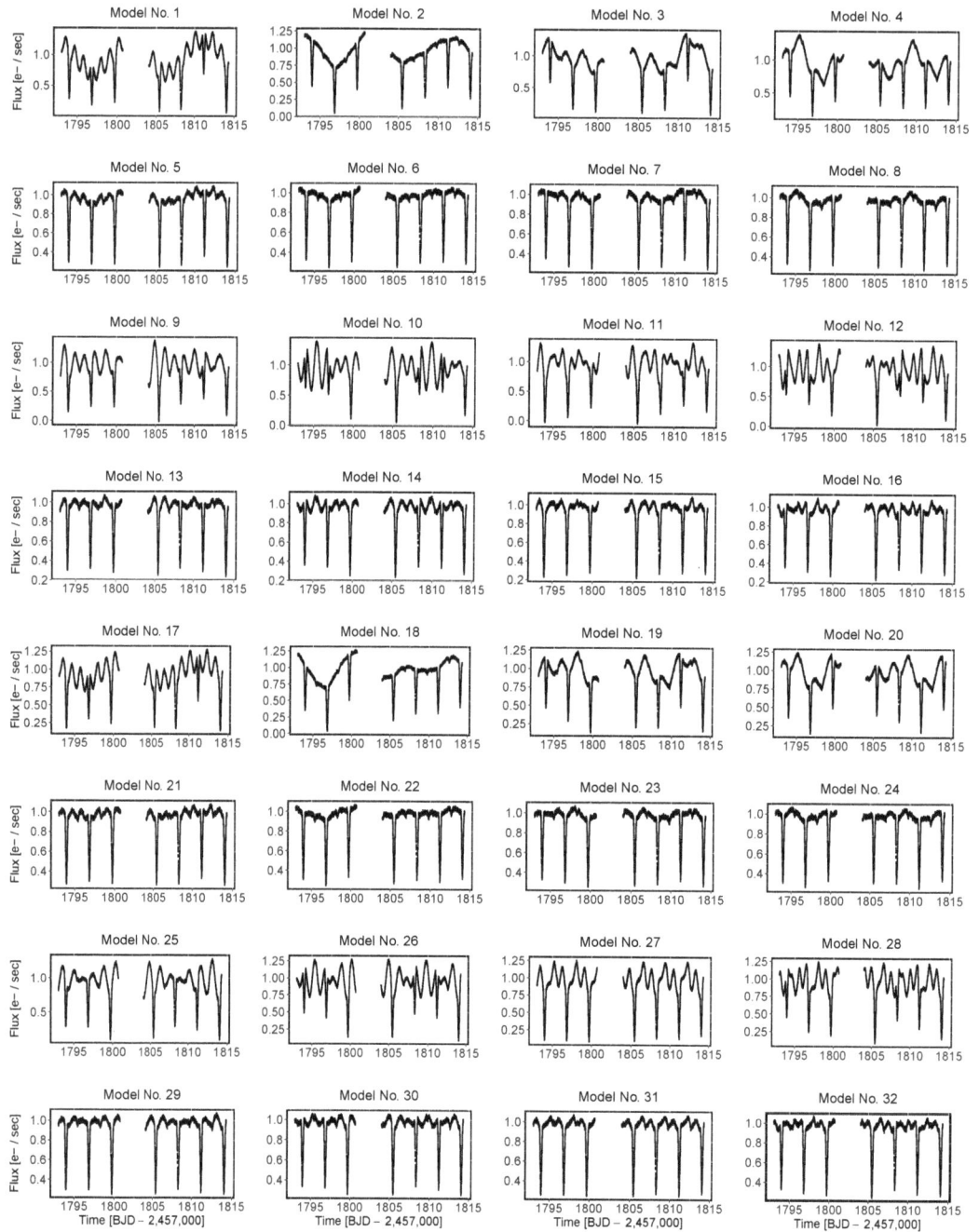

Figure 3. Synthetic light curves obtained from the simulation design. The model number is placed according to the list in Table 1.

Table 2. Comparison of Models No. 0, 1, 5, 17 and 21 with similar properties. The independent periods detected with wavelet (w) and Fourier (F) analysis are provided and their corresponding errors. For the periods obtained with the wavelet analysis, the weighted average with respect to the amount of data for each part of the curve is listed.

Model	Input Period [Days]	Amplitude [%]	Detected Period (w) [Days]	Detected Period (F) [Days]
No. 0	2.87	...	2.790 ± 0.051	2.878 ± 0.003
No. 1	10	20	...	12.911 ± 0.001
			...	9.262 ± 0.013
	8.96		7.852 ± 0.036	...
	3e		2.787 ± 0.061	2.878 ± 0.003
	2.87		1.147 ± 0.105	1.106 ± 0.002
No. 5		4	...	9.467 ± 0.003
			7.865 ± 0.046	...
			2.812 ± 0.048	2.859 ± 0.003
			...	1.100 ± 0.001
No. 17	10	20	...	10.144 ± 0.004
			7.928 ± 0.061	...
	3e		2.809 ± 0.067	2.859 ± 0.001
	2.87		2.506 ± 0.058	...
			1.139 ± 0.103	1.106 ± 0.002
No. 21		4	...	10.144 ± 0.018
			7.924 ± 0.062	...
			2.803 ± 0.053	2.878 ± 0.004
			...	1.103 ± 0.001

For Model No. 0, the wavelet and Fourier analysis recognises the binary's orbital period as a fundamental frequency and finds its harmonics. The wavelet scalogram for both parts of the curve shows a structure (seen as "faces") at the position of the minima of the light curve, also reproducing the eclipses of the binary (compare the timeline and structures of Figures 1 and 4). However, the period detected by the wavelet tool is systematically slightly shorter than the original orbital period. This shift could be related to the fact that the time scale, adopted from a TESS light curve observation, is not completely equispaced.

When examining the wavelet power diagrams for Model No. 1, we noticed some differences from Model No. 0, particularly in the features that identify the eclipses. For example, it loses detail in finding harmonics with periods 1.43 and 0.95 days, maybe due to the size of pulsation amplitude (20%). Even so, it finds the binary 2.87 days period with an error of 0.09, although in the second part of the curve and with much lower power than in Model No. 0. Additionally, a period of about 1.15 days is found, and its trace shows a sinusoidal modulation in the scalogram. Although it could be a spurious value, we associated this period as an independent value, likely with a harmonic of 8.96 or 10 days ($8 \times 1.15 = 9.2$), where the value 8.96 days were artificially introduced to obtain the beating. However, this shift could be due to three near pulsation periods: 3e, 8.96, and 10 days. Furthermore, a period of 7.852 days is also present with high power and could be related to the 3e days period. Even though the wavelet analysis does not find those exact values, the software likely detects the presence of significant periods in that range.

Model No. 5 is affected by the same pulsation periods as Model No. 1 but with a smaller amplitude. Therefore, the model's scalogram and power diagram show a lower significance, but the same periods and scalogram features are maintained. As shown in Figure 2, the wavelet analysis performed with the WaveletComp package presents complications in differentiating nearby periods, particularly cases with a beating, such as Models No. 1 and 5, resulting in an intermediate value.

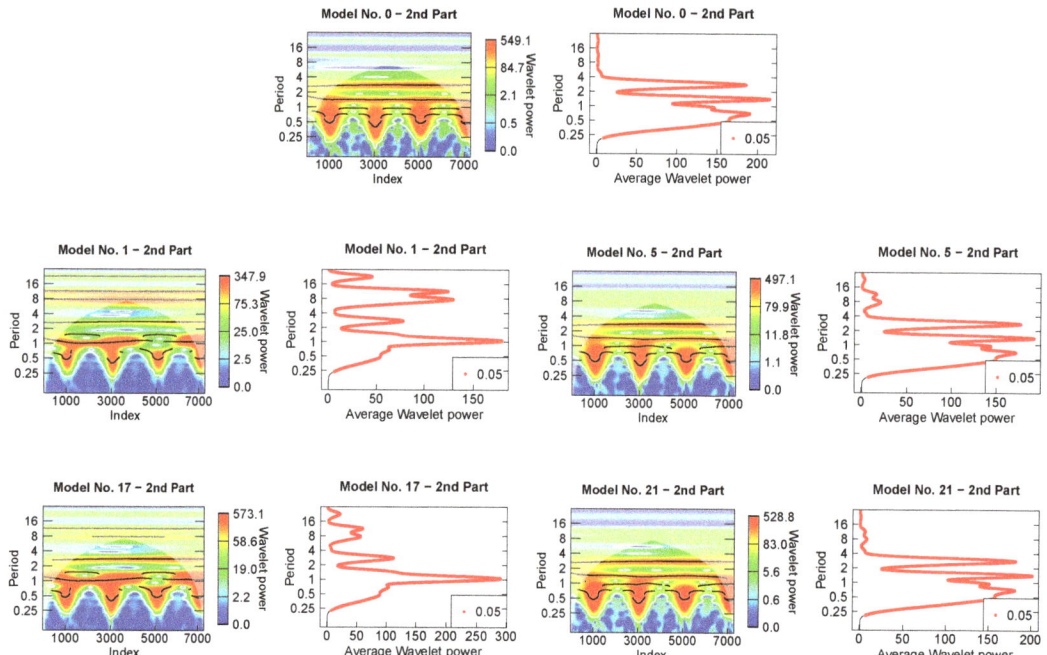

Figure 4. Results of the wavelet analysis for Model No. 0, 1, 5, 17, and 21. From left to right, the wavelet power spectrum and the average wavelet power for the second part of each synthetic light curve.

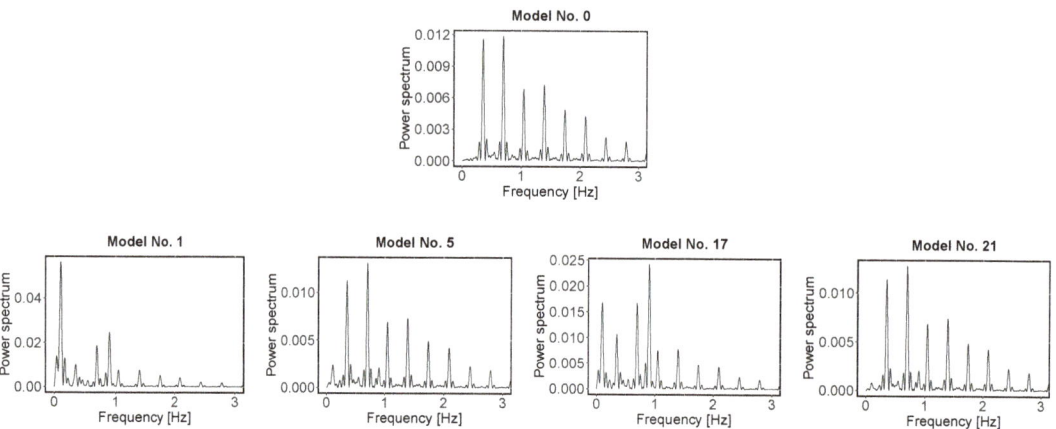

Figure 5. Periodograms obtained for the models No. 0, 1, 5, 17, and 21.

When analysing Model No. 17, we observed that the structures present in the scalogram are very similar to those of Model No. 1, both with a pulsation amplitude of 20% with respect to the depth of the primary eclipse. Since they share two periods, the 7.9 days period and the harmonic of 1.15 days, they could be related to the periods of 3e and 10 days.

Finally, the scalogram and average wavelet power of Model No. 21 is very similar to that found in Model No. 5 since they share two pulsations with amplitudes of 4%. However, in this case, it detects the binary period and another with a 7.92 days period, but with lower power than in Model No. 5.

The periodograms resulting from the Fourier analysis show remarkable similarities between the models affected by low-amplitude pulsations regardless of the periods involved. This is particularly seen in models No. 5 and 21. At the same time, slight variations are observed with respect to model No. 0. The pattern changes markedly for models No. 17 and 21, as seen in Figure 5). A similar situation occurs with models No. 6–8.

Quasi-periodic oscillations decrease the power of detection of the Fourier Transform method, tending not to detect them. The Period04 tool accurately sees the fundamental frequency of the binary orbital motion ($P_{\rm orb} = 2.87 \pm 0.01$ days) and its harmonics in all cases. However, it observes two pulsation periods with a significant error close to $P = 10$ days, detected as 9.26 days and 12.9 days in Model No. 1 and 9.47 days in Model No. 5. The 9.26 days (or 9.47 days) period could be likely related to the beating phenomenon with $P = 8.96$ or the pulsation with $P = 10$ days, or a mix of both. This is supported because, for example, the period 9.26 days also appears in Model No. 2 (see Appendix A) that has regular oscillations (with $P = 7$ days and $P = 10$ days, and a beating of 7.7 days), but in this case, the period 7 days is inexactly seen as $P = 6.657$ days (see also Model No. 6). Similarly to the wavelet analysis, a period of 1.1 days ($f = 0.9036$ days^{-1}) is present in Models No. 1 and 17. Models with the same set of parameters but low amplitude oscillations (i.e., Models No. 5 and 21) behave similarly. Model No. 21, similar to Model No. 17, detects an accurate period for $P = 10$ days.

Generally, the orbital period is well-detected in most models with Fourier analysis; only a few detect half of the period (i.e., Models No. 4, 11, 20, and 27). These models present regular oscillations with short periods or have a quasi-regular period $\sqrt{2}$. The latter is similar to the half-orbital period. Models with quasi-regular pulsations show more dispersion of values, mainly when the time light effect is included. The Fourier analysis finds all periods in 14 models (Models No. 10, 12, 14, 16–19, 21–23, 26, 28, 30 and 32), 9 of which have low pulse amplitude (4%).

On the other hand, the wavelet analysis found the binary's period satisfactorily, although it underestimated the value with an average of 2.81 days (of 31 models), which suggests a period shift of 0.06 days. Concerning the rest of the periods, this method found an equal or smaller number of periods than the Fourier analysis. Although the mean value has a systematic error, its variance is similar to that obtained with the Fourier method. It detected a few periods, mainly in models with low pulsation amplitude and where the period values coincide with a harmonic of the binary's orbital period. However, it found all periods in four models (Models No. 17, 18, 26 and 28). These have a high pulsation amplitude (20%) and absence of beating phenomenon.

A comparison of the methods is shown in Figure 6. The boxplots of the hit ratios of all the prefixed periods obtained by wavelet and Fourier analysis are in the left and right panels, respectively. We see that the median percentages (thick black line inside the box) are much lower for the wavelet analysis (<0.4) than for the Fourier analysis (about 0.8).

Figure 7 presents a boxplot comparison for both techniques, of the hit ratio segmented by the categories of five of the variables included in the simulation design: beating phenomenon (X_1), period of the pulsation (X_2), amplitude of the pulsation (X_3), number of pulsations periods (X_4), quasi-regular pulsations (X_6), calculated with a wavelet and Fourier analysis, respectively. Since we were dealing with the half-model design, the result does not provide information about one variable.

Regarding the results obtained for the wavelet analysis, we can observe that the variable that influences its performance the most is the pulsation amplitude. At low amplitude, it is difficult to detect the intrinsic period of the model. The presence of the beating phenomenon and the fact that the period of the pulsation is quasi-regular also have some influence. The hit ratio increases when the pulsation period is greater than the orbital period and when there is more than one pulsation (see Figure 7). The analysis of the variance of the percentage of hits of a quasi-periodic oscillation with a large pulsation amplitude (i.e., 20%) finds the period only the 55% of the time. While for a regular oscillation, it detects the period 85% of the time. Instead, if the pulsation amplitude is low, it finds the period about 30% of the time in both cases. Quasi-regular or regular period detection is also affected by their interaction with the orbital period, depending if they are shorter or longer than it. The regular period detection rate varies from 60% (for shorter periods) to 50% (for longer periods) and for quasi-regular periods from 65% to 30%.

Unlike wavelet analysis, Fourier analysis is unaffected by the pulsation amplitude. The proportion of detected periods (an approximate value) increases if the period pulsation is longer than the orbital one, but they are more accurate when it is shorter (see Figure 7). The presence of beating and quasi-period decreases the detection capability a little. In both cases, the medians (thick black line in the box) is about 75%, which changes to 100% without the phenomenon. From an analysis of the variance of the hit ratio, we obtain that if the pulsation period is shorter than the orbital period, the presence of the beat phenomenon improves period detection (70% vs 80%). Otherwise, if the pulsation period is longer, the presence of this effect worsens it (95% vs 75%). Quasi-regular or regular period detection is also affected by their interaction with the orbital period, depending if they are shorter or longer than it. The period detection rate for regular oscillations varies from 100% (for shorter periods) to 80% (for longer periods) and for quasi-regular periods from 50% to 85%. For pulsation periods longer than orbital periods, the performance is similar.

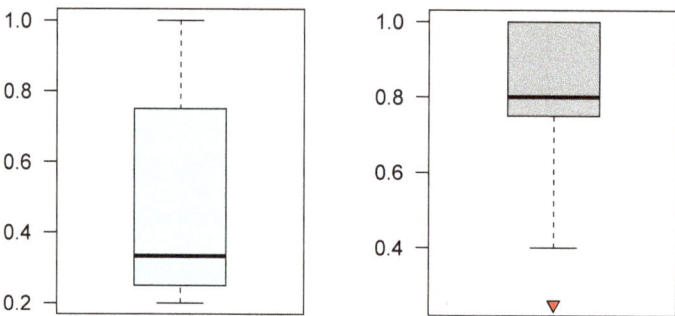

Figure 6. Comparison between the proportion of correctness of the original period obtained by wavelet (cyan) and Fourier (grey) analysis. The thick solid black line represents the median of the data, and the red symbol stands for an outlier or extreme value.

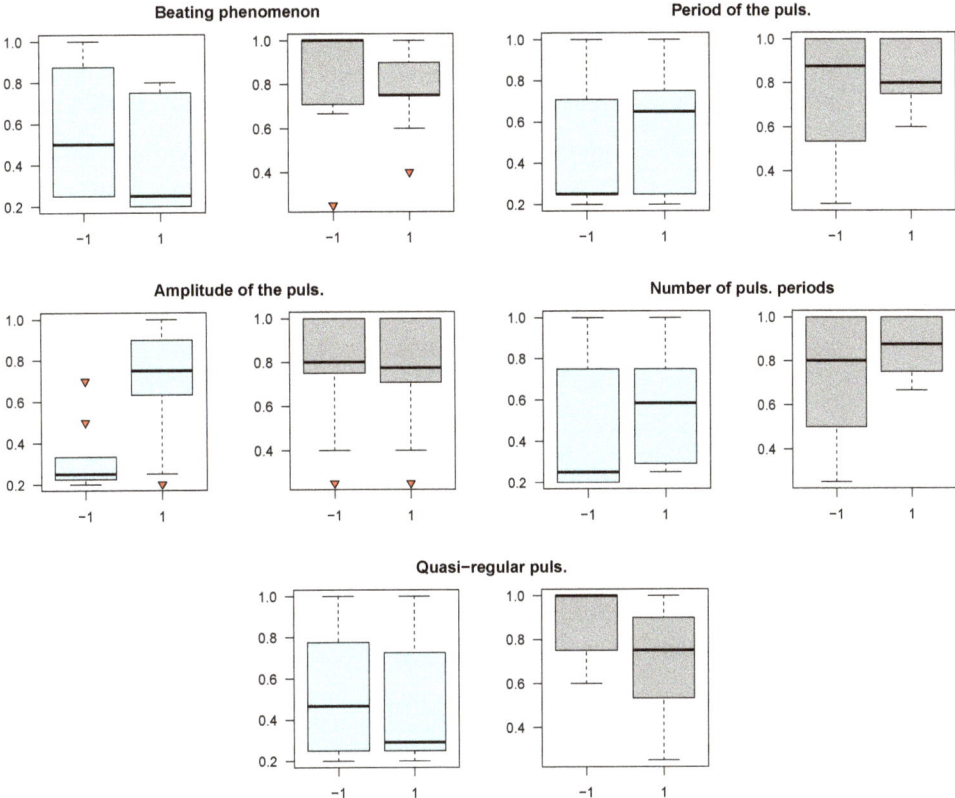

Figure 7. Boxplots of hit ratio of the original periods with the wavelet (cyan) and Fourier (grey) analysis for each variable involved. The thick solid black line represents the median of the data, and the red symbol stands for an outlier or extreme value. The hit rate is on the y-axis and the two categories (indicated by -1 and 1) for each variable are shown on x-axis: beating phenomenon (absence/presence), period of the pulsation (minor/major), amplitude of the pulsation (4%, 20%), number of pulsations periods (1/2), and quasi-regular pulsations (absence/presence).

5. Discussion and Conclusions

It is widely known that power spectra can be misleading and frequently lead to inaccurate conclusions. For this reason, many works present a comparative analysis of popular algorithms applicable to period searching. Sometimes the study of one particular light curve with various numerical methods resulted in a high dispersion or undetected periods as reported by Graham et al. [26] and Distefano et al. [59].

The objective of the current work was to evaluate the response of two methods, based on the Morlet wavelet function and continuous Fourier transform, in the production of bias, alias, or shifts in period detection using a simulation design with pre-fixed and known periods. For this purpose, we performed a simulation design of synthetic light curves of a pulsating star in an eclipsing binary system with an orbital period of 2.87 days. It is a 2^{k-p}-statistical design with $k = 6$ variables, in which the beating phenomenon and the light-time effect were considered. Of the total design method, only one-half was executed. The 32 time series models were analysed using the WaveletComp and Period04 tools. Although many time series analysis tools are available, the application of such a large simulation design to more than two techniques would be unattainable.

The obtained synthetic light curves, shown in Figure 3, present similarities with the light curves of pulsating binary stars obtained with the TESS mission (see Figure 8) in shape and cadence. Among them, we can highlight the similarity between HD 152,248 with model No. 9, TIC 80,042,405 with model No. 27, and TIC 97,467,902 with models with low amplitudes with respect to the depth of the primary eclipse (models No. 5–8, 13–16, 21–24, 29–32).

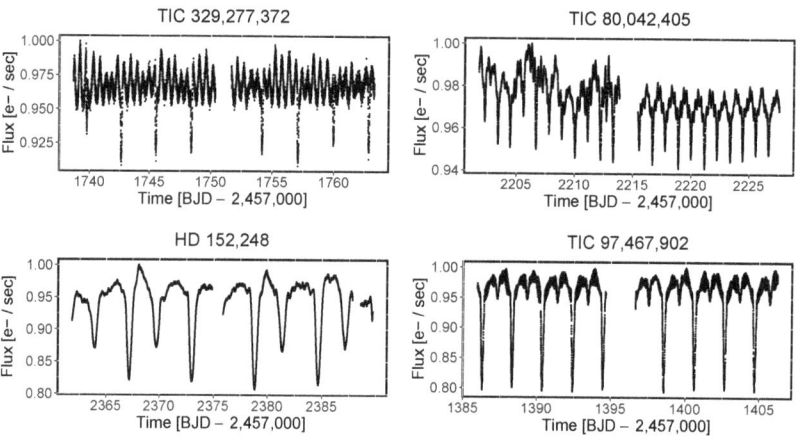

Figure 8. Light curves obtained by TESS of pulsating binary stars.

Comparing both applied methods, we observed that the orbital period and its harmonics were well detected in most models. Even though the Fourier method provided more accurate period detection, the wavelet analysis found it more times. When estimating the orbital period, the WaveletComp package trends to provide a low period value or period shift of approximately 0.06 days in all the models, but this has a similar dispersion to the orbital period obtained from the Fourier analysis. In agreement with this finding, Foster [22] also reported through numerical simulations that for a simple sinusoid, the maximum frequency obtained with wavelet analysis has an offset with respect to the signal frequency, which can be compensated. This author also commented that under irregular sampling, the detection of the period by the wavelet analysis varies with time, making it difficult to give an exact value of the same, especially with the increase in missing data in the sampling, which can reach up to 20% at high frequencies or find spurious periods at low frequencies. Although we considered "almost" a regular sampling, it may be that some of the still sampling irregularity present in the timeline scale selected from TESS observations has affected the performance of the WaveletComp package.

As shown in the previous section, pulsation amplitude is a very powerful variable in period detection. In the case of the wavelet analysis, more satisfactory results are found recovering the periods of the pulsations with large amplitudes. On the other hand, the Fourier analysis did not show difficulties in recognising the periods when the amplitude is low, but they were detected with significantly lower power than models with large amplitudes.

Neither tool was able to distinguish accurately the periods leading to a beating phenomenon when they were longer than the orbital period, resulting in both cases in an intermediate value. However, when the beating was formed by periodic pulsations shorter than the orbital period, the Fourier analysis found it in all cases, but quasi-regular periods remained unresolved. Concerning the beating period $\sqrt{2}$, it is necessary to remember that this period coincided (unfortunately, but possible in a real scenario) with the first harmonic of the selected orbital period of 1.43 days. Therefore, when a value of 1.4 days was detected, it was considered the harmonic of the orbital period. If the values set in the simulation design were unknown, it would be impossible to distinguish between them.

The same holds for the 0.94 days period. By analogy, wavelet analysis does not detect these periods either.

Finally, the wavelet power spectrum revealed in all models a particular structure located at the position of the primary eclipses of the binary in the synthetic light curve, even showing the instant of time in which they occur. Then, if a similar structure is found in a real data set, it could indicate binarity. In addition, as mentioned before, several harmonics of the orbital period were detected by both tools, which is also an indicator of binarity.

In general, the Fourier analysis by means Period04 tool had a higher proportion of periods correctly and accurately detected than the analysis performed with the WaveletComp package. However, the discrepancy could be attributed to the less information available in the latter when performing the period analysis since splitting the light curve into two parts was necessary. Despite the high cadence, the TESS data outside the large gap also present a slightly irregular sampling, which affects the analysis. Furthermore, this package uses a somewhat coarse resolution level, preventing it from accurately determining the periods encountered. Foster [22] also found some kind of asymmetry in period detections, with more values at low frequencies and suppression at high frequencies.

Considering the period hit rate analysed using synthetic light curves for eclipsing binaries, a bias in detecting some pulsation periods (beating phenomenon and light time effects) in multi-periodic light curves would be expected. The hit rate depends on the amplitude and frequency of the pulsation with respect to eclipse depth and orbital period. The situation would be more unfavourable for purely elliptical double stars, partial eclipses, or systems influenced by the gravitational attraction of a third body that might behave similarly to a sinusoidal pulsation. In some cases, only a light modulation could likely be acting because the binary orbit is not aligned with the observer's direction. In these cases, the period hit rate would be lower, masking the binary detection, for example, among B supergiants.

Although the recovery of all pre-fixed periods was not feasible using WaveletComp and Period04, valuable insights were obtained regarding how each technique responds to the various modelled situations. For instance, we observed that WaveletComp can only detect an intermediate value when confronted with two close periods due to the poor spectral resolution of this package. Additionally, we have noted that Period04 fails to recognise quasi-regular periods, but has good accuracy in detecting high-frequency oscillations. Furthermore, we found with both tools structures that indicate binarity (in the periodograms and scalograms), information that will be very useful when performing these analyses on the light curves of binary star candidates, among other findings.

Considering the advantages and disadvantages of the techniques analysed, we recommend using at least two diagnosis tools to conduct a detailed analysis of time series data to obtain confident results. Additionally, a second criterion should be considered to test the accuracy of the found frequencies or periods since the methods might find a harmonic, inaccurate period, or spurious value. Particularly, the wavelet method might predict only a harmonic of the true period. Since approximated values were detected with at least one method, a fine-tuning of trial periods around a certain percentage of its value, by applying phase diagrams, would help to recover accurate values.

Evaluating the behaviour of other tools based on wavelets that do not have this weakness (e.g., Weighted wavelet Z-transform (WWZ) [22]) is one of the goals to follow. This method also enables an improvement in the resolution of the period searches, allowing a high degree of accuracy. In addition, considering even more realistic models or observed light curves that present a gap and time-varying periods would allow us to take full advantage of the main feature of wavelet analysis.

Author Contributions: Formal analysis, A.A.A., G.F.A.M., A.C. and L.S.C.; Investigation, A.A.A., G.F.A.M., A.C. and L.S.C.; Methodology, A.A.A., G.F.A.M., A.C. and L.S.C.; Writing—original draft, A.A.A.; Writing—review & editing, G.F.A.M., A.C. and L.S.C. All authors have read and agreed to the published version of the manuscript.

Funding: This project has received funding from the European Union's Framework Programme for Research and Innovation Horizon 2020 (2014-2020) under the Marie Skłodowska-Curie Grant Agreement No. 823734. AA and LC thanks financial support from CONICET (PIP 1337) and the Universidad Nacional de La Plata (Programa de Incentivos 11/G160), Argentina. AC and GA acknowledges support from Centro de Estudios Atmosféricos y Astroestadística (CEAAS), Universidad de Valparaíso.

Data Availability Statement: Not applicable.

Conflicts of Interest: The authors declare no conflict of interest.

Abbreviations

The following abbreviations are used in this manuscript:

TESS   Transiting Exoplanet Survey Satellite
PHOEBE PHysics Of Eclipsing BinariEs

Appendix A. Fourier Analysis Results

For each synthetic light curve, the frequencies with a signal/noise ratio ≥ 5 and their errors, are listed in Table A1. The fixed listed periods of the design models are listed in column 2. It does not include the period of the binary system (2.87 days) present in all the models, except for model No. 0, where is the only period present. Columns 3 and 4 list the frequency and the corresponding period in days. Column 5 indicates the independent frequencies found and the linear relationship among them. The respective periodograms are shown in Figures A1 and A2.

Table A1. Fourier analysis results reporting the significant frequencies and harmonics. The detect independent frequencies are highlighted in blue. The symbol e is the Euler number.

Model	Fixed Period Values [Days]	Frequencies [Days^{-1}]	Indep. Period [Days]	Comment
No. 0	2.87	$0.6970 \pm 4 \times 10^{-4}$		$\sim 2 f_1$
		$0.3473 \pm 4 \times 10^{-4}$	$2.878 \pm 3 \times 10^{-3}$	f_1
		$1.3941 \pm 3 \times 10^{-4}$		$\sim 4 f_1$
		$1.0467 \pm 6 \times 10^{-4}$		$\sim 3 f_1$
		$2.0912 \pm 7 \times 10^{-4}$		$\sim 6 f_1$
		$1.7419 \pm 5 \times 10^{-4}$		$\sim 5 f_1$
		$2.4386 \pm 1 \times 10^{-4}$		$\sim 7 f_1$
		$2.7883 \pm 2 \times 10^{-4}$		$\sim 8 f_1$
		$3.1380 \pm 1 \times 10^{-4}$		$\sim 9 f_1$
		$3.4824 \pm 3 \times 10^{-4}$		$\sim 10 f_1$
No. 1	3e–8.96–10	$0.1079 \pm 2 \times 10^{-4}$	$9.262 \pm 1.3 \times 10^{-2}$	f_1
		$0.9036 \pm 3 \times 10^{-4}$	$1.106 \pm 2 \times 10^{-3}$	f_2
		$0.3473 \pm 4 \times 10^{-4}$	$2.878 \pm 3 \times 10^{-3}$	f_3
		$0.6970 \pm 4 \times 10^{-4}$		$\sim 2 f_3$
		$1.3941 \pm 5 \times 10^{-4}$		$\sim 4 f_3$
		$1.0467 \pm 6 \times 10^{-4}$		$\sim 3 f_3$
		$2.0935 \pm 7 \times 10^{-4}$		$\sim 6 f_3$
		$1.7438 \pm 7 \times 10^{-4}$		$\sim 5 f_3$
		$2.4409 \pm 9 \times 10^{-4}$		$\sim 7 f_3$
		$2.7883 \pm 1.1 \times 10^{-3}$		$\sim 8 f_3$
		$0.0774 \pm 1.3 \times 10^{-3}$	$12.911 \pm 1 \times 10^{-3}$	f_4
		$3.1380 \pm 1.6 \times 10^{-3}$		$\sim 9 f_3$
No. 2	7–7.70–10	$0.1079 \pm 2 \times 10^{-4}$	$9.262 \pm 1 \times 10^{-3}$	f_1
		$0.6970 \pm 3 \times 10^{-4}$		$\sim 2 f_2$
		$0.3473 \pm 3 \times 10^{-4}$	$2.878 \pm 1 \times 10^{-3}$	f_2

Table A1. Cont.

Model	Fixed Period Values [Days]	Frequencies [Days^{-1}]	Indep. Period [Days]	Comment
		$1.3941 \pm 4 \times 10^{-4}$		$\sim 4 f_2$
		$1.0467 \pm 5 \times 10^{-4}$		$\sim 3 f_2$
		$2.0935 \pm 6 \times 10^{-4}$		$\sim 6 f_2$
		$1.7438 \pm 6 \times 10^{-4}$		$\sim 5 f_2$
		$0.1502 \pm 6 \times 10^{-4}$	$6.657 \pm 2 \times 10^{-3}$	f_3
		$2.4409 \pm 8 \times 10^{-4}$		$\sim 7 f_2$
		$2.7883 \pm 9 \times 10^{-4}$		$\sim 8 f_2$
No. 3	3e–2.12–4.42–8.96	$0.1173 \pm 2 \times 10^{-4}$	$8.521 \pm 1 \times 10^{-3}$	f_1
		$0.3497 \pm 3 \times 10^{-4}$	$2.859 \pm 1 \times 10^{-3}$	f_2
		$0.2182 \pm 3 \times 10^{-4}$	$4.581 \pm 1 \times 10^{-3}$	f_3
		$0.6970 \pm 3 \times 10^{-4}$		$\sim 2 f_2$
		$0.4717 \pm 3 \times 10^{-4}$	$2.179 \pm 2 \times 10^{-3}$	f_4
		$1.3941 \pm 4 \times 10^{-4}$		$\sim 4 f_2$
		$1.0467 \pm 5 \times 10^{-4}$		$\sim 3 f_2$
		$2.0935 \pm 6 \times 10^{-4}$		$\sim 6 f_2$
		$1.7438 \pm 6 \times 10^{-4}$		$\sim 5 f_2$
		$2.4385 \pm 8 \times 10^{-4}$		$\sim 7 f_2$
		$2.7883 \pm 9 \times 10^{-4}$		$\sim 8 f_2$
		$0.0821 \pm 1.2 \times 10^{-3}$	$12.173 \pm 1 \times 10^{-3}$	f_5
No. 4	7–2.03–4.86–7.70	$0.1338 \pm 2 \times 10^{-4}$	$7.475 \pm 1 \times 10^{-3}$	f_1
		$0.6971 \pm 5 \times 10^{-4}$	$1.435 \pm 1 \times 10^{-3}$	f_2
		$0.2793 \pm 4 \times 10^{-4}$		$\sim 2 f_1$
		$0.4228 \pm 4 \times 10^{-4}$	$2.367 \pm 2 \times 10^{-3}$	f_3
		$1.3942 \pm 5 \times 10^{-4}$		$\sim 2 f_2$
		$1.0468 \pm 5 \times 10^{-4}$		$\sim 3/2 f_2$
		$1.7438 \pm 6 \times 10^{-4}$		$\sim 5/2 f_2$
		$2.0912 \pm 6 \times 10^{-4}$		$\sim 3 f_2$
		$0.0962 \pm 6 \times 10^{-4}$	$10.391 \pm 6 \times 10^{-3}$	f_4
		$0.5632 \pm 1 \times 10^{-4}$		$\sim 4 f_1$
		$2.4385 \pm 1 \times 10^{-4}$		$\sim 7/2 f_2$
		$2.7883 \pm 1 \times 10^{-4}$		$\sim 4 f_2$
No. 5	3e–8.96–10	$0.6970 \pm 3.8 \times 10^{-3}$		$\sim 2 f_1$
		$0.3497 \pm 3.6 \times 10^{-3}$	$2.859 \pm 3 \times 10^{-3}$	f_1
		$1.3941 \pm 4.7 \times 10^{-3}$		$\sim 4 f_1$
		$1.0468 \pm 5 \times 10^{-3}$		$\sim 3 f_1$
		$2.0912 \pm 6.2 \times 10^{-3}$		$\sim 6 f_1$
		$1.7439 \pm 1 \times 10^{-4}$		$\sim 5 f_1$
		$0.1056 \pm 7 \times 10^{-3}$	$9.467 \pm 3 \times 10^{-3}$	f_2
		$2.4409 \pm 8.6 \times 10^{-3}$		$\sim 7 f_1$
		$2.7883 \pm 9.6 \times 10^{-3}$		$\sim 8 f_1$
		$0.9083 \pm 1.35 \times 10^{-2}$	$1.100 \pm 1 \times 10^{-3}$	f_3
		$3.1380 \pm 1.49 \times 10^{-2}$		$\sim 9 f_1$
No. 6	7–7.70–10	$0.6970 \pm 3.7 \times 10^{-3}$		$\sim 2 f_1$
		$0.3497 \pm 3 \times 10^{-4}$	$2.859 \pm 3 \times 10^{-3}$	f_1
		$1.3941 \pm 4 \times 10^{-4}$		$\sim 4 f_1$
		$1.0467 \pm 4 \times 10^{-4}$		$\sim 3 f_1$
		$2.0912 \pm 6 \times 10^{-4}$		$\sim 6 f_1$
		$1.7438 \pm 6 \times 10^{-4}$		$\sim 5 f_1$
		$2.4409 \pm 8 \times 10^{-4}$		$\sim 7 f_1$
		$2.7883 \pm 9 \times 10^{-4}$	$9.468 \pm 3 \times 10^{-3}$	f_2
		$3.1380 \pm 1.4 \times 10^{-3}$		$\sim 9 f_1$
		$0.1502 \pm 3.1 \times 10^{-3}$	$6.657 \pm 1 \times 10^{-3}$	f_3

Table A1. *Cont.*

Model	Fixed Period Values [Days]	Frequencies [Days^{-1}]	Indep. Period [Days]	Comment
No. 7	3e–2.12–4.42–8.96	$0.3497 \pm 3 \times 10^{-4}$	$2.859 \pm 3 \times 10^{-3}$	f_1
		$0.6970 \pm 3 \times 10^{-4}$		$\sim 2\, f_1$
		$1.3941 \pm 4 \times 10^{-4}$		$\sim 4\, f_1$
		$1.0467 \pm 5 \times 10^{-4}$		$\sim 3\, f_1$
		$2.0912 \pm 6 \times 10^{-4}$		$\sim 6\, f_1$
		$1.7438 \pm 6 \times 10^{-4}$		$\sim 5\, f_1$
		$2.4409 \pm 8 \times 10^{-4}$		$\sim 7\, f_1$
		$2.7883 \pm 9 \times 10^{-4}$		$\sim 8\, f_1$
		$0.1150 \pm 1.1 \times 10^{-3}$	$8.695 \pm 8 \times 10^{-3}$	f_2
		$3.1380 \pm 1 \times 10^{-4}$		$\sim 9\, f_1$
		$0.2206 \pm 3 \times 10^{-4}$	$4.532 \pm 7 \times 10^{-3}$	f_3
		$0.4717 \pm 3 \times 10^{-4}$	$2.119 \pm 3 \times 10^{-3}$	f_4
No. 8	7–2.03–4.86–7.70	$0.6970 \pm 3 \times 10^{-4}$		$\sim 2\, f_1$
		$0.3473 \pm 3 \times 10^{-4}$	$2.859 \pm 3 \times 10^{-3}$	f_1
		$1.3941 \pm 4 \times 10^{-4}$		$\sim 4\, f_1$
		$1.0467 \pm 5 \times 10^{-4}$		$\sim 3\, f_1$
		$2.0912 \pm 6 \times 10^{-4}$		$\sim 6\, f_1$
		$1.7438 \pm 6 \times 10^{-4}$		$\sim 5\, f_1$
		$2.4409 \pm 8 \times 10^{-4}$		$\sim 7\, f_1$
		$2.7883 \pm 9 \times 10^{-4}$		$\sim 8\, f_1$
		$0.1361 \pm 1.2 \times 10^{-3}$	$7.345 \pm 6 \times 10^{-3}$	f_2
		$3.1380 \pm 1.4 \times 10^{-3}$		$\sim 9\, f_1$
		$0.4905 \pm 3.1 \times 10^{-3}$	$2.038 \pm 4 \times 10^{-3}$	f_3
		$0.2112 \pm 3 \times 10^{-4}$	$4.734 \pm 3 \times 10^{-3}$	f_4
No. 9	$\sqrt{2}$–1.27–1.6	$0.7017 \pm 4 \times 10^{-4}$		$\sim 2\, f_3$
		$0.6196 \pm 8 \times 10^{-4}$	$1.613 \pm 2 \times 10^{-3}$	f_1
		$0.7909 \pm 8 \times 10^{-4}$	$1.264 \pm 1 \times 10^{-3}$	f_2
		$0.3473 \pm 1 \times 10^{-3}$	$2.878 \pm 8 \times 10^{-3}$	f_3
		$1.3941 \pm 1.3 \times 10^{-3}$		$\sim 4\, f_3$
		$1.0468 \pm 1.4 \times 10^{-3}$		$\sim 3\, f_3$
		$2.0936 \pm 1.7 \times 10^{-3}$		$\sim 6\, f_3$
		$1.7439 \pm 1.7 \times 10^{-3}$		$\sim 5\, f_3$
		$2.4386 \pm 2.4 \times 10^{-3}$		$\sim 7\, f_3$
		$2.7883 \pm 2.7 \times 10^{-3}$		$\sim 8\, f_3$
		$3.1380 \pm 4.1 \times 10^{-3}$		$\sim 9\, f_3$
No. 10	1.2–1.32–1.6	$0.8308 \pm 4 \times 10^{-4}$	$1.203 \pm 1 \times 10^{-3}$	f_1
		$0.3473 \pm 7 \times 10^{-4}$	$2.878 \pm 6 \times 10^{-3}$	f_2
		$0.6243 \pm 5 \times 10^{-4}$	$1.601 \pm 1 \times 10^{-3}$	f_3
		$1.3941 \pm 9 \times 10^{-4}$		$\sim 4\, f_2$
		$0.7416 \pm 5 \times 10^{-4}$	$1.348 \pm 1 \times 10^{-3}$	f_4
		$1.0468 \pm 1 \times 10^{-4}$		$\sim 3\, f_2$
		$2.0912 \pm 1 \times 10^{-4}$		$\sim 6\, f_2$
		$1.7415 \pm 1 \times 10^{-4}$		$\sim 5\, f_2$
		$2.7883 \pm 1 \times 10^{-4}$		$\sim 8\, f_2$
		$2.4386 \pm 1 \times 10^{-4}$		$\sim 7\, f_2$
No. 11	$\sqrt{2}$–0.94–2.78–1.27	$0.6970 \pm 4 \times 10^{-4}$	$1.434 \pm 1 \times 10^{-3}$	$\sim 2\, f_1$
		$0.3544 \pm 4 \times 10^{-4}$	$2.821 \pm 3 \times 10^{-3}$	f_1
		$1.0515 \pm 5 \times 10^{-4}$	$0.951 \pm 1 \times 10^{-3}$	$\sim 3\, f_1$
		$0.7886 \pm 5 \times 10^{-4}$	$1.268 \pm 1 \times 10^{-3}$	f_2
		$1.3941 \pm 9 \times 10^{-4}$		$\sim 4\, f_1$
		$2.0936 \pm 1 \times 10^{-4}$		$\sim 6\, f_1$
		$1.7415 \pm 1 \times 10^{-4}$		$\sim 5\, f_1$
		$2.4386 \pm 1 \times 10^{-4}$		$\sim 7\, f_1$
		$2.7883 \pm 2 \times 10^{-4}$		$\sim 8\, f_1$

Table A1. *Cont.*

Model	Fixed Period Values [Days]	Frequencies [Days^{-1}]	Indep. Period [Days]	Comment
No. 12	1.2–0.84–2.06–1.32	$0.8355 \pm 7 \times 10^{-4}$	$1.196 \pm 1 \times 10^{-3}$	f_1
		$0.3473 \pm 8 \times 10^{-4}$	$2.878 \pm 6 \times 10^{-3}$	f_2
		$0.4858 \pm 8 \times 10^{-4}$	$2.058 \pm 3 \times 10^{-3}$	f_3
		$1.1759 \pm 8 \times 10^{-4}$	$0.850 \pm 1 \times 10^{-3}$	f_4
		$1.3918 \pm 1 \times 10^{-4}$		$\sim 4 f_2$
		$1.0468 \pm 1 \times 10^{-4}$		$\sim 3 f_2$
		$0.7011 \pm 1 \times 10^{-4}$		$\sim 2 f_2$
		$2.0912 \pm 1 \times 10^{-4}$		$\sim 6 f_2$
		$1.7415 \pm 1 \times 10^{-4}$		$\sim 5 f_2$
		$2.4409 \pm 2 \times 10^{-4}$		$\sim 7 f_2$
		$2.7883 \pm 2 \times 10^{-4}$		$\sim 8 f_2$
		$0.7440 \pm 2 \times 10^{-4}$	$1.344 \pm 1 \times 10^{-3}$	f_5
		$3.1380 \pm 3 \times 10^{-4}$		$\sim 9 f_2$
No. 13	$\sqrt{2}$–1.27–1.6	$0.6970 \pm 4 \times 10^{-4}$		$\sim 2 f_1$
		$0.3473 \pm 4 \times 10^{-4}$	$2.878 \pm 3 \times 10^{-3}$	f_1
		$1.3941 \pm 5 \times 10^{-4}$		$\sim 4 f_1$
		$1.0468 \pm 6 \times 10^{-4}$		$\sim 3 f_1$
		$2.0912 \pm 7 \times 10^{-4}$		$\sim 6 f_1$
		$1.7439 \pm 7 \times 10^{-4}$		$\sim 5 f_1$
		$2.4386 \pm 1 \times 10^{-4}$		$\sim 7 f_1$
		$2.7883 \pm 1 \times 10^{-4}$		$\sim 8 f_1$
		$3.1380 \pm 1 \times 10^{-4}$		$\sim 9 f_1$
		$0.6219 \pm 1 \times 10^{-4}$	$1.607 \pm 4 \times 10^{-3}$	f_2
		$3.4854 \pm 2 \times 10^{-4}$		$\sim 10 f_1$
		$0.7909 \pm 1 \times 10^{-4}$	$1.264 \pm 2 \times 10^{-3}$	f_3
No. 14	1.2–1.32–1.6	$0.3473 \pm 4 \times 10^{-4}$	$2.878 \pm 3 \times 10^{-3}$	f_1
		$1.3941 \pm 5 \times 10^{-4}$		$\sim 4 f_1$
		$0.6994 \pm 4 \times 10^{-4}$		$\sim 2 f_1$
		$1.0467 \pm 5 \times 10^{-4}$		$\sim 3 f_1$
		$2.0912 \pm 7 \times 10^{-4}$		$\sim 6 f_1$
		$1.7438 \pm 7 \times 10^{-4}$		$\sim 5 f_1$
		$2.4409 \pm 1 \times 10^{-4}$		$\sim 7 f_1$
		$0.8332 \pm 1 \times 10^{-4}$	$1.200 \pm 1 \times 10^{-3}$	f_2
		$2.7883 \pm 1 \times 10^{-4}$		$\sim 8 f_1$
		$3.1380 \pm 1 \times 10^{-4}$		$\sim 9 f_1$
		$3.4853 \pm 2 \times 10^{-4}$		$\sim 10 f_1$
		$0.6290 \pm 1 \times 10^{-4}$	$1.589 \pm 4 \times 10^{-3}$	f_3
		$0.7557 \pm 2 \times 10^{-4}$	$1.323 \pm 3 \times 10^{-3}$	f_4
No. 15	$\sqrt{2}$–0.94–2.78–1.27	$0.6970 \pm 4 \times 10^{-4}$		$\sim 2 f_1$
		$0.3497 \pm 4 \times 10^{-4}$	$2.8594 \pm 3 \times 10^{-4}$	f_1
		$1.3941 \pm 6 \times 10^{-4}$		$\sim 4 f_1$
		$1.0468 \pm 6 \times 10^{-4}$		$\sim 3 f_1$
		$2.0912 \pm 9 \times 10^{-4}$		$\sim 6 f_1$
		$1.7439 \pm 8 \times 10^{-4}$		$\sim 5 f_1$
		$2.4409 \pm 1 \times 10^{-4}$		$\sim 7 f_1$
		$2.7883 \pm 1 \times 10^{-4}$		$\sim 8 f_1$
		$3.1380 \pm 2 \times 10^{-4}$		$\sim 9 f_1$
		$0.7886 \pm 2 \times 10^{-4}$	$1.268 \pm 1 \times 10^{-3}$	f_2
		$3.4854 \pm 2 \times 10^{-4}$		$\sim 10 f_1$
No. 16	1.2–0.84–2.06–1.32	$0.3473 \pm 5 \times 10^{-4}$	$2.878 \pm 3 \times 10^{-3}$	f_1
		$0.6970 \pm 5 \times 10^{-4}$		$\sim 2 f_1$
		$1.3941 \pm 6 \times 10^{-4}$		$\sim 4 f_1$
		$1.0467 \pm 7 \times 10^{-4}$		$\sim 3 f_1$
		$2.0912 \pm 8 \times 10^{-4}$		$\sim 6 f_1$

Table A1. Cont.

Model	Fixed Period Values [Days]	Frequencies [Days^{-1}]	Indep. Period [Days]	Comment
		$1.7438 \pm 8 \times 10^{-4}$		$\sim 5 f_1$
		$2.4409 \pm 1 \times 10^{-4}$		$\sim 7 f_1$
		$2.7883 \pm 1 \times 10^{-4}$		$\sim 8 f_1$
		$0.8308 \pm 2 \times 10^{-4}$	$1.203 \pm 1 \times 10^{-3}$	f_2
		$3.1380 \pm 2 \times 10^{-4}$		$\sim 9 f_1$
		$3.4853 \pm 2 \times 10^{-4}$		$\sim 10 f_1$
		$0.4881 \pm 2 \times 10^{-4}$	$2.048 \pm 5 \times 10^{-3}$	f_3
		$1.1805 \pm 2 \times 10^{-4}$	$0.847 \pm 1 \times 10^{-3}$	f_4
		$0.7510 \pm 2 \times 10^{-4}$	$1.331 \pm 3 \times 10^{-3}$	f_5
No. 17	3e–10	$0.9036 \pm 4 \times 10^{-4}$	$1.106 \pm 2 \times 10^{-3}$	f_1
		$0.0985 \pm 4 \times 10^{-4}$	$10.144 \pm 4 \times 10^{-3}$	f_2
		$0.6970 \pm 6 \times 10^{-4}$		$\sim 2 f_3$
		$0.3497 \pm 6 \times 10^{-4}$	$2.859 \pm 1 \times 10^{-3}$	f_3
		$1.3941 \pm 7 \times 10^{-4}$		$\sim 4 f_3$
		$1.0468 \pm 8 \times 10^{-4}$		$\sim 3 f_3$
		$2.0912 \pm 1 \times 10^{-4}$		$\sim 6 f_3$
		$1.7439 \pm 1 \times 10^{-4}$		$\sim 5 f_3$
		$2.4409 \pm 1 \times 10^{-4}$		$\sim 7 f_3$
		$2.7883 \pm 2 \times 10^{-4}$		$\sim 8 f_3$
		$3.1380 \pm 2 \times 10^{-4}$		$\sim 9 f_3$
No. 18	7–10	$0.1502 \pm 4 \times 10^{-4}$	$6.657 \pm 1 \times 10^{-3}$	f_1
		$0.1009 \pm 4 \times 10^{-4}$	$9.908 \pm 4 \times 10^{-3}$	f_2
		$0.6979 \pm 6 \times 10^{-4}$		$\sim 2 f_3$
		$0.3473 \pm 5 \times 10^{-4}$	$2.878 \pm 1 \times 10^{-3}$	f_3
		$1.3941 \pm 7 \times 10^{-4}$		$\sim 4 f_3$
		$1.0468 \pm 8 \times 10^{-4}$		$\sim 3 f_3$
		$2.0912 \pm 9 \times 10^{-4}$		$\sim 6 f_3$
		$1.7439 \pm 9 \times 10^{-4}$		$\sim 5 f_3$
		$2.4409 \pm 1 \times 10^{-4}$		$\sim 7 f_3$
		$2.7883 \pm 1 \times 10^{-4}$		$\sim 8 f_3$
		$3.1380 \pm 2 \times 10^{-4}$		$\sim 9 f_3$
No. 19	3e–2.12–4.42	$0.3497 \pm 6 \times 10^{-4}$	$2.878 \pm 3 \times 10^{-3}$	f_1
		$0.2182 \pm 7 \times 10^{-4}$	$4.581 \pm 1 \times 10^{-3}$	f_2
		$0.4717 \pm 7 \times 10^{-4}$	$2.119 \pm 3 \times 10^{-3}$	f_3
		$0.6970 \pm 7 \times 10^{-4}$		$\sim 2 f_1$
		$0.1173 \pm 7 \times 10^{-4}$	$8.521 \pm 5 \times 10^{-3}$	f_4
		$1.3941 \pm 8 \times 10^{-4}$		$\sim 4 f_1$
		$1.0467 \pm 9 \times 10^{-4}$		$\sim 3 f_1$
		$2.0935 \pm 1 \times 10^{-4}$		$\sim 6 f_1$
		$1.7438 \pm 1 \times 10^{-4}$		$\sim 5 f_1$
		$2.4385 \pm 1 \times 10^{-4}$		$\sim 7 f_1$
		$2.7883 \pm 2 \times 10^{-4}$		$\sim 8 f_1$
		$3.1380 \pm 2 \times 10^{-4}$		$\sim 9 f_1$
No. 20	7–2.03–4.86	$0.1408 \pm 1 \times 10^{-4}$	$7.101 \pm 5 \times 10^{-3}$	f_1
		$0.6970 \pm 1 \times 10^{-4}$	$1.434 \pm 2 \times 10^{-3}$	f_2
		$0.2792 \pm 1 \times 10^{-4}$		$\sim 2 f_1$
		$0.4224 \pm 9 \times 10^{-4}$		$\sim 3 f_1$
		$1.3941 \pm 1 \times 10^{-4}$		$\sim 2 f_2$
		$1.0467 \pm 1 \times 10^{-4}$		$\sim 3/2 f_2$
		$1.7438 \pm 1 \times 10^{-4}$		$\sim 5/2 f_2$
		$2.0912 \pm 2 \times 10^{-4}$		$\sim 3 f_2$
		$0.5609 \pm 1 \times 10^{-4}$		$\sim 4 f_1$
		$2.4409 \pm 2 \times 10^{-4}$		$\sim 7/2 f_2$
		$2.7883 \pm 2 \times 10^{-4}$		$\sim 4 f_2$
		$0.2041 \pm 1 \times 10^{-4}$	$4.897 \pm 3 \times 10^{-3}$	f_3

Table A1. Cont.

Model	Fixed Period Values [Days]	Frequencies [Days^{-1}]	Indep. Period [Days]	Comment
No. 21	3e–10	$0.6970 \pm 5 \times 10^{-4}$		$\sim 2 f_1$
		$0.3473 \pm 4 \times 10^{-4}$	$2.878 \pm 4 \times 10^{-3}$	f_1
		$1.3941 \pm 6 \times 10^{-4}$		$\sim 4 f_1$
		$1.0467 \pm 6 \times 10^{-4}$		$\sim 3 f_1$
		$2.0912 \pm 8 \times 10^{-4}$		$\sim 6 f_1$
		$1.7438 \pm 8 \times 10^{-4}$		$\sim 5 f_1$
		$2.4409 \pm 1 \times 10^{-4}$		$\sim 7 f_1$
		$2.7883 \pm 1 \times 10^{-4}$		$\sim 8 f_1$
		$0.9059 \pm 1 \times 10^{-4}$	$1.103 \pm 1 \times 10^{-3}$	f_2
		$3.1380 \pm 2 \times 10^{-4}$		$\sim 9 f_1$
		$0.0985 \pm 1 \times 10^{-4}$	$10.144 \pm 1.8 \times 10^{-3}$	f_3
No. 22	7–10	$0.6970 \pm 5 \times 10^{-4}$		$\sim 2 f_1$
		$0.3473 \pm 4 \times 10^{-4}$	$2.878 \pm 4 \times 10^{-3}$	f_1
		$1.3941 \pm 6 \times 10^{-4}$		$\sim 4 f_1$
		$1.0467 \pm 6 \times 10^{-4}$		$\sim 3 f_1$
		$2.0912 \pm 8 \times 10^{-4}$		$\sim 6 f_1$
		$1.7438 \pm 8 \times 10^{-4}$		$\sim 5 f_1$
		$2.4409 \pm 1 \times 10^{-4}$		$\sim 7 f_1$
		$2.7883 \pm 1 \times 10^{-4}$		$\sim 8 f_1$
		$0.1478 \pm 1 \times 10^{-4}$	$6.762 \pm 8 \times 10^{-3}$	f_2
		$3.1380 \pm 2 \times 10^{-4}$		$\sim 9 f_1$
		$0.1009 \pm 1 \times 10^{-4}$	$9.908 \pm 1.7 \times 10^{-3}$	f_3
No. 23	3e–2.12–4.42	$0.3473 \pm 4 \times 10^{-4}$	$2.878 \pm 4 \times 10^{-3}$	f_1
		$0.6970 \pm 5 \times 10^{-4}$		$\sim 2 f_1$
		$1.3941 \pm 6 \times 10^{-4}$		$\sim 4 f_1$
		$1.0467 \pm 6 \times 10^{-4}$		$\sim 3 f_1$
		$2.0912 \pm 8 \times 10^{-4}$		$\sim 6 f_1$
		$1.7438 \pm 8 \times 10^{-4}$		$\sim 5 f_1$
		$2.4409 \pm 1 \times 10^{-4}$		$\sim 7 f_1$
		$2.7883 \pm 1 \times 10^{-4}$		$\sim 8 f_1$
		$3.1380 \pm 2 \times 10^{-4}$		$\sim 9 f_1$
		$0.2182 \pm 2 \times 10^{-4}$	$4.581 \pm 5 \times 10^{-3}$	f_2
		$0.4717 \pm 3 \times 10^{-4}$	$2.119 \pm 1 \times 10^{-3}$	f_3
		$0.1173 \pm 3 \times 10^{-4}$	$8.521 \pm 1.8 \times 10^{-3}$	f_4
No. 24	7–2.03–4.86	$0.6970 \pm 7 \times 10^{-4}$		$\sim 2 f_1$
		$0.3473 \pm 7 \times 10^{-4}$	$2.878 \pm 1 \times 10^{-3}$	f_1
		$1.3941 \pm 9 \times 10^{-4}$		$\sim 4 f_1$
		$1.0467 \pm 1 \times 10^{-4}$		$\sim 3 f_1$
		$2.0912 \pm 1 \times 10^{-4}$		$\sim 6 f_1$
		$1.7438 \pm 1 \times 10^{-4}$		$\sim 5 f_1$
		$2.4409 \pm 2 \times 10^{-4}$		$\sim 7 f_1$
		$2.7883 \pm 2 \times 10^{-4}$		$\sim 8 f_1$
		$3.1380 \pm 3 \times 10^{-4}$		$\sim 9 f_1$
		$0.1408 \pm 3 \times 10^{-4}$	$7.101 \pm 1.6 \times 10^{-3}$	f_2
		$0.4881 \pm 3 \times 10^{-4}$	$2.048 \pm 2 \times 10^{-3}$	f_3
No. 25	$\sqrt{2}$–1.6	$0.7064 \pm 2 \times 10^{-4}$		$\sim 2 f_2$
		$0.6243 \pm 4 \times 10^{-4}$	$1.601 \pm 1 \times 10^{-3}$	f_1
		$0.3473 \pm 5 \times 10^{-4}$	$2.878 \pm 3 \times 10^{-3}$	f_2
		$1.3941 \pm 7 \times 10^{-4}$		$\sim 4 f_2$
		$1.0444 \pm 7 \times 10^{-4}$		$\sim 3 f_2$
		$2.0912 \pm 9 \times 10^{-4}$		$\sim 6 f_2$
		$1.7438 \pm 9 \times 10^{-4}$		$\sim 5 f_2$
		$2.4409 \pm 1 \times 10^{-4}$		$\sim 7 f_2$
		$2.7883 \pm 1 \times 10^{-4}$		$\sim 8 f_2$

Table A1. *Cont.*

Model	Fixed Period Values [Days]	Frequencies [Days^{-1}]	Indep. Period [Days]	Comment
No. 26	1.2–1.6	$0.8332 \pm 4 \times 10^{-4}$	$1.200 \pm 1 \times 10^{-3}$	f_1
		$0.3473 \pm 6 \times 10^{-4}$	$2.878 \pm 4 \times 10^{-3}$	f_2
		$0.6219 \pm 4 \times 10^{-4}$	$1.607 \pm 1 \times 10^{-3}$	f_3
		$1.3941 \pm 7 \times 10^{-4}$		$\sim 4 f_2$
		$0.6970 \pm 6 \times 10^{-4}$		$\sim 2 f_2$
		$1.0467 \pm 8 \times 10^{-4}$		$\sim 3 f_2$
		$2.0912 \pm 1 \times 10^{-4}$		$\sim 6 f_2$
		$1.7438 \pm 1 \times 10^{-4}$		$\sim 5 f_2$
		$2.4409 \pm 1 \times 10^{-4}$		$\sim 7 f_2$
		$2.7883 \pm 1 \times 10^{-4}$		$\sim 8 f_2$
		$3.1380 \pm 2 \times 10^{-4}$		$\sim 9 f_2$
No. 27	$\sqrt{2}$–0.94–2.78	$0.7017 \pm 5 \times 10^{-4}$		$\sim 2 f_1$
		$0.3544 \pm 4 \times 10^{-4}$	$2.821 \pm 3 \times 10^{-3}$	f_1
		$1.0514 \pm 6 \times 10^{-4}$	$0.951 \pm 1 \times 10^{-3}$	$\sim 3 f_1$
		$1.3941 \pm 1 \times 10^{-4}$		$\sim 4 f_1$
		$2.0912 \pm 1 \times 10^{-4}$		$\sim 6 f_1$
		$1.7438 \pm 2 \times 10^{-4}$		$\sim 5 f_1$
		$2.4409 \pm 2 \times 10^{-4}$		$\sim 7 f_1$
		$2.7883 \pm 2 \times 10^{-4}$		$\sim 8 f_1$
		$3.1380 \pm 3 \times 10^{-4}$		$\sim 9 f_1$
No. 28	1.2–0.84–2.06	$0.3473 \pm 6 \times 10^{-4}$	$2.878 \pm 4 \times 10^{-3}$	f_1
		$0.8355 \pm 6 \times 10^{-4}$	$1.196 \pm 1 \times 10^{-3}$	f_2
		$1.1782 \pm 6 \times 10^{-4}$	$0.848 \pm 1 \times 10^{-3}$	f_3
		$0.4834 \pm 6 \times 10^{-4}$	$2.068 \pm 2 \times 10^{-3}$	f_4
		$0.6970 \pm 6 \times 10^{-4}$		$\sim 2 f_1$
		$1.3941 \pm 8 \times 10^{-4}$		$\sim 4 f_1$
		$1.0467 \pm 8 \times 10^{-4}$		$\sim 3 f_1$
		$2.0912 \pm 1 \times 10^{-4}$		$\sim 6 f_1$
		$1.7438 \pm 1 \times 10^{-4}$		$\sim 5 f_1$
		$2.4409 \pm 1 \times 10^{-4}$		$\sim 7 f_1$
		$2.7883 \pm 2 \times 10^{-4}$		$\sim 8 f_1$
		$3.1380 \pm 3 \times 10^{-4}$		$\sim 9 f_1$
No. 29	$\sqrt{2}$–1.6	$0.3473 \pm 4 \times 10^{-4}$	$2.878 \pm 3 \times 10^{-3}$	f_1
		$0.6994 \pm 4 \times 10^{-4}$		$\sim 2 f_1$
		$1.3941 \pm 6 \times 10^{-4}$		$\sim 4 f_1$
		$1.0468 \pm 6 \times 10^{-4}$		$\sim 3 f_1$
		$2.0912 \pm 8 \times 10^{-4}$		$\sim 6 f_1$
		$1.7439 \pm 8 \times 10^{-4}$		$\sim 5 f_1$
		$2.4409 \pm 1 \times 10^{-4}$		$\sim 7 f_1$
		$2.7883 \pm 1 \times 10^{-4}$		$\sim 8 f_1$
		$3.1380 \pm 2 \times 10^{-4}$		$\sim 9 f_1$
		$0.6243 \pm 1 \times 10^{-4}$	$1.601 \pm 4 \times 10^{-3}$	f_2
No. 30	1.2–1.6	$0.3473 \pm 4 \times 10^{-4}$	$2.878 \pm 3 \times 10^{-3}$	f_1
		$0.6970 \pm 5 \times 10^{-4}$		$\sim 2 f_1$
		$1.3941 \pm 6 \times 10^{-4}$		$\sim 4 f_1$
		$1.0467 \pm 6 \times 10^{-4}$		$\sim 3 f_1$
		$2.0912 \pm 8 \times 10^{-4}$		$\sim 6 f_1$
		$1.7438 \pm 8 \times 10^{-4}$		$\sim 5 f_1$
		$2.4409 \pm 1 \times 10^{-4}$		$\sim 7 f_1$
		$2.7883 \pm 1 \times 10^{-4}$		$\sim 8 f_1$
		$0.8332 \pm 2 \times 10^{-4}$	$1.200 \pm 2 \times 10^{-3}$	f_2
		$3.1380 \pm 2 \times 10^{-4}$		$\sim 9 f_1$
		$0.6243 \pm 2 \times 10^{-4}$	$1.601 \pm 3 \times 10^{-3}$	f_3

Table A1. Cont.

Model	Fixed Period Values [Days]	Frequencies [Days^{-1}]	Indep. Period [Days]	Comment
No. 31	$\sqrt{2}$–0.94–2.78	$0.3497 \pm 4 \times 10^{-4}$	$2.859 \pm 3 \times 10^{-3}$	f_1
		$0.6994 \pm 4 \times 10^{-4}$		$\sim 2 f_1$
		$1.3941 \pm 6 \times 10^{-4}$		$\sim 4 f_1$
		$1.0468 \pm 6 \times 10^{-4}$		$\sim 3 f_1$
		$2.0912 \pm 8 \times 10^{-4}$		$\sim 6 f_1$
		$1.7439 \pm 8 \times 10^{-4}$		$\sim 5 f_1$
		$2.4409 \pm 1 \times 10^{-4}$		$\sim 7 f_1$
		$2.7883 \pm 1 \times 10^{-4}$		$\sim 8 f_1$
		$3.1380 \pm 2 \times 10^{-4}$		$\sim 9 f_1$
No. 32	1.2–0.84–2.06	$0.3473 \pm 5 \times 10^{-4}$	$2.878 \pm 4 \times 10^{-3}$	f_1
		$0.6947 \pm 5 \times 10^{-4}$		$\sim 2 f_1$
		$1.3941 \pm 6 \times 10^{-4}$		$\sim 4 f_1$
		$1.0467 \pm 7 \times 10^{-4}$		$\sim 3 f_1$
		$2.0912 \pm 9 \times 10^{-4}$		$\sim 6 f_1$
		$1.7438 \pm 8 \times 10^{-4}$		$\sim 5 f_1$
		$2.4409 \pm 1 \times 10^{-4}$		$\sim 7 f_1$
		$2.7883 \pm 1 \times 10^{-4}$		$\sim 8 f_1$
		$3.1380 \pm 2 \times 10^{-4}$		$\sim 9 f_1$
		$3.4853 \pm 2 \times 10^{-4}$		$\sim 10 f_1$
		$1.1805 \pm 3 \times 10^{-4}$	$0.847 \pm 1 \times 10^{-3}$	f_2
		$0.4834 \pm 3 \times 10^{-4}$	$2.068 \pm 1 \times 10^{-2}$	f_3
		$0.8332 \pm 3 \times 10^{-4}$	$1.200 \pm 3 \times 10^{-3}$	f_4

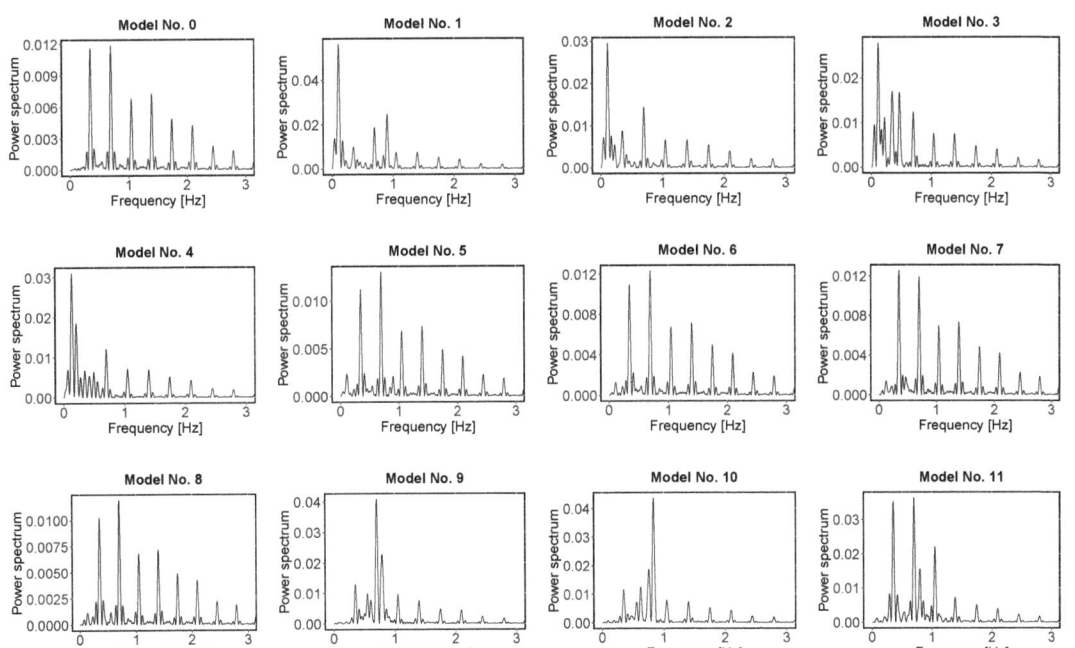

Figure A1. Cont.

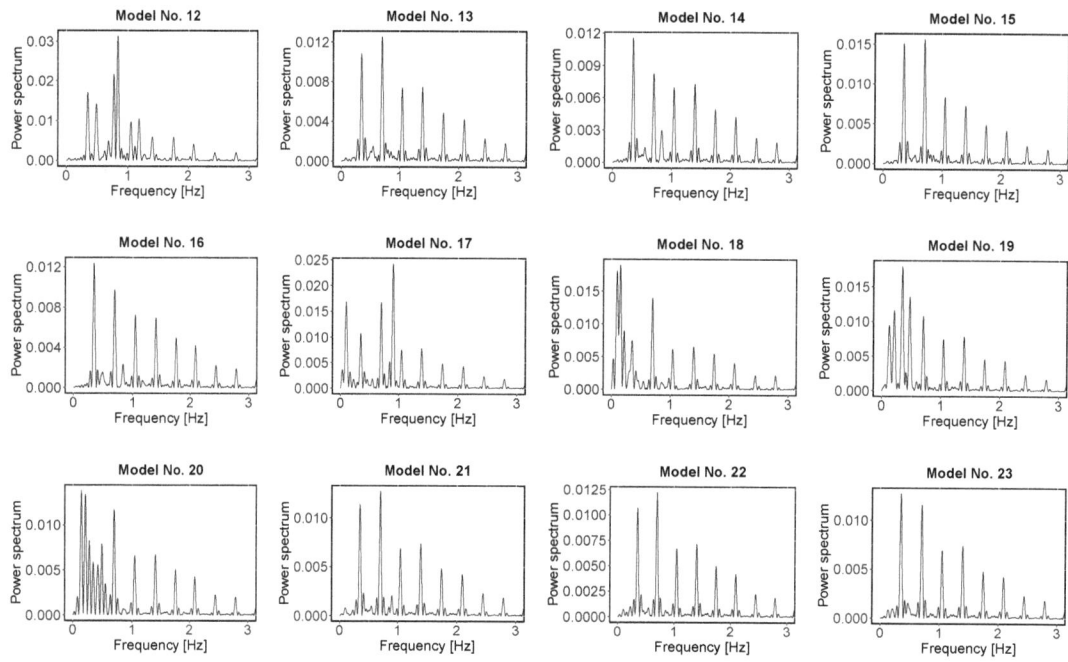

Figure A1. Periodograms obtained from the Fourier analysis for models No. 0 to No. 23.

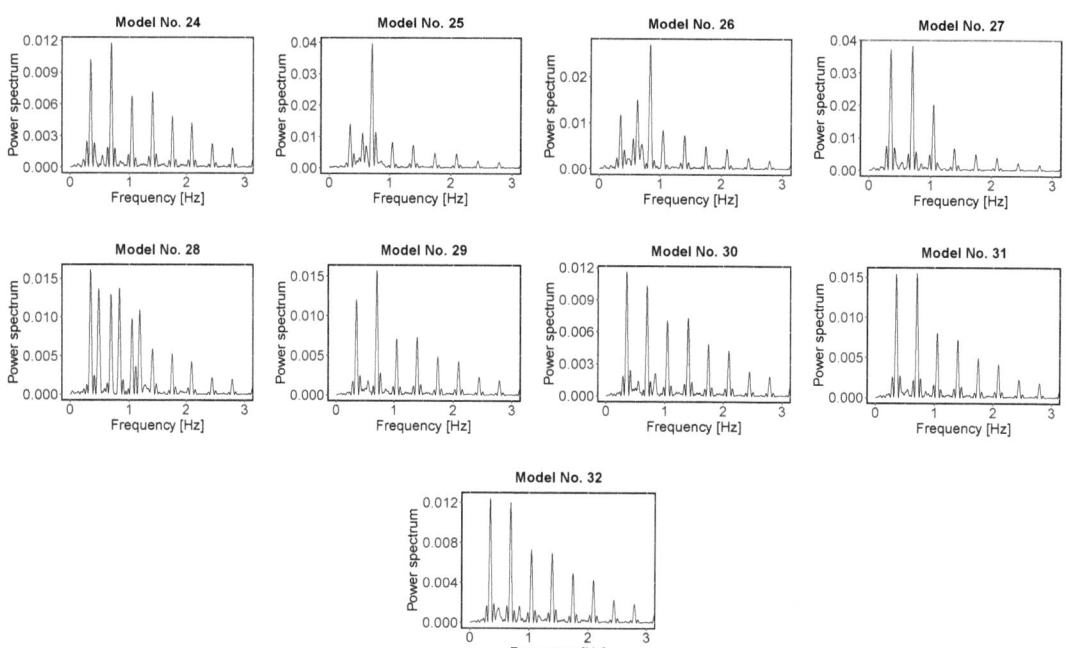

Figure A2. Periodograms obtained from the Fourier analysis for models No. 24 to No. 32.

Appendix B. Wavelet Analysis Results

The periods detected for each synthetic light curve with the wavelet analysis, and their errors, are listed in Table A2. The list of set periods in the simulation design is listed in column 2. It does not include the period of the binary system (2.87 days) present in all the models, except for model No. 0, where is the only period present. Columns 3 and 4 lists the periods present in the first and second parts of the curve. Column 5 indicates the independent periods found and the linear relationship among them. The wavelet power spectrum and the average wavelet power are illustrated in Figures A3–A8.

Table A2. Wavelet analysis results reporting the significant periods and harmonics. The detected independent periods are highlighted in blue. The symbol e is the Euler number.

Model	Fixed Period Values [Days]	Periods [Days] First Part	Periods [Days] Second Part	Comment
No. 0	2.87	$0.548 \pm 2.9 \times 10^{-2}$	$0.539 \pm 7.3 \times 10^{-2}$	$\sim P_1/5$
		$0.728 \pm 1.62 \times 10^{-1}$	$0.72 \pm 5.4 \times 10^{-2}$	$\sim P_1/4$
		$0.979 \pm 1.52 \times 10^{-2}$	$0.943 \pm 2.3 \times 10^{-2}$	$\sim P_1/3$
		$1.43 \pm 2.28 \times 10^{-1}$	$1.442 \pm 3.8 \times 10^{-2}$	$\sim P_1/2$
		$2.785 \pm 2.4 \times 10^{-2}$	$2.811 \pm 7.2 \times 10^{-2}$	P_1
No. 1	3e–8.96–10	$0.549 \pm 1.9 \times 10^{-2}$	$0.52 \pm 5.9 \times 10^{-2}$	$\sim P_1/5$
		$0.740 \pm 1.5 \times 10^{-1}$	$0.726 \pm 1.74 \times 10^{-1}$	$\sim P_1/4$
		$1.168 \pm 2.8 \times 10^{-2}$	$1.132 \pm 1.83 \times 10^{-1}$	P_2
		$1.403 \pm 5.4 \times 10^{-2}$	$1.472 \pm 1.34 \times 10^{-1}$	$\sim P_1/2$
		$2.347 \pm 1.23 \times 10^{-1}$	...	$2P_2$
		...	$2.787 \pm 6.1 \times 10^{-2}$	P_1
		$5.719 \pm 8 \times 10^{-3}$	...	$\sim 2P_1$
		...	$7.852 \pm 3.6 \times 10^{-2}$	P_3
		$11.435 \pm 0 \times 10^{0}$	$11.551 \pm 2.3 \times 10^{-2}$	$\sim 4P_1$
		...	$22.998 \pm 0 \times 10^{0}$	$\sim 8P_1$
No. 2	7–7.70–10	$0.380 \pm 1.5 \times 10^{-2}$	...	P_3
		...	$0.522 \pm 4.4 \times 10^{-2}$	$\sim P_1/5$
		$0.644 \pm 1.18 \times 10^{-1}$	$0.681 \pm 1.0 \times 10^{-1}$	$\sim P_1/4$
		$0.999 \pm 4.0 \times 10^{-2}$	$0.951 \pm 3.5 \times 10^{-2}$	$\sim P_1/3$
		$1.400 \pm 3.7 \times 10^{-2}$	$1.419 \pm 9.6 \times 10^{-2}$	$\sim P_1/2$
		$2.380 \pm 1.20 \times 10^{-1}$	...	P_2
		...	$2.867 \pm 2.1 \times 10^{-2}$	P_1
		...	$4.464 \pm 8.5 \times 10^{-2}$	P_4
		$5.712 \pm 0 \times 10^{0}$	...	$\sim 2P_1$
		...	$8.094 \pm 1.11 \times 10^{-1}$	P_5
		$11.426 \pm 0 \times 10^{0}$	$11.572 \pm 1.5 \times 10^{-2}$	$\sim 4P_1$
		...	$23.190 \pm 0 \times 10^{0}$	$\sim 8P_1$
No. 3	3e–2.12–4.42–8.96	$0.554 \pm 4.4 \times 10^{-2}$	$0.541 \pm 6.7 \times 10^{-2}$	$\sim P_1/5$
		$0.754 \pm 2.18 \times 10^{-1}$	$0.731 \pm 1.86 \times 10^{-1}$	$\sim P_1/4$
		$0.950 \pm 7.4 \times 10^{-2}$	$0.937 \pm 6.6 \times 10^{-2}$	$\sim P_1/3$
		$1.416 \pm 2.71 \times 10^{-1}$	$1.464 \pm 8.2 \times 10^{-2}$	$\sim P_1/2$
		...	$2.158 \pm 1.91 \times 10^{-1}$	P_3
		$2.372 \pm 7.0 \times 10^{-2}$	...	P_2
		...	$2.825 \pm 8.6 \times 10^{-2}$	P_1
		$3.730 \pm 1.83 \times 10^{-1}$	...	P_4
		...	$6.362 \pm 5.21 \times 10^{-1}$	$\sim 3P_3$
		...	$7.360 \pm 7.74 \times 10^{-1}$	$\sim 2P_4$
		$11.435 \pm 0 \times 10^{0}$	$11.513 \pm 4.5 \times 10^{-2}$	$\sim 4P_1$
		...	$23.190 \pm 0 \times 10^{0}$	$\sim 8P_1$

Table A2. *Cont.*

Model	Fixed Period Values [Days]	Periods [Days] First Part	Periods [Days] Second Part	Comment
No. 4	7–2.03–4.86–7.70	$0.557 \pm 3.3 \times 10^{-2}$	$0.556 \pm 1.03 \times 10^{-1}$	$\sim P_1/5$
		$0.723 \pm 1.83 \times 10^{-1}$	$0.696 \pm 1.0 \times 10^{-1}$	$\sim P_1/4$
		$1.003 \pm 1.99 \times 10^{-1}$	$0.940 \pm 7.6 \times 10^{-2}$	$\sim P_1/3$
		$1.422 \pm 5.6 \times 10^{-2}$	$1.440 \pm 1.08 \times 10^{-1}$	$\sim P_1/2$
		$2.027 \pm 1.73 \times 10^{-1}$	...	P_3
		...	$2.257 \pm 4.85 \times 10^{-1}$	P_2
		...	$2.892 \pm 2.61 \times 10^{-1}$	P_1
		$4.038 \pm 5.30 \times 10^{-1}$	...	$\sim 2P_3$
		$5.709 \pm 8 \times 10^{-3}$	$5.590 \pm 2.16 \times 10^{-1}$	$\sim 2P_1$
		$11.419 \pm 0 \times 10^{0}$	$11.470 \pm 5.8 \times 10^{-2}$	$\sim 4P_1$
No. 5	3e–8.96–10	$0.582 \pm 6.1 \times 10^{-2}$	$0.520 \pm 4.5 \times 10^{-2}$	$\sim P_1/5$
		$0.751 \pm 2.04 \times 10^{-1}$	$0.697 \pm 4.3 \times 10^{-2}$	$\sim P_1/4$
		$0.989 \pm 5.3 \times 10^{-2}$	$0.989 \pm 1.64 \times 10^{-1}$	$\sim P_1/3$
		$1.370 \pm 8 \times 10^{-2}$	$1.441 \pm 5.6 \times 10^{-2}$	$\sim P_1/2$
		$2.761 \pm 6 \times 10^{-2}$	$2.812 \pm 4.8 \times 10^{-2}$	P_1
		$5.680 \pm 3.8 \times 10^{-2}$	...	$\sim 2P_1$
		...	$7.865 \pm 4.6 \times 10^{-2}$	P_2
		$11.419 \pm 0 \times 10^{0}$	$11.532 \pm 3.3 \times 10^{-1}$	$\sim 4P_1$
No. 6	7–7.70–10	$0.577 \pm 6.1 \times 10^{-2}$	$0.565 \pm 1.08 \times 10^{-1}$	$\sim P_1/5$
		$0.722 \pm 1.34 \times 10^{-1}$	$0.688 \pm 6.1 \times 10^{-2}$	$\sim P_1/4$
		$0.972 \pm 1.12 \times 10^{-1}$	$0.941 \pm 2.5 \times 10^{-2}$	$\sim P_1/3$
		$1.405 \pm 3.7 \times 10^{-2}$	$1.436 \pm 2 \times 10^{-2}$	$\sim P_1/2$
		$2.780 \pm 5.2 \times 10^{-2}$	$2.823 \pm 5.1 \times 10^{-2}$	P_1
No. 7	3e–2.12–4.42–8.96	$0.571 \pm 5.9 \times 10^{-2}$	$0.547 \pm 8.3 \times 10^{-2}$	$\sim P_1/5$
		$0.729 \pm 1.96 \times 10^{-1}$	$0.692 \pm 6.9 \times 10^{-2}$	$\sim P_1/4$
		$0.952 \pm 4.3 \times 10^{-2}$	$0.938 \pm 3.1 \times 10^{-2}$	$\sim P_1/3$
		$1.397 \pm 4.8 \times 10^{-2}$	$1.438 \pm 2.9 \times 10^{-2}$	$\sim P_1/2$
		$2.767 \pm 6.6 \times 10^{-2}$	$2.794 \pm 9.2 \times 10^{-2}$	P_1
No. 8	7–2.03–4.86–7.70	$0.551 \pm 2.6 \times 10^{-2}$	$0.546 \pm 8.9 \times 10^{-2}$	$\sim P_1/5$
		$0.706 \pm 1.64 \times 10^{-1}$	$0.689 \pm 5.9 \times 10^{-2}$	$\sim P_1/4$
		$0.960 \pm 9.0 \times 10^{-2}$	$0.939 \pm 3.3 \times 10^{-2}$	$\sim P_1/3$
		$1.407 \pm 3.5 \times 10^{-2}$	$1.437 \pm 2.7 \times 10^{-2}$	$\sim P_1/2$
		$2.825 \pm 3.0 \times 10^{-2}$	$2.823 \pm 8.4 \times 10^{-2}$	P_1
No. 9	$\sqrt{2}$–1.27–1.6	$0.524 \pm 4.1 \times 10^{-2}$	$0.554 \pm 1.22 \times 10^{-1}$	$\sim P_1/5$
		$0.735 \pm 1.52 \times 10^{-1}$	$0.681 \pm 5.7 \times 10^{-2}$	$\sim P_1/4$
		$1.346 \pm 5.8 \times 10^{-2}$	$1.410 \pm 1.19 \times 10^{-1}$	$\sim P_1/2$
		$2.831 \pm 5.94 \times 10^{-1}$	$2.773 \pm 1.25 \times 10^{-1}$	P_1
No. 10	1.2–1.32–1.6	$0.535 \pm 3.1 \times 10^{-2}$	$0.500 \pm 3.0 \times 10^{-2}$	$\sim P_1/5$
		$0.766 \pm 1.75 \times 10^{-1}$	$0.671 \pm 4.6 \times 10^{-2}$	$\sim P_1/4$
		$1.246 \pm 3.2 \times 10^{-2}$	$1.219 \pm 5.9 \times 10^{-2}$	P_3
		...	$1.702 \pm 3.8 \times 10^{-2}$	P_2
		$2.745 \pm 1.84 \times 10^{-1}$	$2.781 \pm 1.00 \times 10^{-1}$	P_1
No. 11	$\sqrt{2}$–0.94–2.78–1.27	$0.512 \pm 2.3 \times 10^{-2}$	$0.505 \pm 2.6 \times 10^{-2}$	$\sim P_1/5$
		$0.651 \pm 1.6 \times 10^{-2}$	...	$\sim P_1/4$
		$0.959 \pm 4.7 \times 10^{-2}$	$0.949 \pm 3.9 \times 10^{-2}$	$\sim P_1/3$
		$1.380 \pm 4.6 \times 10^{-2}$	$1.406 \pm 8.1 \times 10^{-2}$	$\sim P_1/2$
		$2.789 \pm 8.1 \times 10^{-2}$	$2.763 \pm 1.37 \times 10^{-1}$	P_1

Table A2. *Cont.*

Model	Fixed Period Values [Days]	Periods [Days] First Part	Periods [Days] Second Part	Comment
No. 12	1.2–0.84–2.06–1.32	$0.380 \pm 1.6 \times 10^{-2}$	...	P_4
		$0.524 \pm 1.5 \times 10^{-2}$	$0.502 \pm 3.8 \times 10^{-2}$	$\sim P_1/5$
		$0.953 \pm 7.2 \times 10^{-2}$	$0.900 \pm 6.8 \times 10^{-2}$	$\sim P_1/3$
		$1.211 \pm 5.3 \times 10^{-2}$	$1.249 \pm 9.7 \times 10^{-2}$	P_3
		$2.099 \pm 2.2 \times 10^{-1}$	$2.136 \pm 1.5 \times 10^{-1}$	P_2
		$2.751 \pm 1.9 \times 10^{-1}$	$2.797 \pm 1.33 \times 10^{-1}$	P_1
No. 13	$\sqrt{2}$–1.27–1.6	$0.546 \pm 2.8 \times 10^{-2}$	$0.512 \pm 3.3 \times 10^{-2}$	$\sim P_1/5$
		$0.696 \pm 3.9 \times 10^{-2}$	$0.687 \pm 4.7 \times 10^{-2}$	$\sim P_1/4$
		$0.952 \pm 6.0 \times 10^{-3}$	$0.949 \pm 6.0 \times 10^{-3}$	$\sim P_1/3$
		$1.397 \pm 6.5 \times 10^{-2}$	$1.409 \pm 3.7 \times 10^{-2}$	$\sim P_1/2$
		$2.831 \pm 2.8 \times 10^{-2}$	$2.805 \pm 7.9 \times 10^{-2}$	P_1
No. 14	1.2–1.32–1.6	$0.547 \pm 2.5 \times 10^{-2}$	$0.514 \pm 3.7 \times 10^{-2}$	$\sim P_1/5$
		$0.704 \pm 5.0 \times 10^{-3}$	$0.689 \pm 5.3 \times 10^{-2}$	$\sim P_1/4$
		$0.976 \pm 5.1 \times 10^{-2}$	$0.971 \pm 3.8 \times 10^{-2}$	$\sim P_1/3$
		$1.385 \pm 8.0 \times 10^{-3}$	$1.404 \pm 5.3 \times 10^{-2}$	$\sim P_1/2$
		$2.811 \pm 4.4 \times 10^{-2}$	$2.784 \pm 1.19 \times 10^{-1}$	P_1
No. 15	$\sqrt{2}$–0.94–2.78–1.27	$0.551 \pm 2.9 \times 10^{-2}$	$0.520 \pm 4.3 \times 10^{-2}$	$\sim P_1/5$
		$0.738 \pm 1.91 \times 10^{-1}$	$0.706 \pm 1.52 \times 10^{-1}$	$\sim P_1/4$
		$0.974 \pm 1.72 \times 10^{-1}$	$0.964 \pm 1.15 \times 10^{-1}$	$\sim P_1/3$
		$1.408 \pm 3.0 \times 10^{-2}$	$1.438 \pm 1.52 \times 10^{-1}$	$\sim P_1/2$
		$2.809 \pm 4.1 \times 10^{-2}$	$2.796 \pm 9.7 \times 10^{-2}$	P_1
No. 16	1.2–0.84–2.06–1.32	$0.544 \pm 2.4 \times 10^{-2}$	$0.527 \pm 6.5 \times 10^{-2}$	$\sim P_1/5$
		$0.686 \pm 1.26 \times 10^{-1}$	$0.788 \pm 2.25 \times 10^{-1}$	$\sim P_1/4$
		$1.028 \pm 2.33 \times 10^{-1}$	$0.946 \pm 7.6 \times 10^{-2}$	$\sim P_1/3$
		$1.436 \pm 2.1 \times 10^{-1}$	$1.429 \pm 7.2 \times 10^{-2}$	$\sim P_1/2$
		$2.803 \pm 7.0 \times 10^{-2}$	$2.791 \pm 1.19 \times 10^{-1}$	P_1
No. 17	3e–10	$0.558 \pm 5.6 \times 10^{-2}$	$0.535 \pm 1.08 \times 10^{-1}$	$\sim P_1/5$
		$0.792 \pm 1.88 \times 10^{-1}$	$0.704 \pm 1.13 \times 10^{-1}$	$\sim P_1/4$
		$1.150 \pm 2.3 \times 10^{-2}$	$1.131 \pm 1.84 \times 10^{-1}$	P_3
		...	$1.472 \pm 7.6 \times 10^{-2}$	$\sim P_1/2$
		$2.506 \pm 5.8 \times 10^{-2}$	...	P_2
		...	$2.809 \pm 6.7 \times 10^{-2}$	P_1
		$5.712 \pm 0 \times 10^{0}$	...	$\sim 2P_1$
		...	$7.928 \pm 6.1 \times 10^{-2}$	P_4
		$11.435 \pm 0 \times 10^{0}$	$11.553 \pm 2.4 \times 10^{-2}$	$\sim 4P_1$
		...	$22.998 \pm 0 \times 10^{0}$	$\sim 8P_1$
No. 18	7–10	$0.388 \pm 2.7 \times 10^{-2}$	...	P_3
		...	$0.498 \pm 3.3 \times 10^{-2}$	$\sim P_1/5$
		$0.634 \pm 9.7 \times 10^{-2}$	$0.674 \pm 5.6 \times 10^{-2}$	$\sim P_1/4$
		$0.983 \pm 2.1 \times 10^{-2}$	$0.954 \pm 7.2 \times 10^{-2}$	$\sim P_1/3$
		$1.401 \pm 3.3 \times 10^{-2}$	$1.411 \pm 4.9 \times 10^{-2}$	$\sim P_1/2$
		$2.361 \pm 1.31 \times 10^{-1}$	...	P_2
		...	$2.983 \pm 6.1 \times 10^{-2}$	P_1
		$5.697 \pm 0 \times 10^{0}$	$5.065 \pm 5.8 \times 10^{-2}$	$\sim 2P_1$
		$11.413 \pm 0 \times 10^{0}$	$11.590 \pm 0 \times 10^{0}$	$\sim 4P_1$
		...	$23.192 \pm 0 \times 10^{0}$	$\sim 8P_1$
No. 19	3e–2.12–4.42	$0.551 \pm 1.17 \times 10^{-1}$	$0.514 \pm 2.70 \times 10^{-2}$	$\sim P_1/5$
		$0.722 \pm 1.15 \times 10^{-1}$	$0.696 \pm 9.80 \times 10^{-2}$	$\sim P_1/4$
		$0.931 \pm 4.60 \times 10^{-2}$	$0.937 \pm 6.40 \times 10^{-2}$	$\sim P_1/3$
		$1.390 \pm 7.80 \times 10^{-2}$	$1.460 \pm 7.00 \times 10^{-2}$	$\sim P_1/2$

Table A2. *Cont.*

Model	Fixed Period Values [Days]	Periods [Days] First Part	Periods [Days] Second Part	Comment
		...	$2.151 \pm 8.90 \times 10^{-2}$	P_2
		$2.710 \pm 2.91 \times 10^{-1}$	$2.840 \pm 4.70 \times 10^{-1}$	P_1
		$3.772 \pm 8.60 \times 10^{-2}$	...	P_3
		$5.703 \pm 3.70 \times 10^{-2}$	$6.108 \pm 2.04 \times 10^{1}$	$\sim 2P_1$
		$11.442 \pm 0 \times 10^{0}$	$11.418 \pm 1.40 \times 10^{-2}$	$\sim 4P_1$
No. 20	7–2.03–4.86	$0.570 \pm 5.3 \times 10^{-2}$	$0.506 \pm 3.1 \times 10^{-2}$	$\sim P_1/3$
		$0.707 \pm 1.67 \times 10^{-1}$	$0.679 \pm 1.04 \times 10^{-1}$	$\sim P_1/2$
		$1.020 \pm 2.10 \times 10^{-1}$	$0.960 \pm 1.0 \times 10^{-2}$	$\sim 2P_1/3$
		$1.422 \pm 4.9 \times 10^{-2}$	$1.455 \pm 7.5 \times 10^{-2}$	P_1
		$2.056 \pm 1.49 \times 10^{-1}$	$2.149 \pm 1.29 \times 10^{-1}$	P_2
		$3.772 \pm 2.11 \times 10^{-1}$	$3.351 \pm 1.78 \times 10^{-1}$	P_3
		$5.702 \pm 1.4 \times 10^{-2}$	$5.050 \pm 6.25 \times 10^{-1}$	$\sim 4P_1$
		$11.442 \pm 0.0 \times 10^{-1}$	$11.427 \pm 1.6 \times 10^{-2}$	$\sim 8P_1$
No. 21	3e–10	$0.572 \pm 5.7 \times 10^{-2}$	$0.512 \pm 2.0 \times 10^{-2}$	$\sim P_1/5$
		$0.745 \pm 1.87 \times 10^{-1}$	$0.680 \pm 1.14 \times 10^{-1}$	$\sim P_1/4$
		$1.033 \pm 1.95 \times 10^{-1}$	$1.083 \pm 3.44 \times 10^{-1}$	$\sim P_1/3$
		$1.382 \pm 6.3 \times 10^{-2}$	$1.438 \pm 4.0 \times 10^{-2}$	$\sim P_1/2$
		$2.789 \pm 4.7 \times 10^{-2}$	$2.814 \pm 5.9 \times 10^{-2}$	P_1
		...	$7.924 \pm 6.2 \times 10^{-2}$	P_2
No. 22	7–10	$0.524 \pm 8 \times 10^{-3}$	$0.518 \pm 2.6 \times 10^{-2}$	$\sim P_1/5$
		$0.736 \pm 2.23 \times 10^{-1}$	$0.682 \pm 1.03 \times 10^{-1}$	$\sim P_1/4$
		$0.987 \pm 1.75 \times 10^{-1}$	$0.955 \pm 1.37 \times 10^{-1}$	$\sim P_1/3$
		$1.405 \pm 3.4 \times 10^{-2}$	$1.434 \pm 2.3 \times 10^{-2}$	$\sim P_1/2$
		$2.775 \pm 5.2 \times 10^{-2}$	$2.845 \pm 4.5 \times 10^{-2}$	P_1
		...	$4.772 \pm 1.23 \times 10^{-1}$	P_2
No. 23	3e–2.12–4.42	$0.542 \pm 2.3 \times 10^{-2}$	$0.520 \pm 4.2 \times 10^{-2}$	$\sim P_1/5$
		$0.707 \pm 1.09 \times 10^{-1}$	$0.690 \pm 7.4 \times 10^{-2}$	$\sim P_1/4$
		$0.954 \pm 2.3 \times 10^{-2}$	$0.947 \pm 1.7 \times 10^{-2}$	$\sim P_1/3$
		$1.412 \pm 1.75 \times 10^{-1}$	$1.437 \pm 2.7 \times 10^{-2}$	$\sim P_1/2$
		$2.791 \pm 5.3 \times 10^{-2}$	$2.790 \pm 1.16 \times 10^{-1}$	P_1
No. 24	7–2.03–4.86	$0.557 \pm 3.5 \times 10^{-2}$	$0.520 \pm 4.2 \times 10^{-2}$	$\sim P_1/5$
		$0.726 \pm 2.05 \times 10^{-1}$	$0.689 \pm 6.7 \times 10^{-2}$	$\sim P_1/4$
		$0.962 \pm 6.4 \times 10^{-2}$	$0.947 \pm 1.15 \times 10^{-1}$	$\sim P_1/3$
		$1.413 \pm 2.3 \times 10^{-2}$	$1.435 \pm 2.5 \times 10^{-2}$	$\sim P_1/2$
		$2.823 \pm 3.1 \times 10^{-2}$	$2.835 \pm 8.9 \times 10^{-2}$	P_1
No. 25	$\sqrt{2}$–1.6	$0.519 \pm 3.3 \times 10^{-2}$	$0.526 \pm 7.5 \times 10^{-2}$	$\sim P_1/5$
		...	$0.670 \pm 5.8 \times 10^{-2}$	$\sim P_1/4$
		$0.836 \pm 2.61 \times 10^{-1}$	...	P_3
		$0.978 \pm 3.4 \times 10^{-2}$	...	$\sim P_1/3$
		$1.380 \pm 3.8 \times 10^{-2}$	$1.411 \pm 4.5 \times 10^{-2}$	$\sim P_1/2$
		$2.061 \pm 2.84 \times 10^{-1}$	...	P_2
		$2.808 \pm 5.14 \times 10^{-1}$	$2.840 \pm 5.85 \times 10^{-1}$	P_1
No. 26	1.2–1.6	$0.522 \pm 3.6 \times 10^{-2}$	$0.507 \pm 2.7 \times 10^{-2}$	$\sim P_1/5$
		$0.742 \pm 1.79 \times 10^{-1}$	$0.751 \pm 2.59 \times 10^{-1}$	$\sim P_1/4$
		$1.182 \pm 1.04 \times 10^{-1}$	...	P_3
		...	$1.347 \pm 4.32 \times 10^{-1}$	$\sim P_1/2$
		$1.894 \pm 7.0 \times 10^{-1}$	$1.637 \pm 7.8 \times 10^{-2}$	P_2
		$2.724 \pm 2.19 \times 10^{-1}$	$2.748 \pm 1.57 \times 10^{-1}$	P_1

Table A2. *Cont.*

Model	Fixed Period Values [Days]	Periods [Days] First Part	Periods [Days] Second Part	Comment
No. 27	$\sqrt{2}$–0.94–2.78	$0.555 \pm 1.09 \times 10^{-1}$	$0.533 \pm 9.8 \times 10^{-2}$	$\sim P_1/5$
		$1.045 \pm 2.74 \times 10^{-1}$	$0.967 \pm 1.4 \times 10^{-1}$	$\sim P_1/3$
		$1.407 \pm 6.1 \times 10^{-2}$	$1.417 \pm 6.6 \times 10^{-2}$	$\sim P_1/2$
		$2.793 \pm 5.3 \times 10^{-2}$	$2.779 \pm 1.11 \times 10^{-1}$	P_1
No. 28	1.2–0.84–2.06	$0.56 \pm 9.4 \times 10^{-2}$	$0.519 \pm 8.7 \times 10^{-2}$	$\sim P_1/5$
		...	$0.903 \pm 8 \times 10^{-2}$	$\sim P_1/3$
		$1.141 \pm 3.67 \times 10^{-1}$	...	P_4
		...	$1.553 \pm 5.44 \times 10^{-1}$	$\sim P_1/2$
		$1.74 \pm 6.92 \times 10^{-1}$	...	P_3
		...	$2.237 \pm 4.93 \times 10^{-1}$	P_2
		$2.629 \pm 4.37 \times 10^{-1}$	$2.813 \pm 1.16 \times 10^{-1}$	P_1
No. 29	$\sqrt{2}$–1.6	$0.542 \pm 2.4 \times 10^{-2}$	$0.535 \pm 7.2 \times 10^{-2}$	$\sim P_1/5$
		$0.742 \pm 1.95 \times 10^{-1}$	$0.685 \pm 6.6 \times 10^{-2}$	$\sim P_1/4$
		$0.989 \pm 1.87 \times 10^{-1}$	$0.971 \pm 1.58 \times 10^{-1}$	$\sim P_1/3$
		$1.466 \pm 2.65 \times 10^{-1}$	$1.425 \pm 2.5 \times 10^{-2}$	$\sim P_1/2$
		$2.799 \pm 4.2 \times 10^{-2}$	$2.796 \pm 9.5 \times 10^{-2}$	P_1
No. 30	1.2–1.6	$0.542 \pm 2.4 \times 10^{-2}$	$0.526 \pm 5.9 \times 10^{-2}$	$\sim P_1/5$
		$0.714 \pm 1.64 \times 10^{-1}$	$0.715 \pm 3.6 \times 10^{-2}$	$\sim P_1/4$
		$0.969 \pm 1.44 \times 10^{-1}$	$0.959 \pm 1.17 \times 10^{-1}$	$\sim P_1/3$
		$1.409 \pm 1.70 \times 10^{-1}$	$1.403 \pm 4.2 \times 10^{-2}$	$\sim P_1/2$
		$2.804 \pm 4.2 \times 10^{-2}$	$2.791 \pm 1.07 \times 10^{-1}$	P_1
No. 31	$\sqrt{2}$–0.94–2.78	$0.545 \pm 2.2 \times 10^{-2}$	$0.526 \pm 5.9 \times 10^{-2}$	$\sim P_1/5$
		$0.723 \pm 1.67 \times 10^{-1}$	$0.705 \pm 1.18 \times 10^{-1}$	$\sim P_1/4$
		$0.984 \pm 1.64 \times 10^{-1}$	$0.962 \pm 1.28 \times 10^{-1}$	$\sim P_1/3$
		$1.405 \pm 4.0 \times 10^{-2}$	$1.436 \pm 4.3 \times 10^{-2}$	$\sim P_1/2$
		$2.810 \pm 3.7 \times 10^{-2}$	$2.801 \pm 8.5 \times 10^{-2}$	P_1
No. 32	1.2–0.84–2.06	$0.543 \pm 2.3 \times 10^{-2}$	$0.525 \pm 6.2 \times 10^{-2}$	$\sim P_1/5$
		$0.742 \pm 2.2 \times 10^{-1}$	$0.735 \pm 1.31 \times 10^{-1}$	$\sim P_1/4$
		$0.967 \pm 4.1 \times 10^{-2}$	$0.941 \pm 4.1 \times 10^{-2}$	$\sim P_1/3$
		$1.488 \pm 3.29 \times 10^{-1}$	$1.435 \pm 7.1 \times 10^{-2}$	$\sim P_1/2$
		$2.798 \pm 6.5 \times 10^{-2}$	$2.791 \pm 1.25 \times 10^{-1}$	P_1

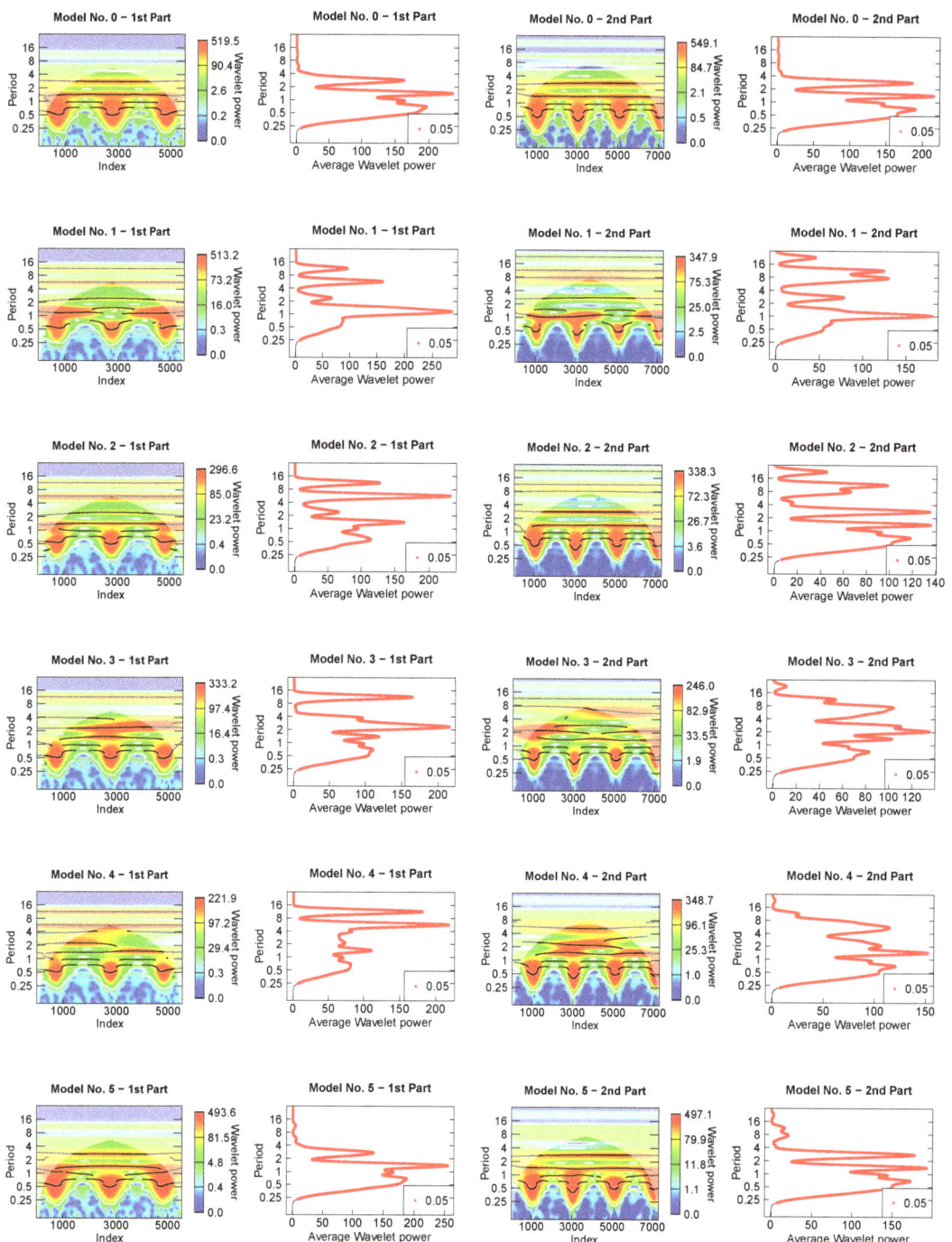

Figure A3. Results of the wavelet analysis for models No. 0 to No. 5. From left to right the wavelet power spectrum and the average wavelet power for the first and second part of each synthetic curve.

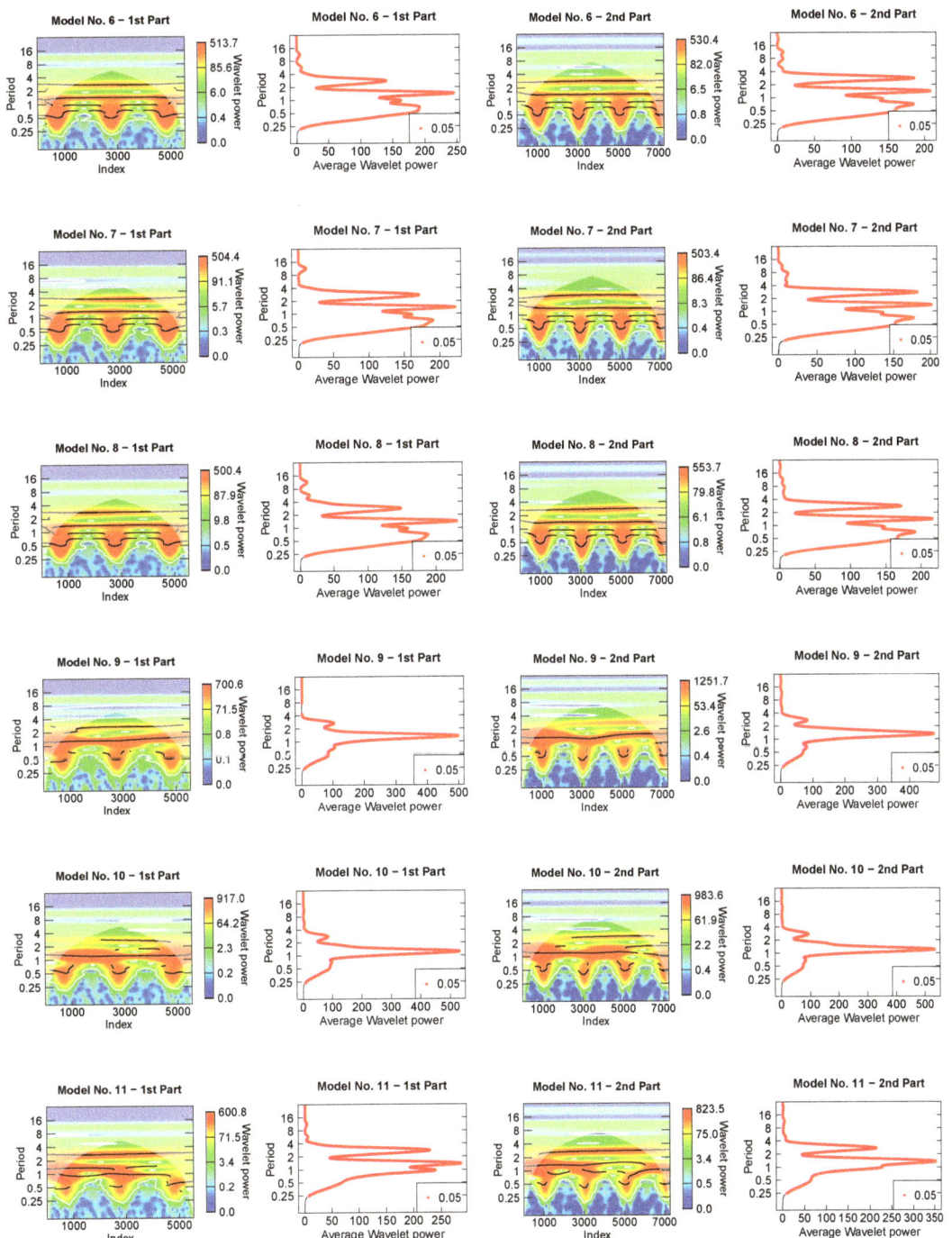

Figure A4. Results of the wavelet analysis for models No. 6 to No. 11. From left to right the wavelet power spectrum and the average wavelet power for the first and second part of each synthetic curve.

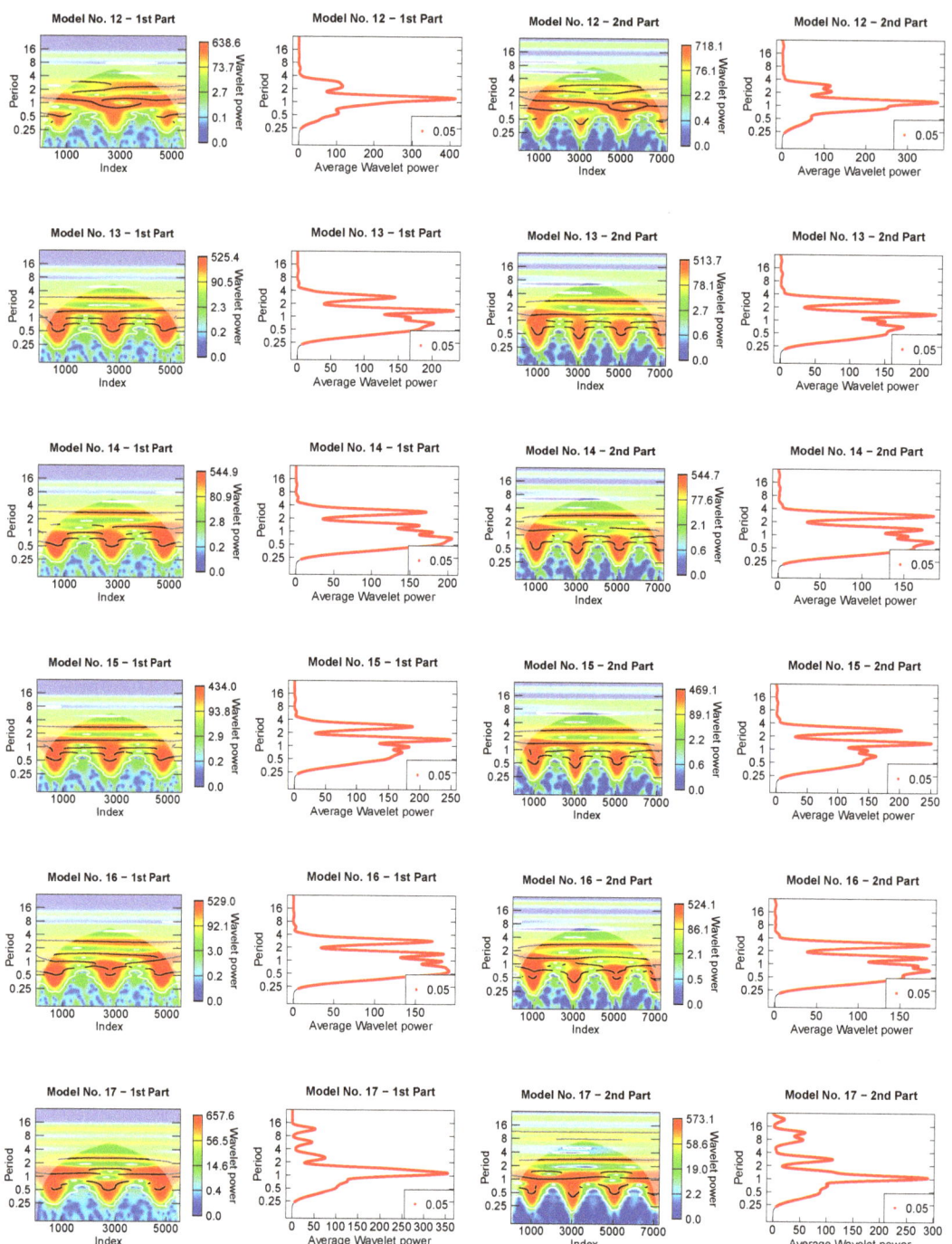

Figure A5. Results of the wavelet analysis for models No. 12 to No. 17. From left to right the wavelet power spectrum and the average wavelet power for the first and second part of each synthetic curve.

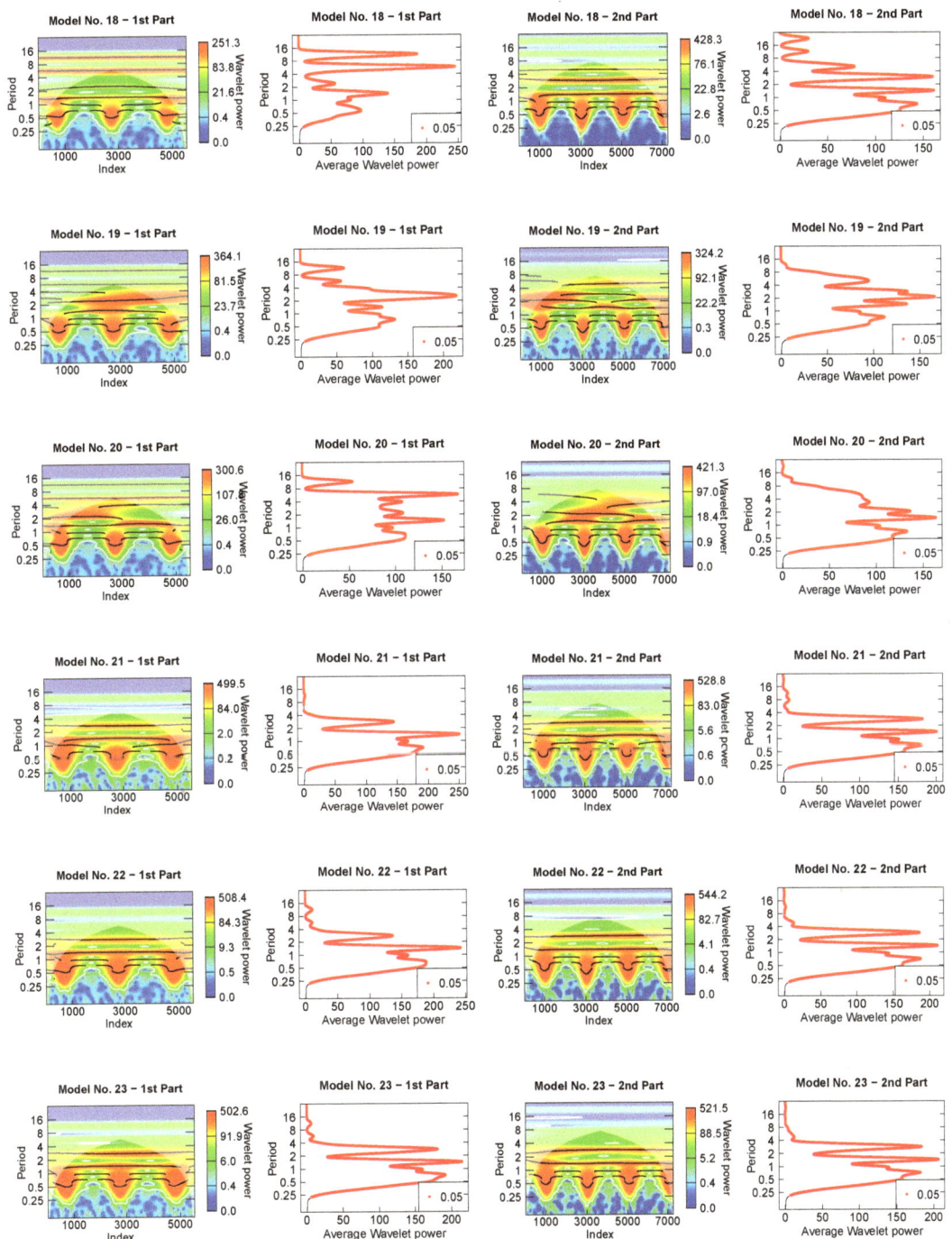

Figure A6. Results of the wavelet analysis for models No. 18 to No. 23. From left to right the wavelet power spectrum and the average wavelet power for the first and second part of each synthetic curve.

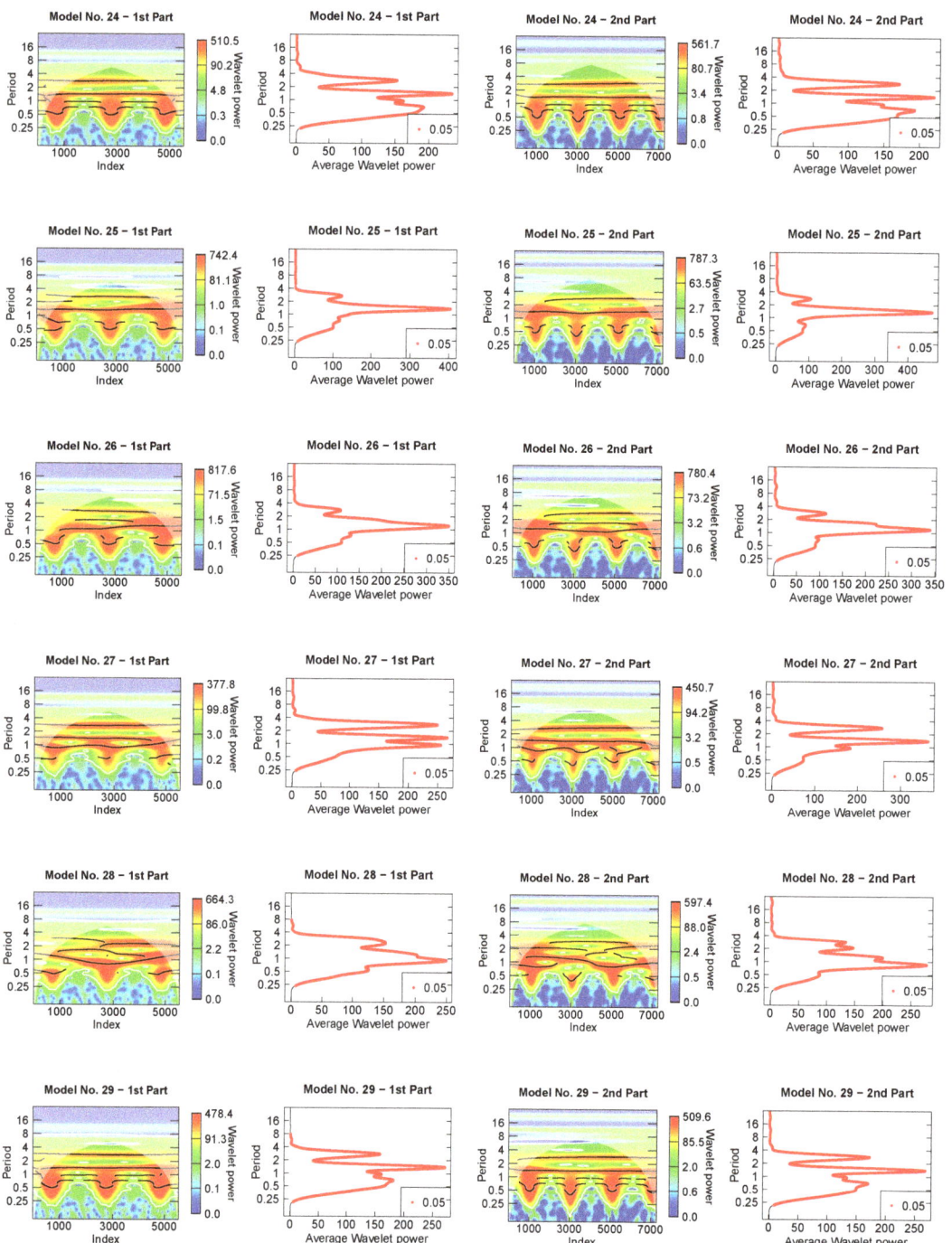

Figure A7. Results of the wavelet analysis for models No. 24 to No. 29. From left to right the wavelet power spectrum and the average wavelet power for the first and second part of each synthetic curve.

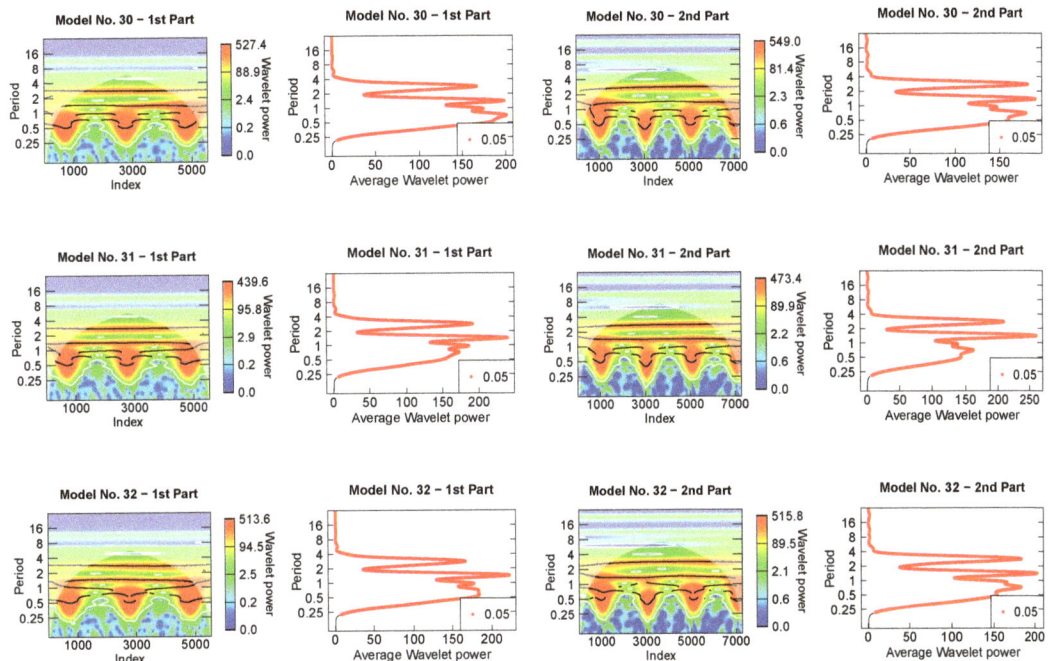

Figure A8. Results of the wavelet analysis for models No. 30 to No. 32. From left to right the wavelet power spectrum and the average wavelet power for the first and second part of each synthetic curve.

Note

1   The search for the most important factors that may be varied in a simulated system.

References

1. Sana, H.; de Mink, S.E.; de Koter, A.; Langer, N.; Evans, C.J.; Gieles, M.; Gosset, E.; Izzard, R.G.; Le Bouquin, J.B.; Schneider, F.R.N. Binary Interaction Dominates the Evolution of Massive Stars. *Science* **2012**, *337*, 444. [CrossRef]
2. Mason, B.D.; Hartkopf, W.I.; Henry, T.J.; Jao, W.C.; Subasavage, J.; Riedel, A.; Winters, J. Nearby Dwarf Stars: Duplicity, Binarity, and Masses. *Noao Propos.* **2009**, *2009B-0044*, 44.
3. Sana, H.; Le Bouquin, J.B.; Lacour, S.; Berger, J.P.; Duvert, G.; Gauchet, L.; Norris, B.; Olofsson, J.; Pickel, D.; Zins, G.; et al. Southern Massive Stars at High Angular Resolution: Observational Campaign and Companion Detection. *Astrophys. J. Suppl. Ser.* **2014**, *215*, 15. [CrossRef]
4. Dunstall, P.R.; Dufton, P.L.; Sana, H.; Evans, C.J.; Howarth, I.D.; Simón-Díaz, S.; de Mink, S.E.; Langer, N.; Maíz Apellániz, J.; Taylor, W.D. The VLT-FLAMES Tarantula Survey. XXII. Multiplicity properties of the B-type stars. *Astron. Astrophys.* **2015**, *580*, A93. [CrossRef]
5. Villaseñor, J.I.; Taylor, W.D.; Evans, C.J.; Ramírez-Agudelo, O.H.; Sana, H.; Almeida, L.A.; de Mink, S.E.; Dufton, P.L.; Langer, N. The B-type binaries characterization programme I. Orbital solutions for the 30 Doradus population. *Mon. Not. R. Astron. Soc.* **2021**, *507*, 5348–5375. [CrossRef]
6. Barbá, R.H.; Gamen, R.; Arias, J.I.; Morrell, N.I. OWN Survey: A spectroscopic monitoring of Southern Galactic O and WN-type stars. In *Proceedings of the The Lives and Death-Throes of Massive Stars*; Eldridge, J.J., Bray, J.C., McClelland, L.A.S., Xiao, L., Eds.; Cambridge University Press: Cambridge, UK, 2017; Volume 329, pp. 89–96. [CrossRef]
7. Godart, M.; Grotsch-Noels, A.; Dupret, M.A. Pulsations in hot supergiants. In *Proceedings of the Precision Asteroseismology*; Guzik, J.A., Chaplin, W.J., Handler, G., Pigulski, A., Eds.; Cambridge University Press: Cambridge, UK, 2014; Volume 301, pp. 313–320. [CrossRef]
8. Saio, H.; Kuschnig, R.; Gautschy, A.; Cameron, C.; Walker, G.A.H.; Matthews, J.M.; Guenther, D.B.; Moffat, A.F.J.; Rucinski, S.M.; Sasselov, D.; et al. MOST Detects g- and p-Modes in the B Supergiant HD 163899 (B2 Ib/II). *Astrophys. J.* **2006**, *650*, 1111–1118. [CrossRef]

9. Mennickent, R.E.; Kołaczkowski, Z.; Soszyński, I.; Cabezas, M.; Garrido, H.E. The SMC B-type supergiant AzV322: A g-mode pulsator with a circumstellar disc. *Mon. Not. R. Astron. Soc.* **2018**, *473*, 4055–4063. [CrossRef]
10. Glatzel, W. Nonlinear strange-mode pulsations. *Commun. Asteroseismol.* **2009**, *158*, 252.
11. Godart, M.; Dupret, M.A.; Noels, A.; Aerts, C.; Simón-Díaz, S.; Lefever, K.; Puls, J.; Montalban, J.; Ventura, P. Pulsations in massive stars: Effect of the atmosphere on the strange mode pulsations. In *Proceedings of the Active OB Stars: Structure, Evolution, Mass Loss, and Critical Limits*; Neiner, C., Wade, G., Meynet, G., Peters, G., Eds.; Cambridge University Press: Cambridge, UK, 2011; Volume 272, pp. 503–504. [CrossRef]
12. Kraus, M.; Haucke, M.; Cidale, L.S.; Venero, R.O.J.; Nickeler, D.H.; Németh, P.; Niemczura, E.; Tomić, S.; Aret, A.; Kubát, J.; et al. Interplay between pulsations and mass loss in the blue supergiant 55 Cygnus = HD 198 478. *Astron. Astrophys.* **2015**, *581*, A75. [CrossRef]
13. Yadav, A.P.; Glatzel, W. Stability analysis, non-linear pulsations and mass loss of models for 55 Cygni (HD 198478). *Mon. Not. R. Astron. Soc.* **2016**, *457*, 4330–4339. [CrossRef]
14. Sterken, C. Some thoughts on pulsation and binarity. *Astrophys. Space Sci.* **2006**, *304*, 139–143. [CrossRef]
15. Simón-Díaz, S.; Herrero, A.; Uytterhoeven, K.; Castro, N.; Aerts, C.; Puls, J. Observational Evidence for a Correlation Between Macroturbulent Broadening and Line-profile Variations in OB Supergiants. *Astrophys. J. Lett.* **2010**, *720*, L174–L178. [CrossRef]
16. Aerts, C.; Puls, J.; Godart, M.; Dupret, M.A. Collective pulsational velocity broadening due to gravity modes as a physical explanation for macroturbulence in hot massive stars. *Astron. Astrophys.* **2009**, *508*, 409–419. [CrossRef]
17. Shibahashi, H.; Kurtz, D.W. FM stars: A Fourier view of pulsating binary stars, a new technique for measuring radial velocities photometrically. *Mon. Not. R. Astron. Soc.* **2012**, *422*, 738–752. [CrossRef]
18. Murphy, S. Pulsating stars in binary systems: A review. *arXiv* **2018**. [CrossRef]
19. Lomb, N.R. Least-Squares Frequency Analysis of Unequally Spaced Data. *Astrophys. Space Sci.* **1976**, *39*, 447–462. [CrossRef]
20. Scargle, J.D. Studies in astronomical time series analysis. II. Statistical aspects of spectral analysis of unevenly spaced data. *Astrophys. J. Part 1* **1982**, *263*, 835–853. [CrossRef]
21. Hübner, M.; Huppenkothen, D.; Lasky, P.D.; Inglis, A.R. Pitfalls of Periodograms: The Nonstationarity Bias in the Analysis of Quasiperiodic Oscillations. *Astrophys. J. Suppl. Ser.* **2022**, *259*, 32. [CrossRef]
22. Foster, G. Wavelets for period analysis of unevenly sampled time series. *Astron. J.* **1996**, *112*, 1709–1729. [CrossRef]
23. López-Santiago, J. On the use of wavelets to reveal oscillatory patterns in stellar flare emission. *Philos. Trans. R. Soc. Lond. Ser. A* **2018**, *376*, 20170253. [CrossRef]
24. Zhou, A.Y.; Hintz, E.G.; Schoonmaker, J.N.; Rodríguez, E.; Costa, V.; Lopez-Gonzalez, M.J.; Smith, H.A.; Sanders, N.; Monninger, G.; Pagel, L. A Pulsational Time-evolution Study for the δ Scuti Star AN Lyncis. *arXiv* **2017**. [CrossRef]
25. Bravo, J.P.; Roque, S.; Estrela, R.; Leã o, I.C.; Medeiros, J.R.D. Wavelets: A powerful tool for studying rotation, activity, and pulsation in Kepler and CoRoT stellar light curves. *Astron. Astrophys.* **2014**, *568*, A34. [CrossRef]
26. Graham, M.J.; Drake, A.J.; Djorgovski, S.G.; Mahabal, A.A.; Donalek, C.; Duan, V.; Maker, A. A comparison of period finding algorithms. *Mon. Not. R. Astron. Soc.* **2013**, *434*, 3423–3444. [CrossRef]
27. Frescura, F.A.M.; Engelbrecht, C.A.; Frank, B.S. Significance of periodogram peaks and a pulsation mode analysis of the Beta Cephei star V403 Car. *Mon. Not. R. Astron. Soc. Mon. Not. R. Astron. Soc.* **2008**, *388*, 1693–1707. [CrossRef]
28. Votruba, V.; Koubský, P.; Korčáková, D.; Hroch, F. False periods in complex chaotic systems. *Astron. Astrophys.* **2009**, *496*, 217–222. [CrossRef]
29. Baluev, R.V. Distinguishing between a true period and its alias, and other tasks of model discrimination. *Mon. Not. R. Astron. Soc.* **2012**, *422*, 2372–2385. [CrossRef]
30. Misiti, M.; Misiti, Y.; Oppenheim, G.; Poggi, J.M. *Wavelets and Their Applications*; Iste: London, UK, 2007.
31. Stéphane, M. Discrete Revolution. In *A Wavelet Tour of Signal Processing*; Elsevier: Amsterdam, The Netherlands, 2009; pp. 59–88.
32. Daubechies, I. *Ten Lectures on Wavelets*; CBMS-NSF Regional Conference Series in Applied Mathematics, Society for Industrial and Applied Mathematics; Society for Industrial and Applied Mathematics: Philadelphia, PA, USA, 1992.
33. Montgomery, D.C. *Design and Analysis of Experiments*; John Wiley and Sons (WIE): Brisbane, Australia, 2000.
34. Box, G.E.P.; Hunter, J.S.; Hunter, W.G. *Statistics for Experimenters*, 2nd ed.; Wiley Series in Probability and Statistics; Wiley-Blackwell: Chichester, UK, 2005.
35. Ricker, G.R.; Winn, J.N.; Vanderspek, R.; Latham, D.W.; Bakos, G.Á.; Bean, J.L.; Berta-Thompson, Z.K.; Brown, T.M.; Buchhave, L.; Butler, N.R.; et al. *Transiting Exoplanet Survey Satellite (TESS)*; SPIE: Bellingham, WA, USA, 2014; Volume 9143, p. 914320. [CrossRef]
36. Kallrath, J.; Milone, E.F. TheWilson–Devinney Program: Extensions and Applications. In *Eclipsing Binary Stars: Modeling and Analysis*; Springer New York: New York, NY, USA, 2009; pp. 305–330. [CrossRef]
37. Alberici, A.A. Uso Del análisis Wavelet Para La Detección De Períodos En Curvas De Luz. Master's Thesis, Facultad de Ciencias Astronómicas y Geofísicas, Buenos Aires, Argentina, 2022.
38. Wecht, K. Determination of Mass Loss and Mass Transfer Rates of Algol (Beta Persei) from the Analysis of Absorption Lines in the UV Spectra Obtained by the IUE Satellite. Doctoral Dissertation, Cornell University, Bethlehem, PA, USA, 2006.
39. Feigelson, E.D.; Babu, G.J.; Caceres, G.A. Autoregressive times series methods for time domain astronomy. *Front. Phys.* **2018**, *6*, 80. [CrossRef]

40. Caceres, G.A.; Feigelson, E.D.; Babu, G.J.; Bahamonde, N.; Christen, A.; Bertin, K.; Meza, C.; Curé, M. Autoregressive Planet Search: Application to the Kepler Mission. *Astron. J.* **2019**, *158*, 58. [CrossRef]
41. Whittle, P. *Hypothesis Testing in Time Series Analysis*; Almqvist & Wiksell: Stockholm, Sweden, 1951.
42. Caceres, G.A.; Feigelson, E.D.; Babu, G.J.; Bahamonde, N.; Christen, A.; Bertin, K.; Meza, C.; Curé, M. Autoregressive Planet Search: Methodology. *Astron. J.* **2019**, *158*, 57. [CrossRef]
43. Shi, X.; Qian, S.; Li, L. New Pulsating Stars Detected in EA-type Eclipsing-binary Systems Based on TESS Data. *Astrophys. J.* **2022**, *259*, 50. [CrossRef]
44. McLaughlin, J.A.; Nakariakov, V.M.; Dominique, M.; Jelínek, P.; Takasao, S. Modelling Quasi-Periodic Pulsations in Solar and Stellar Flares. *Space Sci. Rev.* **2018**, *214*, 45. [CrossRef]
45. van Leeuwen, F.; van Genderen, A.M.; Zegelaar, I. HIPPARCOS photometry of 24 variable massive stars (alpha Cygni variables). *Astron. Astrophys. Suppl. Ser.* **1998**, *128*, 117–129. [CrossRef]
46. Saio, H.; Georgy, C.; Meynet, G. Strange-Mode Instability for Micro-Variations in Luminous Blue Variables. *arXiv* **2013**. [CrossRef]
47. Nieto, N.; Orozco, D.M. El uso de la transformada wavelet discreta en la reconstrucción de señales senosoidales. *Sci. Tech.* **2008**, *14*, 381–386.
48. Castro, L.; Castro, S. *Wavelets y Sus Aplicaciones*; Universidad Nacional de Colombia: Bogota, Colombia, 1995.
49. Lenz, P.; Breger, M. Period04 User Guide. *Commun. Asteroseismol.* **2005**, *146*, 53–136. [CrossRef]
50. Contreras, J.M.; Molina Portillo, E.; Arteaga Cezón, P. Introducción a La Programación Estadística Con R Para Profesores. 2010. Available online: https://1library.co/document/z14pk5dz-introducci%C3%B3n-programaci%C3%B3n-estad%C3%ADstica-profesores-contreras-garc%C3%ADa-portillo-arteaga.html (accessed on 1 May 2023).
51. Roesch, A.; Schmidbauer, H. WaveletComp: Computational Wavelet Analysis. R Package Version 1.1. 2018. Available online: https://cran.r-project.org/web/packages/WaveletComp/index.html (accessed on 1 May 2023).
52. Polanco-Martinez, J.M. Análisis espectral de wavelet: Una concisa revisión. *Rev. Climatol.* **2022**, *22*, 51–95.
53. Torrence, C.; Compo, G.P. A Practical Guide to Wavelet Analysis. *Bull. Am. Meteorol. Soc.* **1998**, *79*, 61–78. [CrossRef]
54. Meyers, S.D.; Kelly, B.G.; O'Brien, J.J. An Introduction to Wavelet Analysis in Oceanography and Meteorology: With Application to the Dispersion of Yanai Waves. *Mon. Weather Rev.* **1993**, *121*, 2858–2866. [CrossRef]
55. Deeming, T.J. Fourier Analysis with Unequally-Spaced Data. *Astrophys. Space Sci.* **1975**, *36*, 137–158. [CrossRef]
56. Grinsted, A.; Moore, J.C.; Jevrejeva, S. Application of the cross wavelet transform and wavelet coherence to geophysical time series. *Nonlinear Process. Geophys.* **2004**, *11*, 561–566. [CrossRef]
57. Baran, A.S.; Koen, C. A Detection Threshold in the Amplitude Spectra Calculated from TESS Time-Series Data. *Acta Astron.* **2021**, *71*, 113–121. [CrossRef]
58. Bognár, Z.; Lampens, P.; Frémat, Y.; Southworth, J.; Sódor, Á.; De Cat, P.; Isaacson, H.T.; Marcy, G.W.; Ciardi, D.R.; Gilliland, R.L.; et al. KIC3489: A genuine adusScuti Kepler hybrid pulsator with transit events. *Astron. Astrophys.* **2015**, *581*, A77. [CrossRef]
59. Distefano, E.; Lanzafame, A.C.; Lanza, A.F.; Messina, S.; Korn, A.J.; Eriksson, K.; Cuypers, J. Determination of rotation periods in solar-like stars with irregular sampling: The Gaia case. *Mon. Not. R. Astron. Soc.* **2012**, *421*, 2774–2785. [CrossRef]

Disclaimer/Publisher's Note: The statements, opinions and data contained in all publications are solely those of the individual author(s) and contributor(s) and not of MDPI and/or the editor(s). MDPI and/or the editor(s) disclaim responsibility for any injury to people or property resulting from any ideas, methods, instructions or products referred to in the content.

Article

A Mini Atlas of H-Band Spectra of Southern Symbiotic Stars

Paula Esther Marchiano [1,2,*], María Laura Arias [1,2,†], Michaela Kraus [3], Michalis Kourniotis [3], Andrea Fabiana Torres [1,2,†], Lydia Sonia Cidale [1,2,†] and Marcelo Borges Fernandes [4]

1 Departamento de Espectroscopía, Facultad de Ciencias Astronómicas y Geofísicas, Universidad Nacional de La Plata, Paseo del Bosque S/N, La Plata B1900FWA, Argentina
2 Instituto de Astrofísica La Plata, CCT La Plata, CONICET-UNLP, Paseo del Bosque S/N, La Plata B1900FWA, Argentina
3 Astronomical Institute, Czech Academy of Sciences, Fričova 298, 251 65 Ondřejov, Czech Republic
4 Observatório Nacional, Rua General José Cristino 77, São Cristóvão, Rio de Janeiro 20921-400, Brazil
* Correspondence: paulam@fcaglp.unlp.edu.ar
† Member of the Carrera del Investigador Científico y Tecnológico, CONICET, Argentina.

Abstract: Symbiotic stars are interacting binary systems composed of an evolved star (generally a late-type red giant) and a degenerate or dwarf companion in orbit close enough for mass transfer to occur. Understanding the status of the late-type star is important for developing binary models for the symbiotic systems as it affects the transfer of matter needed to activate the hot component. Infrared observations have been very useful in probing the nature of late-type stars in symbiotic systems. This work presents a set of symbiotic stars observed with SOAR/OSIRIS ($R\sim3000$) in the H-band. We aimed to search for possible molecular circumstellar emission, to characterize the cool companion in these systems, and to confront the new findings with those obtained from the previous K-band classifications. We detected molecular emission from just one object, BI Cru, which displays the second-overtone CO-bands. To fit the observed photospheric CO absorption bands, we used the MARCS atmosphere models. We present our results as a mini atlas of symbiotic stars in the near-infrared region to facilitate the comparison among different observed symbiotic systems.

Keywords: techniques: spectroscopic; stars: symbiotic, binaries; stars: infrared

Citation: Marchiano, P.E.; Arias, M.L.; Kraus, M.; Kourniotis, M.; Torres, A.F.; Cidale, L.S.; Borges Fernandes, M. A Mini Atlas of H-Band Spectra of Southern Symbiotic Stars. *Galaxies* **2023**, *11*, 80. https://doi.org/10.3390/galaxies11040080

Academic Editor: Oleg Malkov

Received: 29 April 2023
Revised: 4 June 2023
Accepted: 12 June 2023
Published: 22 June 2023

Copyright: © 2023 by the authors. Licensee MDPI, Basel, Switzerland. This article is an open access article distributed under the terms and conditions of the Creative Commons Attribution (CC BY) license (https://creativecommons.org/licenses/by/4.0/).

1. Introduction

Symbiotic stars display a combined spectrum that shows high excitation emission lines superposed on the absorption lines of two stellar components: a late-type giant and a hot compact star. This characteristic led to defining symbiotic stars as strongly interacting binaries consisting of a cool giant of spectral type M (sometimes K- or G-type giant) as a donor, which transfers material to an accretor compact star, most commonly a hot white dwarf (or, in some systems, a neutron star). Such a binary system is surrounded by a circumstellar nebula enriched by the material lost from both components [1,2]. In addition, there are other contributions to the spectral energy distribution (SED) of a symbiotic star, such as emission from heated dust and jets or collimated bipolar outflows. Therefore, the entire SED of symbiotic systems is a superposition of all these contributions, which dominate the emission spectrum in different wavelength regions. With additional processes such as mass loss, accretion, and ionization, the spectrum of each symbiotic object is unique, and as a group, they appear rather heterogeneous. In this sense, Reference [3] proposed a physical definition where "*a symbiotic system is a binary in which a red giant transfers enough material to a compact companion to produce an observable signal at any wavelength*".

Regarding the near-infrared region, symbiotic stars have been divided into two types [4]. If they display an infrared (IR) excess emission that results from dust, they are called D-type (dust). In these systems, the cool giant is a very evolved Mira variable surrounded by warm dust. If they display an IR continuum emission from the cool companion, the symbiotics belong to the S-type (stellar), in which the cool star is a regular red giant, often filling its Roche lobe. The orbital periods of S-type symbiotics are about 2–3 years, while those of the D-types are at least an order of magnitude longer [5]. There is a third type of symbiotic system introduced by Allen et al. [6]: D'-type systems, which are a sub-type of the classical D-types. In these symbiotics, a significantly hotter star of spectral types F and G constitutes the cool component. Furthermore, like D-type symbiotics, D'-type systems also exhibit IR dust emission. However, their SEDs reveal distinct characteristics, with a nearly flat profile indicative of a cool companion and a dusty shell. This is in contrast to D- and S-types, which exhibit intensity peaks at wavelengths between 2 and 4 µm and between 0.8 and 1.7 µm, respectively [7].

As previously mentioned, the accretion object for most symbiotic stars is a white dwarf. However, in a small number of systems, it is believed to be a neutron star. These symbiotic X-ray binaries (SyXBs) represent a special class of symbiotic binaries that have garnered significant interest among astronomers. One notable example is IGR J16194-2810, which was observed in the low/hard state using the Suzaku X-ray satellite. This particular SyXB has been classified as a low-mass X-ray binary [8,9] since the system is composed of an M-type giant and probably a neutron star [10].

Symbiotic systems show a complex emission spectrum with plenty of lines of different ionization degrees and forbidden transitions and also show a different phenomenology, including stellar pulsations (semi-regular and Mira variables), orbital variations, and slow changes due to dust. These systems are interesting objects to study the formation and evolution of gaseous and dusty circumstellar or circumbinary environments, providing clues to investigate other related objects, such as those showing the B[e] phenomenon. Symbiotic stars may also help to solve one of the timeliest problems in modern astrophysics, the missing progenitors of Type Ia supernovae (SNe Ia).

Observations in the IR are particularly important for those systems in which the cold companion is hidden in the optical spectral range. For example, the water vapor and carbon monoxide absorption bands observed in the 1–4 µm spectral range allow us to describe the characteristics of the late-type stars in these systems. It is interesting to note that, in systems with extremely dense environments, the CO-bands might turn into emissions. So far, BI Cru has been known for a long time to show intense CO-bands in emission [11]. Furthermore, H 1-25 (Hen 2-251) and RX Pup were reported by Schmidt and Mikołajewska [12] as symbiotic stars whose spectra show evidence of weak CO emission instead of absorption.

In a previous work, the medium-resolution K-band spectra of a sample of eight symbiotic objects were presented [13]. These observations were used to characterize the cool companions based on their CO first-overtone absorption bands along with measured equivalent widths of Na I and Brγ. Now, in order to obtain a more complete picture of each source, we present the H-band spectra of the same objects and follow a different approach. We derived the effective temperature of the late-type components using synthetic spectra computed with the MARCS atmosphere model' code [14]. However, considering the significant deviation of effective temperatures in red giant stars from temperatures derived using blackbody models, we propose to implement the correction recommended by Akras et al. [7], thereby obtaining what we will refer to as an equivalent temperature T_E. These results were used to estimate a more solid spectral type determination of the red giant.

The paper is structured as follows: Section 2 provides a description of the observations and data reduction process. Section 3 presents the fitting with the MARCS model spectra, the determination of the equivalent temperature T_E, the estimation of the spectral type for each symbiotic object, along with a concise overview of each object. In Section 4, we analyze the advantages and disadvantages of the spectral classification in different spectral bands. Finally, Section 5 ends the paper with our conclusions.

2. Observations and Data Reduction

On 8 June 2014, a selection of nine southern symbiotic stars was observed with the Ohio State InfraRed Imager/Spectrometer (OSIRIS), coupled to the 4.1 m telescope of the

Southern Observatory for Astrophysical Research (SOAR, Cerro Pachón, Chile) under the proposal ID: SO2014A-009. The selected instrumental configuration consisted of a single-order long-slit spectrometer and the camera f/7 to obtain medium-resolution ($R \sim 3000$) spectra in the H- and K-bands. The spectral coverage reached $\lambda\lambda$ 1.50–1.77 μm and $\lambda\lambda$ 2.0–2.4 μm for the H- and K-bands, respectively. The data of eight symbiotic stars obtained in the K-band were published by Marchiano et al. [13].

During all observations, several offset patterns following an ABBA cycle were taken, and the AB pairs were subtracted to remove the sky emission. Each spectrum was flat-fielded, telluric-corrected, and wavelength-calibrated. The reduction process was carried out with the IRAF [1] software package.

Relevant information on the observed targets is given in Table 1. There, we list the star names, the stellar coordinates, the H and K magnitudes, and the IR classification type (D or S) [7,15], and in the last six columns, we show the spectral type of the cool component determined from near-IR spectra (using wavelengths between 7000 and 10,000 Å), the effective temperature T_{cool}, the references where we obtained these data, the spectral type of the cool component derived from the CO-bands at 2.3 μm by Marchiano et al. [13], the spectral type of the cool component obtained in this work, and the equivalent temperature T_E, which we will explain in the next section.

Table 1. Stellar sample data with literature classifications.

Object	α (h:m:s)	δ (° ′ ″)	H (mag)	K (mag)	Type	Sp.T.	T_{cool} (K)	Ref.	Sp.T. Ref. [13]	Sp.T. this work	T_E (K)
BI Cru	12:23:26	−62:38:16	6.187	5.064	D	M0–M1	-	1	-	-	-
RS Oph	17:50:13	−06:42:28	6.858	6.5	S	M2	4100	3, 4	K1–K2	G8	4760
Hen 3-1341	18:08:37	−17:26:39	7.892	7.479	S	M4	3500	2, 3	K3	K3	4170
CL Sco	16 54 51	−30 37 18	-	7.86	S	M5	3400	3	M0	M5	3430
SY Mus	11:32:10	−65:25:11	8.14	4.593	S	M4.5	3400	2, 3	K4–K5	M4	3490
RT Cru	12:25:56	−61:30:28	5.583	5.185	D	M6	-	5	M3.5	M3	3660
V347 Nor	16:14:01	−56:59:28	5.811	4.943	D	M7–M8.5	-	2	M2.5–M3.5	M0	3970
KX TrA	16:44:35	−62:37:14	6.409	5.979	S	M6	3300	2, 3	M2	M3	3600
V694 Mon	07:25:51	−07:44:08	5.471	5.069	S	M6	3300	2, 3	-	M2	3720

References: 1. Schulte-Ladbeck ([16]), 2. Mürset and Schmidt ([17]), 3. Gałan et al. ([18]), 4. Zamanov et al. ([19]), 5. Pujol et al. ([20]).

3. Results

The H-band spectral observation of each symbiotic system listed in Table 1 is presented in Figure 1. This figure shows the complete set of spectra, which were normalized and vertically shifted to ease inspection and comparison. The spectra are first displayed according to the presence of CO-bands in emission and then according to the strength of the atomic absorption lines.

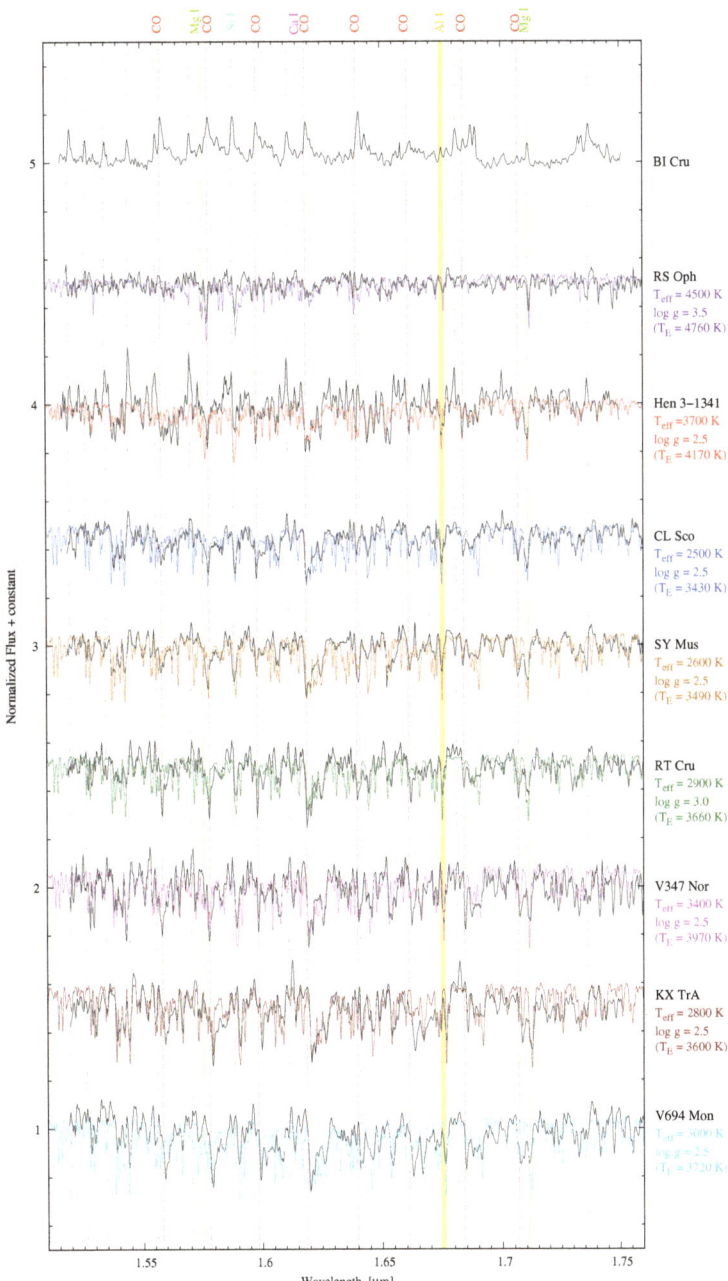

Figure 1. Comparison between the H-band spectra of our sample stars observed with SOAR/OSIRIS (black) and the best synthetic spectra computed with the MARCS model atmospheres (in different colors). Brackett series are indicated in gray, emission/absorption of CO in red, Mg I in green, and Si I in light blue. Ca I lines (1.6150 µm, 1.6157 µm, and 1.6197 µm) are indicated with a pink band and Al I lines (1.67504 µm, 1.67505 µm, 1.67189 µm and 1.67633 µm) with a yellow band. The stellar parameters of each model are also provided.

Our observations showed that all lines in BI Cru were in emission, and the remaining objects all showed CO-bands in absorption. More accurately, carbon monoxide molecules were observed in strong absorption in all stars, confirming the typical atmosphere of cool giants [21].

As a special case, the same as what was seen in the K-band [13], was RS Oph, which showed a weaker CO-band intensity than expected for the spectral type of the cool component. This weakening might be a consequence of the presence of dust with a temperature of ~1000 K. Such a warm, dusty envelope veils the stellar photospheres and contributes to the total SED with emission of the absorbed and scattered stellar light at wavelengths longer than 2 μm [22].

Brackett series emission was significant in the spectra of BI Cru and Hen 3-1341. In addition to the lines of the hydrogen series, we also identified lines from Mg I, Ca I, Si I, and Al I. The Mg I line at λ 1.7109 μm was present in eight of the nine stars of our sample in absorption and in emission only in BI Cru. Note that the Si I line at λ 1.5888 μm was blended with a line of the Brackett series; it is important to take this into account in future observations with higher resolution to better distinguish the contributions of each line.

We fit the observations with the MARCS model spectra [14] to obtain the effective temperature (T_{eff}). A grid was computed for M giant stars, covering the temperature range 2500 K $\leq T_{eff} \leq$ 4600 K, with a step of 100 K. Models were generated with a power resolution of $R = 3000$, solar metallicity [M/H] = 0, $v \sin i = 8$ km s^{-1}, and 30 km s^{-1} and log $g = 3.0, 2.5$, and 2.0. For temperatures lower than 3500 K, the microturbulence was taken with a value of 4 km s^{-1} and the macroturbulence equal to 8 km s^{-1}. For a hotter grid (temperatures higher than 3500 K), lower values for microturbulence (1 km s^{-1}) and macroturbulence (3 km s^{-1}) were considered, based on typical values adopted by studies of red giants and symbiotic stars [23,24]. On the one hand, it is important to note that the projected rotational velocity $v \sin i = 8$ km s^{-1}, or 30 km s^{-1}, taken as a fixed parameter, does not represent the majority of late-type giants in a symbiotic binary system ([25–27]). On the other hand, due to the instrument's spectral resolution used, the minimum detectable velocity would be approximately 100 km s^{-1}. Therefore, this particular instrument would not resolve any velocity change below 100 km s^{-1}.

In order to achieve a more accurate fit, we took into account not only the intensity of the CO absorptions, but also the relative relationships between these lines and other absorptions observed in each spectrum. The estimated values of T_{eff} and log g of each symbiotic star are shown on the right side of Figure 1. The error of T_{eff} was estimated to be around 100–200 K. Temperatures T_E between parentheses refer to the correction made to the effective temperature of red giant stars with $T_{eff} < 4000$ K according to the polynomial functions applied in the work of Akras et al. [7] and van Belle et al. [28]. We used this correction to the T_{eff} values derived from the MARCS model to enable their comparison or equivalence with temperatures calculated via the blackbody approximation. Once this correction is obtained, which we refer to as the equivalent temperature T_E for convenience, we applied the empirical relationships for red giants from the work of van Belle et al. [28] to determine the corresponding spectral type.

A brief summary of each individual symbiotic system is provided below with a description of the emission and absorption characteristics observed in this sample. It is worth clarifying that all systems belong to the catalog of symbiotic stars published by Belczynski et al. [15], which was used for the selection of the nine observed objects. Furthermore, there exist more recent catalogs of symbiotic stars, such as those compiled by Akras et al. [7] and Merc et al. [29].

BI Cru (= Hen 3-782 = WRAY 15-967):

BI Cru is a D-type symbiotic object. It is one out of three currently known symbiotic stars in which CO molecular emission was detected in the near-infrared at 2.3 μm [30], which seems to be stable over long time intervals. From the modeling of the CO-band emission detected in a high-resolution (R~45,000) K-band spectrum, Marchiano et al. [31] derived the kinematics and the physical properties of the molecular circumstellar medium.

This CO emission is associated with a dense disk, which would form after one or more episodes of intense mass loss from the red giant.

McCollum et al. [32] reported an IR shell of BI Cru more than five-times larger in arc size than the star's optical lobe. The temperature of this IR dust emission associated with our object was estimated to be 1300 K [33,34].

RS Oph (= HD 162214 = MWC 414):

RS Ophiuchi is a symbiotic recurrent nova consisting of a white dwarf with a mass near the Chandrasekhar limit that orbits inside the outer wind of a red giant. The system has had numerous outbursts recorded in 1898, 1933, 1958, 1967, 1985, and 2006, and more recently, in August 2021, it underwent its seventh optical eruption. Brandi et al. [35] re-determined the spectroscopic orbit of this star based on the optical spectra over the decade of 1998–2008.

The spectral type of the red giant and its temperature were estimated by Ribeiro et al. [36] and Pavlenko et al. [37]. In Table 1, we mark the reference of Zamanov et al. [19] because these authors applied a more precise method to classify the star. However, the spectral types obtained both by Marchiano et al. [13] and through this work (see Table 1) indicated an earlier spectral type than Zamanov et al. [19].

Hen 3-1341 (= V2523 Oph = SS73 75):

Hen 3-1341 is one of about ten symbiotics that shows hints of jets. According to Stute et al. [38], its optical and ultraviolet spectra show strong emissions in the Balmer continuum from N V, [Fe VII], and He II and the band at 6830 Å due to Raman scattering by neutral hydrogen. Furthermore, the object's infrared colors are appropriate for a cool giant (Spectral Type M4) without circumstellar dust. In quiescence, Hen 3-1341 resembles Z And, the prototype of symbiotic stars [39].

So far, no H-band observations have been reported. Our K-band observations [13] revealed an earlier spectral type for the red giant of this symbiotic star (see Table 1). From our fitting of the spectral features in the H-band, we found $T_{\text{eff}} = 3700$ K and $T_E = 4170$ K if we applied the temperature scale correction [7,28]. With this correction, we obtained the same spectral type found in the K-band. It is important to note that CO-bands in the spectrum of this star (see Figure 1) could be contaminated with emission lines. Furthermore, Hen 3-1341 presents emission lines that resemble those of BI Cru. New observations of this star in this band will be necessary to better confine its temperature.

CL Sco (= Hen 3-1286):

Infrared radial velocities were used by Fekel et al. [40] to compute the orbital parameters of this star. They obtained physical parameters adopting a period $P = 625$ days for a circular solution, which agrees with Kenyon and Webbink [41]. Besides, the orbital solution for both components was presented by Montané et al. [42].

In the near-IR, Gałan et al. [18] measured the photospheric chemical abundances of this object from high-resolution ($R\sim$50,000) spectra. They presented the observed K- and H-band spectra of CL Sco together with synthetic spectra calculated using their own abundance estimates. According to our fitting in the H-band, the obtained T_E value would indicate a later spectral type than that obtained by Marchiano et al. [13] from the analysis of the K-band; however, this coincides with that obtained by Gałan et al. [18].

SY Mus (= HD 100336 = Hen 3-667):

Dumm et al. [43] obtained the orbital parameters of the M star in this eclipsing symbiotic system and observed an asymmetry in the UV continuum light curve, which they explained as being caused by an asymmetric distribution of the wind from the cool component.

SY Mus was observed at high resolution ($R\sim$50,000) in the HA-, K-, and K_r-bands using the Phoenix cryogenic echelle spectrometer on the Gemini South telescope by Mikołajewska et al. [23]. All the spectra cover a narrow spectral range of $\sim$100 Å. Spectrum synthesis employing standard local thermal equilibrium analysis and atmosphere models was used to perform an analysis of the photospheric abundance of CNO for the red giant component. They obtained an isotopic ratio of $^{12}C/^{13}C \sim 6-10$, which indicates that the giant has experienced the first dredge-up.

In this work, we found that SY Mus has an effective temperature $T_{eff} = 2600$ K. Applying the temperature scale correction, we obtained $T_E = 3490$ K. This result is close to that published by Mürset and Schmid [17] and Gałan et al. [18], who indicated that the cold companion has a spectral type M4.5. However, this differs from what was obtained by Marchiano et al. [13], who found an earlier spectral type (see Table 1).

RT Cru (= HV 1245):

RT Crucis was classified as a symbiotic star by Cieslinski et al. [44], and the same authors reported the presence of rapid variations in its brightness. Gromadzki et al. [45] using optical and IR observations, detected two periodicities in the light curves: $P_o = 325 \pm 9$ days and $P_p = 63 \pm 1$ days, which corresponded to the orbital and pulsation periods of the red giant, respectively.

RT Cru also attracted attention due to the discovery of hard X-ray emission detected in 2003–2004 with INTEGRAL/IBIS, which promoted the study of how accumulated material advances through a disk and reaches the surface of the white dwarf [46].

Using the first-overtone band of CO in the K-band, Marchiano et al. [13] found that the spectral type of the red giant companion was earlier than the one published by Pujol et al. [20]. A result in agreement with Marchiano et al. [13] was obtained in this work using synthetic spectra computed by MARCS model atmospheres in the H-band, as can be seen with the best fit in Figure 1 and in Table 1.

V347 Nor (= Hen 2-147 = WRAY 16-208):

This is a symbiotic Mira star. It has a pulsation period of 373 days and lacks the obscuration events typical of many other D-type symbiotics [47]. Its extended nebula was first modeled as an expanding ring inclined with respect to the plane of the sky by Corradi et al. [48]. Utilizing Doppler shift measurements from VLT integral-field spectroscopy in combination with the HST images, Santander-García et al. [47] demonstrated that the intrinsic geometry of the nebula is indeed that of a circular, knotty ring of ionized gas. This ring was found to be inclined by 68 degrees with respect to the line of sight and to expand with a velocity of ~ 90 km s^{-1}.

We present here the first H-band spectrum for this symbiotic system. Figure 1 also shows that the best-fitting spectrum gave a spectral type not only earlier than the one found by Mürset and Schmid [17], but also earlier than the one found by Marchiano et al. [13] (see Table 1).

KX TrA (= Hen 3-1242):

KX TrA is a high-excitation S-type symbiotic binary. Its optical and near-IR spectra were studied by Ferrer et al. [49] and Marchiano et al. [50]. With observations taken between 1995 and 2007, they determined an orbital solution to the absorption lines of the red giant through the radial velocity curve. Since 2004, the system has entered into a period of activity.

Gałan et al. [18,51] presented high-resolution ($R \sim 50,000$) near-IR spectra for a sample of symbiotic systems. They employed stellar atmosphere models using a standard local thermal equilibrium analysis and derived the chemical abundance of several systems including KX TrA. Adopting the spectral type shown in the seventh column in Table 1, their analysis revealed a slightly subsolar metallicity ([Fe/H] ~ -0.3) for KX TrA. Based on our analysis (see Figure 1 and Table 1), the spectral type was slightly later than the spectral type reported by Marchiano et al. [13] and was earlier than the one adopted by Gałan et al. [51] cited above.

V694 Mon (= MWC 560):

The most-spectacular features in the optical spectrum of V694 Mon are the broad absorption lines, most prominently at the Balmer transitions. V694 Mon is a symbiotic star in which the accretion disk drives a powerful high-velocity jet during outbursts, producing broad, blue-shifted, variable absorption lines from atomic transitions that extend up to thousands of km s^{-1}, not only in the optical range, but also in the IR, near-ultraviolet, and far-ultraviolet [52,53].

By fitting the observations of the donor star, we derived for the first time its effective temperature of $T_{\rm eff} = 3000$ K (see Figure 1). We also obtained, through the black body approximation, the $T_{\rm E}$ value, which allowed us to estimate the spectral type of the cool companion (see Table 1).

4. Discussion

Analysis of the near-infrared range in symbiotic systems is key to their study since it has been used to classify these in S- and D-type stars based on their division into color–color diagrams, (J-H) versus (H-K), and (H-K) vs. (K-L) [54]. In addition, the techniques used in this region of the electromagnetic spectrum have proven to be the most-suitable for finding the spectral type of the late-type component in these binary systems. Moreover, the near-infrared is typically not contaminated by the nebula and the emission from the hot component, which both strongly affect the spectra at optical wavelengths.

Keenan and Hynek [55] introduced the spectral classification for the cool component of symbiotic binaries using the red TiO-bands, which increase in strength with decreasing temperature. Sharpless [56] solved the spectral classification using the characteristics of the most-important molecular bands between 7500 and 8900 Å. His temperature classification was also based on the growth of the TiO- and VO-bands, but he found that the CN-bands at 7916, 7941, 7878, and 8068 Å are useful luminosity discriminators in this spectral range as well. According to Schulte-Ladbeck [16], numerous observations of symbiotic systems are subject to contamination by emission lines from the nebula. This represents a source of error because these lines weaken the absorption of molecular bands in relation to the continuum. The longer wavelength bands are expected to be less affected since the IR brightness of the red giant increases rapidly in that range.

The K-band has been widely employed in most spectroscopic studies of cool or obscured objects, due to their higher brightness in this band compared to the J- or H-bands. Moreover, for the spectral classification of cool stars, several atomic features of Mg I, Ca I, and Na I are utilized, along with the band heads of the first-overtone CO-bands, which dominate the K-band spectra. Nevertheless, observations have revealed that stars of spectral types K3-M5 also exhibit significant temperature and luminosity-sensitive features in the H-band, including the lines of Mg I, Al I, and OH, and the band heads of the CO second overtone bands [57].

The circumstellar dust in symbiotic systems (more noticeable in D-type than S-type stars) often causes a continuum emission excess in the IR, which can reach down to wavelengths of around 2 μm. If this is the case, the excess emission can significantly hinder the extraction of precise information about the photosphere of the cool component. In such systems, shorter-wavelength spectra are required to identify and characterize the underlying star.

After considering the pros and cons of performing spectral classification in different near-infrared spectral bands, we propose that the H-band can provide valuable complementary information that can help to better constrain the parameters in symbiotic systems, in particular the temperature of the red giants in dusty symbiotic systems whose K-band spectra might be contaminated by the dust emission. It is also important to clarify that the differentiation of the spectral types obtained in the H- and K-band for each symbiotic object is based on observations that have been taken with the same instrument and within the same observing run. The studies themselves, however, employed different methods. The spectral classification found from the K-band spectra is based on measurements of equivalent line widths [13], unlike in this study, where we used a more robust method. Thus, an estimate of the similarities or differences between the spectral types obtained in each band is not entirely accurate if a different methodology is used, but we are confident that the presented analysis for the H-band spectra of symbiotic objects provides more decent results regarding the classification of their cool components.

5. Conclusions

To summarize, the IR spectra of symbiotic stars are a very useful tool to perform spectral classification of the cool components of these systems [16]. They are also excellent sources of information on the physical conditions of the disks from which molecular emission (if detected) originates. On the other hand, for many years, different studies, dedicated to the near-IR spectra of symbiotic stars have utilized low-resolution data, and in general, many of those studies were restricted to a small sample of stars, some examples being Schulte-Ladbeck [16], Kenyon and Gallagher [58], Schild et al. [59]. In the last two decades, new instruments have given access to the IR spectral range and have made it possible to obtain high-quality spectra with the necessary spectral resolution to improve the understanding of these complex objects.

With the aim to enlarge the sample of symbiotic objects with near-IR classifications, we presented the medium-resolution H-band spectra of nine symbiotic systems obtained with the OSIRIS spectrograph. We summarize our results in two points:

- We observed CO emission only for BI Cru; the rest of the stars of our H-band sample showed CO absorption lines. We identified some pronounced emissions of the hydrogen Brackett series in BI Cru and Hen 3-1341; in the other seven observed stars, these lines were weaker or absent. We also identified the lines of Mg I, Si I, Al I, and Ca I.
- In the spectra with CO-bands in absorption, we fit these features with synthetic spectra obtained using MARCS atmosphere models, and we estimated the values of T_{eff} and $\log g$ of the cool companion of each observed symbiotic system (see the right side of Figure 1). Based on the scheme by van Belle et al. [28] and Akras et al. [7] for correcting the temperatures of red giants, we derived an approximate spectral type of the cool giant. The results are presented in the penultimate column of Table 1.

Author Contributions: Conceptualization, P.E.M., M.L.A., M.K. (Michaela Kraus) and L.S.C.; methodology, P.E.M., M.L.A, M.K. (Michaela Kraus), A.F.T. and L.S.C.; software, P.E.M., M.K. (Michalis Kourniotis); investigation P.E.M., M.L.A, M.K. (Michaela Kraus), M.K. (Michalis Kourniotis), A.F.T., L.S.C. and M.B.F.; resources, P.E.M., M.L.A, M.K. (Michaela Kraus), M.K. (Michalis Kourniotis), A.F.T., L.S.C. and M.B.F.; data curation, M.L.A.; writing—original draft preparation, P.E.M.; writing—review and editing, P.E.M., M.L.A, M.K. (Michaela Kraus), M.K. (Michalis Kourniotis), A.F.T., L.S.C. and M.B.F.; funding acquisition, A.F.T., M.L.A, M.K. (Michaela Kraus) and M.B.F. All authors have read and agreed to the published version of the manuscript.

Funding: A.F.T. and M.L.A. acknowledge financial support from the Universidad Nacional de La Plata (Programa de Incentivos 11/G160) and CONICET (PIP 1337), Argentina. M.K. (Michaela Kraus) acknowledges financial support from the Czech Science Foundation (GAČR, Grant Number 20-00150S). The Astronomical Institute of the Czech Academy of Sciences Ondřejov is supported by the project RVO:67985815. This project has received funding from the European Union's Framework Programme for Research and Innovation Horizon 2020 (2014-2020) under the Marie Skłodowska-Curie Grant Agreement No. 823734. M.B.F. acknowledges financial support from the National Council for Scientific and Technological Development (CNPq) Brazil (grant number: 307711/2022-6)

Institutional Review Board Statement: Not applicable.

Informed Consent Statement: Not applicable.

Data Availability Statement: The data underlying this article will be shared upon reasonable request to the corresponding author.

Acknowledgments: This research made use of the NASA Astrophysics Data System (ADS) and of the SIMBAD database, operated at CDS, Strasbourg, France. This paper is based on observations obtained at the Southern Astrophysical Research (SOAR) telescope, which is a joint project of the Ministério da Ciência, Tecnologia, e Inovação (MCTI) da República Federativa do Brasil, the U.S. National Optical Astronomy Observatory (NOAO), the University of North Carolina at Chapel Hill (UNC), and Michigan State University (MSU) under Program ID SO2014A-009.

Conflicts of Interest: The authors declare no conflict of interest.

Notes

1. IRAF is distributed by the National Optical Astronomy Observatory, which is operated by the Association of Universities for Research in Astronomy (AURA) under cooperative agreement with the National Science Foundation.

References

1. Allen, D. Symbiotic stars. *Astrophys. Space Sci.* **1984**, *99*, 101–125. [CrossRef]
2. Mikołajewska, J. Symbiotic Stars: Continually Embarrassing Binaries. *Balt. Astron.* **2007**, *16*, 1–9.
3. Luna, G.J.M.; Sokoloski, J.L.; Mukai, K.; Nelson, T. Symbiotic stars in X-rays. *Astron. Astrophys.* **2013**, *559*, A6.
4. Webster, B.; Allen, D. Symbiotic stars and dust. *Mon. Not. R. Astron. Soc.* **1975**, *171*, 171–180. [CrossRef]
5. Schmid, H.; Schild, H. Orbital motion in symbiotic Mira systems. *Astron. Astrophys.* **2002**, *395*, 117–127. [CrossRef]
6. Allen, D.; Friedjung, M.; Viotti, R. The Nature of Symbiotic Stars. In *Proceedings of the IAU Colloquium No. 70 Held at the Observatoire De Haute Provence, France, 26–28 August 1981*; Springer: Dordrecht, The Netherlands, 1982.
7. Akras, S.; Guzman-Ramirez, L.; Leal-Ferreira, M.L.; Ramos-Larios, G. A Census of Symbiotic Stars in the 2MASS, WISE, and Gaia Surveys. *Astrophys. J. Suppl. Ser.* **2019**, *240*, 21.
8. Deng, Z.L.; Gao, Z.F.; Li, X.D.; Shao, Y. On the Formation of PSR J1640+2224: A Neutron Star Born Massive? *Astrophys. J.* **2020**, *892*, 4. [CrossRef]
9. Deng, Z.L.; Li, X.D.; Gao, Z.F.; Shao, Y. Evolution of LMXBs under Different Magnetic Braking Prescriptions. *Astrophys. J.* **2021**, *909*, 174. [CrossRef]
10. Kitamura, Y.; Takahashi, H.; Fukazawa, Y. Suzaku observation of the symbiotic X-ray binary IGR J16194-2810. *Publ. Astron. Soc. Jpn.* **2014**, *66*, 6.
11. Whitelocke, P.; Feast, M.; Roberts, G.; Carter, B.; Catchpole, R. Circumstellar CO emission at 2.3 μm in BI Cru, He 3-1138 and He 3-1359. *Mon. Not. R. Astron. Soc.* **1983**, *205*, 1207–1214. [CrossRef]
12. Schmidt, M.R.; Mikołajewska, J. Near-Infrared Spectra of a Sample of Symbiotic Stars. In *Proceedings of the Symbiotic Stars Probing Stellar Evolution, La Palma, Spain, 27–31 May 2002*; Corradi, R.L.M., Mikolajewska, J., Mahoney, T.J., Eds.; Astronomical Society of the Pacific Conference Series; Astronomical Society of the Pacific: San Francisco, CA, USA, 2003; Volume 303, p. 163.
13. Marchiano, P.E.; Cidale, L.S.; Arias, M.L.; Borges Fernandes, M.; Kraus, M. A mini atlas of K-band spectra of southern symbiotic stars. *Bol. Asoc. Argent. Astron.* **2015**, *57*, 87–89.
14. Gustafsson, B.; Edvardsson, B.; Eriksson, K.; Jørgensen, U.G.; Nordlund, Å.; Plez, B. A grid of MARCS model atmospheres for late-type stars. I. Methods and general properties. *Astron. Astrophys.* **2008**, *486*, 951–970. [CrossRef]
15. Belczynski, K.; Mikolajewska, J.; Munari, U.; Ivison, R.; Friedjung, M. A catalogue of symbiotic stars. *Astron. Astrophys. Suppl. Ser.* **2000**, *146*, 407–435. [CrossRef]
16. Schulte-Ladbeck, R.E. Near-infrared spectral classification of symbiotic stars. *Astron. Astrophys.* **1988**, *189*, 97–108.
17. Mürset, U.; Schmid, H.M. Spectral classification of the cool giants in symbiotic systems. *Astron. Astrophys. Suppl. Ser.* **1999**, *137*, 473–493. [CrossRef]
18. Gałan, C.; Mikołajewska, J.; Hinkle, K.H.; Joyce, R.R. Chemical abundance analysis of symbiotic giants - III. Metallicity and CNO abundance patterns in 24 southern systems. *Mon. Not. R. Astron. Soc.* **2016**, *455*, 1282–1293.
19. Zamanov, R.K.; Boeva, S.; Latev, G.Y.; Martí, J.; Boneva, D.; Spassov, B.; Nikolov, Y.; Bode, M.F.; Tsvetkova, S.V.; Stoyanov, K.A. The recurrent nova RS Oph: Simultaneous B- and V- band observationsof the flickering variability. *Mon. Not. R. Astron. Soc.* **2018**, *480*, 1363–1371.
20. Pujol, A.; Luna, G.J.M.; Mukai, K.; Sokoloski, J.L.; Kuin, N.P.M.; Walter, F.M.; Angeloni, R.; Nikolov, Y.; Lopes de Oliveira, R.; Nuñez, N.E.; et al. Taking a break: Paused accretion in the symbiotic binary RT Cru. *Astron. Astrophys.* **2023**, *670*, A32.
21. Rayner, J.; Cushing, M.; Vacca, W. The Infrared Telescope Facility (IRTF) Spectral Library: Cool Stars. *Astrophys. J. Suppl. Ser.* **2009**, *185*, 289–432.
22. Hinkle, K.; Fekel, F.; Joyce, R.; Wood, P. Infrared spectroscopy of symbiotic stars. IX. D-type symbiotic novae. *Astrophys. J.* **2013**, *770*, 28. [CrossRef]
23. Mikołajewska, J.; Gałan, C.; Hinkle, K.H.; Gromadzki, M.; Schmidt, M.R. Chemical abundance analysis of symbiotic giants—I. RW Hya and SY Mus. *Mon. Not. R. Astron. Soc.* **2014**, *440*, 3016–3026.
24. Kondo, S.; Fukue, K.; Matsunaga, N.; Ikeda, Y.; Taniguchi, D.; Kobayashi, N.; Sameshima, H.; Hamano, S.; Arai, A.; Kawakita, H.; et al. Fe i Lines in 0.91–1.33 μm Spectra of Red Giants for Measuring the Microturbulence and Metallicities. *Astrophys. J.* **2019**, *875*, 129. [CrossRef]
25. Fekel, F.C.; Hinkle, K.H.; Joyce, R.R. Rotational Velocities of S-Type Symbiotic Stars. In *Proceedings of the Stellar Rotation, Proceedings of IAU Symposium No. 215, Cancun, Yucatan, Mexico, 11–15 November, 2002*; Maeder, A., Eenens, P., Eds.; Astronomical Society of the Pacific: San Francisco, CA, USA, 2004; Volume 215, p. 168.
26. Schmutz, W.; Schild, H.; Muerset, U.; Schmid, H.M. High resolution spectroscopy of symbiotic stars I. SY Muscae: Orbital elements, M giant radius, distance. *Astron. Astrophys.* **1994**, *288*, 819–828.
27. Zamanov, R.K.; Bode, M.F.; Melo, C.H.F.; Stateva, I.K.; Bachev, R.; Gomboc, A.; Konstantinova-Antova, R.; Stoyanov, K.A. Rotational velocities of the giants in symbiotic stars—III. Evidence of fast rotation in S-type symbiotics. *Mon. Not. R. Astron. Soc.* **2008**, *390*, 377–382.

28. van Belle, G.T.; Lane, B.F.; Thompson, R.R.; Boden, A.F.; Colavita, M.M.; Dumont, P.J.; Mobley, D.W.; Palmer, D.; Shao, M.; Vasisht, G.X.; et al. Radii and Effective Temperatures for G, K, and M Giants and Supergiants. *Astron. J.* **1999**, *117*, 521–533. [CrossRef]
29. Merc, J.; Gális, R.; Wolf, M. New online database of symbiotic variables: Symbiotics in X-rays. *Astron. Nachrichten* **2019**, *340*, 598–606. [CrossRef]
30. McGregor, P.; Hyland, A.; Hillier, D. Atomic and molecular line emission from early-type high-luminosity stars. *Astrophys. J.* **1988**, *324*, 1071–1098. [CrossRef]
31. Marchiano, P.E.; Kraus, M.; Arias, M.L.; Torres, A.F.; Cidale, L.S.; Vallverdú, R. Molecular emission of CO in BI Cru with high resolution spectroscopy. *Bol. Asoc. Argent. Astron.* **2022**, *63*, 101–103.
32. McCollum, B.; Bruhweiler, F.; Wahlgren, G.; Eriksson, M. A Large Infrared Shell Associated with BI Crucis. *Astrophys. J.* **2008**, *682*, 1087–1094. [CrossRef]
33. Marchiano, P.E.; Cidale, L.S.; Brandi, E.; Muratore, M.F. Spectral Energy Distribution in the symbiotic system BI Cru. *Bol. Asoc. Argent. Astron.* **2013**, *56*, 163–166.
34. Henize, K.G.; Carlson, E.D. BI CRU: A new symbiotic star. *Publ. Astron. Soc. Pac.* **1980**, *92*, 479–483. [CrossRef]
35. Brandi, E.; Quiroga, C.; Mikolajewska, J.; Ferrer, O.E.; García, L.G. Spectroscopic orbits and variations of RS Ophiuchi. *Astron. Astrophys.* **2009**, *497*, 815–825. [CrossRef]
36. Ribeiro, V.A.R.M.; Bode, M.F.; Darnley, M.J.; Harman, D.J.; Newsam, A.M.; O'Brien, T.J.; Bohigas, J.; Echevarría, J.M.; Bond, H.E.; Chavushyan, V.H.; et al. The Expanding Nebular Remnant of the Recurrent Nova RS Ophiuchi (2006). II. Modeling of Combined Hubble Space Telescope Imaging and Ground-Based Spectroscopy. *Astrophys. J.* **2009**, *703*, 1955. [CrossRef]
37. Pavlenko, Y.V.; Evans, A.; Kerr, T.; Yakovina, L.; Woodward, C.E.; Lynch, D.; Rudy, R.; Pearson, R.L.; Russell, R.W. Metallicity and effective temperature of the secondary of RS Ophiuchi. *Astron. Astrophys.* **2008**, *485*, 541–545. [CrossRef]
38. Stute, M.; Luna, G.J.M.; Pillitteri, I.F.; Sokoloski, J.L. Detection of X-rays from the jet-driving symbiotic star Hen 3-1341. *Astron. Astrophys.* **2013**, *554*, A56.
39. Kenyon, S.J. *The Symbiotic Stars*; Cambridge Astrophysics; Cambridge University Press: Cambridge, UK, 1986. [CrossRef]
40. Fekel, F.C.; Hinkle, K.H.; Joyce, R.R.; Wood, P.R.; Lebzelter, T. Infrared Spectroscopy of Symbiotic Stars. V. First Orbits for Three S-Type Systems: Henize 2-173, CL Scorpii, and AS 270. *Astron. J.* **2006**, *133*, 17. [CrossRef]
41. Kenyon, S.J.; Webbink, R.F. The nature of symbiotic stars. *Astrophys. J.* **1984**, *279*, 252–283. [CrossRef]
42. Montané, B.; Quiroga, C.; Brandi, E. Estudio espectroscópico de la binaria simbiótica CL Scorpii. *Bol. Asoc. Argent. Astron.* **2013**, *56*, 295–298.
43. Dumm, T.; Schmutz, W.; Schild, H.; Nussbaumer, H. Circumstellar matter around M-giants in symbiotic binaries: SY MUSCAE and RW Hydrae. *Astron. Astrophys.* **1999**, *349*, 169–176.
44. Cieslinski, D.; Elizalde, F.; Steiner, J.E. Observations of suspected symbiotic stars. *Astron. Astrophys. Suppl.* **1994**, *106*, 243–251.
45. Gromadzki, M.; Mikołajewska, J.; Soszyński, I. Light Curves of Symbiotic Stars in Massive Photometric Surveys II. S and D'-Type Systems. *Acta Astron.* **2013**, *63*, 405–428.
46. Luna, G.J.M.; Mukai, K.; Sokoloski, J.L.; Lucy, A.B.; Cusumano, G.; Segreto, A.; Jaque Arancibia, M.; Nuñez, N.E.; Puebla, R.E.; Nelson, T.; et al. X-ray, UV, and optical observations of the accretion disk and boundary layer in the symbiotic star RT Crucis. *Astron. Astrophys.* **2018**, *616*, A53. [CrossRef]
47. Santander-García, M.; Corradi, R.L.M.; Whitelock, P.A.; Munari, U.; Mampaso, A.; Marang, F.; Boffi, F.; Livio, M. HST and VLT observations of the symbiotic star Hen 2-147. Its nebular dynamics, its Mira variable and its distance. *Astron. Astrophys.* **2007**, *465*, 481–491. [CrossRef]
48. Corradi, R.L.M.; Ferrer, O.E.; Schwarz, H.E.; Brandi, E.; García, L. The optical nebulae around the symbiotic Miras He 2-147, HM Sagittae and V1016 Cygni. *Astron. Astrophys.* **1999**, *348*, 978–989.
49. Ferrer, O.; Quiroga, C.; Brandi, E.; García, L.G. The Symbiotic System KX TrA. In *Proceedings of the Symbiotic Stars Probing Stellar Evolution, La Palma, Spain, 27–31 May 2002*; Corradi, R.L.M., Mikolajewska, J., Mahoney, T.J., Eds.; Astronomical Society of the Pacific Conference Series; Astronomical Society of the Pacific: San Francisco, CA, USA, 2003; Volume 303, p. 117.
50. Marchiano, P.E.; Brandi, E.; Quiroga, C.; Garcia, L.G.; Ferrer, O.E. Parámetros orbitales de KX TrA. *Bol. Asoc. Argent. Astron.* **2008**, *51*, 117–120.
51. Gałan, C.; Mikołajewska, J.; Hinkle, K.H. Chemical abundance analysis of symbiotic giants - II. AE Ara, BX Mon, KX TrA, and CL Sco. *Mon. Not. R. Astron. Soc.* **2015**, *447*, 492–502.
52. Ando, K.; Fukuda, N.; Sato, B.; Maehara, H.; Izumiura, H. Optical spectroscopic observations of a symbiotic star MWC 560 in the mass accumulation phase. *Publ. Astron. Soc. Jpn.* **2021**, *73*, L37–L41. [CrossRef]
53. Lucy, A.B.; Sokoloski, J.L.; Munari, U.; Roy, N.; Kuin, N.P.M.; Rupen, M.P.; Knigge, C.; Darnley, M.J.; Luna, G.J.M.; Somogyi, P.; et al. Regulation of accretion by its outflow in a symbiotic star: The 2016 outflow fast state of MWC 560. *Mon. Not. R. Astron. Soc.* **2020**, *492*, 3107–3127.
54. Allen, D.A.; Glass, I.S. Infrared photometry of southern emission-line stars. *Mon. Not. R. Astron. Soc.* **1974**, *167*, 337–350. [CrossRef]
55. Keenan, P.C.; Hynek, J.A. The Use of Infrared Spectra for the Determination of Absolute Magnitudes. *Astrophys. J.* **1945**, *101*, 265. [CrossRef]
56. Sharpless, S. The Infrared Spectral Classification of M-Type Stars. *Astrophys. J.* **1956**, *124*, 342. [CrossRef]

57. Meyer, M.R.; Edwards, S.; Hinkle, K.H.; Strom, S.E. Near-Infrared Classification Spectroscopy: H-Band Spectra of Fundamental MK Standards. *Astrophys. J.* **1998**, *508*, 397. [CrossRef]
58. Kenyon, S.; Gallagher, J. Infrared spectroscopy of symbiotic stars and the nature of their cool components. *Astron. J.* **1983**, *88*, 666–673. [CrossRef]
59. Schild, H.; Boyle, S.J.; Schmid, H.M. Infrared spectroscopy of symbiotic stars: Carbon abundances and 12C/13C isotopic ratios. *Mon. Not. R. Astron. Soc.* **1992**, *258*, 95–102.

Disclaimer/Publisher's Note: The statements, opinions and data contained in all publications are solely those of the individual author(s) and contributor(s) and not of MDPI and/or the editor(s). MDPI and/or the editor(s) disclaim responsibility for any injury to people or property resulting from any ideas, methods, instructions or products referred to in the content.

Article

Revealing the Binarity of HD 36030—One of the Hottest Flare Stars

Olga Maryeva [1,*], Péter Németh [1,2] and Sergey Karpov [3,4]

1. Astronomical Institute of the Czech Academy of Sciences, Fričova 298, 25165 Ondřejov, Czech Republic
2. Astroserver.org, Főtér 1, 8533 Malomsok, Hungary
3. Institute of Physics of the Czech Academy of Sciences, Na Slovance 1999/2, 18200 Prague, Czech Republic
4. Special Astrophysical Observatory of the Russian Academy of Sciences, 369167 Nizhnii Arkhyz, Russia
* Correspondence: olga.maryeva@asu.cas.cz

Abstract: The Kepler and TESS space missions significantly expanded our knowledge of what types of stars display flaring activity by recording a vast amount of super-flares from solar-like stars, as well as detecting flares from hotter stars of A-F spectral types. Currently, we know that flaring occurs in the stars as hot as B-type ones. However, the structures of atmospheres of hot B-A stars crucially differ from the ones of late types, and thus the occurrence of flaring in B-A type stars requires some extension of our theoretical views of flare formation and therefore a detailed study of individual objects. Here we present the results of our spectral and photometric study of HD 36030, which is a B9 V star with flares detected by the *TESS* satellite. The spectra we acquired suggest that the star is in a binary system with a low-mass secondary component, but the light curve lacks any signs of periodic variability related to orbital motion or surface magnetic fields. Because of that, we argue that the flares originate due to magnetic interaction between the components of the system.

Keywords: stars flares; stars activity; stars binaries spectroscopic; early-type stars; stars individual; HD 36030

Citation: Maryeva, O.; Németh, P.; Karpov, S. Revealing the Binarity of HD 36030—One of the Hottest Flare Stars. *Galaxies* **2023**, *11*, 55. https://doi.org/10.3390/galaxies11020055

Academic Editors: Lydia Sonia Cidale, Michaela Kraus and María Laura Arias

Received: 14 March 2023
Revised: 1 April 2023
Accepted: 10 April 2023
Published: 12 April 2023

Copyright: © 2023 by the authors. Licensee MDPI, Basel, Switzerland. This article is an open access article distributed under the terms and conditions of the Creative Commons Attribution (CC BY) license (https://creativecommons.org/licenses/by/4.0/).

1. Introduction

Stars classified as B-type span a very wide range of temperatures (from 25,000 to 11,000 K) and masses (from 50 $M_\odot$ for B0 Ia0 down to 3.5 $M_\odot$ for B9 V). They usually display the variability related to changes of the photospheric structure, e.g., pulsations of different varieties [1], such as: slowly pulsating B-type stars (SPB) [2,3], variables of the β Cephei type [4], variables of the α Cygni type [5], or classical Be (CBe) stars [6]. Less frequently, B-type stars show variability related to extended atmospheres and outflows, usually observed in peculiar B-type stars, such as B[e] [7,8] or luminous blue variables (LBVs) [9,10]. Quite unexpectedly, some B stars have also been shown to manifest very energetic stellar flares (e.g., [11] and references therein)—unpredictable dramatic increases in brightness for a few dozen minutes.

Such flaring activity is a well-known effect that is often seen both on the Sun and on so-called flare stars. Energetic flaring events are attributed to the release of the energy stored in coronal magnetic fields in magnetic reconnection events that accelerate the particles (electrons and ions) to high energies, thus heating dense regions of the stellar atmosphere and generating the flare-like emission in a wide range of frequencies, from radio to gamma-rays [12,13].

Although the majority of flare stars were found among objects of late spectral types, primarily cool K-M type red dwarfs [14,15], such outbursts were also detected in some hot B-A type stars, both in the optical range [16–18] and in X-rays [19,20]. In cool stars (F5 and later), the flares are linked with strong magnetic fields produced by a dynamo mechanism working in sufficiently deep outer convection zones [21]. On the other hand, the interiors of hot early-type stars are mostly radiative, with no convective motions to power the dynamo, and therefore magnetic field production is not expected in them, and other explanations of the eruptive activity are needed.

Pedersen et al. [22] suggested that all flaring hot stars are parts of binary systems and that the flares occur on unresolved cool components. However, Švanda and Karlický [23] and Bai and Esamdin [24] found that the frequency distribution of flare energies for A-type stars is steeper (with more flares having large energies) than the one for stars of later spectral types. It may suggest that the nature of flares on hot stars is indeed different, and they are not occurring on their colder companions. Balona [17,18] studied the variability of flaring B-A type stars and argued that rotational modulation seen in them confirms the presence of strong surface magnetic fields. One more possible mechanism for the flares may be related to magnetic interaction between the components in a close binary system, as was suggested by Yanagida et al. [20] for the pre-main sequence stars HD 47777 and HD 261902.

Currently, more than one hundred hot flaring B-A type stars are known from observations of the *Kepler* and *TESS* missions [11,17,18]. Van Doorsselaere et al. [25] demonstrated that about 2.45% of all B-A stars display detectable flares. However, the nature of their flares is still elusive. Thus, the study of these stars, and their possible binarity, is of utmost importance.

This paper is devoted to the spectral and photometric study of HD 36030[1], classified as B9 V [26] and belonging to the galactic open cluster ASCC 21 (or [KPR2005] 21 [27,28]). For the first time, HD 36030 was mentioned as a flare star in Maryeva et al. [11], where the authors searched for flare stars among confirmed members of galactic open clusters using high-cadence photometry from the *TESS* mission. Maryeva et al. [11] found eight B-A type flaring objects; among them, HD 36030 was the hottest. Independently, Balona [18] and Yang et al. [29] also selected HD 36030 as a flare star while searching for flares in *TESS* data. Despite being included in a couple dozens of catalogs, and being detected as an X-ray source by the Swift X-ray Telescope (XRT) [30]), to date HD 36030 lacks any specific studies devoted to it.

In Section 2, we present new spectral data, their analysis, and the results of atmospheric modeling using the TLUSTY code. In Section 3, we perform the analysis of photometric data on different time scales, and in Section 4, we discuss the results, the parameters of the second component, and the nature of flares. Finally, short conclusions are given in Section 5.

2. Spectroscopy of HD 36030

2.1. Observations

To clarify the nature of HD 36030 flaring activity, we initiated its spectral monitoring at the Perek 2 m telescope of the Ondřejov observatory of the Astronomical Institute of the Czech Academy of Sciences. High-resolution spectra of HD 36030 were obtained with the Ondřejov echelle spectrograph (OES) [31,32] during the autumn of 2021 and spring of 2022 (Table 1). The OES provides a wavelength range of 3750–9200 Å and a spectral resolving power of 50,000. The spectra were reduced using a dedicated IDL-based package. After primary reduction, all spectra were normalized and corrected for barycentric velocity.

Table 1. Log of HD 36030 observations at Perek 2 m telescope.

Date	Exposure [s]	Sp. Range [ÅÅ]	Instrument
19 Sep 2020	3600	6250–6740	Coudé
24 Oct 2021	5200	3750–9200	OES
29 Oct 2021	5200	3750–9200	OES
6 Jan 2022	5200	3750–9200	OES
2 Feb 2022	3 × 5200	3750–9200	OES
10 Mar 2022	2 × 5200	3750–9200	OES
11 Mar 2022	2 × 5200	3750–9200	OES
18 Mar 2022	5200	3750–9200	OES

Hydrogen lines in the acquired OES spectra have profiles with broad wings, with the width of the lines comparable to an echelle spectral order. For this reason, for the spectral normalization in the Hα region we used the medium resolution spectrum of HD 36030 covering a 6250–6740 Å range acquired using the Coudé spectrograph ($R \simeq 13000$; Slechta and Skoda [33]), which was also obtained at the Perek 2-m telescope on 20 September 2020. The spectrum was reduced using an IDL-based package, in the same way as described in Maryeva et al. [34]. Unfortunately, we have medium-resolution spectra only for the Hα region, so we could not perform the same routine for the orders containing other hydrogen lines. For them, we performed manual continuum drawing, and thus the normalization in these regions is less robust.

2.2. Spectral Analysis

In the spectrum of HD 36030, there is a strong Mg II λ4481 line, Ca II λ3933, 3968 doublet, Mg I λ5167, 5172, 5183 triplet, as well as a number of Fe II lines. All the lines in the spectrum besides the hydrogen lines are narrow. As it was noted above, the hydrogen lines show prominent Voigt profiles, which are a signature of the Stark effect (Figure 1). It is possible to also detect the lines of He I λ4471, 5876, which is a piece of evidence that the star is hotter than A0 [35]. Figure 2 shows a comparison of the HD 36030's spectrum with ones of B9 V and A0 V stars.

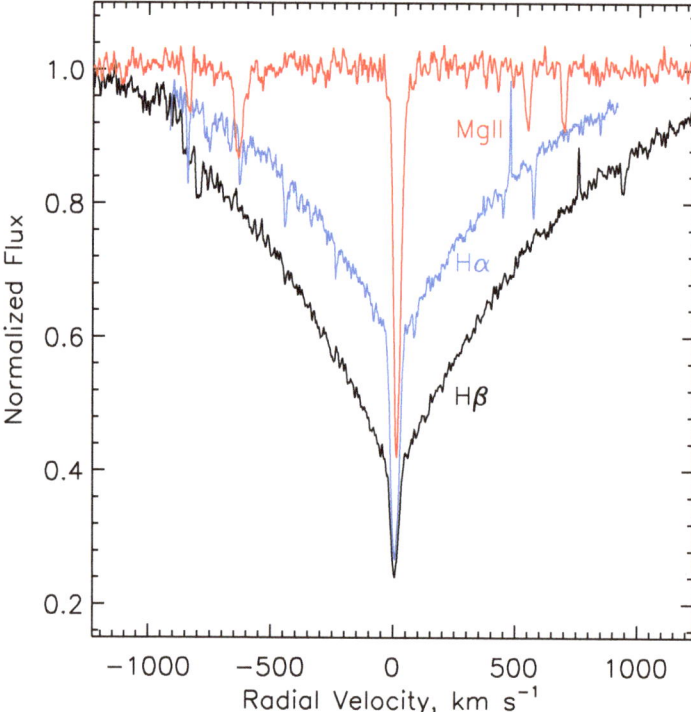

Figure 1. Comparison of broad wings hydrogen lines and narrow of Mg II λ4481 observed in the spectrum of HD 36030.

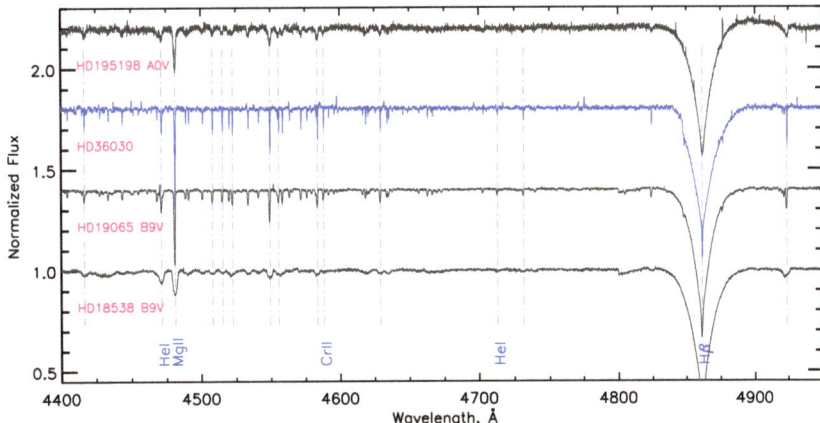

Figure 2. Comparison of HD 36030 spectrum with the ones of B9 V and A0 V stars. Spectra of HD 18538 and HD 19065 are taken from IACOB database [36–38] (spectral resolution is R = 85,000), whereas the spectrum of HD 195198 is from [39] (R = 42,000). Unnamed lines correspond to Fe II.

We estimated the reddening by comparing the photometric data from the Tycho-2 catalogue [40] ($B = 8.95 \pm 0.02$ mag and $V = 8.96 \pm 0.01$ mag) with the intrinsic color for a B9 star[2] (($B - V)_0 = -0.07$ mag; Pecaut and Mamajek [41]) as $E(B - V) = 0.06$ mag. Then, assuming the distance to its host cluster ASCC 21 to be $d = 345.5^{+12.3}_{-11.6}$ pc [28], the absolute magnitude of HD 36030 is $M_V = 1.09 \pm 0.07$ mag. It is fainter than the expected absolute magnitude of a B9 dwarf ($M_{V\text{table}} = 0.5$ mag from the table of Pecaut and Mamajek [41]). However, the location of HD 36030 in the Hertzsprung–Russell (HR) diagram for ASCC 21 cluster (Figure 3) accurately corresponds to the Main Sequence, and therefore we consider HD 36030 to be a B9 dwarf (B9 V), in agreement with Houk and Swift [26].

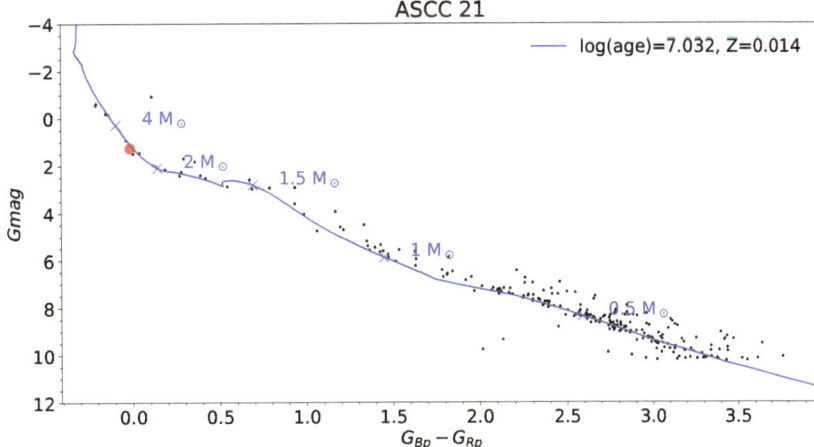

Figure 3. Hertzsprung–Russell diagram for the ASCC 21 cluster based on *Gaia* photometric data and compared with PARSEC isochrones [42] for the solar metallicity Z = 0.014. Age of the cluster (10.8 Myr) is taken from Bossini et al. [43]. Red circle shows the position of HD 36030.

Lines in the spectrum of HD 36030 clearly display a shift from night to night. We measured radial velocities (RVs) of selected lines (collected in Table 2). For the RV measurements, we used Gaussian profile fittings, and, for the case of hydrogen lines, we measured the positions of the narrow central absorptions. As Table 2 shows, the lines of different elements and ions differ in RV. RVs of Mg II $\lambda 4481$ and O I $\lambda 8446$ lines are higher than those of other lines, but all the lines demonstrate the same pattern of variability. Figure 4 shows average RVs, and those of the interstellar Na I $\lambda 5890, 5896$, measured in the same way as a reference. The scatter of the RV values is from 6 up to 45 km s^{-1}, and they clearly show that HD 36030 is a binary system. However, we do not see any lines of the secondary component in the spectrum, so we classified HD 36030 as a single-lined spectroscopic binary (SB1) system.

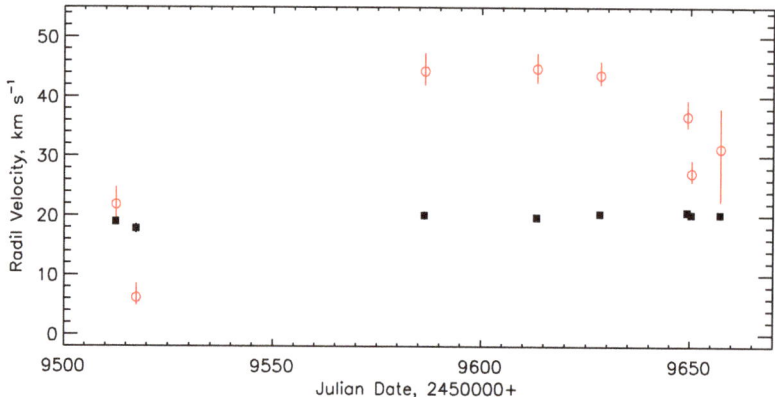

Figure 4. Radial velocity curve for HD 36030. Red dots display average values for radial velocity for different lines listed in Table 2 for a given night. Error bars represent the spread of individual line velocities. For comparison, black dots show the average velocity of the NaI $\lambda 5890, 5896$ interstellar doublet.

2.3. Spectral Analysis with XTGRID

To determine the atmospheric parameters of HD 36030 in local thermodynamic equilibrium (LTE), we fitted its OES observations with synthetic spectra calculated from TLUSTY (v207) model atmospheres [44–46]. The models include opacities from H, He, C, N, O, Ne, Mg, Si, P, S, and Fe. The spectral analysis was done with a steepest-descent spectral analysis procedure, implemented in the XTGRID code [47]. The procedure is a global fitting method that simultaneously reproduces all line profiles with a single atmosphere model. XTGRID starts with an input model and, by successive approximations, it calculates new model atmospheres and their corresponding synthetic spectra iteratively in the direction of decreasing χ^2. The procedure adjusts the atomic data and microphysics to the changing conditions in the atmosphere as the iterations move across the parameter space. Once the fitting procedure reaches the global minimum, statistical errors are calculated by changing each parameter in one dimension until the χ^2 variation corresponds to the 60% confidence. To avoid local minima, the procedure returns to the descent part if a better fit is found during the error analysis. Our models evaluated the conditions for convection, but the convective gradients indicated there are no convective layers in the atmosphere. We have made attempts to calculate non-LTE models and evaluate departures from the LTE approximation; however, all such models met numerical instabilities and failed. A non-LTE analysis will require the latest version of TLUSTY (v208) and will be reported in a forthcoming publication.

The spectroscopic parameters obtained from TLUSTY LTE models are summarized in Table 3, and the best fit is shown in Figures 5 and 6. The large error of log g measurement is mostly due to uncertainties in the continuum normalization of the broad hydrogen lines. Such big uncertainties in log g and $T_{\rm eff}$ give us a broad range of a possible mass of the star, $M = 2.1 \div 9$ M$_\odot$. However, the position of HD 36030 in the HR diagram (Figure 3) is consistent with $M_* = 3$ M$_\odot$.

To check the validity of TLUSTY models, we have repeated the analysis with interpolated ATLAS9 LTE models from the BOSZ spectral library [48]. The ATLAS9 models confirmed the effective temperature and surface gravity within error bars, but the BOSZ library was calculated for scaled solar metallicities; therefore, it is not suitable to derive individual element abundances for further comparisons.

Table 2. Radial velocity measurements for different lines on different nights; Avg is average values for radial velocity. The colons after values mark the data with lower accuracy of position determination.

	Time, JD-2450000							
	9512.463	9517.464	9586.371	9613.234	9628.234	9649.242	9650.239	9657.250
SiII 4128.05	21.26	7.35	44.53	45.61	42.53	37.62	27.39	35.31
SiII 4130.89	21.38	6.08	42.73	42.74	42.78	36.76	25.79	38.09
FeII 4173.45	24.71	6.17	43.38	46.39	45.07	37.99	29.35	—
FeII 4233.17	19.55	5.53	43.25	45.08	43.40	36.20	27.51	30.84
Hγ 4340.46	20.90	4.89	41.99	42.87	43.36	38.13	27.30	22.44
MgII 4481.13	27.91	12.49	50.22	50.94	49.62	42.80	33.23	38.11
FeII 4508.29	20.93	5.05	44.73	44.03	42.33	35.62	26.79	—
FeII 4515.34	21.28	5.71	44.66	42.45	44.85	36.23	26.38	32.61
FeII 4522.63	23.46	7.39	46.31	46.10	44.30	36.18	26.59	—
FeII 4549.47	24.81	8.55	47.36	47.38	46.01	39.46	29.39	36.36
FeII 4555.89	22.38	5.94	43.32	45.34	43.85	37.01	26.06	31.53
FeII 4583.83	22.67	6.07	43.86	44.88	43.46	36.80	27.50	33.81
CrII 4588.20	20.39	6.64	45.12	44.18	44.73	36.67	26.42	30.59
FeII 4629.33	20.62	5.47	44.98	43.82	42.56	36.69	26.55	29.68
Hβ 4861.32	21.70	5.63	44.51	45.16	44.30	36.38	27.23	25.98
FeII 4923.92	22.13	5.95	45.11	45.21	43.15	36.97	27.57	30.14
FeII 5018.44	22.00	6.52	44.92	44.76	44.00	36.79	27.83	31.05
SiII 5041.03	20.91	6.32	42.70	44.29	42.67	36.25	26.98	30.38
SiII 5055.98	26.04	10.63	47.98	50.36	47.87	41.70:	32.30	35.37
FeII 5100.73	21.90	6.55	44.71	45.12	44.78	36.76	27.72	33.09
FeII 5169.03	21.08	5.51	43.81	44.09	43.33	37.20	26.70	30.81
MgI 5172.68	21.90	7.20	45.38	45.56	44.21	36.98	28.01	31.29
MgI 5183.60	22.55	5.94	44.42	45.90	44.59	36.94	28.05	31.03
FeII 5197.57	22.46	5.67	43.52	43.66	42.24	36.93	26.73	30.18
FeII 5275.99	23.09	6.38	42.87	46.52:	42.09	34.90	26.02	32.96
FeII 5316.62	22.64	7.90	45.74	45.18	44.90	37.40	28.14	33.12
FeII 5362.87	21.55	5.46	43.33	45.05	43.59	36.20	27.05	30.11
SiII 6347.09	20.44	5.18	42.55	42.56	42.20	35.66	25.80	29.80
SiII 6371.36	21.61	6.10	—	44.44	43.67	36.53	27.15	30.10
Hα 6562.79	21.68	6.14	46.75	45.92	43.88	37.15	27.34	31.29
MgII 7877.13	—	0.89	39.54	—	41.64	34.30	23.38	24.77
MgII 7896.37	—	−2.87	43.00	44.15	41.05	35.66	24.90	—
OI 8446.35	28.33	12.08	49.48	48.55	49.32	40.71	32.85	34.61
Avg	$22.40^{5.01}_{-2.85}$	$6.72^{5.77}_{-1.84}$	$44.58^{5.63}_{-5.05}$	$45.15^{5.79}_{-3.51}$	$44.0^{5.6}_{-2.96}$	$37.14^{5.66}_{-2.34}$	$25.52^{5.71}_{-4.14}$	$31.43^{+6.60}_{-8.99}$
NaI 5889.95	18.60	17.03	19.38	19.47	20.10	20.36	20.03	20.51
NaI 5895.92	19.32	18.56	20.69	19.92	20.56	20.90	20.38	19.89

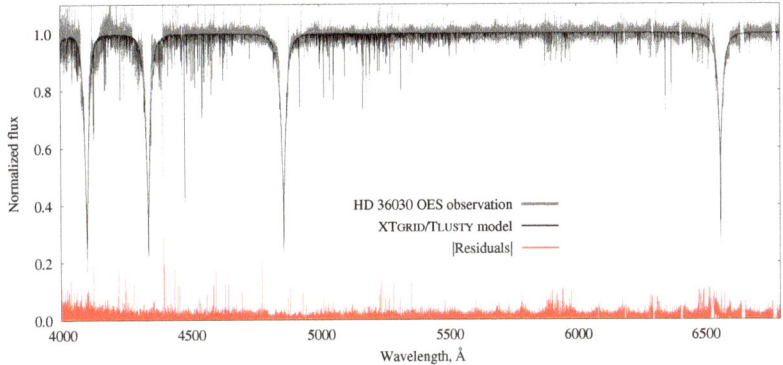

Figure 5. Normalized spectrum (grey line) of HD 36030 taken on 29 October 2021 with OES spectrograph compared with the best-fitting TLUSTY model (black line). The absolute values of the residuals between observed and synthetic spectra are shown in red.

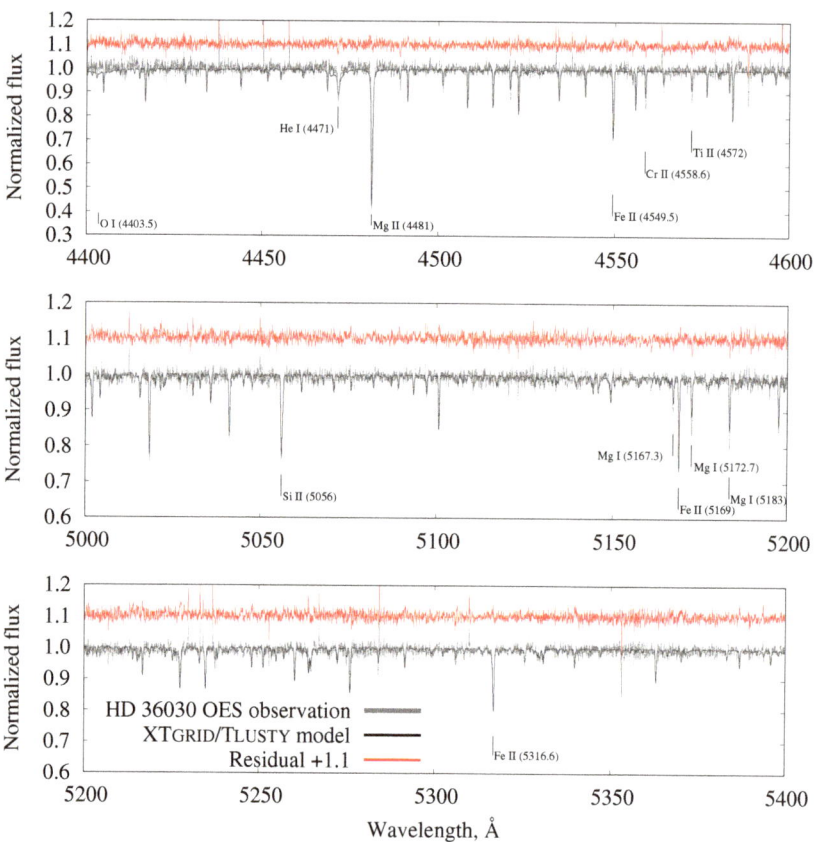

Figure 6. Selected intervals from Figure 5.

Table 3. Surface parameters of HD 36030 from the analysis of the OES spectra.

Parameter		Value	
T_{eff} (K)		$11{,}900 \pm 1100$	
$\log g$ (cm s^{-2})		4.69 ± 0.15	
$v \sin i$ (km s^{-1})		15.00 ± 0.94	
Abundance	by number $\log(nX/nH)$	mass fraction	solar fraction $\log(\epsilon/\epsilon_\odot)$
H (reference)	1	9.16×10^{-1}	0
He	-1.75 ± 0.32	6.49×10^{-2}	-0.68
C	<-2.3	$<4.60 \times 10^{-4}$	<-0.8
N	<-1.9	$<1.36 \times 10^{-3}$	<-0.2
O	-3.25 ± 0.02	8.11×10^{-3}	0.06
Ne	<-2	$<1.50 \times 10^{-3}$	<-0.03
Na	-5.42 ± 73	7.89×10^{-5}	0.34
Mg	-4.39 ± 0.04	9.00×10^{-4}	0.01
Al	-5.89 ± 0.49	3.19×10^{-5}	-0.34
Si	-4.21 ± 0.08	1.59×10^{-3}	0.28
Ca	<-5.48	$<7.3 \times 10^{-5}$	<-0.04
Ti	-7.00 ± 0.11	4.38×10^{-6}	0.05
Cr	-6.11 ± 0.21	3.67×10^{-5}	0.25
Mn	-6.94 ± 0.88	5.73×10^{-6}	-0.37
Fe	<-4.13	$<3.74 \times 10^{-3}$	<0.37
Ni	<-5	$<5.34 \times 10^{-4}$	<0.78

3. Photometry

In order to investigate the short time scale photometric variability of HD 36030, we used the data from the Transiting Exoplanet Survey Satellite (*TESS*; Ricker et al. [49]). HD 36030 was observed by *TESS* twice during its ongoing operation, in Sector 6 (11 December 2018–7 January 2019) of the prime mission, and Sector 32 (19 November 2020–17 December 2020) of the extended mission. We downloaded the data products for these observations produced by the *TESS* Science Processing Operations Center pipeline (SPOC; Jenkins et al. [50]) and available at the Barbara A. Mikulski Archive for Space Telescopes (MAST)[3] public data archive. We used the data products for the Long Cadence data, having 1800 s effective exposure during the prime mission, and 600 s—during the extended mission. In order to minimize the instrumental effects that are abundant in *TESS* data, we decided to use pre-search data conditioning simple aperture photometry (PDCSAP) light curves, which are corrected for instrumental trends using singular value decomposition, as well as for flux contributions from nearby objects in crowded fields [50]. We also filtered out the measurements marked with bad quality flags. We did not apply any additional detrending or pre-processing to the data, and the resulting light curves are shown in Figure 7.

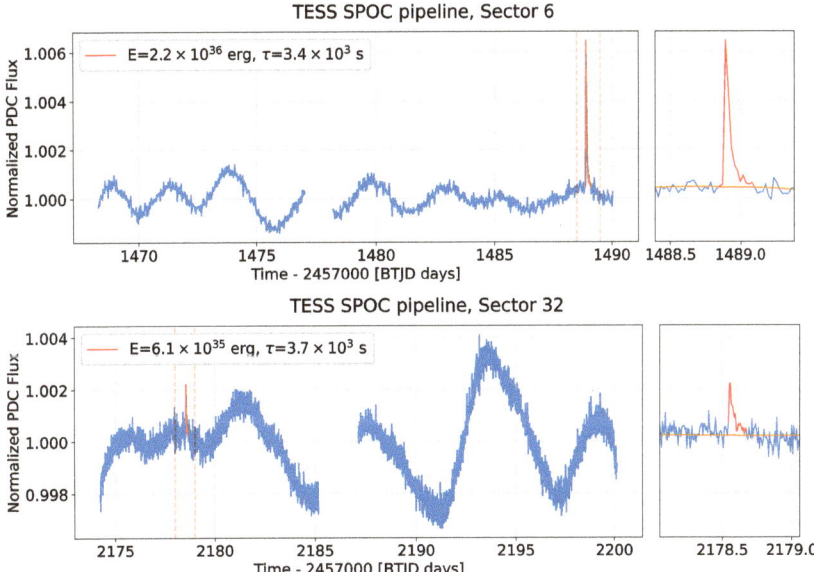

Figure 7. *TESS* light curves (blue lines) of HD 36030 for two sectors when the object was observed. Visually selected flares in both sectors are highlighted in red and also shown in the separate panels on the right. The total energies of the flares as well as their durations are also shown. Orange lines represent the smooth interpolation of a quiescent light curve behaviour during the flare, used for removing background contribution from its energy.

The light curve from Sector 6 shows a strong flare that has already been reported by Maryeva et al. [11] and Balona [18]. Visual inspection of Sector 32 also revealed a flare of smaller amplitude but similar duration. Both flares are shown in Figure 7. We characterized them by subtracting the smooth quiescent emission and integrating the flare profile to get its total fluence and fitted the fading part with an exponential function to get the characteristic duration, which we define as an e-folding time.

We also estimated the total energetics of the flares by integrating the flare profile after subtraction of a smooth trend of a star' quiescent emission. In order to correct for the fact that the majority of both the star and flare emission is outside of the *TESS* sensitivity range, we applied the bolometric corrections by convolving the spectrum of the star as well as flare spectrum with the sensitivity curve of *TESS*. As the temperatures of flares on the hot stars are not known yet, as a conservative estimate, following Shibayama et al. [51] and Günther et al. [15], we assumed the flares to have a blackbody spectrum with $T = 9000$ K, whereas for the star we used the spectral energy distribution of the best fitting TLUSTY model derived in Section 2.3. Then we used these bolometric corrections to convert the relative amplitude of the flare to its relative total fluence, and, knowing the luminosity of HD 36030, we converted them to absolute values. The resulting flare energies are shown in Figure 7.

In mid-October 2022, we also initiated a series of photometric observations of HD 36030 on FRAM-ORM, which is a 10-inch Meade f/6.3 Schmidt–Cassegrain telescope with custom Moravian Instruments G2-1600 CCD installed in the Roque de Los Muchachos Observatory, La Palma. The data were acquired in Johnson–Cousins *B* (20 s exposures), *V*, and *R* filters (10 s exposures) to avoid saturation due to the brightness of the star and then automatically processed by a dedicated Python pipeline based on the STDPIPE package [52], which includes bias and dark current subtraction, flat-fielding, cosmic ray removal, astrometric calibration, aperture (with a 5-pixel radius) photometry, and photometric calibration using the catalog of synthetic photometry based on Gaia DR3 low-resolution XP spectra [53]. The resulting light curve is shown in the upper panel of Figure 8.

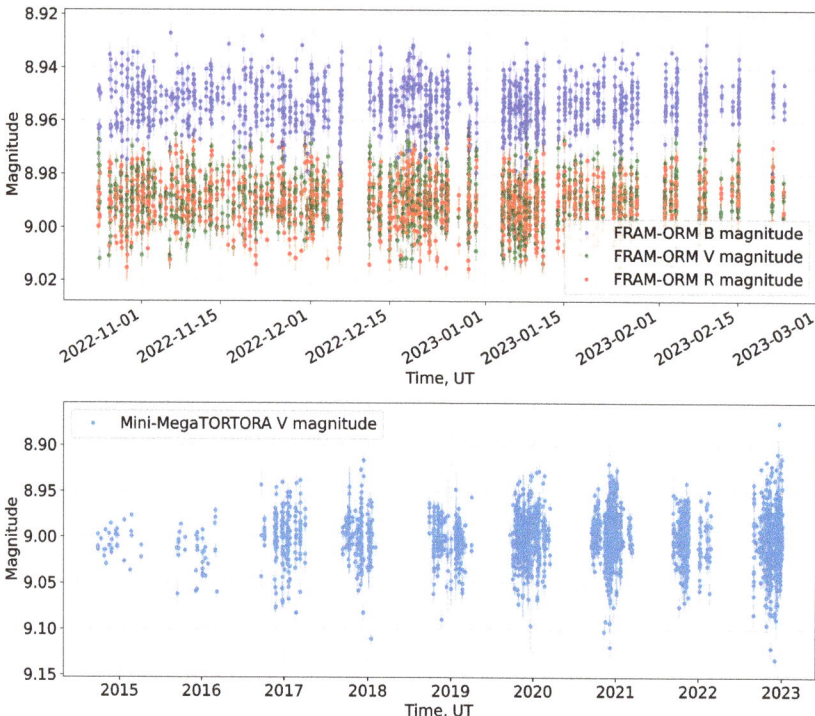

Figure 8. Light curves of HD 36030 from FRAM-ORM (**upper panel**) and Mini-MegaTORTORA (**lower panel**). FRAM-ORM observed the star in three Johnson–Cousins photometric filters, whereas Mini-MegaTORTORA data are in white light and re-calibrated to Johnson *V* bandpass. Both light curves lack any signs of systematic variability or statistically significant flares.

Finally, in order to assess the longer time scale variability of HD 36030, we acquired its photometry from the data archive of Mini-MegaTORTORA [54], which is a nine-channel wide-field optical monitoring system with high temporal resolution, operated since mid-

2014 and located at the Special Astrophysical Observatory, Nizhny Arkhyz, Russia. As part of its systematic observations of the northern sky, primarily targeted at the detection and characterization of optical transients on a sub-second time scale, it acquires deeper "survey" images with 20 to 60 s exposures in white light, covering every point of the northern sky on average several times per night. These images are processed by a dedicated pipeline that, apart from standard calibration steps, determines the effective photometric system of every frame and then employs this information to derive the $(B - V)$ colors of every star and re-calibrate the measurements to a Johnson V filter [55]. The resulting measurements are published online on the dedicated portal[4]. The data for HD 36030 have been extracted from the Mini-MegaTORTORA archive and passed through quality cuts in order to filter out the points corresponding to bad weather intervals and images where photometric calibration was too noisy. This resulted in more than 2300 points with good V magnitudes for the star, spanning more than 8 years since mid-2014.

The light curves from both *TESS* sectors display prominent oscillating patterns, but with sufficiently different amplitudes and characteristic time scales. The upper panel of Figure 9 shows the Lomb–Scargle periodogram [56,57] of both light curves. There are no common peaks visible. Thus, we may attribute the oscillations to residual instrumental effects not fully corrected by the data processing, or probably to some complex pulsational pattern of the star. We also computed the periodograms of the FRAM-ORM and Mini-MegaTORTORA light curves. For the former, we used the approach of VanderPlas and Ivezić [58] to get the combined periodogram of the measurements in all three colors. The periodograms are also shown in Figure 9. They display no prominent peaks or structures common to all light curves, and so we may conclude that we see no signs of any photometric periodicity in HD 36030.

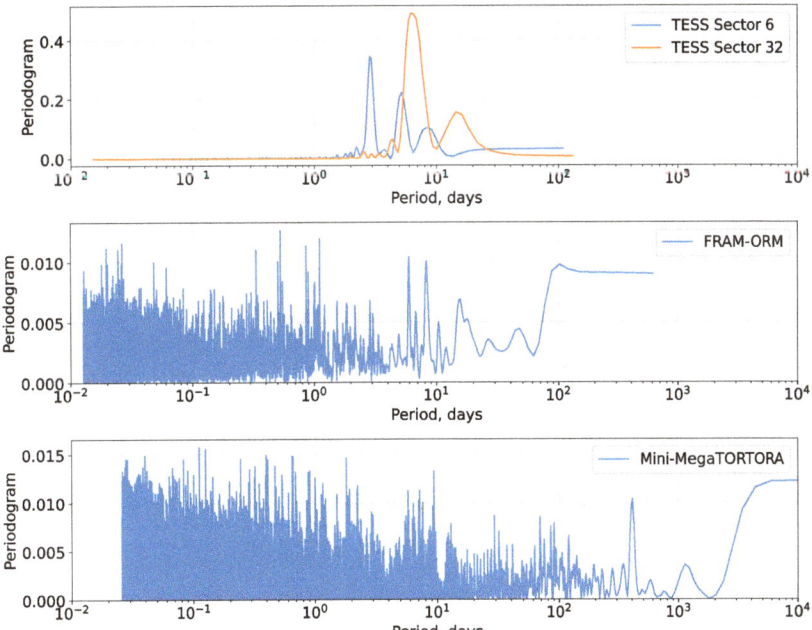

Figure 9. Lomb–Scargle periodograms of HD 36030 light curves from *TESS*, FRAM-ORM, and Mini-MegaTORTORA. For *TESS*, the data for both sectors are shown in the **upper panel**. The peaks there supposedly correspond to uncorrected instrumental effects in the light curves. The **middle panel** shows the multiband periodogram [58] for the FRAM-ORM light curve, and the **lower panel** shows the periodogram of Mini-MegaTORTORA data. There are no high-significance peaks visible in both of the latter panels.

4. Discussion

Our new spectroscopic data clearly display changes in the RV of HD 36030 across different epochs of observations. It directly testifies to the binarity of the star, which is thus an SB1 binary, as there are no spectral signatures of a second component in the spectrum. Although our data are too scarce to measure the orbital period, from Figure 4 we may constrain it as follows. The fastest change we see it is about $10\,\mathrm{km\,s^{-1}}$ between two consecutive nights, with a total spread of velocities of about $38\,\mathrm{km\,s^{-1}}$. Thus, the period should not exceed about 8 days, which gives the mass function of a second component $f = M_2^3 \sin^3 i / (M_1 + M_2)^2 < 0.006 M_\odot$. Thus, it favors a low-mass companion on an orbit with low to moderate inclination, which is consistent with the absence of eclipses in the light curve. The semi-major axis of the orbit is $a < 1.7 \cdot 10^{12}$ cm assuming the total system mass of $3 M_\odot$ and the period shorter than 8 days. Further spectral monitoring is necessary in order to better constrain the parameters of the second component of the system, as well as orbital parameters.

On the other hand, we do not see any signs of periodic variations in the light curves (see Figures 7 and 9) on these time scales, apart from quasi-periodic structures in the *TESS* data, with both periods and amplitudes varying both within the spans of individual observations and between different epochs, which we cannot attribute to the binary period. Their frequencies are also lower than the ones of rotational modulation seen in B-type stars in both *Kepler* [59] and *TESS* [60] data, so we assume that these oscillations are also not a signature of stellar rotation that might appear due to, e.g., starspots on the surface of HD 36030, but most probably they are a signature of uncorrected systematic effects in the data (see, e.g., Hattori et al. [61]).

We detected two flares on HD 36030 in the *TESS* data from two sectors, i.e., one in each 27-day series of continuous observations separated by nearly two years, meaning that these flares are not some extreme events but occur regularly. It is consistent with the detection of repeating flares on hot stars by Yang et al. [29]. The energies of these flares[5], $2.2 \cdot 10^{36}$ and $6.1 \cdot 10^{35}$ erg, place them among the most energetic ones detected by Balona [18] and Yang et al. [29] (see Figure 10). This fact makes it highly improbable that the flares originate from the low-mass companion[6] of HD 36030, which may only be a late-type, low-luminosity star due to the absence of any spectroscopic signature from it and a very low value of its mass function, which implies its low mass for reasonable range of values of the system' inclination angle. On the other hand, our modeling of HD 36030 suggests that it lacks the convective layers that may produce magnetic fields. There are no conclusive signs of surface magnetic fields—no strict periodicity in the light curve disfavors the presence of starspots; however, quasi-periodic patterns seen in Figure 7, if they are not instrumental in nature, may in principle be the manifestation of short-lived spots on HD 36030 surface—but in the absence of convection, the formation of such spots may be unrealistic. The star also lacks any signs of peculiarities in its spectra, such as additional emission components in hydrogen lines that might be a sign of a circumstellar outflow or disk that might help generate the magnetic fields due to star–disk interaction. However, the presence of a low-mass companion star, most probably possessing significant magnetic fields, on a close orbit around HD 36030 may be the source of magnetic fields spanning across the system and both storing the energy enough for powering the flares and maybe inducing the starspots on HD 36030 surface.

Thus, we may conclude that the only mechanism that may produce the magnetic fields necessary for powering the energetic flares may be the interaction between components inside the binary system.

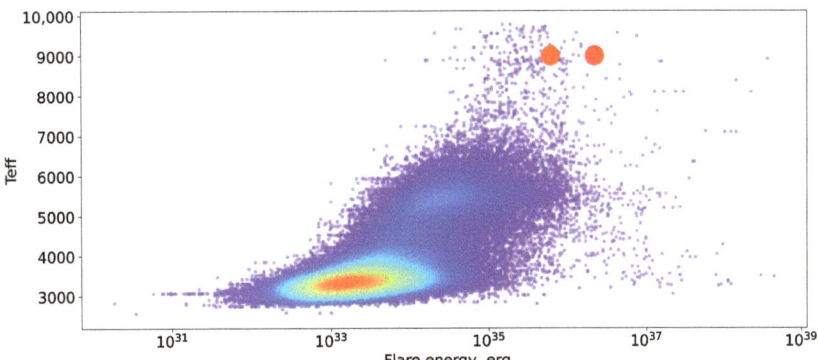

Figure 10. Energies of the flares of HD 36030 (red circles) in comparison with the ones detected by Yang et al. [29], as a function of the effective temperatures of the stars. The temperatures are originating from Gaia DR2 [63] and are underestimated for the hottest objects. The colors of the dots represent the density of points.

5. Conclusions

We performed a dedicated study of HD 36030, which was the hottest flare star detected by Maryeva et al. [11]. We found one more flare in the *TESS* data that confirmed the repeating nature of the flaring. We initiated a spectroscopic monitoring of the star in order to better understand the physics behind these flares. The spectra did not display any peculiarities and suggest that HD 36030 is a normal main-sequence B9 V star. On the other hand, we clearly detected radial velocity changes between different epochs of observations, thus confirming the binarity of the star. We did not detect any coherent variability in the light curve of the star, so we could not estimate the period of the orbital motion of the star or confirm the presence of the spots on its surface. The latter may be a sign of the absence of strong surface magnetic fields on HD 36030. Thus, the question of the origin of magnetic field powering the strong flares from HD 36030 is still open, and we favor the magnetic interaction with a second low-mass component in a binary system as their cause.

Author Contributions: O.M. proposed the concept of the study, performed observations on the Perek 2-m telescope, reduced the spectroscopic material, and performed spectral analysis; P.N. performed numerical modeling of the stellar atmosphere using TLUSTY code; S.K. performed photometric monitoring, reduction, and analysis of photometric data, and estimated the properties of flares. All authors participated in the discussion of the results and preparation of the manuscript. All authors have read and agreed to the published version of the manuscript.

Funding: This research received funding from the European Union's Framework Programme for Research and Innovation Horizon 2020 (2014–2020) under the Marie Skłodowska-Curie Grant Agreement No. 823734 (POEMS project). The Astronomical Institute in Ondřejov is supported by the project RVO:67985815. P.N. acknowledges support from the Grant Agency of the Czech Republic (GAČR 22-34467S) and from the Polish National Science Centre under projects No. UMO-2017/26/E/ST9/00703 and UMO-2017/25/B/ST9/02218. S.K. acknowledges support from the European Structural and Investment Fund and the Czech Ministry of Education, Youth and Sports (Project CoGraDS—CZ.02.1.01/0.0/0.0/15_003/0000437). This research has used the services of www.Astroserver.org. The operation of the robotic telescope FRAM-ORM is supported by the grant of the Ministry of Education of the Czech Republic LM2018102. The operation of the Mini-MegaTORTORA was supported under the Ministry of Science and Higher Education of the Russian Federation grant 075-15-2022-262 (13.MNPMU.21.0003). This paper includes data collected by the *TESS* mission, which are publicly available from the Mikulski Archive for Space Telescopes (MAST). Funding for the *TESS* mission is provided by NASA's Science Mission directorate. This research was made by using of the SIMBAD database and the VizieR catalogue access tool, both operated at CDS, Strasbourg, France.

Institutional Review Board Statement: Not applicable.

Informed Consent Statement: Not applicable.

Data Availability Statement: The data presented in this study are available on request from the corresponding author, or available in publicly accessible data archives as specified in the manuscript text.

Acknowledgments: The authors acknowledge the help from Suryani Guha with observations with Coudé spectrograph at the Perek 2 m telescope.

Conflicts of Interest: The authors declare no conflict of interest.

Abbreviations

The following abbreviations are used in this manuscript:

IDL	Interactive data language
LBV	Luminous blue variable
LTE	Local thermodynamic equilibrium
OES	Ondřejov echelle epectrograph
RV	Radial velocity
SB1	single-lined spectroscopic binary
TESS	Transiting Exoplanet Survey Satellite

Notes

1. RA $05^h28^m58^s \cdot 526$; Dec $+03°38'49'' \cdot 28$.
2. Up to date table of colors and effective temperatures of stars from Pecaut and Mamajek [41] is available at https://www.pas.rochester.edu/~emamajek/EEM_dwarf_UBVIJHK_colors_Teff.txt accessed on 10 March 2023.
3. https://archive.stsci.edu accessed on 10 March 2023.
4. http://survey.favor2.info/ accessed on 10 March 2023.
5. We must note that our estimation of flare energies is a conservative one, as it assumes their temperatures to be 9000 K, which may not be a good approximation, as the effective temperature of the star itself is above that. Moreover, there are signatures that the temperatures of superflares even on cool stars may be significantly larger than that [62]. Thus, the actual energies of the flares we detected may also be significantly larger.
6. The same argument also holds against any other similar source in the field of HD 36030 that may occasionally pollute the large aperture of *TESS*.

References

1. McNamara, B.J.; Jackiewicz, J.; McKeever, J. The Classification of Kepler B-star Variables. *Astron. J.* **2012**, *143*, 101. [CrossRef]
2. Waelkens, C. Slowly pulsating B stars. *Astron. Astrophys.* **1991**, *246*, 453. [CrossRef]
3. Waelkens, C. Slowly Pulsating B Stars. In *International Astronomical Union Colloquium*; Cambridge University Press: Cambridge, UK, 1993; p. 180.
4. Stankov, A.; Handler, G. Catalog of Galactic β Cephei Stars. *Astrophys. J. Suppl. Ser.* **2005**, *158*, 193–216. [CrossRef]
5. Saio, H.; Georgy, C.; Meynet, G. Evolution of blue supergiants and α Cygni variables: Puzzling CNO surface abundances. *Mon. Not. R. Astron. Soc.* **2013**, *433*, 1246–1257. [CrossRef]
6. Rivinius, T.; Carciofi, A.C.; Martayan, C. Classical Be stars. Rapidly rotating B stars with viscous Keplerian decretion disks. *Astron. Astrophys. Rev.* **2013**, *21*, 69. [CrossRef]
7. Krtičková, I.; Krtička, J. An ultraviolet study of B[e] stars: Evidence for pulsations, luminous blue variable type variations and processes in envelopes. *Mon. Not. R. Astron. Soc.* **2018**, *477*, 236–253. [CrossRef]
8. Kraus, M. A Census of B[e] Supergiants. *Galaxies* **2019**, *7*, 83. [CrossRef]
9. Lobel, A.; Groh, J.H.; Martayan, C.; Frémat, Y.; Torres Dozinel, K.; Raskin, G.; Van Winckel, H.; Prins, S.; Pessemier, W.; Waelkens, C.; et al. Modelling the asymmetric wind of the luminous blue variable binary MWC 314. *Astron. Astrophys.* **2013**, *559*, A16. [CrossRef]
10. Clark, J.S.; Najarro, F.; Negueruela, I.; Ritchie, B.W.; Urbaneja, M.A.; Howarth, I.D. On the nature of the galactic early-B hypergiants. *Astron. Astrophys.* **2012**, *541*, A145. [CrossRef]
11. Maryeva, O.; Bicz, K.; Xia, C.; Baratella, M.; Čechvala, P.; Vida, K. Flare stars in nearby Galactic open clusters based on TESS data. *Contrib. Astron. Obs. Skaln. Pleso* **2021**, *51*, 78–97. [CrossRef]
12. Priest, E.; Forbes, T. *Magnetic Reconnection*; Cambridge University Press: Cambridge, UK, 2000; p. 612.
13. Benz, A.O.; Güdel, M. Physical Processes in Magnetically Driven Flares on the Sun, Stars, and Young Stellar Objects. *Annu. Rev. Astron. Astrophys.* **2010**, *48*, 241–287. [CrossRef]

14. Gershberg, R.E. Time scales and energy of flares on red dwarf stars (a review). *Mem. Della Soc. Astron. Ital.* **1989**, *60*, 263–287.
15. Günther, M.N.; Zhan, Z.; Seager, S.; Rimmer, P.B.; Ranjan, S.; Stassun, K.G.; Oelkers, R.J.; Daylan, T.; Newton, E.; Kristiansen, M.H.; et al. Stellar Flares from the First TESS Data Release: Exploring a New Sample of M Dwarfs. *Astron. J.* **2020**, *159*, 60. [CrossRef]
16. Schaefer, B.E. Flashes from Normal Stars. *Astrophys. J.* **1989**, *337*, 927. [CrossRef]
17. Balona, L.A. Kepler observations of flaring in A-F type stars. *Mon. Not. R. Astron. Soc.* **2012**, *423*, 3420–3429. [CrossRef]
18. Balona, L.A. Spots and flares in hot main sequence stars observed by Kepler, K2 and TESS. *Front. Astron. Space Sci.* **2021**, *8*, 32. [CrossRef]
19. Schmitt, J.H.M.M.; Guedel, M.; Predehl, P. Spatially resolved X-ray and radio observations of Castor A+B+C. *Astron. Astrophys.* **1994**, *287*, 843–851.
20. Yanagida, T.; Ezoe, Y.; Kawaharada, M.; Kokubun, M.; Makishima, K. Large X-ray Flares from B-Type Stars, HD261902 and HD47777, in NGC2264 Observed with CHANDRA. In Proceedings of the Active OB-Stars: Laboratories for Stellar and Circumstellar Physics, ASP Conference Series, Sapporo, Japan, 29 August–2 September 2005; Okazaki, A.T., Owocki, S.P., Stefl, S., Eds.; Astronomical Society of the Pacific: San Francisco, CA, USA, 2007; Volume 361, p. 533.
21. Rosner, R.; Vaiana, G.S. Stellar Coronae from Einstein: Observations and Theory. In *X-ray Astronomy*; Giacconi, R., Setti, G., Eds.; NATO Advanced Study Institutes Series; Springer: Dordrecht, The Netherlands, 1980; Volume 60, p. 129. [CrossRef]
22. Pedersen, M.G.; Antoci, V.; Korhonen, H.; White, T.R.; Jessen-Hansen, J.; Lehtinen, J.; Nikbakhsh, S.; Viuho, J. Do A-type stars flare? *Mon. Not. R. Astron. Soc.* **2017**, *466*, 3060–3076. [CrossRef]
23. Švanda, M.; Karlický, M. Flares on A-type Stars: Evidence for Heating of Solar Corona by Nanoflares? *Astrophys. J.* **2016**, *831*, 9. [CrossRef]
24. Bai, J.Y.; Esamdin, A. Flare Properties of A-type Stars in Kepler Data. *Astrophys. J.* **2020**, *905*, 110. [CrossRef]
25. Van Doorsselaere, T.; Shariati, H.; Debosscher, J. Stellar Flares Observed in Long-cadence Data from the Kepler Mission. *Astrophys. J. Suppl. Ser.* **2017**, *232*, 26. [CrossRef]
26. Houk, N.; Swift, C. Michigan catalogue of two-dimensional spectral types for the HD Stars, Vol. 5. *Mich. Spectr. Surv.* **1999**, *5*.
27. Kharchenko, N.V.; Piskunov, A.E.; Röser, S.; Schilbach, E.; Scholz, R.D. 109 new Galactic open clusters. *Astron. Astrophys.* **2005**, *440*, 403–408. [CrossRef]
28. Cantat-Gaudin, T.; Jordi, C.; Vallenari, A.; Bragaglia, A.; Balaguer-Núñez, L.; Soubiran, C.; Bossini, D.; Moitinho, A.; Castro-Ginard, A.; Krone-Martins, A.; et al. A Gaia DR2 view of the open cluster population in the Milky Way. *Astron. Astrophys.* **2018**, *618*, A93. [CrossRef]
29. Yang, Z.; Zhang, L.; Meng, G.; Han, X.L.; Misra, P.; Yang, J.; Pi, Q. Properties of flare events based on light curves from the TESS survey. *Astron. Astrophys.* **2023**, *669*, A15. [CrossRef]
30. Evans, P.A.; Page, K.L.; Osborne, J.P.; Beardmore, A.P.; Willingale, R.; Burrows, D.N.; Kennea, J.A.; Perri, M.; Capalbi, M.; Tagliaferri, G.; et al. 2SXPS: An Improved and Expanded Swift X-Ray Telescope Point-source Catalog. *Astrophys. J. Suppl. Ser.* **2020**, *247*, 54. [CrossRef]
31. Koubský, P.; Mayer, P.; Čáp, J.; Žďárský, F.; Zeman, J.; Pína, L.; Melich, Z. Ondřejov Echelle Spectrograph—OES. *Publ. Astron. Inst. Czechoslov. Acad. Sci.* **2004**, *92*, 37–43.
32. Kabáth, P.; Skarka, M.; Sabotta, S.; Guenther, E.; Jones, D.; Klocová, T.; Šubjak, J.; Žák, J.; Špoková, M.; Blažek, M.; et al. Ondřejov Echelle Spectrograph, Ground Based Support Facility for Exoplanet Missions. *Publ. Astron. Soc. Pac.* **2020**, *132*, 035002. [CrossRef]
33. Slechta, M.; Skoda, P. 2-meter telescope devices: Coudé slit spectrograph and HEROS. *Publ. Astron. Inst. Czechoslov. Acad. Sci.* **2002**, *90*, 1–4.
34. Maryeva, O.V.; Karpov, S.V.; Kniazev, A.Y.; Gvaramadze, V.V. How long can luminous blue variables sleep? A long-term photometric variability and spectral study of the Galactic candidate luminous blue variable MN 112. *Mon. Not. R. Astron. Soc.* **2022**, *513*, 5752–5765. [CrossRef]
35. Gray, R.O.; Corbally, C.J. *Stellar Spectral Classification*; Princeton University Press: Princeton, NJ, USA, 2009.
36. Simón-Díaz, S.; Castro, N.; Garcia, M.; Herrero, A.; Markova, N. The IACOB spectroscopic database of Northern Galactic OB stars. *Bull. De La Soc. R. Des Sci. De Liege* **2011**, *80*, 514–518.
37. Simón-Díaz, S.; Garcia, M.; Herrero, A.; Maíz Apellániz, J.; Negueruela, I. The IACOB Project: Synergies for the Gaia Era. *arXiv* **2011**, arXiv:1109.2665.
38. Simón-Díaz, S.; Negueruela, I.; Maíz Apellániz, J.; Castro, N.; Herrero, A.; Garcia, M.; Pérez-Prieto, J.A.; Caon, N.; Alacid, J.M.; Camacho, I.; et al. The IACOB spectroscopic database: Recent updates and first data release. *arXiv* **2015**, arXiv:1504.04257.
39. Prugniel, P.; Soubiran, C. A database of high and medium-resolution stellar spectra. *Astron. Astrophys.* **2001**, *369*, 1048–1057. [CrossRef]
40. Høg, E.; Fabricius, C.; Makarov, V.V.; Urban, S.; Corbin, T.; Wycoff, G.; Bastian, U.; Schwekendiek, P.; Wicenec, A. The Tycho-2 catalogue of the 2.5 million brightest stars. *Astron. Astrophys.* **2000**, *355*, L27–L30.
41. Pecaut, M.J.; Mamajek, E.E. Intrinsic Colors, Temperatures, and Bolometric Corrections of Pre-main-sequence Stars. *Astrophys. J. Suppl. Ser.* **2013**, *208*, 9. [CrossRef]
42. Chen, Y.; Girardi, L.; Bressan, A.; Marigo, P.; Barbieri, M.; Kong, X. Improving PARSEC models for very low mass stars. *Mon. Not. R. Astron. Soc.* **2014**, *444*, 2525–2543. [CrossRef]

43. Bossini, D.; Vallenari, A.; Bragaglia, A.; Cantat-Gaudin, T.; Sordo, R.; Balaguer-Núñez, L.; Jordi, C.; Moitinho, A.; Soubiran, C.; Casamiquela, L.; et al. Age determination for 269 Gaia DR2 open clusters. *Astron. Astrophys.* **2019**, *623*, A108. [CrossRef]
44. Hubeny, I.; Lanz, T. Non-LTE Line-blanketed Model Atmospheres of Hot Stars. I. Hybrid Complete Linearization/Accelerated Lambda Iteration Method. *Astrophys. J.* **1995**, *439*, 875. [CrossRef]
45. Lanz, T.; Hubeny, I. A Grid of NLTE Line-blanketed Model Atmospheres of Early B-Type Stars. *Astrophys. J. Suppl. Ser.* **2007**, *169*, 83–104. [CrossRef]
46. Hubeny, I.; Lanz, T. TLUSTY User's Guide III: Operational Manual. *arXiv* **2017**. [CrossRef]
47. Németh, P.; Kawka, A.; Vennes, S. A selection of hot subluminous stars in the GALEX survey—II. Subdwarf atmospheric parameters. *Mon. Not. R. Astron. Soc.* **2012**, *427*, 2180–2211. [CrossRef]
48. Bohlin, R.C.; Mészáros, S.; Fleming, S.W.; Gordon, K.D.; Koekemoer, A.M.; Kovács, J. A New Stellar Atmosphere Grid and Comparisons with HST/STIS CALSPEC Flux Distributions. *Astron. J.* **2017**, *153*, 234. [CrossRef]
49. Ricker, G.R.; Winn, J.N.; Vanderspek, R.; Latham, D.W.; Bakos, G.Á.; Bean, J.L.; Berta-Thompson, Z.K.; Brown, T.M.; Buchhave, L.; Butler, N.R.; et al. Transiting Exoplanet Survey Satellite (TESS). *J. Astron. Telesc. Instru. Syst.* **2015**, *1*, 014003. [CrossRef]
50. Jenkins, J.M.; Twicken, J.D.; McCauliff, S.; Campbell, J.; Sanderfer, D.; Lung, D.; Mansouri-Samani, M.; Girouard, F.; Tenenbaum, P.; Klaus, T.; et al. The TESS science processing operations center. In Proceedings of the Software and Cyberinfrastructure for Astronomy IV, Edinburgh, UK, 26–30 June 2016; Chiozzi, G., Guzman, J.C., Eds.; Society of Photo-Optical Instrumentation Engineers (SPIE): Edinburgh, UK, 2016; Volume 9913, p. 99133E. [CrossRef]
51. Shibayama, T.; Maehara, H.; Notsu, S.; Notsu, Y.; Nagao, T.; Honda, S.; Ishii, T.T.; Nogami, D.; Shibata, K. Superflares on Solar-type Stars Observed with Kepler. I. Statistical Properties of Superflares. *Astrophys. J. Suppl. Ser.* **2013**, *209*, 5. [CrossRef]
52. Karpov, S. STDPipe: Simple Transient Detection Pipeline. Astrophysics Source Code Library, Record ascl:2112.006. 2021. Available online: https://ascl.net/2112.006 (accessed on 10 March 2023).
53. Gaia Collaboration; Montegriffo, P.; Bellazzini, M.; De Angeli, F.; Andrae, R.; Barstow, M.A.; Bossini, D.; Bragaglia, A.; Burgess, P.W.; Cacciari, C.; et al. Gaia Data Release 3: The Galaxy in your preferred colours. Synthetic photometry from Gaia low-resolution spectra. *arXiv* **2022**. [CrossRef]
54. Beskin, G.M.; Karpov, S.V.; Biryukov, A.V.; Bondar, S.F.; Ivanov, E.A.; Katkova, E.V.; Orekhova, N.V.; Perkov, A.V.; Sasyuk, V.V. Wide-field optical monitoring with Mini-MegaTORTORA (MMT-9) multichannel high temporal resolution telescope. *Astrophys. Bull.* **2017**, *72*, 81–92. [CrossRef]
55. Karpov, S.; Beskin, G.; Biryukov, A.; Bondar, S.; Ivanov, E.; Katkova, E.; Orekhova, N.; Perkov, A.; Sasyuk, V. Photometric calibration of a wide-field sky survey data from Mini-MegaTORTORA. *Astron. Nachrichten* **2018**, *339*, 375–381. [CrossRef]
56. Lomb, N.R. Least-Squares Frequency Analysis of Unequally Spaced Data. *Astrophys. Space Sci.* **1976**, *39*, 447–462. [CrossRef]
57. Scargle, J.D. Studies in astronomical time series analysis. II. Statistical aspects of spectral analysis of unevenly spaced data. *Astrophys. J.* **1982**, *263*, 835–853. [CrossRef]
58. VanderPlas, J.T.; Ivezić, Ž. Periodograms for Multiband Astronomical Time Series. *Astrophys. J.* **2015**, *812*, 18. [CrossRef]
59. Balona, L.A. Rotational modulation in B stars observed by the Kepler K2 mission. *Mon. Not. R. Astron. Soc.* **2016**, *457*, 3724–3731. [CrossRef]
60. Balona, L.A.; Handler, G.; Chowdhury, S.; Ozuyar, D.; Engelbrecht, C.A.; Mirouh, G.M.; Wade, G.A.; David-Uraz, A.; Cantiello, M. Rotational modulation in TESS B stars. *Mon. Not. R. Astron. Soc.* **2019**, *485*, 3457–3469. [CrossRef]
61. Hattori, S.; Foreman-Mackey, D.; Hogg, D.W.; Montet, B.T.; Angus, R.; Pritchard, T.A.; Curtis, J.L.; Schölkopf, B. The unpopular Package: A Data-driven Approach to Detrending TESS Full-frame Image Light Curves. *Astron. J.* **2022**, *163*, 284. [CrossRef]
62. Howard, W.S.; Corbett, H.; Law, N.M.; Ratzloff, J.K.; Galliher, N.; Glazier, A.L.; Gonzalez, R.; Vasquez Soto, A.; Fors, O.; del Ser, D.; et al. EvryFlare. III. Temperature Evolution and Habitability Impacts of Dozens of Superflares Observed Simultaneously by Evryscope and TESS. *Astrophys. J.* **2020**, *902*, 115. [CrossRef]
63. Gaia Collaboration; Brown, A.G.A.; Vallenari, A.; Prusti, T.; de Bruijne, J.H.J.; Babusiaux, C.; Bailer-Jones, C.A.L.; Biermann, M.; Evans, D.W.; Eyer, L.; et al. Gaia Data Release 2. Summary of the contents and survey properties. *Astron. Astrophys.* **2018**, *616*, A1. [CrossRef]

Disclaimer/Publisher's Note: The statements, opinions and data contained in all publications are solely those of the individual author(s) and contributor(s) and not of MDPI and/or the editor(s). MDPI and/or the editor(s) disclaim responsibility for any injury to people or property resulting from any ideas, methods, instructions or products referred to in the content.

MDPI
St. Alban-Anlage 66
4052 Basel
Switzerland
www.mdpi.com

Galaxies Editorial Office
E-mail: galaxies@mdpi.com
www.mdpi.com/journal/galaxies

Disclaimer/Publisher's Note: The statements, opinions and data contained in all publications are solely those of the individual author(s) and contributor(s) and not of MDPI and/or the editor(s). MDPI and/or the editor(s) disclaim responsibility for any injury to people or property resulting from any ideas, methods, instructions or products referred to in the content.

www.ingramcontent.com/pod-product-compliance
Lightning Source LLC
LaVergne TN
LVHW070246100526
838202LV00015B/2184